I0063135

Fire Regimes: Spatial and Temporal Variability and Their Effects on Forests

Special Issue Editors

Yves Bergeron
Sylvie Gauthier

MDPI

Special Issue Editors

Yves Bergeron
Forest Research Institute
Canada

Sylvie Gauthier
Laurentian Forestry Centre
Canada

Editorial Office
MDPI AG
St. Alban-Anlage 66
Basel, Switzerland

This edition is a reprint of the Special Issue published online in the open access journal *Forests* (ISSN 1999-4907) from 2016–2017 (available at: http://www.mdpi.com/journal/forests/special_issues/fire_variability).

For citation purposes, cite each article independently as indicated on the article page online and as indicated below:

Author 1; Author 2; Author 3 etc. Article title. *Journal Name*. **Year**. Article number/page range.

ISBN 978-3-03842-390-4 (Pbk)
ISBN 978-3-03842-391-1 (PDF)

Articles in this volume are Open Access and distributed under the Creative Commons Attribution license (CC BY), which allows users to download, copy and build upon published articles even for commercial purposes, as long as the author and publisher are properly credited, which ensures maximum dissemination and a wider impact of our publications. The book taken as a whole is © 2017 MDPI, Basel, Switzerland, distributed under the terms and conditions of the Creative Commons license CC BY-NC-ND license (http://creativecommons.org/licenses/by-nc-nd/4.0/).

Table of Contents

Section 1: Fire Regimes

Section 2: Impact on Vegetation

Section 3: Impact on Ecosystems

About the Guest Editors

Yves Bergeron is a professor of Forest Ecology at the Université du Québec à Montréal and at the Université du Québec en Abitibi-Témiscamingue. Dr. Bergeron holds the NSERC Industrial Chair and the Canadian Research Chair in Ecology and Sustainable Forest Management, and he is known internationally for the exceptional quality of his research on boreal forest. He has authored more than 400 publications and received numerous awards. He has been amember of the Royal Society of Canada since 2010.

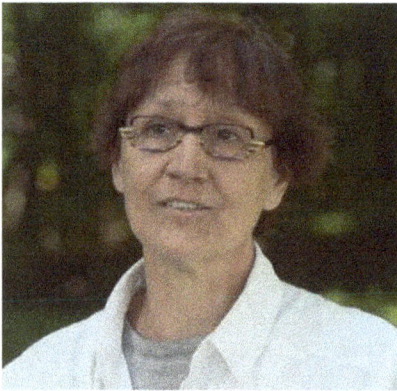

Sylvie Gauthier joined the Canadian Forest Service of Natural Resources Canada in 1993, initially with the Petawawa National Forestry Institute, and then with the Laurentian Forestry Centre in Québec. She is an associate member of the Centre for Forest Research (CFR), and an adjunct professor: in biology at the Université du Québec à Montréal (UQAM); in forestry at Université Laval; and in applied science at Université du Québec en Abitibi-Témiscamingue (UQAT). She has authored more than 100 scientific articles. Her work focuses on: (1) the characterization of disturbance regimes and the factors affecting them; (2) the effect of disturbance regimes on forest dynamics; and (3) the translation of this knowledge into sustainable forest management and climate change adaptation strategies. She recently published an invited review on the health of the circumboreal forest in *Science*.

Preface to "Fire Regimes: Spatial and Temporal Variability and Their Effects on Forests"

Fire regimes (occurrence, cycle, severity, size, etc.) are key factors in many forest ecosystems, as they are often critical drivers of forest composition, dynamics, and ecosystem processes. Fire regimes vary in space and time according to climatic, physical and biological factors. A better understanding of the interacting factors controlling fire regimes may contribute to improving fire and forest management and their future projection in the context of global change. Knowledge of how fire regimes affect natural landscapes is also used in forestry as a template to manage the forest for wood production. This book aims to synthesize current understanding of factors affecting fire regime characteristics, to present recent research on fire regimes and their effects on forest ecosystems, and to illustrate how this knowledge could be translated into forest or fire management strategies. The book is divided into three sections. The first one (Fire Regime) includes papers that described the different parameters of fire regime and how fire regime is controlled by climate, landscape features, fuels, and human impact. The second section includes papers on how fire regimes control vegetation composition and its spatial distribution. Finally the third section includes papers that discussed how fire regimes may influence forest health, forest regeneration, ecosystems processes and carbon emissions. The last paper discusses a wildfire mitigation strategy. Overall the papers present a large spectrum of current research on fire regime and their impacts on vegetation and ecosystems. It provides background information for maintaining forest productivity and resilience in the face of climate change.

Yves Bergeron and Sylvie Gauthier

Guest Editors

Section 1:
Fire Regimes

![forests logo] *forests*

MDPI

Article

Quantifying Fire Cycle from Dendroecological Records Using Survival Analyses

Dominic Cyr [1,*], Sylvie Gauthier [1], Yan Boulanger [1] and Yves Bergeron [2]

1   Canadian Forest Service-Laurentian Forestry Centre, Quebec, QC G1V 4C7, Canada;
    sylvie.gauthier2@canada.ca (S.G.); yan.boulanger@canada.ca (Y.B.)
2   Forest Research Institute, Industrial Chair NSERC-UQAT-UQAM in Sustainable Forest Management,
    Université du Québec en Abitibi-Témiscamingue, Rouyn-Noranda, QC J9X 5E4, Canada;
    yves.bergeron@uqat.ca
*   Correspondence: cyr.dominic@gmail.com; Tel.: +1-418-648-5834

Academic Editor: Eric J. Jokela
Received: 12 May 2016; Accepted: 16 June 2016; Published: 28 June 2016

Abstract: Quantifying fire regimes in the boreal forest ecosystem is crucial for understanding the past and present dynamics, as well as for predicting its future dynamics. Survival analyses have often been used to estimate the fire cycle in eastern Canada because they make it possible to take into account the censored information that is made prevalent by the typically long fire return intervals and the limited scope of the dendroecological methods that are used to quantify them. Here, we assess how the true length of the fire cycle, the short-term temporal variations in fire activity, and the sampling effort affect the accuracy and precision of estimates obtained from two types of parametric survival models, the Weibull and the exponential models, and one non-parametric model obtained with the Cox regression. Then, we apply those results in a case area located in eastern Canada. Our simulation experiment confirms some documented concerns regarding the detrimental effects of temporal variations in fire activity on parametric estimation of the fire cycle. Cox regressions appear to provide the most accurate and robust estimator, being by far the least affected by temporal variations in fire activity. The Cox-based estimate of the fire cycle for the last 300 years in the case study area is 229 years (CI95: 162–407), compared with the likely overestimated 319 years obtained with the commonly used exponential model.

Keywords: accuracy; boreal forest; coverage; Cox regression; dendrochronology; exponential; fire cycle; precision; survival analyses; time since fire; Weibull

1. Introduction

Fire is a fundamental process in the natural dynamics of the boreal forest of North America. Quantifying fire regime characteristics is therefore crucial to the understanding of past and present dynamics of the boreal ecosystem as well as for predicting its future dynamics. It has also become a necessary step for many important forest management planning issues. Over the last few decades, there has been a shift from a commodity production-oriented approach to an ecosystem-based approach in forest management that aims to maintain ecological integrity of the boreal forest [1,2]. Fire regimes are central to this paradigm shift as new management strategies are being developed that are better aligned with the characteristics and relative importance of stand-initiating disturbances (fire) and other types of natural disturbances that usually occur at finer spatial scales or that are more selective in nature (e.g., wind and insects). In addition, a total exclusion of fire and replacement by harvesting is economically unrealistic and ecologically questionable [3]. Consequently, there is a growing interest in incorporating fire a priori in strategic forest management planning to prevent eventual shortage in timber supplies caused by fire [4–6]. Furthermore, fire is a key element of the carbon cycle in the boreal forest, which makes the quantification of the fire regime necessary to the assessment of the carbon budget of countries and industrial sectors related to the boreal forest [7]. Such assessments can directly affect the effort and levels of mitigation needed to reach atmospheric CO_2 stabilization targets in a carbon economy.

An important characteristic of the fire regime is the fire cycle (FC), defined as the number of years required to burn an area equal to the total area surveyed. The FC is equivalent to the mean fire return interval when used from a point-based perspective, and both concepts are also reciprocal to the mean burn rate or fire frequency, depending on whether the term is used from an area-based or point-based perspective [8]. In this study, we will mainly use the term fire cycle.

The FC is generally difficult to accurately and precisely quantify in eastern Canada for several reasons: First, typical fire return intervals are relatively long [9–15] compared with the maximum longevity of most tree species (generally between 100 and 300 years). In addition, the area burned annually is associated with a large interannual and interdecadal variability as well as subject to observed or suspected long-term trends. Those difficulties are exacerbated by the fact that burned areas only started to be identified, delineated and organized in provincial or national databases in the last few decades [16]. Moreover, aging and "overburning" of the forest gradually erase traces of past fire events. In eastern Canada, most fires typically are stand-replacing, and multiple fire scars are only exceptionally encountered The most reliable estimates of FC currently available for the eastern boreal forests of Canada are those obtained from dendroecological reconstructions, in which the age of the first post-fire tree cohort (most commonly) or fire scars (seldom) are sought to infer the time elapsed since the last fire (TSF) and are then processed in survival analyses (SA). Representative surveys of large landscapes are also sometimes made difficult because of limited road access, which reduces the number of points that can be visited for random or systematic sampling. Aside from allowing us to quantify the FC and relate it to environmental covariates that create spatiotemporal heterogeneity, SA allows us to consider the fire intervals that are known only to a limited extent. That happens when no trace of past fire events can be detected or dated with a sufficient level of confidence, in which case only a minimum amount of TSF can be attributed to the site. The probability of observing such a minimum interval increases with TSF and, consequently, the proportion of such stands across the landscape increases with the length of the FC. They usually become dominant in the landscape when FC reaches or exceeds 200–300 years. It has been shown that those minimum intervals, which are referred to as censored observations in SA literature [17–19], represent incomplete but meaningful information that cannot be dismissed as missing observations without introducing major biases [20]. However, an important proportion of censored observations decreases the precision of FC estimates in a way that is similar to decreasing the sampling effort [19], and it may affect accuracy if censoring does not occur in all stands in the same fashion.

Several analytical approaches involving SA have previously been described and used to quantify FC or other closely related metrics [8,21] without necessarily yielding the same results. The general objective of this paper is to assess and compare the accuracy and precision of three different SA-based methods for estimating FC from dendroecological reconstructions of fire history by means of a modeling experiment designed to simulate a range of conditions that are representative of what is typically encountered in such studies [10,11,14,22–24]. More specifically, we compare the robustness of three methods in the face of three sources of uncertainty that interfere or limit the estimation of historical FCs: (1) the length of the FC, which, when put in relationship with longevity of the main three species, directly affects the proportion of censored observations; (2) the sampling effort, which is a ubiquitous constraint in empirical field studies; and (3) past variation in fire activity, which has been recognized as a potential bias and source of assumption violation for some analytical approaches [25–28]. Most of these criticisms are related to the assumption that the disturbance process is stationary, which has been suggested to be unrealistic. We try to assess the extent to which this affects the validity of FC estimation in eastern Canada. The first two methods assessed consist in fitting the negative exponential [29] and the Weibull [30] distributions to the TSF distribution obtained from a representative sample of an area of interest. The third one consists in using the baseline, non-parametric hazard function of a Cox regression model [31] to quantify the length of the FC. Although we provide a brief description of these three methods, the interested reader is invited to refer to specialized literature for a more complete description of the mathematical grounds as well as a wider spectrum of applications of these methods [17–19]. Here, our main focus is on the practical implications of using these methods to estimate FC from dendroecological studies. The second objective is to apply the resulting outcomes of this modeling experiment to a case study of a 1.5 Mha boreal landscape of eastern North America.

Before presenting the modeling approach used in this study, we first present some general aspects of SA applied to fire ecology, and the three SA-based methods used to estimate FC. Then, we present the modeling approach and a case study in the Côte-Nord region of Quebec, in eastern Canada.

2. Materials and Methods

2.1. Theoretical Background

2.1.1. Time Intervals, Censoring and Truncation in the Context of FC Estimation Based on the TSF Dataset

Generally, SA involves the modeling of time to event data. Ideally, the date of "birth" and "death" of a subject are both known, in which case the lifespan is unambiguous. However, it is relatively common in survival data to know the length of the lifespan only to a limited extent, i.e., it is at least a certain duration (when the birth is known but not the death, or the reverse situation). This is called a censored observation. The most common type of censoring is right-censoring, and it typically occurs when an experiment is terminated before the event of interest could be observed (the time axis is represented from left to right). Left-censoring is also possible, and it occurs when the lifespan of an individual began prior to the beginning of the experiment with no way of retrospectively determining the moment at which it began. In the present situation, the event of interest is a stand-replacing fire and, in an ideal dataset, two successive fire events would have to be known to observe a closed interval, i.e., uncensored information, e.g., [28,32]. Most of the time, however, it is not possible to obtain such complete information with landscape-scale dendroecologically reconstructed fire history data, especially in regions where fire-free intervals are long. In fact, in a stand-replacing fire context, only one past fire event can usually be dated on each site using archives or dendroecological methods. Fire history is reconstructed by conducting a cross-sectional field survey, sometimes combined with archival data, to obtain a "snapshot" or static image at the time of sampling based on a representative sample or a complete fire map. This approach is mostly applicable to fire regimes that are dominated by stand-replacing fires. A TSF distribution is then obtained and used as survival data.

In eastern Canada, it has been shown that, in the absence of fire for more than 100–150 years, the post-fire cohort is gradually replaced by new trees in the canopy [33,34], a process that erases all traces of the initial post-fire cohort whose age is used to infer TSF. This causes left censoring because the beginning of the lifespan is unknown except for the fact that it occurred prior to the age of the oldest trees, assuming they did not survive the last fire. In the case of cross-sectional field surveys, the entire TSF distribution can also be considered to be censored at the time of sampling because no other fire event has yet to occur. This right-censoring affects all observations. With the type of datasets that are available for the boreal forest of North America, where there is already a relatively high percentage of left-censored observations because of the typically long fire-free intervals and rare complete intervals, we usually consider the time of sampling as time $t = 0$ to avoid the issue of right censoring, mainly because of the lack of alternatives. In fact, at least one complete time-to-event observation is needed for survival function fitting. Then, we work in reverse time [35,36] to transform cases of left censoring into right censored observations, which is relatively straightforward using standard SA, but it may have important repercussions as it will be discussed below.

2.1.2. Survival Analyses

Survival datasets can be described using several interrelated functions. In statistical terms, the three most important are the cumulative density function, the probability density function and the hazard function. Following the terminology used by Johnson and Gutsell [8], who put these functions in the context of stand-replacing fire regimes, the cumulative density function corresponds to the survivorship distribution function $A(t)$, which is the probability of having gone without fire (survived) longer than time t, that is:

$$A(t) = P(T < t), t \geqslant 0 \tag{1}$$

The probability density function corresponds to the fire interval distribution $f(t)$, which is the probability of having a fire in the interval t to $t + \Delta t$:

$$f(t) = \frac{dA(t)}{dt} \tag{2}$$

Finally, the hazard function, or hazard of burning function, is the chance that each element survived to time t and burned in the interval t to $t + \Delta t$:

$$\lambda(t) = \frac{f(t)}{A(t)} \tag{3}$$

2.1.3. Estimating FC Using Parametric Survival Models

Survival data can be fit using parametric models based on many distributions, upon which the most commonly used for FC estimations are the negative exponential and the two-parameter Weibull distribution, the former being a special case of the latter. When fit to the more general Weibull distribution, the survivorship distribution is:

$$A(t) = e^{-(tb)^c} \tag{4}$$

where e is the Napierian base, t is time, b is dimensioned in time and is the scale parameter, and c is dimensionless and is known as the shape parameter. The negative exponential is a special case of the Weibull distribution where $c = 1$. Modeled using a Weibull distribution (or negative exponential). The fire interval (probability density) distribution is:

$$f(t) = \frac{c\,t^{c-1}}{b^c} e^{-(tb)^c} \tag{5}$$

and the hazard of burning function is:

$$\lambda(t) = \frac{c\,t^{c-1}}{b^c} \tag{6}$$

Here, we can easily see the special case of the negative exponential where $c = 1$ makes the hazard constant through time ($1/b$).

From these distributions, it is possible to estimate FC (area-based), or the average fire interval (element-based), which is:

$$FC = b\Gamma\left(\frac{1}{c} + 1\right) \tag{7}$$

where Γ is the gamma function. Then again, the equation is considerably simplified when $c = 1$ as FC equals b. When not fixed to 1, the shape parameter c is adjusted in order to allow the hazard of burning to change over time. The hazard increases with time when $c > 1$, and decreases with time when $c < 1$.

In this study, we use the *survreg* function in R package "survival" [37], which provides maximum likelihood estimates of Weibull and negative exponential parameters.

2.1.4. Estimating FC Using Non-Parametric Models

The Cox regression [31] is the most widely used type of survival regression in general, but it is only recently that it has been more broadly applied to forest fire studies [13,14,24,25,38,39]. The Cox regression is a semi-parametric survival model. One interesting feature of the Cox regression that explains its enormous popularity in other fields is that it does not require that we choose some particular probability distribution to represent survival times. In other words, the underlying hazard function is left unspecified at first, except that it cannot be negative, and it can take any form depending on the empirical observations [19]. The hazard function is the non-parametric portion of the model. Although the baseline hazard (null model) is unspecified, the covariates' influence, which is the parametric portion of the model, can still be estimated by the method of partial likelihood developed by Cox [31] and presented in the same paper in which he introduced the Cox regression model. Even though the resulting estimates are not as efficient as maximum-likelihood estimates for a correctly specified parametric hazard regression model, not having to make arbitrary (and possibly incorrect) assumptions about the form of the baseline hazard is a compensating feature of Cox's specification. It is also relatively straightforward to fit a Cox model using the *coxph* function in R package "survival" and extract the baseline hazard function using the *basehaz* function, also part of the R package "survival" [37]. By default, *basehaz* yields Nelson–Aalen cumulative hazard estimates [37]. To estimate FC, we simply identify the time t at which the cumulative hazard reaches its maximum value, which makes the calculation of FC quite simple:

$$FC = \frac{\Lambda(t)}{t} \tag{8}$$

where t is the time at which the cumulative hazard reaches its maximum value and $\Lambda(t)$ is the corresponding cumulative hazard.

2.2. Modeling Approach

To assess and compare the three above-mentioned SA-based methods for estimating FC, we simulated the aging and burning of a virtual landscape. This allowed us to record a "true" fire history and the associated FC. At the end of each simulation, we simulated the censoring process that naturally occurs around the stand break-up of a post-fire cohort, sampled it, and conducted SA to estimate FC in conditions similar to those that prevail in a real field sampling. Then, we analyzed the differences between SA-based estimates of FC and true FC statistics as a function of the length of the simulated FC, the sampling effort and long-term trends in fire activity.

2.2.1. Aging and Burning

In this study, we simulated a forest landscape as a two-dimensional grid of square pixels that aged at a yearly time step. All pixels (stands) were considered equally susceptible to burn as long as they were at least 1 year old, i.e., they could not burn twice during a single time step. Fires were simulated using simple cellular automata that "ignited" randomly within the simulated landscape and that could spread following queen's case (eight adjacent unburned neighbors of each ignited cells) until a predefined fire size was reached. We paid no particular attention to the realism of the fire shape, although we set the probability of fire spreading from an ignited cell to an adjacent unburned one to 0.32 to produce irregular shapes that are typical of forest fires.

Virtual landscapes were initiated by randomly assigning to each pixel an age (time since fire; TSF) drawn from an exponential distribution of mean equal to the length of the simulated FC. Simulations ran for 300 years, and TSF values were reset to zero in pixels affected by fire.

The virtual landscape was 2.1025 Mha in size (145 km $\times$ 145 km) with a 1-km resolution (100-ha pixels). A 10-cells (10 km) band was cropped out at the end of simulations on the periphery to avoid edge effects that make marginal cells less susceptible to burning if ignition is made random.

Consequently, only the 1.5625 Mha central portion of the landscape was considered in downstream sampling and analyses, which roughly corresponds to the size of our case example.

The size of individual fire events was modeled as a serially independent draw from a log-normal probability distribution. The log-normal distribution that we used to approximate burned area size distribution [40] has the following probability density function:

$$f(x; \mu, \sigma) = \frac{1}{x\sigma\sqrt{2\pi}} e^{\left[\frac{-(\ln(x)-\mu)^2}{2\sigma^2}\right]}, x > 0 \tag{9}$$

where x is the fire size, μ is the mean of the natural logarithm of the fire size, and σ is the standard deviation of the natural logarithm of the fire size. The parameters were estimated from the empirical size distribution of all fires ignited by lightning within a 100-km distance from the outermost boundaries of the case study area ($n = 123$) during 1959–1999 obtained from the Large Fire Database [16].

We modeled different FCs to assess their influence on subsequent survival-based estimators. FC can be calculated as:

$$FC = \frac{T \cdot A}{S \cdot N} \tag{10}$$

where A is the size of the study area, T is the length of the simulation (years), S is the mean fire size, and N is the number of fires. Isolating N allowed us to determine the number of fire events that must be drawn from the log-normal fit of the fire size distribution in order to stimulate each FC scenario.

The fire series was then distributed in time following different temporal patterns, random or structured, allowing for multiple fires in the same years, in which case individual fire sizes were summed up to obtain the annual area burned.

More details about the different fire regimes simulated can be found below, in the Experimental Design section, while additional details about the implementation of the experiment in R can be found in the in the online repository [41].

2.2.2. Censoring and Sampling

At the end of each simulation, the virtual landscape was randomly sampled with varying efforts in order to assess the impact of the sampling effort on survival-based estimates of the FC. To model the loss of information that is caused by the aging of the forest, when traces of the initial post-fire cohort gradually disappear, thus making it impossible to date the last fire event from the within-stand age structure, we applied a linear censoring function (Figure 1). This function determines the probability that a sample of particular TSF will be censored. It was decided from the authors' personal experience in reconstructing fire history in eastern North America that a linearly increasing probability of censorship between TSF = 100 and TSF = 300 was a valid approximation of what is generally observed in the boreal forest of eastern North America. Stand break-up usually occurs between 100 and 150 years, depending on the species [34], but a successful determination of TSF is often possible for several decades after the initial post-fire cohort breaks up as veteran stems are targeted in dendroecological sampling. Fires that occurred over 250 to 300 years earlier are almost never dated as this period of time roughly corresponds to the longevity of the longest-living and most common species in the area covered by our case study, i.e., black spruce (*Picea mariana* Mill. BSP).

The implementation of the censoring function is straightforward: for each TSF observation between 100 and 300 years, one value is drawn from a uniform set ranging between the same values, indicating the time at which TSF would become censored. If the true TSF value is lower than the drawn value, it remains unchanged and uncensored. On the contrary, if the drawn value is equal or lower than the true TSF, it replaces it as the minimum measurable TSF (censored observations).

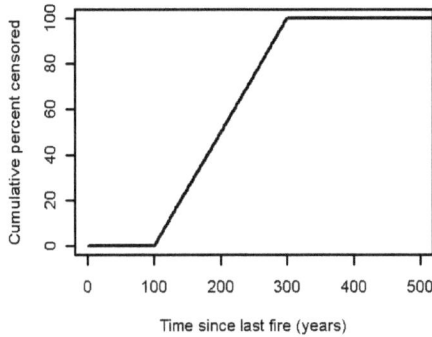

Figure 1. Cumulative probability function of time since fire (TSF) censorship.

2.2.3. Experimental Design

We simulated FCs of four different lengths (62.5, 125, 250 and 500 years) to assess their influence on subsequent survival-based estimators. The range of FCs covered was partly based on the range of conditions that can be found in fire-driven North American boreal forests, but also on the range within which SA can actually be useful considering the inherent limits of dendroecological fire history reconstruction methods in those systems.

Because considerable temporal variations in fire activity have been documented in most regions, we also wanted to assess the impacts of such patterns on the survival-based estimation of FC. Thus, for each FC length, we simulated three different fire regimes that followed distinct temporal trends in fire activity. First, constant fire regimes were simulated for each FC, i.e., the fire series drawn from the log-normal fire size distribution were distributed randomly during the course of the 300-year simulations. We also generated fire series that were correlated with the simulated years, i.e., some simulations were subject to increasing fire activity, while others were subject to decreasing fire activity, although they presented the same fire activity when averaged over the entire simulation. To implement that, we imposed a 0.5 and a -0.5 correlation coefficient between annual area burned and simulated years for regimes with increasing and decreasing fire activity, respectively.

A total of 12 distinct fire regimes were thus simulated (4 distinct FCs $\times$ 3 distinct temporal patterns).

For each of those treatments, we conducted 100 independent simulations, at the end of which we drew 100 independent samples for each predefined sample size (Table 1).

Table 1. Summary of simulation parameters.

Total area/Pixel size of simulated landscape (ha)	1,562,500 ha*/100 ha
Spatial resolution	1 km
Simulation duration	300 years
Fire size distribution	Log-normal (fit from empirical records)
Length of the FC (years, averaged over the entire simulation)	62.5, 125, 250 and 500 years
Temporal trends in fire activity	Increasing, constant or decreasing
Number of independent simulation replications	100
Number of independent samples for each simulation	100
Simulated sampling effort (random points)	25, 50, 75, 94*, 250 and 500

* Similar or equal to the empirical dataset from the case study area.

2.3. Analysis

2.3.1. Assessment of Accuracy and Precision of the Estimated FC

To assess the accuracy of each survival-based method of estimation, we first recorded the exact true FC over the course of the entire 300 years, which varied around the target FC because of the stochasticity of the fire series drawn. Then, we computed the three survival-based estimates (Cox, Weibull and Exponential) with varying sampling efforts. By analyzing how the residuals (estimated minus true value) were distributed as a function of the fire regimes simulated and the sampling effort, we produced a first general assessment of the accuracy and precision of each method and its relative sensitivity to the conditions under which estimates were obtained.

We also wanted to assess whether it was possible to report valid quantification of the uncertainty associated with the estimated FC. Thus, we computed 95% confidence intervals (CI95) by bootstrapping (10,000 resamplings). The CI95 were computed using the *"boot.ci"* function of the R package *"boot"* [42], which implements various types of non-parametric confidence interval. Here, we present four of those: Basic bootstrap intervals (*basic*), first-order normal approximations (*norm*), bootstrap percentile intervals (*perc*), and adjusted bootstrap percentiles (BC_a). More details about each method can be found in the package documentation, and the formulae on which the calculations are based on can be found in Davison and Hinkley [43]. We used those CI95 and verified whether they achieved the advertised nominal level of coverage of 95%, i.e., whether the reported CI95 indeed included the true FC 95% for the time period. Higher coverage (> 95%) would indicate that the reported CI95 are too wide (conservative), and inversely.

2.3.2. Case Study

We also aimed to quantify FC in a coniferous boreal landscape of eastern Canada in light of the information that was sought about the accuracy and precision of the three above-mentioned SA-based methods. The case study area covers 15,961 km^2 (1.5961 Mha) of boreal forest in eastern Quebec, specifically in the Côte-Nord region, between longitudes 67.00° W and 69.00° W and between latitudes 49.00° N and 50.25° N (Figure 2). This region has a cold, maritime climate with an average annual temperature of 1.7 °C and average precipitation of 1001 mm, measured in Baie-Comeau, in the southwest corner of the study area. Precipitation is evenly distributed during the year, and is about 70% rain [44]. The topography is moderately uneven with high hills with rounded summits and many rocky escarpments. The highest hills, located in the northeastern part of the area, are just over 700 m high, while other sparsely distributed hills are above 500 m high. The average elevation ranges within landscape subunits vary between 150 m and 200 m. Three of these landscape subunits, as described by Robitaille and Saucier [45], make up for almost the totality of the case study area. The average slope is 15%. The hydrography is complex with numerous small lakes and rivers of varying sizes, some of them very large. The configuration of the topography produces a drainage system with a mainly north–south orientation [45]. There are rocky outcrops on slightly more than a third of the total land area, while the rest of the land area consists mainly of shallow tills on sloping areas and deep tills at the bottom of slopes. To a lesser extent, there are glaciofluvial deposits on valley floors [45].

Black spruce and balsam fir (*Abies balsamea* (L.) Mill.) are the dominant species, making up a very well connected coniferous matrix, along with white spruce (*Picea glauca* (Moench) Voss) and white birch (*Betula papyrifera* Marsh.). Also found in the region, but more sporadically, are trembling aspen (*Populus tremuloides* Michx.) and jack pine (*Pinus banksiana* Lamb.), mainly in recently burned areas. Tamarack (*Larix laricina* (Du Roi) K. Koch) can also be found along with black spruce in a few rare hydric stations in the region.

Information from various sources has been used to compile the fire history of this area. All fires affecting a surface area of one or more hectares that have occurred since 1941 are listed in Quebec's Ministère des Forêts, de la Faune et des Parcs's archives (MFFPQ-Direction de l'environnement et de la protection des forêts), and aerial photographs dating back to 1931–1932 were interpreted in order to map two older fires (1923 and 1896). In some areas, dendroecological surveys conducted for decadal forest inventories of the MFFPQ were used to assess the amount of time elapsed since the most recent fires. In order to take full advantage of these available data and focus our efforts on ground sampling in areas where the fire history was less known, a preliminary TSF map of the area was created. This rough demarcation included recently mapped fires and sections of the study area covered mainly by even-aged forests of black spruce, which were determined with the help of dendroecological surveys carried out during the MFFPQ's forest inventory campaigns.

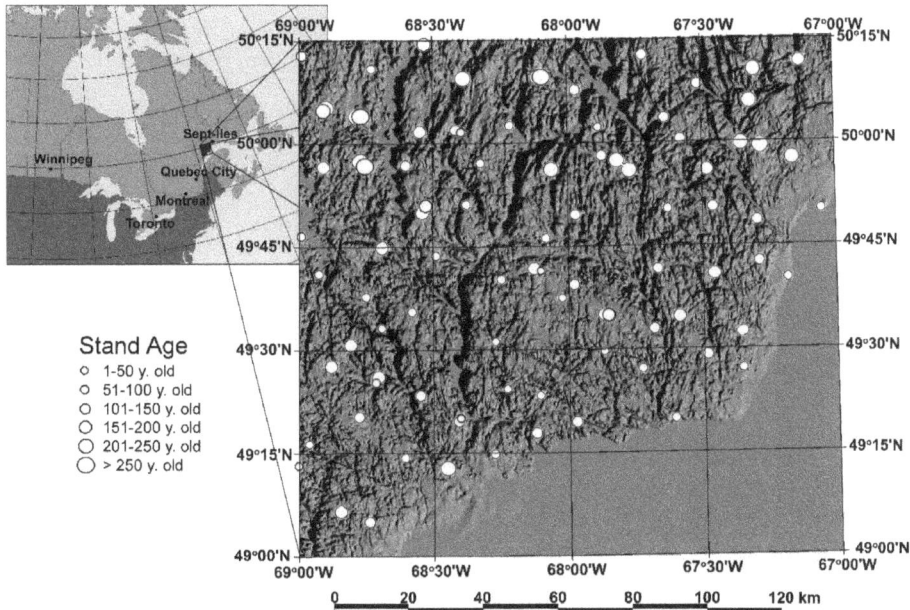

Figure 2. Case study area and empirical dendroecological survey (*n* = 94). Stand age can be associated with either a known TSF (uncensored) or a minimum TSF (censored). Topography is amplified 10 times.

A total of 94 points made up the final sample. The MFFPQ map archives were used to directly assign a fire date to some recent fires, i.e., 12.8% of cases. Dendroecological analyses were used to estimate the time interval elapsed since the most recent fires for the rest of the sample, based on data gathered in the MFFPQ's decadal inventories (37.2%) and during the sample-gathering campaign carried out for this study (50%). The time intervals since the most recent fires, estimated with the help of the dendroecological surveys, were inferred based on the conventional methods of Arno and Sneck [46]. At each site, between 10 and 15 dominant trees were cut at the root collar and dated. Fire dates were deemed sufficiently reliable if they concerned even-aged stands of a pioneer species that commonly establishes itself after fire. A minimum age (censored data) was assigned to uneven-aged stands, i.e., where a 20-year interval included less than 60% of the sampled dominant trees. However, a visual examination of each stand's age structure suggested that this 20-year interval was too restrictive in the case of some of the older stands that seemed in fact even-aged, probably because of an increasing imprecision of dating with stand age, and we thus chose to extend it to 30 years for stands older than 200 years. A minimum age was also assigned to stands consisting primarily of one species that usually does not establish itself after fire, such as balsam fir, independently of their age structure.

In that data set, no points where two distinct fire dates are known were present. Thus, there were no closed fire intervals.

The median fire-free interval was estimated at 191 years in a previous study [47], and a preliminary estimate of FC based on a negative exponential fitting of the survival data was 295 years, with 53.2% of the observed TSF censored.

Fire behavior in the study area is known to be characterized by large, intense, stand-replacing fires. The largest fire that occurred in the area in recorded history was a little over 200,000 ha in size and partly affected the southwestern portion of the case study area [16]. Some exceptional sites show multiple-scarred jack pines that suggest an alternative, more frequent and less severe fire regime. However, this possible alternative fire regime will be ignored in the present study because the scarcity of evidence suggests that it does not play a major role in structuring the landscape in general. However, it might explain in part the distribution of fire-dependent jack pine in that area, which is otherwise characterized by generally long fire-free intervals.

2.3.3. Estimation of FC in the Case Study Area

The empirical survival data obtained from the field survey in the case study area was submitted to the three above-mentioned methods to estimate FC (negative exponential fitting, Weibull fitting, and Cox regression). Associated CI95 were computed by means of bootstrapping consisting of 10,000 resamplings of the original empirical sample using the same methods as those assessed in the simulation experiment presented here, i.e., basic, normal approximations, percentiles, and adjusted percentiles.

3. Results

3.1. Empirical Fire Size Distribution

Fire size distribution is the only part of the simulation experiment that directly depends on empirical data. The mean size of recorded fires ignited by lightning within a 100-km radius of the case study area is 6365 ha. The maximum likelihood estimates of the log-normal distribution that was fit to the empirical size distribution of individual fire sizes are $\mu = 7.406$ (SE = 0.136) and $\sigma = 1.511$ (SE = 0.096) (units: ln (ha); Figure 3). Individual fire sizes were drawn from this log-normal distribution for all simulations.

Fire size cumulative probability distribution

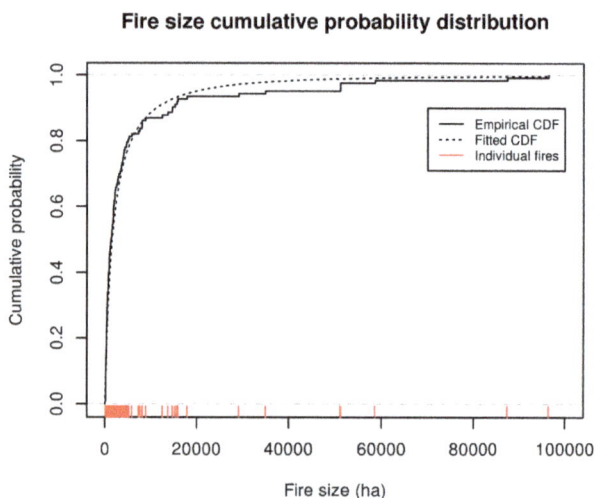

Figure 3. Empirical and fitted cumulative fire size probability distribution ($n = 123$).

3.2. General Description of Simulation Dataset

A total of 1200 simulations made up the dataset on which subsequent processing and analyses were conducted. In these simulations, aging and burning occurred in a relatively simple manner that produced a pixelated mosaic of "stands" to each one of which was attributed a TSF value. Even though we aimed at producing specific fire regimes, the resulting FC varied because of the stochastic approach we adopted for simulating fires (Figure 4). Nevertheless, all treatments were contrasting enough to nicely cover the range of landscape conditions that we wanted to simulate and submit to simulated dendroecological samplings and FC estimation.

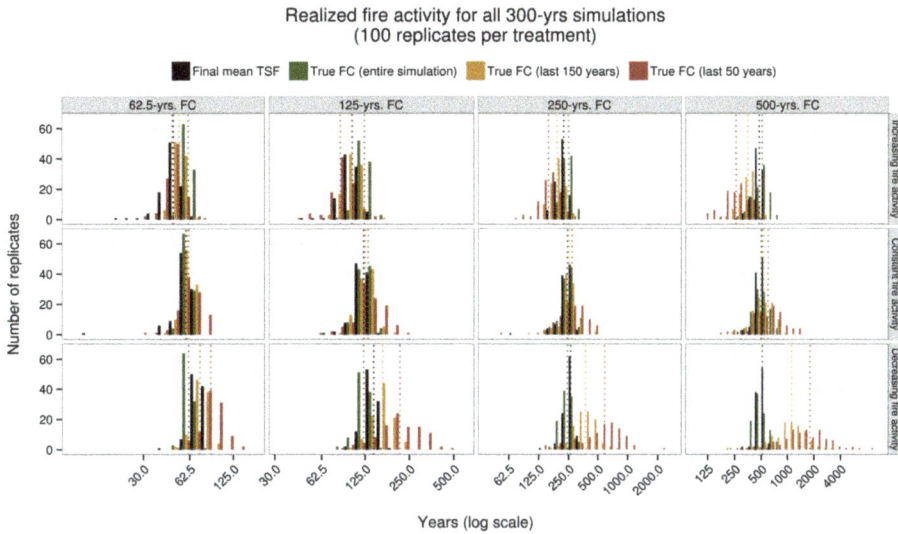

Figure 4. Simulated fire activity for all treatments (length of fire cycle and temporal patterns). Final mean TSF is computed at time $t = 300$. The true FC is computed over the entire course of the simulations, and is also computed for shorter periods of time toward the end of the simulations (last 150 and 50 years) to illustrate temporal patterns. Dotted lines indicate average values.

Generally, mean TSF roughly equals the length of the FC when the fire regime is constant. Alternatively, the shorter the FC is, the more mean TSF is linked to recent fire activity, i.e., increasing or decreasing trends in fire activity will affect the age of the simulated mosaic in a similar fashion. Moreover, the link between recent fire activity and mean TSF will be stronger when the length of the FC is shorter. A given landscape can be made drastically younger by more frequent large fires whatever its age previous to the burn, and while it can only get older one year at a time, one year is proportionally more important in younger landscapes, all of which is reflected in Figure 4.

Once FC statistics had been recorded, we mainly worked with the TSF raster produced at the final time-step of our simulations, i.e., at time $t = 300$ (e.g., Figure 5a). On those rasters, we randomly sampled true TSF values of varying efforts and applied a censoring function to simulate the loss of information available through dendroecological field surveys. Although we only applied censoring to our samples in order to minimize the computing time and data storage requirements, we applied censoring to an entire landscape for the sake of an example (Figure 5b), in which we can visualize how youngest patches can still be easily identified while older patches get increasingly more difficult to decipher.

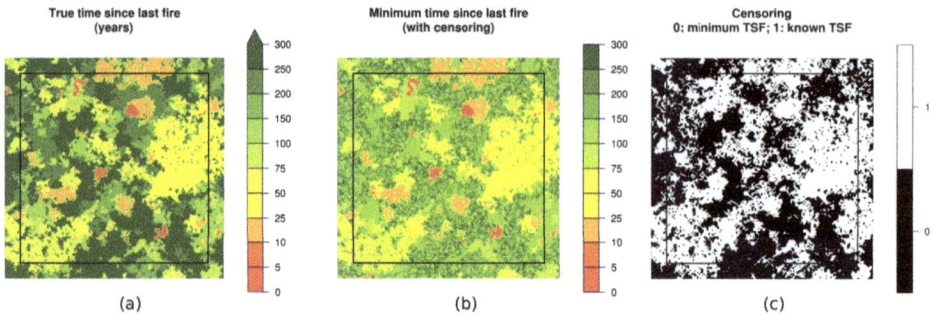

Figure 5. Example of a simulated landscape subject to a constant fire regime of 250 years: (**a**) true TSF at time *t* = 300; (**b**) minimum TSF at time *t* = 300 after the censoring function was applied; and (**c**) location of pixels where only a minimum age can be inferred from simulated dendroecological sampling. For all panels, the inner rectangles indicate the region that was subject to simulated samplings. (Patch contours were accentuated for the sake of illustration and may appear to be of an intermediate age. This does not reflect the true TSF values that were considered in our simulations.)

3.3. Accuracy and Precision of the Three Survival-Based Estimators of FC

For all three methods, FC estimates that were applied to samples on which the censoring function was applied appear relatively unbiased when fire activity is constant as there is no notable departure from 0 for the mean residuals (Figure 6).

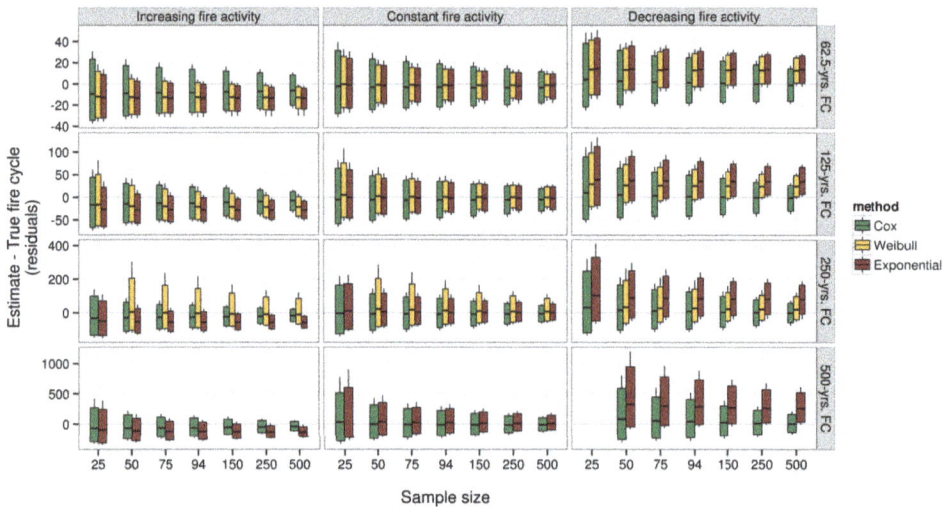

Figure 6. Fire cycle estimation error from simulated forest fire history reconstruction affected by censoring. Boxes contain 95% of the variations, while whiskers extend to cover 99% of the variations. Black lines within boxes indicate mean residuals.

The picture changes considerably when simulated fire regimes present increasing or decreasing fire activity. All methods of estimation may become biased, to different extents, toward the most recent fire activity, i.e., longer FC when fire activity is decreasing, and inversely. However, there are some differences, with exponential- and Weibull-based estimates being the most sensitive by far. While exponential-based estimates seem to simply become more representative of the most recent fire activity instead of that affecting the entire statistical population, Weibull-based estimates react in a more complex manner. In fact, they react similarly to the exponential-based estimates for the shorter FCs. However, when global FC becomes greater than or equal to 250 years, estimates become very unstable, with many values so much higher than the true ones that we could not plot them without making other patters impossible to visualize. Cox regression-based estimates are much less sensitive to temporal variations in fire regimes and seem to describe in the most accurate way the fire regime affecting the entire statistical population under analysis.

Sampling effort mainly affects how variable the estimates are around the true values (Figure 6). The decrease in variability with an increase in sampling effort is fairly gradual, presents no notable thresholds, and appears to follow a similar pattern for all three means of estimation. The narrower distribution of residuals with a greater sampling effort combined with the biases observed for parametric methods, especially the exponential one, resulted in the entire ranges covered by 99% of the estimates often not even including the true FC when fire activity was not constant, a result that is even better illustrated below, in the section describing the performances of the CI95.

3.4. Performance of the Non-Parametric CI95

Although the different types of confidence intervals that we computed differ in terms of performance, i.e., by their coverage being different from the nominal value of 95%, those differences are subordinate to those associated with the fire regime simulated, mostly the temporal patterns, and to the survival model used to estimate the FC (Figure 7).

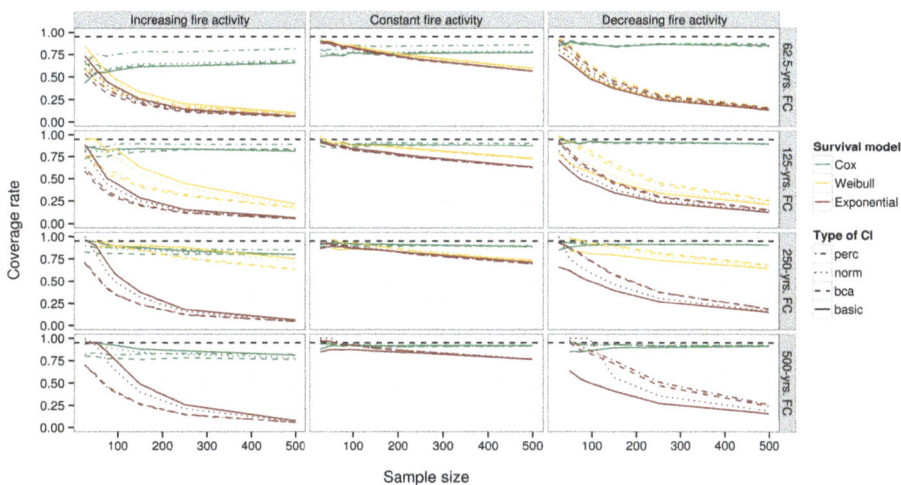

Figure 7. Coverage rate of bootstrap CI95 (10,000 resamplings). Black dashed lines indicate the nominal coverage rate of 95%.

When estimating FC in a context of increasing or decreasing fire activity, all types of confidence intervals perform very badly when based on parametric survival models (exponential or Weibull), mainly because of the considerable biases that are associated with those methods. The CI95 associated with the estimates obtained using Cox regressions actually are the only ones that behave in a consistent manner when increasing the sampling effort or under different temporal patterns of fire regimes.

However, even Cox-based CI95 provide coverage that is narrower than the nominal one in almost every situation, although they approach the nominal value when recent fire activity is not too high, i.e., when it is generally decreasing or when it is constant and FC $\geqslant$ 125 years. In those situations, all types of CI95 are roughly equivalent in terms of coverage performances for Cox-based estimates.

15

3.5. Fire Cycle Estimation in the Case Study Area

The Weibull fit and Cox regression estimates of FC for the case study area are very similar and are considerably shorter than those obtained from the negative exponential fit (Figure 3), although their CI95 overlap. Overall, the CI95 are narrower for Weibull estimates, but all CI95, irrespective of the survival method or type of CI, are considerably wide relative to central estimates.

Among the different types of non-parametric CI95 that were computed, two differ notably. First, the bias-corrected and accelerated (BC_a) method tends to extend the CI95 by 50 years or more in the upper range compared to the others, while the percentile-based method (*perc*) does the same, to a lesser extent, in the lower range (Figure 8). The other two methods, i.e., the *basic* and normal approximation (*norm*) methods, report CI95 that are very similar and intermediate to the two others.

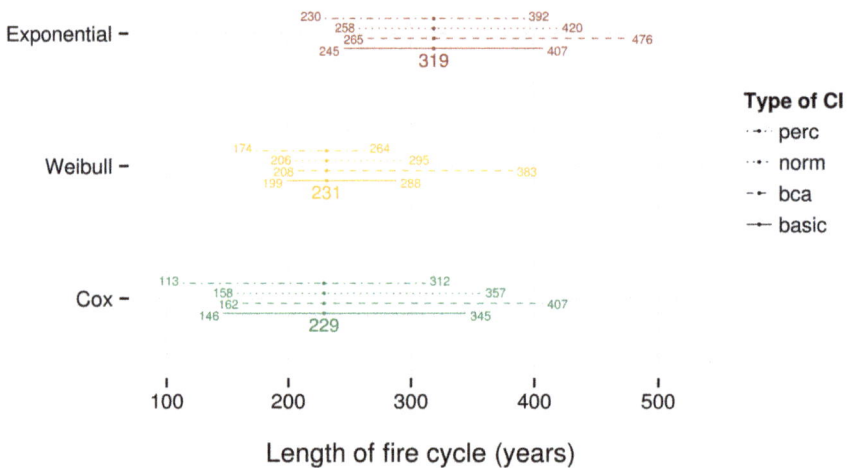

Figure 8. Fire cycle estimates and CI95 for the Côte-Nord case study area, eastern Canada.

4. Discussion

In our simulation experiment, we tried to incorporate the most important sources of uncertainty that affect estimation of FC based on dendroecologically reconstructed fire history. In the context of fire history reconstruction obtained from cross-sectional field surveys, such as those simulated in the present study for our case study, sources of uncertainty in the estimation of FC are multiple and include: (1) the variation inherent to sampling; (2) the spatial stochasticity of the phenomenon, which involves the random burning of a landscape in which stands with various TSF values are affected; and (3) temporal variations in fire activity, which interfere with the stationarity assumption that is needed to substitute observed TSF for complete fire intervals [28]. By integrating those sources of uncertainty and by testing a variety of estimation methods in a range of conditions that are representative of the field conditions for which those reconstructions can be useful, we managed to identify the most accurate and robust method as well as the factors that limit its application and reliability.

We defined the scope of this experiment so that the most commonly raised issues would be addressed, but we acknowledge that other factors may be of interest to consider in some particular contexts. That is why we made the experiment entirely reproducible and tried to make it easily adaptable.

With no surprise, all methods are affected by the quality and quantity of information available as input. However, they are not equal, as Cox regression clearly comes out as the most reliable method for the estimation of FC. It is in fact much more robust in the face of the most common sources of uncertainty and is therefore less likely to be affected by the potentially misleading biases that often affect parametric methods such as Weibull- and exponential-based estimates.

4.1. Assessment of Estimators and Their Reported CI

Our simulations allowed us to single out temporal variations in fire activity as the most important factor interfering with estimation of FC from cross-sectional sampling, which confirms previous criticism made in that regard [25,28]. That is because survival analyses are based on TSF data and not on complete fire-return intervals, which can be treated interchangeably only if the failure process (fire) is assumed to be stationary [28]. For all methods, increases in fire activity are associated with an underestimation of the length of the FC, and vice versa, although this bias is almost negligible for the FC obtained using Cox regressions. Those biases are related to the gradual loss of information that is caused by overburning and/or censoring, which create conditions where the most recent fire activity gets a better representation in the samples that are used for estimating FC. This appears to be particularly problematic when using parametric methods, with the exception of Weibull-based estimates in some very specific situations, an aspect that we will develop below. Substantial increases in fire activity mean that records of lower amounts of area burned in previous periods (older stands) get "erased" from the landscape. In the opposite situation, when fire activity decreases, it is censoring that gradually erases records of the higher proportions of stands that were initiated in earlier decades when fire activity was higher, when the interval between fires allows it.

Compared with parametric methods, which appear to have biases that are comparable in magnitude in situations where fire activity increases or decreases, Cox regression-based estimation does considerably better in situations of decreasing fire activity. This suggests that although the a priori unspecified hazard function of the Cox regression, i.e., its non-parametric component, deals very well with varying "failure" rates, it is still sensitive to trends induced by the most intense, recent burning, which simply "erases" traces of past fire activity on the landscape.

The Weibull-based estimates show some distinct patterns that are worth discussing a little further. In situations of low fire activity, the large fires events that actually occur once in a while strongly affect the estimation of the shape parameter c of the Weibull distribution (see equation 4), which determines how the hazard of burning increases or decreases along the time axis. Weibull fit on survival samples affected by a recent increase in fire activity, or even by a punctual peak in an otherwise relatively constant fire activity, will typically yield a survival model associated with a decreasing hazard of burning along the time axis (in reverse time), a trend that is perpetuated to infinity. This is of course unrealistic as it implies that older forests see their vulnerability to fire decrease in a monotonous manner down to zero at infinity, and it thus generates overestimates of FC. Instability in the estimation of the shape parameter explains why we had to simply discard all Weibull-based estimates for FC $\geqslant 500$ as well as those with smaller sampling sizes ($n \leqslant 25$) when FC $\geqslant 250$. In other set-ups, the actual length of the FC at which this becomes an issue could vary as a function of the size of the area being analyzed compared with the average size of fires, the age at which censoring occurs and starts erasing traces of past fire events, and the sampling effort. The distribution of Weibull-based estimates becomes well-centered around true values at FC = 250 with an increasing fire activity, likely because the shape parameter c fortuitously fits well that specific temporal trend in fire activity.

One interesting feature of Weibull-based survival models in fire ecology is related to their ability to formally test the age-dependency of hazard of burning within a parametric framework [30,48]. This has been a commonly used approach to test the effect of fuel loading as an endogenous source of variation in hazard of burning, especially in fire prone Mediterranean types of ecosystems [27,32,49]. In the context of the boreal forest of eastern Canada, however, typical fire return intervals are usually long enough that exogenous factors affecting the hazard of burning, such as climate, may change concurrently with potential endogenous factors, which considerably complicates the interpretation of the shape parameter that defines how hazard of burning changes along the time axis. Some ecological limits where fuel can become insufficient to support a higher level of fire activity are possible in the boreal forest of eastern Canada, although it has only been observed north of the limit of commercial forestry [50]. However, it appears unrealistic that any relationship between age and fuel loading would be monotonically increasing (or decreasing) on the entire typical lifespan of a boreal stand [51]. In fact, one would rather expect to find thresholds related to the delay in establishment of a fuel load sufficient to carry fire, as well as to succession from broadleaved species to coniferous ones, or the reverse [52,53]. In this context, the potential advantages related to the use of Weibull-based survival models to estimate FC compared with the negative exponential model are likely outweighed by the aforementioned artifacts that were observed in our simulations. We also suggest that it might be more appropriate to incorporate such empirically and independently developed relationships as the one between age and fuel loading [51] within a Cox regression model as time-dependent covariates, which would allow exogenous factors such as climate and land use to be treated separately by the unspecified hazard function. To sum up, for this reason and because the Cox regression does not require that we choose some particular probability distribution to represent survival time, we recommend the use of Cox regression to estimate FC from cross-sectional field surveys such as those modeled in our study. In our opinion, the slight underestimation of FC by Cox regression, which is very small compared with the width of the CI95, is a minor drawback considering its vastly superior robustness in terms of accuracy and CI95 performance.

Regardless of the method used, it is important to highlight the fact that a wide confidence interval must be associated with estimates of FC for the last 300 years. In many cases, the type of bootstrap CI95 does not seem to make any difference in terms of their coverage performance. However, here we suggest the use of the bias-corrected and accelerated bootstrap CI95 (BC_a) for the following reasons: (1) the estimate bootstrap distribution is moderately biased and skewed toward higher values (data not shown), which corresponds exactly to the situation for which BC_a bootstrapping has been developed in the first place [54]; and (2) all other methods, i.e., *basic*, *percentile* and *normal approximation* bootstrap CI, have been shown not to perform well in those types of situation, especially the percentile bootstrap, which, by inverting the lower and higher percentiles in the calculation of the confidence interval [43] may create artifacts such as the considerably lower limit produced from the present case study. For all three of those methods, biased and skewed bootstrap distribution may cause confidence interval coverage to be lower than the nominal value. We have not investigated why some of those latter types of confidence interval seem to perform better when the landscape being analyzed has been subject to an increase in fire activity, but we suspect this may be caused by some fortuitously convenient bias similar to that related to the surprisingly good performance of Weibull under very specific conditions that cannot a priori be assumed with empirical data.

4.2. Estimation of the FC in the Study Area

The three estimates of FC for the last 300 years in the case study area of eastern Canada range between 229 and 319 years, with CI95 that considerably extend the range of possible values. However, given the results of our simulations, the Cox regression-based estimates of FC for the last 300 years is by far the most credible, which implies that the true FC is more likely closer to 229 years than to the 319 years the exponential model yielded. Moreover, a decreasing trend in fire activity has been observed at the scale of the last 300 years in areas of the boreal forest of eastern Canada where fire history has been reconstructed (see, [12]). According to the results of our simulation experiment for a global FC of 250 years, which is the treatment that is the most similar to our case study, this trend over such a time frame is associated with considerable overestimation of the length of the FC by the negative exponential model, thus supporting the use of Cox regression-based estimation.

In the most likely and empirically supported situation where fire activity has been decreasing during the last 300 years, all types of CI95 that we tested seem to perform similarly in terms of coverage. Because of the mathematical construct and properties of BC_a confidence intervals that have been discussed above, we believe that the best available CI95 for the case study area are covered by values ranging from 162 to 407 years. Our simulations suggest that it should probably be even wider since the actual coverage of all the tested bootstrap intervals is in fact slightly less than the nominal value of 95%.

4.3. Implications

The TSF distributions obtained from natural fire regimes that make up boreal landscapes are often suggested as "guidelines" for forest management because they constitute a very integrative characteristic of boreal landscapes that also directly and indirectly determines the relative influence of finer-scale types of disturbances [2]. Reproducing these natural patterns through harvesting is therefore believed to act as a coarse filter for the conservation of biodiversity [2,55] and other known and unknown ecosystem services [56,57]. With such an approach, FC is the key parameter defining the respective proportions of young and old forests, as well as the relative importance of silvicultural practices that better emulate their dynamics and maintain important processes and functions. However, the age structure at the landscape level from which the survival sample is obtained is only one result of the stochastic process [41] and, for this reason, some criticisms arose that argued that using FC as one strict reference may be too restrictive and does not adequately recognize the changing nature of fire-regulated boreal forests [26,58]. Our results show that the estimation of FC itself is subject to considerable imprecision. However, in the boreal forest of eastern North America, the idea that stands with TSF exceeding the length of typical harvesting rotations under low-retention harvesting regimes (60–100 years) made up the vast majority of large productive landscapes remains unchallenged. Even with a FC at the lower limit of the CI95 ($\approx$ 160 years), the expected proportion of stands older than 100 years would be around 54%. Considering the central measure of the estimated FC for the last 300 years (229 years), the expected proportion of old forests would be around 65%. Of course, these proportions vary more at finer spatiotemporal scales, but there are no indications whatsoever from any kind of proxy (e.g., dendroecological or paleoecological) that boreal landscapes made up by a majority of young stands (TSF < 100 years) are the norm rather than the exception in eastern North America, south of what is today considered as the northern limit of commercial forestry [59]. Moreover, our results illustrate that FC estimated from periods of reference longer than what is available in provincial archives and that fully takes advantage of dendroecologically reconstructed fire history, i.e., a minimum of 150 years and, ideally, 300 years, is more closely related to the true mean TSF of the forest, especially in eastern Canada where fire free intervals are typically long. The use of mean TSF to set targets for age structure and composition has indeed previously been suggested [12,60], as it better summarizes the overall age-class distribution at the landscape level than FC estimated from short reference periods and because it takes the inertia of large landscapes into better consideration.

The uncertainty associated with FC estimates, however, may impose more restrictive limits to other applications of the knowledge of the length of the FC that require a more precise quantification. For instance, it may considerably complicate the detection of changes in fire regime in the past and near future because of the great variability inherent to the process [61], which is an important issue on many levels in the context of climate change. Average annual area burned has shown long-term trends in the past in relation to climate change and is expected to increase in most of the boreal forest, but these changes will very likely need to be quite dramatic to be statistically discernible. Consequently, not only is the baseline contribution of fire disturbance to the global carbon cycle difficult to assess, but recent and future changes in fire regime that might need to be documented are also considerably uncertain [62]. This also applies to the effect of fire on annual allowable cut predictions in strategic forest management planning [6]. This is why we stress the importance of integrating uncertainty into the management decision-making process by using a wide array of scenarios.

There are other sources of uncertainty or potential issues related to the use of survival-based estimates of FC. First, we did not assess how temporal partitioning of dendroecologically reconstructed fire history [12,35] could temper the bias associated with the parametric estimation of FC. Nevertheless, such temporal partitioning must generally be made a priori and consequently generates a certain level of subjectivity. This increases the risk of highly "volatile" period-specific estimates becoming mostly linked to a few spikes in fire activity, although it is generally agreed upon that even relatively constant fire regimes are characterized by considerable temporal variability.

In this study, we basically use null models to obtain estimates of FC for entire landscapes. It is often relevant from an ecological perspective as well as from a risk management one, i.e., for timber supply, infrastructures and human lives, to test for the influence of covariates that may influence fire hazard in forested landscapes. Several studies suggest that the power of survival models is somewhat limited when applied to dendroecological reconstructions of fire history obtained from cross-sectional surveys [13,63]. Based on the present simulation experiment, it would be possible to actually quantify their power and experiment procedures that aim to improve it.

5. Conclusions

The simulation approach used in this study is a transparent way to incorporate multiple types of uncertainty and to provide an integrative assessment of the accuracy and precision of survival model-based estimates of FC. We have shown that even with an increased sampling effort, large confidence intervals are obtained.

Our study confirms some of the statistical issues that were put forward with the use of survival analyses for estimating FC from a cross-sectional sample of the landscape, mainly the influence of temporal variations in fire activity [25,28]. However, the length of the FC remains important information for many practical management issues, and sources of information are limited, especially where typical fire return intervals are long. Using a modeling approach allowed us to better quantify the impact of current methodological limitations and showed that SA conducted on cross-sectional sampling of large boreal landscapes can provide valid estimates of FC even though they are associated with large confidence intervals.

Although all methods remain subject to irreducible levels of uncertainty, Cox regressions comes out as a clear winner as it is by far the least sensitive to the most common sources of error, to the point where a reassessment of many existing estimations of FC obtained using parametric methods could be justified.

Supplementary Materials: All code, input data, and future development directly based on the present study can be found in DC's personal Github repository: https://github.com/dcyr/survFire.

Acknowledgments: We thank Jeanne Poirier for helpful comments and Isabelle Lamarre for linguistic revision.

Author Contributions: D.C., S.G. and Y.Be. provided the original ideas. D.C., S.G., Y.Be. and Y.Bo. wrote the paper. D.C. designed and performed the simulation experiment and analyzed the results. All authors participated in the interpretation of the results.

Conflicts of Interest: The authors declare no conflict of interest.

Abbreviations

The following abbreviations are used in this manuscript:

CI95	95% confidence intervals
FC	Fire cycle
SA	Survival analyses
TSF	Time since fire

References

1. Perera, A.H.; Buse, L.J.; Weber, M.G. *Emulating Natural Forest Landscape Disturbances*; Columbia University Press: New York, NY, USA, 2004.
2. Gauthier, S.; Vaillancourt, M.-A.; Leduc, A.; De Grandpré, L.; Kneeshaw, D.D.; Morin, H.; Drapeau, P.; Bergeron, Y. *Ecosystem Management in the Boreal Forest*; Les Presses de l'Université du Québec: Québec, QC, Canada, 2009.

3. Hirsch, K.; Kafka, V.; Tymstra, C.; McAlpine, R.; Hawkes, B.; Stegehuis, H.; Quintilio, S.; Gauthier, S.; Peck, K. Fire-smart forest management: A pragmatic approach to sustainable forest management in fire-dominated ecosystems. *For. Chron.* **2001**, *77*, 357–363. [CrossRef]

4. Armstrong, G. Sustainability of timber supply considering the risk of wildfire. *For. Sci.* **2004**, *50*, 626–639.

5. Bureau du forestier en chef. *Manuel de Détermination des Possibilités Forestière 2013–2018*; Gouvernement du Québec: Roberval, QC, Canada, 2013.

6. Leduc, A.; Bernier, P.Y.; Mansuy, N.; Raulier, F.; Gauthier, S.; Bergeron, Y. Using salvage logging and tolerance to risk to reduce the impact of forest fires on timber supply calculations. *Can. J. For. Res.* **2014**, *45*, 480–486. [CrossRef]

7. Amiro, B.D.; Cantin, A.; Flannigan, M.D.; de Groot, W.J. Future emissions from Canadian boreal forest fires. *Can. J. For. Res.* **2009**, *39*, 383–395. [CrossRef]

8. Johnson, E.A.; Gutsell, S.L. Fire frequency models, methods and interpretations. *Adv. Ecol. Res.* **1994**, *25*, 239–287.

9. Bergeron, Y.; Gauthier, S.; Kafka, V.; Lefort, P.; Lesieur, D. Natural fire frequency for the eastern Canadian boreal forest: Consequences for sustainable forestry. *Can. J. For. Res.* **2001**, *31*, 384–391. [CrossRef]

10. Lesieur, D.; Gauthier, S.; Bergeron, Y. Fire frequency and vegetation dynamics for the south-central boreal forest of Quebec, Canada. *Can. J. For. Res.* **2002**, *32*, 1996–2009. [CrossRef]

11. Drever, C.R.; Messier, C.; Bergeron, Y.; Doyon, F. Fire and canopy species composition in the Great Lakes-St. Lawrence forest of Témiscamingue, Québec. *For. Ecol. Manag.* **2006**, *231*, 27–37. [CrossRef]

12. Bergeron, Y.; Cyr, D.; Drever, C.R.; Flannigan, M.; Gauthier, S.; Kneeshaw, D.; Lauzon, È.; Leduc, A.; Le Goff, H.; Lesieur, D.; et al. Past, current, and future fire frequencies in Quebec's commercial forests: Implications for the cumulative effects of harvesting and fire on age-class structure and natural disturbance-based management. *Can. J. For. Res.* **2006**, *36*, 2737–2744. [CrossRef]

13. Cyr, D.; Gauthier, S.; Bergeron, Y. Scale-dependent determinants of heterogeneity in fire frequency in a coniferous boreal forest of eastern Canada. *Landsc. Ecol.* **2007**, *22*, 1325–1339. [CrossRef]

14. Bélisle, A.C.; Gauthier, S.; Cyr, D.; Bergeron, Y.; Morin, H. Fire regime and old-growth boreal forests in central Quebec, Canada: An ecosystem management perspective. *Silva. Fenn.* **2011**, *45*, 889–908.

15. Bergeron, Y.; Fenton, N.J. Boreal forests of eastern Canada revisited: Old growth, nonfire disturbances, forest succession, and biodiversity. *Botany* **2012**, *90*, 509–523. [CrossRef]

16. Stocks, B.J.; Mason, J.A.; Todd, J.B.; Bosch, E.M.; Wotton, B.M.; Amiro, B.D.; Flannigan, M.D.; Hirsch, K.G.; Logan, K.A.; Martell, D.L.; et al. Large forest fires in Canada, 1959–1997. *J. Geophys. Res.* **2002**, *107*, 5-1–5-12. [CrossRef]

17. Hosmer, D.W., Jr.; Lemeshow, S. *Applied Survival Analysis-Regression Modeling of Time to Event Data*; John Wiley & Sons, Inc.: New York, NY, USA, 1999.

18. Lawless, J.F. Observation schemes, censoring, and likelihood. In *Statistical Models and Methods for Lifetime Data*, 2nd ed.; John Wiley & Sons, Inc.: New York, NY, USA, 2002; pp. 49–78.

19. Allison, P.D. *Survival Analysis Using SAS: A Practical Guide*; SAS Institute Inc.: Cary, NC, USA, 2005.

20. Moritz, M.A.; Moody, T.J.; Miles, L.J.; Smith, M.M.; de Valpine, P. The fire frequency analysis branch of the pyrostatistics tree: Sampling decisions and censoring in fire interval data. *Environ. Ecol. Stat.* **2009**, *16*, 271–289. [CrossRef]

21. Reed, W.J. Estimating historical forest-fire frequencies from time-since-last-fire-sample data. *Math. Med. Biol.* **1997**, *14*, 71–83. [CrossRef]

22. Bergeron, Y.; Flannigan, M.; Gauthier, S.; Leduc, A.; Lefort, P. Past, current and future fire frequency in the Canadian boreal forest: Implications for sustainable forest management. *Ambio* **2004**, *33*, 356–360. [CrossRef] [PubMed]

23. Lauzon, È.; Kneeshaw, D.; Bergeron, Y. Reconstruction of fire history (1680–2003) in Gaspesian mixedwood boreal forests of eastern Canada. *For. Ecol. Manag.* **2007**, *244*, 41–49. [CrossRef]

24. Senici, D.; Chen, H.Y.H.; Bergeron, Y.; Cyr, D. Spatiotemporal variations of fire frequency in central boreal forest. *Ecosystems* **2010**, *13*, 1227–1238. [CrossRef]

25. Clark, J.S. Ecological disturbance as a renewal process: Theory and application to fire history. *Oikos* **1989**, *56*, 17–30. [CrossRef]

26. Boychuk, D.; Perera, A.H. Modeling temporal variability of boreal landscape age-classes under different fire disturbance regimes and spatial scales. *Can. J. For. Res.* **1997**, *27*, 1083–1094. [CrossRef]

27. Polakow, D.A.; Dunne, T.T. Modelling fire-return interval T: Stochasticity and censoring in the two-parameter Weibull model. *Ecol. Model.* **1999**, *121*, 79–102. [CrossRef]
28. Polakow, D.A.; Dunne, T.T. Numerical recipes for disaster: Changing hazard and the stand-origin-map. *For. Ecol. Manag.* **2001**, *147*, 183–196. [CrossRef]
29. Van Wagner, C.E. Age-class distribution and the forest fire cycle. *Can. J. Restor.* **1978**, *8*, 220–227. [CrossRef]
30. Johnson, E.A. Fire recurrence in the subarctic and its implications for vegetation composition. *Can. J. Bot.* **1979**, *57*, 1374–1379. [CrossRef]
31. Cox, D.R. Regression models and life-tables. *J. R. Stat. Soc. Ser. B* **1972**, *34*, 187–220.
32. Moritz, M.A. Spatiotemporal analysis of controls on shrubland fire regimes: Age dependency and fire hazard. *Ecology* **2003**, *84*, 351–361. [CrossRef]
33. Bergeron, Y. Species and stand dynamics in the mixed woods of Quebec's southern boreal forest. *Ecology* **2000**, *81*, 1500–1516. [CrossRef]
34. Gauthier, S.; Boucher, D.; Morissette, J.; De Grandpré, L. Fifty-seven years of composition change in the eastern boreal forest of Canada. *J. Veg. Sci.* **2010**, *21*, 772–785. [CrossRef]
35. Reed, W.J.; Larsen, C.P.S.; Johnson, E.A.; MacDonald, G.M. Estimation of temporal variations in historical fire frequency from time-since-fire map data. *For. Sci.* **1998**, *44*, 465–475.
36. Reed, W.J.; Johnson, E.A. Reply-reverse cumulative standing age distributions in fire-frequency analysis. *Can. J. For. Res.* **1999**, *29*, 1812–1815. [CrossRef]
37. Therneau, T.M. Package 'survival'. Available online: https://cran.r-project.org/web/packages/survival/survival.pdf (accessed on 21 April 2016).
38. Bouchard, M.; Pothier, D.; Gauthier, S. Fire return intervals and tree species succession in the North Shore region of eastern Quebec. *Can. J. For. Res.* **2008**, *38*, 1621–1633. [CrossRef]
39. Portier, J.; Gauthier, S.; Leduc, A.; Arseneault, D.; Bergeron, Y. Fire regime variability along a latitudinal gradient of continuous to discontinuous coniferous boreal forests in Eastern Canada. *Ann. Stat.* **1981**, *9*, 93–108.
40. Armstrong, G.W. A stochastic characterisation of the natural disturbance regime of the boreal mixedwood forest with implications for sustainable forest management. *Can. J. For. Res.* **1999**, *29*, 424–433. [CrossRef]
41. Cyr, D. Accuracy and precision of three survival-based methods for estimating fire cycle from dendroecological data. Available online: https://github.com/dcyr/survFire (accessed on 21 April 2016).
42. Canty, A.; Ripley, B. boot: Boostrap functions. Available online: https://cran.r-project.org/web/packages/boot/index.html (accessed on 21 April 2016).
43. Davison, A.C.; Hinkley, D.V. *Bootstrap Methods and Their Application*; Cambridge University Press: Cambridge, UK, 1997.
44. Environment Canada. Climate normal and averages, 1981–2010. Available online: http://climate.weather.gc.ca/climate_normals/ (accessed on 3 January 2016).
45. Robitaille, A.; Saucier, J. *Paysages rÉgionaux du Québec Méridional*; Les Publications du Québec: Sainte-Foy, QC, Canada, 1998.
46. Arno, S.F.; Sneck, K.M. *A Method for Determining Fire History in Coniferous Forests of the Mountain West*; USDA Forest Service: Ogden, UT, USA, 1977.
47. Cyr, D.; Gauthier, S.; Bergeron, Y. The influence of landscape-level heterogeneity in fire frequency on canopy composition in the boreal forest of eastern Canada. *J. Veg. Sci.* **2012**, *23*, 140–150. [CrossRef]
48. Johnson, E.A.; Van Wagner, C.E. The theory and use of two fire history models. *Can. J. For. Res.* **1985**, *15*, 214–220. [CrossRef]
49. Moritz, M.A.; Keeley, J.E.; Johnson, E.A.; Schaffner, A.A. Testing a basic assumption of shrubland fire management: How important is fuel age? *Front. Ecol. Environ.* **2004**, *2*, 67–72. [CrossRef]
50. Héon, J.; Arseneault, D.; Parisien, M.-A. Resistance of the boreal forest to high burn rates. *Proc. Natl. Acad. Sci. USA* **2014**, *111*, 13888–13893.
51. Schoenberg, F.F.P.; Peng, R.; Huang, Z.; Rundel, P. Detection of non-linearities in the dependence of burn area on fuel age and climatic variables. *Int. J. Wildl. Fire* **2003**, *12*, 1–6. [CrossRef]
52. Terrier, A.; Girardin, M.P.; Périé, C.; Legendre, P.; Bergeron, Y. Potential changes in forest composition could reduce impacts of climate change on boreal wildfires. *Ecol. Appl.* **2013**, *23*, 21–35. [CrossRef] [PubMed]

53. Girardin, M.P.; Ali, A.A.; Carcaillet, C.; Blarquez, O.; Hély, C.; Terrier, A.; Genries, A.; Bergeron, Y. Vegetation limits the impact of a warm climate on boreal wildfires. *New Phytol.* **2013**, *199*, 1001–1011. [CrossRef] [PubMed]

54. Efron, B. Better bootstrap confidence intervals. *J. Am. Stat. Assoc.* **1987**, *82*, 171–185. [CrossRef]

55. Hunter, M.L. Natural fire regimes as spatial models for managing boreal forests. *Biol. Conserv.* **1993**, *65*, 115–120. [CrossRef]

56. Attiwill, P.M. The disturbance of forest ecosystems: the ecological basis for conservative management. *For. Ecol. Manag.* **1994**, *63*, 247–300. [CrossRef]

57. Christensen, N.L.; Bartuska, A.M.; Brown, J.H.; Carpenter, S.; D'Antonio, C.; Francis, R.; Franklin, J.F.; MacMahon, J.A.; Noss, R.F.; Parsons, D.J.; et al. The report of the Ecological Society of America committee on the scientific basis for ecosystem management. *Ecol. Appl.* **1996**, *6*, 665–691. [CrossRef]

58. Armstrong, G.; Adamowicz, W.; Beck, J. Coarse filter ecosystem management in a nonequilibrating forest. *For. Sci.* **2003**, *49*, 209–223.

59. Jobidon, R.; Bergeron, Y.; Robitaille, A.; Raulier, F.; Gauthier, S.; Imbeau, L.; Saucier, J.-P.; Boudreault, C. A biophysical approach to delineate a northern limit to commercial forestry: The case of Quebec's boreal forest. *Can. J. For. Res.* **2015**, *45*, 515–528. [CrossRef]

60. Gauthier, S.; Lefort, P.; Bergeron, Y.; Drapeau, P. *Time since Fire Map, Age-Class Distribution and Forest Dynamics in the Lake Abitibi Model Forest*; Natural Resources Canada: Québec, QC, Canada, 2002.

61. Metsaranta, J.M. Potentially limited detectability of short-term changes in boreal fire regimes: A simulation study. *Int. J. Wildl. Fire* **2010**, *19*, 1140–1146. [CrossRef]

62. Kurz, W.A.; Stinson, G.; Rampley, G.J.; Dymond, C.C.; Neilson, E.T. Risk of natural disturbances makes future contribution of Canada's forests to the global carbon cycle highly uncertain. *Proc. Natl. Acad. Sci. USA* **2008**, *105*, 1551–1555. [CrossRef] [PubMed]

63. Cyr, D.; Bergeron, Y.; Gauthier, S.; Larouche, A.C. Are the old-growth forests of the Clay Belt part of a fire-regulated mosaic? *Can. J. For. Res.* **2005**, *73*, 65–73. [CrossRef]

© 2016 by the authors. Licensee MDPI, Basel, Switzerland. This article is an open access article distributed under the terms and conditions of the Creative Commons Attribution (CC BY) license (http://creativecommons.org/licenses/by/4.0/).

![forests logo] *forests*

MDPI

Article

Fire Regime along Latitudinal Gradients of Continuous to Discontinuous Coniferous Boreal Forests in Eastern Canada

Jeanne Portier [1],*, Sylvie Gauthier [2], Alain Leduc [3], Dominique Arseneault [4] and Yves Bergeron [5]

[1] Département des Sciences Biologiques, Université du Québec à Montréal and Centre for Forest Research, Case postale 8888, Succursale Centre-ville Montréal, QC H3C 3P8, Canada

[2] Natural Resources Canada, Canadian Forest Service, Laurentian Forestry Centre, 1055 du PEPS, P.O. Box 10380, Stn. Sainte-Foy, Québec, QC G1V 4C7, Canada; sylvie.gauthier2@canada.ca

[3] Département des Sciences Biologiques, Chaire Industrielle CRSNG UQAT-UQAM en Aménagement Forestier Durable, Université du Québec à Montréal and Centre for Forest Research, Case postale 8888, Succursale Centre-ville Montréal, QC H3C 3P8, Canada; leduc.alain@uqam.ca

[4] Département de Biologie, Chimie et Géographie, Centre for Northern Studies, Université du Québec à Rimouski, 300, Allée des Ursulines, Rimouski, QC G5L 3A1, Canada; dominique_arseneault@uqar.ca

[5] Forest Research Institute, Université du Québec en Abitibi-Témiscamingue and Université du Québec à Montréal, 445, boul. de l'Université, Rouyn-Noranda, QC J9X 5E4, Canada; Yves.Bergeron@uqat.ca

* Correspondence: portier.jeanne@courrier.uqam.ca; Tel.: +1-514-987-3000 (ext. 7608)

Academic Editor: Timothy A. Martin
Received: 6 June 2016; Accepted: 8 September 2016; Published: 24 September 2016

Abstract: Fire is the main disturbance in North American coniferous boreal forests. In Northern Quebec, Canada, where forest management is not allowed, the landscape is gradually constituted of more opened lichen woodlands. Those forests are discontinuous and show a low regeneration potential resulting from the cumulative effects of harsh climatic conditions and very short fire intervals. In a climate change context, and because the forest industry is interested in opening new territories to forest management in the north, it is crucial to better understand how and why fire risk varies from the north to the south at the transition between the discontinuous and continuous boreal forest. We used time-since-fire (TSF) data from fire archives as well as a broad field campaign in Quebec's coniferous boreal forests along four north-south transects in order to reconstruct the fire history of the past 150 to 300 years. We performed survival analyses in each transect in order to (1) determine if climate influences the fire risk along the latitudinal gradient; (2) fractionate the transects into different fire risk zones; and (3) quantify the fire cycle—defined as the time required to burn an area equivalent to the size of the study area—of each zone and compare its estimated value with current fire activity. Results suggest that drought conditions are moderately to highly responsible for the increasing fire risk from south to north in the three westernmost transects. No climate influence was observed in the last one, possibly because of its complex physical environment. Fire cycles are shortening from south to north, and from east to west. Limits between high and low fire risk zones are consistent with the limit between discontinuous and continuous forests, established based on recent fire activity. Compared to the last 40 years, fire cycles of the last 150–300 years are shorter. Our results suggest that as drought episodes are expected to become more frequent in the future, fire activity might increase significantly, possibly leading to greater openings within forests. However, if fire activity increases and yet remains within the range of variability of the last 150–300 years, the limit between open and closed forests should stay relatively stable.

Keywords: fire history reconstruction; fire cycle; fire risk; black spruce–moss forests; lichen woodlands; boreal ecosystems; fire weather; survival analyses

1. Introduction

By controlling structural and compositional attributes, fire is the main disturbance shaping the North American boreal forest [1,2]. Fires affect the forest's structure by creating a mosaic of stands of different ages and sizes [3,4], thus constantly rejuvenating stands and landscapes. Fire cycles, defined as the time required to burn an area equivalent to that of the study area [5,6], determine the age structure of forest stands [7,8] across the landscape. Fires also influence stands' composition by controlling succession patterns, for instance, by favoring fire-adapted species such as jack pine (*Pinus banksiana*) or black spruce (*Picea mariana*) [9–11]. Fire regimes are highly variable in space as a result of various environmental factors acting on different scales [12–14]. Climate acts as a top-down factor from regional to continental scales. In Canada, for instance, the increasing gradient of fire activity observed from east to west is caused by the spatial variability in the frequency of drought events [15,16]. However, topography [14,17], surficial deposits and drainage [18], or fuel type and availability [19] are bottom-up factors which act from stand to regional scales. Fire regimes also vary in time; for example, the end of the Little Ice Age that occurred around 1850 represents a well-known transition to lower fire cycles in eastern Canada [8,20]. Temporal variations in fire activity are mainly driven by climatic factors such as shifting air masses responsible for dry conditions [20–22].

The coniferous boreal forest of Quebec, eastern Canada, experiences a gradient of dense, continuous forests to the south that transition to discontinuous, less productive forests [23,24], and finally to the forest tundra in the north [2]. The northern open forests are mainly constituted of lichen woodlands resulting from numerous factors such as limited post-fire regeneration due to low seed production [25], unfavorable climate, short intervals between successive fires [26–29], and high severity of large fires [30].

Transition ecosystems are known to be extremely vulnerable to climate change [31], and particularly so for the boreal forest where fire activity is expected to increase [32]. Because the opening of these forests is closely related to fire activity, studying their fire regime is crucial. In Quebec, there is evidence that current fire activity is higher in northern discontinuous forests than in the commercial boreal forest further south [24]. However, it is not clear whether climate is responsible for this latitudinal gradient, or if the underlying climate factors are constant.

Moreover, from a forest management perspective, it is important to understand how fire regimes vary depending on the latitude. In Quebec, northern discontinuous forests are protected from commercial forest harvesting by the legal limit of the commercial forest. It is thought that forest management could worsen the problem of regeneration failures at high northern latitudes under climatic influence. However, the spatial and temporal variability of fire regimes along the latitudinal gradient is still poorly known. The zonation of fire activity is also of interest, as zones with high annual area burned can jeopardize forests' post-fire recovery [29], although recent studies have shown that the boreal forest could express a certain resistance toward high burn rates [33]. Learning more about the spatial consistency of high fire risk zones is particularly important in the context of the northern limit of the commercial forest because their expansion to the south could lead to a reduction in the area available for forest management.

The objective of this research is to assess the latitudinal variability of fire regime at the transition between continuous and discontinuous coniferous boreal forests in Quebec over the last 150–300 years, and its relation to climatic conditions. Even if a zonation of fire activity had been developed in previous studies based on current fire regimes [24], the spatiotemporal consistency of the fire zones over a longer temporal scale has not been explored. The first step of this study was to reconstruct the fire history along four north-south roads almost equally distributed over the black spruce forest of Quebec, using fire archives and dendroecological surveys. We used survival models to assess whether climate was influencing fire risk—defined here as the relative hazard of burning compared with the road average—along each road. The latitudinal distribution of fire risk was then used to delimit homogeneous fire risk zones for each road. Then, the fire cycle of each fire risk zone was calculated allowing for an assessment of fire risk variability along the longitudinal gradient. Finally, fire cycles were compared to previous estimates based on the recent fire history in order to highlight the temporal variability of the fire regime in the study area.

2. Materials and Methods

2.1. Study Area

The study area is located in the boreal zone of Quebec and lies between latitudes 49.5° N and 53° N. The region is mainly coniferous and dominated by black spruce. It covers a gradient from closed, dense forests in the spruce-moss domain to the south, to more open and fragmented forests in the spruce-lichen domain to the north (Figure 1b). The limit of the commercial forest crosses the study area separating managed forests in the south from unmanaged ones in the north.

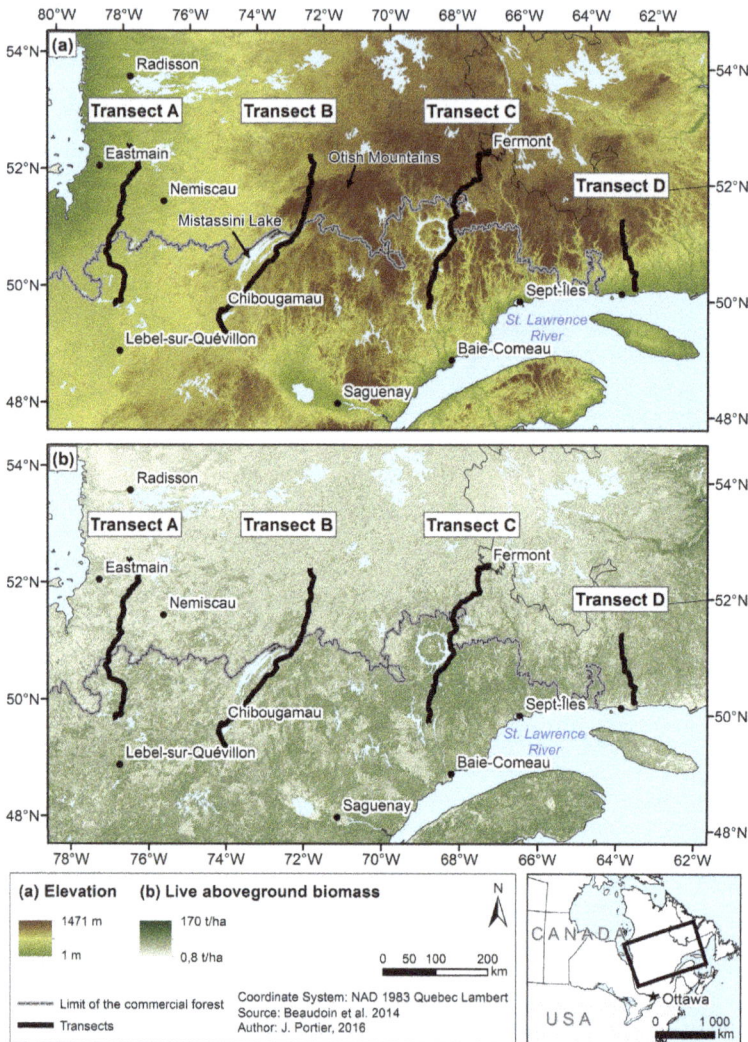

Figure 1. Maps of the study area showing the four transects being analyzed, along with (**a**) the elevation profile and (**b**) the live aboveground biomass in tons per hectare from Beaudoin et al. [34] of the study area. The northern limit of the commercial forest in Quebec is also shown.

In order to cover the latitudinal gradient and because access is difficult in the north of the study area, four north-south roads that are almost evenly distributed from west to east were chosen as a means in which to reach forest stands and served as a basis for our sampling design. Those roads were divided into consecutive 2500 ha-cells (5 km long by 5 km wide) and will hereafter be referred to as transects (Figures 1 and 2).

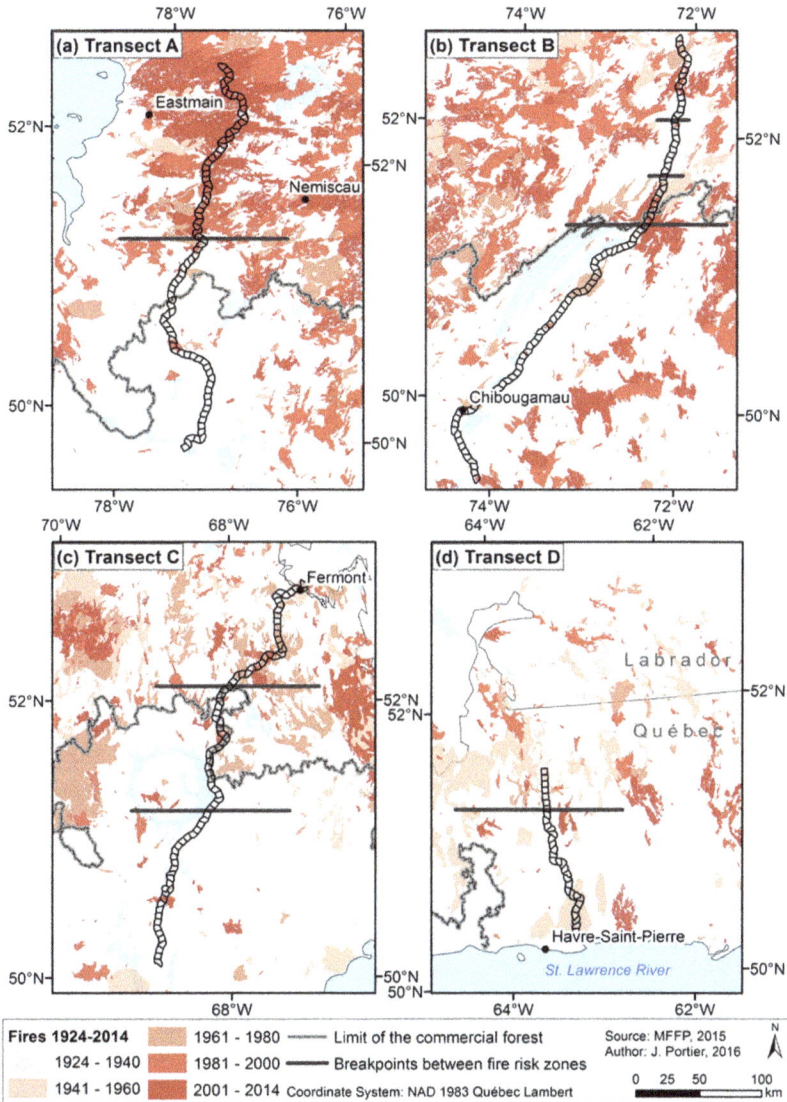

Figure 2. Detailed maps of the four transects showing the corresponding 2500 ha-cells. Fire archives (from 1924 to 2014) and the northern limit of the commercial forest in Quebec are also shown. Each cell that is not overlapped by a recent fire has been subject to a dendroecological survey at a randomly located point in order to determine its time-since-fire (TSF). Latitudinal breakpoints between the fire risk zones are also presented; the two smaller lines in panel (**b**) delimit the mountainous area excluded for the calculation of fire cycles.

The transects are located in the James Bay region (A), the Chibougamau area (B), the North Shore (C), and along the Romaine River (D). They are 81, 94, 83, and 34-cell-long, respectively. Transect D comprises less cells because of a shorter road. However, 28 cells were added to the original dataset using plots previously sampled for the Northern Forest Inventory Program of the Ministère des Forêts, de la Faune et des Parcs (MFFP) (NFIP; 2005 to 2009). These additional cells are located at a maximum distance of 45 km on either side of the transect.

2.2. Time-Since-Fire Data

As the study area is remote and not easily accessible, we developed a data collection strategy which attempts to maximize the use of fire archive data available for the area between 1924 and 2014 and to complement it with field sampling.

2.2.1. Fire Archive Data (1924–2014)

First, fire archives for the 1924–2014 period obtained from the MFFP were used to reconstruct the most recent fire history (Figure A1). However, this database may gradually become less complete or lose dating precision with time, particularly towards the northern areas, as it becomes more likely that some fires were either overlapped by most recent fires, or not detected, reported, and archived. The database precision is excellent after 1972 [24], very good after 1940, but some small fires can be missing or not perfectly delimited between 1924 and 1940. In some cases—mainly in the north and for the oldest fires recorded—the fire date is noted to the nearest five- to ten year interval. For those fires, the middle year of the range was used in the analyses.

Using ArcGIS 10.0, fires at least partially overlapping a cell were identified for each transect. The corresponding time-since-fire (TSF) was then assigned to the cells regardless of the relative importance of the area burned in the cell. When more than one fire partly overlapped a cell, only the most recent TSF was kept. This last situation concerned mostly areas where fires are recurrent (Figure 2).

2.2.2. Field Sampling Design

The field campaign took place in 2013 for transects A, B, and C, and in 2014 for transect D. In the north of transect A, from the top of the transect to Broadback River, the TSF data from Héon et al. and Erni et al. [33,35] collected along a 200 km long transect was used. The data was adapted to our study design and sampling effort by rescaling their original 2 km by 1 km sampling cells.

In each cell where no TSF was assigned from the fire archives, a point corresponding to a sampling plot was randomly generated from 100 m to 750 m on either side of the road. We assumed that those random points are representative of their corresponding cells in terms of TSF because in our study area, the mean fire size—calculated over the 1924–2014 period—is 5200 ha while our cells are 2500 ha. For this reason, a random point is very likely to capture the most representative fire that occurred in a cell.

In our effort to compare our results with those of Gauthier et al. [24] who studied the fire regime between 1972 and 2009, all cells that burned between 1972 and 2014 needed to be assigned with a pre-1972 TSF. Therefore, an additional point was sampled outside the polygon of the post-1972 fire in those cells if no pre-1972 fire date was recorded in the fire archives.

For each plot, 10-dominant trees were sampled (section or core taken as close to the ground as possible), with priority given first to jack pine (*Pinus blanksiana*), and then to black spruce (*Picea mariana*), paper birch (*Betula papyfera*), trembling aspen (*Populus tremuloides*) and, lastly, to balsam fir (*Abies balsamea*). This order of priority was determined based on the rapidity of these species to regenerate after a fire in order to better approximate the real TSF. The trees' age was determined by counting their annual growth rings. In a given cell, if all trees were the same age plus or minus 20 years, the TSF of the cell was considered equal to the age of the oldest tree. Otherwise, the age of the oldest tree was considered a minimum TSF, and therefore right-censored. Because post-fire trees are eventually replaced by new ones as succession proceeds, assessing the exact TSF becomes more difficult as a stand ages.

2.2.3. Relative Importance of Recent Fires

Road layout depends on different physical attributes of the landscape as they are usually built on specific surficial deposits. They are also more used by humans than the rest of the landscape. Both these particularities of our transect roads, which have all been built after 1924, can bias the TSF distribution toward more recent fires, thereby affecting estimations of burn rate. Moreover, in some areas, fires can be rare and yet very large. When roads cross such large fires, it can lead to an overestimation of the number of burned cells compared with the average burn rate of the corresponding region. For all those reasons, we computed for each transect the proportion of recent fires (after 1924) in the cells as well as in a 45 km-wide buffer around the transect. Because recent fires were slightly overestimated in all transects comparatively to the 45 km-wide buffers, the cells that burned after 1924 were down-weighted in the analyses. Weights were calculated in each transect in order to match the relative frequency of recent fires with the relative area recently burned in the surrounding landscape. Weights of cells burned between 1924 and 2014 of transects A, B, C, and D are 0.57, 0.41, 0.36, and 0.46, respectively; while all other cells were assigned a weight of 1.

2.3. Climate Data

The Fire Weather Index (FWI) System consists of six indices derived from meteorological observations—namely temperature, relative humidity, wind speed, and 24-h rainfall—which provide numeric ratings of relative potential for wildland fire. The Fine Fuel Moisture Code (FFMC), the Duff Moisture Code (DMC), and the Drought Code (DC) constitute the fuel moisture codes, and the Initial Spread Index (ISI), the Buildup Index (BUI), and the Fire Weather Index (FWI) constitute the fire behaviour indices. We extracted the value of these indices for each cell within each transect using the BioSIM 9 software [36]. BioSIM allows the user to compensate for the scarcity of weather stations in a study area by interpolating climate data from nearby weather stations, with adjustments for elevation, latitude, and longitude [36]. We extracted the mean value of each index in the FWI system over the period of 1971 to 2000 at the cell's centroids. Means of each index in each cell were calculated for spring months (April to June), summer months (July to September), and fire season months (May to August).

2.4. Statistical Analyses

Survival analyses are often used in fire studies because of their ability to examine the time required for an event to occur, which in our case refers to the TSF, and its relationship with one or more covariates. They produce a survival distribution function corresponding to the probability of having gone without fire at each time *t*, from which a fire cycle can be calculated. Not only are survival analyses adapted to time to event data, but they also allow for censored observations in the modelling process. This is a major advantage compared to regular fire cycle analyses, which strictly assume that TSF is given by the stand age, with no distinction in stands that have been attributed a minimum age. This often leads to an underestimation of the fire cycle that can be attenuated with survival analyses [37].

Among the different types of survival analyses available, we selected a semi-parametric model known as the Cox proportional hazard regression [38]. Although this model is one of the most commonly used for survival analyses in other fields—mainly in medicine—still very few fire studies have explored its potential (e.g., [39,40]).

The Cox proportional hazard model is made of two distinct parts. The first part corresponds to the baseline hazard function (cumulative hazard function when all covariates values are set to zero), i.e., the non-parametric portion of the model, which is initially left unspecified. This has great advantage over parametric models (such as the Weibull distribution, which has been widely used in fire cycle studies) because it avoids making arbitrary and possibly incorrect assumptions about the form of the baseline hazard function. Instead, it is derived from the empirical TSF distribution. In terms of fire history, it means that since the model does not assume a constant risk of burning through time, it allows for variations in the fire regime that could have happened in the past, resulting, for example, from human activities or climate change. Those variations are therefore taken into account while calculating the fire cycle, giving a more precise fire history estimate [37]. The second part of the Cox model is parametric and is estimated using the method of partial likelihood [38]. It is used to evaluate the relationship between the tested covariates and survival. Our survival models were built using the *coxph* function of the *survival* R package [41].

Cox proportional hazard models were used all along the analysis process. They were built either with covariates in order to test the effect of climate on fire risk and delimitate fire risk zones along the transects, or as null models in order to calculate the fire cycle of each fire risk zone previously determined. Analyses were performed for each transect separately, as four independent entities, each representative of their surrounding region. Indeed, the four transects are under very different climatic regimes, and merging them into the same analysis process would make impossible the estimation of the climate effect at the scale of one particular transect. Moreover, we wanted to identify variables affecting the fire risk independently for each transect. Although analyses are realized per transect and allow for the latitudinal assessment of the fire risk variability, calculating fire cycles provides a means of assessing the longitudinal variability by comparing fire activity among transects.

2.4.1. Climate Influence on Fire Risk

For each transect, survival models were built in order to examine the influence of the different FWI indices—hereafter referred to as climate variables—on TSF. A supervised forward model selection was conducted in order to select the climate variables that best explained the fire risk. This multi-step process was conducted using the Akaike Information Criterion (AIC). Figure 3 summarizes the different steps of the model selection process. First, univariate models were built in order to test for each climate variable individually. For each model, the AIC and Δ_{AIC} (i.e., the difference from the model having the lowest AIC) were calculated. Models with a Δ_{AIC} higher than 6 from the best univariate model were discarded [42]. The second step consisted in adding a second variable to each model selected. Only variables that were not collinear with the first one (threshold: correlation coefficient of Pearson < 0.7) were tested as second variables. A second variable was kept only if the model with two variables showed a lower AIC value by at least 2 than the AIC of the corresponding univariate model, in which case the univariate model was discarded. The same process of adding variables was repeated until the model could not be improved by any additional variable. The AIC of all selected models were compared and those having a Δ_{AIC} value higher than 2 from the best model were discarded. Among all models having a Δ_{AIC} lower than 2, only the most parsimonious ones were retained, and the one with the lowest AIC value was kept as the final model. The AIC of this model was finally compared with the AIC of a null model to ensure the overall improvement. Bootstrap was then applied by randomly sampling with replacement (1000 iterations) the original dataset containing TSF and climate variables to extract a 95% confidence interval on the variables' estimates using the lower and upper percentiles.

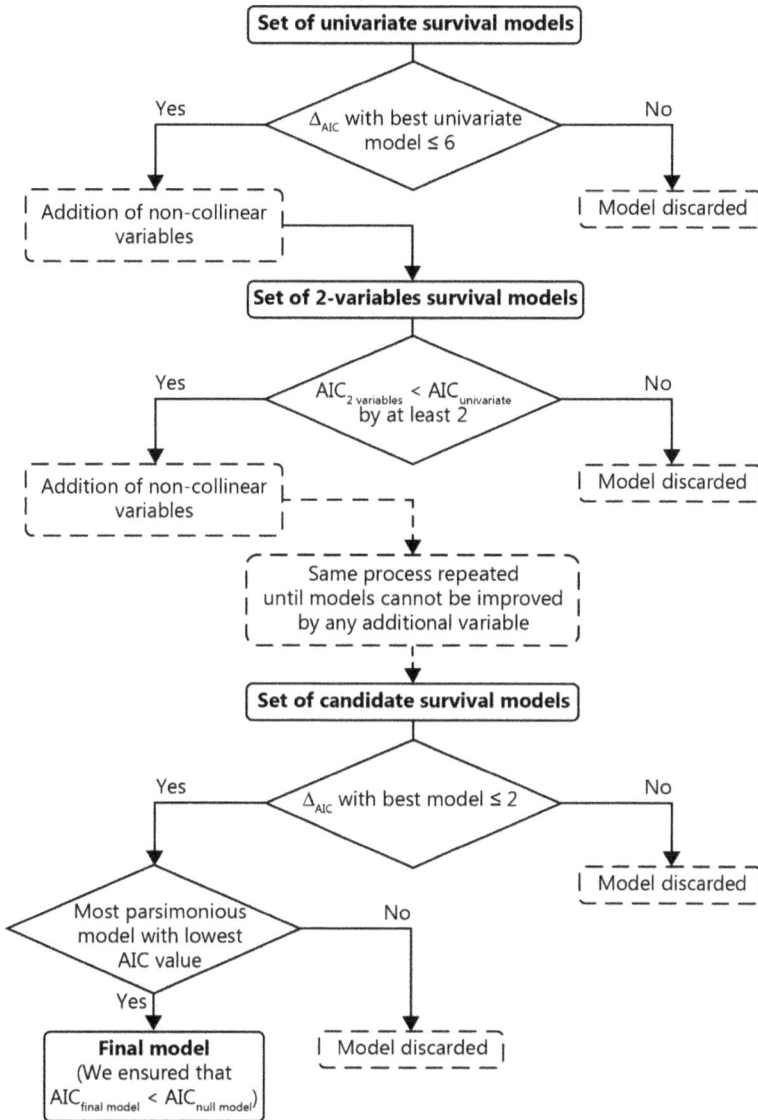

Figure 3. Diagram summarizing the different steps of the model selection process using Akaike Information Criterion (AIC). This procedure is applied to each transect individually. The set of univariate survival models is built using each climate variable separately.

2.4.2. Relative Fire Risk and Latitudinal Risk Zonation

For each transect, the predicted fire risk of each cell was extracted from the final best-fitted model containing the selected climate variables (Figure 3). A 95% confidence interval on the fire risk was calculated using the same bootstrapping process detailed in the previous section. Because the Cox model is a relative risk model, the predicted risk is relative to the sample used in the model, so it can only be interpreted within a transect. The mean risk of a transect is set to one, and is associated with the mean value of the variables used in the model. The value of the risk can then take any positive value and show how many times the risk equals the mean risk of the transect. We chose to graphically represent the results using the log-transformed values of the predicted risk. This scale indeed implies the same range of risk values on both sides of the mean risk value, which on this scale equals zero. For each transect, the log-transformed predicted fire risk variations along the latitudinal gradient allowed us to identify fire risk zones where the fire risk was diverging significantly from the mean risk of the transect. Each transect was thereby separated into different zones, where each was attributed either a low, moderate, or high fire risk relative to the mean risk of the entire transect.

2.4.3. Fire Cycle

Calculating the fire cycle of each transect zone allows for the comparison of fire activity between transects, as we are no longer dealing with relative estimates within transects. Fire cycles can therefore be used to assess the fire activity variability along both latitudinal and longitudinal gradients. Moreover, in order to compare our results with those of Gauthier et al. [24] who regionalized the entire coniferous boreal vegetation domain based on fire cycles over the period 1972–2009, fire cycles were calculated for two different periods (i.e., previous to 2014 and to 1972). Calculating fire cycles with and without the 1973–2014 years also allows for the ability to highlight the impact of recent years (post-1972) on past fire regimes, and therefore to assess the temporal variability of fire activity over these two periods.

In order to calculate the observed fire cycle of each transect zone, a stratified null Cox model was built for each transect. No variables were added to the models in order to capture the observed fire cycle per zone, as opposed to a predicted fire cycle, based on the prevailing climate conditions. A special *strata* term specifying which cell belongs to which transect zone was added to the models in order to take into account how transects were split into different zones. The estimated cumulative hazard of burning (baseline hazard function) could then be extracted for each transect zone [41], representing the accumulated hazard of burning through time. The time it takes for the cumulative hazard to reach 1 is equivalent to the fire cycle [37,39]. To estimate the fire cycle, the time at which the cumulative hazard reached or exceeded 1 was then divided by its associated cumulative hazard. In case the cumulative hazard never reached 1, the fire cycle was calculated as the time at which the cumulative hazard reached its maximum value, divided by this maximum cumulative hazard value. A 95% bootstrap confidence interval on the fire cycle was calculated using 1000 randomizations with replacement of the original TSF dataset. The confidence interval was computed using the lower and upper percentiles.

3. Results

The frequency distributions of TSF (Figure 4) show that whereas most recent fires are dated to the year, they are mostly dated with a minimum TSF beyond 90–100 years. Transects A, B, C, and D show 26%, 32%, 43%, and 44% of minimum TSF data, respectively, thus underlining the importance of considering censored data in survival analyses. Transects C and D are located in the North Shore region of Quebec where the proportion of balsam fir, a fire-sensitive species, is much greater than in the other transects, suggesting that these stands did not establish themselves immediately after a fire event. In these old stands, it is usually difficult to date the TSF precisely, which explains the higher percentage of censored data in these two transects. The minimum dates we recorded for transects A, B, C, and D are 1719, 1703, 1731, and 1663, respectively.

Peaks of TSF can sometimes correspond to single fires. For example, the most recent peak in transect A results at 80% from a very large 2013 fire, although immense fires are common in this region [33,35]. In transect D, a large fire occurred in the 1940s in its southernmost part that covered close to 26% of the entire transect. Unlike in transect A, this fire appears as an exceptional event when compared to the surrounding landscape. It is not only the largest, it also covers more than 31% of the area burned since 1924 within a 200 km-wide area centered on the transect and delimited in the north by the breakpoint in the latitudinal zonation section (see below), and in the south, by the bottom of the transect. Moreover, the fire is about 16 times larger than the mean size of all fires that have occurred in this area since 1924. For this reason, the cells associated to this particular fire were either removed or re-associated with a previous TSF (obtained from field data or fire archives). We will refer to this configuration as transect D2. This approach also allows for a clearer demonstration of how this fire is influencing our results, as it could either lead to an overestimation of the fire risk or to a misinterpretation of the climate's influence on the fire risk.

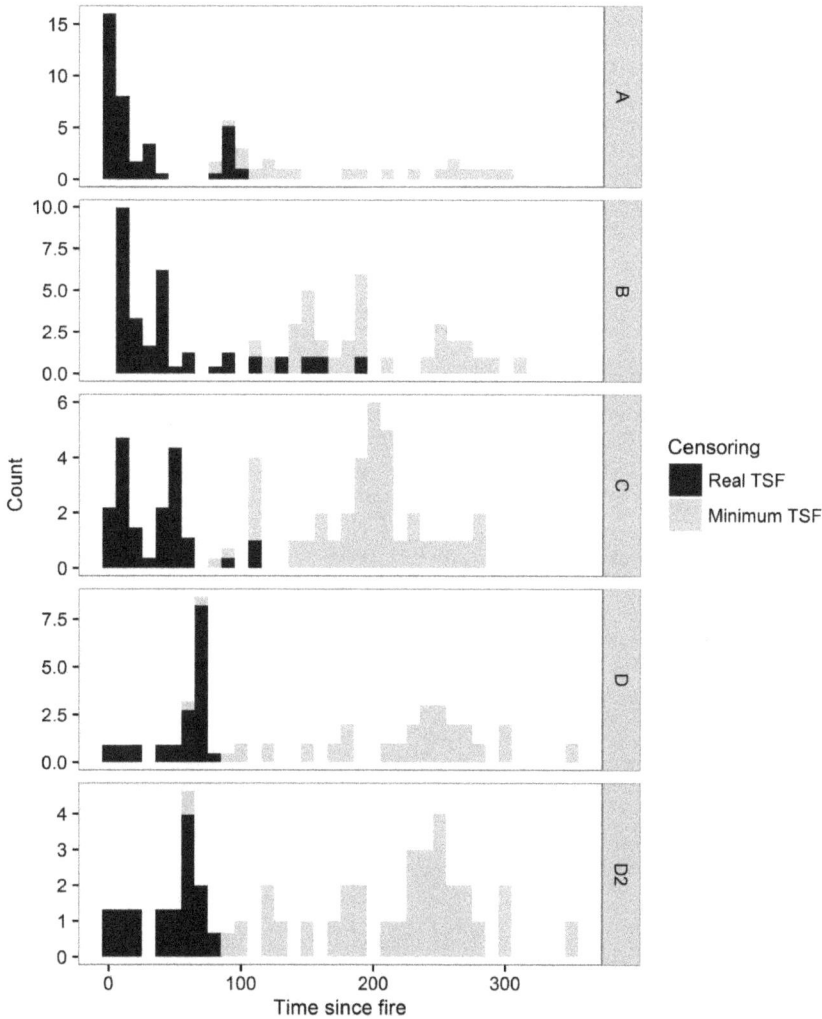

Figure 4. Decadal weighted frequency distribution of TSF for each transect. Weights are calculated in the same way that they were in the survival analyses in order to compensate for the over-representation of recent fires (after 1924) in transects. The non-weighted frequency distribution of TSF is presented in Figure A2. The proportion of right-censored data (minimum TSF) and real TSF are shown in grey and black, respectively.

3.1. Climate Influence on Fire Risk

In all transects, the selected models have a lower AIC than the null models with Δ_{AIC} values higher than 7 (Table 1), meaning that null models can be discounted [43]. Moreover, in all transects except B, the Δ_{AIC} values with null models are higher than 10, a threshold indicating with high certainty that the selected models are highly plausible [44]. Pseudo-R^2 are all above 0.35, except for transect B.

Table 1. Best models according to the supervised forward model selection procedure for each transect. AIC of the best and null models are given as well as their difference (Δ_{AIC}). Cox and Snell's pseudo-R^2 of best models, their associated maximum value, and the corresponding pseudo-R^2 value for max pseudo-R^2 = 1 are also shown.

Transect	Best Model	AIC	AIC Null Model	Δ_{AIC}	Pseudo-R^2 (Max Pseudo-R^2)	Pseudo-R^2 for Max Pseudo-R^2 = 1
A	~DC fire season	206.11	262.03	55.92	0.51 (0.96)	0.53
B	~DC max spring + DMC fire season	213.11	220.49	7.38	0.16 (0.91)	0.18
C	~DC spring + DC fire season	110.34	133.34	23.00	0.28 (0.80)	0.35
D	~FWI fire season + FFMC fire season	90.78	111.73	20.95	0.39 (0.82)	0.48
D2	~FFMC fire season	55.46	96.12	40.66	0.43 (0.81)	0.53

DC, Drought Code; DMC, Duff Moisture Code; FWI, Fire Weather Index; FFMC, Fine Fuel Moisture Code.

All variables in AIC selected models have significant effects on the fire risk (Table 2). In transect A, the Drought Code (DC) during the fire season increases the fire risk. In transects B and C, models with two variables were selected. The first variables (lowest *p*-value) with the most important positive effect on the fire risk, are maximum DC and DC during spring, respectively. The second variables selected show a slight negative effect on the fire risk for both transects, suggesting an adjustment to the positive effects of the first variables. In transects D and D2, the main climatic factor selected is the Fine Fuel Moisture Code (FFMC) during fire season. In contrast to the other transects, FFMC decreases the fire risk even though it is an indicator of sustained flaming ignition and fire spread [45,46].

Table 2. Coefficients of models presented in Table 1. The 95% confidence interval (CI95) for each coefficient was obtained after 1000 randomizations with replacement of the original dataset. *p*-values and exponentiated coefficients with their CI are also shown.

Transect	Variables	Coefficient (CI95)	exp (Coefficient) exp (CI95))	*p*-value
A	DC fire season	0.16 (0.13; 0.19)	1.17 (1.14; 1.21)	$3.61e^{-11}$
B	DC max spring	0.30 (0.21; 0.40)	1.35 (1.23; 1.49)	$9.32e^{-5}$
	DMC fire season	−0.61 (−0.89; −0.35)	0.54 (0.41; 0.70)	$5.67e^{-4}$
C	DC spring	0.75 (0.55; 1.00)	2.12 (1.73; 2.72)	$8.59e^{-6}$
	DC fire season	−0.23 (−0.38; −0.08)	0.79 (0.68; 0.92)	$3.08e^{-2}$
D	FFMC fire season	−3.97 (−5.42; −2.88)	0.02 (0.00; 0.06)	$1.14e^{-5}$
	FWI fire season	4.11 (1.66; 6.80)	60.95 (5.26; 897.85)	$1.25e^{-2}$
D2	FFMC fire season	−2.73 (−3.62; −2.14)	0.07 (0.03; 0.12)	$2.10e^{-6}$

In the Cox proportional hazard model, the relevant estimates are the exponentiated coefficients, which represent the multiplicative effect on the risk of burning. Thus, if we take the example of transect C (Table 2), when holding the DC fire season constant, an increase of 1 in the DC spring value increases the risk of burning by an average factor of 2.12. Likewise, an increase of 1 in the DC fire season value decreases the risk of burning by a factor of 0.79 on average.

3.2. *Relative Fire Risk and Latitudinal Risk Zonation*

For each transect, we defined homogeneous fire risk zones based on whether or not the predicted risk diverged from the mean fire risk of the transect (Figures 2 and 5). Because the predicted fire risks extracted from the models are relative to each transect, relative risk values cannot be compared from one transect to another.

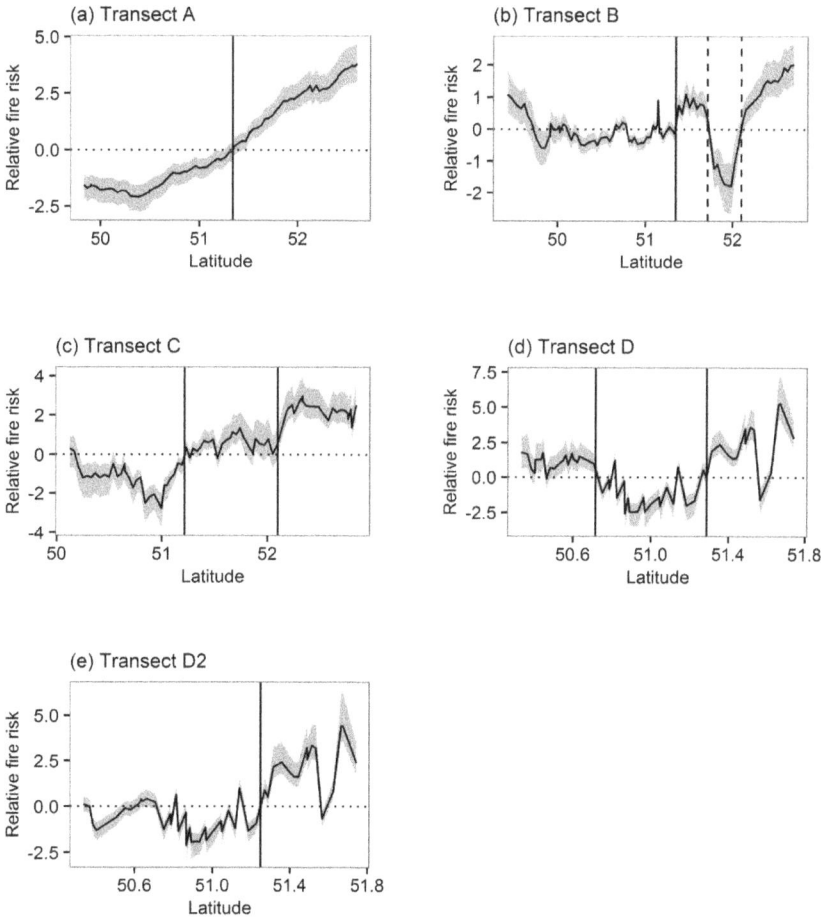

Figure 5. Log-Transformed predicted fire risk according to latitude for each transect. Zero represents the mean fire risk of each transect. Vertical lines separate fire risk zones where for each transect the relative fire risk differs from the mean fire risk of the transect. 95% bootstrap confidence intervals (CI95) are represented by shaded areas. CI95 were obtained after 1000 randomizations with replacement of the original dataset. The two vertical dashed lines in panel (**b**) delimit the non-representative mountainous area excluded for the calculation of fire cycles.

In transect A, the fire risk gradually increases from south to north, thus allowing two zones (north and south) to emerge. On average, the northern zone shows a fire risk 15.61 times higher than the mean risk of the transect, while the southern zone shows a risk 3.33 times lower. 95% confidence intervals (CI95) for the average relative fire risk of each zone can be found in Table A1.

In transect B, the predicted fire risk increases from moderate (not significantly different from the mean fire risk of the transect) in the south to high in the north, except within the zone between latitudes 51.707° N and 52.105° N (coinciding with the plateau located between the Otish Mountains to the east and the Tichegami Mountains to the west, Figure 1a), where the risk abruptly drops (dashed lines in Figure 5b). The fire regimes of high elevation areas are often idiosyncratic because hilltops and upperslopes can be subject to lower fire frequency [40] due to shorter fire seasons and lower temperatures. Because those mountains are not representative of the regions they cross, this section has been removed from the rest of the analyses. Transect B was divided into a southern zone with an average fire risk similar to the mean risk of the transect (1.05 times higher), and a northern zone with a fire risk 3.70 times higher than the mean risk of the transect (excluding the plateau near the Otish Mountains).

In transect C, the predicted fire risk increases stepwise from south to north. This transect was therefore split into a high risk zone in the north, with a risk 10.48 times higher than the mean risk of the transect, a moderate risk zone in the center, with a risk 1.92 times higher than the mean risk, and a low risk zone in the south, showing a risk 2.63 times lower than the mean risk.

In transect D, the fire risk increases from south to north, although the southernmost portion was highly influenced by the 1940s fire (Figure 5d,e). The transect was thus split into three zones when this fire is included in the analysis, with the southernmost zone—almost exclusively inside the 1940s fire area—showing a moderate to high fire risk 3.75 times higher than the mean risk of the transect. The rest of the transect includes a high fire risk zone in the north and a low fire risk zone in the center, showing fire risks 48.14 times higher and 1.75 times lower than the mean risk of the transect, respectively. When removing the 1940s fire from the analysis, transect D2 could be split into two zones (north and south) with risks 21.30 times higher and 1.45 times lower than the mean risk of the transect, respectively. The breakpoint separating the northern and southern zones in D2 was located 4.4 km south of the northern breakpoint in D. This slight difference results from the increased mean fire risk of the transect due to the 1940s fire in transect D.

3.3. Fire Cycles

In all transects, fire cycles (Table 3) calculated over the whole period are seen to lengthen from north to south. Globally, there is also a lengthening from west to east. Fire cycles were calculated over two different periods, prior to 1972 and prior to 2014, in order to check for recent changes in fire regimes. Except for zones A north and B north, the fire cycles calculated for the period before 1972 are shorter in all zones than those calculated over the whole period (prior to 2014). The temporal variability in fire activity in each transect zone (Figure A3) also shows that over the last 300 years most fire activity recorded in each transect zone occurred before 1972, except for A north and B north.

Table 3. Fire cycle (FC) of each transect zone calculated from the cumulative baseline survival function of stratified null models. For each transect zone, we assume that the FC calculated is representative of a period starting with the 10th percentile of the TSF data (i.e., from that date on, 90% of the cells in that particular transect zone were burned). The 95% bootstrap confidence interval (CI95) was obtained after 1000 randomizations with replacement of the original dataset and computation of the FC for each transect using the upper and lower percentiles. FC are calculated for the entire period (prior to 2014) and for the period prior to 1972. FC calculated over the 1972–2009 period by Gauthier et al. [24] in the corresponding regions are also given.

Transect	Zone	Starting Date of Period Covered	FC (CI95) (<2014)	FC (CI95) (<1972)	FC (CI95) (1972–2009) [24]
A	North	1994	5 * (2; 9)	44 (23; 57)	94 (85; 105)
	South	1756	168 (104; 263)	125 (71; 219)	712 (636; 816)
B	North	1836	33 (11; 71)	38 (9; 102)	
	Plateau	1739	408 (101; 1544)	145 (37; 264)	183 (155; 221)
	South	1822	233 (144; 354)	154 (109; 218)	

Table 3. *Cont.*

Transect	Zone	Starting Date of Period Covered	FC (CI95) (<2014)	FC (CI95) (<1972)	FC (CI95) (1972–2009) [24]
	North	1957	37 (22; 50)	8 (4; 10)	183 (155; 221)
C	Center	1800	183 (29; 396)	143 (26; 361)	712 (636; 816)
	South **	1763	720 (326; 1515)	361 (71; 1014)	712 (636; 816)/ 1668 (1286; 2380)
	North	1924	60 (35; 112)	20 (11; 32)	272 (239; 312)
D	Center	1725	785 (290; 1970)	170 (72; 404)	712 (636; 816)
	1940s fire	1929	57 (26; 98)	18 (10; 29)	1668 (1286; 2380)
	North	1927	53 (32; 83)	19 (10; 29)	272 (239; 312)
D2	South **	1737	732 (201; 1747)	177 (65; 543)	712 (636; 816)/ 1668 (1286; 2380)

* 89% of the cells in this zone have a TSF of <13 years; the remaining 11% have a TSF of <40. Moreover, an important fire that occurred in 2013 burned 51% of the cells, also highly influencing the FC estimate. As large fires are common in this region, the size of the transect zone is small when taking the full variability of the fire regime of that region into account when using TSF data. We thus consider the five-years FC observed here to be highly underestimated. ** As both C South and D2 South cross two fire regions delimited by Gauthier et al. [24], two corresponding FC values are shown for each of these zones, respectively.

4. Discussion

4.1. Climate Influence on Fire Risk

The Drought Code (DC) significantly increased the fire risk in the three westernmost transects (i.e., A, B, and C) while no climate influence was detected in transect D. This result is consistent with other studies that have shown a similar effect of the DC on fire risk [16,21,47] and on the number of fires and annual area burned [48] in the Canadian boreal forest. The DC is a numeric rating of the average moisture content of deep, compact organic layers and is an indicator of seasonal droughts [45,46]. This suggests that drought conditions are partly responsible for the variability of the fire risk along the latitudinal gradient over most of our study area. However, the better model fit for transects A and C compared with transect B may indicate that climate is not the only factor influencing the fire risk. In particular, the large Mistassini Lake to the west and the Tichégami and Otish Mountains to the southern and northern zones (Figure 1a), respectively, may have acted as large firebreaks. Indeed, in Quebec, prevailing winds blow predominantly from west to east. These geographical features may thus be accompanied by large fire shadows that attenuate the effect of climate in this region.

As Cox survival models from both transects D and D2 show good fits (pseudo-$R^2 \geq 0.48$), the negative, counterintuitive effect of the FFMC on the fire risk could reflect the influence of some bottom-up drivers. For instance, this region is known for its complex topography, from plains on the edge of St. Lawrence River to mountains toward the north [49] (Figure 1a). The region is also characterized by an important variability of surficial deposits, from organic deposits and bedrock to till dominance from south to north [49]. Topography and surficial deposits are two significant factors of fire risk because they influence the drying potential of the forest floor as well as fuel composition and structure [14,18,40]. Well-drained stands are more likely to burn [18], and the slopes found in the northern portion of the transect could, for example, help with draining and thus drying the forest floor, thereby facilitating fire spread. Moreover, a limited fire ignition due to a very low occurrence of lightning strikes [50] is an additional factor that may explain the low fire activity of this region, as well as the difficulty to detect any climate effect on fire risk. When lightning strikes happen in conjunction with weather favorable to fire spread, the accumulated fuel can, therefore, allow for very large fires to occur. This control over fire activity could therefore prevail over other factors in this transect.

4.2. Fire Cycle

The computation of the fire cycle of each fire risk zone allowed the assessment of the spatiotemporal variability of fire activity within the study area. Although recent fires (after 1924) were over-represented in our transects, we trust the down-weighting applied to recently burned cells significantly reduced this bias and allowed for the calculation of realistic fire cycles. On a broad scale, other studies have shown the same gradient of increasing fire activity from south to north [24,48] and from east to west [21,24,48,51] in eastern Canada.

On a narrower spatial scale, our fire cycle estimates are consistent with values calculated for similar regions over the same time period. In the southern zone of transect A, our fire cycle is consistent with previous estimates made in the commercial forest further south [51–53]. In the northern zone of the same transect, one of the most fire-prone regions of boreal Canada where very large fires occur at a high frequency [33,35], we estimated a fire cycle of five years. However, compared to the size of those fires, our transect zone is relatively small, meaning that individual fires can intersect a large fraction of the transect (Figure 2a). In this particular region where very large fires occur, thus regularly erasing marks of older fires, a method using TSF data over a relatively short transect in order to estimate fire cycles does not appear to be well suited. In this situation, the use of archive data to compute annual area burned, or a different sampling design covering a larger area better adapted to accounting for immense fires could be used instead. Based on fire interval data for the same transect zone, Héon et al. [33] estimated a fire cycle of 42 years for the time period 1910–2013. With such a short fire cycle, shifts of vegetation from black spruce- to jack pine- dominated stands can occur in this area [54,55]. Short fire intervals are also likely to limit stand regeneration and consequently lead to an opening of the forest [55,56], possibly into lichen-woodlands [55]. However, recent studies have shown that a negative feedback between fuel availability and fire activity has strongly limited the occurrence of these short intervals during the last two centuries [33,57]. The whole landscape could nonetheless burn regardless of the fuel continuity if either the number of ignition points or the frequency of extremely severe weather events are high enough. The northern zone of transect A thus constitutes a very interesting area to monitor in the future in order to better understand how strong the forest's resilience to high fire activity is in this boreal ecosystem [35].

Our fire cycle estimates along transect B are similar to values obtained in previous studies of the same region. Indeed, in the southern zone and in the plateau near the Otish Mountains, Mansuy et al. [18] calculated fire cycles of 205 (CI95: 128; 502) and 237 (CI95: 136; 929), respectively. Southeast of the transect, Bélisle et al. [39] also found a similar fire cycle (247 (CI95: 187; 309)) to our southern zone. Mansuy et al. [18] did estimate a longer fire cycle (129 (CI95: 86; 257)) in the region corresponding to the northern zone of our transect. However, our estimate may have been highly influenced by the most recent decades as 60% of the cells in this zone have a TSF of <14 years.

Transects C and D are both located in the North Shore region where fire cycles have been shown to lengthen on an eastward course. Previous studies have estimated fire cycles varying from 250 years around the southern and center zones of transect C [9] to between 295 (40,51,58) and 600 years [9] further east, toward transect D. Those values are in agreement with our results as they are included within the confidence intervals of our fire cycle estimates of these regions. Furthermore, our study is one of the first that analyzes the fire regime from empirical data in the area covered by the northern zones of transects C and D, thus making it difficult to compare our results with others. However, we assume our fire cycle estimates are realistic based on the overall consistency between our results and those of other studies as well as the reliability of the estimates produced by the Cox analyses [37].

4.3. Fire Risk Zonation and Temporal Variability

Our study indicates that the fire risk increases from south to north, either gradually as in transects A, B, and D2, or stepwise as in transect C. In all transects, high and low fire risk zones could be delimited in the north and south, respectively. The localization of the breakpoints between fire risk zones is generally consistent with the regional boundaries set by Gauthier et al. [24] based on the recent fire regime (1972–2009), except for transect B which they consider to be more homogeneous.

The fire cycles estimated by Gauthier et al. [24] are longer than ours for all transect zones (Table 3). Temporal variations in the fire regime can explain these differences. Indeed, our fire cycle estimates for the time period prior to 1972 are generally shorter than estimates for the entire study period, suggesting a decrease in fire activity during the last four decades. Similar shifts were previously observed in Quebec around the middle of the 20th century, thus f the decrease in fire activity experienced since the end of the Little Ice Age that occurred around 1850 [51,58,59]. Moreover, previous studies targeting the fire activity of the last 150–300 years (e.g., [9,18,39,40,53]) have estimated fire cycles similar to ours. In the northern zone of transect A however, the difference between our estimates and the ones calculated by Gauthier et al. [24] seems to results from the most recent years (i.e., after 2009) when most of this zone was burned (Figure A3). This is consistent with previous studies showing that fire activity has been increasing since 1980 in this area [35].

All indices of the FWI, particularly the DC, are expected to increase in the future in response to climate change [60,61]. During the last few decades, the northern zone of transect A, which is the driest sector of the study area, has experienced very high fire activity (Figure 2a and Figure A3). With climate change, this phenomenon could propagate over the whole study area, leading to a large scale increase of fire activity in the near future [48], possibly returning to the fire regime levels of the last 150–300 years.

The relative stability of fire zone boundaries in the past may have resulted either from top-down or bottom-up processes. The climatic zonation may have remained somewhat constant with proportional changes among regions. Alternatively, bottom-up factors, such as fuel availability or surficial deposits, may have determined the observed spatial variability. As these factors are spatially stationary, they could account for the inertia of the limit between fire risk zones regardless of any changes in climate, provided that future fire risks remain in the range of past ones. In both cases, this could suggest that if climate and fire regimes are predicted to change in the future, the limits between fire risk zones might remain stable. This has great implications for forest management planners, as if they have to adjust for future changes in fire activity they will nonetheless be able to rely on the stability of their management unit layout in regards to fires. However, this should be accepted with caution as Boulanger et al. [48] have shown that in the future slight changes in fire regions could occur based on the projected area burned and the number of fires.

5. Conclusions

Considered as a whole, the latitudinal breakpoints separating our fire risk zones are largely consistent with the recent evaluation of the northern limit of the commercial forest. This limit has been drawn across the coniferous boreal forest of Quebec to delineate the more opened forests to the north from the tall and dense forests that are suitable for forest management to the south [23,62]. As there is a direct link between high fire activity and the opening of forests [26–29], as we have shown that fire risk appears to be higher in the northern zones over the last 150–300 years, and as these zones have been relatively stable through time, it seems reasonable to conclude that the limit between open and closed forests has also been somewhat stable. If the expected climate change leads to a fire activity level that remains in the same range of variability as the last 150–300 years, which Girardin et al. [63] consider a plausible scenario, this limit may also remain stable in the future. Indeed, boreal forests south of the northern limit of the commercial forest seem to be well-adapted to large changes in fire activity [63,64]. However, if fire activity increases beyond its range of variability in the south, dense forests could start opening [2,56] and thus eventually change the location of the limit.

Acknowledgments: We are grateful to André Robitaille (MFFP), Alain Tremblay (Hydro-Québec), and Danielle Charron (UQAM) for their help with field logistic. We thank Mélanie Desrochers (CFR), Aurélie Terrier (UQAM), Annie Claude Bélisle (UQAT), and Dominic Cyr (CFS) for their valuable technical advices. We heartily thank Sylvain Larouche, Alix Daguzan, Dave Gervais, Aurélie Terrier, and Joannie Hébert for their assistance in the field and Evick Mestre in the lab. We finally thank two anonymous reviewers for their helpful comments on an earlier version of the paper. We also acknowledge the MFFP for providing us with forest inventory and fire archive data. This work was supported by a Natural Sciences and Engineering Research Council of Canada strategic partnership grant awarded to Y.B. and S.G. and by a Mitacs Accelerate grant awarded to J.P. in partnership with Hydro-Quebec.

Author Contributions: J.P., S.G., and Y.B. conceived and designed the study; J.P. and D.A. provided TSF data; J.P., S.G., and A.L. performed the statistical analyses; J.P., S.G., A.L., and Y.B. interpreted the results; J.P. wrote the paper and all authors revised it.

Conflicts of Interest: The authors declare no conflict of interest.

Abbreviations

The following abbreviations are used in this manuscript:

AIC	Akaike Information Criterion
BUI	Buildup Index
CI	Confidence Interval
DC	Drought Code
DMC	Duff Moisture Code
FC	Fire Cycle
FFMC	Fine Fuel Moisture Code
FWI	Fire Weather Index
ISI	Initial Spread Index
MFFP	Ministère des Forêts, de la Faune et des Parcs (Québec)
TSF	Time Since Fire

Appendix A

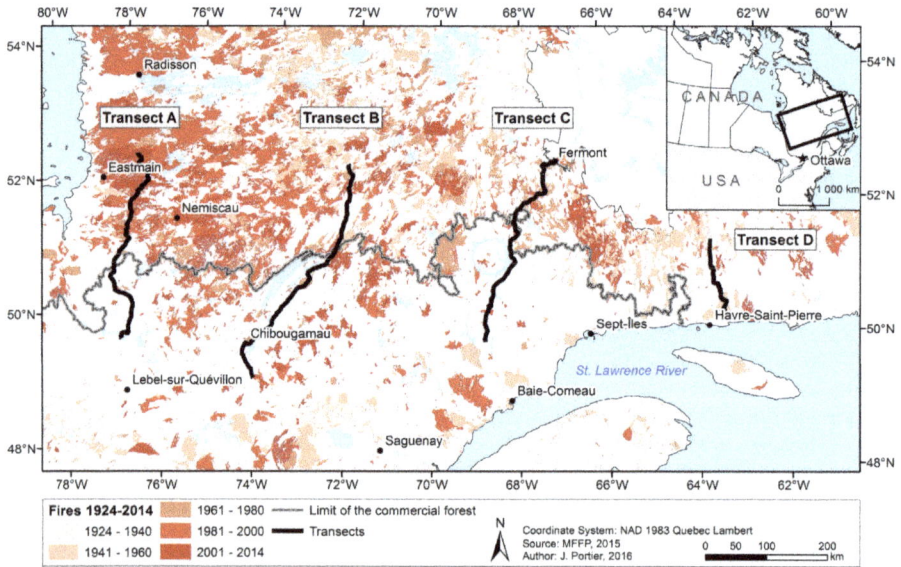

Figure A1. Map of the study area showing the four transects being analyzed, along with the recent fires from 1924 to 2014. The northern limit of the commercial forest in Quebec is also shown.

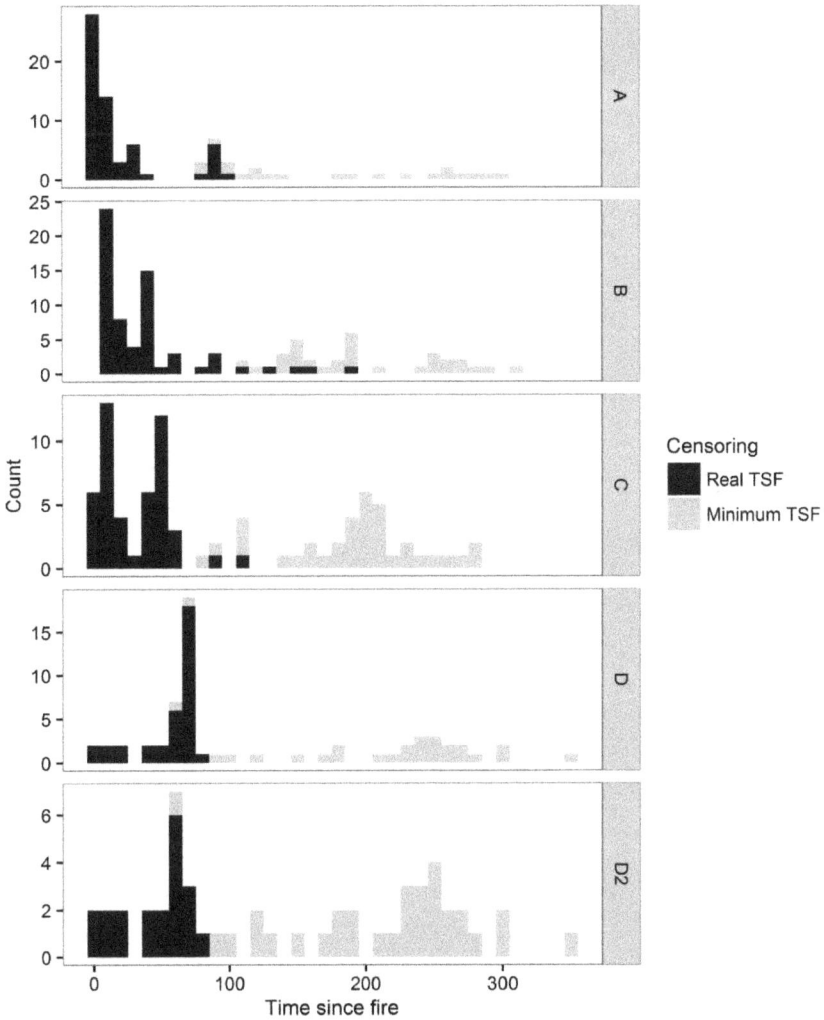

Figure A2. Decadal frequency distribution of TSF for each transect. No weights are applied to cells where a fire occurred between 1924 and 2014. The proportion of right-censored data (minimum TSF) and real TSF are shown in grey and black, respectively.

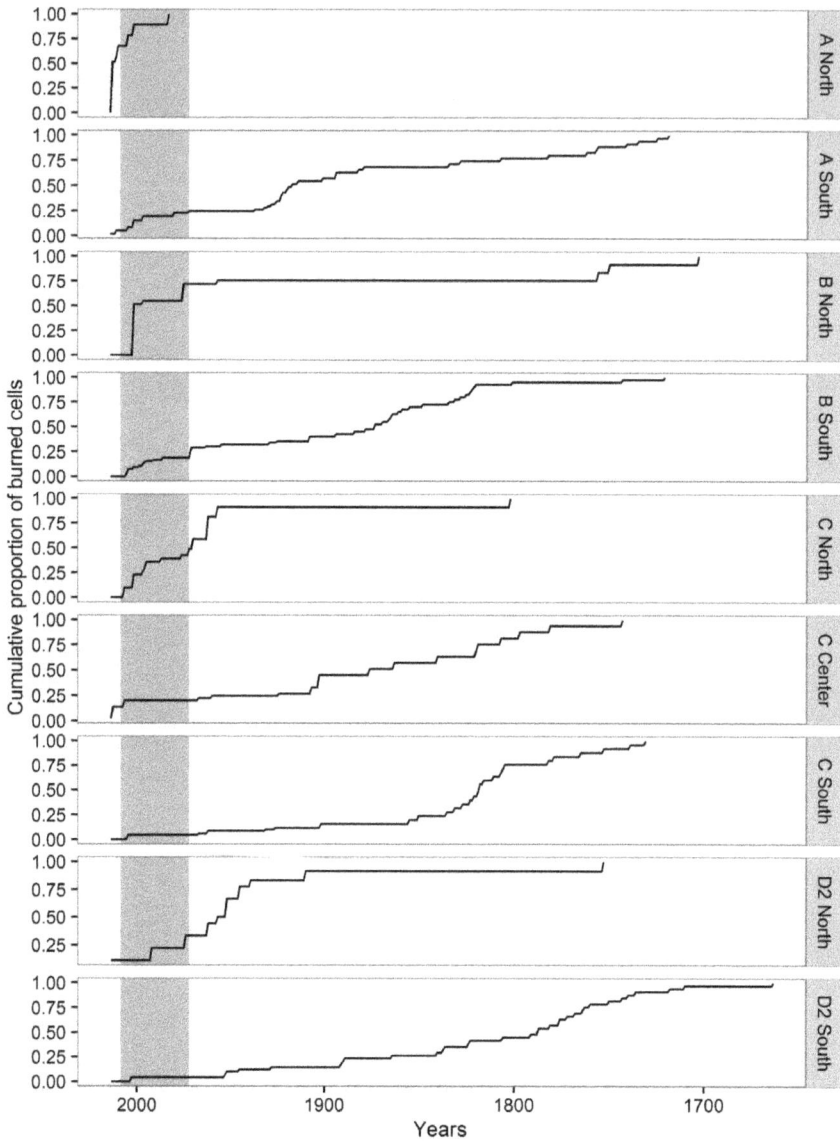

Figure A3. Reverse weighted cumulative proportion of cells that burned per decade in each transect zone. Weights are calculated in the same way that they were in the survival analyses in order to compensate for the over-representation of recent fires (after 1924) in transects. The shaded area shows the 1972–2009 period covered by the fire cycle estimates calculated by Gauthier et al. [24].

Table A1. Mean relative fire risk of transect zones compared to mean risk of the transect with 95% bootstrap confidence intervals (CI95) (e.g., zone north of transect A has a risk 15.61 times higher than the mean risk of the transect, while zone south of the same transect has a risk 3.33 times lower or equals to 0.30 times the mean risk of the transect). Relative fire risk values take into account the down-weighting of recently burned cells.

Transect	Zone	Mean Relative Risk	CI95
A	North	15.61	(7.96; 34.91)
	South	0.30 (−3.33)	(0.18; 0.47)
B	North	3.70	(2.16; 6.76)
	Plateau	0.33 (−3.03)	(0.18; 0.59)
	South	1.05	(0.75; 1.56)
C	North	10.48	(5.29; 24.09)
	Center	1.92	(1.00; 4.09)
	South	0.38 (−2.63)	(0.19; 0.70)
D	North	48.14	(14.66; 306.15)
	South	0.57 (−1.75)	(0.29; 1.06)
	1940s fire	3.75	(1.59; 11.64)
D2	North	21.30	(8.40; 98.19)
	South	0.69 (−1.45)	(0.43; 1.16)

References

1. Johnson, E.A. *Fire and Vegetation Dynamics: Studies from the North American Boreal Forest*; Cambridge: Cambridge, UK, 1992.

2. Payette, S. Fire as a controlling process in the North American boreal forest. In *A System Analysis of the Global Boreal Forest*; Shugart, H.H., Leemans, R., Bonan, G.B., Eds.; Cambridge University Press: Cambridge, UK, 1992; pp. 144–169.

3. Gauthier, S.; Leduc, A.; Bergeron, Y. Forest dynamics modelling under natural fire cycles: A tool to define natural mosaic diversity for forest management. *Environ. Monit. Assess.* **1996**, *39*, 417–434. [CrossRef] [PubMed]

4. Johnson, E.A.; Miyanishi, K.; Weir, J.M.H. Wildfires in the western Canadian boreal forest: Landscape patterns and ecosystem management. *J. Veg. Sci.* **1998**, *9*, 603–610. [CrossRef]

5. Johnson, E.A.; Gutsell, S.L. Fire frequency models, methods and interpretations. *Adv. Ecol. Res.* **1994**, *25*, 239–287.

6. Li, C. Estimation of fire frequency and fire cycle: A computational perspective. *Ecol. Modell.* **2002**, *154*, 103–120. [CrossRef]

7. Van Wagner, C.E. Age-Class distribution and the forest fire cycle. *Can. J. For. Res.* **1978**, *8*, 220–227. [CrossRef]

8. Cyr, D.; Gauthier, S.; Bergeron, Y.; Carcaillet, C. Forest management is driving the eastern North American boreal forest outside its natural range of variability. *Front. Ecol. Environ.* **2009**, *7*, 519–524. [CrossRef]

9. Bouchard, M.; Pothier, D.; Gauthier, S. Fire return intervals and tree species succession in the North Shore region of eastern Quebec. *Can. J. For. Res.* **2008**, *38*, 1621–1633. [CrossRef]

10. Gauthier, S.; De Grandpré, L.; Bergeron, Y. Differences in forest composition in two boreal forest ecoregions of Quebec. *J. Veg. Sci.* **2000**, *11*, 781–790. [CrossRef]

11. Girard, F.; Payette, S.; Gagnon, R. Origin of the lichen-spruce woodland in the closed-crown forest zone of eastern Canada. *Glob. Ecol. Biogeogr.* **2009**, *18*, 291–303. [CrossRef]

12. Senici, D.; Chen, H.Y.H.; Bergeron, Y.; Cyr, D. Spatiotemporal variations of fire frequency in central boreal forest. *Ecosystems* **2010**, *13*, 1227–1238. [CrossRef]

13. Parisien, M.-A.; Parks, S.A.; Miller, C.; Krawchuk, M.A.; Heathcott, M.; Moritz, M.A. Contributions of ignitions, fuels, and weather to the spatial patterns of burn probability of a boreal landscape. *Ecosystems* **2011**, *14*, 1141–1155. [CrossRef]

14. Bélisle, A.; Leduc, A.; Gauthier, S.; Desrochers, M.; Mansuy, N.; Morin, H.; Bergeron, Y. Detecting local drivers of fire cycle heterogeneity in boreal forests: A scale issue. *Forests* **2016**, *7*, 1–21. [CrossRef]

15. Flannigan, M.D.; Harrington, J.B. A study of the relation of meteorological variables to monthly provincial area burned by wildfire in Canada (1953–80). *J. Appl. Meteorol.* **1988**, *27*, 441–452. [CrossRef]

16. Girardin, M.P.; Wotton, B.M. Summer moisture and wildfire risks across Canada. *J. Appl. Meteorol. Climatol.* **2009**, *48*, 517–533. [CrossRef]

17. Van Wagner, C.E. Effect of slope on fires spreading downhill. *Can. J. For. Res.* **1988**, *18*, 818–820. [CrossRef]

18. Mansuy, N.; Gauthier, S.; Robitaille, A.; Bergeron, Y. The effects of surficial deposit-drainage combinations on spatial variations of fire cycles in the boreal forest of eastern Canada. *Int. J. Wildl. Fire* **2010**, *19*, 1083–1098. [CrossRef]

19. Hély, C.; Fortin, C.M.-J.; Anderson, K.R.; Bergeron, Y. Landscape composition influences local pattern of fire size in the eastern Canadian boreal forest: Role of weather and landscape mosaic on fire size distribution in mixedwood boreal forest using the Prescribed Fire Analysis System. *Int. J. Wildl. Fire* **2010**, *19*, 1099–1109. [CrossRef]

20. Ali, A.A.; Carcaillet, C.; Bergeron, Y. Long-Term fire frequency variability in the eastern Canadian boreal forest: The influences of climate vs. *local factors. Glob. Chang. Biol.* **2009**, *15*, 1230–1241. [CrossRef]

21. Girardin, M.P.; Tardif, J.C.; Flannigan, M.D.; Bergeron, Y. Forest fire-conducive drought variability in the southern Canadian boreal forest and associated climatology inferred from tree rings. *Can. Water Resour. J.* **2006**, *31*, 275–296. [CrossRef]

22. Girardin, M.P.; Tardif, J.; Flannigan, M.D.; Wotton, B.M.; Bergeron, Y. Trends and periodicities in the Canadian Drought Code and their relationships with atmospheric circulation for the southern Canadian boreal forest. *Can. J. For. Res.* **2004**, *34*, 103–119. [CrossRef]

23. Jobidon, R.; Bergeron, Y.; Robitaille, A.; Raulier, F.; Gauthier, S.; Imbeau, L.; Saucier, J.-P.; Boudreault, C. A biophysical approach to delineate a northern limit to commercial forestry: The case of Quebec's boreal forest. *Can. J. For. Res.* **2015**, *45*, 515–528. [CrossRef]

24. Gauthier, S.; Raulier, F.; Ouzennou, H.; Saucier, J.-P. Strategic analysis of forest vulnerability to risk related to fire: An example from the coniferous boreal forest of Quebec. *Can. J. For. Res.* **2015**, *45*, 553–565. [CrossRef]

25. Sirois, L. Spatiotemporal variation in black spruce cone and seed crops along a boreal forest-tree line transect. *Can. J. Bot.* **2000**, *30*, 900–909. [CrossRef]

26. Jasinki, J.P.P.; Payette, S. The creation of alternative stable states in the southern boreal forest, Québec, Canada. *Ecol. Monogr.* **2005**, *75*, 561–583. [CrossRef]

27. Payette, S.; Bhiry, N.; Delwaide, A.; Simard, M. Origin of the lichen woodland at its southern range limit in eastern Canada: The catastrophic impact of insect defoliators and fire on the spruce-moss forest. *Can. J. For. Res.* **2000**, *30*, 288–305. [CrossRef]

28. Girard, F.; Payette, S.; Gagnon, R. Rapid expansion of lichen woodlands within the closed-crown boreal forest zone over the last 50 years caused by stand disturbances in eastern Canada. *J. Biogeogr.* **2008**, *35*, 529–537. [CrossRef]

29. Jayen, K.; Leduc, A.; Bergeron, Y. Effect of fire severity on regeneration success in the boreal forest of northwest Québec, Canada. *Ecoscience* **2006**, *13*, 143–151. [CrossRef]

30. Arseneault, D. Impact of fire behavior on postfire forest development in a homogeneous boreal landscape. *Can. J. For. Res.* **2001**, *31*, 1367–1374. [CrossRef]

31. IPCC Climate Change 2014. *Synthesis Report: Contribution of Working Groups I, II and III to the Fifth Assessment Report of the Intergovernmental Panel on Climate Change*; Core Writing Team, Pachauri, R.K., Meyer, L.A., Eds.; IPCC: Geneva, Switzerland, 2014.

32. Flannigan, M.D.; Logan, K.A.; Amiro, B.D.; Skinner, W.R.; Stocks, B.J. Future area burned in Canada. *Clim. Change* **2005**, *72*, 1–16. [CrossRef]

33. Héon, J.; Arseneault, D.; Parisien, M.-A. Resistance of the boreal forest to high burn rates. *Proc. Natl. Acad. Sci. USA* **2014**, *111*, 13888–13893. [CrossRef] [PubMed]

34. Beaudoin, A.; Bernier, P.Y.; Guindon, L.; Villemaire, P.; Guo, X.J.; Stinson, G.; Bergeron, T.; Magnussen, S.; Hall, R.J. Mapping attributes of Canada's forests at moderate resolution through kNN and MODIS imagery. *Can. J. For. Res.* **2014**, *44*, 521–532. [CrossRef]

35. Erni, S.; Arseneault, D.; Parisien, M.-A.; Bégin, Y. Spatial and temporal dimensions of fire activity in the fire-prone eastern Canadian taiga. *Glob. Chang. Biol.* **2016**. [CrossRef] [PubMed]

36. Régnière, J.; Saint-Amant, R. *BioSIM 9-User's Manual, Information Report LAU-X-134*; Natural Resources Canada: Quebec, Canada, 2008.

37. Cyr, D.; Gauthier, S.; Boulanger, Y.; Bergeron, Y. Quantifying fire cycle from dendroecological records using survival analyses. *Forests* **2016**, *7*, 1–21. [CrossRef]
38. Cox, D.R. Regression models and life-tables. *J. R. Stat. Soc. Ser. B* **1972**, *34*, 187–220.
39. Bélisle, A.C.; Gauthier, S.; Cyr, D.; Bergeron, Y.; Morin, H. Fire regime and old-growth boreal forests in central Quebec, Canada: An ecosystem management perspective. *Silva Fenn.* **2011**, *45*, 889–908. [CrossRef]
40. Cyr, D.; Gauthier, S.; Bergeron, Y. Scale-Dependent determinants of heterogeneity in fire frequency in a coniferous boreal forest of eastern Canada. *Landsc. Ecol.* **2007**, *22*, 1325–1339. [CrossRef]
41. Therneau, T.M. Package "Survival". Available online: https://cran.r-project.org/web/packages/survival/survival.pdf (accessed on 12 July 2016).
42. Symonds, M.R.E.; Moussalli, A. A brief guide to model selection, multimodel inference and model averaging in behavioural ecology using Akaike's information criterion. *Behav. Ecol. Sociobiol.* **2011**, *65*, 13–21. [CrossRef]
43. Richards, S.A. Testing ecological theory using the information-theoretic approach: Examples and cautionary results. *Ecology* **2005**, *86*, 2805–2814. [CrossRef]
44. Burnham, K.P.; Anderson, D.R. *Model Selection and Multimodel Inference: A Practical Information-Theoretic Approach*, 2nd ed.; Springer: New York, NY, USA, 2002.
45. Amiro, B.D.; Logan, K.A.; Wotton, B.M.; Flannigan, M.D.; Todd, J.B.; Stocks, B.J.; Martell, D.L. Fire weather index system components for large fires in the Canadian boreal forest. *Int. J. Wildl. Fire* **2004**, *13*, 391–400. [CrossRef]
46. Van Wagner, C.E. *Development and Structure of the Canadian Forest Fire Weather Index System*; Canadian Forestry Service: Ottawa, Canada, 1987.
47. Girardin, M.P.; Ali, A.A.; Carcaillet, C.; Mudelsee, M.; Drobyshev, I.; Hély, C.; Bergeron, Y. Heterogeneous response of circumboreal wildfire risk to climate change since the early 1900s. *Glob. Chang. Biol.* **2009**, *15*, 2751–2769. [CrossRef]
48. Boulanger, Y.; Gauthier, S.; Gray, D.R.; Le Goff, H.; Lefort, P.; Morissette, J. Fire regime zonation under current and future climate over eastern Canada. *Ecol. Appl.* **2013**, *23*, 904–923. [CrossRef] [PubMed]
49. Robitaille, A.; Saucier, J.-P.; Chabot, M.; Côté, D.; Boudreault, C. An approach for assessing suitability for forest management based on constraints of the physical environment at a regional scale. *Can. J. For. Res.* **2015**, *45*, 529–539. [CrossRef]
50. Morissette, J.; Gauthier, S. Study of cloud-to-ground lightning in Quebec: 1996–2005. *Atmosphere-Ocean* **2008**, *46*, 443–454. [CrossRef]
51. Bergeron, Y.; Cyr, D.; Drever, C.R.; Flannigan, M.; Gauthier, S.; Kneeshaw, D.; Lauzon, È.; Leduc, A.; Le Goff, H.; Lesieur, D.; et al. Past, current, and future fire frequencies in Quebec's commercial forests: Implications for the cumulative effects of harvesting and fire on age-class structure and natural disturbance-based management. *Can. J. For. Res.* **2006**, *36*, 2737–2744. [CrossRef]
52. Bergeron, Y.; Gauthier, S.; Kafka, V.; Lefort, P.; Lesieur, D. Natural fire frequency for the eastern Canadian boreal forest: Consequences for sustainable forestry. *Can. J. For. Res.* **2001**, *31*, 384–391. [CrossRef]
53. Bergeron, Y.; Gauthier, S.; Flannigan, M.; Kafka, V. Fire regimes at the transition between mixedwood and coniferous boreal forest in northwestern Quebec. *Ecology* **2004**, *85*, 1916–1932. [CrossRef]
54. Le Goff, H.; Sirois, L. Black spruce and jack pine dynamics simulated under varying fire cycles in the northern boreal forest of Quebec, Canada. *Can. J. For. Res.* **2004**, *34*, 2399–2409. [CrossRef]
55. Lavoie, L.; Sirois, L. Vegetation changes caused by recent fires in the northern boreal forest of eastern Canada. *J. Veg. Sci.* **1998**, *9*, 483–492. [CrossRef]
56. Brown, C.D.; Johnstone, J.F. Once burned, twice shy: Repeat fires reduce seed availability and alter substrate constraints on Picea mariana regeneration. *For. Ecol. Manag.* **2012**, *266*, 34–41. [CrossRef]
57. Parisien, M.-A.; Parks, S.A.; Krawchuk, M.A.; Little, J.M.; Flannigan, M.D.; Gowman, L.M.; Moritz, M.A. An analysis of controls on fire activity in boreal Canada: Comparing models built with different temporal resolutions. *Ecol. Appl.* **2014**, *24*, 1341–1356. [CrossRef]
58. Gauthier, S.; Leduc, A.; Bergeron, Y.; Le Goff, H. La fréquence des feux et l'aménagement forestier inspiré des perturbations naturelles. In *Aménagement Écosystémique en Forêt Boréale*; Gauthier, S., Vaillancourt, M.-A., Leduc, A., De Grandpré, L., Kneeshaw, D.D., Morin, H., Drapeau, P., Bergeron, Y., Eds.; Presses de l'Université du Québec: Québec, QC, Canada, 2009; pp. 61–78. (In French)

59. Le Goff, H.; Flannigan, M.D.; Bergeron, Y.; Girardin, M.P. Historical fire regime shifts related to climate teleconnections in the Waswanipi area, central Quebec, Canada. *Int. J. Wildl. Fire* **2007**, *16*, 607–618. [CrossRef]

60. Wang, X.; Thompson, D.K.; Marshall, G.A.; Tymstra, C.; Carr, R.; Flannigan, M.D. Increasing frequency of extreme fire weather in Canada with climate change. *Clim. Chang.* **2015**, *130*, 573–586. [CrossRef]

61. Flannigan, M.D.; Wotton, B.M.; Marshall, G.A.; de Groot, W.J.; Johnston, J.; Jurko, N.; Cantin, A.S. Fuel moisture sensitivity to temperature and precipitation: Climate change implications. *Clim. Chang.* **2016**, *134*, 59–71. [CrossRef]

62. Ministère des Ressources Naturelles du Québec. *Rapport du Comité Scientifique Chargé d'Examiner la Limite Nordique Des Forêts Attribuables*; Gouvernement du Québec Secteur des Forêts: Québec, Canada, 2013. (In French)

63. Girardin, M.P.; Ali, A.A.; Carcaillet, C.; Gauthier, S.; Hély, C.; Le Goff, H.; Terrier, A.; Bergeron, Y. Fire in managed forests of eastern Canada: Risks and options. *For. Ecol. Manag.* **2013**, *294*, 238–249. [CrossRef]

64. Bergeron, Y.; Cyr, D.; Girardin, M.P.; Carcaillet, C. Will climate change drive 21st century burn rates in Canadian boreal forest outside of its natural variability: Collating global climate model experiments with sedimentary charcoal data. *Int. J. Wildl. Fire* **2010**, *19*, 1127–1139. [CrossRef]

© 2016 by the authors. Licensee MDPI, Basel, Switzerland. This article is an open access article distributed under the terms and conditions of the Creative Commons Attribution (CC BY) license (http://creativecommons.org/licenses/by/4.0/).

forests

MDPI

Review

Spatiotemporal Variability of Wildland Fuels in US Northern Rocky Mountain Forests

Robert E. Keane

Missoula Fire Sciences Laboratory, Rocky Mountain Research Station, US Forest Service, 5775 Highway 10 West, Missoula, MT 59808, USA; rkeane@fs.fed.us; Tel.: +1-406-329-4846

Academic Editors: Yves Bergeron and Sylvie Gauthier
Received: 3 May 2016; Accepted: 12 June 2016; Published: 27 June 2016

Abstract: Fire regimes are ultimately controlled by wildland fuel dynamics over space and time; spatial distributions of fuel influence the size, spread, and intensity of individual fires, while the temporal distribution of fuel deposition influences fire's frequency and controls fire size. These "shifting fuel mosaics" are both a cause and a consequence of fire regimes. This paper synthesizes results from two major fuel dynamics studies that described the spatial and temporal variability of canopy and surface wildland fuel characteristics found in US northern Rocky Mountain forests. Eight major surface fuel components—four downed dead woody fuel size classes (1, 10, 100, 1000 h), duff, litter, shrub, and herb—and three canopy fuel characteristics—loading, bulk density and cover—were studied. Properties of these fuel types were sampled on nested plots located within sampling grids to describe their variability across spatiotemporal scales. Important findings were that fuel component loadings were highly variable (two to three times the mean), and this variability increased with the size of fuel particles. The spatial variability of loadings also varied by spatial scale with fine fuels (duff, litter, 1 h, 10 h) varying at scales of 1 to 5 m; coarse fuels at 10 to 150 m, and canopy fuels at 100 to 600 m. Fine fuels are more uniformly distributed over both time and space and decayed quickly, while large fuels are rare on the landscape but have a high residence time.

Keywords: range; semi-variogram; fuel deposition; decomposition; fuel component; vegetation development

1. Introduction

Fire regimes are created by the interaction of bottom–up and top–down controls [1]; bottom–up controls, such as vegetation, topography, and disturbance history, often dictate fire spread, intensity, and severity at fine scales, while coarse scale, top–down controls, such as climate and weather, dictate fire frequency, duration, and synchrony [2]. Of all bottom–up controls, wildland fuels are important because they govern most of fire's combustion processes [3]. The spatial and temporal variability of wildland fuel directly impacts fire regimes which, in turn, has major implications for fire management [4]. Landscape patches that have minimal fuels, such as recently treated or burned areas, form fuel breaks that may limit growth, reduce intensity, and minimize severity for future fires [5]. This self-organizational property of wildland fire is incredibly important in predicting future fire dynamics under climate change [6,7]. Fire and fuel management should use the changing fuel mosaic to develop management plans that effectively integrate wildfires, controlled wildfires, prescribed fires, and fuel treatments to minimize firefighting costs and maximize ecosystem resilience while still protecting homes and people [8].

Wildland fuels are live and dead organic matter called *biomass* [9]. The forest fuelbed is vertically stratified into three fuel layers—*ground, surface, and canopy* fuels (Figure 1). *Surface fuels* are all biomass within 2 m above the ground surface. *Ground fuels* are all organic matter below the litter and above the mineral soil, which is called duff in most upland forests. *Canopy fuels* are the biomass above the surface fuel layer. Fuelbed layers are composed of finer-scale elements called *fuel components*, which are fuel types that are defined for specific purposes, mostly for fire behavior and effects prediction. A woody fuel type, for example, might be defined as a fuel component based on particle diameter size range (Table 1). The finest scale of fuelbed description is the fuel *particle*, which is a general term that defines a specific piece of fuel that is part of a fuel type or component of a fuelbed (Figure 1); a fuel particle can be an intact or fragmented stick, grass blade, shrub leaf, or pine needle. There are many physical properties that can describe fuel particles, such as specific gravity, heat content, weight, and shape, and statistical summaries of these particle properties are often used to quantify coarser fuel component properties. However, the fuel property most used in fire management is *loading* or amount of biomass per unit area (kg· m^{-2} in this paper).

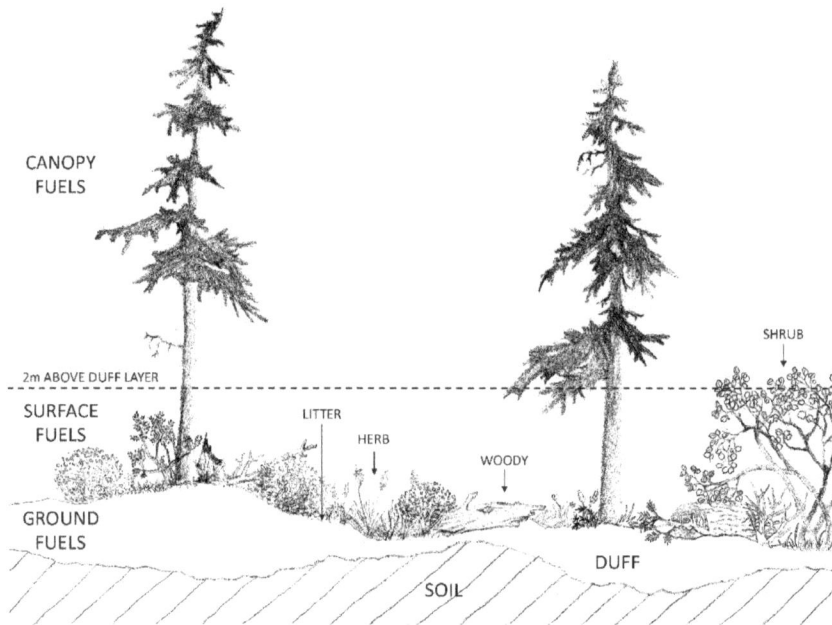

Figure 1. Illustration of a wildland forest fuelbed showing the three major strata: ground, surface, and canopy fuels from Keane [9] and drawn by Ben Wilson.

An often overlooked feature of wildland fuelbeds is that they are always changing in space and in time; live and dead biomass are constantly being added, modified, and removed by various ecological processes, thereby changing particle, component, and layer properties [10]. The annual shed of leaves and small woody twigs from trees, for example, creates significantly different spatial distributions than the infrequent toppling of tree boles to create logs or coarse woody debris (CWD). As a result, wildland fuel landscapes can be thought of as shifting mosaics of hierarchically intersecting fuel characteristics [9]. This dynamic and complex character of fuelbeds across space and time is responsible for the great variability found in wildland fuel characteristics [11–13] and is perhaps the single most important concept to understand in fire management today because it influences strategic fuel management considerations such as fuel treatment longevity and effectiveness, fire return intervals, and smoke potential [9].

Numerous ecological processes influence fuel dynamics, but four are particularly important in controlling spatial and temporal distributions of wildland fuels—vegetation development, deposition, decomposition, and disturbance—the four "Ds" [9]. Wildland fuels accumulate as a result of the establishment, growth, and mortality of vegetation (*development*). Rates of biomass accumulation, often called productivity, are dictated by the interactions of the plant species available to occupy a site and the site's physical environment (e.g., climate, soils, and topography) [14]. Over time, portions of living biomass are shed or die and get *deposited* on the ground to become dead surface fuels or necromass. Below- and above-ground necromass is eventually *decomposed* by microbes and soil macrofauna. *Disturbances*, such as fire, insects, and disease, act on living and dead biomass to change the magnitude, trend, and direction of fuel dynamics in space and time. These four fuel processes interact, but they are often influenced by different environmental factors depending on the ecosystem. In wildland fuel science, for example, many assume fuels are closely related to vegetation characteristics [15], but this would be true only if the first two processes (development and deposition) were considered thereby ignoring the role of decomposition and disturbance in fuelbed development.

The landscape ecology of wildland fuels is the interaction of the above processes across multiple space and time scales to create the shifting mosaics of fuel conditions [16]. Understanding the spatial and temporal dynamics of fuels may provide a better grasp of the impact of various wildland fuel management activities on fuel properties [17] and it also might help explain unexpected fire behaviors and effects (e.g., [18]). It may also aid in developing effective fuel applications that integrate spatial variability in their design such as new fuel classifications, sampling methods, and geospatial data [9]. Patterns of fuel characteristics will be important inputs to the fire effects and behavior models of the future [18,19].

Few have explored spatial and temporal relationships of the wildland fuels. Reich et al. [20] evaluated the spatial variability of several fuel components over a large landscape in the US Black Hills and found that the variability was governed by topography and vegetation. Hiers et al. [21] measured small-scale variations in surface fuel using LiDAR and found that fuelbed depths become spatially independent after small distances (~0.5 m²). Spatial variability of grasslands have been described in the context of population dynamics and restoration potential but have not been related to fuel characteristics [22]. Theobald [23] found that while fine scale variation in fuels dictated fire behavior, the distribution of CWD dictated germination in longleaf pine ecosystems. Kreye et al. [24] described the spatial structure of duff near tree boles using Moran's I and found duff depth had high spatial correlations at short distances (<1 m). While some studies described fuel distributions across landscapes [25,26], few have actually quantified the variability of fuel properties across space [19,27,28]. And, while many have identified fuel continuity as an important spatial characteristic of wildland fuels [29,30], few studies have addressed the structure of fuel spatial variation at landscape scales.

This paper is a synthesis of two long-term projects that were designed to understand the spatial and temporal dynamics of wildland fuelbeds. Spatial fuel characteristics were measured for eight fuel components (Table 1) in an extensive study called FUELVAR that assessed fuel component properties at various distances in a 1 km sampling grid installed in six US northern Rocky Mountain stands [11,16]. Temporal fuel dynamics were measured on 28 plots across the northern US Rocky Mountains over a period of 10–12 years to assess deposition and decomposition rates for five fuel components in the FUELDYN study (Table 1) [31,32]. While these two studies were not directly linked, many methods and analyses overlap, and as a result, the findings can be described in a similar context. In this paper, the two studies are used to demonstrate the variability and complexity of wildland forest fuelbeds over time and space.

Table 1. Descriptions of the three canopy fuel characteristics, eight surface fuel components, and one ground fuel components included in this paper. FWD is fine woody debris, a term often given to wood fuel particles less than 8 cm in diameter. CWD is coarse woody debris, a term used to woody fuel particles greater than 8 cm in diameter. Those fuel components in bold indicate that they were included in the FUELDYN project, while all fuel components were included in the FUELVAR project.

Fuel Type	Fuel Component/Attribute	Common Name	Size	Description
Canopy Fuels				
Canopy	Canopy bulk density $(kg \cdot m^{-3})$	CBD	<3 mm diameter	All canopy material less than 3 mm diameter
	Canopy fuel loading $(kg \cdot m^{-2})$	CFL	<3 mm diameter	All canopy material less than 3 mm diameter
	Canopy cover (%)	CC	All material	Vertically projected canopy cover
Surface Fuels				
Downed Dead Woody	**1 h woody**	Twigs, FWD	<0.6 cm (0.25 inch) diameter	Detached woody fuel particles on the ground
	10 h woody	Branches, FWD	0.6–2.5 cm (0.25–1.0 inch) diameter	Detached woody fuel particles on the ground
	100 h woody	Large Branches, FWD	2.5–8 cm (1–3 inch) diameter	Detached woody fuel particles on the ground
	1000 h woody	Logs, CWD	8+ cm (3+ inch) diameter	Detached woody fuel particles on the ground
Shrubs	Shrub	Shrubby	All shrubby material less than 5 cm diameter	All burnable shrubby biomass with branch diameters less than 5 cm
Herbaceous	Herb	Herbs	All sizes	All live and dead grass, forb, and fern biomass
Litter	**Litter**	Litter	All sizes, excluding woody	Freshly fallen non-woody material which includes leaves, cones, pollen cones
Ground Fuels				
Duff	Duff	Duff	All sizes	Partially decomposed biomass whose origins cannot be determined

2. Methods

2.1. Spatial Methods

In the FUELVAR study, a nested grid design within a square 1 km^2 area was installed in the center of six selected study sites [11] (Figure 2a). Sites sampled included a second-growth dry mixed conifer stand of ponderosa pine (*Pinus ponderosa*), Douglas-fir (*Pseudotsuga menziesii*), and western larch (*Larix occidentalis*) that had been thinned, a ponderosa pine-Douglas-fir stand that had been prescribed burned, a lodgepole pine (*Pinus contorta*) stand with a history of non-lethal surface fires, a pinyon pine (*Pinus edulis*) and juniper (*Juniperus occidentalis*) woodland, a ponderosa pine savanna, and a sagebrush grassland. Sides of the sampling grid were oriented along the four cardinal directions. Transects were established at each corner and at 100-m intervals along each grid side (Figure 2a). Sampling was intensified around four central grid points to increase the number of distances between sample points by installing a nested sampling grid of 16 additional sampling points centered around one of the four grid points using a 100-m square (eight sampling points) and 50-m square (another eight points) design (Figure 2b). These additional sampling points were placed at the corners and side mid-points for the two nested squares. This intensive grid provided the additional distances, including 25, 35, 50, and 100 m.

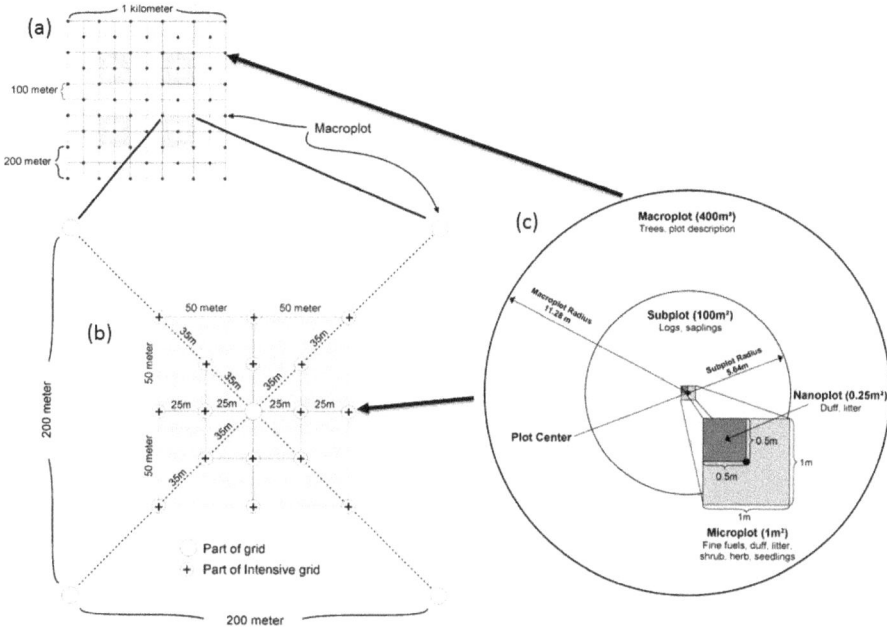

Figure 2. The sampling design of the FUELVAR study showing (**a**) the 1 km^2 sampling grid, (**b**) the intensification of the grid, and (**c**) the nested plot sampling scheme in the upper right.

A 400-m^2 circular *macroplot* was established at each grid sample point for sampling trees greater than 10 cm DBH (diameter breast height) and canopy cover (Figure 2c). Using the same sample point as plot center, we installed a 100-m^2 circular *subplot* on which we sampled logs (woody fuel particles greater than 8 cm diameter) and sapling trees (trees greater than 1.37 m tall and less than 10 cm DBH). We then centered a 1-m^2 square *microplot* over the grid sampling point, within which we measured shrub, herb, and fine woody (wood fuel particles less than 8 cm diameter; twigs and branches) fuel characteristics. Last, a 0.25-m^2 (50 by 50 cm) square *nanoplot* was installed in the northwest corner of the microplot to measure duff and litter fuels.

Spatial variability of the measured fuel loadings of each component (Table 1) was described using semi-variograms; a descriptive technique that graphically represents the spatial continuity and spatial autocorrelation of a spatial data set [33,34]. Semi-variogram *range*, the distance where the variance curve first flattens, is important in landscape ecology because it represents the spatial scale at which the entity of concern is best described in space, often called the inherent patch size [35].

2.2. Temporal Methods

In 1993, a set of litter traps were installed on four plots on each of two sites in western Montana to parameterize and validate two ecosystem models [31]. Then, in 1995, four new sites were established along elevational and aspect gradients within the larger US Northern Rockies study area (Figure 3a). Plots were established only in mature stands that had no evidence of disturbance. Forest types represented by these sites include stands dominated by ponderosa pine, Douglas-fir), western red cedar (*Thuja plicata*), subalpine fir (*Abies lasiocarpa*), and whitebark pine (*Pinus albicaulis*). In 1996, a seventh site was established in the ubiquitous lodgepole pine ecosystem that occurs east of the Continental Divide (Figure 3a).

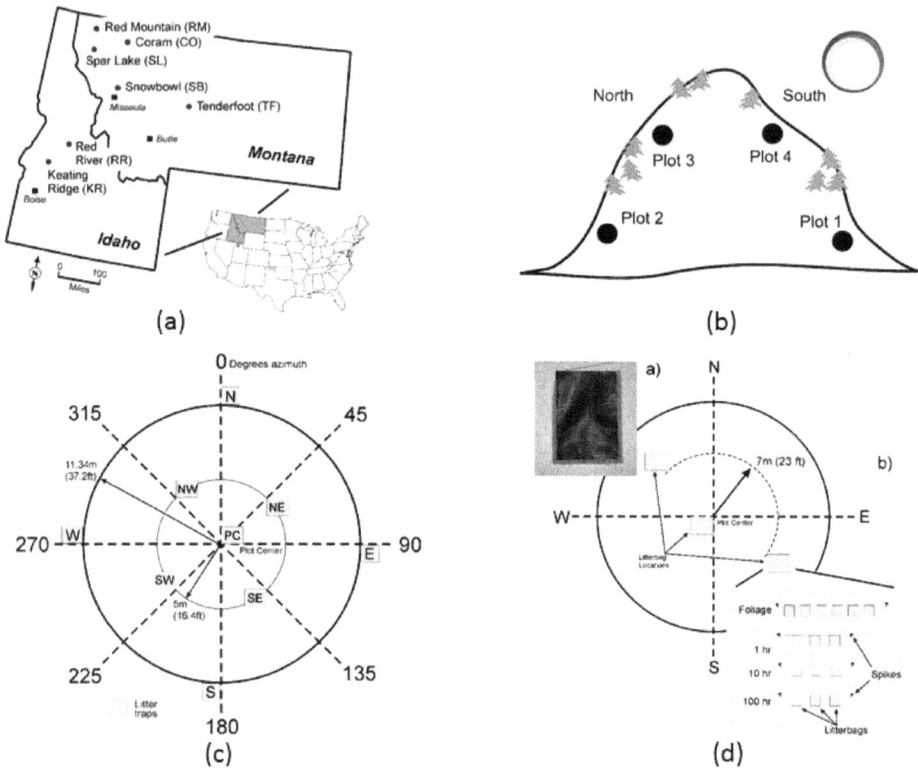

Figure 3. The sampling design of the FUELDYN study showing the (**a**) seven sampling sites; (**b**) the four plots per sampling site placed at high and low elevations and north and south aspects; (**c**) the littertrap placement within the 400 m^2 plot; and (**d**) the decomposition bag placement on the 400 m^2 plot.

Four plots were established at each site established along major topographic gradients of elevation and aspect to capture the diversity of the important direct environmental gradients such as productivity, moisture, and temperature (Figure 3b). A number of topographic, vegetation, and ecosystem characteristics were measured on the 0.1 acre (0.04 ha) circular plots, and then an inventory of all trees within plot boundaries was conducted. A network of 30 m fuel transects to estimate fuel loadings for five fuel components was also used in this study.

At each plot, we installed nine 1 m by 1 m litter traps in the star pattern to collect fallen biomass (Figure 3c) but two were removed (NW, SE) after statistical analysis revealed they weren't needed. Each plot was visited once a month during the snow-free periods of the year and all material in each trap was placed into heavy paper bags, transported to the laboratory, and the labeled bags were placed in an oven set at 90 °C for two to three days. The weight of each fuel component was recorded to the nearest 0.01 g along with the date, site, plot, and trap information written on the bag. A small sample of the dried material was set aside for the decomposition experiment. We measured litterfall in these traps for 10–12 years depending on the site.

Litter bags were used to estimate the rate of decay for four fuel components of freshly fallen foliage, twigs, branches, and large branches [31]. Approximately 100–150 g of the material taken from the litter traps was put into each bag and then the bag was sewn closed. Decomposition rates for logs and other canopy material were not measured because of limited time, lack of appropriate equipment, and incompatible methods. At each plot, three sets of three bags for the three fine woody fuel components (1, 10, and 100 h) and three sets of six bags for the foliage material were installed in the pattern shown in Figure 3d. Decomposition was measured over three years by taking one foliage bag from each wire set every 6 months and one woody bag from each woody fuel set every 12 months. The litter bags were then placed in paper bags, dried at 80 °C for 3 days, and then weighed to the nearest 0.01 g with the weight.

Two estimates of decomposition were calculated. A mass loss rate (percent year^{-1}) was calculated from differences in bag weights over the three year period. Then, we estimated the decomposition parameter k in the Olson [36] equation using a linear mixed effects model whose form is as follows:

$$\ln\left(\frac{x_{ij}}{x_{i0}}\right) = (-k + b_i)t_j + \varepsilon_{ij} \tag{1}$$

where x_{ij} is the weight of the ith trap at time j (t_j) and x_{i0} is the initial weight of the ith trap; b_i is the random effect of trap i representing the deviation of the slope from the fixed effect for trap i; and ε_{ij} are the random errors assumed to be independently distributed with a normal distribution.

3. Results

3.1. Spatial Dynamics

Using semi-variogram analysis, Keane, Gray and Bacciu [11] estimated the spatial scale of individual canopy and surface fuel components and found that the smaller the fuel component, the finer the scale of spatial distribution. Fine woody debris (FWD, Table 1) varied at scales of 1–5 m, depending on the size of fuel particle; CWD varied at 22–160 m; and canopy fuel characteristics varied at 120–600 m scales (Table 2). In fact, Keane, Gray, BacciuandLeirfallom [16] related fuel particle diameter with semi-variogram range and found that an increase in 1 cm in fuel particle diameter resulted in an increase of the range (inherent patch size) by 4.6 m (Figure 4). Results from the FUELVAR study showed that each fuel component has its own inherent spatial scale and that this scale varies by biophysical environment, vegetation structure and composition, and time since disturbance.

Table 2. Semi-variogram range statistics for all surface fuel components and three canopy fuel variables across the six FUELVAR sites. Empty cells indicate no spatial model could be fit to the data. Canopy fuel attributes: CBD-Canopy bulk density, CFL-Canopy fuel load, and CC-Canopy cover.

Fuel Component	Sagebrush Grassland	Pinyon Juniper	Ponderosa Pine Savanna	Ponderosa Pine-Fir	Pine-Fir-Larch	Lodgepole Pine
			Range (m)			
1 h	4.7	2.5	2.8	16.3	8.9	6.0
10 h	6.6	2.4	0.9	4.9	2.2	11.1
100 h	No 100 h	2.5	2.5	4.6	2.4	4.1
1000 h	No Logs	No Logs	84.0	22.0	87.3	157.0
Shrub	2.4	15.1	0.9	1.8	3.6	2.7
Herb	0.7	1.1	0.8	3.5	0.5	1.8
Litter + Duff	0.5	1.4	2.5	1.3	0.5	0.9
CBD	No trees	440.0	-	412.0	100.0	120.0
CFL	No trees	560.0	-	600.0	310.0	560.0
CC	No trees	407.0	-	-	230.0	300.0

There were several other findings of the FUELVAR study that warrant mention. First Keane, Gray and Bacciu [11] found high variability in a number of fuel properties both across and within sites; coefficients of variation (variation expressed as a percent of the mean) exceeded 200% for loading of most woody fuel components and that variation was correlated with particle size ($R^2 = 0.6$). They also found that this variability was not normally distributed, but instead, highly skewed (skewness statistic >2.0 for most components) because many microplots were missing fuel components. Next, Keane, GrayandBacciu [11] found that none of the surface fuel components were correlated with each other; correlation coefficients were <0.4 for the loading of all combinations of surface fuel components. Even more important was the fact that none of the surface fuel components correlated with canopy fuel components or numerous other stand characteristics, such as tree density, basal area, and tree diameter. Each fuel component was distributed independently of all other components.

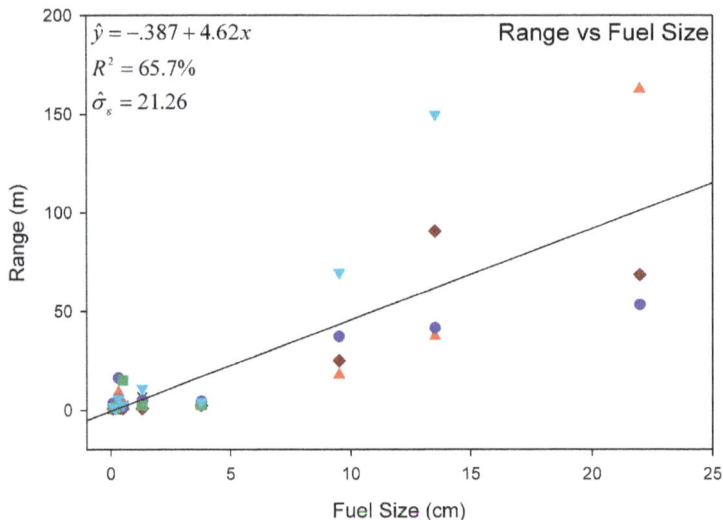

Figure 4. Relationship of semi-variogram range (m) with the size of fuel particle represented by diameter (cm).

3.2. Temporal Dynamics

3.2.1. Deposition

In the FUELDYN study, Keane [31] found that rates of deposition (litterfall) varied by fuel component with foliage having the greatest deposition rates (0.05–0.15 kg$\cdot$m$^{-2}\cdot$year^{-1}) and the largest woody fuel components (100 and 1000 h) having the slowest rates (<0.03 kg$\cdot$m$^{-2}\cdot$year^{-1}) (Figure 5a). Moreover, fuel deposition varied greatly by site, from less than 0.05 kg$\cdot$m$^{-2}\cdot$year^{-1} on unproductive sites (hot, dry and cold, wet) to ten times more (over 0.5 kg$\cdot$m$^{-2}\cdot$year^{-1}) on warm mesic sites (Figure 5b). These deposition rates were more closely correlated to vegetation characteristics than to environmental or climatic factors (e.g., temperature, elevation, and precipitation). In addition, deposition rates of each fuel component differed by site (Figure 5b); woody fuel deposition rates varied widely across all sites (an order of magnitude for 10, 100 h).

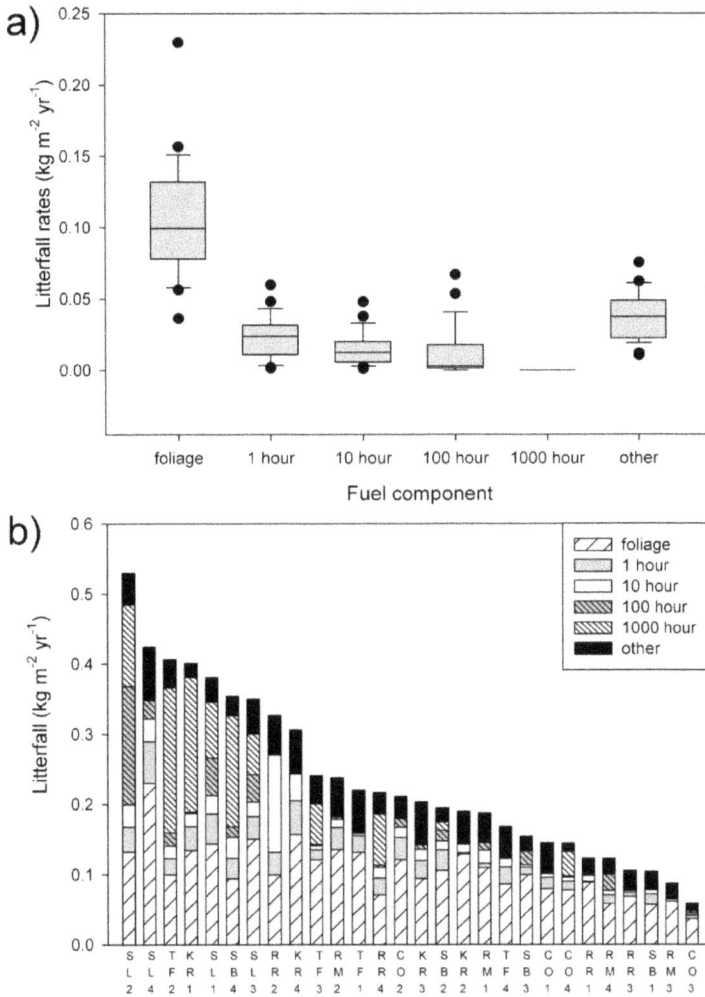

Figure 5. The amount of fuel deposited each year (litterfall) by (**a**) fuel component and (**b**) the 28 sites in the Keane [31] study. Site descriptions ID numbers on the X axis can be found in Keane [31]. Sites are arranged in order of productivity from highest to lowest.

Keane [31] also found that the temporal pattern (i.e., annual variation) of deposited biomass differed by fuel component (Figure 6). It appeared that approximately the same amount of litter and 1 h woody fuels were deposited each year with little year-to-year variability, but the deposition of coarser woody fuels (10, 100 h) was more highly variable in time, with most coarse fuels deposited in one year (Figure 6). Keane [31] also found that only the foliage material was evenly deposited across the littertraps; all woody fuel components were unevenly distributed across both space and time.

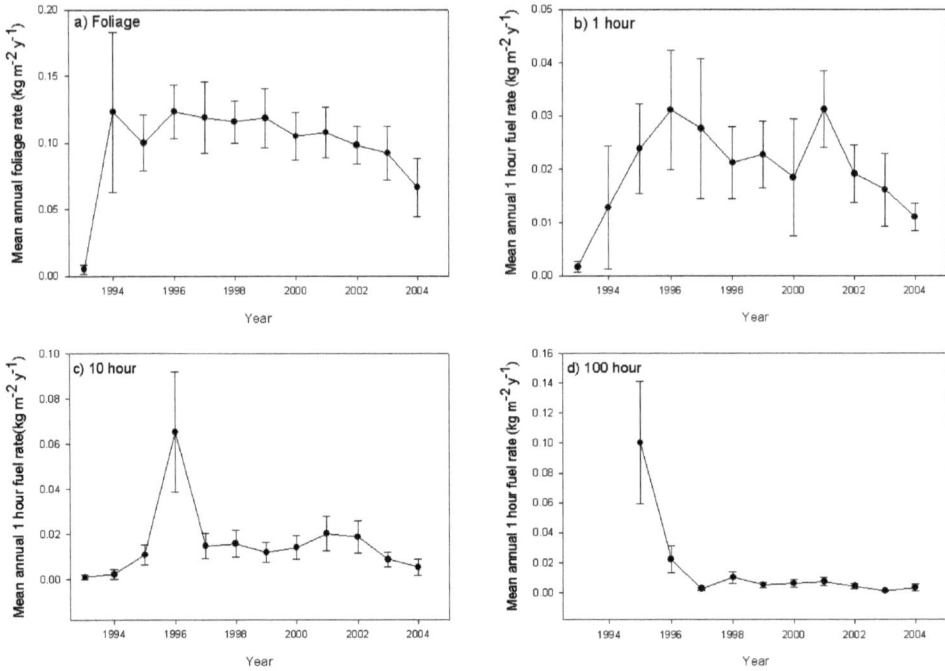

Figure 6. Coarser fuels have higher interannual variability. The amount of deposited biomass by fuel component on the Coram site over 12 years of collection (**a**) foliage (litter); (**b**) 1 h; (**c**) 10 h; and (**d**) 100 h.

3.2.2. Decomposition

Similar to litterfall, decomposition also varies by fuel component and site (Figure 7). The smallest fuel particles (foliage) decayed at the fastest rates, often losing more than a fifth of their mass per year, while the coarser woody particles lost less than 5% of their weight each year (Figure 7a,d). Also, the rate of decomposition was greatly driven by an elevation gradient with the high elevation sites having the fastest decomposition (k values > 0.3 year^{-1}) and the dry, hot low elevation sites had the slowest decomposition (k > 0.1 year^{-1}). In fact, Keane [32] found decomposition rates were best correlated with climate gradients of temperature and moisture. To add to the complexity, measured decomposition rates for each fuel component also differed by site with rates of woody fuel decomposition often unrelated to foliar decomposition rates, and some larger woody fuels having k values greater than smaller woody fuels (Figure 7c).

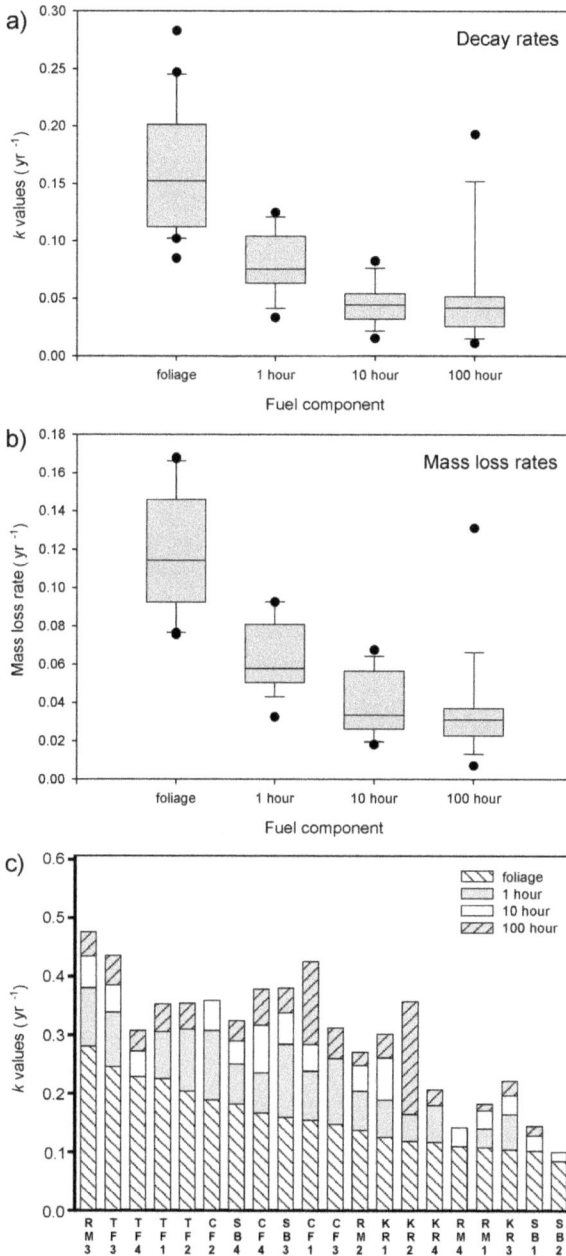

Figure 7. Rates of decomposition expressed as (**a**) k values for each fuel component, (**b**) mass loss rates for each fuel component; and (**c**) *k* values for each site by fuel component. Site descriptions ID numbers on the X axis can be found in Keane [31]. Sites are arranged in order of elevation from highest to lowest. Some sites did not have decomposition estimates [31].

4. Discussion

4.1. The Wildland Fuel Mosaic

Results from these two studies indicate that the landscape mosaic of wildland fuel is much more complex than we once thought [9]. Fuel components are distributed differently in space with fine fuels changing over spatial scales that are an order of magnitude smaller than coarse woody fuels and canopy fuels (Table 2). Fine fuel components are more uniformly distributed in space with lower variabilities than the high variability coarse fuels. The primary reasons for this complexity is that fuel components are deposited on the ground at different rates and patterns (Figure 4), and after they land on the ground, they decompose at different rates (Figure 7), and this deposition and decomposition are governed by different factors (vegetation controls deposition, climate controls decomposition). Fine fuels are added to the surface fuel layer at the greatest rates, but they may decompose quicker, whereas CWD particles are more scattered across space, and it may take decades to centuries for it to decompose depending on site factors.

A framework for understanding spatiotemporal surface and canopy fuel dynamics can be constructed from findings of these two studies [9,31]. First, the spatial locations of the plants that contribute both live and dead fuels provide the coarse template for spatial fuel dynamics. Existing plants grow and die, and new plants are always becoming established. The phenological, physiological, and morphological characteristics of the plants dictate the types of fuels (i.e., live and dead components) available for combustion. Dead material from these plants gets deposited on the ground in patterns based on particle size and weight, height of release, intercepting vegetation, and wind dynamics [25]; small particles may be blown great distances while large heavy particles tend to drop straight down. A greater amount of smaller particles, such as foliage and twigs, are deposited than the large particles, such as branches and logs (Figure 4). Once on the ground, the process of decay acts on these particles to further fracture and decompose organic components [37,38]. Rates of decomposition also depend on the size of the particle with smaller particles decomposing faster than large materials (Figure 6). So, while large particles are rarer than small particles, they have a longer residence time and they tend to be found near the parent plants [39]. Disturbance alters development and deposition rates by killing and maybe consuming whole plants and plant parts, and they can also change the pattern of surface dead fuel by consuming necromass and altering decomposition. The interactions of disturbance and decomposition processes with vegetation development and deposition controls wildland fuelbed properties thereby creating the dynamic fuel mosaic, and this changing fuel mosaic ultimately controls fire regimes.

Two factors that control fuel dynamics were not directly assessed in these two studies—vegetation development and disturbance. Rates of live fuel growth were not included in FUELDYN study because of time and cost considerations, and disturbance was ignored because FUELDYN sites were selected based on their lack of disturbance [31]. Had these two factors been studied, they probably would have revealed even greater complexity and corresponding variability in wildland fuel dynamics. In fact, the two sites in the FUELVAR study that were thinned had twice the variability in loading as the other sites. Live plant biomass accumulates at rates much greater than decompositional losses, and the diversity of species, sizes, and coverage of plants contributes to a diverse and complex set of canopy and surface fuel characteristics. And while biomass accumulation is slow and gradual, the effects of disturbance on wildland fuelbed dynamics may be immediate and extensive. Wildland fire, the most pervasive disturbance in the US northern Rocky Mountains [40,41], causes major changes in wildland fuelbeds through fuel consumption and fire-caused plant mortality. Disturbance impacts are also manifest differently by fuel component; most fine fuels are consumed in wildfires, for example, while the coarse fuels may only partially be consumed [42].

4.2. Implications

Findings from the FUELVAR and FUELDYN studies have sweeping implications for all of fire management. Traditional methods and sampling designs that are used to describe and inventory fuels may be fundamentally limited if they don't recognize the high variability of fuel attributes across space and time [43]. Planar intersect sampling of woody fuel loadings, for example, may be inappropriate for describing spatial variability of woody fuel loading [44]. Conventional fuel inventory techniques, such as photo series [45], may be inappropriate because fuels vary at scales different from the scales represented by the picture in the series; the limited area of evaluation, for example, may be too small to accurately assess loadings for CWD [44]. Inappropriate sampling methods may result in higher uncertainty in fuel loadings that may mask subtle treatment effects [46]. Remotely sensed products used to map fuels, such as Landsat Thematic Mapper, may have resolutions that are inappropriate for capturing the spatial distribution of fuels, especially FWD [47]. Fuel classifications may be ineffective because fuel components vary independently of each other and the high variability of fuel characteristics within a site may overwhelm unique fuelbed identification across sites [43]. Keane et al. [48], for example, found that the high variability of fuel loadings within a classification category resulted in the inability of that category to be accurately mapped. Many fire hazard and risk analyses assume fuels do not change over time [49–51], yet temporal changes in surface and canopy fuels can be large enough to influence fire behavior predictions (Figure 7). High deposition rates, coupled with disturbance effects, may rapidly change fuelbed characteristics and quickly render fuel maps out-dated [47]. Each year, the US spends millions of dollars on fuel treatments that may fast become ineffective because the design failed to account for temporal fuel dynamics [52].

These findings also provide valuable insight into why it is so difficult to create accurate fuel applications and products for fire behavior modeling—complex ecological interactions create high spatiotemporal variability of fuel component characteristics that may compromise predictions. This variability is different for each fuel component and it is so high that it often overwhelms most statistical analyses and classification schemes [48]. Traditional approaches of over-simplifying fuel descriptions for fire simulations may rarely be appropriate, if they ever worked at all [9,46]. Fire behavior fuel models, for example, may provide acceptable results in a one-dimensional application, such as the BEHAVE model [53], but three-dimensional fire behavior models may need more realistic inputs of spatial fuel distributions [18].

Future wildland fuel applications and tools must account for high spatiotemporal variability to be effective in fire management. Instead of managing for average fuel conditions, the ranges and variations must be used to better approximate the highly variable stand conditions. Fire hazard and risk models must have the ability to simulate changes in the fuelbed due to the 4 D's described above [54]. Fuel inventory and monitoring methods must have the ability to sample fuel components at their inherent spatial scale of distribution [44]. Fuel classifications must be designed to account for high variability in loading over multiple scales; Lutes et al. [55], for example, created a fuel classification of potential fire effects using field sampled fuel component loadings but failed to include measures of variability. Fuel maps must be designed to successfully link fuel component spatial scales with mapping methods, and these maps should be linked to fuel dynamics models to simulate potential changes that may render the map quickly outdated.

5. Summary

The ecological processes of vegetation development (succession), deposition, decomposition, and disturbance interact to create complex wildland fuel mosaics in the forests of the US northern Rocky Mountains. This fuel mosaic provides a template for future fire patterns, and its shifts over time influence the intrinsic fire regime that eventually emerges from the interactions of the fuel mosaic, fire ignition, and climate [41,56]. The fuel mosaic is constantly changing at rates dictated by the biophysical environment, the vegetation, and disturbance interactions. This dynamic character of fuel mosaics contributes to high spatiotemporal variability that confounds development and use of wildland fuel applications in fire management. Future wildland fuel applications must address both the spatial and temporal variability of wildland fuels to ensure accurate fuel inputs to fire management applications.

Acknowledgments: This work was partially funded by the USGS National Biological Service and Glacier National Park's Global Change Research Program under Interagency Agreements 1430-1-9007 and 1430-3-9005 and the USGS CLIMET project. It was also funded using National Fire Plan Research monies.

Conflicts of Interest: The author declares no conflict of interest.

Abbreviations

The following abbreviations are used in this manuscript:

FUELVAR	A comprehensive project aimed at quantifying the spatial variability of fuels
FUELDYN	A 12 years study quantifying litterfall and decomposition rates for US northern Rocky Mountain undisturbed stands
1 h	downed, dead woody fuel particles less than 6 mm in diameter
10 h	downed, dead woody fuel particles less than 25 mm in diameter
100 h	downed, dead woody fuel particles less than 76 mm in diameter
1000 h	downed, dead woody fuel particles greater than 76 mm in diameter
FWD	all dead fuel particles below 76 mm in diameter
CWD	all dead fuel particles greater than 76 mm in diameter

References

1. Heyerdahl, E.K.; Brubaker, L.B.; Agee, J.K. Spatial controls of historical fire regimes: A multiscale example from the Interior West, USA. *Ecology* **2001**, *82*, 660–678. [CrossRef]
2. Swetnam, T.W.; Betancourt, J.L. Fire-southern oscillation relations in the southwestern United States. *Science* **1990**, *249*, 1017–1020. [CrossRef] [PubMed]
3. Weise, D.R.; Wright, C.S. Wildland fire emissions, carbon and climate: Characterizing wildland fuels. *For. Ecol. Manag.* **2014**, *317*, 26–40. [CrossRef]
4. Rocca, M.E.; Brown, P.M.; MacDonald, L.H.; Carrico, C.M. Climate change impacts on fire regimes and key ecosystem services in Rocky Mountain forests. *For. Ecol. Manag.* **2014**, *327*, 290–305. [CrossRef]
5. Agee, J.K.; Skinner, C.N. Basic principles of forest fuel reduction treatments. *For. Ecol. Manag.* **2005**, *211*, 83–96. [CrossRef]
6. McKenzie, D.; Kennedy, M.C. Scaling laws and complexity in fire regimes. In *The Landscape Ecology of Fire*; McKenzie, D., Miller, C., Falk, D.A., Eds.; Springer Ltd.: Dordrecht, The Netherlands, 2011.
7. McKenzie, D.; Shankar, U.; Keane, R.E.; Stavros, E.N.; Heilman, W.E.; Fox, D.G.; Riebau, A.C. Smoke consequences of new wildfire regimes driven by climate change. *Earth's Future* **2014**, *2*, 35–59. [CrossRef]
8. Reinhardt, E.D.; Keane, R.E.; Calkin, D.E.; Cohen, J.D. Objectives and considerations for wildland fuel treatment in forested ecosystems of the interior western United States. *For. Ecol. Manag.* **2008**, *256*, 1997–2006. [CrossRef]
9. Keane, R.E. *Wildland Fuel Fundamentals and Applications*; Springer: New York, NY, USA, 2015.
10. Harmon, M.E.; Franklin, J.F.; Swanson, F.J.; Sollins, P.; Gregory, S.V.; Lattin, J.D.; Anderson, N.H.; Cline, S.P.; Aumen, N.G.; Sedell, J.R.; et al. Ecology of coarse woody debris in temperate ecosystems. *Adv. Ecol. Res.* **1986**, *15*, 133–302.
11. Keane, R.E.; Gray, K.; Bacciu, V. *Spatial Variability of Wildland Fuel Characteristics in Northern Rocky Mountain Ecosystems*; Research Paper RMRS-RP-98; USDA Forest Service Rocky Mountain Research Station: Fort Collins, CO, USA, 2012; p. 56.
12. Kalabokidis, K.; Omi, P. Quadrat analysis of wildland fuel spatial variability. *Int. J. Wildland Fire* **1992**, *2*, 145–152. [CrossRef]
13. Brown, J.K.; Bevins, C.D. *Surface Fuel Loadings and Predicted Fire Behavior for Vegetation Types in the Northern Rocky Mountains*; INT-358; U.S. Department of Agriculture, Forest Service, Intermountain Forest and Range Experiment Station: Ogden, UT, USA, 1986; p. 9.
14. Waring, R.H.; Running, S.W. *Forest Ecosystems: Analysis at Multiple Scales*, 2nd ed.; Academic Press, Inc.: San Diego, CA, USA, 1998; p. 370.

15. Menakis, J.P.; Keane, R.E.; Long, D.G. Mapping ecological attributes using an integrated vegetation classification system approach. *J. Sustain. For.* **2000**, *11*, 245–265. [CrossRef]

16. Keane, R.; Gray, K.; Bacciu, V.; Leirfallom, S. Spatial scaling of wildland fuels for six forest and rangeland ecosystems of the northern Rocky Mountains, USA. *Landsc. Ecol.* **2012**, *27*, 1213–1234. [CrossRef]

17. Stephens, S.L.; McIver, J.D.; Boerner, R.E.; Fettig, C.J.; Fontaine, J.B.; Hartsough, B.R.; Kennedy, P.L.; Schwilk, D.W. The effects of forest fuel-reduction treatments in the United States. *BioScience* **2012**, *62*, 549–560.

18. Parsons, R.A.; Mell, W.E.; McCauley, P. Linking 3D spatial models of fuels and fire: Effects of spatial heterogeneity on fire behavior. *Ecol. Model.* **2010**, *222*, 679–691. [CrossRef]

19. King, K.J.; Bradstock, R.A.; Cary, G.J.; Chapman, J.; Marsden-Smedley, J.B. The relative importance of fine-scale fuel mosaics on reducing fire risk in south-west Tasmania, Australia. *Int. J. Wildland Fire* **2008**, *17*, 421–430. [CrossRef]

20. Reich, R.M.; Lundquist, J.E.; Bravo, V.A. Spatial models for estimating fuel loads in the Black Hills, South Dakota, USA. *Int. J. Wildland Fire* **2004**, *13*, 119–129. [CrossRef]

21. Hiers, J.K.; O'Brien, J.J.; Mitchell, R.J.; Grego, J.M.; Loudermilk, E.L. The wildland fuel cell concept: An approach to characterize fine-scale variation in fuels and fire in frequently burned longleaf pine forests. *Int. J. Wildland Fire* **2009**, *18*, 315–325. [CrossRef]

22. Peters, D.P.C.; Mariotto, I.; Havstad, K.M.; Murray, L.W. Spatial variation in remnant grasses after a grassland-to-shrubland state change: implications for restoration. *Rangel. Ecol. Manag.* **2006**, *59*, 343–350. [CrossRef]

23. Theobald, D. A general model to quantify ecological integrity for landscape assessments and US application. *Landsc. Ecol.* **2013**, *28*, 1859–1874. [CrossRef]

24. Kreye, J.K.; Varner, J.M.; Dugaw, C.J. Spatial and temporal variability of forest floor duff characteristics in long-unburned Pinus palustris forests. *Can. J. For. Res.* **2014**, *44*, 1477–1486. [CrossRef]

25. Ferrari, J.B. Fine-scale patterns of leaf litterfall and nitrogen cycling in an old-growth forest. *Can. J. For. Res.* **1999**, *29*, 291–302. [CrossRef]

26. Jin, S. Computer classification of four major components of surface fuel in Northeast China by image: The first step towards describing spatial heterogeneity of surface fuels by imags. *For. Ecol. Manag.* **2004**, *203*, 395–406. [CrossRef]

27. Jia, G.J.; Burke, I.C.; Goetz, A.F.H.; Kaufmann, M.R.; Kindel, B.C. Assessing spatial patterns of forest fuels using AVIRIS data. *Remote Sens. Environ.* **2006**, *102*, 318–327. [CrossRef]

28. Miller, C.; Urban, D.L. Connectivity of forest fuels and surface fire regimes. *Landsc. Ecol.* **2000**, *15*, 145–154. [CrossRef]

29. Knapp, E.E.; Keeley, J.E. Heterogeneity in fire severity within early season and late season prescribed burns in a mixed-conifer forest. *Int. J. Wildland Fire* **2006**, *15*, 37–45. [CrossRef]

30. Jenkins, M.J.; Page, W.G.; Hebertson, E.G.; Alexander, M.E. Fuels and fire behavior dynamics in bark beetle-attacked forests in Western North America and implications for fire management. *For. Ecol. Manag.* **2012**, *275*, 23–34. [CrossRef]

31. Keane, R.E. *Surface Fuel Litterfall and Decomposition in the Northern Rocky Mountains, USA*; Research Paper RMRS-RP-70; USDA Forest Service Rocky Mountain Research Station: Fort Collins, CO, USA, 2008; p. 22.

32. Keane, R.E. Biophysical controls on surface fuel litterfall and decomposition in the northern Rocky Mountains, USA. *Can. J. For. Res.* **2008**, *38*, 1431–1445. [CrossRef]

33. Bellehumeur, C.; Legendre, P. Multiscale sources of variation in ecological variables: Modeling spatial dispersion, elaborating sampling designs. *Landsc. Ecol.* **1998**, *13*, 15–25. [CrossRef]

34. Townsend, D.E.; Fuhlendorf, S.D. Evaluating Relationships between Spatial Heterogeneity and the Biotic and Abiotic Environments. *Am. Midl. Nat.* **2010**, *163*, 351–365. [CrossRef]

35. Fortin, M.-J. Spatial statistics in landscape ecology. In *Landscape Ecological Analysis: Issues and Applications*; Klopatek, J.M., Gardner, R.H., Eds.; Springer-Verlag, Inc.: New York, NY, USA, 1999; pp. 253–279.

36. Olson, J. Energy storage and the balance of the producers and decomposers in ecological systems. *Ecology* **1963**, *44*, 322–331. [CrossRef]

37. Harmon, M.E.; Krankina, O.N.; Sexton, J. Decomposition vectors: A new approach to estimating woody detritus decomposition dynamics. *Can. J. For. Res.* **2000**, *30*, 76–84. [CrossRef]

38. Kaarik, A.A. Decomposition of wood. In *Biology of Plant Litter Decomposition*; Dickinson, C.H., Pugh, G.J.F., Eds.; Academic Press: London, UK, 1974; Volume 1, pp. 129–174.

39. Harmon, M.E.; Hua, C. Coarse woody debris dynamics in two old-growth ecosystems. *Bioscience* **1991**, *41*, 604–610. [CrossRef]

40. McKenzie, D.; Miller, C.; Falk, D.A. *The Landscape Ecology of Fire*; Springer Ltd.: Dordrecht, The Netherlands, 2011.

41. Scott, A.C.; Bowman, D.M.J.S.; Bond, W.J.; Pyne, S.J.; Alexander, M.E. *Fire on Earth: An Introduction*; John Wiley and Sons Ltd.: Chichester, UK, 2014; p. 413.

42. Brown, J.K.; Reinhardt, E.D.; Fischer, W.C. Predicting duff and woody fuel consumption in northern Idaho prescribed fires. *For. Sci.* **1991**, *37*, 1550–1566.

43. Keane, R.E. Describing wildland surface fuel loading for fire management: A review of approaches, methods and systems. *Int. J. Wildland Fire* **2013**, *22*, 51–62. [CrossRef]

44. Keane, R.E.; Gray, K. Comparing three sampling techniques for estimating fine woody down dead biomass. *Int. J. Wildland Fire* **2013**, *22*, 1093–1107. [CrossRef]

45. Fischer, W.C. *Photo Guide for Appraising Downed Woody Fuels in Montana Forests: Interior Ponderosa Pine, Ponderosa Pine-Larch-Douglas-Fir, Larch-Douglas-Fir, and Interior Douglas-Fir Cover Types*; INT-97; USDA Forest Service Intermountain Research Station: Ogden, UT, USA, 1981; p. 133.

46. Sikkink, P.G.; Keane, R.E. A comparison of five sampling techniques to estimate surface fuel loading in montane forests. *Int. J. Wildland Fire* **2008**, *17*, 363–379. [CrossRef]

47. Keane, R.E.; Burgan, R.E.; Wagtendonk, J.V. Mapping wildland fuels for fire management across multiple scales: Integrating remote sensing, GIS, and biophysical modeling. *Int. J. Wildland Fire* **2001**, *10*, 301–319. [CrossRef]

48. Keane, R.E.; Herynk, J.M.; Toney, C.; Urbanski, S.P.; Lutes, D.C.; Ottmar, R.D. Evaluating the performance and mapping of three fuel classification systems using Forest Inventory and Analysis surface fuel measurements. *For. Ecol. Manag.* **2013**, *305*, 248–263. [CrossRef]

49. Hall, S.A.; Burke, I.C. Considerations for characterizing fuels as inputs for fire behavior models. *For. Ecol. Manag.* **2006**, *227*, 102–114. [CrossRef]

50. Finney, M.A. Design of regular landscape fuel treatment patterns for modifying fire growth and behavior. *For. Sci.* **2001**, *47*, 219–228.

51. Finney, M.A. A Computational Method for Optimizing Fuel Treatment Locations. *In Fuels management-how to measure success*, Proceedings of the Rocky Mountain Research Station Conference, Portland, OR, USA, 28–30 March 2006; Andrews, P.L., Butler, B.W., Eds.; U.S. Department of Agriculture, Forest Service, RMRS-P-41: Portland, OR, USA, 2006; pp. 107–123.

52. Parks, S.A.; Miller, C.; Holsinger, L.M.; Baggett, L.S.; Bird, B.J. Wildland fire limits subsequent fire occurrence. *Int. J. Wildland Fire* **2016**, *25*, 182–190. [CrossRef]

53. Andrews, P.L. *BehavePlus Fire Modeling System, Version 4.0: Variables*; RMRS-GTR-213WWW; USDA Forest Service Rocky Mountain Research Station: Fort Collins, CO, USA, 2008; p. 107.

54. Beukema, S.J.; Greenough, J.A.; Robinson, D.C.E.; Kurtz, W.A.; Reinhardt, E.D.; Crookston, N.L.; Brown, J.K.; Hardy, C.C.; Stage, A.R. An introduction to the Fire and Fuels Extension to FVS. In Proceedings of the Forest Vegetation Simulator Conference, Ft. Collins, CO, USA, 3–7 February 1997; Teck, R., Moeur, M., Adams, J., Eds.; United States Department of Agriculture, Forest Service, Intermountain Forest and Range Experiment Station: Ft. Collins, CO, USA, 1997; pp. 191–195.

55. Lutes, D.C.; Keane, R.E.; Caratti, J.F. A surface fuels classification for estimating fire effects. *Int. J. Wildland Fire* **2009**, *18*, 802–814. [CrossRef]

56. Bowman, D.M.J.S.; Balch, J.K.; Artaxo, P.; Bond, W.J.; Carlson, J.M.; Cochrane, M.A.; D'Antonio, C.M.; DeFries, R.S.; Doyle, J.C.; Harrison, S.P.; et al. Fire in the Earth System. *Science* **2009**, *324*, 481–484. [CrossRef] [PubMed]

© 2016 by the author. Licensee MDPI, Basel, Switzerland. This article is an open access article distributed under the terms and conditions of the Creative Commons Attribution (CC BY) license (http://creativecommons.org/licenses/by/4.0/).

forests

MDPI

Article

Detecting Local Drivers of Fire Cycle Heterogeneity in Boreal Forests: A Scale Issue

Annie Claude Bélisle [1,*], Alain Leduc [2], Sylvie Gauthier [3], Mélanie Desrochers [4], Nicolas Mansuy [3], Hubert Morin [5] and Yves Bergeron [6]

[1] Institut de Recherche sur les Forêts, Université du Québec en Abitibi-Témiscamingue and Centre d'étude de la forêt, Case Postale 8888, Succursale Centre-Ville Montréal, Quebec, QC H3C 3P8, Canada

[2] Département des Sciences Biologiques, Chaire Industrielle CRSNG UQAT-UQAM en Aménagement Forestier Durable, Université du Québec à Montréal and Centre d'étude de la forêt, Case Postale 8888, succursale Centre-Ville Montréal, Quebec, QC H3C 3P8, Canada; leduc.alain@uqam.ca

[3] Natural Resources Canada, Canadian Forest Service, Laurentian Forestry Centre, 1055 du PEPS, PO Box 10380, Stn. Sainte-Foy, Québec, QC G1V 4C7, Canada; sylvie.gauthier2@canada.ca (S.G.); nicolas.mansuy@canada.ca (N.M.)

[4] Centre d'étude de la forêt, Case Postale 8888, Succursale Centre-Ville Montréal, Montreal, QC H3C 3P8, Canada; desrochers.melanie@uqam.ca

[5] Département des Sciences Fondamentales, Université du Québec à Chicoutimi, 555 Boulevard de l'Université, Chicoutimi, QC G7H 2B1, Canada; hubert_morin@uqac.ca

[6] Chaire Industrielle CRSNG UQAT-UQAM en Aménagement Forestier Durable, Université du Québec en Abitibi-Témiscamingue and Université du Québec à Montréal, 445, boul. de l'Université, Rouyn-Noranda, QC J9X 5E4, Canada; Yves.Bergeron@uqat.ca

* Correspondence: AnnieClaude.Belisle@uqat.ca; Tel.: +1-877-870-8728 (ext. 4353)

Academic Editors: Timothy A. Martin and Eric J. Jokela
Received: 30 May 2016; Accepted: 29 June 2016; Published: 12 July 2016

Abstract: Severe crown fires are determining disturbances for the composition and structure of boreal forests in North America. Fire cycle (FC) associations with continental climate gradients are well known, but smaller scale controls remain poorly documented. Using a time since fire map (time scale of 300 years), the study aims to assess the relative contributions of local and regional controls on FC and to describe the relationship between FC heterogeneity and vegetation patterns. The study area, located in boreal eastern North America, was partitioned into watersheds according to five scales going from local (3 km^2) to landscape (2800 km^2) scales. Using survival analysis, we observed that dry surficial deposits and hydrography density better predict FC when measured at the local scale, while terrain complexity and slope position perform better when measured at the middle and landscape scales. The most parsimonious model was selected according to the Akaike information criterion to predict FC throughout the study area. We detected two FC zones, one short (159 years) and one long (303 years), with specific age structures and tree compositions. We argue that the local heterogeneity of the fire regime contributes to ecosystem diversity and must be considered in ecosystem management.

Keywords: forest ecology; fire cycle; multi-scale models; survival analysis; physiographic drivers

1. Introduction

Patterns and processes of boreal forests are highly determined by large-scale disturbance regimes [1,2]. In North America, stand-replacing crown fires [3–5] shape landscapes in patchworks where forest stands differ in their age, internal structure and composition [6,7]. Disturbance regime parameters are used as benchmarks by forest managers concerned with ecosystem management and the natural range of variability [8,9]. Understanding fire regime drivers and their consequences on forest landscapes is hence a key issue for boreal landscape ecology, resilience and management [10–13].

Fire frequency, or rate, is a parameter of fire regime expressing the mean proportion of area that is burned yearly in the region of interest [14]. It can also be expressed by the concept of fire cycle (FC), having both area- and point-based meanings and interpretations [15]. From the area-based perspective, it is defined as the time required (year) to burn an area equivalent to the region of interest and is mathematically the inverse of annual burn rate [15]. It thus assumes homogeneity over the entire region of interest. From the point-based perspective, FC is the mean time interval (year) between fires during a certain period of time (e.g., the Holocene) for a single location. FC in a region drives landscape age-structure [16], species composition [17] and succession pathways [18–20]. Easily transferable to forest ecosystem management, FC and associated landscape age-structure are useful to compare harvested and natural landscapes [10,21]. For example, FC was used to describe old-growth forests' natural range of abundance and to highlight their depletion in managed landscapes of eastern Canada [22]. In this research, we stress the homogeneity assumption and propose a way to deal with a heterogeneous area.

FC spatial variability depends on the interactions between top-down controls, such as weather (regional to global scales), and bottom-up controls, such as fuel availability, quality and connectivity (local to landscape scale) [23–25]. In boreal North America, large fires, even if infrequent compared to small fires, are responsible for most of the burned area (3% of the fires account for 97% of burned area in Canada) [3]. Their occurrence and distribution are generally assumed to be driven by weather conditions [26,27]. FC hence follows a continental climate gradient, going from the dry inland continent where fires are recurrent to the wet East Coast where fires are rare [28]. Additional spatial variability in FC is observed at the regional scale and may be explained by soil drainage and regional fire weather [28,29]. It is a more complicated task to identify local drivers of FC because of large and severe fires, making it impossible to compare for instance the number of fires or the area burned in localities. Researchers interested in bottom-up drivers of FC need to turn towards historical fire data collected with dendrochronology or soil and lake charcoal deposits [30–33]. Local variability may also be captured with simulations based on fire behavior models [24,34]. The first presents the advantage of describing an empirical reality, but requires greater effort for historical data acquisition. The second allows covering greater spatial extents and taking into account the mechanisms involved, but relies mostly on simulations.

In eastern North America, FC is generally assumed to be regionally homogeneous. It ranges from ~100 years–>500 years for the last 300 years and from 388–645 for the recent period (1940–2003) [29,35,36]. However, vegetation patterns are sometimes in apparent contradiction with the estimated FC within the homogeneous FC assumption. For instance, FC reported for jack pine (*Pinus banksiana* Lamb.) stands is often shorter than regional FC, partly because of low-severity fire occurrence (e.g., [33,37]. Moreover, jack pine should theoretically be excluded from zones where fire frequency is low because of its serotinous cones, which need extreme heat to open [18,32,38]. Otherwise, its distribution extends on sandy soils of regions where historical FC exceeds species longevity [36,39]. Driving factors of FC recently appeared to be highly scale dependent, making studies undertaken on a single scale unable to capture FC spatial complexity [40–43]. A multi-scale approach is necessary to investigate the influence of both local and regional factors on FC. Measuring and understanding this relation is important for fire ecology, modeling and forest management.

The objective of this research is to assess the relative contributions of local versus regional controls on FC and to describe the relation between FC and vegetation succession patterns in boreal forests of Eastern Canada. We hypothesize that: (1) FC is primarily influenced by bottom-up processes driven by coarse topographic features, especially valleys where dry soils in conjunction with wind corridors may enhance fire frequency [44,45] and hills that may act as fire breaks [46]; (2) FC is locally shorter where soils are well drained [29,47] and longer where fire breaks are abundant [48]; and (3) the vegetation succession pathway spatial distribution goes along with the FC spatial distribution.

In this paper, we assess the relationship between FC and physiographic variables for a boreal region of eastern Canada. We used survival analysis to model FC with time since fire data covering the last 300 years. We addressed the scale issue by partitioning the study area into five scale classes, from local to landscape, based on watershed Strahler order. All physiographic variables were thus measured for five scales in order to identify the most accurate to predict FC. Model selection based on the Akaike information criterion was performed to choose the "best" variable scale combinations.

2. Materials and Methods

2.1. Study Area

The study area (Figure 1) covers 540,300 ha in the continuous boreal forest of eastern North America (71°15′ W–72°45′ W, 49°36′ N–50°59′ N). Black spruce (*Picea mariana* Mill.) feather moss forest is the most abundant ecological type [49]. Jack pine, paper birch (*Betula papyrifera* Marsh.) and trembling aspen (*Populus tremuloides* Michx.) are present in some young forests. Balsam fir (*Abies balsamea* Mill.) is co-dominant with black spruce in some old-growth forests [39]. Elevation ranges between 120 m and 870 m, striking a contrast between the typical rounded hills of the Boreal Shield and the valleys of the deeply embedded rivers [50,51]. Hydrography is structured by two north-south-oriented rivers (Mistassibi and Mistassibi Nord-Est). Surficial materials are mainly composed of glacial tills with rocky outcrops and fluvioglacial sands following river beds. Some depletion moraines are located on the northern part of the study area; organic surficial materials are also common [51,52]. Mean temperature ranges between a minimum of 6 °C–19 °C and a maximum of 21 °C–25 °C in July and between a minimum of −29 °C–−25 °C and a maximum of −14 °C–−10 °C in January (1971–2001). Average annual precipitation ranges between 900 mm and 1200 mm, 30%–35% of which falls as snow [50,53].

As elsewhere in the boreal forest of eastern North America, the landscape is subject to severe crown fires [4]. FC in the region is estimated at 247 years for the last 300 years and at 375 years for the recent period (1949–2009). The mean fire size was 10,000 ha between 1970 and 2010 [13]. Most of the burned area is attributable to four fire decades: the 1820s, 1860s, 1920s and 2000s [13]. The study area is located in a homogeneous fire region according to [28]. No influence of latitude and longitude on FC has been observed [39].

Located on public tenures, the study area has been partly, but extensively managed for wood supply since the 1970s [54,55]. Before 1950, the landscape remained almost free from human activities, with a native population density estimated at 0.005 individuals/km^2 [56]. The closest villages are located 70 km south and were settled between 1870 and 1930 [57].

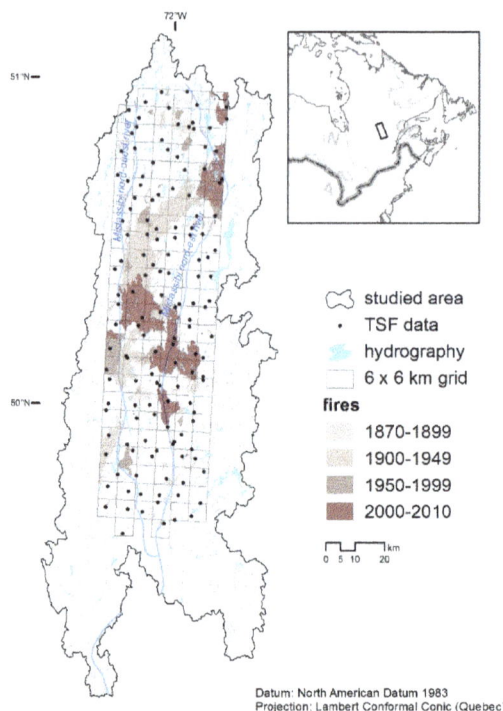

Figure 1. Stand origin map [21] within the study area, including fire delineation (1900–2010) from aerial photos and time elapsed since last fire (TSF) sampling points.

2.2. Environment Delineation and Scaling

We designed environment delineation and scaling according to the drainage network and associated watersheds. Watersheds rely on water bodies with a hierarchical drainage network, making them embedded and scalable. Strahler order [58] is used to classify rivers and their catchment areas according to their importance (see Appendix A Figure A1). Moreover, contrary to administrative borders or the circular buffers generally used (e.g., [30,40]), they rely on real physiographic features [59].

Watersheds were delineated using the ArcGIS 10.1 (Esri, Redlands, CA, USA) Hydrology toolbox (see Table A1). Strahler orders were computed using a 20-m flow accumulation raster in which a cell was considered a river when its flow accumulation was greater than 16 ha (400 cells, as suggested by the ArcGIS 10.1 help module). Order 1 rivers were fuzzy and did not fit well with previously mapped rivers, so we only considered Orders 2–6, reported as 1–5. We delineated watersheds for every river (manual pour-point distribution, Watershed tool), defined by their hierarchical importance. We consider Orders 1 and 2 as local scale, Order 3 as middle scale and Orders 4 and 5 as landscape scale.

The number of watersheds exponentially decreases when the scale gets coarser, ranging from 3736 watersheds of Order 1 to 6 of Order 5 (Figure 2). Conversely, mean watershed size increases exponentially with order, ranging from 2.9 km^2 (Order 1) to 1782.5 km^2 (Order 5). Watersheds generally present an elongated shape and follow a north-south axis, especially at the middle and landscape scales.

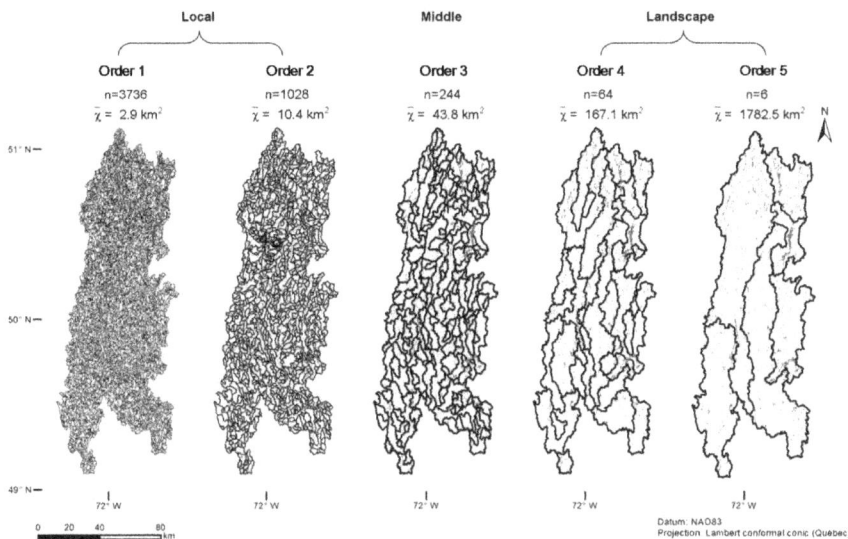

Figure 2. Watershed delineation for Strahler Orders 1–5, grouped into local (Orders 1 and 2), middle (Order 3) and landscape (Orders 4 and 5) scales. *n* = number of watersheds, $\bar{x}$ = mean area (km^2).

2.3. Data Collection

The method used to study the spatial distribution of FC uses historical time since fire data. When assessing FC, spatial extent can be traded for temporal depth and vice versa. For example, to assess FC for a precise location (point-based), paleoclimatic data are needed. Alternatively, to assess FC for a whole region (area-based), annual burn rate (area burned/year) by itself allows calculating FC. This research is at the mid-point between both perspectives. Using historical data allowed us to divide the study area into smaller zones according to their specific FC.

We used a stand-origin map to derive FC from the time elapsed since last fire (TSF) distribution [60]. This methodology has proven to be useful for regimes of severe and infrequent fires where fire scars are rare and long time series are needed [31,32,36,61]. TSF point data (*n* = 144, density = 1 point every 36 km^2) are derived from government fire registers (Direction de la protection des forêts, 1950–2010), vegetation databases [62], aerial photographs (1948) and tree ring analysis (see [21] for details on data collection) (Figure 1). TSF data are scattered on a grid (resolution: 6 km) with one random point in every cell. TSF is considered real when traces of the last fire make it possible to determine a fire year ($\pm$ 10 years) (from aerial photos or even-aged tree distribution); otherwise, a minimum and censored TSF, corresponding to the age of the oldest tree sampled, is attributed [62].

In order to test the effect of physiographic factors on FC, we built scale-dependent metrics of soil drying potential, fire break density, terrain complexity and slope position (Table 1). This means that for a single location in the study area, those metrics take on different values depending on the scale at which they are measured. Spatial analyses were processed on ArcGIS 10.0 using a 20-m raster from which water bodies were excluded (except for fire break density calculation).

$$\ln h\,(t) = \ln h_0\,(t) + \beta_1 X_1 + \ldots + \beta_p X_p \;(1) \tag{1}$$

h(t), Hazard to survive (not to burn) to time t; h$_0$ (t), Unspecified baseline hazard; P, Number of predictor variables; β, Coefficient estimate; X, Predictor value.

Table 1. Definition of explanatory (predictor) variables (*p* in the Cox proportional hazard model, see Equation (1)).

Symbol	Variable	Definition	Units
DRY	Dry surficial deposits density *	The ratio between dry deposit area and watershed area	%
HD	Hydrographic density	The ratio between lake and river area and watershed area	%
ESD	Elevation standard deviation		m
TPI	Topographic position index	Standardized local elevation deviation from the mean elevation	-

* Dry materials of the study area were mainly composed of sandy textured fluvioglacial and juxtaglacial deposits, disintegration moraines and rocky outcrops.

We hypothesized that well-drained soils enhance combustible dryness and inflammability and thus shorten FC. We used a surficial materials classification [29] to make the distinction between high and low soil drying potential. Dry materials of the study area were mainly composed of sandy textured fluvioglacial and juxtaglacial deposits, disintegration moraines and rocky outcrops. Dry surficial material density (DRY) is thereby defined as the proportion of an area occupied by dry deposits on a given area. Hydrology density (HD) is defined as the proportion of an area occupied by water bodies (lakes and rivers) and is used as an indicator of fire break abundance [48]. HD was computed using a water body map (1:50,000) provided by the Government of Canada [50]. Literally defined as the elevation standard deviation, Elevation standard deviation (ESD) is an index of topography ruggedness and may also be associated with fire break density. The more the terrain presents slopes and hills, the higher the ESD. Finally, the topographic position index (TPI) is a measure of slope position [63] and is defined as the difference between local elevation and mean environment elevation, divided by environment ESD. High slope position locations may be less prone to fire because of a colder and wetter climate creating a persistent snow cover during spring [46]. A TPI below -1 indicates valleys and down-slopes; a TPI between -1 and 1 indicates mid-slopes; and a TPI greater than 1 indicates up-slopes and mountain tops.

The variable distribution changes with scale, so for a single point in space, the same variable can take on different values according to the scale. DRY, HD, ESD and TPI distributions throughout the study area are presented for all five scales (Figure 3). The range of DRY and HD narrows when the scale gets coarser, so their distribution follows small patches in the study area. ESD range increases with scale, so the topography is characterized by coarse features. TPI range remains the same as expected because this metric is normalized. For TPI, HD and ESD, watershed is described by a unique value, so for Order 5 watersheds, they can take on only six possible values.

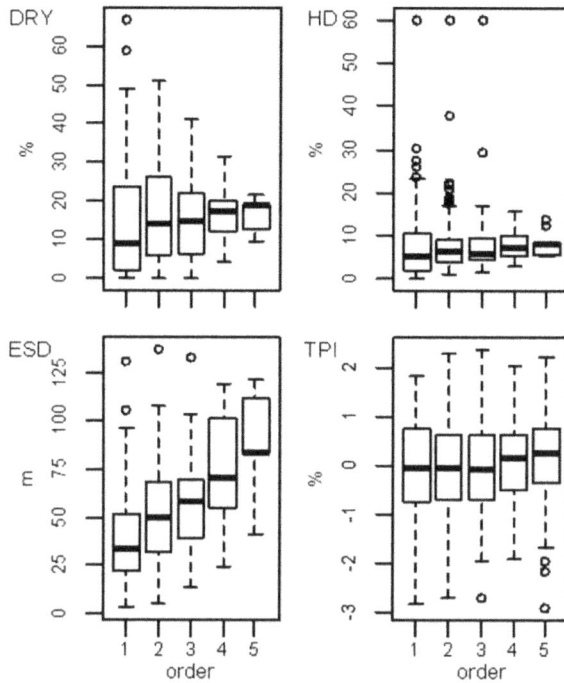

Figure 3. Dry surface deposits density (DRY) (%), Hydrographic density (HD) (%), Elevation Standard Deviation (ESD) (m) and Topographic position index (TPI) (%) distributions (see Table 1) for Orders 1–5 (see Figure 2). Horizontal lines represent the 25th, 50th and 75th percentiles, and whiskers indicate the range. Dots represent values for which the distance from the box is 1.5-times greater than the length of the box.

2.4. Survival Analysis

Survival analysis tests the relation between two or more variables and includes censored data that are only partially known [64]. TSF is censored when the year of the last fire is unknown so the only information we have is a minimum time elapsed without fire. We used the Cox proportional hazard model (Equation (1)), a semi-parametric survival analysis, to model the influence of physical environment on fire frequency [30,31,64]. Statistical analyses were computed in the software R (version 2.15.1, The R Foundation for Statistical Computing, Vienna, Austria), package survival [65].

We first tested the effect of environmental variables and scales on fire hazard using and comparing single-variable survival analysis (see Table 2 for the FC modeling main steps). We tested 20 proportional hazard models (function coxph) (4 variables × 5 scales) [66]. DRY was log-transformed to meet the proportional hazard model assumption, i.e., the relative risk for each stratum must remain the same over time. Models were performed with a single dataset of 144 TSF points.

Table 2. Main steps for the model selection and projection.

1.	Univariate survival models	Figure 5
2.	Selection of variable scales (ΔAICcNull < 2)	Figure 5
3.	Removal of correlated variable scales	-
4.	Modeling all possible combinations of variables and scales	-
5.	Analysis of variable scales contributions using model averaging	Table 3
6.	Selection of the model used for FC prediction based on: (a) ΔAICc (b) Simplicity criteria (number of variable-scales involved)	Table A1
7.	Projection of FC for the whole study area	Figure 6
8.	Classification of the study area according to FC zones	Table 5, Figure 7
9.	Validation with independent vegetation data	Table 6

Model accuracy and performance were compared on the basis of their second-order Akaike information criterion (AICc), informing on the balance between model complexity and predictivity [67]. Wald test and log-likelihood ratio test scores ($p < 0.05$) inform on the goodness of fit compared to the null model [64]. Linear predictors indicate the sense and the relative magnitude of the relation. Confidence intervals (95%) on linear predictors were computed with bootstrapping (1000 resamplings with replacement from the original dataset, $n = 144$). We also suspected an overwhelming effect of recent fires, so we sequentially excluded fire years (2009, 2007, 2005) to check their influence on results [68]. Only relevant variable-scale combinations, with an AICc difference with null model (ΔAICcNull) > 2 [69], were considered for multi-scale models. If two variable scales were correlated (Pearson's coefficient >0.7), one of them was removed in further analysis. After selecting significant and uncorrelated variables and scales, we then tested existing combinations of variables and scales (CRAN R package MuMin) [70]). The potential best models were identified by an AICc difference with the lowest AICc (ΔAICc) <2, indicating a balance between model complexity and predictivity [67,69]. Variables relative contributions, or cumulated weights, were calculated by model averaging (Σ (linear predictor × model weight)) [67].

2.5. FC Prediction and Distribution

In accordance with the parsimony principle, we selected the simplest model (i.e., involving the lower number of variables and scales) among the set of potential best models. For instance, Fictive Model 1 (DRY1 + HD1) would be chosen over Model 2 (DRY1 + HD3) or Model 3(DRY1 + HD1 + ESD1). Mean annual hazard was estimated for the 200 first years (1810–2010) of the baseline hazard function (basehaz in package survival [66]) and multiplied by the hazard ratio:

$$\frac{h_i(t)}{h_m(t)} = \frac{e^{X_i\beta}}{e^{X_m\beta}} \tag{2}$$

($h_m(t)$ = mean hazard at time t, X_m = mean predictor) to predict annual burn rate, which is the inverse of FC. We predicted FC with the previous equation both for the original dataset and for the whole study area raster. For interpretation and validation purposes, FC was grouped into classes in accordance with the shape and the extent of FC distribution. The number of classes was dictated by the number of modes and their limits by the breakpoints between modes. FC was calculated for each class of the TSF dataset using their specific basehaz functions. We calculated a 95% confidence interval on FC by bootstrapping (1000 resamplings with replacement from the original dataset, $n = 144$).

Landscape age structure directly depends on FC [16]. We validated the model by testing if observed vegetation age-structures in the study area were consistent with the predicted FC. We used the photo-interpreted vegetation database (SIFORT) [52], which provides information on stands' time since disturbance and composition with a resolution of 14 ha [71]. Although inventories exist for the decades of 1980, 1990, 2000 and 2010, we restricted analysis to the 1980s data to minimize the influence of forest management. Early succession forests (TSF < ~100 years) were recognized by delineated fires or by the presence (or dominance) of post-fire species (jack pine, paper birch (*Betula papyrifera* Marsh.), trembling aspen) [17,39]. Late succession forests were recognized by the presence of old-growth species, such as balsam, and by traces of insect outbreaks affecting specifically this species [72]. Pure black spruce stands, which can grow in all succession stages, were considered early-succession when the photo-interpreted age was below 90 years; otherwise, their status remained unknown. We expect a higher abundance of young forests where FC is short and a higher abundance of old-growth forests where FC is long. Succession stage abundances were compared between short and long FC with contingency table analysis (chi-square and Freeman–Tukey deviate).

2.6. Vegetation Composition

We hypothesized that forest composition is influenced by FC, especially in young forests. We identified post-fire cohorts throughout the study area according to the fire map described previously (1900–2010). We classified young forests into composition groups, corresponding to different succession pathways: jack pine (presence), broadleaf (presence) and black spruce (pure) [39]. Young forest succession pathway distributions were compared between short and long FC with contingency table analysis using chi-square and Freeman–Tukey deviate. Jack pine should theoretically be restricted to young forests in areas where FC is <200 years [18].

3. Results

3.1. Time since Fire Distribution

TSF is estimated for 144 locations throughout the study area. We were able to date the last fire (±10 years) for 60% of these locations, while a minimum TSF, generally based on the age of the oldest tree, was determined for the remaining 40% (Figure 4). TSF (including both real and minimum age) ranges from 0–340 years.

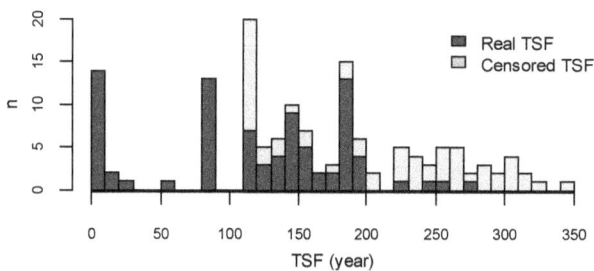

Figure 4. Real and censored (minimum) TSF distributions, grouped in 10-year age classes (*n* = number of TSF data).

3.2. FC Modeling

Single variable models are compared to the difference of AICc with the null model (ΔAICcNull) (Figure 5). ln(DRY + 1) at local and middle scales has the best predictive power (ΔAICcNull between -11.5 and -11.8). HD performs better at the local scale with ΔAICcNull of -4.5. For ESD at local and middle scales, ΔAICc ranges between -2.5 and -3.0. Finally, TPI has not shown any explicative power except at the landscape scale (Order 4), where ΔAICc is -1.9. Coefficients, confidence intervals, significance tests and controls for bias associated with recent fires are available in the Supplementary Material (Figure A2).

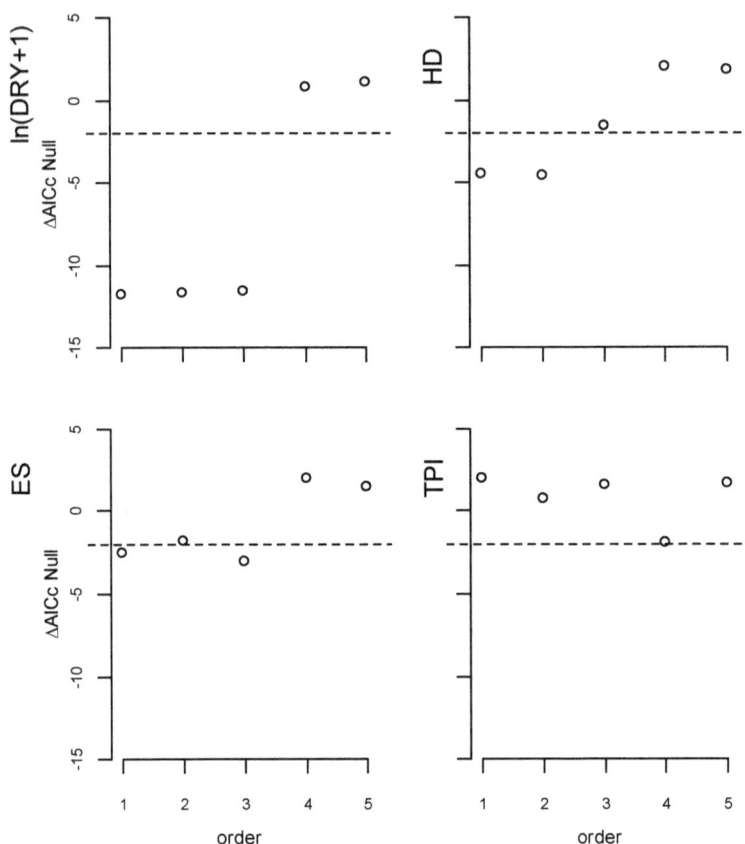

Figure 5. AICc difference with the null model (ΔAICcNull) for the Cox univariate models with four variables (ln(DRY + 1), HD, ES, TPI) (see Table 1) and scale Orders 1–5 (see Figure 2). The dotted lines indicate ΔAICc = −2. Dots under the line are better than the null model [67,69], and full dots indicate variables qualified for model selection (Table 3).

Significant variable-scale associations according to ΔAICcNull are $ln(DRY + 1)_1$, $ln(DRY + 1)_3$, HD_1, ESD_1 and ESD_3. We also kept HD3 and TPI4 in the process, even if their performance was under the ΔAICcNull to avoid missing significant interactions or patterns while knowing that they would be excluded from further analyses if no pattern emerged. Despite good individual performances, we excluded $ln(DRY + 1)_2$, HD2 and ESR2 because they were too correlated with other variables. From all possible combinations, fifteen had a ΔAICc < 2 and were thus considered as potential best models (Supplementary Material). Model averaging shows that the parameters that contribute the most are $HD1$, $ln(DRY + 1)_1$ and $ln(DRY + 1)_3$, which all had a cumulated weight greater than 65% (Table 3). From the 15 potentially best models, we selected the simplest one to predict FC, in accordance with the parsimony principle. The selected model thus includes the proportion of dry surficial deposits and hydrologic density, both at the local scale, to predict FC (Table 4).

Table 3. Relative importance of parameters (cumulated weight), coefficient averaged estimate and 95% confidence interval from multi-model inference for the DRY, HD, ESD and TRI selected orders.

Explanatory Variables	Watershed Order	Relative Importance (Cumulated Weight)	Model-Averaged Estimate	95% Confidence Interval	
				Lower	Upper
ln(DRY + 1)	1	0.73	0.1940	−0.0073	0.5097
	3	0.68	0.2789	−0.0315	0.7927
HD	1	0.83	−0.0337	−0.0830	−0.0020
	3	0.32	−0.0043	−0.1039	0.0471
ESD	1	0.45	0.0026	−0.0029	0.0143
	3	0.38	0.0014	−0.0038	0.0109
TRI	4	0.36	−0.0426	−0.3868	0.0543

Table 4. Coefficients and confidence intervals (CI), Wald test score (z) and *p*-values for the selected model.

	Coefficients	CI (2.5%)	CI (97.5%)	z	Pr (>|z|)
ln(DRY + 1)$_1$	0.355	0.170	0.546	3.798	0.000146
HD$_1$	−0.0430	−0.0827	−0.0154	−2.510	0.012063
ΔAICc	0.66				

Likelihood ratio test = 21.94 on 2 *df*, *p* = 1.724^{-5}. Wald test = 18.82 on 2 *df*, *p* = 8.177^{-5}.

3.3. FC Distribution

For the original TSF dataset (n = 144), predicted FC ranges between 74 and 2041 years with a mean of 241 years (Figure 6a). For the entire study area, it ranges from 60–8000 years with a mean of 224 years (Figure 6b). The FC distribution has a bimodal shape for both the TSF dataset and the whole study area, so FC is classified into two groups, short and long. The break between modes is estimated to be located at 275 years according to the entire study area, so 73% belongs to the short FC zone while 27% belongs to the long FC zone. The difference between the FC of each zone is confirmed when FC and confidence intervals are estimated from the TSF distribution (Table 5).

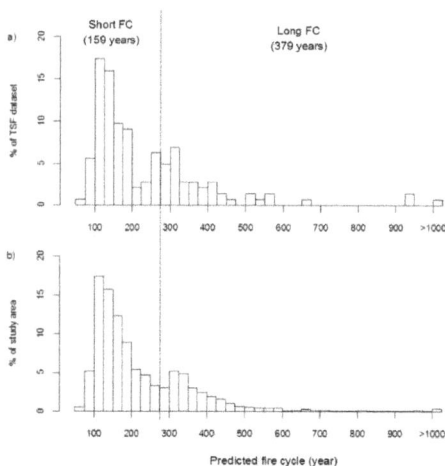

Figure 6. Predicted fire cycle (FC) (years) distribution according to the model detailed in Table 4 for (**a**) the TSF dataset (n = 144) and (**b**) the whole study area. The dotted line shows the division between short and long FC zones estimated with the whole study area (b). FC for each zone is calculated using the TSF dataset (see Table 5 for details).

Table 5. FC (year) and confidence intervals (CI) for the period 1810–2010 for the short and long FC zones, calculated from Cox regressions using the TSF dataset.

	n	FC	CI (2.5%)	CI (97.5%)
Short	100	159	126	198
Long	44	379	238	590

3.4. Model Validation

Pearson's chi-square test confirms that age-structure is younger where FC is short. The Freeman–Tukey deviate shows that each observed deviance is significant ($p < 0.001$) (Table 6a). Young forests compose 38% of short FC zones, which is 11.8% more than expected from a homogeneous distribution. They compose 26% of long FC zones, which is 20% less abundant than expected from a homogeneous distribution. We did not perform statistics on old-growth forest distribution because for certain types of vegetation composition, it was impossible to distinguish mature from old-growth forests. However, both old-growth and unknown age classes were underrepresented in short FC zones (respectively 15% and 46%) and overrepresented in long FC zones (respectively 21% and 46%).

3.5. FC and Succession Pathways

Pearson's chi-square test confirms that there are differences in the composition of young forests according to FC (Table 6b). In short FC zones, 59% of young forests are pure black spruce stands; broadleaf species are present in 12% and jack pine in 29%. In long FC zones, 67% of young forests are pure black spruce stands; broadleaf species are present in 11% and jack pine in 22%. Jack pine is especially rare (-20% less than expected for a random distribution) among young forests in the long FC zones.

Table 6. Contingency tables and chi-square (X^2) test of observed frequencies and their deviance from expected frequencies (in parentheses) for (a) young, unknown (mature or old-growth) and old-growth forest types and (b) young forest compositions in short and long FC areas.

	Predicted FC (Year)	Forest Age		
		Young	Unknown	Old-Growth
(a)	Short (159)	16 301 (+11.8%)	19 874 (−5.6%)	6575 (−8.1%)
	Long (379)	4 093 (−21.2%)	8 544 (+8.8%)	3389 (+12.2%)
	$X^2 = 873.5$		$p < 0.001$	
	df = 2		Critical distance = 13.82	
	Predicted FC (Year)	Young Forest Composition		
		Black Spruce	Broadleaf	Jack Pine
(b)	Short (159)	2864 (−1.9%)	609 (+1.6%)	1408 (+3.4%)
	Long (379 year)	555 (+11.2%)	93 (−9.2%)	186 (−20.0%)
	$X^2 = 19.33$		$p < 0.001$	
	df = 2		Critical distance = 13.82	

4. Discussion

4.1. FC Physical Drivers and Scales

Results show that physical environment influences local FC length. As hypothesized, FC is shorter where soil drainage is locally high (Orders 1–3) and river density is low (Orders 1, 2). Contrary to the original hypothesis, landscape terrain complexity (Orders 4, 5) is positively associated with short FC. In the study area, terrain complexity arises along rivers, which are both associated with sandy deposits and rocky outcrops. Contrary to the initial hypothesis, topographic features may not act as fire breaks, but rather, as fire corridors. Riverbeds are embedded in rocky outcrops and covered by sandy and dry soils, which may contribute to combustible flammability and fire propagation in wind corridors [44,45]. We also observed that landscape slope position (Order 4) is associated with long FC, meaning valleys burn more than hills, which may act as fire breaks [25,73].

Our results show the importance of bottom-up controls on FC in the study area. Similar phenomena are observed in many fire-prone ecosystems (e.g., [48,74–76]), but findings remain scarce for the boreal forest. Some studies highlight the association between FC and local soil drying potential [47,48,68,77], fire-break configuration [34] and topography [30,46,78], while others did not find any or very low bottom-up controls on fire frequency (e.g., [31,79]). As proposed by [76], the lack of consistency between results may be due to different sensitivities depending on FC. The longer FC is, the more influence bottom-up controls should have [29,80]. In another vein, we observed that we can miss some variable effects if we measure them at the wrong scale. We thus argue that single-scale approaches may not be suitable for observing both bottom-up and top-down controls [81,82]. The study area division according to watershed and river orders performed well with identifying FC drivers and their optimal measurement scales.

4.2. A Local FC Model

We built a spatially-explicit and scale-relevant FC model, driven by local physical factors. Model selection based on AICc allowed us to order models from the best to the worst with little ambiguity. We first expected to build a multi-scale model whose performance would include different drivers measured at their optimal scale. However, from the pool of possible models, ΔAICc indicated a good performance from a model based solely on the local scale for both HD and DRY variables. We identified two possible explanations for the better performance at the local scale. First, local FC could be bottom-up controlled, primarily influenced by local environment. This idea is however not in accordance with the generally accepted idea that fire regimes are driven by top-down, regional factors [78,83]. In that sense, we acknowledge that top-down drivers provide regional trends of FC, but we can argue that local physiography creates non-negligible variability. The second explanation concerns a suspected low performance of landscape-scale watersheds to fit landscape physiography compared to local-scale watersheds. We observe that short and long FC local-scale watersheds are aggregated in patches (Figure 7). We thus understand that local measurement of variables is the best to predict FC according to our scaling design, but this is not interpreted as the absence of coarser scale patterns.

Two fire regimes co-occur in the study area, as shown by the bimodal distribution of predicted FC. Heterogeneity in FC distribution can be caused either by temporal or spatial variability in FC. If only temporal variations occurred, FC would be randomly distributed throughout the study area, and no spatial patterns could be observed. However, we observe that zones where fires are frequent follow latitudinal valleys, while zones where fires are rare are mainly concentrated in inter-valley corridors (Figure 7). As observed by [44], valleys in this region may be prone to fire because dry deposits follow riverbeds, whose orientation goes along with wind direction, enhancing combustible drought and fire propagation.

4.3. Model Validation

Predicted zones of short and long FC were validated with independent vegetation data. Landscape age-structure theoretically fits a negative exponential distribution whose scaling depends on FC [16]. Following this rule, young forests (TSF < 100 years) should account for 47% of the short FC zone (FC = 159 years) and for 23% of the long FC zone (FC = 379 years). We have observed more abundant young forests where predicted FC is short (vegetation maps from the 1980s [52]), but young forests are still less abundant than expected (38%). We explain this gap by the fire occurrence temporal pattern. Large fires are not equally distributed in time, but occur in some decades of fire-prone weather [21]. Landscape age-structure thus changes in steps, facing significant rejuvenation following fire decades and getting gradually older until the next one. Considering that when vegetation data were collected, there had been no large fires in the study area since the 1920s, we assume that the 1980s landscape was more at the end than at the beginning of this continuum.

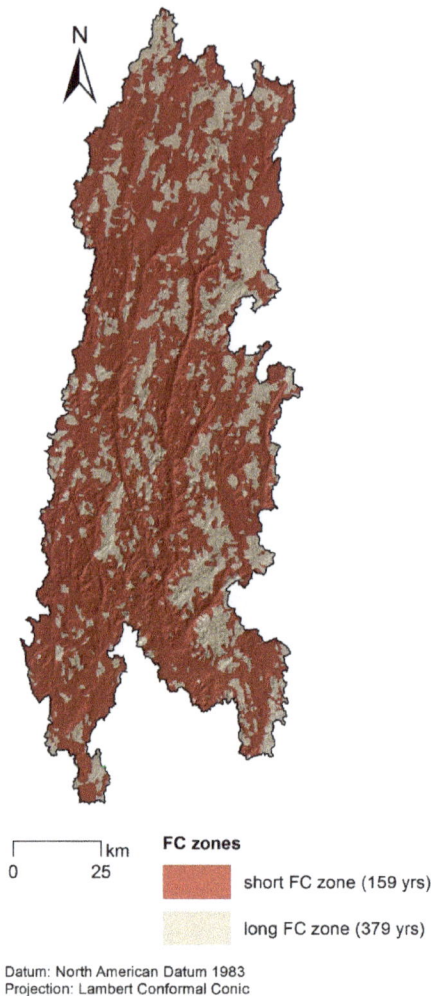

Figure 7. Predicted FC spatial distribution as presented in Figure 6b, grouped into short (159 years) and long (379 years) FC zones. Topography is presented in background hillshade.

4.4. FC and Vegetation

The initial hypothesis that jack pine, a pioneer fire-dependent species [38,84], is restricted to zones of short FC is only partly verified. Jack pine is more abundant where FC cycle is short, but is not excluded from zones where FC is long. Theory says jack pine should be restricted to sectors where FC < 220 years [18]. In the whole study area, mean FC is 224 years, and most burned areas are attributable to rare and large fires (>10,000 ha), so the time lapse between fires may be too long to sustain an abundant jack pine population. We thus suspect short FC zones to be a reservoir of jack pine during prolonged periods without fire. When large fires occur, jack pine may spread outside short FC zones, helped by post-fire colonization adaptations such as seed abundance and dispersal, fast growth and serotinous cones [38,85,86]. Paleo-ecological researchers [33,87] observed a stable jack pine and black spruce local co-occurrence during the Holocene. Their relative abundance depends on time elapsed since last fire. We suspect this stable co-occurrence is specific to short FC zones. Further research is needed to investigate local composition changes during the Holocene along a local FC gradient.

Our results suggest disturbance regime heterogeneity allows different habitats to co-occur in the study area and thus contribute to species diversity. Jack pine is adapted to short FC [86]. It is abundant in western Quebec where FC is generally short and rare in eastern Quebec where FC is long [88]. On the other hand, balsam fir is a shade-tolerant late-succession species [17,89] generally associated with long FC, such as in eastern Quebec [90,91]. Both species are abundant throughout the study area, located in central Quebec [39].

Broadleaf species distribution, although associated with fire frequency [92], was not analyzed because it is mainly restricted to the southern extreme of the study area.

4.5. Implications for Forest Management

FC spatial heterogeneity challenges the scale at which forest management is planned. Ecosystem-based management inspired by natural disturbance regimes aligns harvest rates with historical FC and associated age-structures [12,35,93]. While we showed that fire regime is locally variable, forest management and harvest rates are mostly planned at the landscape or management unit scale [94]. Moreover, depending on available data, FC may be only available for an even wider scale or for a neighboring area (e.g., [61,95]). We found that FC is complex and heterogeneous at a smaller scale than management units.

We argue for the consideration of local FC variability in forest ecosystem management. Different issues affect different locations in a single management unit. On the one hand, zones where FC is short are especially vulnerable to the expected increase in fire frequency due to climate change (e.g., [17,96,97]). The resilience of these forests should be a main management concern, as successive disturbances lead closed forests to turn into open lichen woodlands [98–101]. On the other hand, zones where FC is long face issues associated with old-growth forest depletion, due to a shorter time lapse between harvests than between natural fires [21,22,102]. Strategies for old-growth preservation, such as conservation zones [103], extended harvest rotation time [104] and adapted cutting designs for old-growth structure maintenance [105], should thus be concentrated in long FC zones.

5. Conclusions

FC is widely used for the basic description of boreal forest ecology and has a crucial place in ecosystem management. However, FC relies on the assumption that fire hazard is homogeneous throughout the region of interest. In this research, we demonstrated that homogeneity should not be systematically presumed, and we adapted the FC concept to make it relevant also for heterogeneous regions.

Ecological functions of the patchwork landscape structure and interactions between short and long FC zones remain unknown, although we expect them to contribute to forest diversity and resilience. Each zone may act as a reservoir of species adapted to specific fire intervals, such as jack pine in short FC zones and balsam fir in long FC zones. This may enhance the adaptation potential to changes in the fire regime and, subsequently, to climate change.

Finally, our results suggest that FC drivers are scale dependent, so we argue for further consideration of scale issues in fire regime studies. Research in the area of FC spatial heterogeneity is needed to address the question of the mechanisms involved. In that sense, comparing the results from statistical models as we did and from fire behavior models is an interesting avenue to explore. Assessing the heterogeneity of other parameters of fire regimes, such as ignition patterns and fire severity, could also provide useful insights.

Acknowledgments: We greatly thank Élisabeth Turcotte, Myriam Jourdain, David Gervais, Léa Langlois, Alexandre Turcotte, Nicolas Fauvart, Jean-Guy Girard, as well as Université du Québec à Chicoutimi plant ecology lab members for their assistance in the lab and in the field. We thank Aurélie Terrier for revising the manuscript and the anonymous reviewers for their helpful comments. We also acknowledge the Natural Sciences and Engeneering Research Council of Canada-UQAT-UQAM Industrial Chair in Sustainable Forest Management for material and scientific support. This project was funded by the Fonds de la Recherche Forestière du Saguenay-Lac-Saint-Jean and the Fonds de Recherche Nature et Technologies du Québec. We used forest inventory and fire data from the ministère des Ressources naturelles et de la Faune du Québec. We also acknowledge the financial contribution of the Natural Sciences and Engineering Research Council of Canada. Finally, we thank Resolute Forest Products for their partnership in the project and accommodation during field work.

Author Contributions: H.M., S.G. and Y.B. conceived of and coordinated the research project. A.C.B. and S.G. designed data collection. A.C.B. and M.D. conceived of the cartography tools. N.M. provided support for the surficial deposits classification. A.C.B. and A.L. analyzed the data. A.C.B. wrote the paper. All authors revised it.

Conflicts of Interest: The authors declare no conflicts of interest.

Abbreviation

The following abbreviations are used in this manuscript:

FC	Fire cycle
TSF	Time since last fire
DRY	Dry surficial deposit density
HD	Hydrographic density
ESD	Elevation standard deviation
TPI	Topographic position index

Appendix

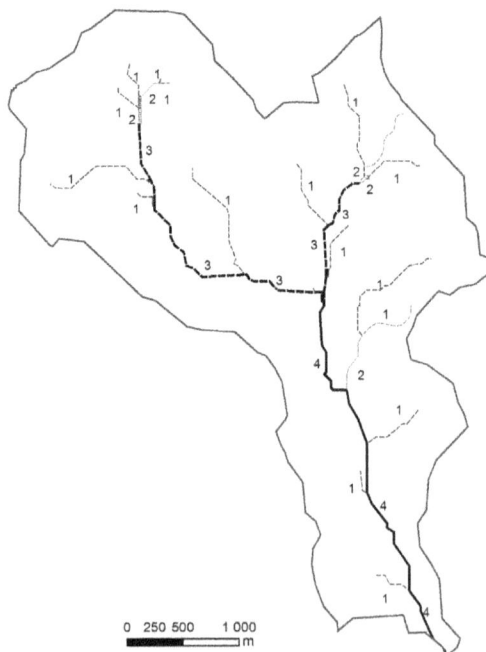

Figure A1. Rivers' Strahler order (1–4) inside an Order 4 watershed. Order 1 rivers connecting together make an Order 2 river; Order 2 rivers connecting together make an Order 3 river, etc.

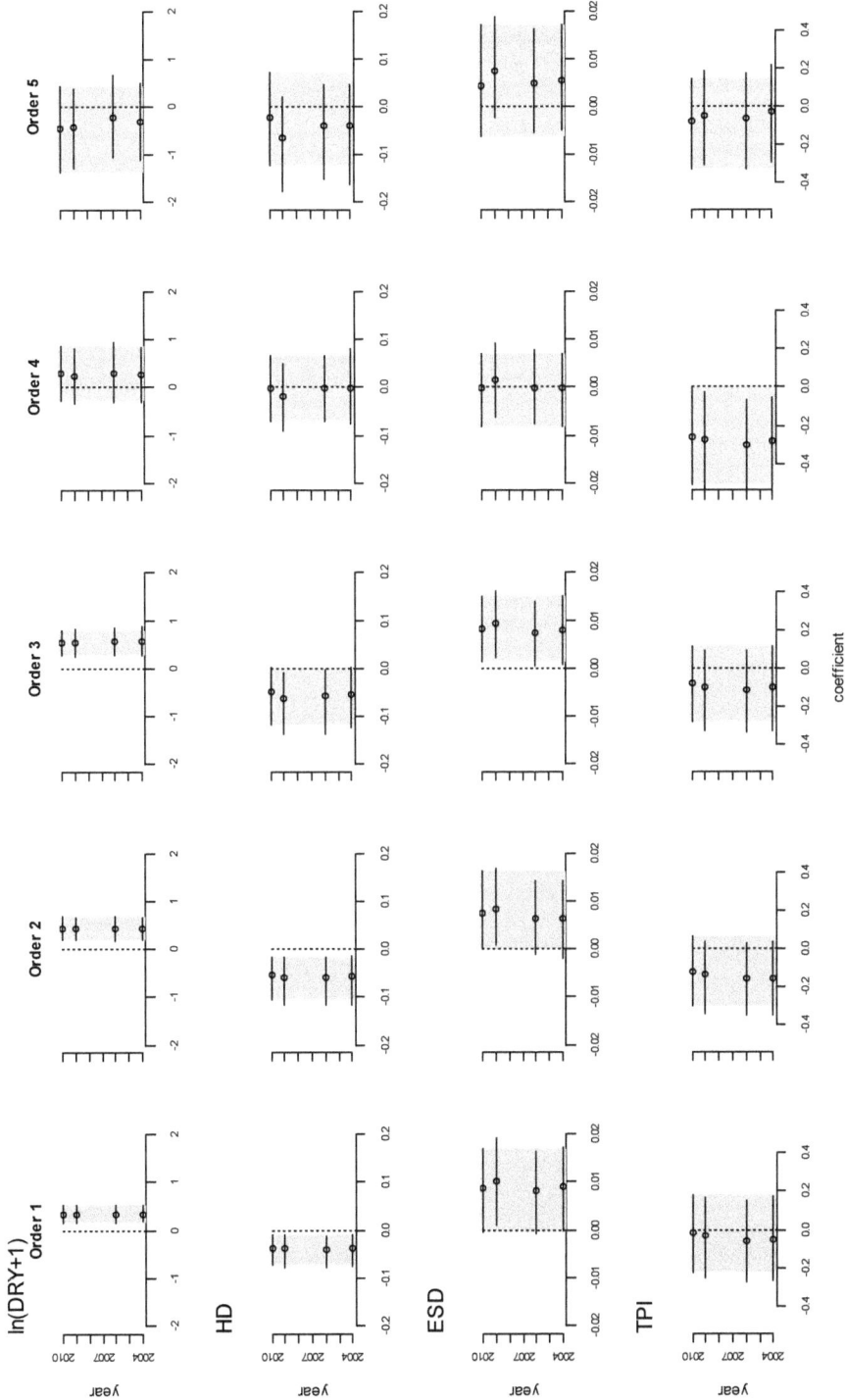

Figure A2. Coefficient and confidence intervals (calculated with bootstrap) plots of variables ln(DRY + 1), HD, ESD and TPI, Orders 1–5, for univariate Cox regressions. TSF data describe the study area age-structure as it was in 2010 (shaded rectangle, as used for analysis), 2009, 2006 and 2004.

Table A1. Model comparisons based on AICc. Results for the 15 candidate models (ΔAICc < 2) are presented. The selected model is shaded.

	Model	Degrees of Freedom	Loglikelihood	AICc	ΔAICc	Weight
1	log(DRY1) + log(DRY3) + HD1	3	−357.8589	721.8892	0.0000	0.0575
2	log(DRY3) + HD1 + ESD1	3	−357.8718	721.9151	0.0259	0.0568
3	log(DRY1) + log(DRY3)3 + HD1 + ESD1	4	−356.8615	722.0108	0.1217	0.0541
4	log(DRY1) + HD1 + ESD3	3	−358.1568	722.4850	0.5958	0.0427
5	log(DRY1) + HD1	2	−359.2333	722.5517	0.6626	0.0413
6	log(DRY1) + log(DRY3) + HD1 + ESD3	4	−357.1722	722.6322	0.7430	0.0397
7	log(DRY1) + log(DRY3) + HD1 + TPI4	4	−357.3373	722.9624	1.0732	0.0336
8	log(DRY3) + HD1 + ESD1 + TPI4	4	−357.4987	723.2851	1.3959	0.0286
9	log(DRY1) + HD1 + TPI4	3	−358.5942	723.3597	1.4706	0.0276
10	log(DRY1) + log(DRY3) + HD1 + ESD1 + TPI4	5	−356.5770	723.5888	1.6997	0.0246
11	log(DRY1) + HD1 + ESD3 + TPI4	4	−357.7217	723.7312	1.8421	0.0229
12	log(DRY3) + HD1 + HD3 + ESD1	4	−357.7719	723.8315	1.9423	0.0218
13	log(DRY1) + HD1 + ESD1	3	−358.8381	723.8476	1.9584	0.0216
14	log(DRY3) + HD1 + ESD1 + ESD3	4	−357.7942	723.8762	1.9870	0.0213
15	log(DRY1) + log(DRY3) + HD1 + HD3	4	−357.8120	723.9117	2.0225	0.0209

Steps for watershed computation from a digital elevation model (20 m resolution). Hydrology tools in ArcGIS are indicated in italics. 1. *Fill*; 2. *Flow direction*; 3. Identify streams: *Reclass*, >16 ha = 1; 4. *Snap Pour points* + edit pour points by visual inspection; 5. *Stream order* (Strahler); Order 1 is excluded for fuzziness; 6. Compute *Watersheds* separately for all Strahler orders; 7. Visual inspection and corrections if needed (e.g., one cell watersheds).

References

1. Bonan, G.B.; Shugart, H.H. Environmental factors and ecological processes in boreal forests. *Annu. Rev. Ecol. Syst.* **1989**, *20*, 1–28. [CrossRef]
2. Payette, S. Fire as a controlling process in the North American boreal forest. *Syst. Anal. Glob. Boreal For.* **1992**, 144–169. [CrossRef]
3. Stocks, B.J.; Mason, J.A.; Todd, J.B.; Bosch, E.M.; Wotton, B.M.; Amiro, B.D.; Flannigan, M.D.; Hirsch, K.G.; Logan, K.A.; Martell, D.L.; et al. Large forest fires in Canada, 1959–1997. *J. Geophys. Res.* **2003**, *108*, 1–12. [CrossRef]
4. Johnson, E.A. *Fire and Vegetation Dynamics: Studies from the North American Boreal Forest*; Cambridge University Press: Cambridge, UK, 1992; p. 129.
5. Bergeron, Y.; Gauthier, S.; Kafka, V.; Lefort, P.; Lesieur, D. Natural fire frequency for the eastern Canadian boreal forest: Consequences for sustainable forestry. *Can. J. For. Res.* **2001**, *31*, 384–391. [CrossRef]
6. Wu, J.; Loucks, O.L. From balance of nature to hierarchical patch dynamics: A paradigm shift in ecology. *Q. Rev. Biol.* **1995**, *70*, 439–466. [CrossRef]
7. Dix, R.L.; Swan, J.M.A. The roles of disturbance and succession in upland forest at Candle Lake, Saskatchewan. *Can. J. Bot.* **1971**, *49*, 657–676. [CrossRef]
8. Landres, P.B.; Morgan, P.; Swanson, F.J. Overview of the use of natural variability concepts in managing ecological systems. *Ecol. Appl.* **1999**, *9*, 1179–1188.
9. Gauthier, S.; Vaillancourt, M.-A.; Kneeshaw, D.D.; Drapeau, P.; de Grandpré, L.; Claveau, Y.; Paré, D. Forest Ecosystem Management: Origins and Foundations. In *Ecosystem Management in the Boreal Forest*; Gauthier, S., Vaillancourt, M.-A., Leduc, A., de Grandpré, L., Kneeshaw, D.D., Morin, H., Drapeau, P., Bergeron, Y., Eds.; Les Presses de l'Université du Québec: Quebec, QC, Canada, 2009; pp. 13–38.
10. Gauthier, S.; Leduc, A.; Bergeron, Y. Forest dynamics modelling under natural fire cycles: A tool to define natural mosaic diversity for forest management. *Environ. Monit. Assess.* **1996**, *39*, 417–434. [CrossRef] [PubMed]
11. Kuuluvainen, T. Natural variability of forests as a reference for restoring and managing biological diversity in boreal Fennoscandia. *Silva Fenn.* **2002**, *36*, 97–125. [CrossRef]
12. Franklin, J.F.; Spies, T.A.; Pelt, R.V.; Carey, A.B.; Thornburgh, D.A.; Berg, D.R.; Lindenmayer, D.B.; Harmon, M.E.; Keeton, W.S.; Shaw, D.C.; et al. Disturbances and structural development of natural forest ecosystems with silvicultural implications, using Douglas-fir forests as an example. *For. Ecol. Manag.* **2002**, *155*, 399–423. [CrossRef]
13. Gauthier, S.; Raulier, F.; Ouzennou, H.; Saucier, J.-P. Strategic analysis of forest vulnerability to risk related to fire: An example from the coniferous boreal forest of Quebec. *Can. J. For. Res.* **2015**, *45*, 553–565. [CrossRef]

14. Vaillancourt, M.-A.; de Grandpré, L.; Gauthier, S.; Leduc, A.; Kneeshaw, D.D.; Claveau, Y.; Bergeron, Y. How can natural disturbances be a guide for forest ecosystem management? In *Ecosystem Management in the Boreal Forest*; Gauthier, S., Vaillancourt, M.A., Leduc, A., de Grandpré, L., Kneeshaw, D.D., Morin, H., Drapeau, P., Bergeron, Y., Eds.; Les Presses de l'Université du Québec: Quebec, QC, Canada, 2009; pp. 39–56.
15. Li, C. Estimation of fire frequency and fire cycle: A computational perspective. *Ecol. Model.* **2002**, *154*, 103–120. [CrossRef]
16. Van Wagner, C.E. Age-class distribution and the forest fire cycle. *Can. J. For. Res.* **1978**, *8*, 220–227. [CrossRef]
17. Bergeron, Y. Species and stand dynamics in the mixed woods of Quebec's southern boreal forest. *Ecology* **2000**, *81*, 1500–1516. [CrossRef]
18. Le Goff, H.; Sirois, L. Black spruce and jack pine dynamics simulated under varying fire cycles in the northern boreal forest of Quebec, Canada. *Can. J. For. Res.* **2004**, *34*, 2399–2409. [CrossRef]
19. Frelich, L.E.; Reich, P.B. Spatial patterns and succession in a Minnesota southern-boreal forest. *Ecol. Monogr.* **1995**, *65*, 325–346. [CrossRef]
20. Mansuy, N.; Gauthier, S.; Robitaille, A.; Bergeron, Y. Regional patterns of postfire canopy recovery in the northern boreal forest of Quebec: Interactions between surficial deposit, climate, and fire cycle. *Can. J. For. Res.* **2012**, *42*, 1328–1343. [CrossRef]
21. Bélisle, A.C.; Gauthier, S.; Cyr, D.; Bergeron, Y.; Morin, H. Fire regime and old-growth boreal forests in central Quebec, Canada: An ecosystem management perspective. *Silva Fenn.* **2011**, *45*, 889–908. [CrossRef]
22. Cyr, D.; Gauthier, S.; Bergeron, Y.; Carcaillet, C. Forest management is driving the eastern North American boreal forest outside its natural range of variability. *Front. Ecol. Environ.* **2009**, *7*, 519–524. [CrossRef]
23. Krawchuk, M.A.; Moritz, M.A. Constraints on global fire activity vary across a resource gradient. *Ecology* **2011**, *92*, 121–132. [CrossRef] [PubMed]
24. Parks, S.A.; Parisien, M.-A.; Miller, C. Spatial bottom-up controls on fire likelihood vary across western North America. *Ecosphere* **2012**, *3*. [CrossRef]
25. Heyerdahl, E.K.; Brubaker, L.; Agee, J.K. Spatial controls of historical fire regimes: A multiscale example from the interior west, USA. *Ecology* **2001**, *82*, 660–678. [CrossRef]
26. Flannigan, M.D.; Harrington, J.B. A study of the relation of meteorological variables to monthly provincial area burned by wildfire in Canada (1953–1980). *J. Appl. Meteorol.* **1988**, *27*, 441–452. [CrossRef]
27. Van Wagner, C.E. *Development and Structure of the Canadian Forest Fire Weather Index System*; Canadian Forestry Service: Ottawa, ON, Canada, 1987; p. 37.
28. Boulanger, Y.; Gauthier, S.; Burto, P.J. A refinement of models projecting future Canadian fire regimes using homogeneous fire regime zones. *Can. J. For. Res.* **2014**, *44*, 365–376. [CrossRef]
29. Mansuy, N.; Gauthier, S.; Robitaille, A.; Bergeron, Y. The effects of surficial deposit-drainage combinations on spatial variations of fire cycles in the boreal forest of eastern Canada. *Int. J. Wildland Fire* **2010**, *19*, 1083–1098. [CrossRef]
30. Cyr, D.; Gauthier, S.; Bergeron, Y. Scale-dependent determinants of heterogeneity in fire frequency in a coniferous boreal forest of eastern Canada. *Landsc. Ecol.* **2007**, *22*, 1325–1339. [CrossRef]
31. Senici, D.; Chen, H.Y.H.; Bergeron, Y.; Cyr, D. Spatiotemporal variations of fire frequency in central boreal forest. *Ecosystems* **2010**, *13*, 1227–1238. [CrossRef]
32. Larsen, C.P.S. Spatial and temporal variations in boreal forest fire frequency in northern Alberta. *J. Biogeogr.* **1997**, *24*, 663–673. [CrossRef]
33. Frégeau, M.; Payette, S.; Grondin, P. Fire history of the central boreal forest in eastern North America reveals stability since the mid-Holocene. *Holocene* **2015**, *25*, 1912–1922. [CrossRef]
34. Parisien, M.A.; Parks, S.A.; Miller, C.; Krawchuk, M.A.; Heathcott, M.; Moritz, M.A. Contributions of ignitions, fuels, and weather to the spatial patterns of burn probability of a boreal landscape. *Ecosystems* **2011**, *14*, 1141–1155. [CrossRef]
35. Gauthier, S.; Leduc, A.; Bergeron, Y.; le Goff, H. Fire frequency and forest management based on natural disturbances. In *Ecosystem Management in the Boreal Forest*; Gauthier, S., Vaillancourt, M.A., Leduc, A., de Grandpré, L., Kneeshaw, D.D., Morin, H., Drapeau, P., Bergeron, Y., Eds.; Les Presses de l'Université du Québec: Quebec, QC, Canada, 2009; pp. 39–56.
36. Bouchard, M.; Pothier, D.; Gauthier, S. Fire return intervals and tree species succession in the North Shore region of eastern Quebec. *Can. J. For. Res.* **2008**, *38*, 1621–1633. [CrossRef]

37. Smirnova, E.; Bergeron, Y.; Brais, S. Influence of fire intensity on structure and composition of jack pine stands in the boreal forest of Quebec: Live trees, understory vegetation and dead wood dynamics. *For. Ecol. Manag.* **2008**, *255*, 2916–2927. [CrossRef]

38. Yarranton, M.; Yarranton, G. Demography of a jack pine stand. *Can. J. Bot.* **1975**, *53*, 310–314. [CrossRef]

39. Bélisle, A.C. Régime des feux, dynamique forestière et aménagement de la pessière à mousses au nord du Lac St-Jean. Master's Thesis, Université du Québec à Montréal, Quebec, QC, Canada, September 2012.

40. Cavard, X.; Boucher, J.-F.; Bergeron, Y. Vegetation and topography interact with weather to drive the spatial distribution of wildfires in the eastern boreal forest of Canada. *Int. J. Wildland Fire* **2015**, *24*, 391–406. [CrossRef]

41. Falk, D.A.; Heyerdahl, E.K.; Brown, P.M.; Farris, C.; Fulé, P.Z.; McKenzie, D.; Swetnam, T.W.; Taylor, A.H.; van Horne, M.L. Multi-scale controls of historical forest-fire regimes: New insights from fire-scar networks. *Front. Ecol. Environ.* **2011**, *9*, 446–454. [CrossRef]

42. Kennedy, M.C.; McKenzie, D. Using a stochastic model and cross-scale analysis to evaluate controls on historical low-severity fire regimes. *Landsc. Ecol.* **2010**, *25*, 1561–1573. [CrossRef]

43. Miller, C. The spatial context of fire: A new approach for predicting fire occurrence. In Proceedings of the Fire Conference 2000: The First National Congress on Fire Ecology, Prevention, and Management, San Diego, CA, USA, 27 November–1 December 2000; Galley, K.E.M., Klinger, R.C., Sugihara, N.G., Eds.; Tall Timbers Research Station: Tallahassee, FL, USA, 2003; pp. 27–34.

44. Mansuy, N.; Boulanger, Y.; Terrier, A.; Gauthier, S.; Robitaille, A.; Bergeron, Y. Spatial attributes of fire regime in eastern Canada: Influences of regional landscape physiography and climate. *Landsc. Ecol.* **2014**, *29*, 1157–1170. [CrossRef]

45. Barros, A.M.; Pereira, J.M.; Lund, U.J. Identifying geographical patterns of wildfire orientation: A watershed-based analysis. *For. Ecol. Manag.* **2012**, *264*, 98–107. [CrossRef]

46. De Lafontaine, G.; Payette, S. The origin and dynamics of subalpine white spruce and balsam fir stands in boreal Eastern North America. *Ecosystems* **2010**, *13*, 932–947. [CrossRef]

47. Harden, J.W.; Meier, R.; Silapaswan, C.; Swanson, D.K.; McGuire, A.D. Soil drainage and its potential for influencing wildfires in Alaska. *Stud. US Geol. Surv. Alas.* **2001**, *1678*, 139–144.

48. Hellberg, E.; Niklasson, M.; Granatröm, A. Influence of landscape structure on patterns of forest fires in boreal forest landscapes in Sweden. *Can. J. For. Res.* **2004**, *34*, 332–338. [CrossRef]

49. Saucier, J.-P.; Bergeron, J.-F.; Grondin, P.; Robitaille, A. *Les Régions Écologiques du Québec Méridional (3e Version): Un des Éléments du Système Hiérarchique de Classification Écologique du Territoire mis au Point par le Ministère des Ressources Naturelles du Québec*; Ministère des Ressources naturelles du Québec: Québec, QC, Canada, 1998.

50. Ressources naturelles Canada, L'Atlas du Canada. Available online: http://atlas.nrcan.gc.ca/site/index.html (accessed on 1 February 2012).

51. Robitaille, A.; Saucier, J.-P. *Paysages Régionaux du Québec Méridional*; Gouvernement du Québec: Québec, QC, Canada, 1998; p. 213.

52. Direction des Inventaires Forestiers. *Normes de Cartographie Écoforestière, Troisième Inventaire Écoforestier*, 2e édition; Ministère des Ressources naturelles de la Faune et des Parcs du Québec, Ed.; Gouvernement du Québec: Québec, QC, Canada, 2003; p. 95.

53. Environnement Canada. *Normales et Moyennes Climatiques du Canada 1971–2001*; Environnement Canada: Ottawa, Canada, 2010.

54. Coulombe, G. *Commission d'étude sur la Gestion de la Forêt Publique Québécoise*; MFFP: Quebec, QC, Canada, 2004; p. 307.

55. Boucher, Y.; Arseneault, D.; Sirois, L. Logging history (1820–2000) of a heavily exploited southern boreal forest landscape: Insights from sunken logs and forestry maps. *For. Ecol. Manag.* **2009**, *258*, 1359–1368. [CrossRef]

56. Helm, J. *Handbook of North American Indians*; Smithsonian Institution Press: Washington, DC, USA, 1978; Volume 6, p. 837.

57. Ministère de L'agriculture et de la Colonisation du Québec. *La Contrée du Lac Saint-Jean*; Tremblay Chicoutimi: Chicoutimi, QC, Canada, 1888.

58. Strahler, A.N. Quantitative analysis of watershed geomorphology. *Trans. Am. Geophys. Union* **1957**, *38*, 913–920. [CrossRef]

59. MacMillan, R.; Jones, R.K.; McNabb, D.H. Defining a hierarchy of spatial entities for environmental analysis and modeling using digital elevation models (DEMs). *Comput. Environ. Urban Syst.* **2004**, *28*, 175–200. [CrossRef]

60. Johnson, E.A.; Gutsell, S.L. Fire frequency models, methods and interpretation. *Adv. Ecol. Res.* **1994**, *25*, 239–287.

61. Bergeron, Y.; Cyr, D.; Drever, C.R.; Flannigan, M.; Gauthier, S.; Kneeshaw, D.; Lauzon, È.; Leduc, A.; LeGoff, H.; Lesieur, D.; et al. Past, current, and future fire frequencies in Quebec's commercial forests: Implications for the cumulative effects of harvesting and fire on age-class structure and natural disturbance-based management. *Can. J. For. Res.* **2006**, *36*, 2737–2744. [CrossRef]

62. Direction des Inventaires Forestiers. *Placettes-Échantillons Temporaires, Peuplements de 7 m et Plus de Hauteur*; Forêt Québec, ministère des Ressources naturelles du Québec, Ed.; Gouvernement du Québec: Québec, QC, Canada, 2002; p. 194.

63. Tagil, S.; Jenness, J. GIS-based automated landform classification and topographic, landcover and geologic attributes of landforms around the Yazoren Polje, Turkey. *J. Appl. Sci.* **2008**, *8*, 910–921. [CrossRef]

64. Hosmer, D.W.; Lemeshow, S.; May, S. *Applied Survival Analysis: Regression Modeling of Time-to-Event Data*; Wiley-Interscience: Hoboken, NJ, USA, 2008.

65. R Core Team. *R: A Language and Environment for Statistical Computing*; R Core Team: Vienna, Austria, 2012.

66. Therneau, T. *Survival Analysis, including Penalised Likelihood*; CRAN R: Vienna, Austria, 2011.

67. Burnham, K.P.; Anderson, D.R. *Model Selection and Multimodel Inference: A Practical Information-Theoretic Approach*; Springer: New York, NY, USA, 2002.

68. Bergeron, Y.; Gauthier, S.; Flannigan, M.; Kafka, V. Fire regimes at the transition between mixedwood and coniferous boreal forest in northwestern Quebec. *Ecology* **2004**, *85*, 1916–1932. [CrossRef]

69. Mazerolle, M.J. Improving data analysis in herpetology: Using Akaike's Information Criterion (AIC) to assess the strength of biological hypotheses. *Amphib. Reptil.* **2006**, *27*, 169–180. [CrossRef]

70. Package "MuMIn". Available online: https://cran.r-project.org/web/packages/MuMIn/MuMIn.pdf (accessed on 30 June 2016).

71. Pelletier, G.; Dumont, Y.; Bédard, M. *Système d'Information FORestière par Tesselle, Manuel de l'usager Québec*; ministère des Ressources naturelles et de la Faune du Québec: Québec, QC, Canada, 2007; p. 125.

72. Blais, J. Trends in the frequency, extent, and severity of spruce budworm outbreaks in eastern Canada. *Can. J. For. Res.* **1983**, *13*, 539–547. [CrossRef]

73. Kasischke, E.S.; Williams, D.; Barry, D. Analysis of the patterns of large fires in the boreal forest region of Alaska. *Int. J. Wildland Fire* **2002**, *11*, 131–144. [CrossRef]

74. Beaty, R.M.; Taylor, A.H. Spatial and temporal variation of fire regimes in a mixed conifer forest landscape, Southern Cascades, California, USA. *J. Biogeogr.* **2001**, *28*, 955–966. [CrossRef]

75. Mermoz, M.; Kitzberger, T.; Veblen, T.T. Landscape influences on occurrence and spread of wildfires in Patagonian forests and shrublands. *Ecology* **2005**, *86*, 2705–2715. [CrossRef]

76. Flatley, W.T.; Lafon, C.W.; Grissino-Mayer, H.D. Climatic and topographic controls on patterns of fire in the southern and central Appalachian Mountains, USA. *Landsc. Ecol.* **2011**, *26*, 195–209. [CrossRef]

77. Syrjanen, K.; Kalliola, R.; Puolasmaa, A.; Mattsson, J. Landscape structure and forest dynamics in subcontinental Russian European taiga. *Ann. Zool. Fenn.* **1994**, *31*, 19–34.

78. Kasichke, E.S.; Turetsky, M.R. Recent changes in the fire regime across the North American boreal region—Spatial and temporal patterns of burning across Canada and Alaska. *Geophys. Res. Lett.* **2006**, *33*. [CrossRef]

79. Lesieur, D.; Gauthier, S.; Bergeron, Y. Fire frequency and vegetation dynamics for the south-central boreal forest of Quebec, Canada. *Can. J. For. Res.* **2002**, *32*, 1996–2009. [CrossRef]

80. Bessie, W.C.; Johnson, E.A. The relative importance of fuels and weather on fire behavior in subalpine forests. *Ecology* **1995**, *76*, 747–762. [CrossRef]

81. Kushla, J.D.; Ripple, W.J. The role of terrain in a fire mosaic of a temperate coniferous forest. *For. Ecol. Manag.* **1997**, *95*, 97–107. [CrossRef]

82. Levin, S.A. The problem of pattern and scale in ecology: The Robert H. MacArthur award lecture. *Ecology* **1992**, *73*, 1943–1967. [CrossRef]

83. Portier, J.; Gauthier, S.; Leduc, A.; Arsenault, D.; Bergeron, Y. Fire regime variability along a latitudinal gradient of continuous to discontinuous coniferous boreal forests in Eastern Canada. *Forests* **2016**, submitted.

84. Cayford, J.; McRae, D. The ecological role of fire in jack pine forests. In *The Role of Fire in Northern Circumpolar Ecosystems*; Wein, R.W., MacLean, D.A., Eds.; John Wiley and Sons Ltd.: New York, NY, USA, 1983; pp. 183–199.

85. Gauthier, S.; Bergeron, Y.; Simon, J.-P. Cone serotiny in jack pine: Ontogenetic, positional, and environmental effects. *Can. J. For. Res.* **1993**, *23*, 394–401. [CrossRef]
86. Rudolph, T.D.; Laidly, P.R. *Pinus banksiana* Lamb. *Silvics of North America: Conifers*; Burns, R.M., Honkala, B.H., Technical Coordinators, Eds.; U.S. Department of Agriculture, Forest Service: Washington, DC, USA, 1990; Volume 1.
87. Sirois, L. Distribution and dynamics of balsam fir (*Abies balsamea* L. Mill.) at its northern limit in the James Bay area. *Ecoscience* **1997**, *4*, 340–352.
88. Beaudoin, A.; Bernier, P.; Guindon, L.; Villemaire, P.; Guo, X.; Stinson, G.; Bergeron, T.; Magnussen, S.; Hall, R. Mapping attributes of Canada's forests at moderate resolution through *k*NN and MODIS imagery. *Can. J. For. Res.* **2014**, *44*, 521–532. [CrossRef]
89. Chen, H.Y.H.; Popadiouk, R.V. Dynamics of North American boreal mixedwoods. *Environ. Rev.* **2002**, *10*, 137–166. [CrossRef]
90. De Grandpré, L.; Morissette, J.; Gauthier, S. Long-term post-fire changes in the northeastern boreal forest of Quebec. *J. Veg. Sci.* **2000**, *11*, 791–800. [CrossRef]
91. Gauthier, S.; Boucher, D.; Morissette, J.; de Grandpré, L. Fifty-seven years of composition change in the eastern boreal forest of Canada. *J. Veg. Sci.* **2010**, *21*, 772–785. [CrossRef]
92. Terrier, A.; Girardin, M.P.; Périé, C.; Legendre, P.; Bergeron, Y. Potential changes in forest composition could reduce impacts of climate change on boreal wildfires. *Ecol. Appl.* **2013**, *23*, 21–35. [CrossRef] [PubMed]
93. Grumbine, R.E. What is ecosystem management? *Conserv. Biol.* **1994**, *8*, 27–38. [CrossRef]
94. Kneeshaw, D.; Leduc, A.; Messier, C.; Drapeau, P.; Paré, D.; Gauthier, S.; Carignan, R.; Doucet, R.; Bouthillier, L. Developing biophysical indicators of sustainable forest management at an operational scale. *For. Chron.* **2000**, *76*, 482–493. [CrossRef]
95. Boucher, Y.; Bouchard, M.; Grondin, P.; Tardif, P. *Le Registre des États de Référence: Intégration des Connaissances sur la Structure, la Composition et la Dynamique des Paysages Forestiers Naturels du Québec Méridional*; Direction de la recherche forestière, Ed.; Ministère des Ressources naturelles et de la Faune du Québec: Québec, QC, Canada, 2011; p. 21.
96. Amiro, B.; Cantin, A.; Flannigan, M.; de Groot, W. Future emissions from Canadian boreal forest fires. *Can. J. For. Res.* **2009**, *39*, 383–395. [CrossRef]
97. Girardin, M.P.; Ali, A.A.; Carcaillet, C.; Gauthier, S.; Hély, C.; Le Goff, H.; Terrier, A.; Bergeron, Y. Fire in managed forests of eastern Canada: Risks and options. *For. Ecol. Manag.* **2013**, *294*, 238–249. [CrossRef]
98. Girard, F.; Payette, S.; Gagnon, R. Rapid expansion of lichen woodlands within the closed-crown boreal forest zone over the last 50 years caused by stand disturbances in eastern Canada. *J. Biogeogr.* **2008**, *35*, 529–537. [CrossRef]
99. Payette, S.; Delwaide, A. Shift of conifer boreal forest to lichen-heath parkland caused by successive stand disturbances. *Ecosystems* **2003**, *6*, 540–550. [CrossRef]
100. Rapanoela, R.; Raulier, F.; Gauthier, S. Regional instability in the abundance of open stands in the boreal forest of Eastern Canada. *Forests* **2016**, *7*, 103–120. [CrossRef]
101. Van Bogaert, R.; Gauthier, S.; Raulier, F.; Saucier, J.-P.; Boucher, D.; Robitaille, A.; Bergeron, Y. Exploring forest productivity at an early age after fire: A case study at the northern limit of commercial forests in Quebec. *Can. J. For. Res.* **2015**, *45*, 579–593. [CrossRef]
102. Shorohova, E.; Kneeshaw, D.; Kuuluvainen, T.; Gauthier, S. Variability and dynamics of old-growth forests in the circumboreal zone: Implications for conservation, restoration and management. *Silva Fenn.* **2011**, *45*, 785–806. [CrossRef]
103. Bengtsson, J.; Angelstam, P.; Elmqvist, T.; Emanuelsson, U.; Folke, C.; Ihse, M.; Moberg, F.; Nyström, M. Reserves, resilience and dynamic landscapes. *Ambio* **2003**, *32*, 389–396. [CrossRef] [PubMed]
104. Burton, P.J.; Kneeshaw, D.D.; Coates, K.D. Managing forest harvesting to maintain old growth in boreal and sub-boreal forests. *For. Chron.* **1999**, *75*, 623–631. [CrossRef]
105. Harvey, B.D.; Leduc, A.; Gauthier, S.; Bergeron, Y. Stand-landscape integration in natural disturbance-based management of the southern boreal forest. *For. Ecol. Manag.* **2002**, *155*, 369–385. [CrossRef]

© 2016 by the authors. Licensee MDPI, Basel, Switzerland. This article is an open access article distributed under the terms and conditions of the Creative Commons Attribution (CC BY) license (http://creativecommons.org/licenses/by/4.0/).

forests

MDPI

Article

350 Years of Fire-Climate-Human Interactions in a Great Lakes Sandy Outwash Plain

Richard P. Guyette [1,*], Michael C. Stambaugh [1], Daniel C. Dey [2], Joseph M. Marschall [1], Jay Saunders [3] and John Lampereur [3]

[1]   Missouri Tree-Ring Laboratory, School of Natural Resources, University of Missouri, 203 ABNR Building, Columbia, MO 65211, USA; stambaughm@missouri.edu (M.C.S.); marschallj@missouri.edu (J.M.M.)
[2]   U.S. Forest Service, Northern Research Station, 11 Campus Blvd., Suite 200. Newtown Square, PA 19073, USA; ddey@fs.fed.us
[3]   U.S. Forest Service, Chequamegon-Nicolet National Forest, 500 Hanson Lake Road, Rhinelander, WI 54501, USA; jsaunders@fs.fed.us (J.S.); jlampereur@fs.fed.us (J.L.)
*   Correspondence: guyetter@missouri.edu; Tel.: +01-573-882-7741

Academic Editors: Sylvie Gauthier and Yves Bergeron
Received: 30 June 2016; Accepted: 18 August 2016; Published: 27 August 2016

Abstract: Throughout much of eastern North America, quantitative records of historical fire regimes and interactions with humans are absent. Annual resolution fire scar histories provide data on fire frequency, extent, and severity, but also can be used to understand fire-climate-human interactions. This study used tree-ring dated fire scars from red pines (*Pinus resinosa*) at four sites in the Northern Sands Ecological Landscapes of Wisconsin to quantify the interactions among fire occurrence and seasonality, drought, and humans. New methods for assessing the influence of human ignitions on fire regimes were developed. A temporal and spatial index of wildland fire was significantly correlated ($r = 0.48$) with drought indices (Palmer Drought Severity Index, PDSI). Fire intervals varied through time with human activities that included early French Jesuit missions, European trade (fur), diseases, war, and land use. Comparisons of historical fire records suggest that annual climate in this region has a broad influence on the occurrence of fire years in the Great Lakes region.

Keywords: anthropogenic fire regimes; humans; red pine; ignition; drought

1. Introduction

In much of eastern North America, the quantitative data describing historical fire regimes over the last several centuries hinge on fire-scarred wood and trees. Annual resolution fire scar histories provide data and perspective on past fire intervals, fire extent, fire severity, and forcing factors that inform forest management and restoration [1]. For fire history, northern Wisconsin is a unique study region because of the considerable body of early written human history and the abundance of dateable and fire scarred wood. With these sources, excellent potential exists to study interactions among humans, climate, and fire.

Although droughts tend to occur infrequently in this cool-wet climate region, they have been associated with occurrences of high severity and culturally significant fire events (e.g., Peshtigo Fire of 1871). Attribute data from the 1800s General Land Office survey notes describe the historically complex vegetation patterning of northern Wisconsin ecosystems and the influence of drought and fire [2,3]. Multi-century fire scar records in the Great Lakes Region indicate that increased fire frequency, extent, and severity were associated with both past drought conditions and human activity [4,5]. Predisposing factors such as drought and weather combined with frequent (e.g., daily) human fire use (whether accidental or purposeful) made large areas prone to recurring fires. Approximately fifty percent of Wisconsin's vegetation and landscapes are estimated to have been influenced by Native American fires [6,7].

Depending on the ignition source, fire scars on trees can be viewed as both natural and cultural resources. In many forested ecosystems with hundreds, if not thousands, of years of human activity, a large proportion of basal injuries can result from human ignitions [8,9]. In the case where probability for anthropogenic ignitions is high, ecosystem conditions may be byproducts of human activity. Landscapes cultured by anthropogenic fire are recognized throughout North America [10–12].

Before about 1800 CE, reliable records of early human population, cultures, and fire are difficult to obtain or are unavailable for much of North America. In the Great Lakes region, early and literate travelers (Jesuits missionaries and French traders) provided records of population, culture, and trade [9]. These records are particularly important for understanding historical fire regimes since changes in populations and cultures often coincide with changes in fire frequency [10–12]. Thus, considering human history in analyses of fire regimes and fire ecology is imperative, especially in regions where anthropogenic fire has been deemed an important ignition source [12,13].

During at least the last three and a half centuries, historical records indicate that the Great Lakes region has been influenced by diverse cultures and population densities [14,15]. Generally, significant environmental changes occurred in response to cultures switching from subsistence to market economies [16]. In northern Wisconsin specifically, several Native American populations circa 1650 are estimated at 2000–3000 Menominee, several thousand Ojibwa, with additions by French and Indian 'fur traders' [15,16]. Driven by distant European markets, the trade of animal furs had an early and widespread influence on the number and locations of humans in this region [17]. Fire regimes changed in a multitude of ways during European settlement and development. Increased wildfire activity occurred, particularly along railways (due to increased ignitions embers from wood- and charcoal-burning railroad engines, sparks from railroad tracks, deliberate burning to clear lands by prospective farmers) while decreased fire activity occurred with EuroAmerican settlements (e.g., fuel alterations [6] and fire suppression) and continued to do so due to fire suppression policies [18,19].

The primary objective of this study was to describe and quantify the historical fire regime characteristics in the Northern Sands Ecological Landscapes of Wisconsin [20]. The second objective was to examine the relative influences of climate and humans on fire regimes. We hypothesized that: 1) interactions between human ignitions and drought changed fire frequency and extent, and 2) fluctuating human populations, cultures interacting with 'climate' were major factors influencing fire occurrence over the last three and a half centuries.

2. Data and Methods

2.1. Site Descriptions

Study sites were established in the Chequamegon-Nicolet National Forest (CNNF), northern Wisconsin, USA (Table 1). Three sites were located in the Northeast Sands Ecological Landscape in Oconto County (Lakewood District) and one site was located in the Northwest Sands Ecological Landscape in Bayfield County (Washburn District) (Figure 1). The climate of northern Wisconsin is characterized as humid-continental with cold winters and warm summers [21]. The three eastern-most sites have a mean maximum temperature of 11.7 °C, an average annual temperature of 5.5 °C and an annual precipitation of 83 cm. The northwestern site is slightly cooler with a mean maximum temperature of 10.5 °C, an average temperature of 4.9 °C, and an annual precipitation of 84 cm. Sites occurred on pitted outwash glacial deposits characterized by sandy and gravely soils [22]. The topography at all sites is variable with undulating glacial terrain and sandy outwash flats interspersed with lakes, ponds, and wetlands. Site areas ranged from 0.3 to 3 km^2.

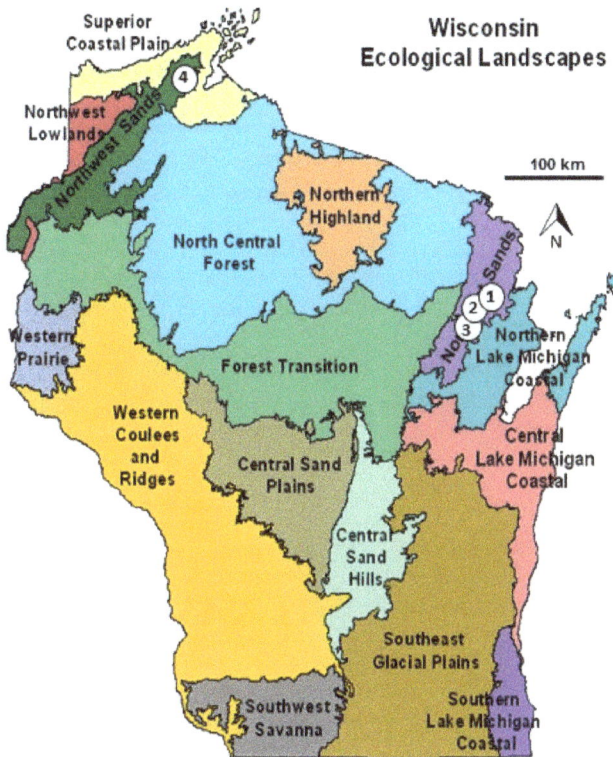

Figure 1. Locations of fire history sites (white circles) in Wisconsin's Ecological Landscapes [20]. Numbers in circles refer to site numbers in Table 1.

Table 1. Descriptions of site locations, vegetation, current dominate tree species, area, topography, and fire history records.

Site name	Grindle Lake	Waubee Lake	Airport Road	Moquah Barrens
Site number	1	2	3	4
Tree species	Red pine	Oaks and MH	MH & red pine	MH & red pine
Community structure	Woodland & Wetland	Forest	Forest, Woodland, Wetlands	Woodland & Forest
Location	N 45°13.89' W 88°21.46'	N 45°21.30' W 88°26.13'	N 45°10.92' W 88°19.57'	N 46°36.17' W 91°17.89'
Site area	1 km^2	0.3 km^2	3 km^2	0.6 km^2
Topography	Gentle slopes and wetlands	Hummocky terrain, steep slopes	Low hills, flatlands, & wetlands	Steep & gentle slopes
Number trees	28	15	23	32
Number scars	141	68	110	146
Number fire years	25	24	23	27
Period of fire record	1650–2008	1638–1873	1650–2009	1585–2009

MH = Mixed Hardwoods.

Three of the four study sites share similar current vegetation and site characteristics with pine and mixed-hardwood forest types growing on well to excessively drained, sandy outwash plains (Table 1). Somewhat differently, the Waubee Lake site has the greatest topographic variability and is presently a northern red oak (*Quercus rubra*) forest type with abundant red pine remnant wood (i.e., stumps, dead trees) but few living red pines. The Grindle Lake and the Airport Road sites have relatively gentle slopes, are interspersed with wetlands, and are more heavily covered in red pine. The Moquah Barrens site has the greatest relief and is adjacent to a large grassy area of jack (*Pinus banksiana*) and red pine barrens.

Historically, the four study sites had varied proximities to human activities associated with trading centers, camps, and villages. The Ojibwa (Chippewa) and other native peoples lived on Madeline Island about 40 km northeast of the Moquah Barrens site. About 1660 CE, French fur traders arrived in the region and by 1690 CE a trade center at La Point, Madeline Island was established. The fur trade likely resulted in increased probability and distribution of anthropogenic ignitions. Although the northeastern study sites were further from Lake Superior trade routes [23], they were also within the trade network associated with Lake Michigan. These three sites (1–3; Table 1) are within 30 km of the present day Menominee Indian Reservation and 60 km from the Menominee River.

2.2. Fire Scar Collection and Data

Cross-sections from ninety-six fire-scarred red pine stumps, snags, and live trees were cut from the four study sites. Location (GPS coordinates), slope, and aspect were recorded for each tree sampled. Surfaces of cross-sections were sanded with successively finer sandpaper (80–1200 grit) until the cellular detail of annual rings and fire scar injuries was revealed. A radius (pith-to-bark tree-ring series) of the cross-section with the least amount of ring-width variability due to fire injuries was chosen for ring-width measurement and crossdating. Ring-width series from each sample were plotted and used for visual crossdating and signature year identification [24]. Visual matching of ring-width patterns allowed for the weighting of important "signature years" over years with low common variability among trees. Ring-width plots aided in identifying errors in measurement, missing rings, and/or false rings which can be associated with injury or drought. The samples and dating chronology were crossdated and verified using a master red pine chronology developed from the Cathedral Pines Natural Area (Oconto County; unpublished data) and live trees within the study sites. Computer program COFECHA [24,25] was used to check the accuracy of crossdated tree-ring series. Fire scars were identified by the presence of callus tissue, charcoal, traumatic resin canals, liquefaction of resin and cambial injury. Calendar years of fire events represented the year of the tree's response to cambial injury. Fire events (year, season based on scar position) were recorded in standard FHX2 format [26]. Fire-free periods associated with red pine survival were estimated when possible.

2.3. Reconstructed Drought Data

Reconstructed Palmer Drought Severity Indices (PDSI) [27] during the period of the fire scar records were collated for use in analysis with fire events. Indices were averaged for the northern Wisconsin region (PDSI grid points: 198, 207, 215, and 216).

2.4. AnthroFire Index (AFI)

An AnthroFire Index (AFI) has been developed to detect changes in the fire regime that reflect human ignition influence [28,29]. The AFI was calculated using reconstructed drought [27] and fire occurrence data. The AFI operates under two assumptions concerning the relationship between drought and anthropogenic fires: 1) the absence of fires during drought years was due to a lack of human ignitions and, 2) the occurrence of fires in wet years resulted from increased human ignitions (accidental or purposeful). Assumption 1 relates to identifying the potential human ignition influence in an ecosystem and that, during drought conditions, fire probability is increased. Assumption 2 relates to fires occurring in 'wet' years; fires occurring in wet years are limited to burning during short, dry periods and require abundant or 'smart ignition' by humans. No human data (population, cultures, or others) are used in the AFI calculation and are limited to periods without modern fire suppression technology.

To calculate the AFI, fires occurring in wet years (PDSI > 0) were assigned increasing values from 1 to 2.5 corresponding with increased wetness. Dry years (PDSI < 0) without fire were assigned decreasing values from −1 to −2.5 corresponding to increased dryness. Fire occurring in dry years (PDSI < 0) and wet years without fire were not assigned values nor used. The time series of AFI values was then smoothed using moving averages of two lengths, 11 and 21 years, due to the irregular occurrence of AFI values through time and the need to include a sufficient number of observations to reduce spurious, short-term (e.g., <4 years) wet-dry variability. We scaled the AFI time series to a mean value of zero by subtracting the series mean. Thus, increasing AFI values were interpreted as increasing human influence ('smart ignitions') and decreasing AFI values were interpreted as decreasing human influence.

2.5. Fire Extent Index (FEI)

We developed a coarse-scale annual Fire Extent Index (FEI) to examine the relationship between drought and the potential sites burned and trees scarred [11,28]. The FEI was not developed for estimating the area burned but rather to estimate a relative spatial extent variable that could be related to PDSI. We calculated the FEI from data of the number of sites scarred in a year (range = 0 to 4, a value reflecting regional spatial extent of fire in a year) and the percent of trees scarred at a site (range = 0–90, a value reflecting the scale of fire severity at a site). The FEI is the product of the number of sites scarred and the percent of trees scarred at all the sites (Equation (1)).

$$\text{FEI for year } x \ = \ \#\text{fire} - \text{scarred sites } x\% \text{ trees fire} - \text{scarred} \tag{1}$$

2.6. Fire Scar Data Analysis

For each site we summarized fire event data including ranges of fire intervals, mean fire return intervals, seasonality of fires, and fire severity (based on percentages of trees scarred) [30,31]. Data summaries were developed for the full periods of records and for sub-periods separated based on known changes in the fire environment (e.g., humans, climate, land use). The quality of the fire scar record can vary by the size of the site area and the number of recorder trees in the record at any time. We estimated the effects of sample size on fire frequency by comparing the decade-to-decade relationship between sample depth and the number of fire years detected. The asymptote of the relationship between sample depth and number of fire years was assumed to be an indicator of the sample depth required to adequately characterize fire events (i.e., quality of the fire scar record at any year).

Correlation analysis was used to determine if PDSI was significantly related to fire frequency, percentage of trees scarred, or the FEI. Regression analysis was conducted to relate PDSI to FEI and develop a model predicting variability in the FEI [31]. FEI was transformed utilizing a natural log prior to regression analysis to adjust for the non-linear relationship with PDSI. SAS/STAT software [32] was used for statistical summaries, analysis of means, and regression and correlation analyses. We collated nine other existing fire scar records from the Great Lakes Region [5,13,29,33] and compared them to the northern Wisconsin fire scar record. Comparisons among sites included regional fire years (years in which many sites burned), percent trees scarred, and the FEI. In addition, we compared the fire frequencies derived from the study sites with those predicted by a model of historic fire frequency based on climate parameters [34].

3. Results

3.1. Fire Scar Data

We dated a total of 465 fire scars on 98 red pine remnant and living trees from the four study sites (Table 1). Ninety percent or more of the fire scar years was captured by as few as 11 trees (Figure 2). Fire scar dates spanned the period 1591–1948 (Figure 3). The number (and extent) of fires was lowest from 1665 to 1718 and highest from 1719 to 1820. Prior to 1780, mean fire intervals ranged from 8.9 to 29 years at all sites (Table 2). The four most frequently occurring fire years were 1664, 1743, 1756, and 1780 (i.e., occurring during the fur trade era and prior to EuroAmerican settlement). The occurrence of fire scars decreased sharply in the 1920s (Figure 3). In the pre-EuroAmerican settlement period (1650–1864), about 25 percent of the years had a fire recorded in at least one study site (Figures 3 and 4). Of these 54 fire years, 30 occurred during drier years (PDSI <0) and 24 occurred during wetter years (PDSI > 0).

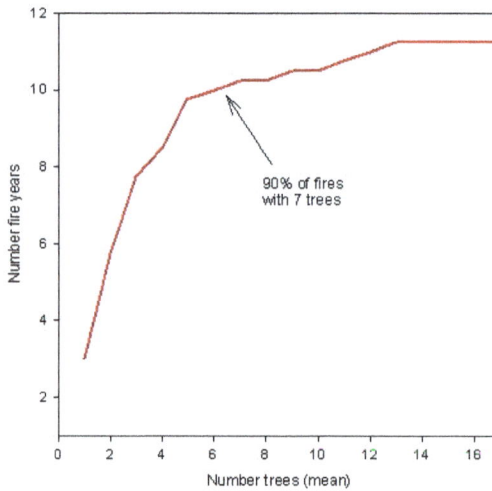

Figure 2. Relationship between the number of trees sampled and new fire years detected based on data collected at the four study sites. The beginning of the asymptote of the curve reflects the minimum sample depth needed in a given area for a high-quality fire scar record.

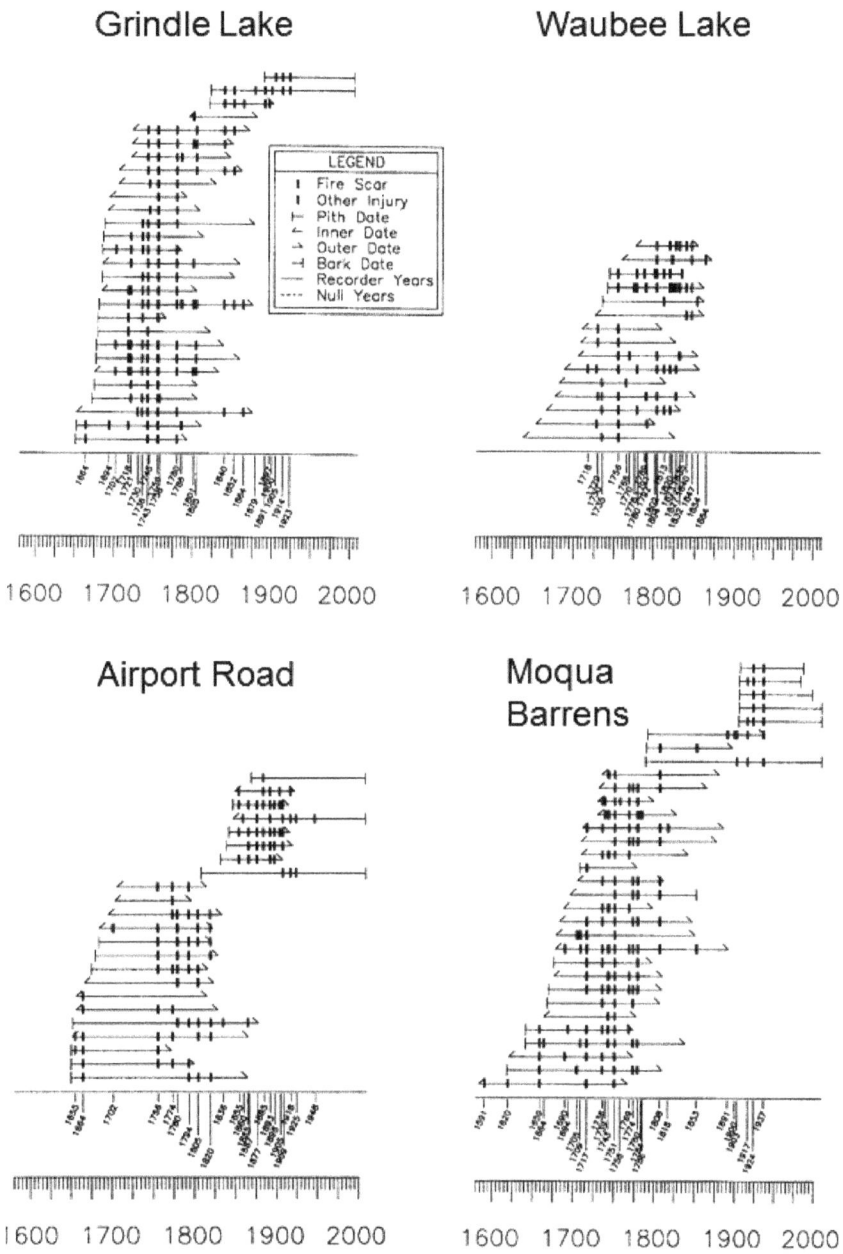

Figure 3. Fire history diagrams showing fire scar dates (vertical short lines) and composite fire scar records (bottom line with dates) at the four study sites. Each horizontal line represents the tree-ring record of a red pine remnant or living tree.

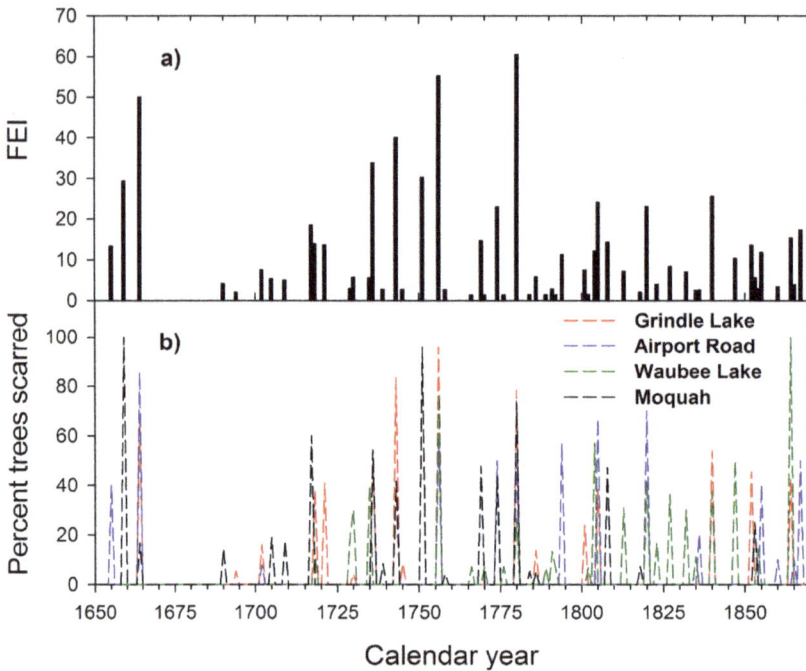

Figure 4. (**a**) Bar plot of the Fire Extent Index (FEI) for Sands Ecological Landscapes of Wisconsin (*n* = 98 red pine trees), (**b**) percentage of trees scarred at each of the four study sites through time.

Table 2. Fire frequency statistics for study sites. Time periods were selected based on characteristics of fire frequency, human history, and fire history record overlap. Precise time periods vary slightly owing to fire intervals that span the boundaries between time periods. Model predicted MFIs (2nd to last row) are from Guyette et al. 2012. Periods not covered by the fire scar record are represented by "na".

Period	Grindle Lake	Waubee Lake	Airport Road	Moquah Barrens
Mean fire interval (~1660–1780)	9.7 years	9.9 years	29.0 years	8.9 years
Mean fire interval (~1780–1864)	12.0 years	5.1 years	14.4	14.6 years
Mean fire interval (~1852–1925)	10.1 years	na	6.4 years	14.2 years
Mean fire interval (~1925–2010)	>87 years	na	>44 years	>43 years
Mean fire interval (start to <1937)	10.4 years 1664–1923	6.3 years 1707–1864	12.9 years 1655–1925	12.8 years 1591–1937
Mean fire interval (climate model predicted MFI [33])	14–18 years	14–18 years	14–18 years	20–22 years
Range of pre-1925 intervals	2–35 years	3–21 years	2–54 years	2–35 years

Fire scar seasonality was determinable for 316 scars (68 percent of scars). Dormant season (between growth rings) fire scars were the most common among sites with the exception of the Airport Road site which had more growing season fires than dormant season fires (Table 3). For all sites combined, 65 percent of the fires occurred during the dormant season (inter-ring and early earlywood scars), whereas 35 percent occurred during the growing season (middle earlywood, late earlywood and latewood scars; Table 3). Mixed seasonality fires (i.e., individuals years with trees recording different seasons of fires at a site) occurred during several years at three of the sites (Table 3). The dramatic fire year of 1780 [4] had the highest variability in fire seasonality both within and between sites.

Table 3. Fire scar positions within annual rings and their related fire seasonality. Season of fires are the percent of scars in each scar class.

	Grindle Lake	Waubee Lake	Airport Road	Moquah Barrens	All Sites & Years
Scar location within annual ring					
Between rings (dormant season)	52%	88%	10%	72%	56%
Early earlywood	24%	0%	7%	6%	9%
Middle earlywood	16%	0%	17%	3%	9%
Late earlywood	5%	0%	22%	1%	7%
Latewood	3%	12%	43%	17%	19%
Season of fire					
Late fall, winter, & early spring scars	76 %	88%	17%	79%	65%
Late spring & summer scars	24 %	12%	83%	21%	35%
Mixed seasons fires at site (percent and number years)	29%, 5	0%, 0	16%, 3	5%, 1	12%, 2

3.2. Tree Age and Recruitment

The youngest ages of red pine trees surviving fires ranged from 6 to over 30 years. The average age of trees at time of first scar was 15 years. Red pine overstory recruitment appeared to have occurred primarily as single trees or in small groups. However, at least two sites had two periods of major regeneration and recruitment, indicative of major disturbance. At Grindle Lake 52 percent of the trees sampled had established circa 1670–1685 and at the Airport Road site 44 percent of the trees were established circa 1660.

3.3. Fire Extent Index and Climate

Reconstructed drought indices [27] were a better predictor of the FEI than fire interval length. The FEI was highest during drought years (PDSI < 0). However, between 1650 and 1864 (i.e., the most replicated period of record) about 70 percent of drought years had no evidence of fire. When fires did occur ($n = 54$ years), drought (PDSI) explained about 24 percent of the variance in the FEI (Equation (2), Figure 5). A model predicting the FEI as a function of drought for fires years between 1650 and 1864 was given as:

$$\mathrm{Ln(FEI)} = 1.84 - (0.32 \times \mathrm{PDSI}) \tag{2}$$

where: ln(FEI) = natural log of (Fire Extent Index), PDSI = reconstructed drought [27], $n= 54$, $r^2 = 0.23$, $p < 0.001$ for model and variable.

Based on the FEI, fires in 1780, 1774, 1756, and 1664 were extensive. These years were dry with the exception of 1756. The year 1756 was incipient wet (PDSI = 0.98) and fire scars were limited to the northeastern sites (#'s 1, 2, 3; Table 1). The average PDSI of the fire years 1780, 1774, and 1664 was −1.9 (mild drought).

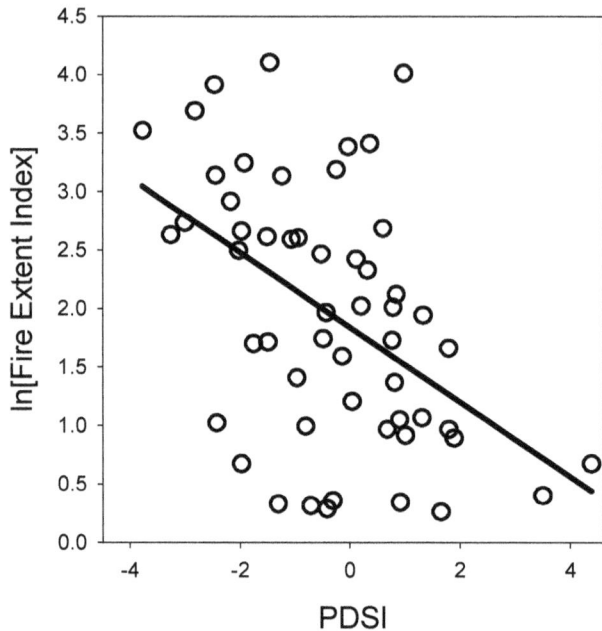

Figure 5. Scatterplot of the Fire Extent Index (FEI) and reconstructed Palmer Drought Severity Index (PDSI) for 54 historical fire years in the Northwest and Northeast Sands Ecological Landscapes of northern Wisconsin (Figure 1). Data are from the period 1660–1864. The regression indicates prediction of FEI from reconstructed drought (ln(FEI) = 1.84 − (0.32 × PDSI), Equation (2), r^2= 0.23). Decreasing values of PDSI relate to increasing dryness.

3.4. Human Population, Culture, and Fire Frequency

Native American populations plummeted soon after European encounters. In 1634, Jean Nicolet [35] wrote 'Already in 1644, the smallpox had reached the Winnebago in Green Bay where the rotting corpses caused great mortality and they could not bury the dead' [36]. Although we identified few fire years before the 1660s, this could be partly due to the reduced sample size in the record. The study fire regime, as described by the AFI (Figure 6), was increasingly influenced by humans from 1660 to 1755 after Native American populations increased upward as the Fox and Sauk immigrated to northern Wisconsin [36]. Continued human land use in the region occurred from 1755 to 1860, but was interrupted with short decadal decreases in circa 1775–1785 and 1810–1825 (Figure 6a). The lowest levels of the AFI occurred during 1770s and 1780s and corresponds with the "Great Smallpox Epidemic of 1775–1782" [37] and the population recovery thereafter.

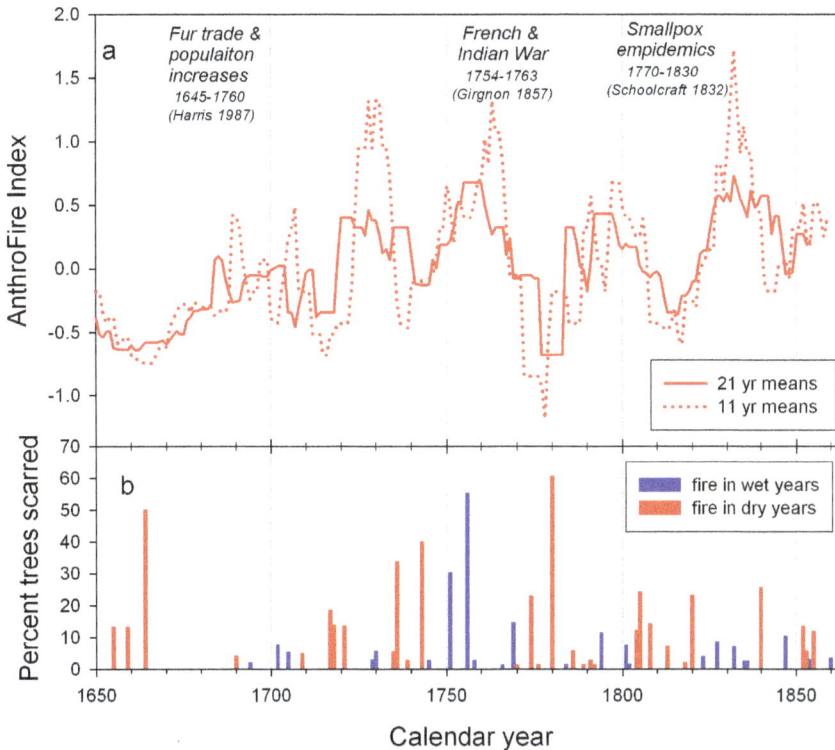

Figure 6. (**a**) Running means (21- and 11-year) of the AnthroFire Index (AFI). Larger AFI values indicate increasing human influence based on the occurrence of fire in 'wet' years and the absence of fire in drought years. The text identifies possible factors and periods of regional human influence in the fire regime of Northeastern Sands Ecological Landscape [23,38,39]. (**b**) Time series of the percent of trees scarred and reconstructed drought conditions (blue for wet years (PDSI > 0 and red for dry years PDSI < 0).

Early trends of increased fire frequency were associated with increased human population density, movement, and trade. Shipping records of the tonnage of trade goods in and out of the Great Lakes Region [23] were associated with fire frequency during the period 1640 and 1785 (Figure 7). The significance of the relationship between fire and trade could not be tested due to autocorrelation in the time series data and the lack of independence of the observations in trade records. Perhaps the most profound change in the fire regime occurred after 1925 when fire frequency was diminished to a rate of only two fires in an 85-year period (<0.6 fires per 25 years). Despite the low number of sample trees during this period, the fire record is fairly accurate because of the abundance of younger trees and trees not sampled (they had no fire scars) and the observation that 11 trees should capture much of the fire record (Figure 3).

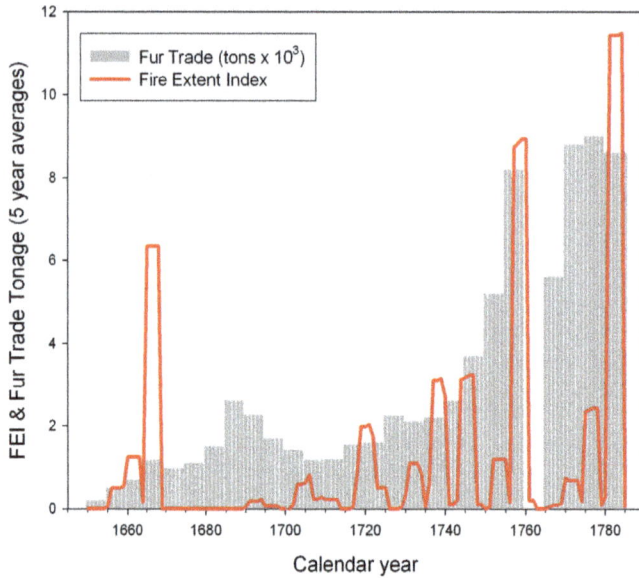

Figure 7. Trends in Great Lakes fur trade per five-year period [23] plotted with a five-year moving average of the number of fires at the study sites. The 1760–1765 trade data is missing from Harris [23]. The similar trends in trade and fire regimes support the 1650–1785 increase in human influence as calculated by the AnthroFire Index.

3.5. Climate-Fire Comparison in Other Fire History Records

We compared the northern Wisconsin fire scar record with nine fire scar records from the Great Lakes Region [5,13,29,33]. The synchrony of years reveals the probable influence of large-scale climate influence on regional fire activity (Figure 8). The temporal and spatial synchrony of percentage of trees scarred is significant ($r = 0.55$, $p < 0.01$). Curiously, fire scar records outside of northern Wisconsin, but in the Great Lakes region, are a stronger predictor of fire in Wisconsin than reconstructed drought [27].

Estimates of historical fire extents were based on the FEI (Figure 7); an index developed from a spatially limited (<6 km^2) collection area. In an attempt to overcome this limitation and to add support or refute extrapolating the characteristics of Wisconsin fire record to larger spatial extents, we compared the Wisconsin fire records to historical fire records in the Midwest and Great Lakes regions. Many of the years calculated as having large FEI values in northern Wisconsin were also extensive fire years in more northerly and eastern regions. For example, in 1780 all sites in Wisconsin burned and three of seven sites (43 percent) were burned in the Huron Mountains of Michigan (175 km northeast) [28]. Similarly, in 1664 fires also coincided in the Huron Mountains and in Wisconsin.

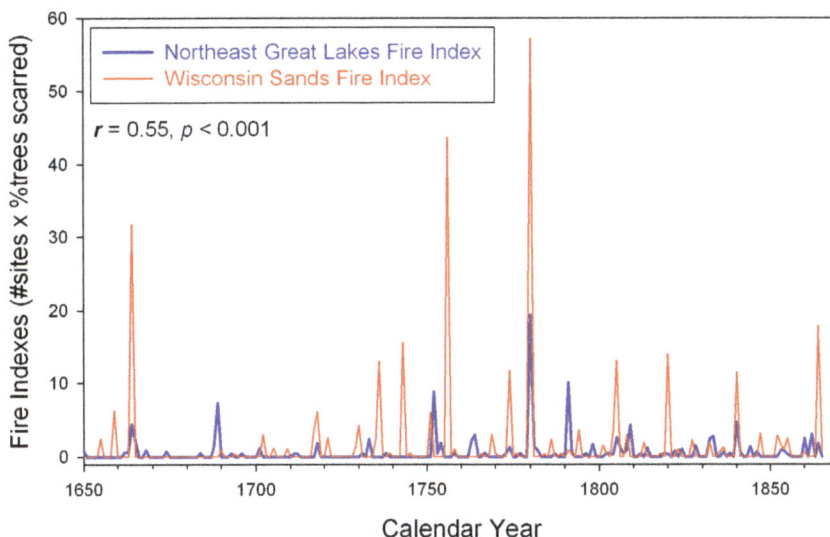

Figure 8. Comparison of the fire records from the Northwest and Northeast Sands Ecological Units of northern Wisconsin (98 trees, this study) plotted with separate fire records in the Great Lakes region including Ontario, Upper Peninsula of Michigan, and Minnesota (239 trees). Correlation analysis includes all years. Removal of no fire years increases the correlation coefficient.

In 1780, 9300 hectares were estimated to have burned in the Ontario's Algonquin Park Barron Township along with other fires in the eastern Algonquin Highlands [4,5,32]. The 1780 fire year is also represented 150 km northwest in northern Minnesota [40] and other areas [41]. Remarkably, 1000 kilometers to the south, sites across Missouri, Arkansas, and Oklahoma (an estimated 56 percent (122,000 km^2) of the Ozarks Highlands) burned in 1780 [11]. Fires in these studies are hypothesized to have resulted from the interactions between human ignitions and drought.

Mean fire intervals from northern Wisconsin sand ecosystems (Table 2, data rows 1 and 2) tend to be shorter than climate modeled fire intervals for the region (PC2FM [33], Table 2, data row 6). Two of the Wisconsin fire history records (Grindle and Waubee Lakes) were included in a dataset of 170 North American sites used to construct and calibrate this model [33]. Generally, mean fire intervals in this study (Table 2, data rows 1 and 2) are from 5 to 11 years shorter than this climate model estimates. This difference may result from more frequent human ignitions or landscape influences such as reduced topographic roughness.

4. Discussion

4.1. Fire Intervals

Red pine stumps, snags, and trees provided a fire scar record spanning three centuries. The length and quality of fire records provided by this wood was comparable to fire histories from Ontario, Michigan, Minnesota, and Wisconsin [18,41–45]. The fire scar record that the northern Wisconsin remnant wood provides is decaying and will probably be unavailable in a few decades. Much of the remnant wood used in this study was sampled near ground level with decay advancing from the stump top. Because sample trees were not randomly located and represent a small portion of the landscape, the detection of small fires may be less well represented than that of larger fires.

Fire intervals were more temporally variable within sites (Figure 3) than between them (Table 2). We hypothesize that this indicates strong local fire control (probably ignition abundance) on the study sites. For example, the Airport Road site, with the exception of a single fire scar (1702, growing season) had a 92-year fire interval (1664–1756). Also, drought explained only about one quarter (23 percent) of the variance in fire extent. This could be due to strong anthropogenic ignitions during windy and dry weather. The short length of the fire intervals at the Waubee Lake site may also be due to relatively high human activity. As is the case in many areas of the U.S., the greatest change in MFI at the sites resulted from fire suppression after about 1910. This is probably the most radical change from the historical fire regime in the record. Although this period appears low on fire scar data, during field collection we observed an absence of fire scars or charcoal on thousands of younger live trees (<100 years in age) across all sites indicating that few if any recent fires had occurred.

Most fires were likely single event fires with short durations (a few days as opposed to months). This conclusion is based on the consistency of fire scar positions within the dated annual rings when compared among different trees at the study sites. Little is known about the causes of fire years with mixed seasonality of fire scar positions. We surmise that this indicates either variability in the timing of fires or a difference among trees in the timing of wood formation and dormancy (Table 3). Scar seasonality differences within a year at a single site may indicate that fire burned very slowly, burned as ground fires (the study sites often included small areas of peat) that "flared up" again throughout the dormant and growing seasons, or there were multiple fires within a site within the year.

Often, intra-ring fire scar seasonality data has limitations due to the unknown differences in the timing of wood formation among trees and through time. Additionally, the effects of site conditions on the timing of cambial dormancy and physiology account for differences in the timing and rates of wood formation among trees. The timing of historical fires at Waubee Lake appears unique based on the preponderance of dormant season fires (88 percent, Table 3). Several factors unique to this site may be important. First, the fire history data are from red pine remnants scattered within a forest currently dominated by northern red oak (*Quercus rubra*). If this site was dominated by oak during the period of fire record, then deciduous leaf litter may have promoted dormant season fires. However, the composition of the prior stand is in question as northern red oak has benefitted greatly in northern Wisconsin since European settlement [19]. A second hypothesis is that Native Americans were actively using the site and conducted dormant season burns for their benefit such as harvesting acorns in the fall [46]. Dormant season and frequent burning could also have been used to promote the growth of nearby extensive areas of blueberries [47]. A trading post located at Waubee Lake in the mid-1800s and an early Native American presence in the region make it probable that abundant anthropogenic ignitions existed which would have been necessary to maintain frequent fire in the hummocky topography of the site.

4.2. Climate, Humans, and Fire

During fire years, the best predictor of fire severity (i.e., percentage of trees scarred during fire events) and extent of fire at the study sites was drought [27]. However, reconstructed drought was a poor predictor (<0.06 percent variance explained) of percentages of trees scarred alone. From this, we infer that between 1650 and 1864 the fire regimes of the study region were in an ignition-limited stage [10] and that fire frequency was primarily controlled by non-climatic factors such as ignition frequency.

Through time, the fire scar record defines a fire regime that had strong but variable anthropogenic influence. Compared to fire frequencies of other eastern and southern fire regimes [33], study sites had a higher frequency of fires than might be expected based on climate conditions (temperature and precipitation). As has been found elsewhere, anthropogenic influences on the fire regimes are expected to have been caused by multiple factors such as changes in human population density, cultures, and values. In northern Wisconsin, changes in commerce, land use, war, conflict, material culture, and even spirituality likely influenced the fire regime. Ignitions from flint and steel were brought to northern Wisconsin by Jesuit missionaries in the early 17th century as documented by Jean De Brebeuf in 1633 [48]: "You should love the Indians like brothers, with whom you are to spend the rest of your life. Never make them wait for you in embarking. Take a flint and steel to light their pipes and kindle their fire at night...".

The French fur trade and material culture developed quickly with the European religious missions. Material culture soon increased in influence over the religious mission, the fur trade soon dominated, and the Jesuits were expelled for a time (1712–1738) during the Fox Wars [36]. The material culture driven by the economics of the fur trade (Figure 8) and fire technology of Europe (e.g., flint and steel, rifles, metal pots) modified the fire regime and resulted in new human trails beyond the traditional waterways and into the "wilderness". During the most replicated period of our fire history data (1650–1864), commerce in the northern Wisconsin economy was centered on the fur trade. For over 200 years, the European demand for furbearers, especially beaver, likely affected ignitions (i.e., rates, seasons). In summary, in northern Wisconsin this period of the fire regime may have been influenced by climate conditions and commercial activity including the politics and material demands from Europe.

Early commercial activities influencing the fire regime include the increased movement of humans (and potential ignitions) during the fur trade. In this case, ignitions may have occurred in remote regions (i.e., low human population density) where beaver and their habitats were abundant. Fires may have been set to alter habitat for beaver and other furbearers [49,50] and encourage an increase small woody (e.g., willow, birch, aspen) and herbaceous vegetation that are a primary source of food for beaver and other wildlife. By the 1820s, overharvesting of beaver in northern Wisconsin began to reduce the fur trade and perhaps ignitions. Beginning in the 1850s, land use in northern Wisconsin changed to include extensive commercial activities (i.e., 'The Great Cutover') [51] of logging and agriculture. Vegetation conversion with fire and axe altered the fuel environment (fuel types, fuel fragmentation, and fuel loading) while human population density continued to increase.

Movements of human ignitions can create complex, abrupt, and potentially rapidly changing characteristics in fire regimes and vegetation [11]. Although many native cultures had remarkable abilities to travel, few had the transport technology of the Ojibwa birch-bark canoe [52]. Canoes allowed travel for trade and the migration of human populations. Native American oral histories indicate long-distance movements by water of tribal groups from east to west during the early period of European contact. Ojibwa tradition indicates that they lived near the Saint Lawrence River and about 1660 C.E. the Ojibwa migrated westward in the Great Lakes region [53]. Early population shifts of humans with 'intelligent ignitions' (i.e., ignited through a thought process) may have altered the vegetation of the study region with a design of subsistence fire use.

Primary among Ojibwa was the use of paper birch (*Betula papyrifera*) bark. The bark was incorporated into shelter (i.e., hogans) and canoe construction, thus, incentive for burning existed to increase this critical resource when and where it was needed [52]. Paper birch is a shade intolerant early successional tree species that requires bare mineral or organic soil for seedling reproduction [54,55], conditions that are enhanced by fire. Forest canopy and litter disturbance such as that from fire are accountable for the establishment of many stands. In addition, Native Americans had many other reasons for manipulating vegetation for their use and benefit.

4.3. Human Conflict, War, and Fire

The Potawatomi, Ojibwa, Sauk, Fox, and Ottawa groups moved into Wisconsin as early as the 1600s, near the beginning of our fire history. The early fire history of Wisconsin cannot be separated from war and conflict among the cultural groups, commercial enterprises, and the societies that lived there. Fire has long been a tool of war and conflict [56,57]. Groups that vied for power, land, furs, influence, and 'souls' included the Menominee, Sioux, Ojibwa, French, English, Americans, and Jesuits. Many conflicts occurred in Wisconsin during the Iroquois-French wars (~1701), Fox Wars (1712–1738), the French and Indian War (1754–1763), and the War of 1812 [36]. Generally, with the exception of the French and Indian War, these conflict periods do not appear to be associated with changes in fire regime characteristics, even despite new Native American groups moving into Wisconsin as a result.

Characteristics of fire scars in 1756 suggest that it may be pronounced due to the 'French and Indian War' (1754–1763) or 'Seven Years War'. During the War, the Menominee and others were allied with the French while the Iroquois were allied with the British. The year 1756 is an extreme outlier because of a high proportion of trees scarred (55 percent), sites with fire (3), and occurrence during incipient wet conditions (PDSI +0.98) [27]. Over the full period of the fire scar record, 1756 ranked third in numbers of fire scars. In the Northeastern Sands Ecological Landscape of Wisconsin (near present Menominee tribal lands), 81 percent of the trees were scarred in 1756. Alternative hypotheses for the magnitude of the 1756 fire year are 1) weather, not summer season drought conditions, was the major and sole factor for the pronounced fire year, and/or 2) the interaction of weather and abundant human ignitions were predisposing and inciting factors.

4.4. Diseases and Population

Perhaps a greater factor affecting anthropogenic fire regimes than wars or conflicts was the devastation of Native American populations by diseases introduced from Europe. Disease has been linked to depopulations and changes in fire activity at multiple sites [5,10]. Jean Nicolet visited thousands of the Ho-Chunk in Wisconsin in 1634 [57]. Twenty years later, French fur traders found only 700 people remaining. In northern Wisconsin, smallpox was introduced by Europeans in the mid-1600s and continued as a serious episodic epidemic until at least the 1830s [34,58,59]. The later medical history of smallpox among the Ojibwa indicates that this disease was brought by British fur traders in 1770. We observed declines in the AnthroFire Index (Figure 6) circa 1780 and 1830 coincident with documented smallpox out breaks further implicating humans in regional fire activity.

Although written history rarely provides the data needed for systematic analysis, it does provide important context to the fire scar data. Hunting and gathering activities often utilize landscape-scale fires. Studies on hunting patterns and fur trade of Native Americans document human locations, timing, and activities [36]. Kay [36] references European's first encounter with the Menominee Indians at the mouth of the Menominee River (south of sites 1–3). By 1720, sites 1–3 may have been influenced by the Menominee, Fox, and Sauk (see Figure 1 in [36]).

4.5. Management Implications

Knowledge of fire history is often a first step in ecosystem restoration activities that incorporate fire management. Data of historical fire intervals, severity, and extent provides a wealth of information for consideration in management. Paired with knowledge of species and ecosystem ecology, past vegetation composition, structure and change may be realized. The historical range of variability in fire disturbance provides a sense for landscape dynamics through space and time [1]. Coupled with other historical and ecological information, such as Public Land Survey notes [60], county records, aerial photography, soil surveys, and ecological land type maps, it is possible to set targets such as forest patch sizes, structure diversity, and stem density. In red pine and oak forest ecosystems, fire intervals and tree regeneration and successful recruitment may be linked to fire. Identification of extreme fire events and their inciting factors (e.g., drought, humans) can be important examples beyond the potential future management opportunities.

5. Conclusions

Fire histories developed from fire scars are capable of revealing numerous and varied details of the long-term changes in fire, forests, and land use that are otherwise unobtainable by other methods. We expect these fire history data and analyses provide land managers, fire practitioners, policymakers, and the public a greater perspective on the ecology and historical role of fire in northern Wisconsin. In northern Wisconsin Sands Ecological Landscapes, fires were highly variable through time, ranging from uncommon to frequent and spatially limited to extensive. A temporal understanding of fire regimes aids in realizing not only factors that influence fire but also the full range of potential fire conditions. In the Lake States, historical fire events are among the most powerful and dangerous (rate of spread, energy release) experienced in the U.S. (e.g., 1871 Peshtigo Fire, 1980 Mack Lake Fire) [61]. From the current network of fire history data in the Great Lakes region and expansive unstudied areas, high potential exists to further describe the landscape variability in historical fire frequency and, specifically, the climatological conditions associated with high severity fire event years (ex. 1664, 1756, and 1780).

Acknowledgments: The Northern Research Station of the USDA Forest Service provided funding for this project. Support for location, collection, and historical information was provided by the Chequamegon-Nicolet National Forest. Greg Nowacki provided helpful comments on previous versions of the manuscript.

Author Contributions: R.P. conceived the study, aided in sample collection, analyzed the data; M.S. aided in sample collection and manuscript writing; D.D. supported analysis and manuscript writing; J.M. processed samples and crossdated tree-rings and fire scars; J.S. located study sites, aided in sample collection and provided management information; J.L. located study sites, provided forest information, and contributed to manuscript writing and review.

Conflicts of Interest: The authors declare no conflict of interest.

References

1. Swetnam, T.W.; Allen, C.D.; Betancourt, J.L. Applied historical ecology: Using the past to manage for the future. *Ecol. Appl.* **1999**, *9*, 1189–1206. [CrossRef]
2. Schulte, L.A.; Mladenoff, D.J. Severe wind and fire regimes in northern Wisconsin (USA) forests: Historical variability at the regional scale. *Ecology* **2005**, *86*, 431–445. [CrossRef]
3. Schulte, L.A.; Mladenoff, D.J.; Burrows, S.N.; Sickley, T.A.; Nordheim, E.V. Spatial controls of Pre-EuroAmerican wind and fire in Wisconsin (USA) forests: A multiscale assessment. *Ecosystems* **2005**, *8*, 73–94. [CrossRef]
4. McMurry, E.R.; Stambaugh, M.C.; Guyette, R.P.; Dey, D.C. Fire scars reveal the source of New England's 1780 Dark Day. *Int. J. Wildland Fire* **2007**, *16*, 266–270. [CrossRef]
5. Dey, D.C.; Guyette, R.P. Anthropogenic fire history and red oak forests in south-central Ontario. *For. Chron.* **2000**, *76*, 339–347. [CrossRef]
6. Curtis, J.T. *The Vegetation of Wisconsin*; University of Wisconsin Press: Madison, WV, USA, 1959.
7. Rohe, R.; Miller, S.; Eisele, T. *One Hundred Years of Wisconsin Forestry*; Trails Custom Publishing: Black Earth, WI, USA, 2004.
8. Stewart, O.C. Barriers to understanding the influence of use of fires by aborigines on vegetation. *Proc. Second Annu. Tall Timbers Fire Ecol. Conf.* **1963**, *2*, 117–126.
9. Wisconsin Historical Society. Available online: http://www.wisconsinhistory.org/ (accessed on 18 August 2016).
10. Guyette, R.P.; Muzika, R.M.; Dey, D.C. Dynamics of an anthropogenic fire regime. *Ecosystems* **2002**, *5*, 472–486.
11. Guyette, R.P.; Spetich, M.A.; Stambaugh, M.C. Historic fire regime dynamics and forcing factors in the Boston Mountains, Arkansas, USA. *For. Ecol. Manag.* **2006**, *234*, 293–304. [CrossRef]
12. Stambaugh, M.C.; Guyette, R.P.; Marschall, J.M. Fire history in the Cherokee Nation of Oklahoma. *Human Ecol.* **2013**, *41*, 749–758. [CrossRef]
13. Torretti, R.L. Traditional Stories from Non-Traditional Stories: Tree-Rings Reveal Historical Use of Fire by Native Americans on Lake Superior's Southern Shore. Master's Thesis, Northern Michigan University, Marquette, MI, USA, 2003.
14. Loope, W.L.; Anderton, J.B. Human versus lightning ignition of presettlement surface fires in coastal pine forests of the upper Great Lakes. *Am. Midland Nat.* **1998**, *140*, 206–218. [CrossRef]
15. Mooney, J. The Aboriginal population of America north of Mexico. In *Smithsonian Miscellaneous Collections*; Swanton, J., Ed.; Smithsonian Institution: Washington, DC, USA, 1928; Volume 80, pp. 1–40.
16. Swanton, J.R. *Bureau of American Ethnology*; The Indian tribes of North America Smithsonian Institution: Washington, DC, USA, 1952.
17. Terrell, J.U. *American Indian Almanac*; The World Publishing Co.: New York, NY, USA, 1971; p. 494.

18. Sands, B.A.; Abrams, M.D. A 183-year history of fire and recent fire suppression impacts in select pine and oak forest stands of the Menominee Indian Reservation, Wisconsin. *Am. Midland Nat.* **2011**, *166*, 325–338. [CrossRef]

19. Nowacki, G.L.; Abrams, M.D.; Lorimer, C.G. Composition, structure, and historical development of northern red oak stands along an edaphic gradient in north-central Wisconsin. *For. Sci.* **1990**, *36*, 276–292.

20. Ecological Landscapes of Wisconsin. State of Wisconsin. Available online: http://dnr.wi.gov/topic/landscapes/book.html (accessed on 26 August 2016).

21. Peel, M.C.; Finlayson, B.L.; McMahon, T.A. Updated world map of Koppen-Geiger climate classification. *Hydrol. Earth Syst. Sci.* **2007**, *11*, 1633–1644. [CrossRef]

22. Kotar, J.; Kovach, J.A.; Locey, C.T. *Field Guide to Forest Habitats Types of Northern Wisconsin*; Wisconsin Department of Natural Resources: Madison, WI, USA, 1998.

23. Harris, R.C.; Matthews, G.J. *Historical Atlas of Canada*; University of Toronto Press: Toronto, ON, 1987; Volume 1.

24. Stokes, M.A.; Smiley, T.L. *An Introduction to Tree-Ring Dating*; University of Arizona Press: Tucson, AZ, USA, 1996; p. 73.

25. Holmes, R.L.; Adams, R.; Fritts, H.C. Quality Control of Crossdating and Measuring: A User's Manual for Program COFECHA. In *Tree-Ring Chronologies of Western North America: California, Eastern Oregon and Northern Great Basin*; University of Arizona, Laboratory of Tree-Ring Research: Tucson, AZ, USA, 1986; pp. 41–49.

26. Grissino-Mayer, H.D.; Holmes, R.L.; Fritts, H.C. *International Tree-Ring Data Bank Program Library Version 2.0 User's Manual*; University of Arizona, Laboratory of Tree-Ring Research: Tucson, AZ, USA, 1996; p. 106.

27. Cook, E.R.; Meko, D.M.; Stahle, D.W.; Cleaveland, M.K. *North American Summer PDSI Reconstructions*; NOAA/NGDC, Paleoclimatology Program: Boulder, CO, USA, 2004.

28. Muzika, R.M.; Guyette, R.P.; Stambaugh, M.C.; Marschall, J.M. Fire, drought, and humans in a heterogeneous Lake Superior landscape. *J. Sustain. For.* **2015**, *34*, 49–70. [CrossRef]

29. Guyette, R.P.; Stambaugh, M.C. A quantitative method for estimating long-term influence of human ignitions on fire regimes. In Uniting Research, Education, and Management, Proceedings of 5th International Fire Ecology and Management Congress, Portland, OR, USA, 3–7 December 2012.

30. Dietrich, J.H. *The Composite Fire Interval—A Tool for more Accurate Interpretation of Fire History*; USDA Forest Service General Technical Report RM-81; Rocky Mountain Forest and Range Experiment Station Forest Service US, Department of Agriculture: Tucson, AZ, USA, 1980; pp. 8–14.

31. SAS/STAT. *SAS User's Guide: Statistics*, 5th ed.; SAS Institute: Cary, NC, USA, 2002; p. 955.

32. Cwynar, L.C. The recent fire history of Barron Township, Algonquin Park. *Can. J. Bot.* **1977**, *55*, 1524–1538. [CrossRef]

33. Guyette, R.P.; Stambaugh, M.C.; Dey, D.C.; Muzika, R.M. Estimating fire frequency with the chemistry of climate. *Ecosystems* **2012**, *15*, 322–335. [CrossRef]

34. Schlesier, K.H. Epidemics and Indian Middlemen: Rethinking the Wars of the Iroquois, 1609–1653. *Ethnohistory* **1976**, *23*, 129–145. [CrossRef] [PubMed]

35. Blair, E.H. *The Indian tribes of the Upper Mississippi and Great Lakes Region*; University of Nebraska Press: Lincoln, MT, USA, 1911.

36. Kay, J. The fur trade and Native American population growth. *Ethnohistory* **1984**, *31*, 265–287. [CrossRef]

37. Fenn, E.A. *PoxAmerica: The Great Smallpox Epidemic of 1775-82*; Cahners Business Information, Inc.: New York, NY, USA, 2001.

38. Seventy-two years' recollections of Wisconsin. Available online: http://www.wisconsinhistory.org/turningpoints/search.asp?id=28 (accessed on 26 August 2016).

39. Schoolcraft, H.R. *Narrative of an expedition through the upper Mississippi to Itasca Lake, 1832*; Harper & Brothers: New York, NY, USA, 1834.

40. Heinselman, M.L. Fire in the virgin forests of the Boundary Waters Canoe Area, Minnesota. *Quat. Res.* **1973**, *3*, 329–382. [CrossRef]

41. Guyette, R.P.; Palik, B.; Dey, D.C. Three Centuries of Burning on the Chippewa National Forest and the Leech Lake Band of Ojibwa's near Cass Lake, Minnesota. Unpublished work. 2015.

42. Guyette, R.P.; Dey, D.C. *A Pre-Settlement Fire History of an Oak-Pine Forest near Basin Lake, Algonquin Park, Ontario*; Forest Research Report No. 132; Ontario Forest Research Institute: Marie, Canada, 1995.

43. Guyette, R.P.; Dey, D.C. *A Dendrochronological Fire History of Opeongo Lookout in Algonquin Park, Ontario*; Forest Research Report No. 134; Ontario Forest Research Institute: Ontario, Canada, 1995.

44. McEwan, R.; Aldrich, S.; Bauer, J.; Gentry, C.; Kernan, J.; Lusteck, R.; Martinez, P.; Shapiro, L.; Sprenger, C.; Vining, M.; et al. A tree-ring based fire history reconstruction of red pine (*Pinus resinosa*) forests, Lake Itasca, Minnesota. Ohio Academy of Science 114th Annual Meeting Abstracts. *Ohio J. Sci.* **2005**, *105*, A-36.

45. Drobyshev, I.; Goebel, P.C.; Hix, D.M.; Corace, R.G., III; Semko-Duncan, M.E. Pre- and post-European settlement fire history of red pine dominated forest ecosystems of Seney National Wildlife Refuge, Upper Michigan. *Can. J. For. Res.* **2008**, *38*, 2497–2514. [CrossRef]

46. Use of acorns for food in California: past, present and future. Available online: http://timeless-environments. blogspot.com/p/use-of-acorns-for-food-in-california.html (accessed on 18 August 2016).

47. Kautz, E.W. Prescribed fire in blueberry management. *Fire Manag. Notes* **1987**, *USDA FS 48*, 9–12.

48. Parkman, F. *The Jesuits in North America in the Seventeenth Century*; Little, Brown and Company: Boston, MA, USA, 1983.

49. Baskin, Y. *The Work of Nature: How The Diversity Of Life Sustains Us*; Island Press: Washington, DC, USA, 1997.

50. Hood, G.A.; Bayley, S.E.; Olson, W. Effects of prescribed fire on habitat of beaver (*Castor canadensis*) in Elk Island National Park, Canada. *For. Ecol. Manage.* **2007**, *239*, 200–209. [CrossRef]

51. Schulte, L.A.; Mladenoff, D.J.; Crow, T.R.; Merrick, L.; Cleland, D.T. Homogenization of northern USA Great Lakes forests as a result of land use. *Landsc. Ecol.* **2007**, *22*, 1089–1103. [CrossRef]

52. Matile, R. *Those Marvelous Ojibwa Built Birch Bark Canoes*; The Canadian Canoe Museum: Oswego, IL, USA, 2013.

53. Swenson, J.C. Canoe Passages: Cross-Cultural Conveyance in USA and Canadian Literature. Ph.D. Thesis, University of Iowa, Iowa City, IA, USA, 2007.

54. Burns, R.M.; Honkala, B.H. *Silvics of North America: Volume 2. Hardwoods*; USDA Forest Service: Washington, DC, USA, 1990; pp. 158–171.

55. Fowells, H.A. *Silvics of Forest Trees of the United States/Prepared by the Division of Timber Management Research, Forest Service, 1965*; Agricultural Handbook USDA no. 271; USDA Forest Service: Washington, DC, USA, 1965.

56. Pyne, S.J. *Fire in America*; University of Washington Press: Princeton, NJ, USA, 1982; p. 655.

57. Kellogg, L.P. *Early Narratives of the Northwest, 1917, 1634-1699*; Charles Scribner's Sons: New York, NY, USA, 1917.

58. Houghton, D. Vaccination of the Indians. In *Henry Schoolcraft, Summary Narrative of an Exploratory Expedition to the Sources of the Mississippi River*; Philadelphia, Lippincott, Grambo, and Co.: Philadelphia, PA, USA, 1855; pp. 574–581.

59. Fenn, E.A. *PoxAmerica: The Great Smallpox Epidemic of 1775-82*; Hill & Wang: New York, NY, USA, 2001.

60. Schulte, L.A.; Mladenoff, D.J. The original USA Public Land Survey Records: Their use and limitations in reconstructing presettlement vegetation. *J. For.* **2001**, *99*, 5–10.

61. Simard, A.J.; Haines, D.A.; Blank, R.W.; Frost, J.S. *The Mack Lake Fire*; USDA Forest Service: St, Paul, MN, USA, 1983.

© 2016 by the authors. Licensee MDPI, Basel, Switzerland. This article is an open access article distributed under the terms and conditions of the Creative Commons Attribution (CC BY) license (http://creativecommons.org/licenses/by/4.0/).

forests

MDPI

Article

Mapping Local Effects of Forest Properties on Fire Risk across Canada

Pierre Y. Bernier [1],*, Sylvie Gauthier [1], Pierre-Olivier Jean [1,2], Francis Manka [1], Yan Boulanger [1], André Beaudoin [1] and Luc Guindon [1]

[1]   Canadian Forest Service, Natural Resources Canada, Quebec, QC G1V 4C7, Canada; sylvie.gauthier2@canada.ca (S.G.); peo.jean@gmail.com (P.-O.J.); francismanka@hotmail.com (F.M.); yan.boulanger@canada.ca (Y.B.); andre.beaudoin@canada.ca (A.B.); luc.guindon@canada.ca (L.G.)
[2]   Currently at Premier Tech Horticulture, 1 Avenue Premier, Rivière-du-Loup, QC G5R 6C1, Canada
*   Correspondence: pierre.bernier2@canada.ca; Tel.: +1-418-648-4524

Academic Editor: Eric J. Jokela
Received: 9 May 2016; Accepted: 19 July 2016; Published: 27 July 2016

Abstract: Fire is a dominant mechanism of forest renewal in most of Canada's forests and its activity is predicted to increase over the coming decades. Individual fire events have been considered to be non-selective with regards to forest properties, but evidence now suggests otherwise. Our objective was therefore to quantify the effect of forest properties on fire selectivity or avoidance, evaluate the stability of these effects across varying burn rates, and use these results to map local fire risk across the forests of Canada. We used Canada-wide MODIS-based maps of annual fires and of forest properties to identify burned and unburned pixels for the 2002–2011 period and to bin them into classes of forest composition (% conifer and broadleaved deciduous), above-ground tree biomass and stand age. Logistic binomial regressions were then used to quantify fire selectivity by forest properties classes and by zones of homogeneous fire regime (HFR). Results suggest that fire exhibits a strong selectivity for conifer stands, but an even stronger avoidance of broadleaved stands. In terms of age classes, fire also shows a strong avoidance for young (0 to 29 year) stands. The large differences among regional burn rates do not significantly alter the overall preference and avoidance ratings. Finally, we combined these results on relative burn preference with regional burn rates to map local fire risks across Canada.

Keywords: boreal forest; Canada; fire selectivity; MODIS; wildfires

1. Introduction

Fire is a dominant disturbance across boreal forests, with surface fires and stand-replacing canopy fires dominating in the Eurasian and North-American boreal forests respectively [1]. In Canada, the rate at which boreal and Montane forest burn, and the inter-annual variability in area burned affect all ecological processes, from landscape diversity to carbon sequestration [2]. Wildfires are also a threat to communities and infrastructures, and their frequency is predicted to increase during this century with climate change [3]. Because of this, improved maps of local fire risks are now needed to help inform management actions for risk mitigation.

Fire was initially studied in terms of its behaviour based on the detailed analysis of individual fires, with results showing the importance of weather conditions, and tree species on variables such as crowning, rate of spread and fuel consumption, e.g., [4,5]. This knowledge base was later extended to fire frequency using regional studies. In general, fire frequency or fire risk was found to be related to a host of factors, including longitude and latitude, topography, surficial deposits and forest properties [6,7], often dependant on the scale at which the study was conducted. In particular, fire was shown to strongly select conifer stands as compared to stands with a large deciduous component [8,9].

Finally, over the past decade, efforts have included national evaluations of fire risk and projections of such risks over time as a function of climate scenarios [10,11]. Because of their national coverage, these studies have been based exclusively on historical fire records and regional climate data and projections.

In the studies and approaches described above, the shift from understanding behavior to mapping risks at the national level has entailed the use of a data at coarse resolution, and an associated shift from local to regional drivers of fire dynamics. However, the mapping of fire risks across Canada's forests at a scale relevant to management decisions requires a merging of regional climatic drivers of burn rates and local drivers of fire risk. Possibly the only study that tackled this scale issue was that of Parisien and colleagues [12] who looked at drivers of fire dynamics across Canada's forests within 10,000-km^2 hexagonal pixels. In that study, the authors extracted both regional (climate) and local (forest type) drivers of burn rates in the same analysis, and concluded that the climate variables largely dominate the explained variation in regional burn rates.

Recent developments described below now enable us to use a different approach for getting at a much finer evaluation of local forest fire risks across Canada. Regional fire risks, in terms of annual area burned and burn rate have been evaluated by Boulanger and colleagues [11] across 16 zones of homogeneous fire regimes (HFR) across Canada's vegetated landscape. At the same time, Canada-wide maps of forest properties for base year 2001 and of fire for years 2001 to 2011 were created by Beaudoin and colleagues [13] and by Guindon and colleagues [14] respectively at a 250 m (6.25 ha) resolution corresponding to the pixel grid of the space-borne MODIS sensor. These last two products offer the possibility to quantify vegetation controls on fire risk with a massive dataset of fire and forest property observations. The goal of this study was therefore, by capitalising on this new information, to create a map of local (pixel-level) fire risk by combining quantified effect of vegetation properties on fire risk and the regional, climate-driven burn rates.

More specifically, the objectives of this study were (1) to quantify the effect of forest properties on fire selectivity; and (2) to assess the stability of this selectivity by forest property class across the range of regional burn rates found in Canada's forests; and (3) to combine regional burn rates and fire selectivity by forest property class into a map of current fire risk at a 250 m resolution. We chose forest age, biomass and composition as forest properties, with age and biomass used as equivalent proxies to fuel loads, and composition used as a proxy to fuel flammability. In this analysis, forest composition is expressed as the percent composition in either coniferous evergreen (including *Larix* species) or broadleaved deciduous tree species. Across Canada, coniferous species are mostly represented by the genus *Picea*, *Abies* and *Pinus*, while broadleaved deciduous tree species in fire-prone areas are mostly represented by species from the genus *Populus* and *Betula*, while forest zones in south-eastern Canada with very limited fire activity have additional and often dominant components in trees of the genus *Acer*, with a large number of other broadleaved deciduous genus at various level of representation.

2. Materials and Methods

2.1. Region of Interest and Source of Data

The study was conducted across all forests of Canada using the 16 Homogeneous Fire Regime (HFR) zones from which we selected the percent annual area burned, or burn rate, as our regional fire risk metric [11] (Figure 1). These historical (1959–1999) annual burn rates range from less than 0.1% in the broadleaved-deciduous dominated forests of the Eastern Temperate zone, to nearly 1.5% in the conifer-dominated Lake Athabaska zone, for a 50-fold difference across the HFR zones.

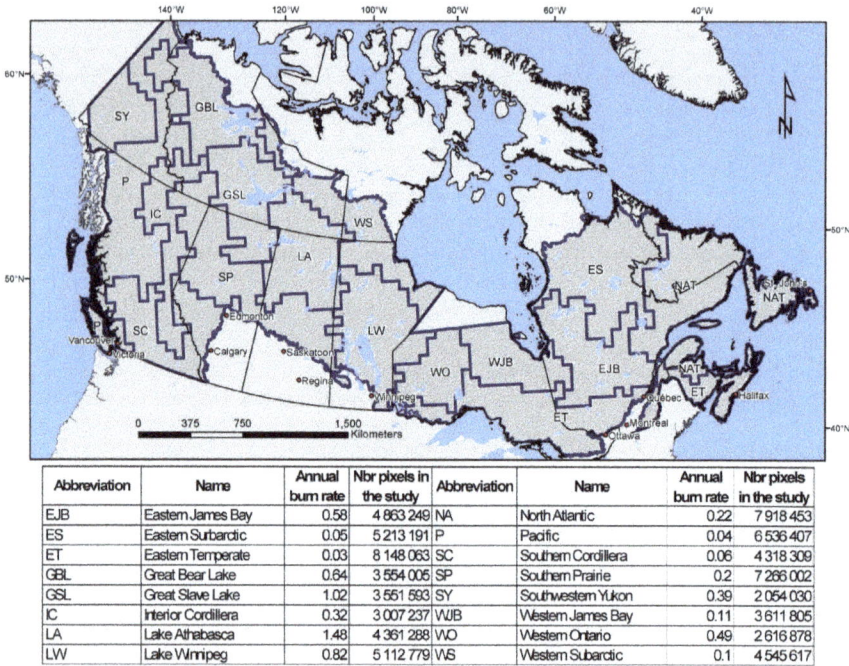

Abbreviation	Name	Annual burn rate	Nbr pixels in the study	Abbreviation	Name	Annual burn rate	Nbr pixels in the study
EJB	Eastern James Bay	0.58	4 863 249	NA	North Atlantic	0.22	7 918 453
ES	Eastern Subarctic	0.05	5 213 191	P	Pacific	0.04	6 536 407
ET	Eastern Temperate	0.03	8 148 063	SC	Southern Cordillera	0.06	4 318 309
GBL	Great Bear Lake	0.64	3 554 005	SP	Southern Prairie	0.2	7 266 002
GSL	Great Slave Lake	1.02	3 551 593	SY	Southwestern Yukon	0.39	2 054 030
IC	Interior Cordillera	0.32	3 007 237	WJB	Western James Bay	0.11	3 611 805
LA	Lake Athabasca	1.48	4 361 288	WO	Western Ontario	0.49	2 616 878
LW	Lake Winnipeg	0.82	5 112 779	WS	Western Subarctic	0.1	4 545 617

Figure 1. Homogeneous fire regime zones of Canada [11] with their associated burn rate, expressed as the mean annual area burned, in percent of their total area, and with the number of pixels from each homogeneous fire regime (HFR) used in the present analysis.

Forest composition, age and biomass were obtained at a pixel resolution of 250 m × 250 m (6.25 ha) from the Canada-wide maps of forest properties of Beaudoin and colleagues [13]. These maps were created for base year 2001 using the non-parametric *k*-nearest-neighbour as the statistical estimation method, the National Forest Inventory photoplot data [15] as reference information, and a set of Canada-wide variables including MODIS spectral reflectance at 250 m (6.25 ha) pixel resolution as predictors. For our analysis, pixels in which the forest was identified as recently disturbed in 2001 and pixels with <80% of vegetation cover were eliminated. The analyses were carried out on the remaining 76,678,906 pixels, referred below as the "full set".

Yearly Canada-wide maps of fire and harvest were produced for years 2001–2011 by Guindon and colleagues [14], also on the 250 m MODIS grid (available as supplementary material b of the original publication). These maps were created by analysing the pixel-level change in spectral properties over a moving 4-year window with the use of models trained on a large representative set of pixels that had been either harvested or burned. We used these maps to extract from our full set of pixels those that had burned in years 2002 to 2011. This led to the identification of 2,739,728 pixels distributed across all the HFR zones in which a fire had taken place in those years, with the fire extending across the full extent of the pixel in nearly 80% of the cases. The pre-burn composition, age and biomass of these burned pixels had been extracted earlier from the 2001 forest properties maps. The fire map product was specifically designed to minimize commission errors [14], and all pixels thus identified as having burned were retained for the analysis.

All pixels within the full set and the burned subset were binned into four composition classes, three age classes and three biomass classes based on their 2001 properties. Composition classes were: pure conifers (more than 75% conifers), conifer-dominated mixtures (between 50% and 75% of conifers), deciduous-dominated mixtures (between 25% and 50% of conifers), and pure deciduous (less than 25%

of conifers). Age classes were: young (0 to 29 years), mature (30 to 89) and old (90+). Biomass classes were divided so as to contain the same number of pixels across the full dataset: low: (less than 19 tons/ha), medium: (19–55 tons/ha) and high (more than 55 tons/ha). The analyses were to be carried out on two combinations of independent variables: "composition and age", and "composition and biomass". Preliminary tests showed no significant correlation between composition and age or biomass across the HFR zones.

2.2. Analytical Methods

We used logistic binomial regressions to examine the pixel-level fire selectivity to biomass or age and composition. The full set of pixels was used as the trial set from which fire could select whereas the burned subset of pixels was used as the event set that was selected by the fire. Pixels in the full set and in the burned subset were processed by HFR zone using logistic regressions in which "composition and age" or "composition and biomass" were the independent class variables, and tested with or without interaction among them. Calculations were done using the *glm* function as programmed in the *R* open software [16]. Models were compared based on the smallest area under the Receiver Operating Characteristic (ROC) curve obtained using the *pROC* procedure of [17] programmed within the *R* open software.

Probabilities derived from the logistic models inherently reflect the 10-year burn rate in each individual HFR zones, which is thought to be in large part driven by climatic conditions (top-down factors). This effect had to be removed in order to compare the relative selectivity of fire for each "composition × age" or "composition × biomass" class across all HFR zones. Within each HFR zone, we therefore divided the burn probability of each class by the mean probability of the 12 classes, thereby transforming the probabilities into a set of normalised selection ratios in which a value of 1 represents a random selection process [18]. This normalisation enabled us to extract the effect of forest composition and age (or biomass) on fire selectivity within each HFR zone, as required by the first objective of this study.

For the second objective, two approaches were used to test whether or not the selectivity of fire with respect to forest composition and age or biomass changed with the regional burn rate. In the first approach, we used the hypothesis from [9] that an increased burn rate would decrease fire selectivity, and thus reduce the variability in selection ratios among the 12 "composition × age" or "composition × biomass". To test this, we determined the significance of both a linear and a non-linear relationship between the burn rate (R_b) and the standard deviation among the 12 selection ratios (S_{sr}) across the 16 HFR zones. The linear and non-linear functions tested were:

$$S_{sr} = aR_b + b \tag{1}$$

$$S_{sr} = \frac{1}{aR_b + b} \tag{2}$$

where *a* and *b* are parameters to be adjusted, using the *nls* function of the *R* open software.

In a second approach, we evaluated the relative contribution of composition, age (or biomass) and HFR zones in classifying the selection ratios within a regression tree. Since mean yearly burn rates vary by about 50-fold across the 16 HFR zones, we hypothesized that the HFR zones would be a strong classifying variable should burn rates influence the selection ratios of the 12 different forest classes. The selection ratios were used as response variable, and age or biomass, composition type and HFR zone were included as predictor variables. The analysis was carried out using the *party* procedure of [19] within the *R* open software with Bonferonni correction applied at *p* < 0.01.

As we will see in the results section, selection ratios were unaffected by fire regime and were therefore averaged by forest property class across Canada into national selection ratios. We combined these national selection ratios and the HFR zone burn rates of [11] to map pixel-level fire risk as a function of pixel-level forest properties. The national selection ratios provide a relative measure of fire selectivity among forest property classes, but their application at the pixel level must account for the relative abundance of these classes within each HFR zone (Table S1a,b). For this reason, we performed the following calculations by HFR zone. We first calculated a "fire risk adjustment factor" by forest property class as the national selection ratio of that class divided by the weighed mean of the 12 national selection ratios (Table S2), where the weights are the relative abundances (proportion of HFR zone pixels) of their respective forest property class. The fire risk of each forest property class was then computed as the product of its "fire risk adjustment factor" and the burn rate of [11] (Table S3a,b), and mapped by pixel across each HFR zone based on the forest property maps of Beaudoin and colleagues [13].

3. Results

In each of the 16 HFR zone, the interaction between the independent variables (i.e., "age × composition" or "biomass × composition") were non-significant, and the logistic regression model with both independent variables performed better than the model with only one variable (Table S4a,b). The exceptions to this were two HFR zones, the low-fire conifer-dominated North Atlantic (NA) and the sub-arctic Western James Bay (WBJ) (see Figure 1 for location), in which models with age only were the best. The pixel-level probability of burning increased consistently with age class (or biomass) within each HFR zone, and with the proportion of conifers in all HFR zones other than NA and WBJ. Predicted 10-year probability of burning and corresponding standard errors for each of the 12 "composition × age" (or "composition × biomass") classes for the 2002 to 2011 period are shown by HFR zone in Table S5a,b. The probability values were normalized into selection ratios by HFR zone (Table S6a,b) as described in the Methods section so as to express the relative effect of composition and age (or biomass) while controlling for the HFR annual burn rate on fire selectivity.

Overall, we found no strong evidence of a relationship between selection ratios and fire regime. In spite of a 50-fold span of burn rates across the 16 HFR zones, we found no significant linear or non-linear relationship ($p > 0.55$) between the standard deviation of the mean in selection ratios and the burn rate (Figure 2), thereby invalidating our hypothesis of a regional burn rate effect on fire selectivity. Furthermore, only composition and age were necessary to classify the selection ratios of forest composition and ages classes within a regression tree analysis, with $p < 0.01$ (Figure 3; note that NA and WJB were not included in these analyses). The patterns were the same when using biomass classes instead of the age ones, with the exception of a modest HFR zone effect on selection ratio classification of the combined mixed-deciduous and deciduous composition classes (Figure S1). This effect is very modest in that it affects only a small part of the classification tree for composition types that are less abundant, and was therefore disregarded in the remainder of the analysis.

The lack of a consistent burn rate effect on selection ratios allowed us to compute national selection ratios as the mean of all HFR zone selection ratios by "composition × age" or "composition × biomass" classes (Figure 4A,B, Table 1). Only these national values are referred to hereafter. When averaged across all age classes, pixels of pure conifer forests were selected by fire 1.9 times more, and pixels of pure deciduous avoided (avoidance = 1/selection) 2.6 times more, than what would be expected from a random selection process. Averaged across all composition classes, pixels of old forests were selected by fire 1.6 times more, and young forests avoided by fire 2.5 times more, than would be expected from a random selection process. Results suggest that old conifer forests were selected 2.9 times more and young deciduous forests avoided 6.6 times more than what would be expected from a random fire selection process. Age (Figure 4A) and biomass (Figure 4B) performed nearly interchangeably as controls of fire selectivity or avoidance. Although the regression tree analysis was unable to separate old from mature age classes in the pure conifer composition, and mixed deciduous from deciduous compositions across all three age classes (Figure 3), their respective national selections ratios are reported and were used in the mapping exercise.

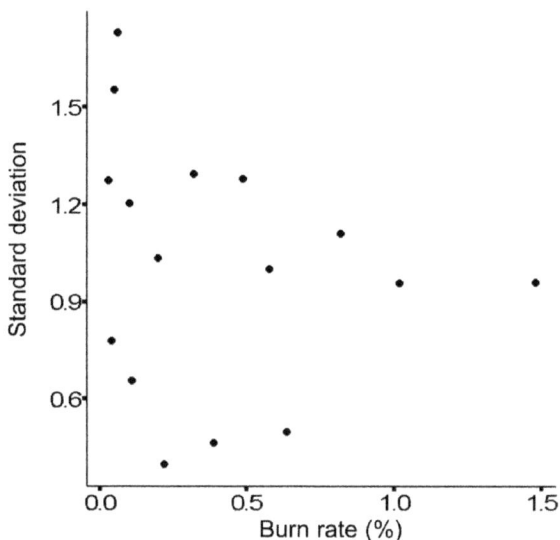

Figure 2. Standard deviations of the selection ratios of the 12 classes of composition and age across the 16 homogeneous fore regime (HFR) zones of [11] as a function of the annual burn rate of each HFR zone.

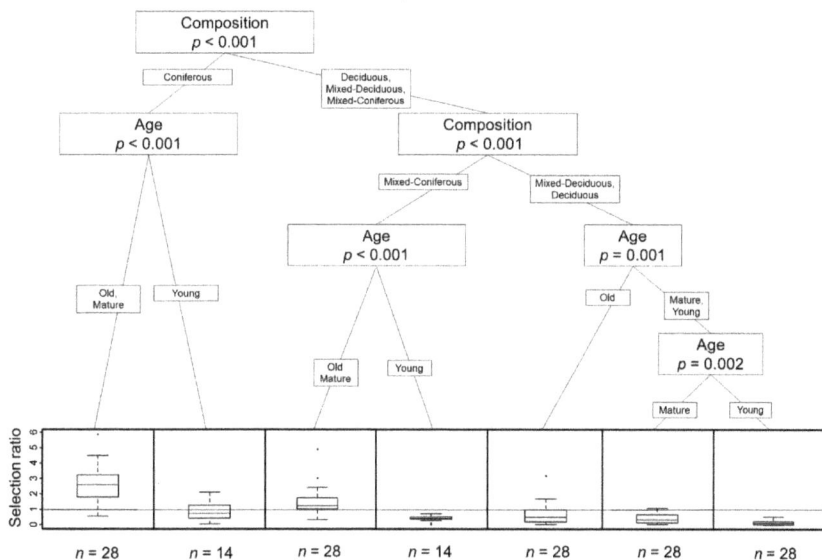

Figure 3. Regression tree classification of selection ratios among the 12 composition × age classes across the 14 homogeneous fire regime zones (HFR), using class variables composition, age and HFR as classifiers. Box-and-whiskers plots show values of the median, first and fourth quartiles and the 95% confidence interval of the selection ratios within each classification. The regression tree in which age is replaced by biomass is shown in Supplementary Material Figure S1. Note that NA and WJB zones were excluded from these analysis.

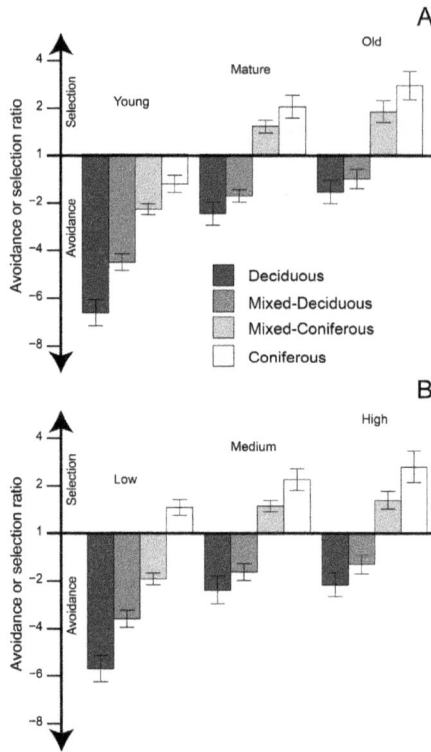

Figure 4. Values of the national selection ratios and avoidance ratios for (**A**) classes of "composition × age"; and (**B**) classes of "composition × biomass". A value of 1 is expected if fire selectivity is random. Selection ratios < 1 indicate fire avoidance and are therefore expressed as avoidance ratios (1/selection ratio) for this graphical representation. The national selection ratios are means across the 16 homogeneous fire regime (HFR) zones. Also shown are the standard deviations of the selection and avoidance ratios among HFR zones.

Table 1. Mean selection ratios and corresponding standard errors (SE) of the mean calculated from the 16 homogenous fire regime zones for (**A**) composition × age classes, and (**B**) composition × biomass classes.

A						
	Young		Mature		Old	
Forest Type	**Ratio**	**SE**	**Ratio**	**SE**	**Ratio**	**SE**
Conifer	0.80	0.15	2.00	0.26	2.90	0.35
Mixed-conifer	0.43	0.05	1.16	0.16	1.79	0.26
Mixed-deciduous	0.22	0.04	0.57	0.08	0.96	0.20
Deciduous	0.15	0.04	0.40	0.10	0.63	0.16
B						
	Low		Medium		High	
Forest Type	**Ratio**	**SE**	**Ratio**	**SE**	**Ratio**	**SE**
Conifer	1.10	0.18	2.28	0.27	2.82	0.37
Mixed-conifer	0.52	0.07	1.16	0.14	1.41	0.20
Mixed-deciduous	0.28	0.05	0.61	0.11	0.77	0.16
Deciduous	0.18	0.05	0.42	0.12	0.46	0.12

The fire risk map resulting from the incorporation of local vegetation effect on regional burn rates shows the extent to which burn rates by HFR zone still dominate the national picture, with abrupt transitions between contrasting HFRs (Figure 5). However, the details shown in the cut-outs emphasize the extent to which accounting for pixel-level forest properties creates significant within-HFR zone modulation of local fire risk. The most striking within-HFR zone features are the lower fire risks (longer return intervals) areas associated with the recent fire scars in the western cut-out, but also with the agricultural areas around the near-circular Lac St-Jean in the eastern cut-out.

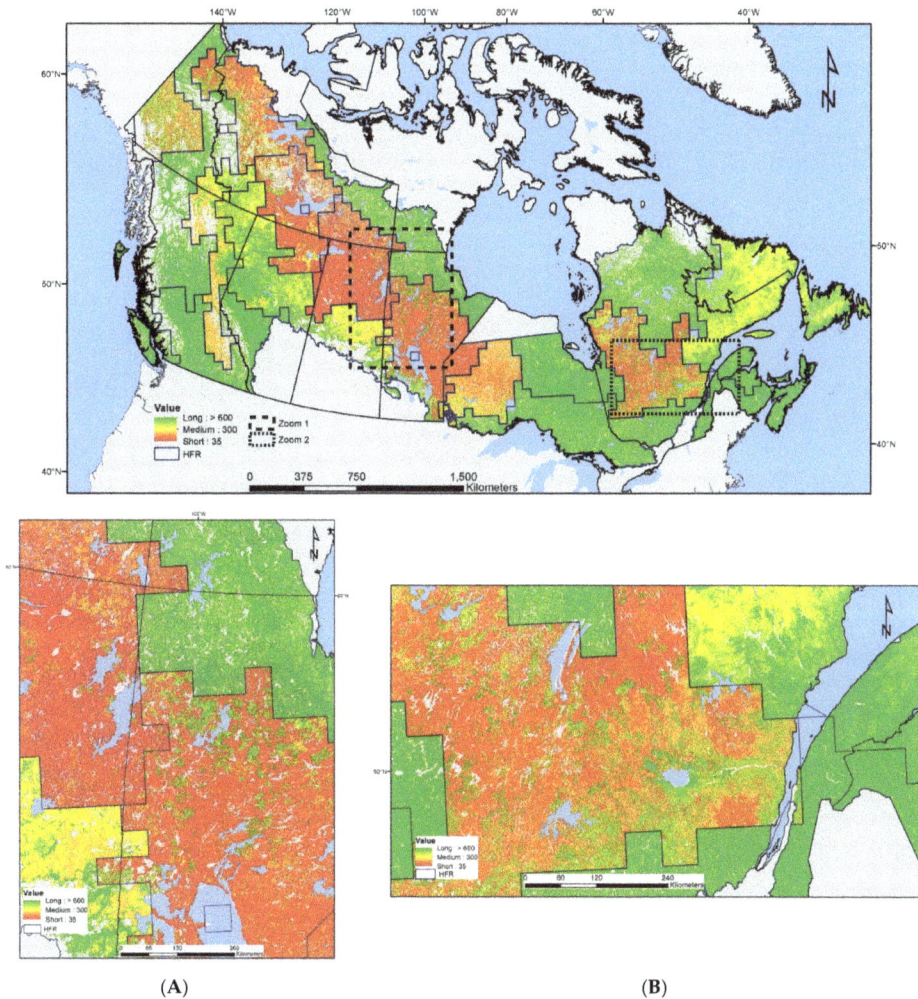

(A)   **(B)**

Figure 5. Map of local (pixel-level) fire risks in which the HFR-level fire risks of [11] have been modulated at the pixel level by fire selectivity given the pixel's composition and age. For the sake of clarity, fire risk is presented in terms of fire return interval (1/burn rate), and where long intervals represent a low fire risk. Cut-outs show in greater details the within-HFR degree of modulation as well as the abrupt transitions at HFR zone borders for regions across (**A**) western and (**B**) eastern forest regions of Canada.

4. Discussion

Our results demonstrate that fire selectivity or avoidance with respect to stand properties can be described in terms of forest composition and age (or biomass). Composition in terms of conifers or deciduous species was used as a proxy for flammability, while age or biomass was used as a proxy for fuel load. Both proxies proved to be powerful and independent classifiers of fire selectivity. These results agree with studies that found lower fire risks in young or deciduous stands [9,20–22] in regional analyses of fire statistics and forest composition. Our results also complement those of [23] on the relationship between forest composition and regional burn rates, by providing an evaluation at a local level and as a function of both composition and age (or biomass).

The absence of interaction between composition and age (or biomass) with respect to fire risk in our analysis contrasts with results from studies that have highlighted such interactions [8] or shown differences in fire risk among different coniferous species [24,25]. These studies were apparently able to capture subtleties in fire behavior induced by local vegetation composition and age at the within-fire scale. Such effects may simply be too local to be captured in our national-scale analysis.

The absence of interaction between burn rates and the forest property-based selection ratios is in line with the regional results of [9] and suggests that our results capture the general flammability and loading within our generic species classification. Conifer species dominance across Canada ranges from *Abies balsamea*, (L.) Mill. a fire avoider (*sensu* [1]) in the wet North Atlantic HFR zone to the fire-embracing *Pinus banksiana* Lamb. in the high-burn rate Lake Athabaska HFR zone. Such a difference in species dominance does not seem to affect the selection ratios, and may suggest a low impact of the dominant conifer species on the regional burn rate. By contrast, [1] linked the prevalence of canopy versus ground fires in North America versus Eurasia to trait differences between dominant conifer tree species. It is apparent that the fundamental differences in tree architecture linked to fire types in the intercontinental comparison do not seem to be prevalent within Canada's forests.

Many studies have found small scale differences in fire behavior among conifer species, with results for Canada's forests encapsulated in the 16 fuel types of the Canadian Forest Fire Behavior Prediction (FBP) System [26]. Studies have also shown the temporally or locally dominant effects of weather and of fine-scale variables such as topography or surficial deposits [7,27] on fire occurrence. Our analysis is based on empirical "realized" fire occurrence, and, as in [28], yields results that may differ from those from process-based analysis of fire potential or behavior. Such a difference, however, is likely a matter of scale and optics as we do not attempt to capture elements fire behavior such as the rate of spread or the energy of the fire front, but rather assess the fire risk of pixel-level forest types independently of their neighbors. Also, our choice to perform this analysis at the scale of Canada and at the resolution of 250 m pixels means a near impossibility to reconcile our national-level results with highly local effects other than forest properties. Nevertheless, the quantitative evaluation of forest selectivity or avoidance based on our independent variables provides a level of information and spatial resolution that can be used to support management decisions.

5. Conclusions

The projections of future regional fire risk as performed by [11] and others assume that the influence of vegetation is stationary, based on the assumption of complete fire randomness within a fire regime zone. However, it is now apparent that there is a negative feedback process between fire regime and vegetation dynamics, as suggested by [21] and [29], on account of the avoidance of regenerating stands by fire, and even more so if hardwood regeneration is enhanced. Although not perfect, our results should help improve the evaluation of landscape-level flammability [30], the design of fire management activities and the development of adaptation measures to an increased fire risk.

Supplementary Materials: The following are available online at www.mdpi.com/1999-4907/7/8/157/s1, Figure S1: Regression tree classification of selection ratios among the 12 composition × biomass classes across the 14 homogeneous fire regime zones (HFR), using class variables composition, age and HFR as classifiers. Box-and-whiskers plots show values of the median, first and fourth quartiles and the 95% confidence interval of the selection ratios within each classification. Note that the homogeneous fire zones NA and WJB were not included in the analysis, Figure S2: Map of the homogeneous fire regime (HFR) zones of [11] in which the HFR-level fire risk has been modulated at the pixel level by fire selectivity given the pixel's composition and biomass. The legend refers to the fire return interval, in years, which is equal to 1/(fractional annual burn rate), Table S1a: Relative abundance of each composition and age class, expressed as a fraction of total pixel numbers, within each of the 16 homogeneous fire regime (HFR) zones, Table S1b: Relative abundance of each composition and biomass class, expressed as a fraction of total pixel numbers, within each of the 16 homogeneous fire regime (HFR) zones, Table S2: Abundance-weighed mean of the 12 national selection ratios for Age and composition and for Biomass and composition Age × Composition forest property classes, Table S3a: Adjusted burn rate for each forest property classes of each HFR region for Age × composition forest property classes, Table S3b: Adjusted burn rate for each forest property classes of each HFR region for Biomass × composition forest property classes, Table S4a: Values of "Receiver Operating Characteristic" (ROC) used to compare regression models to explain the pixel-level occurrence of fire within the homogeneous fire regime zones of [11], and using forest composition (conifer, mixed-conifer, mixed-deciduous and deciduous) and age (0–30, 31–90, 91+) as independent variables. Values in bold indicate the model used to compute the regional selection ratio based on the logistic regression results obtained with *glm*, Table S4b: Values of "Receiver Operating Characteristic" (ROC) used to compare regression models to explain the pixel-level occurrence of fire within the homogeneous fire regime zones of [11], and using forest composition (conifer, mixed-conifer, mixed-deciduous and deciduous) and biomass (low: <19 tons/ha; medium: 19–55 tons/ha; high: >55 tons/ha) as independent variables. Values in bold indicate the model used to compute the regional selection ratio based on the logistic regression results obtained with *glm*, Table S5a: Predicted 10-year probabilities of burning (2002–2011) and their corresponding standard error for each combination of composition and age from logistic binomial regression models. Homogeneous fire regime (HFR) zones and their annual burn rates (BR; %) are from [11]. HRF zone names are detailed in Figure 1 of the main text, Table S5b: Predicted 10-year probabilities of burning (2002–2011) and their corresponding standard error for each combination of composition and biomass from logistic binomial regression models. Homogeneous fire regime (HFR) zones and their annual burn rates (BR; %) are from [11]. HRF zone names are detailed in Figure 1 of the main text, Table S6a: Selection ratios by homogeneous fire regime (HFR) zone for the 12 composition and age classes, Table S6b: Selection ratios by homogeneous fire regime (HFR) zone for the 12 composition and biomass classes.

Acknowledgments: The authors would like to acknowledge the Forest Change program within the Canadian Forest Service for its support to the development of the underlying spatial databases and to the current analysis.

Author Contributions: P.Y.B. and S.G. contributed equally to the conceptual development of the project, the interpretation of results and the writing of the manuscript. P.O.J. and F.M. carried out the analyses. Y.B., A.B. and L.G. provided expertise and interpretation related to the spatial databases of forest properties and fire dynamics that underpinned the analyses.

Conflicts of Interest: The authors declare no conflict of interest.

References

1. Rogers, B.M.; Soja, A.J.; Goulden, M.L.; Randerson, J.T. Influence of tree species on continental differences in boreal fires and climate feedbacks. *Nat. Geosci.* **2015**, *8*, 228–234. [CrossRef]

2. Kurz, W.A.; Stinson, G.; Rampley, G.J.; Dymond, C.C.; Neilson, E.T. Risk of natural disturbances makes future contribution of Canada's forests to the global carbon cycle highly uncertain. *Proc. Nat. Acad. Sci. USA* **2008**, *105*, 1551–1555. [CrossRef] [PubMed]

3. Flannigan, M.D.; Logan, K.A.; Amiro, B.D.; Skinner, W.R.; Stocks, B.J. Future area burned in Canada. *Clim. Chang.* **2005**, *72*, 1–16. [CrossRef]

4. Van Wagner, C.E. Age class distribution and the forest fire cycle. *Can. J. For. Res.* **1978**, *8*, 220–227. [CrossRef]

5. Stocks, B.J. Fire behaviour in mature jack pine. *Can. J. For. Res.* **1989**, *19*, 783–790. [CrossRef]

6. Cyr, D.; Gauthier, S.; Bergeron, Y. Scale-dependent determinants of heterogeneity in fire frequency in a coniferous boreal forest of eastern Canada. *Landsc. Ecol.* **2007**, *22*, 1325–1339. [CrossRef]

7. Mansuy, N.; Gauthier, S.; Robitaille, A.; Bergeron, Y. The effects of surficial deposit-drainage combinations on spatial variations of fire cycles in the boreal forest of eastern Canada. *Int. J. Wildland Fire* **2010**, *19*, 1083–1098. [CrossRef]

8. Kafka, V.; Gauthier, S.; Bergeron, Y. Fire impacts and crowning in the boreal forest: Study of a large wildfire in western Quebec. *Int. J. Wildland Fire* **2001**, *10*, 119–127. [CrossRef]

9. Cumming, S.G. Forest type and wildfire in the Alberta boreal mixedwood: What do fire burn? *Ecol. Appl.* **2001**, *11*, 97–110. [CrossRef]

10. Jiang, Y.; Zhuang, Q.; Flannigan, M.D.; Little, J.M. Characterization of wildfire regimes in Canadian boreal terrestrial ecosystems. *Int. J. Wildland Fire* **2009**, *18*, 992–1002. [CrossRef]

11. Boulanger, Y.; Gauthier, S.; Burton, P.J. A refinement of models projecting future Canadian fire regimes using homogeneous fire regime zones. *Can. J. For. Res.* **2014**, *44*, 365–376. [CrossRef]

12. Parisien, M.-A.; Parks, S.A.; Krawchuk, M.A.; Flannigan, M.D.; Bowman, L.M.; Moritz, M.A. Scale-dependent controls on the area burned in the boreal forest of Canada, 1980–2005. *Ecol. Appl.* **2011**, *21*, 789–805. [CrossRef] [PubMed]

13. Beaudoin, A.; Bernier, P.Y.; Guindon, L.; Villemaire, P.; Guo, X.J.; Stinson, G.; Bergeron, T.; Magnussen, S.; Hall, R.J. Mapping attributes of Canada's forests at moderate resolution through *k*NN and MODIS imagery. *Can. J. For. Res.* **2014**, *44*, 521–532. [CrossRef]

14. Guindon, L.; Bernier, P.Y.; Beaudoin, A.; Pouliot, D.; Villemaire, P.; Hall, R.J.; Latifovic, R.; St-Amant, R. Annual mapping of severe forest disturbances across Canada's forests using 250 m MODIS imagery from 2000 to 2011. *Can. J. For. Res.* **2014**, *44*, 1545–1554. [CrossRef]

15. Gillis, M.D.; Omule, A.Y.; Brierley, T. Monitoring Canada's forests: The national forest inventory. *For. Chron.* **2005**, *81*, 214–221. [CrossRef]

16. R Core Team. R: A language and environment for statistical computing. R Foundation for statistical Computing: Vienna, Austria. Available online: https://www.R-project.org/.

17. Robin, X.; Turck, N.; Hainard, A.; Tiberti, N.; Lisacek, F.; Sanchez, J.-C.; Müller, M. pROC: An open-source package for R and S+ to analyze and compare ROC curves. *BMC Bioinform.* **2011**, *12*, 77. [CrossRef] [PubMed]

18. Manly, B.; McDonald, L.; Thomas, D. *Resource Selection by Animals: Statistical Design and Analysis for Field Studies*; Chapman and Hall: New York, NY, USA, 1993; p. 221.

19. Hothorn, T.; Hornik, K.; Strobl, C.; Zeileis, A. Package 'party': A Laboratory for Recursive Partytioning. 2015. Available online: https://cran.r-project.org/web/packages/party/party.pdf (accessed on 22 July 2016).

20. Niklasson, M.; Granström, A. Numbers and sizes of fires: Long-term spatially explicit fire history in a Swedish boreal landscape. *Ecology* **2000**, *81*, 1484–1499. [CrossRef]

21. Héon, J.; Arseneault, D.; Parisien, M.-A. Resistance of the boreal forest to high burn rates. *Proc. Natl. Acad. Sci. USA.* **2014**, *111*, 13888–13893. [CrossRef] [PubMed]

22. Krawchuk, M.A.; Cumming, S.G. Effects of biotic feedback and harvest management on boreal forest fire activity under climate change. *Ecol. Appl.* **2011**, *21*, 122–136. [CrossRef] [PubMed]

23. Girardin, M.P.; Terrier, A. Mitigating risks of future wildfires by management of the forest composition: An analysis of the offsetting potential through boreal Canada. *Clim. Chang.* **2015**, *130*, 587–601. [CrossRef]

24. Leduc, A.; Bergeron, Y.; Gauthier, S. Relationships between prefire composition, fire impact, and postfire legacies in the boreal forest of eastern Canada. In Proceedings of the RMRS-P-46CD, Fort Collins, CO, USA, 26–30 March 2007; Department of Agriculture, Forest Service, Rocky Mountain Research Station: Fort Collins, CO, USA, 2007.

25. Van Wagner, C.E. Prediction of crown fire behavior in two stands of jack pine. *Can. J. For. Res.* **1993**, *23*, 442–449. [CrossRef]

26. Hirsch, K.G. *Canadian Forest Fire Behavior Prediction (FBP) System: User's Guide*; Special Report 7; Natural Resources Canada, Canadian Forest Service/Northern Forestry Centre: Edmonton, AB, Canada, 1996; p. 122.

27. Parisien, M.-A.; Moritz, M.A. Environmental controls on the distribution of wildfire at multiple spatial scales. *Ecol. Monogr.* **2009**, *79*, 127–154. [CrossRef]

28. Wang, X.; Parisien, M.-A.; Flannigan, M.D.; Parks, S.A.; Anderson, K.R.; Little, J.M.; Taylor, S.W. The potential and realized spread of wildfires across Canada. *Glob. Chang. Biol.* **2014**, *20*, 2518–2530. [CrossRef] [PubMed]

29. Parks, S.A.; Holsinger, L.M.; Miller, C.; Nelson, C.R. Wildland fire as a self-regulating mechanism: The role of previous burns and weather in limiting fire progression. *Ecol. Appl.* **2015**, *25*, 1478–1492. [CrossRef] [PubMed]

30. Palma, C.D.; Cui, W.; Martell, D.L.; Robak, D.; Weintraub, A. Assessing the impact of stand-level harvests on the flammability of forest landscapes. *Int. J. Wildland Fire* **2007**, *16*, 584–592. [CrossRef]

© 2016 by the authors. Licensee MDPI, Basel, Switzerland. This article is an open access article distributed under the terms and conditions of the Creative Commons Attribution (CC BY) license (http://creativecommons.org/licenses/by/4.0/).

![forests logo] *forests*

MDPI

Article

Fuel Classes in Conifer Forests of Southwest Sichuan, China, and Their Implications for Fire Susceptibility

San Wang and Shukui Niu *

College of Forestry, Beijing Forestry University, No. 35, Qinghua East Road, Haidian District, Beijing 100083, China; gzwangsan@126.com
* Correspondence: bjniushukui@126.com

Academic Editors: Yves Bergeron and Sylvie Gauthier
Received: 21 October 2015; Accepted: 23 February 2016; Published: 7 March 2016

Abstract: The fuel characteristics that influence the initiation and spread of wildfires were measured in *Keteleeria fortune* forest (FT1), *Pinus yunnanensis* forest (FT2), *P. yunnanensis* and *Platycladus orientalis* (L.) Franco mixed forest (FT3), *P. yunnanensis* Franch and *K. fortunei* (Murr.) Carr mixed forest (FT4), *Tsuga chinensis* forest (FT5), and *P. orientalis* forest (FT6) in southwest Sichuan Province, China. We compared vertical distributions of four fuel classes (active fuel, fine fuel, medium fuel and thick fuel) in the same vertical strata and in different spatial layers, and analyzed the fire potential (surface fire, passive and active crown fires) of the six forest types (FT). We then classified the six forest types into different groups depending on their wildfire potential. By using the pattern of forest wildfire types that burnt the most number of forests, we identified four fire susceptibility groups. The first two groups had the lowest susceptibility of active crown fires but they differed in the proportion of surface and passive crown fires. The third group was positioned in the middle between types with low and extremely high fire susceptibility; while the fourth group had the highest susceptibility of active crown fires. The results of this study will not only contribute to the prediction of fire behavior, but also will be invaluable for use in forestry management.

Keywords: stand characteristics; fire susceptibility; forest types; canonical component analyses

1. Introduction

Wildfire is a primary source of natural disturbance in forest ecosystems, and it plays an important role in determining the landscape structure and plant community composition. Wildfires are classified into ground, surface, and crown fires, based on the strata where the burning occurs [1]. For example, the shrub and small trees stratum act as a "ladder fuel" that forms a continuous ladder fuel from the surface fuel up to the canopy fuel, which can turn low-intensity surface fires into severe canopy fires, potentially resulting in active crown fires [2,3]. Crown fires include active crown fires when a solid flame develops in the crown of trees where the surface and crown phases advances as a linked unit dependent on each other. Passive crown fire is where a fire in the crown of the trees in which a tree or group of trees torch, ignited by the passing front of the fire [1,4,5]. Fire susceptibility is determined by the distribution, type and continuity of fuel [6]. Assessing the potential of wildfires is an important factor for fire prevention and suppression planning. Forest managers are expressing a growing interest in proactively reducing an area's susceptibility to fires, yet fires remain responsible for a large proportion of the annual area burned in fire-prone ecosystems. Our knowledge of wildfire prevention and its relationship with forest type is lacking [7–9].

Fuels, weather, and terrain are key factors influencing the initiation and spread of wildfires [10], and of these three factors, only fuels can be actively managed [11]. The term "fuel" does not stand for a single object but is a complex multi-layered system, which includes the following layers from surface to canopy: a semi-decomposed layer, an active/fine fuel layer, a large fuel layer, a herb layer, a shrub layer and a tree layer [12]. The spatial characteristics of fuel include the quantity, size, and continuity, which mainly affect behavior of the fire, such as the rate of spread and fire intensity [13,14]. It would be prohibitively difficult to maintain an inventory of all fuel characteristics because fuel is structurally complex and varies widely in its physical attributes. Thus, an orderly method of classifying fuels and inferring fuel properties from limited observations is needed.

A classification and characterization method simplifies the complexity to a reasonable degree and does not oversimplify the description of forest fuel [15,16]. Since the 1930s, Hornby [17] classified fuels based on their potential rate of spread in the United States. Rothermel [10] developed a mathematical model that allowed consideration of the intensity and rate of spread of fires among reasonably homogenous fuels types, including fire behavior prediction systems. Increasing characterization of different forest types has allowed for the establishment of 13 fuel models to simulate fire intensity and the spread of surface fires through grass, shrub, timber, and logging slash forest types [18]. The BEHAVE model [19] was developed to provide standardized numerical fuelbed descriptions in order to generate reasonable and accurate fire behavior predictions using Rothermel's spread model. This model is the most widely used to predict fire behavior, today, and it greatly increased the demand for quantitative data of different forest types.

Fire spread models were tentatively established by Chinese scholars using the Rothermel model [9,20,21]. These models incorporate an assessment of surface and crown fire potential and enable the user (the forest manager) to create and catalogue fuel classes. However, the limitations of these fuel models and the frequent change of forest types make it difficult to apply fuel models that were developed for other regions.

Fires are a natural part of the landscapes of Sichuan Province in southwestern China, so the dynamics of burning are part of their natural ecosystems [22,23]. In Sichuan province, the forested area until the Fifth Forest Resources Inventory was 13,301,500 ha, and was 11.6 ha per capita [9]. During the 29-year period, from 1979 to 2008, there were 4774 wildfires; 90 percent of all wildfires in a given year occur from January to May [9]. Therefore, wildfires are an integral part of the region's ecology and botanical diversity, which makes understanding their dynamics essential for forest management in Sichuan. When a fire occurs, it poses a serious threat to property and safety in regions where natural areas are adjacent to urban areas. Thus, the importance of predicting wildfire behavior in assessing fire potential is a prerequisite for evaluating the effectiveness of fuel management during fire prevention planning [24,25].

Since the 1980s, studies on forest fuel in southwestern Sichuan have assessed the main vegetation types according to fire danger classes [26], the influence of extreme climate on fuel distribution [23], the combustion characteristics of *Pinus yunnanensis* Franch [20], and control measures for wildfires [27]. However, previous studies did not take into consideration the different forest types that are found in the region, mainly due to the difficulty of determining the potential fire behaviour by forest managers during daily observations. To improve fuel management and fire suppression planning, there is a clear need for additional information on the spatial distribution and classification of fuel types in Sichuan Province.

In this study, we conducted field measurements of forests to classify coniferous forests into forest types as a function of their fire potential. Our study includes detailed field classifications in every layer where different diameters of fuel were found, and their implications for the potential of fire spread at the stand scale. The study objectives were to: (1) compare the spatial distribution of fuel loadings in different coniferous forest types, (2) analyze the fire potential of different conifer forests, and (3) classify the different forest types into different fuel types depending on the pattern of fire types that burned them.

2. Materials and Methods

2.1. Study Area

The study area was conducted in the southwestern region of Sichuan Province (between 101°08′–103°53′ E and 26°05′–29°27′ N), China, which included two regions, the Liangshan Autonomous Prefecture and Panzhihua City (Figure 1). The region has a dry, subtropical monsoon climate typified by hot summers, warm winters, and slightly more precipitation during the summer than the winter and spring. The annual average temperature of 15 °C is one of the highest in Sichuan Province, with mean daily temperatures ranging from 13 °C in winter to 26 °C in summer. Mean average annual rainfall ranges from 820 to 1160 mm and is concentrated in the summer, with winter rainfall representing only about 7% of the annual total. Average annual sunshine ranges between 2000 and 2700 h, relative humidity ranges from 56% to 71%, while the mean annual average wind speed is 1.75 m·s^{-1}.

The study site contained 74.2% mountain, 10.3% hilly, 8.2% plain and 7.3% plateau regions. The elevation is between 186 and 6511 m, and the average slope is 18° [28]. By the end of 2014, the human population was 81,400,000. Vegetation mainly includes four types, they are cold-temperate zone coniferous forest, temperate broad-leaved coniferous mixed forest, north-subtropical deciduous and evergreen broad-leaved mixed forest, and mid-subtropical deciduous and evergreen broad-leaved mixed forest.

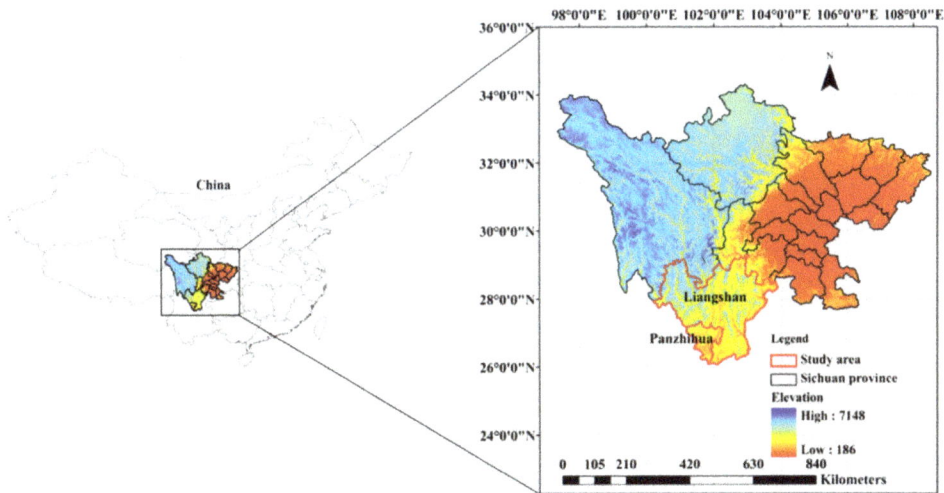

Figure 1. Study area in Sichuan Province, China.

2.2. Site Selection and Measurements

During the initial stage of the research, we evaluated candidate sites where wildfires had occurred more than twenty years prior and where no fuel treatment activities had been carried out within the perimeter of the wildfire. We consulted with officials from the Forest Fire Prevention Headquarters Office of Sichuan Province and reviewed wildfire records to determine the suitability of sites, identifying six forest types (hereafter, FT) that met our selection criteria: *Keteleeria fortune* forest (FT1), *Pinus yunnanensis* forest (FT2), *P. yunnanensis* and *Platycladus orientalis* (L.) Franco mixed forest (FT3), *P. yunnanensis* Franch and *K. fortunei* (Murr.) Carr mixed forest (FT4), *Tsuga chinensis* forest (FT5), and *P. orientalis* forest (FT6). We avoided selecting locations with confounding influences that likely changed the wildfire potential, such as roads, rivers, or constructed firelines. For each selected forest type, three to five transects that spanned the area were placed parallel to each other and 100 meters apart [15]. We aimed to locate different stands with the same species composition and to minimize differences due to topography and local climate by selecting transects that were in close proximity to each other. Three to ten fuel plots were established at 100-m intervals along each transect based on the size of the forested area and variability in landforms. A plot area covering 400 m^2 (20 m × 20 m) was used to measure the fuel composition [14]. A total of 5 transects (45 plots) were sampled in FT1, 4 transects (37 plots) in FT2, 5 transects (44 plots) in FT3, 5 transects (48 plots) in FT4, 4 transects (36 plots) in FT5, and 5 transects (43 plots) in FT6. A total of 28 transects (253 plots) were sampled in the six forest types.

We considered all trees higher than 3 m in each plot and we classified them into three classes: small (3–5 m), medium (5–8 m) and large trees (>8 m). Then, we measured the following: diameter at breast height, height, age, canopy closure, canopy base height and canopy width. The initial classification of stands was based on site specific data including elevation, slope, and aspect. Total height and the height to the live crown base were measured with a hypsometer laser Vertex III [29]. The canopy closure in each plot was estimated using a spherical densitometer [30], and the slope was measured with a forest compass [31]. The elevation was estimated by GPS.

2.3. Forest Fuel Profiles

2.3.1. Crown Layer

We estimated the crown fuel loading using the following steps:

(1) Selection of standard wood: In each plot, two trees in the forests composed of single species and three to five trees in mixed forests were randomly chosen as standard wood (SW) from each representative forest (FT1–FT6). The selection criteria were as follows: (1) the difference was no more than ±10% between the diameter of standard wood and the average diameter; (2) the difference was no more than ±5% between the height of standard wood and average height [14].

(2) Selection of standard branches (SB): We selected three live and three dead standard branches (SB) for each species of SW in each plot, and calculated the average wet weight of an SB for each species. An SB was representative of the size and distribution of other branches. Then, standard branches were divided into six components (see (4)) based on diameter. Determining standard branches provides an advantage by providing information about the amount of fuel per tree without cutting down the trees. For other examples of standard branches used in forestry studies, see Zhang [28], and Niu [14].

(3) The crown fuel was divided into 11 vertical layers: 0–1, 1–2, 2–3, 3–4, 4–5, 5–6, 6–7, 7–8, 8–9, 9–10, and above 10 m. The number of live and dead SBs (N) in every layer of the tree was recorded.

(4) SBs were divided into six components, based on diameter (D): (1) dead fuel, (a) 1-h fuels (D < 0.64 cm), (b) 10-h fuels (woody biomass 0.64 ⩽ D < 2.54 cm), (c) 100-h fuels (2.54 ⩽ D < 7.6 cm), and (d) 1000-h fuels (D ⩾ 7.6 cm); and (2) live fuel, (a) foliage and twig (⩽1 cm), and (b) big branch (1< D ⩽ 10 cm) [32]. All the SB harvested material was weighed by size class to the nearest gram using a digital scale in the field [33]. After that, subsamples of each SB size class were taken back to the laboratory to be weighted, and oven dried at 105 °C [34], then weighed again [14].

(5) Calculations

Formula 1 was used to calculate the ratio of the fresh and dry (R) material.

$$R = \frac{W_D}{W_N} \tag{1}$$

where R is fuel ratio of the wet and dry (g), W_N is the wet weight of the fuel (g), and W_D is the dry weight of the fuel (g).The fuel loading (W) was calculated with Equation (2).

$$W = \frac{N \sum\limits_{i=1}^{n} F_i R}{1000SH} \tag{2}$$

where W is the canopy bulk density (kg·m^{-3}), F_i is the fresh weight of the ith SB fuel classes (g), N is the number of SBs, and S is the area of the site (m^2), H is the height of every layer.

2.3.2. Shrub and Herb Layer

We determined the shrub and herb species, the total number of each species, and their height (cm) and weight (g) in 2 m × 2 m quadrats for shrubs and 1 m × 1 m quadrats for herbs within the plots at each of the four corners and in the center using the clear cut method. All shrub and herb harvested material was weighed by their size classes to the nearest gram using a digital scale [33]. After that, subsamples of each the shrub and herb size class were taken back to the laboratory to be weighed and dried at 105 °C [34], then weighed again [14].

2.3.3. Surface Dead Layer

In 1 m× 1 m sampling quadrats, we investigated the thickness and type of litter layers (leaves, cones, and bark) and dead fuels (1-h, 10-h, 100-h,and 1000-h) [35]. All litter and dead fuel harvested material was weighed by their size classes to the nearest gram using a digital scale [33]. After that, subsamples of each litter size class were taken back to the laboratory to be weighed and dried at 105 °C [34], then weighed again [14].

We calculated the loading of shrubs, herbs and surface dead fuel with Equation (3):

$$W = \frac{F_i R}{1000S} \tag{3}$$

where W is the fuel loading (kg·m^{-2}), S is the area of the quadrats, and F_i is the fresh weight of the shrubs, herbs and dead surface fuel (g), R is the ratio of the fresh to dry material.

2.4. Data Analysis

We separated surface and crown fuel into four fuel classes: (1) active fuel (litter layers and 1-h), (2) fine fuel (herbs, foliage and 10-h), (3) medium fuel (branches ($D \leqslant 1$ cm) and 100-h), and (4) thick fuel (branches ($1 < D \leqslant 10$ cm) and 1000-h) [36–39] (1-h, 10-h, 100-h, 1000-h from the tree or dead surface fuels). There are many factors that affect the combustion characteristics of fuel, such as species, continuity, and moisture content; however, we focused on the diameter of the fuel in this paper.

The homogeneity of variances and the normality of distribution were determined for the six forest types and four fuel classes using the least significant difference (LSD) multiple comparison method. To clearly illustrate the spatial distribution of surface fuel, we analyzed the surface fuel loading as a variable using descriptive statistics—frequencies. The statistical analyses were performed with SPSS software (version 18.0), and differences were considered significant at the level of $\alpha = 0.05$.

2.5. Fire Behavior Simulations

We consulted with officials from the Forest Fire Prevention Headquarters Office of Sichuan Province, reviewed wildfire records, and the meteorological conditions to determine three burning conditions: low (fuel moisture of 15% and a wind speed of 5 km·h^{-1}), moderate (fuel moisture of 10% and a wind speed of 15 km·h^{-1}), and extreme (fuel moisture of 5% and a wind speed of 30 km·h^{-1}). We conducted separate analyses for each plot depending on plot location and the three burning conditions. All of the wind values refer to 10-m open windspeeds. The wind adjustment factor was considered as 0.3, which was applied to all fuel models. Heat content values for all simulations were obtained from Niu's [14] work. Fire intensity and spread rate of a crown fire were evaluated using Rothermel's [10] surface fire spread model and Rothermel's [40] crown fire spread model, respectively, using BehavePlus v. 5.0.4 [41]. We used the custom fuel model and set many parameters (surface fuel loading, crown fuel density of each plot, moisture content, heat content values, canopy base height, wind speed and slope). According to the simulated effects of the fire on the plot and the neighboring plot, we distinguished three fire types (surface fires, passive crown fires, and active crown fires) by Alvarez's [29] criterion.

We used canonical component analyses (CCA) to evaluate the differences among the six forest types in terms of outcome of fire type under extreme fire weather conditions. The environmental dataset included tree age, height, elevation, slope, aspect, density, canopy closure, and diameter at breast height (DBH) in each plot. The environmental data were measured in the field except aspect need to be converted by digitizing in the calculation process [13]. Additionally, the fire type in each plot (under extreme burning conditions) was considered the "species" in the data matrix. From these analyses, we identified the groups of forest types that showed a similar response, yet could be considered different fire types. We did not measure ground fires because in these unmanaged forests, ground fires are commonly absent.

3. Results

3.1. Description of Six Forest Types

We described each forest type with statistics on surface fuel horizontal continuity, canopy closure, and fuel horizontal (which had been measured through the density and canopy width) and vertical continuity (which had been measured through the proportion on trees in different height class and vertical distribution of fuel classes). Results are summarized in Tables 1–3 and Figures 2 and 3.

Forest Type 1 (FT1) had three layers, a high density of stems (Table 1) that contributed to its very high horizontal continuity, vertical continuity and high canopy fuel loading (Figure 3). As the height of trees increased, active fuel declined gradually, while fine fuel, medium fuel, and thick fuel increased gradually. Only fine fuel, medium fuel, and thick fuel were distributed above 7 m (Figure 3). This gap in the canopy also led to a high understory density. The surface continuity had a loading between 0.50 and 0.75 kg· m^2, which accounted for about 80% of the total loading, while more than 0.90 kg· m^{-2} accounted only for 6.7% (Figure 2).

Table 1. Mean of the variables used to describe the six forest types identified in southwest Sichuan Province, China.

Forest types	N	Canopy Closure	Diameter at Breast Height (cm)	Height (m)	Crown Width (m)	Small Trees (%)	Medium Trees (%)	Large Trees (%)	Slope(°)	Elevation (m)	Aspect	Tree Age
FT1	45	0.8	11.8 (4.6)	10.7 (0.7)	3.5 (0.3)	20 (6)	15 (4)	65 (16)	28 (2)	1507 (20)	NW	37 (1)
FT2	37	0.7	16.5 (3.6)	10.9 (1.1)	4.1 (0.9)	0	6 (2)	94 (11)	30 (1)	1598 (21)	SW	30 (2)
FT3	44	0.7	10.9 (3.2)	12.7 (2.1)	2.2 (0.3)	27 (4)	4 (1)	71 (10)	27 (2)	1609 (27)	NE	28 (2)
FT4	48	0.6	14.6 (4.1)	10.3 (1.7)	2.6 (0.5)	26 (3)	23 (3)	51 (12)	35 (2)	1509 (32)	SW	35 (3)
FT5	36	0.8	12.0 (5.3)	12.8 (0.9)	3.2 (1.6)	17 (2)	65 (7)	18 (8)	34 (1)	3014 (25)	NE	30 (1)
FT6	43	0.6	7.4 (2.9)	10.5 (0.8)	1.6 (0.5)	33 (7)	33 (9)	44 (9)	30 (3)	1551 (35)	NW	40 (5)

The number in parentheses is the standard deviation. We classified trees into three classes based on total height: small (3–5m), medium (5–8 m) and large trees (>8 m). N: number of plots.

Forest Type 2 (FT2) was a simple structure that was characterized by a layer of large trees (>90%) with lower values of surface structure, canopy fuel loading, and tree density (Tables 1 and 2 and Figure 3). FT2 had the lowest small-sized trees, the lowest canopy base height, the highest value of DBH, and largest crown width. Therefore, FT2 had low vertical and high horizontal continuity. The total surface fuel loading was significantly lower than the other two forests that contained *P. yunnanensis* forests (Table 2). FT2 contained three fuel classes (active fuel, fine fuel, and thick fuel) on the surface. Thick fuel was significantly different from the other forests at the *p*-value of 0.05. The active fuel loading in FT2 was more than 0.14 kg· m^{-2}, which accounted for 41% of the total loading. The horizontal distributions had mid-continuity (Figure 3). Due to its large crown width and the low light levels at the forest floor, FT2 had the lowest understory density.

Table 2. Loadings of surface fuel classes (active fuel, fine fuel, medium fuel and thick fuel) by six forest types (unit: kg·m^{-2}).

Forest types	Active Fuel	Fine Fuel	Medium Fuel	Thick Fuel	Total
FT1	0.30 (0.07)	0.04 (0.01)	0.08 (0.03)	0.20 (0.07)	0.62 (0.10)
FT2	0.14 (0.04)	0.07 (0.01)	0.00	0.13 (0.05)	0.34(0.08)
FT3	0.63 (0.16)	0.05 (0.01)	0.00	0.36 (0.14)	1.04 (0.15)
FT4	0.25 (0.06)	0.08 (0.01)	0.18 (0.04)	0.16 (0.05)	0.67 (0.07)
FT5	1.11 (0.25)	0.60 (0.17)	0.06 (0.01)	0.57 (0.17)	2.34 (0.72)
FT6	0.16 (0.04)	0.02 (0.01)	0.00	0.05 (0.01)	0.23 (0.07)

The number in parentheses is the standard deviation.

Table 3. Percent of fire types (surface (S), Passive crown fire (P) and Active crown fire (A) from simulation of different fire weather scenarios at 10 m wind speed (U), and fuel moisture content (M_f) for the six forest types. Fire types were derived according to Alvarez [29] fire spread criteria and Rothermel's [10] surface fire spread model and Rothermel's [41] crown fire spread model.

Forest types	Low (U = 5 km·h^{-1}, Mf = 15%)			Moderate (U = 15 km·h^{-1}, Mf = 10%)			Extreme (U = 30 km·h^{-1}, Mf = 5%)		
	S	P	A	S	P	A	S	P	A
FT1	78	22	0	33	38	29	22	45	33
FT2	100	0	0	85	15	0	70	30	0
FT3	80	20	0	32	47	21	0	25	75
FT4	79	21	0	30	35	35	0	39	61
FT5	95	5	0	40	45	15	0	8	92
FT6	87	13	0	27	65	8	0	17	83

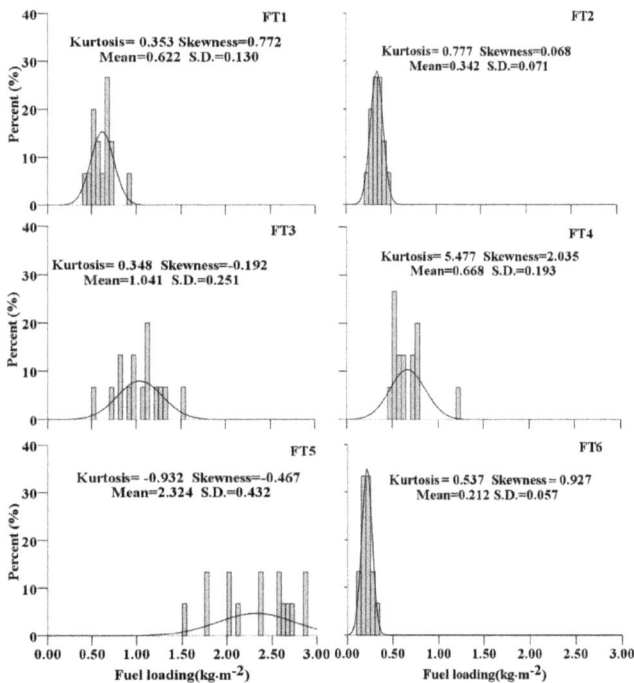

Figure 2. Observed (bar) and fitted (curve) distributions of surface fuel loading by six forest types.

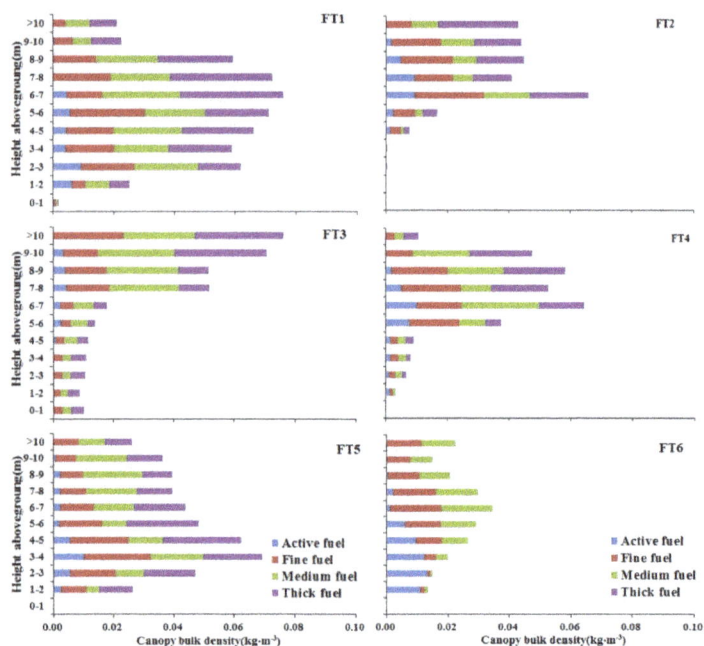

Figure 3. A vertical profile of the aboveground canopy bulk density of four fuel classes in six forest types.

Forest Type 3 (FT3) was a structure with a layer of large trees (percent of large trees was 50%) and a second layer of medium and small trees that contributed to its higher vertical continuity. FT3 had a higher tree density and canopy base height than FT2. The crown distribution of the various fuel classes in each layer was different (Figure 3), but the total fuel loading increased sharply above 10 m. However, there was a small quantity of loading between 6 m to 7 m. Active fuel was only distributed between 4 m and 9 m (Figure 3). Surface fuels were also distributed relatively heterogeneously, though not to the degree as we observed in FT5 (Figure 3).

Forest Type 4 (FT4) had three layers, the first layer was composed of large trees, small trees and medium trees were almost identical (26% *versus* 23%). Almost all fuel classes were distributed in each canopy layer, which reached a maximum between 5 and 8 m; however, there was a small quantity of loading between 1 m and 5 m. The loading of active fuel was 0.25 kg· m^{-2}, which accounted for 37% of the total loading in the type (Table 2). The surface fuel distribution was continuous enough to allow for the spread of a surface fire.

Forest Type 5 (FT5) was an irregular structure that contained three layers (each with a proportion of 17% to 65%), and had a medium tree density (>60%) and high vertical continuity (Figure 3). Structure and loading in the crown were similar in the layers above 1 m. The moss attached to branches acted as a ladder that allowed the fire to climb up to the crown. The surface loading of active fuel (1.11 kg· m^{-2}), fine fuel (0.60 kg· m^{-2}), and thick fuel (0.57 kg· m^{-2}) reached a maximum among the six forests (Table 2). Surface fuels had the lowest horizontal discontinuity compared with the other forest types, and we observed a large area of bare grounds (Figure 3).

Finally, Forest Type 6 (FT6) was also a three-layer structure, with high vertical continuity and a low density of stems that resulted in this forest type having the lowest horizontal continuity and canopy loading. The majority of crown loading was fine fuel and medium fuel (>70%). Active fuel declined gradually as the height of the trees increased, while fine fuel and medium fuel increased gradually with an increase in tree height. Although the high horizontal continuity of surface fuels in these plots would be conducive to fire spread (Figure 2), the low fuel loadings would only support low intensity surface fires.

3.2. Pattern of Fire Types Burning in the Different Forest Types

We compared the pattern of fire types between different forests types under three burning conditions (low, moderate and extreme) in Table 3. In the low burning condition, there were only surface fire and passive crown fires. Under low burning conditions, only surface fires occurred in FT2 because the fuel structure had the lowest vertical continuity and highest horizontal continuity. In FT5, five percent of the fires were passive crown fires and the remaining were surface fires (Tables 1 and 3, Figure 3). FT4 and FT3 had similar proportions of surface fire, while FT1 and FT6 had similar fire types. Compared to low burning conditions, a proportion of passive crown fires occurred in all forest types under moderate burning conditions and extreme burning conditions because of low moisture content and high wind alignment. Under extreme burning conditions, there were high proportions of plots burning with active crown fires in all forests except FT2. There was no active crown fires in FT2 and 70% of the fires remained as surface fires burning under extreme conditions.

3.3. The Differences among the Six Forest Types Using CCA

Figure 4 illustrates the differences in fire type outcome among the six forest types as viewed using CCA. The first axis of this analysis explained 78.9% of the variance in fire type under extreme conditions. Four groups could be differentiated based on our CCA (Figure 4A). FT1 was significantly different from all of the other forest types and was located at the positive end of axis 1, and is likely in this location due to the higher surface fuel (Table 2), canopy fuel loading (Figure 3) and tree density (Table 1).The lower left of the diagram contained FT5 and FT6, showing that the interaction between the environmental factors and stand structure influenced the entire fuel "system". FT4 and FT3 emerged to the right of axis 2, and FT2 was located on the bottom right of the diagram. According to Figure 4B, the first axis contained positive values that were associated with density ($r = 0.865$), diameter at breast height ($r = 0.623$), and canopy closure ($r = 0.616$); the second axis contained positive values that were associated with height ($r = 0.605$) and negative values associated with aspect ($r = -0.668$), The slope, tree age and diameter at breast height had a lesser influence as indicated by the small arrow length compared to height, density, elevation and canopy closure.

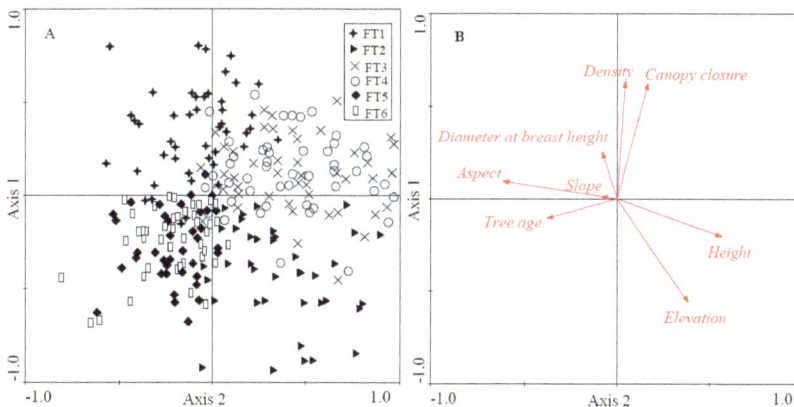

Figure 4. Representation of (A) plots of the six forest types and (B) environmental factor variables on the first two axes of the canonical component analysis. *n* = 253 plots.

4. Discussion

Our results indicated that the distribution and structure of four fuel classes, the total loading and fire potential differed among the six forest types. Using CCA analysis, four groups of FT were differentiated (FT1), (FT2), (FT4 and FT3), and (FT5 and FT6), had characteristics in common that determined the type of fires that burned them under the extreme burning condition. The results of the predicted fire potential indicated that the presence of a particular forest type could increase the susceptibility of active crown fires. Thus, less understory loading at the plot level but continuous distribution and lower fuel moisture could have the same effect in initiating crown fires as a higher amount of understory. It is possible that wind speed and continuity are the key elements for the transition from a surface fire to a passive crown fire and, later, to an active crown fire. Niu [14] suggested that the highest increase in size and severity of fires occur with higher vertical and horizontal fuel continuity. Larger crown width, higher canopy fuel loading, and higher density of trees increase the horizontal continuity and created favorable opportunities for the spread of crown fire. Vertical continuity increased with the presence of three tree layers and the "ladder fuel" that provides continuity from the surface fuel to canopy fuel. The FT1 has an increased susceptibility of having active crown fires due to a high tree density and canopy fuel loading, which result in high horizontal continuity and high density as well as a lower live crown with vertical continuity of branches that result in a high vertical continuity. However, FT1 did not generate the highest percentage of active crown fires (Table 3). This may be due to unsuitable conditions to initiate crown fires, such as the proportion of medium fuel and thick fuel being higher than 68%, which can raise the ignition point [14], or due to a lower wind speed as a consequence of it being a dense forest [42]. Thus, FT1 had a high minefield of potential susceptibility for generating active crown fires, but the conditions for crown fire initiation were lower.

Compared to FT1, the susceptibility of having active crown fires in FT5 can increase by the increment of ladder fuels (represented by medium trees, or small trees, especially with epiphytic moss) (Tables 1 and 3). Thus, irregular understory density and lower fuel moisture can lead to the highest susceptibility of surface fires transiting to active crown fires. According to the historical fire record, surface, ground and crown fires occurred simultaneously, which caused irreversible damage. Only two fires (with burn areas of 948 and 666 ha, both in March 1999) occurred in FT5 in March 1999 when there was little rainfall. In meteorological records, there were 74 consecutive days without rain fall in our study area. Therefore, the identification of FT5 is important for suppression planning, because it generate fires with behavior exceeding the extinction capacity. It would be best to reduce the amount of ladder fuels by thinning from below, as well as small diameter trees, cutting shrubs and herbs [43,44]. Additionally, pruning the crown to a half or third of the height of the tree could help to reduce ladder fuels. It is possible to also immediately fell a few trees to reduce the canopy density and break the horizontal continuity.

FT2 had a single layer with large trees, low surface fuel loading, and a high canopy base height that originated the most surface fires without any active crown fire initiation (Tables 2 and 3, Figure 3). The fuel quantity, structure and continuity of four fuel classes in FT2 makes the transition to crown fire more difficult (Table 3, Figure 3). However, combined with the higher horizontal continuity of the surface in this type, contributes to greater spreading for surface fires (Figure 2). Therefore, cleaning the surface fuel could help suppress the initiation of surface fire.

The type that resulted in the highest instances of passive and active crown fires was FT6 (Table 3), which had a structure with three layers of small trees, medium trees and large trees (Table 1), and higher vertical continuity (Figure 3). In addition, FT6 only included three fuel classes in the crown: active fuel, fine fuel and medium fuel (active fuel + fine fuel > 60%). As a consequence, crown fires can consume more biomass due to the presence of active fuel and fine fuel above a critical rate. This critical spread sustains passive crown fires, which increases the spread, and causes a larger consumption of trees, including active fuel and fine fuel [45]. FT6 would benefit from the removal of some "ladder fuels" to reduce the canopy density. This could be effective in stopping passive crown fires and preventing the start of new fires, especially if it was combined with patchy canopy and vertical layer management [13,14,29].

The increment of fuels confirms that types similar to FT6, but with higher vertical continuity (e.g., FT3 and FT4) caused by ladder fuels and higher canopy loading, can increase the susceptibility of having active crown fires (Figure 3). The two types that caused moderate passive crown fires and active crown fires were FT4 and FT3, which had middle-range canopy fuel loading (Figure 3) and tree density (Table 1). FT3 and FT4 had higher spreading for surface fires compared to FT6, which could be caused by higher continuity (Figure 2) and the large quantity of active fuel and fine fuel, which can cause easy spreading [22]. It had a high spread of surface fire that climbed up to the canopy because of the distribution of active fuel and fine fuel between 0 m and 1 m. FT3 and FT4 are effective in stopping active crown fires and preventing the start of new active crown fires if the canopy continuity is lessened by the removal of small trees.

5. Conclusions

Loading and continuity of surface fuels are an important element in determining the spread of surface fire and the transition. The higher continuity of certain forest types results in higher spreading susceptibility of crown fire spread. Ladder fuels result in vertical continuity, which can increase the susceptibility of having active crown fires; large crown width, high canopy fuel loading and tree density increase horizontal continuity and create favorable conditions for burning.

As a consequence, the six forest types have two effects on fire behavior: on one hand they favor the ignition of an active crown with vertical continuity of fuels classes, which determines crown fire initiation; on the other hand, there could be a point of a new crown fire run because of horizontal continuity through the canopy layer, which determines crown fire spread. Forestry managers should take into consideration fuel management and preventive silviculture to estimate the susceptibility of wildfire for the following two reasons. Firstly, this strategy can be effective to stop passive crown fires and also prevent new fires when combined with surface/crown patchiness, understory management and "ladder fuel" management. Several measures can be taken to achieve this goal, such as reducing surface fuel (needles, leaves, grass, dead and down branch wood) [46], and reducing the amount of ladder fuels (small trees and shrubs) [14,28,29]. The pruned plant material can be masticated or chipped and left on site, which is an option an increasing number of forest managers are using [24,28]. The pruning can increase the gap between the surface fuel and the crown base height up to 3 m, effectively breaking the continuity of the distribution of the surface fuel [14,26]. Secondly, this strategy could be effective in stopping active crown fire if combined with canopy patchiness of low density and canopy fuel loading management. According to Hall and Burke [47], crown fuel structure is a critical factor determining whether crown fires occur. Treatments to reduce crown fire susceptibility often focus on reducing canopy fuels and interrupting the surface-canopy fuel continuum. Combining this with a lower vertical and horizontal continuity makes the transition to crown fire more difficult [29]. Horizontal patchiness of the canopy will reduce the spread of fire within the canopy layer [29], using measures such as cleaning up litter (active fuel and fine fuel) [24], increasing canopy gaps [48], and decreasing tree density [28].

Acknowledgments: This study is part of the Special Research Program for Public-welfare Forestry: Responses of forests to climate change and adaptive strategy of forestry in China, Grant No. 200804001. This research was also supported by National Forest Science and Technology Promotion Project: Medium-long-term and multi-spatio-temporal scales prediction technologies and demonstration of forest fire in Yunnan Province, Grant No. 2015-04. Thanks are given to the Headquarters of Forest Fire Prevention, Sichuan Province, for useful fire data in the statistical analysis. We would also like to thank the Command College of the Armed Police Forces, Beijing, China, for this experiment.

Author Contributions: San Wang and Shukui Niu conceived and designed the study. San Wang performed the experiments and wrote the paper. Shukui Niu reviewed and edited the manuscript. All authors read and approved the manuscript.

Conflicts of Interest: The authors declare no conflict of interest.

References

1. Pyne, S.J.; Andrews, P.L.; Laven, R.D. *Introduction to Wildland Fire Science*; Wiley: New York, NY, USA, 1996.
2. Peterson, D.L.; Johnson, M.C.; Agee, J.K.; Jain, T.B.; McKenzie, D.; Reinhardt, E.D. *Forest Structure and Fire Hazard in Dry Forests of the Western United States*; General Technical Report PNW-GTR-628; United States Department of Agriculture: Washington, DC, USA, 2005; p. 30.
3. Menning, K.M.; Stephens, S.L. Fire climbing in the forest: A semi-qualitative, semiquantitative approach to assessing ladder fuel hazards. *West. J. Appl. For.* **2007**, *22*, 88–93.
4. Alexander, M.E. Help with making crown fire hazard assessments. In Proceedings of the Protecting People and Homes from Wildfire in the Interior West: Proceedings of the Symposium and Workshop, Missoula, MT, USA, 6–8 October 1988; Fischer; Fischer, W.C., Arno, S.F., Eds.; USDA Forest Service: Washington, DC, USA, 1988.
5. Scott, J.H.; Reinhardt, E.D. *Assessing Crown Fire Potential by Linking Models of Surface and Crown Fire Potential*; Research Paper RMRS-29; USDA Forest Service, Rocky Mountain Research Station: Washington, DC, USA, 2001.
6. Xu, L. The freezing weather and forest fire. *For. Fire Prev.* **1992**, *3*, 26–27.
7. Wang, J.H. *Forest Fire Hazardous Areas Based on RS and GIS in Panzhihua*; Beijing Forestry University: Beijing, China, 2012.
8. Li, F.J. *Status of Forest Resources and Sustainable Management Countermeasures in Yanbian Country, Panzhihua*; Sichuan Agricultural Uniersity: Ya'an, China, 2009.
9. Li, D. *The Relationship between Forest Fire and Meterological Factors in the Key Areas of Sichuan Province*; Beijing Forestry University: Beijing, China, 2013.
10. Rothermel, R.C. *A Mathematical Model for Predicting Fire Spread in Wildland Fuels*; Res. Pap. INT-115; USDA Forest Service, Intermountain Forest and Range Experiment Station: Ogden, UT, USA, 1972; p. 40.
11. Wu, Z.W.; He, H.S.; Liu, X.M.; Deng, H.; He, W.; Li, X.; Li, S.; Song, G.; Wang, Q. Relationship between Loading of Dead Forest Fuels in Surface Soil and Environmental Factors in Fenglin Nature Reserve. *J. Northest For. Univ.* **2011**, *39*, 52–55.
12. Sandbergh, D.; Ottmar, R.D.; Cushon, G.H. Characterizing fuels in the 21st century. *Int. J. Wildland Fire* **2001**, *10*, 381–387. [CrossRef]
13. Stephens, S.L.; Moghaddas, J.J.; Edminster, C.; Fiedler, C.E.; Haase, S.; Harrington, M.; Keeley, J.E.; Knapp, E.E.; McIver, J.D.; Metlen, K.; *et al.* Fire treatment effects on vegetation structure, fuels, and potential fire severity in western US Forests. *Ecol. Appl.* **2009**, *19*, 305–320. [CrossRef] [PubMed]
14. Niu, S.K. *Fire Behavior and Fuel Spatial Continuity of Major Forest Types in the Mountainous Area, Beijing*; Beijing Forestry University: Beijing, China, 2013.
15. Stambaugh, M.C.; Dey, D.C.; Guyette, R.P.; He, H.S.; Marschall, J.M. Spatial patterning of fuels and fire hazard across a central US deciduous forest region. *Landsc. Ecol.* **2011**, *26*, 923–935. [CrossRef]
16. Uhl, C.; Kauffman, J.B. Deforestation, fire susceptibility, and potential tree responses to fire in the eastern amazon. *Ecology* **1990**, *71*, 437–449. [CrossRef]
17. Hornby, L.G. *Fire Control Planning in the Northern Rocky Mountain Region*; Progress Report No. 1; USDA Forest Service, Northern Rocky Mountain Forest and Range Experiment Station: Ogden, UT, USA, 1936.
18. Albini, F.A. *Estimating Wildfire Behavior and Effects*; General Technical Report INT-30; USDA Forest Service, Intermountain Forest and Range Experiment Station: Ogden, UT, USA, 1976.
19. Andrews, P.L. *Behave: Fire Behavior Prediction and Fuel Modeling System—BURN Subsystem, Part 1*; USDA Forest Service: Washington, DC, USA, 1986.
20. Zheng, H.N. Study on control measures of forest fires in southwest. *Fire Saf. Sci.* **1996**, *1*, 8–12.
21. Hu, H.; Zhang, Z.; Wu, X. Type Classification of Forest Fuel in Tahe Forestry Bureau Based on Remote Sensing. *J. Northeast For. Univ.* **2007**, *7*, 20–21.
22. Wang, S.; Chen, F.; Li, D. Vertical distribution characteristics of fuels in Pinus yunnanensis and its influence factors. *Chin. J. Applly Ecol.* **2013**, *24*, 331–337.
23. Tian, X.R. Influences of ice storm on fuels in southern Sichuan. *Fire Saf. Sci.* **2011**, *11*, 43–47.
24. Mitsopoulos, I.D.; Dimitrakopoulos, A.P. Canopy fuel characteristics and potential crown fire behavior in Aleppo pine (Pinus halepensis Mill.) forests. *Ann. For. Sci.* **2007**, *64*, 287–299. [CrossRef]

25. Fernandes, P.M. Combining forest structure data and fuel modelling to classify fire hazard in portugal. *Ann. For. Sci.* **2009**, *66*, 415. [CrossRef]

26. Mou, K.H.; Wang, J.X.; Ma, Z.G. Study on combustibility of forest (Pinus Yunnanensis Faranch) in west of Panzhihua. *J. Sichuan For. Sci. Technol.* **1991**, *12(2)*, 28–36.

27. Fan, J.R.; Zhang, Z.Y.; Li, L.H. Mountain demarcation and mountainous area divisions of Sichuan Province. *Geogr. Res.* **2015**, *34*, 65–73.

28. Zhang, J.Q.; Wang, C.L.; Wang, D.X. Study on relationship between crown fire and flammable fuel in interlayer. *Fire Saf. Sci.* **1995**, *4*, 5–9.

29. Alvarez, A.; Gracia, M.; Retana, J. Fuel types and crown fire potential in Pinus halepensis forests. *Eur. J. For. Res.* **2012**, *131*, 463–474. [CrossRef]

30. Lemmon, P.E. A spherical densiometer for estimating forest overstory density. *For. Sci.* **1956**, *1*, 314–320.

31. Liu, S.Q.; Shen, Z.H.; Sun, Y.L. Elementary discussions on the application of forest compass in Geological Profile Survey. *Jiangxi Build. Mater.* **2015**, *2*, 214.

32. Deeming, J.E.; Burgan, R.E.; Cohen, J.D. *The National Fire-Danger Rating System—1978*; General Technical Report INT-39; USDA Forest Service: Washington, DC, USA, 1977; p. 6.

33. Keyser, T.; Smith, F.W. Influence of Crown Biomass Estimators and Distribution on Canopy Fuel Characteristics in Ponderosa Pine Stands of the Black Hills. *For. Sci.* **2010**, *56*, 156–165.

34. Matthews, S. Effect of drying temperature on fuel moisture content measurements. *Int. J. Wildland Fire* **2010**, *19*, 800–802. [CrossRef]

35. Brown, J.K. *Handbook for Inventorying Down Woody Material*; General Technical Report INT-16; USDA Forest Service, Intermountain Forest and Range Experiment Station: Ogden, UT, USA, 1974; p. 24.

36. Lin, Q.Z.; Shu, L.F. *Forest Fire Introduction*; University of Science and Technology of China Press: Hefei, China, 2003; pp. 142–148.

37. Hu, H.Q. *Forest Fire Ecology and Management*; China Forestry Publishing House: Beijing, China, 2005; pp. 17–91.

38. Zhou, Y.F.; Zhou, G.M.; Yu, S.Q.; Xu, X.J.; Jin, W. Spatial distribution of combustible substance of Schima superba stands in Zhejiang Province, Eastern China. *J. Beijing For. Univ.* **2008**, *30*, 99–107.

39. Kucuk, O.; Bilgili, E.; Saglam, B. Estimating crown fuel loading for calabrian pine and Anatolian black pine. *Int. J. Wildland Fire* **2008**, *17*, 147–154. [CrossRef]

40. Rothermel, R.C. *Predicting behavior and size of crown fires in the Northern Rocky Mountains*; Intermountain Forest and Range Experiment Station of USDA Forest Service: Ogden, UT, USA, 1991; pp. 1–46.

41. Andrews, P.L.; Bevins, C.D.; Seli, R.C. *BehavePlus Fire Modeling System User's Guide*; General Technical Report RMRS-GTR-106; USDA Forest Service: Washington, DC, USA, 2008.

42. Graham, J.B.; McCarthy, B.C. Fuel and fire dynamics in eastern mixed-oak forests. In Proceedings of the Conference on Fire in Eastern Oak Forests: Delivering Science to Land Managers, Columbus, OH, USA, 15–17 November 2005; General Technical Report NRS-P-1. USDA Forest Service, Northern Research Station: Newtown Square, PA, USA, 2006; p. 278.

43. Cruz, M.G.; Alexander, M.E.; Wakimoto, R.H. Assessing canopy fuel stratum characteristics in crown fire prone fuel types of western North America. *Int. J. Wildland Fire* **2003**, *12*, 39–50. [CrossRef]

44. Johnson, E.A. *Fire and Vegetation Dynamics: Studies from the North American Boreal Forest*; Cambridge University Press: London, UK, 1992; p. 129.

45. Scott, J.H. *Fuel Reduction in Residential and Scenic Forests: A Comparison of Three Treatments in a Western Montana Ponderosa Pine Stand*; Research Paper RMRS-RP-5; USDA Forest Service, Rocky Mountain Research Station: Ogden, UT, USA, 1998; p. 19.

46. Ascoli, D.; Lonati, M.; Marzano, R.; Bovio, G.; Cavallero, A.; Lombardi, G. Prescribed burning and browsing to control tree encroachment in southern european heathlands. *For. Ecol. Manag.* **2013**, *289*, 69–77. [CrossRef]

47. Hall, S.A.; Burke, I.C. Considerations for characterizing fuels as inputs for fire behavior models. *For. Ecol. Manag.* **2006**, *227*, 102–114. [CrossRef]

48. Albrecht, M.A.; Mccarthy, B.C. Effects of prescribed fire and thinning on tree recruitment patterns in central hardwood forests. *For. Ecol. Manag.* **2006**, *226*, 88–103. [CrossRef]

© 2016 by the authors. Licensee MDPI, Basel, Switzerland. This article is an open access article distributed under the terms and conditions of the Creative Commons Attribution (CC BY) license (http://creativecommons.org/licenses/by/4.0/).

forests

MDPI

Article

Spatio-Temporal Configurations of Human-Caused Fires in Spain through Point Patterns

Sergi Costafreda-Aumedes [1,*], Carles Comas [2] and Cristina Vega-Garcia [1,3]

[1] Department of Agriculture and Forest Engineering, University of Lleida, Alcalde Rovira Roure 191, Lleida 25198, Spain; cvega@eagrof.udl.cat
[2] Department of Mathematics, University of Lleida, Agrotecnio Center, Avinguda Estudi General 4, Lleida 25001, Spain; carles.comas@matematica.udl.cat
[3] Forest Sciences Centre of Catalonia, Ctra. Sant Llorenç de Morunys km 2, Solsona 25280, Spain
* Correspondence: scaumedes@gmail.com; Tel.: +34-973-702-876

Academic Editors: Yves Bergeron and Sylvie Gauthier
Received: 23 June 2016; Accepted: 18 August 2016; Published: 26 August 2016

Abstract: Human-caused wildfires are often regarded as unpredictable, but usually occur in patterns aggregated over space and time. We analysed the spatio-temporal configuration of 7790 anthropogenic wildfires (2007–2013) in nine study areas distributed throughout Peninsular Spain by using the Ripley's K-function. We also related these aggregation patterns to weather, population density, and landscape structure descriptors of each study area. Our results provide statistical evidence for spatio-temporal structures around a maximum of 4 km and six months. These aggregations lose strength when the spatial and temporal distances increase. At short time lags after a wildfire (<1 month), the probability of another fire occurrence is high at any distance in the range of 0–16 km. When considering larger time lags (up to two years), the probability of fire occurrence is high only at short distances (>3 km). These aggregated patterns vary depending on location in Spain. Wildfires seem to aggregate within fewer days (heat waves) in warm and dry Mediterranean regions than in milder Atlantic areas (bimodal fire season). Wildfires aggregate spatially over shorter distances in diverse, fragmented landscapes with many small and complex patches. Urban interfaces seem to spatially concentrate fire occurrence, while wildland-agriculture interfaces correlate with larger aggregates.

Keywords: inhomogeneous spatio-temporal point patterns; Ripley's K-function; spatio-temporal point patterns; wildfires

1. Introduction

Human-caused fires (HCFs) do not occur randomly, they follow spatio-temporal patterns that change depending on the socioeconomic activity linked to the use or misuse of fire triggering ignitions [1]. Ignition points have been proved to show broadly identifiable spatial and temporal patterns [2]. For instance, fire starts have occurred most often near roads [3], near urban- and cropland-forest interfaces [4] and in areas with an extensive presence of shrubs or conifers [5]. Fire starts also showed clustered temporal structures due to the seasonal distribution of the risk of ignitions [6].

The number of HCFs can vary widely between locations and time spans. Thus, the characterization of spatio-temporal patterns of fire ignition can provide important information for optimizing resource allocation in strategic firefighting [7]. Fire management strategies usually focus on the control of potential multiple-fire situations in areas and periods with high risk of fire [8]. Because of budgetary restrictions and rising firefighting costs, it is usually impossible to maintain sufficient resources to cope with all potential multiple-fire occurrences. In fact, under extreme weather conditions, available firefighting resources may be overloaded beyond suppression capacity. In these cases, the ability to anticipate high-risk wildfire conditions and take preventive actions, or to pre-position firefighting resources in advance, can reduce the damages and optimize the use of the suppression resources [7,9].

A number of previous studies have focused on the spatial and/or temporal distribution of wildland fires. For instance, [10] identified the most significant spatial variables for analysing human-caused wildfire occurrences using non-spatially explicit models (autoregressive Poisson and logit processes). Other studies have used spatially explicit models to explain patterns of fire occurrence, for instance, geographically weighted regression models [11], ignition density estimates [12], log-Gaussian Cox processes [13,14], scan statistics permutation [15], or Ripley's K-function [16–18]. A few studies have focused on the temporal pattern of fire ignitions; [19] found temporal aggregations using temporal trajectory metrics of wildfire ignition densities, while [20] found temporal aggregations when analysing the fire weather indices of summer fire ignitions in Finland. In addition, time series of the fire occurrence models of [6] included temporal and spatio-temporal lags lasting up to 2–3 days.

Wildfire occurrences have also been analysed as points placed within a spatio-temporal region using point process statistical tools. These tools include, for instance, analysis of inhomogeneous spatio-temporal structures of wildfire ignitions [21], cluster analysis [15,22], modelling of fire locations by spatio-temporal Cox point processes [23], and spatio-temporal analysis of fire ignition points combined with geographical and environmental variables [2]. For instance, [21] analysed space-time configuration of forest fires assuming spatial tools for each year of study separately, and they did not consider a continuous space-time approach for the fire occurrence.

Here we consider inhomogeneous spatio-temporal point processes to analyse the point pattern configuration of human-caused wildfire ignition points of several data sets in Spain. We applied the inhomogeneous spatio-temporal counterpart version of Ripley's K-function proposed by Gabriel & Diggle [24]. This approach was adopted because of the apparent inhomogeneous structure of the spatio-temporal point patterns suggested by the analysis of available official fire reports from the Spanish Ministry of Environment. The analysis of these point configurations would be valuable for interpreting the space-time dependencies of fire ignition points in order to understand wildfire dynamics.

The expected spatial and temporal aggregation patterns of HCFs should be related to the underlying fire risk factors [10] found in previous work such as weather or population. Land use has been used often as a proxy variable for distribution of vegetation/fuels and the presence and activity of human sources of ignition [25,26]. However, the spatial structure of the land mosaic is rarely considered [26], although its composition, configuration, and length of land use interfaces should be of special interest in spatial processes like this. Advances in landscape ecology provide abundant indices to measure mosaic characteristics [27]. Consequently, we also test linear correlations between spatial and temporal parameters derived from the fire patterns and relevant spatial variables linked to the structure of the fire environment with the Pearson product-moment correlation coefficient [28].

2. Materials and Methods

2.1. Study Area

This study analysed nine regions in windows of 40 km × 40 km distributed over forested areas (at least >20% forest area) in Peninsular Spain (Figure 1). These study areas comprise a wide range of forest environments with different landscape structures, but all have fire use levels conducive to significant fire occurrence (at least 100 fires over the study period).

Most of peninsular Spain is dominated by a Mediterranean climate, and only 15% of the land area, located in the north, has an Atlantic climate. These climatic zones and the complex topography combined with human socio-economic development over millennia have given way to a very uneven spatial distribution of the vegetation, combining the presence of medium-scale farming areas, areas with scarce natural vegetation cover (grasses, rangelands), extensive shrub-lands, park-like open forest structures (*dehesas*) with undergrowth, and high forests of variable densities [29]. Tables 1 and 2 include a subset of the total number of independent variables that were generated to capture weather, socioeconomic, and landscape composition and configuration traits of the nine study areas; these variables were selected for their potential relation to the spatio-temporal aggregation of fires. Population density was derived from the municipal registry of 2014 available on the website of the National Institute of Statistics of Spain (http://www.ine.es) and is weighted by the township area included in each study area. Annual climate data was derived from the Digital Climate Atlas of the Iberian Peninsula (1971–2000) (http://www.opengis.uab.es). Landscape ecology indices (landscape and class levels) [27] were calculated with Patch Analyst 5.2 [30] extension of ArcGis 10.3 over a land use reclassification (Figure 2) of the Forest Map of Spain (digitized at 1:50,000 from 1997 to 2006) from Ruiz de la Torre and available on the website of the Spanish Nature Databank of the Ministry of Agriculture, Food and Environment (http://www.magrama.gob.es). Woodland-urban interfaces (WUI), woodland-agriculture interfaces (WAI), and urban-agriculture interfaces (UAI) were evaluated, firstly, calculating a 100 m-buffer of each land use [31] and intersecting them, and secondly, by dividing the area of each interface by all interface areas.

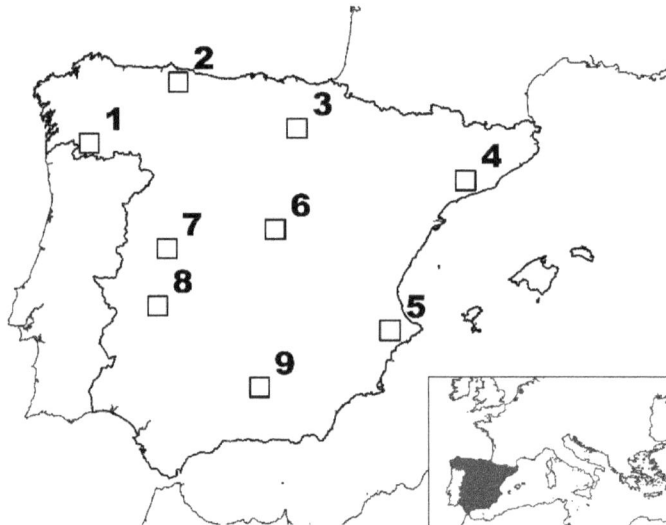

Figure 1. Location of the nine study areas in the Spanish peninsula. 1. Ourense; 2. Asturias; 3. La Rioja; 4. Tarragona; 5. Alicante; 6. Guadalajara; 7. Caceres; 8. Badajoz; 9. Jaen.

Table 1. A subset of independent variables for general characterization of each study area.

Location	Pp	Weather		Land Use			Interfaces				Landscape Metrics			
		Tmax	P	Wil	Agr	Urb	WUI	WAI	UAI	NP	MdPS	MPE	PAR	SDI
Ourense	37.1	17.8	1076	67.0	31.5	1.0	4.3	90.4	2.9	493	21.7	16.1	0.37	0.71
Asturias	275.4	16.9	1169	56.1	39.4	4.3	4.7	82.8	10.9	1516	8.8	10.5	0.66	0.84
La Rioja	112.7	17.4	606	36.9	59.9	2.9	4.2	81.9	10.6	4469	0.6	3.1	1.54	0.80
Tarragona	89.9	18.9	583	45.3	50.9	3.7	3	86.6	4.8	1542	6.3	9.2	0.83	0.83
Alicante	135.9	20.3	541	53.6	42.0	4.0	7.5	65.7	12.2	1401	5.3	8.2	0.55	0.85
Guadalajara	298.3	19.6	478	24.5	68.7	6.6	3.8	85.1	9.8	1023	8.2	10.8	0.54	0.79
Caceres	32.7	18.4	1073	81.0	17.5	0.9	3.9	88.7	6.9	782	7.6	8.4	0.43	0.55
Badajoz	29.6	22.3	580	49.1	49.0	1.3	4.4	80	12.3	441	12.7	12	0.31	0.79
Jaen	78.9	20.4	568	40.9	57.2	1.8	3.4	86.2	8.8	966	5.2	7.5	0.54	0.77

Pp: Population density (inhab/km^2); *Tmax*: Annual maximum temperature (°C); *P*: Annual precipitation (mm); *Wil*: Forest, shrubs and pastures (%); *Agr*: Croplands (%); *Urb*: Urban (%); *WUI*: Wildand-Urban interface (%); *WAI*: Wildland-Agriculture interface (%); *UAI*: Urban-Agriculture interface (%); *NP*: Number of patches; *MdPS*: Median patch size (ha); *MPE*: Mean patch edge (km); *PAR*: Perimeter-Area ratio (km/ha); *SDI*: Shannon's diversity index.

Table 2. Landscape ecology metrics at the land use class level by study area.

Location	Land Use	CA	NP	MPS	MdPS	PSSD	MPE	ED	PAR	MSI
Ourense	Agriculture	31.5	316	159.6	28.7	0.705	11.6	23	324.7	2.896
	Wildland	67	127	843.9	11.2	7.136	30.4	24.2	508.3	2.386
	Urban	1	44	37.7	10.6	0.086	6.5	1.8	284.1	2.434
	Water	0.5	6	124.1	25.2	0.207	16	0.6	438.2	4.157
Asturias	Agriculture	39.4	793	79.5	11.2	0.795	9.4	46.8	773.8	3.015
	Wildland	56.1	422	212.6	9.5	2.594	16.7	44	569.9	2.904
	Urban	4.3	289	23.7	3.8	0.216	4.5	8.1	472.3	2.445
	Water	0.2	12	31.8	19.3	0.030	10.6	0.8	412.0	5.084
La Rioja	Agriculture	59.9	1494	64.1	0.8	0.952	4.3	40.2	916.4	1.814
	Wildland	36.9	2302	25.7	0.4	0.668	2.7	38.9	2198.3	2.047
	Urban	2.9	621	7.4	0.9	0.044	16.4	6.4	679.3	1.860
	Water	0.4	52	11.1	1.5	0.030	4.5	1.5	543.8	2.470
Catalonia	Agriculture	50.9	722	112.8	6.4	0.123	9.4	42.6	580.8	2.565
	Wildland	45.3	616	117.7	5.5	2.206	10.6	40.7	1290.5	2.920
	Urban	3.7	199	30.1	9.4	0.079	3.8	4.8	340.1	2.184
	Water	0.1	5	21.0	20	0.012	7.8	0.2	376.6	4.676
Alicante	Agriculture	42	804	83.6	6.5	0.342	6.7	33.9	487.1	2.369
	Wildland	53.6	352	243.8	3.1	4.228	14	30.7	810.5	2.255
	Urban	4	226	28.5	7.1	0.115	4.3	6	345.3	2.219
	Water	0.3	19	29.0	9.8	0.042	9.6	1.2	620.1	5.406
Guadalajara	Agriculture	68.7	388	283.4	8.3	2.251	13.7	33.2	726.3	2.323
	Wildland	24.5	460	85.4	7.1	0.468	10.7	30.8	501.5	2.833
	Urban	6.6	165	64.4	12.1	0.401	4.6	4.7	195.9	1.861
	Water	0.1	10	13.4	11.6	0.009	7.3	0.5	473.7	5.040
Caceres	Agriculture	17.5	493	56.7	9.5	0.201	6.1	18.7	381.4	2.407
	Wildland	81	138	939.2	4.5	10.875	22.9	19.7	745.2	2.153
	Urban	0.9	110	13.1	6.2	0.022	2.4	1.6	268.5	1.962
	Water	0.6	41	24.7	6.6	0.066	4.3	1.1	354.2	2.846
Badajoz	Agriculture	49	202	387.9	13.4	1.409	11.2	14.1	243.8	2.014
	Wildland	49.1	135	582.0	10.9	4.315	15.8	13.4	421.2	2.507
	Urban	1.3	62	34.3	16.4	0.077	8.3	3.2	239.4	2.911
	Water	0.6	42	22.7	12.3	0.042	8.9	2.3	420.7	4.471
Jaen	Agriculture	57.2	298	306.9	5.1	2.659	11.8	21.9	686.2	2.093
	Wildland	40.9	524	124.8	5.3	2.353	6.2	20.3	513.5	2.273
	Urban	1.8	123	23.8	5.5	0.086	3.6	2.7	323.5	2.043
	Water	0.1	21	9.4	4.7	0.025	2.6	0.3	487.3	2.658

CA: Land use class (%); NP: Number of patches; MPS: Mean patch size (ha); MdPS: Median patch size (ha); PSSD: Patch size standard deviation (ha); MPE: Mean patch edge (km); ED: Edge density (km/ha); PAR: Perimeter-Area ratio (km/ha); MSI: Mean shape index.

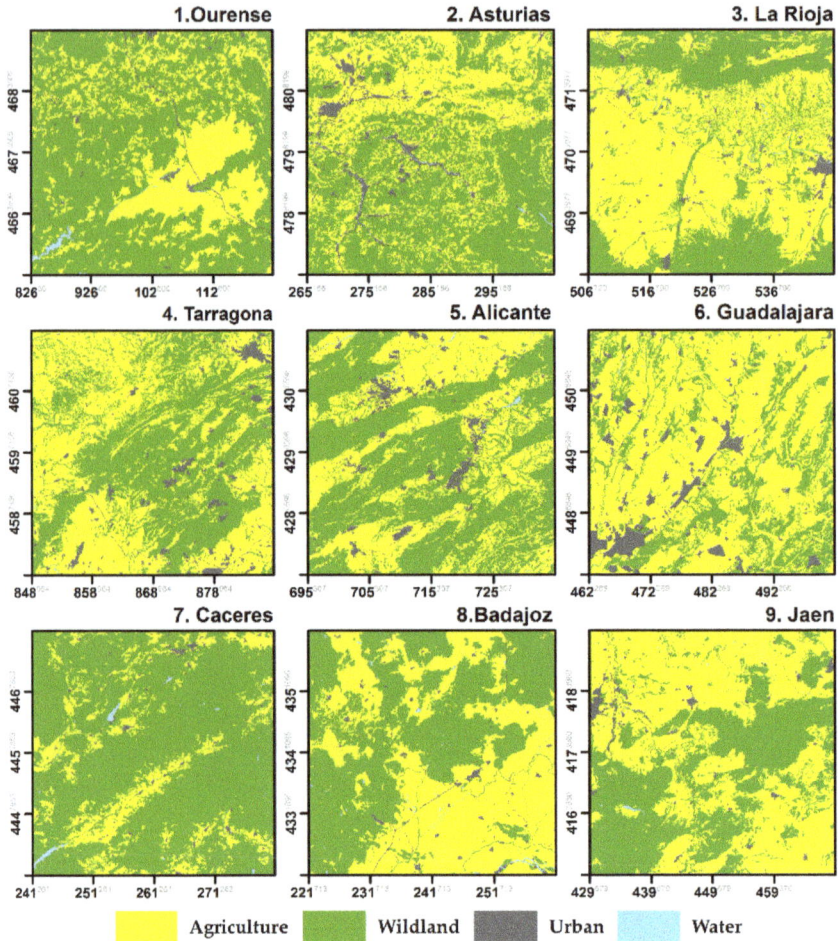

Figure 2. Land use map of the study areas.

2.2. Fire Data

The Spanish Forest Service of the Ministry of Environment and Rural and Marine Affairs (MAGRAMA) provided the fire records for the study. The nine study areas held a sufficient number of fire ignition points to study the spatio-temporal dynamics of fire ignition: at least 100 fires over the study period. Our data sets involved historical records of daily human-caused fire occurrences during the period 2007–2013. The period of study was restricted to seven years due to data availability (precise GPS locations available), but this period was considered appropriate because it surpassed the usual time framework for fire prevention planning in Spain [32]. This period included a variety of weather conditions, with mild years but also years with high risk weather conditions, i.e., 2006 in NW Spain (1900 fires set in just 12 days in August) [33].

The spatio-temporal point pattern analysed consisted of 7790 fire ignition points located in nine square areas of 40 km × 40 km for the seven years, with 877 ignitions in 2007, 1060 in 2008, 1298 in 2009, 1032 in 2010, 1308 in 2011, 1478 in 2012, and 903 in 2013. Figure 3 displays their distribution by study area, monthly.

Figure 4 shows the HCFs spatial pattern of the nine selected study regions. Visual inspection of the point pattern in the nine plots suggests that the point structures are inhomogeneous, with areas of high point intensity juxtaposed to areas of low point intensity. This figure also highlights the presence of point clusters, suggesting that fire events aggregate in space and in time.

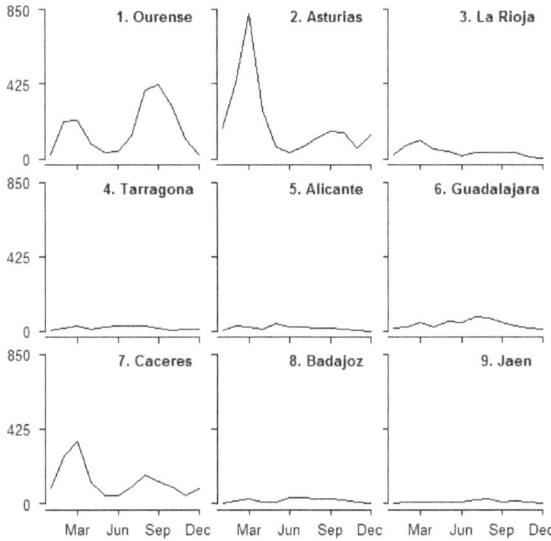

Figure 3. Human-caused wildfire frequency for the period 2007–2013 given by study area and month.

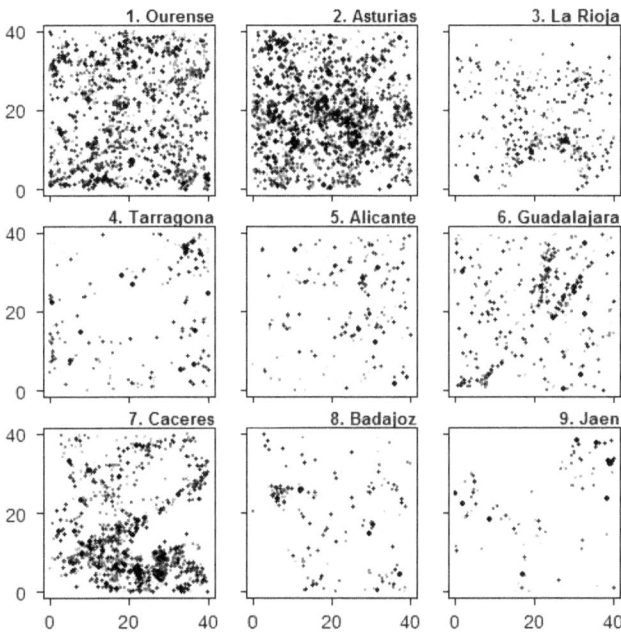

Figure 4. Spatial positions of the 7790 human-caused fires (HCFs) in the study. The biggest and darkest points correspond to recent fires while the lightest points occurred earlier.

2.3. Spatio-Temporal Statistics

To analyse the spatio-temporal structure of inhomogeneous point patterns representing ignition point fires, we used the spatio-temporal counterpart version of Ripley's K-function proposed by [24]. For a review about space-time point processes see [34]. Consider a stationary and anisotropic spatio-temporal point process Φ on $\Re^2 \times \Re$ whose elements form a countable set $S_i = (X_i, t_i)$, for $i = 1, \ldots, n$ and $X_i = (x_i, y_i) \in \Re^2$ and $t_i \in \Re$ in a bounded region $M = W \times T$. This M region contains all the ignition fires for a given planar region W for a time interval $T \in [T_0, T_1]$. Now, the point pattern should be assumed as a set of points in a continuous tridimensional space. The inhomogeneous spatio-temporal Ripley's K-function proposed by [24] assumes that the point pattern under analysis is second-order intensity reweighted stationary and isotropic or, in other words, it assumes a weaker form of stationarity and, therefore, relaxes the hypothesis of homogeneity. A point process is stationary and isotropic if its statistical properties do not change under translation and rotation, respectively. Informally, stationarity implies that one can estimate properties of the process from a single realization on $W \times T$, by exploiting the fact that these properties are the same in different, but geometrically similar, subregions of $W \times T$; isotropy means that there are no directional effects. Function $K_{st}(u, v)$ is the expected number of further points in a spatio-temporal region delimited by a cylinder whose bottom surface area is centred at an arbitrary point of Φ (a point process) with radius u (a spatial distance) and height $2v$ (a time interval). For any inhomogeneous Poisson process (i.e., a Poisson process where the constant intensity is replaced by a spatially varying intensity function) with spatio-temporal intensity function bounded away from zero, $K_{st}(u, v) = 2\pi u^2 v$, and hence $K_{st}(u, v) - 2\pi u^2 v$ (i.e., the empirical spatio-temporal Ripley's K-function minus this function under the hypothesis of no spatial-temporal structure, fire ignitions are independently distributed) can be considered a measure for detecting spatio-temporal point dependences [24]. Values of $K_{st}(u, v) - 2\pi u^2 v < 0$ will indicate regularity, while $K_{st}(u, v) - 2\pi u^2 v > 0$ will suggest spatio-temporal clustering. Moreover, $K_{st}(u, v)$ can also be used to detect the absence of spatio-temporal interaction. In particular, separability of $K_{st}(u, v)$ into purely spatial and temporal components, $K_{st}(u, v) = K_s(u)K_t(v)$, suggests the absence of spatio-temporal dependency [35]. The lack of spatio-temporal interaction indicates that ignition point locations and ignition times are independent, i.e., there is no correlation between where a fire happens and when it happens. However, in real life one may expect these two components to be correlated, so the time occurrence of a fire will depend on the spatial location. An edge-corrected estimator of $K_{st}(u, v)$ can be defined via [36].

$$\hat{K}_{st}(u, v) = \frac{1}{|W \times T|} \sum_{i=1}^{n} \sum_{j \neq i}^{n} \frac{1}{\omega_{ij} v_{ij}} \frac{I(u_{ij} \leq u) I(|t_j - t_i| \leq v)}{\hat{\lambda}(S_i) \hat{\lambda}(S_j)} \tag{1}$$

where n is the total number of points in M, $u_{ij} = \| X_i - X_j \|$, $I(\cdot)$ is the indicator function where $I(F) = 1$ if F is true and $I(F) = 0$ otherwise, $|W \times T|$ denotes the volume of this region and $\hat{\lambda}(\cdot)$ is an estimator of the spatio-temporal intensity function at the location S_i or, in other words, an estimator of expected number of points per unit volume at this exact location. To correct spatial edge effects we use Ripley's factor ω_{ij} [37] and to deal with time-edge effects we consider v_{ij}. This v_{ij} equals 1 if both ends of the interval of length $2|t_j - t_i|$ centred at t_j lie between T, and it equals $1/2$ otherwise [36]. Note that correctors for edge-effects are necessary to deal with window sampling where information outside this space-time window (unobserved points) is lost, introducing, usually, a negative bias for $\hat{K}_{st}(u, v)$. Edge-effect correctors such as the Ripley's factor and the time correction considered here are standard approximations to reduce these bias effects based on mathematical arguments. Usually these arguments consider that the unobserved numbers of points outside the observation windows are proportional to those inside these windows.

In order to obtain (1), we need to obtain an estimator of the spatio-temporal intensity function. Here we adopted a kernel-based estimator for this space-time function. First we need to assume that first-order effects (i.e., the intensity function) are separable from the space and the time domain, as suggested by Gabriel & Diggle [24], i.e.,

$$\lambda(X, t) = m(X)\mu(t) \tag{2}$$

and thus any non-separable effects can be considered as second-order effects (i.e., related to the variance of the process) rather than first-order effects. Then from Equation (2) we can estimate $\lambda(\cdot)$ [38] as:

$$\hat{\lambda}(X, t) = \hat{m}(X)\hat{\mu}(t)/n \tag{3}$$

as:

$$\int_W \hat{m}(X)d(X) = \int_T \hat{\mu}(t)d(t) = n \tag{4}$$

Now, we can estimate both the space and the time intensity function separately. For the space point intensity $m(X)$, we used a Gaussian kernel-based estimator, with bandwidth initially chosen to minimize the estimated mean-square error of $\hat{m}(X)$, as suggested in [39]. In some cases, this optimal bandwidth was slightly increased to provide a good visual fitting to the point patterns. Moreover, for time point intensity, we adopted a Gaussian kernel estimator since we did not consider covariate information related the fire locations.

Note that a kernel-based estimator for the time intensity does not assume any previous knowledge of the time series, while providing a reasonable approximation for the intensity function. After some experimentation, we considered $\sigma_\mu = 10.0$ as it provides a good visual fitting to the data while reproducing quite well some of the outliers observed in the time series.

To test for evidence of spatio-temporal clustering or regular structures, we compared the estimator $\hat{K}_{st}(u, v)$ with estimates obtained for simulations under a suitable null hypothesis. Here the null hypothesis is that the underlying point process is an inhomogeneous Poisson process, and, therefore, the empirical spatio-temporal pattern is compared with a spatio-temporal Poisson process with point intensity (3) based on a Monte Carlo test. This is a space-time Poisson process where the constant intensity is replaced by a spatially varying intensity function estimated by (3).

We simulated 1000 spatio-temporal point patterns under this null hypothesis and for each one an estimator of Equation (1) was obtained. This set of functions was then compared with the resulting estimator for the empirical data under analysis. Under this test, we rejected the null hypothesis (spatio-temporal point independence) if the resulting estimator of this function lay above the 95th percentile of estimates calculated from the 1000 simulations of inhomogeneous Poisson point process with intensity (3). This 95th percentile of estimated values formed the tolerance envelopes for our test. In this case, we should accept spatio-temporal clustering of fire locations.

All the spatio-temporal statistical analyses were computed using the stpp statistical package [36] for the R statistical environment [40].

2.4. Spatio-Temporal Aggregation Trends

In order to explain possible spatio-temporal HCFs aggregations, the Pearson product-moment correlation coefficient was selected to measure the strength of the linear dependence between the spatial and temporal aggregation patterns of HCFs and the independent weather, socioeconomic, and landscape composition and configuration variables (Tables 1 and 2). This estimator ranges from +1 to −1, where the positive and negative values indicate, respectively, positive and negative correlation (data-pairs best regression fit), and 0 indicates no significant correlation between variables [12].

3. Results

3.1. Spatio-Temporal Aggregation of HCFs

In Figure 5, we compare $\hat{K}_{st}(u,v) - 2\pi u^2 v$ and tolerance envelopes suggesting the presence of different spatio-temporal structures for time lags of less than two years and ignition point distances in the range of 0–16 km. In particular, this figure provides results on clustering patterns in the nine regions under analysis; black values indicate spatio-temporal clustering for these space-time scales. We used these maximum time and space intervals to avoid edge effects that may not be corrected by the mathematical assumptions made here. The maximum scale of the spatio-temporal aggregation of HCFs was found to be around 4–4.5 km and 5.5–6 months in Tarragona, and the minimum, less than one month and one km, in La Rioja. The spatio-temporal structures generally lose strength as the time lag increases; at the time lag of six months or higher, these dependencies are only observable for short inter-ignition point distances of less than 3 km.

However, in the NW of Spain with Atlantic climate (Ourense and Asturias), the spatial pattern shows a cyclical aggregation trend of around one year (a hump in the plot around the 12-month value). These Atlantic areas also display aggregation up to the maximum spatial distance considered (16 km) for all time lags under three months. This means that once a fire happens, the probability that another takes place within a wide area around the first one (up to 16 km in radius) persists for a period close to three months. In the other study areas, the probability that an additional fire occurs once one takes place, only persist for the maximum distance (up to 16 km in radius) for short time lags (less than 1–2 months).

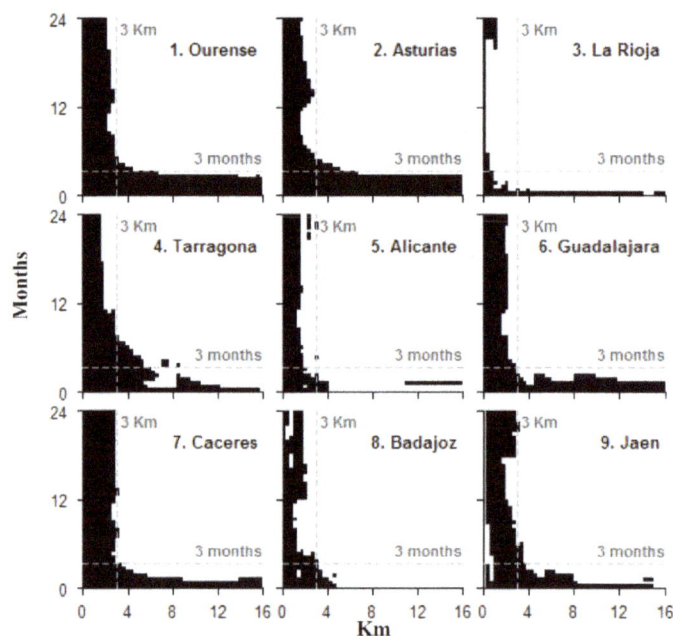

Figure 5. Comparison between $\hat{K}_{st}(u,v) - 2\pi u^2 v$ and tolerance envelopes indicating spatio-temporal clustering (black values) for each study area. Reference of 3 km and three months in grey dotted line.

3.2. Trends of the Spatio-Temporal HCFs Pattern

The values found in each plot at the limits of the maximum intervals considered or axis—the *x* for 24 months, *x*24; the *y* for 16 km, *y*16—were taken as descriptors of spatial and temporal aggregation.

*x*24 is the spatial lag for aggregation at any time lag (spatial lag of aggregation independent of the time lag). *y*16 is the time lag for aggregation at any spatial lag considered (time lag of aggregation independent of the space lag). Table 3 shows the Pearson's correlation between those aggregation parameters, the number of fires of Figure 3 and the descriptive variables of Table 1 at the landscape level. The number of HCFs shows a positive correlation with the time-independent spatial aggregation values *x*24, even higher with the distance-independent temporal aggregation values *y*16. Therefore, the higher the fire occurrence caused by humans, the more likely fires aggregate over longer distances and longer time frames, and vice versa.

Table 3. Pearson product-moment correlation coefficient between each variable of Table 1 and the descriptors of spatial and temporal aggregation. In bold, values over 0.5.

Variable	Spatial *x*24	Temporal *y*16	Variable	Spatial *x*24	Temporal *y*16
FF	**0.609**	**0.813**	WAI	**0.539**	0.192
Pp	−0.218	0.391	UAI	**−0.562**	−0.418
Tmax	−0.123	**−0.688**	NP	**−0.768**	−0.162
P	**0.696**	**0.693**	MdPS	**0.648**	0.488
Wil	**0.683**	0.327	MPE	**0.649**	**0.514**
Agr	**−0.681**	−0.371	PAR	**−0.744**	−0.157
Urb	−0.422	0.198	SDI	**−0.627**	0.035
WUI	−0.221	0.154			

FF: Fire frequency; *Pp*: Population density (inhab/km^2); *Tmax*: Annual maximum temperature (°C); *P*: Annual precipitation (mm); *Wil*: Forest, shrubs and pastures (%); *Agr*: Croplands (%); *Urb*: Urban (%); *WUI*: Wildand-Urban interface (%); *WAI*: Wildland-Agriculture interface (%); *UAI*: Urban-Agriculture interface (%); *NP*: Number of patches; *MdPS*: Median patch size (ha); *MPE*: Mean patch edge (km); *PAR*: Perimeter-Area ratio (km/ha); *SDI*: Shannon's diversity index.

According to the results, higher population density causes distance-aggregation or spatially closer fires and dilates the time lag for wildfire occurrence, but the correlation is not high. Drought weather conditions (higher *Tmax* and lower *P*) influence wildfire aggregates by decreasing the distance and time lags. Weather, mean patch edge (*MPE*), and fire frequency (*FF*, the occurrence of other fires) seem to be the variables mainly related to temporal aggregation of fires.

Proportions of land covers (*Wil*, *Agr*, *Urb*) indicate an effect on spatial aggregation of fires, linked by larger distances in landscapes with higher wildland cover and closer distances in landscapes with a higher relative proportion of agriculture and urban areas. Interfaces between land covers are correlated to the time-independent spatial aggregation of fires, particularly, urban interfaces (*WUI* and *UAI*) seem to spatially concentrate fire occurrence, while *WAI* correlates with larger aggregates.

The time-independent spatial lag for aggregation *x*24 is clearly influenced by landscape composition and configuration, being negatively correlated to *SDI*, *NP* and *PAR* and positively to *MPE* and *MdPS*. In other words, wildfires aggregate over closer distances in diverse, fragmented landscapes with many patches, where patches are small and with complex shapes.

Table 4 shows the Pearson's correlation between the descriptors of spatial and temporal aggregation and the variables in Table 2 for landscape structure analysed at the land use class level. Higher relative proportion of wildland organized in larger patches (*MPS*, *MdPS*), with more edges (*MPE*), and lower complexity (*PAR*) and number of patches (*NP*) favour larger spatial aggregation distances. In general, fire spatial aggregates grow in coarse-grained landscapes, with decreasing number of patches (*NP*) and compact shapes (*PAR*) in all land use classes. The temporal lag for aggregation of fires seems to be positively related to the presence of larger and complex agriculture patches (*MdPS*, *MSI*), and wildland edges (*MPE*).

Table 4. Pearson product-moment correlation coefficient between each variable in Table 2 with the descriptors of spatial and temporal aggregation. In bold, values over 0.5.

	Class	NP	MPS	MdPS	PSSD	MPE	ED	PAR	MSI
	Agr	**−0.696**	0.042	**0.659**	−0.108	0.378	−0.444	**−0.521**	**0.635**
Spatial *x*24	*Wil*	**−0.769**	**0.711**	**0.580**	**0.709**	**0.777**	−0.403	**−0.712**	0.109
	Urb	**−0.761**	0.107	0.200	−0.054	0.245	**−0.597**	**−0.575**	0.204
	Agr	0.009	−0.459	**0.572**	−0.344	0.161	0.421	0.073	**0.915**
Temporal *y*16	*Wil*	−0.257	0.163	0.389	0.145	**0.569**	0.459	−0.241	0.364
	Urb	−0.089	0.174	−0.200	0.304	0.062	0.227	0.040	0.044

Wil: Forest, shrubs and pastures (%); *Agr*: Croplands (%); *Urb*: Urban (%); *NP*: Number of patches; *MPS*: Mean patch size (ha); *MdPS*: Median patch size (ha); *PSSD*: Patch size standard deviation (ha); *MPE*: Mean patch edge (km); *ED*: Edge density (km/ha); *PAR*: Perimeter-Area ratio (km/ha); *MSI*: Mean shape index.

4. Discussion

The methodology used appears to be suitable for identifying differentiated patterns of spatio-temporal aggregation for HCFs in environments with different fire incidence, such as Peninsular Spain, even though the influence of window size (40 km × 40 km) and study period remains to be explored in future research. This method is especially useful in regions with enough observations because its negative simulations' bias decreases as the number of observations increases [24]. The largest spatial and temporal distances for wildfire aggregation were found with increased fire occurrence, which is coherent with higher risk levels that cause more fires over longer time spans and greater distances (i.e., Galicia, Asturias).

Our results provide statistical evidence for spatio-temporal structures around a maximum of 4 km and three months, but these aggregated structures lose strength when the spatial and temporal distances increase. These results agree with previous work [2,17,22,41–43] which detected spatial and temporal structures in wildfire occurrence in Portugal and Spain, and the general state-of-knowledge on fire occurrence in Spain; at short time lags after a wildfire (<1 month), the probability of another fire occurrence is high at any distance in the range of 0–16 km. This is in agreement with the fact that in the short term, weather is the main driver of fire occurrence, and its effects are regional. When considering larger time lags (up to two years, or 24 months), the probability of fire occurrence is high only at short distances, closer than 3 km, which is consistent with the presence of local structural risk factors independent of the season or weather condition (i.e., arson, [13]). These results agree with [15,22], which mention that aggregations between fires are more often at the local level and are not visible in larger distances (15 or 50 km).

Nevertheless, these aggregated patterns vary depending on location in Spain, suggesting the existence of varied spatio-temporal aggregation patterns of HCFs throughout the country, mainly related to fire frequency, weather, and landscape structure variables, and hence, fire regimes.

Patterns in Atlantic (Ourense, Asturias) and Mediterranean Spain (the other areas) differ, which should be expected given their climatic and landscape structure characteristics that determine different fire regimes [5]. The spatial aggregation found (up to the maximum distance considered, 16 km) for all time lags under three months is likely determined by the duration of the bimodal fire season in the milder Atlantic region (February–April, June–August, three months), but also a consequence of a fragmented landscape and a generally high human risk and occurrence all year round. This pattern has also been identified in Portugal [42] linked with the annual cycle of weather and vegetation phenology. Relatively stable conditions with higher rainfall (>1000 mm) and lower maximum temperatures extend risk over longer periods than in the Mediterranean (around 550 mm). Variations of weather events occur gradually in the NW, so the range of variation in temperature and precipitation is low within each Atlantic study area. In areas with higher precipitation there are more rainy days and, therefore, the number of fire-days decreases [44,45], so fires aggregate over longer time lags [24].

Wildfires seem to aggregate within fewer days in warm and dry Mediterranean regions (0–1.5 months). The annual weather cycle [46] favours multiple fires per day or in a few days in the summer fire season [47]. Fire suppression resources sufficient to manage one fire may be challenged on high temperature days with simultaneous occurrences [48], which require exhaustive firefighting personnel management [49]. During high temperature days, the temporal aggregation of HCFs decreases, since the occurrence of new fires is associated to those spells of extreme weather conditions [10,46,50]. Our temporal results are coherent with the occurrence of heat waves (high-temperature days (HTDs) [46]) that combine with more uneven human risk levels over coarser landscapes to render a lower fire occurrence, though these fires may have catastrophic results in terms of burned area.

Previous studies done in the Iberian Peninsula [43,50], found direct relations between population density and HCF occurrence; we found that higher population density causes distance-aggregation or spatially closer fires and dilates the time lag for wildfire occurrence, though the correlations were not very high. We used a single population density value for each study area (40 km × 40 km), but considering mean distances to towns [3,10,13], access by road [3,51,52], and trails [43,53] could also be adequate to account for the combined effect of population and access on HCF spatio-temporal aggregation.

Landscape structure clearly influences spatio-temporal patterns in wildfire occurrence. We found that wildfires aggregate spatially in closer distances in diverse, fragmented landscapes with many patches, where patches are small and have complex shapes. Ignitions have been found before to concentrate in highly fragmented landscapes [4,54] or, in other words, in areas with a larger number of small patches [25,26] in anthropic environments where patches are more compact [25,26] with shorter edges [19], raising doubts about the role of patch shape on ignition. We propose that patch shape is relevant in combination with landscape composition, depending on the class under consideration and its dominance in the landscape. Landscapes with a higher relative proportion of wildland coverage organized in larger patches (*MPS*, *MdPS*), with more edges (*MPE*), and lower complexity (*PAR*) and number of patches (*NP*) favour larger spatial aggregation distances. In general, fire spatial aggregates cover larger distances in coarse-grained landscapes, with decreasing number of patches (*NP*) and compact shapes (*PAR*) in all classes. These forest landscapes are typically created by land abandonment processes [55] which have been expanding in all the Southern European Mediterranean countries for the last 60 years.

Landscape composition and patch shapes determine the presence of interfaces between land use classes. Urban interfaces with wildlands and crops (*WUI* and *UAI*) seem to spatially decrease the distance for fire clustering, while increasing percentages of wildland-agriculture interfaces correlate with larger aggregates, effectively showing a spatial extension in risk. Previous studies in HCF prediction [43,56] have linked wildfires to agricultural cover over wildland [13,45,51] and urban covers [57]. Fires in Spain often occur at the wildland-agriculture interface [52]. The study areas with a lower proportion of wildland-agriculture interface have wildfires clustered at shorter distances and they seem to aggregate on patch perimeters between these classes (*WAI*). This finding agrees with [57,58] who have also associated a higher proportion of wildland-agriculture interface with an increase of fire occurrence in Spain. Interestingly, the temporal lag for aggregation of fires seems to be positively related to the presence of larger and complex agriculture patches (*MdPS*, *MSI*), and wildland edges (*MPE*) pointing again to the importance of *WAI* interfaces in fire occurrence.

Beyond supporting previous findings in the field of fire occurrence prediction related to fire frequency, weather, and landscape structure variables, we would like to point out that our analysis contributes additional information that is useful for fire management. The descriptors of spatial and temporal aggregation (*x*24, *y*16) have different values in different study areas, and may serve as indicators for diverse applications, for instance, fire regimes classification concerning fire occurrence. A better knowledge of factors related to occurrence is useful for prevention and suppression, but the spatial and temporal dimensions added for each window of analysis have direct operational applications. Wildfire suppression performance in the fire season depends on the number and behaviour of active fires [49]; fire managers must make crucial decisions on the amount, type, and allocation of the fire suppression resources required. For instance, risk levels and probability of new fire occurrences remain high in Ourense for up to three months, which allows for less mobility in the positioning of initial attack crews than in Badajoz with no temporal aggregation, or La Rioja (<1 month). Spatial risk at any time lag occurs under 0.75 km distance in La Rioja, but reaches 2.75 km in Caceres or Ourense, with implications for the design of the detection network. This persistent local risk is related to complex socioeconomic factors [6], but can be linked to landscape structure, which can also be used to inform general prevention and land planning to avoid risky structures.

5. Conclusions

This study demonstrates the existence of spatio-temporal aggregation patterns of human-caused fires in Peninsular Spain. This aggregation reaches maximum values around 4 km and six months, but decreases with increasing temporal and spatial distances, and varies in different study areas. The probability of an additional fire is higher at any distance in the range of 0–16 km for short periods after a fire. In the long term, the probability of fire occurrence is higher at distances closer than 3 km from the location of a first fire. Temporal aggregation is mainly related to meteorology (annual rainfall and maximum temperature), while spatial aggregation is mainly linked to the structure and composition of the landscape. Our results suggest that wildfires temporally aggregate in fewer days in warm and dry Mediterranean regions than in milder Atlantic areas; wildfires spatially aggregate in fewer kilometres in highly fragmented wildland and agriculture landscapes with high land use diversity, and spatially disperse comparatively more in forest coarse-grained landscapes resulting from abandonment. Our results also suggest the existence of local risk conditions that persist over time, probably related to land structure and complex socioeconomic factors.

Acknowledgments: We gratefully acknowledge the Spanish Ministry of Environment, Rural and Marine Affairs (MAGRAMA) for allowing us to use the historical wildfire registry (EGIF).

Author Contributions: S. Costafreda-Aumedes, C. Comas and C. Vega-Garcia conceived and designed the experiments; C. Comas performed the experiments; S.Costafreda-Aumedes and C. Vega-Garcia analysed the data; S. Costafreda-Aumedes, C. Comas and C. Vega-Garcia wrote the paper.

Conflicts of Interest: The authors declare no conflict of interest.

References

1. González-Olabarria, J.R.; Mola-Yudego, B.; Coll, L. Different Factors for Different Causes: Analysis of the Spatial Aggregations of Fire Ignitions in Catalonia (Spain). *Risk Anal.* **2015**, *35*, 1197–1209. [CrossRef] [PubMed]
2. Juan, P.; Mateu, J.; Saez, M. Pinpointing spatio-temporal interactions in wildfire patterns. *Stoch. Environ. Res. Risk Assess.* **2012**, *26*, 1131–1150. [CrossRef]
3. Badia-Perpinyà, A.; Pallares-Barbera, M. Spatial distribution of ignitions in Mediterranean periurban and rural areas: The case of Catalonia. *Int. J. Wildland Fire* **2006**, *15*, 187–196. [CrossRef]
4. Martínez, J.; Vega-Garcia, C.; Chuvieco, E. Human-caused wildfire risk rating for prevention planning in Spain. *J. Environ. Manag.* **2009**, *90*, 1241–1252. [CrossRef] [PubMed]
5. Verdú, F.; Salas, J.; Vega-García, C. A multivariate analysis of biophysical factors and forest fires in Spain, 1991–2005. *Int. J. Wildland Fire* **2012**, *21*, 498–509. [CrossRef]
6. Prestemon, J.P.; Chas-Amil, M.L.; Touza, J.M.; Goodrick, S.L. Forecasting intentional wildfires using temporal and spatiotemporal autocorrelations. *Int. J. Wildland Fire* **2012**, *21*, 743–754. [CrossRef]
7. Genton, M.G.; Butry, D.T.; Gumpertz, M.L.; Prestemon, J.P. Spatio-temporal analysis of wildfire ignitions in the St Johns River Water Management District, Florida. *Int. J. Wildland Fire* **2006**, *15*, 87–97. [CrossRef]
8. Gonzalez-Olabarria, J.R.; Brotons, L.; Gritten, D.; Tudela, A.; Teres, J.A. Identifying location and causality of fire ignition hotspots in a Mediterranean region. *Int. J. Wildland Fire* **2012**, *21*, 905–914. [CrossRef]
9. Boychuk, D.; Martell, D.L. A Markov chain model for evaluating seasonal forest fire fighter requirements. *For. Sci.* **1988**, *34*, 647–661.

10. Padilla, M.; Vega-Garcia, C. On the comparative importance of fire danger rating indices and their integration with spatial and temporal variables for predicting daily human-caused fire occurrences in Spain. *Int. J. Wildland Fire* **2011**, *20*, 46–58. [CrossRef]

11. De la Riva, J.; Pérez-Cabello, F.; Lana-Renault, N.; Koutsias, N. Mapping wildfire occurrence at regional scale. *Remote Sens. Environ.* **2004**, *92*, 363–369. [CrossRef]

12. Amatulli, G.; Peréz-Cabello, F.; de la Riva, J. Mapping lightning/human-caused wildfires occurrence under ignition point location uncertainty. *Ecol. Model.* **2007**, *200*, 321–333. [CrossRef]

13. Serra, L.; Saez, M.; Mateu, J.; Varga, D.; Juan, P.; Díaz-Ávalos, C.; Rue, H. Spatio-temporal log-Gaussian Cox processes for modelling wildfire occurrence: The case of Catalonia, 1994–2008. *Environ. Ecol. Stat.* **2014**, 531–563. [CrossRef]

14. Najafabadi, A.T.P.; Gorgani, F.; Najafabadi, M.O. Modeling forest fires in Mazandaran Province, Iran. *J. For. Res.* **2015**, *26*, 851–858. [CrossRef]

15. Vega Orozco, C.; Tonini, M.; Conedera, M.; Kanveski, M. Cluster recognition in spatial-temporal sequences: the case of forest fires. *Geoinformatica* **2012**, *16*, 653–673. [CrossRef]

16. Serra, L.; Juan, P.; Varga, D.; Mateu, J.; Saez, M. Spatial pattern modelling of wildfires in Catalonia, Spain 2004–2008. *Environ. Model. Softw.* **2013**, *40*, 235–244. [CrossRef]

17. Fuentes-Santos, I.; Marey-Pérez, M.F.; González-Manteiga, W. Forest fire spatial pattern analysis in Galicia (NW Spain). *J. Environ. Manag.* **2013**, *128*, 30–42. [CrossRef] [PubMed]

18. Turner, R. Point patterns of forest fire locations. *Environ. Ecol. Stat.* **2009**, *16*, 197–223. [CrossRef]

19. Gralewicz, N.J.; Nelson, T.A.; Wulder, M.A. Spatial and temporal patterns of wildfire ignitions in Canada from 1980 to 2006. *Int. J. Wildland Fire* **2012**, *21*, 230–242. [CrossRef]

20. Tanskanen, H.; Venäläinen, A. The relationship between fire activity and fire weather indices at different stages of the growing season in Finland. *Boreal Environ. Res.* **2008**, *13*, 285–302.

21. Hering, A.S.; Bell, C.L.; Genton, M.G. Modeling spatio-temporal wildfire ignition point patterns. *Environ. Ecol. Stat.* **2009**, *16*, 225–250. [CrossRef]

22. Pereira, M.G.; Caramelo, L.; Orozco, C.V.; Costa, R.; Tonini, M. Space-time clustering analysis performance of an aggregated dataset: The case of wildfires in Portugal. *Environ. Model. Softw.* **2015**, *72*, 239–249. [CrossRef]

23. Møller, J.; Díaz-Avalos, C. Structured Spatio-Temporal Shot-Noise Cox Point Process Models, with a View to Modelling Forest Fires. *Scand. J. Stat.* **2010**, *37*, 2–25. [CrossRef]

24. Gabriel, E.; Diggle, P.J. Second-order analysis of inhomogeneous spatio-temporal point process data. *Stat. Neerl.* **2009**, *63*, 43–51. [CrossRef]

25. Henry, M.; Yool, S. Assessing Relationships between Forest Spatial Patterns and Fire History with Fusion of Optical and Microwave Remote Sensing. *Geocarto Int.* **2004**, *19*, 25–37. [CrossRef]

26. Costafreda-Aumedes, S.; Garcia-Martin, A.; Vega-Garcia, C. The relationship between landscape patterns and human-caused fire occurrence in Spain. *For. Syst.* **2013**, *22*, 71–81. [CrossRef]

27. McGarigal, K.; Cushman, S.A.; Ene, E. FRAGSTATS v4: Spatial Pattern Analysis Program for Categorical and Continuous Maps. Computer Software Program Produced by the Authors at the University of Massachusetts, Amherst. Available online: http://www.umass.edu/landeco/research/fragstats.html (accessed on 11 January 2016).

28. Pearson, K. Notes on the History of Correlation. *Biometrika* **1920**, *13*, 25–45. [CrossRef]

29. EEA (European Environment Agency). *European Forest Types. Categories and Types for Sustainable Forest Management Reporting and Policy*; EEA: Copenhagen, Denmark, 2006.

30. Rempel, R.S.; Kaukinen, D.; Carr, A.P. *Patch Analyst and Patch Grid*; Ontario Ministry of Natural Resources Centre for Northern Forest Ecosystem Research: Thunder Bay, ON, Canada, 2012.

31. Ruiz Cejudo, J.A.; Madrigal Olmo, J. Caracterización de la interfaz-urbano forestal en la provincia de Valencia: implicaciones en la evaluación riesgo y en la prevención de incendios forestales. In Proceedings of the 6° Congreso Forestal Español; Vitoria-Gasteiz, España, 10–14 June 2013; p. 16.

32. Generalitat Valenciana. Conselleria de governació i justicia. Plan de prevención de incendios forestales de la demarcación de Altea. In *Instrucciones para la Redacción de Planes de Prevención de Incendios Forestales en Espacios Naturales Protegidos Distintos de Parques Naturales*; Generalitat Valenciana: Valencia, Spain, 2012.

33. Chas-Amil, M.L.; Touza, J.; Prestemon, J.P. Spatial distribution of human-caused forest fires in Galicia (NW Spain). *Ecol. Environ.* **2010**, *137*, 247–258.

34. Illian, J.; Penttinen, A.; Stoyan, H.; Stoyan, D. *Statistical Analysis and Modelling of Spatial Point Patterns*; John Wiley & Sons, Ltd.: Chichester, UK, 2008.

35. Diggle, P.; Chetwynd, A.G.; Häggkvist, R.; Morris, S.E. Second-order analysis of space-time clustering. *Stat. Methods Med. Res.* **1995**, *4*, 124–136. [CrossRef] [PubMed]
36. Gabriel, E.; Rowlingson, B.; Diggle, P.J. Stpp: An R Package for Plotting, Simulating and Analysing Spatio-Temporal Point Patterns. *J. Stat. Softw.* **2013**, *53*, 1–29. [CrossRef]
37. Ripley, B.D. The Second-Order Analysis of Stationary Point Processes. *J. Appl. Probab.* **1976**, *13*, 255–266. [CrossRef]
38. Ghorbani, M. Testing the weak stationarity of a spatio-temporal point process. *Stoch. Environ. Res. Risk Assess.* **2013**, *27*, 517–524. [CrossRef]
39. Berman, M.; Diggle, P. Estimating Weighted Integrals of the Second-Order Intensity of a Spatial Point Process. *J. R. Stat. Soc. Ser. B* **1989**, *51*, 81–92.
40. R Development Core Team. *R: A Language and Environment for Statistical Computing*; The R Core Team: Auckland, New Zealand, UK, 2005; Volume 1.
41. Alonso-Betanzos, A.; Fontenla-Romero, O.; Guijarro-Berdiñas, B.; Hernández-Pereira, E.; Paz Andrade, M.I.; Iménez, E.; Legido Soto, J.L.; Carballas, T. An intelligent system for forest fire risk prediction and fire fighting management in Galicia. *Expert Syst. Appl.* **2003**, *25*, 545–554. [CrossRef]
42. Telesca, L.; Pereira, M.G. Time-clustering investigation of fire temporal fluctuations in Portugal. *Nat. Hazards Earth Syst. Sci.* **2010**, *10*, 661–666. [CrossRef]
43. Chas-Amil, M.L.; Prestemon, J.P.; Mcclean, C.J.; Touza, J. Human-ignited wild fire patterns and responses to policy shifts. *Appl. Geogr.* **2015**, *56*, 164–176. [CrossRef]
44. Boubeta, M.; Lombardía, M.J.; Marey-Pérez, M.F.; Morales, D. Prediction of forest fires occurrences with area-level Poisson mixed models. *J. Environ. Manag.* **2015**, *154*, 151–158. [CrossRef] [PubMed]
45. Garcia-Gonzalo, J.; Zubizarreta-Gerendiain, A.; Ricardo, A.; Marques, S.; Botequim, B.; Borges, J.G.; Oliveira, M.M.; Tomé, M.; Pereira, J.M.C. Modelling wildfire risk in pure and mixed forest stands in Portugal. *Allg. Forst Jagdztg.* **2012**, *183*, 238–248.
46. Cardil, A.; Molina, D.M.; Kobziar, L.N. Extreme temperature days and their potential impacts on southern Europe. *Nat. Hazards Earth Syst. Sci.* **2014**, *14*, 3005–3014. [CrossRef]
47. De Haan, J.; Icove, D.J. *Kirk's Fire Investigation*, 7th ed.; Pearson Education: New York, NY, USA, 2012.
48. Rachaniotis, N.P.; Pappis, C.P. Scheduling fire-fighting tasks using the concept of "deteriorating jobs". *Can. J. For. Res.* **2006**, *36*, 652–658. [CrossRef]
49. Haight, R.G.; Fried, J.S. Deploying wildland fire suppression resources with a scenario-based standard response model. *INFOR* **2007**, *45*, 31–39. [CrossRef]
50. Barreal, J.; Loureiro, M.L. Modelling spatial patterns and temporal trends of wildfires in Galicia (NW Spain). *For. Syst.* **2015**, *24*, e022.
51. Oliveira, S.; Pereira, J.M.C.; San-Miguel-Ayanz, J.; Lourenço, L. Exploring the spatial patterns of fire density in Southern Europe using Geographically Weighted Regression. *Appl. Geogr.* **2014**, *51*, 143–157. [CrossRef]
52. Rodrigues, M.; de la Riva, J. An insight into machine-learning algorithms to model human-caused wildfire occurrence. *Environ. Model. Softw.* **2014**, *57*, 192–201. [CrossRef]
53. Vasilakos, C.; Kalabokidis, K.; Hatzopoulos, J.; Matsinos, I. Identifying wildland fire ignition factors through sensitivity analysis of a neural network. *Nat. Hazards* **2009**, *50*, 125–143. [CrossRef]
54. Ruiz-Mirazo, J.; Martínez-Fernández, J.; Vega-García, C. Pastoral wildfires in the Mediterranean: Understanding their linkages to land cover patterns in managed landscapes. *J. Environ. Manag.* **2012**, *98*, 43–50. [CrossRef] [PubMed]
55. Vega-García, C.; Chuvieco, E. Applying local measures of spatial heterogeneity to Landsat-TM images for predicting wildfire occurrence in Mediterranean landscapes. *Landsc. Ecol.* **2006**, *21*, 595–605. [CrossRef]
56. Catry, F.X.; Rego, F.C.; Bação, F.L.; Moreira, F. Modeling and mapping wildfire ignition risk in Portugal. *Int. J. Wildland Fire* **2009**, *18*, 921–931. [CrossRef]
57. Gonzalez-Olabarria, J.R.; Mola-Yudego, B.; Pukkala, T.; Palahi, M. Using multiscale spatial analysis to assess fire ignition density in Catalonia, Spain. *Ann. For. Sci.* **2011**, *68*, 861–871. [CrossRef]
58. Martínez-Fernández, J.; Chuvieco, E.; Koutsias, N. Modelling long-term fire occurrence factors in Spain by accounting for local variations with geographically weighted regression. *Nat. Hazards Earth Syst. Sci.* **2013**, *13*, 311–327. [CrossRef]

© 2016 by the authors. Licensee MDPI, Basel, Switzerland. This article is an open access article distributed under the terms and conditions of the Creative Commons Attribution (CC BY) license (http://creativecommons.org/licenses/by/4.0/).

forests

Article

Wildfires Dynamics in Siberian Larch Forests

Evgenii I. Ponomarev [1,2], Viacheslav I. Kharuk [1,2,*] and Kenneth J. Ranson [3]

[1] V.N. Sukachev Institute of Forest, Siberian Branch of Russian Academy of Sciences,
    Krasnoyarsk 660036, Russia; evg@ksc.krasn.ru
[2] Siberian Federal University, Krasnoyarsk 660041, Russia
[3] NASA Goddard Space Flight Center, Greenbelt, Maryland 20771, USA; kenneth.j.ranson@nasa.gov
* Correspondence: kharuk@ksc.krasn.ru; Tel.: +7-391-249-4447; Fax: +7-391-243-3686

Academic Editors: Yves Bergeron and Sylvie Gauthier
Received: 29 February 2016; Accepted: 8 June 2016; Published: 17 June 2016

Abstract: Wildfire number and burned area temporal dynamics within all of Siberia and along a south-north transect in central Siberia (45°–73° N) were studied based on NOAA/AVHRR (National Oceanic and Atmospheric Administration/ Advanced Very High Resolution Radiometer) and Terra/MODIS (Moderate Resolution Imaging Spectroradiometer) data and field measurements for the period 1996–2015. In addition, fire return interval (FRI) along the south-north transect was analyzed. Both the number of forest fires and the size of the burned area increased during recent decades ($p < 0.05$). Significant correlations were found between forest fires, burned areas and air temperature ($r = 0.5$) and drought index (The Standardized Precipitation Evapotranspiration Index, SPEI) ($r = -0.43$). Within larch stands along the transect, wildfire frequency was strongly correlated with incoming solar radiation ($r = 0.91$). Fire danger period length decreased linearly from south to north along the transect. Fire return interval increased from 80 years at 62° N to 200 years at the Arctic Circle (66°33′ N), and to about 300 years near the northern limit of closed forest stands (about 71°+ N). That increase was negatively correlated with incoming solar radiation ($r = -0.95$).

Keywords: wildfires; drought index; larch stands; fire return interval; fire frequency; burned area; climate-induced trends in Siberian wildfires

---

1. Introduction

Siberia is within the region of observed and predicted future accelerated climate change [1]. Increased air temperature may lead to an increase in wildfire frequency and burned area. According to some previous publications, the annual burned area in Russia was estimated as 4 to 20 MHa [2,3]. According to official data, the annual burned area was 0.55–2.4 MHa (http://www.gks.ru; [4]). More than 70% and up to 90% (i.e., 2–14 MHa annually) of the total area burned in Russia occurred in Siberia [3,5]. The majority (>50%) of wildfires in Siberia were observed in larch (*Larix sibirica*, *L. gmelinii*), because it dominated forest communities and due to its low crown closure which spread surface fires. The dense lichen and moss ground cover can support severe groundfires covering up to several million hectares. Due to the shallow root zone (limited by permafrost) those wildfires were mostly stand-replacing fires [6]. Thus, the largest area of stand-replacement fires (0.58 MHa) in the last decade occurred in Yakutia (Northeast Siberia) [7]. Non stand-replacement fires were most common in the forests of southern Siberia. Gauthier et al. (2015) found that at high-latitude areas in Canada and Siberia the mean annual fraction burned was similar and ranges within 2%–2.5% of the forested area [8].

The Siberian taiga is expected to become more prone to forest fires [3,9]. This may result in an increase in both fire frequency and carbon emissions, and may convert this area to a source for greenhouse gases [1]. Surprisingly, the important issue of climate impact on the wildfires and burns dynamics in Siberia has been discussed in only a few papers [3,5,7,10,11]. It was also shown that the occurrence of extreme fire events in Central Siberia and the Trans-Baikal region were related to soil moisture and precipitation anomalies [5,10]. Recently, Ponomarev and Kharuk [11] showed an increase in both fire frequency and size of the burned area in the Altai-Sayan region of southern Siberia [11].

The goal of this paper is an analysis of climate impacts during recent decades on fire frequency and burned area in Siberia. We sought answers to the following questions:

(i)   what is the fire frequency and burned area within (a) all of Siberia and (b) along a "south-north" gradient? How do these relate to climate variables?

(ii) what is the fire return interval dependence on the "south-north" climatic gradient?

2. Materials and Methods

2.1. Study Area

The study covered the whole territory of Siberia. In addition, a south-north transect was selected for detailed analysis as shown in Figure 1. The whole Siberia polygon covers 1000 MHa with a forested area of about 600 MHa. Maps of forest types (with 1000 m spatial resolution) within the area were derived from the forest map of Bartalev et al., 2011 [12].

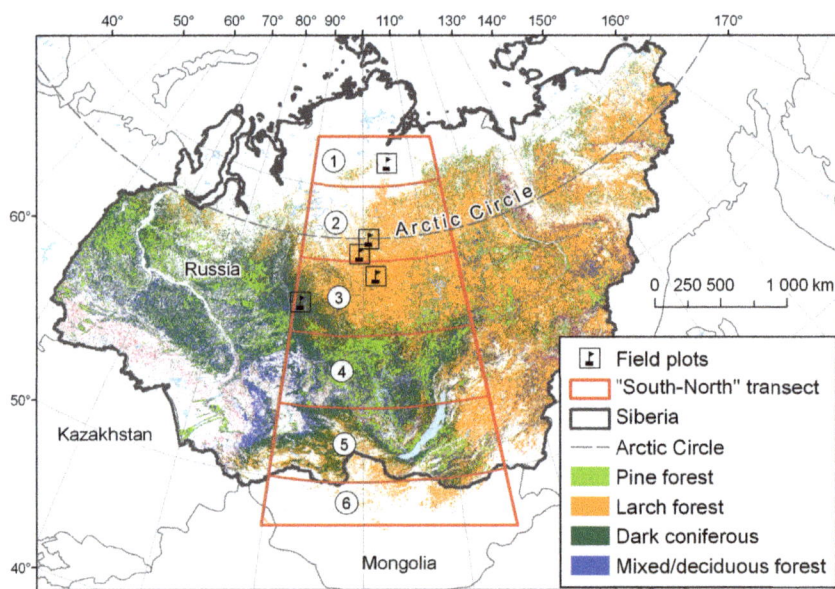

Figure 1. Study area and forest map. 1–6: 5-degree latitudinal zones within the south-north transect. Background: forest map [11].

The territory of Siberia includes the following forest types: Light coniferous taiga composed of larch (50%), Scots pine (*Pinus sylvestris*) (about 18%) and mixed stands; "dark coniferous stands" composed of fir (*Abies sibirica*), spruce (*Picea obovata*), and Siberian pine (*Pinus sibirica*) (area < 17% of total), and deciduous/mixed forests (*Betula sp.*, *Populus tremula*) (about 10% of total area). Forests dominated by larch (*Larix sibirica*, *L. gmelinii*) range over an area 270–300 MHa; the area of Scots pine was 120 MHa, dark coniferous were about 100 MHa and mixed forests about 77 MHa.

Within the "south-north" transect we considered wildfires within larch forests only. This was done for consistency, i.e., excludes the impact of different forests types on the fire patterns along the south-north transect. The transect area is divided into six zones with a width of 5-degrees latitude (Figure 1). The transect length was 2900 km with an area of about 400 MHa. The transect includes the known range larch stands—from the southern border in Mongolia to northern boundary of closed forests (about 72° + N).

2.2. Methods

The Sukachev Institute of Forest wildfires database was used in this study. This database was generated based on National Oceanic and Atmospheric Administration' Advanced Very High Resolution Radiometer (NOAA/AVHRR) (1996–2003) and Terra/Aqua/MODIS (Moderate Resolution Imaging Spectroradiometer) (2003–2015) scenes acquired directly by Sukachev Institute of Forest receiving station. The database contains daily wildfire information over all of Siberia. Both satellites used have similar overlapping characteristics. Thus, we used similar (1000 m) pixel size and wavelength bands. In earlier studies (e.g., Loboda et al., 2004 [13]) it was shown that burned area estimation based on NOAA/AVHRR and Terra/MODIS sensors were highly correlated. We processed satellite scenes using threshold-based software "Fire Processor 4.03" which was elaborated in the Sukachev Forest Institute. A threshold method was used for detecting "active pixels" based on reflectance in near-infrared (0.8–0.9 μm) and emission in medium-infrared (3.5–4.0 μm) and long wave infrared (11–12 μm) spectral bands. The method used enables detection of each wildfire based on only one satellite record. Typically, for large-scale fires, 50–100 satellite records were used in the analysis.

Annual polygonal layers for the years 1996–2015 were obtained using GIS software (ESRI ArcGIS). All active fire pixel data were preprocessed and aggregated into fire polygons based on spatial and temporal thresholds. The resulting wildfire database also included wildfires' coordinates, date, area burned, and energy characteristics.

Landsat/TM/ETM scenes with higher (pixel size = 30 m) resolution were used for burned area correction. For this purpose, a sample size of 5% of the total burned area in Siberia was used. Based on the comparison of Landsat vs. AVHRR/MODIS burned area, the regression equations were generated to correct AVHRR/MODIS estimates. The latter were used for the burned area database correction. Along with this, a geometric correction of the fire polygons was also performed [14].

A larch cover map was obtained from a vegetation map (consisting of areas with larch presence >80%) [12]. The map was used to locate wildfires within larch forests. Every burn polygon was intersected with layer of Larch forest polygons in the GIS. Only part of a burn included in the Larch class was used for further analysis. Then, the burned area was normalized with respect to the larch area within each transect cell (i.e., the burned area was divided by the larch area).

The following parameters were calculated:

(a) Normalized wildfires number (NFN) for areas with larch only, defined as:

$$\text{NFN} = \frac{n}{N} \times 100\% \tag{1}$$

where n—fires number within given latitude range; N—total fires number within transect for areas with larch only.

(b) Relative fire frequency (RFF, number of fires per 10^5 ha per given time interval), defined as:

$$RFF = \frac{n}{S_{Larch} \times t} \times 10^5 \tag{2}$$

where n—number of fires within the latitude zone for areas with larch only; S_{larch}—larch forests area within the latitude zone, t—time interval, 10^5—normalizing coefficient.

(c) Relative burned area (RBA, %):

$$RBA = \frac{S_{burned(i)}}{S_{Larch} \times t} \times 100\% \tag{3}$$

where S_{burned}—total larch burned area within given latitude zone (i); $S_{larch\ (i)}$—area of larch forests within given latitude zone (i), *t*—time interval.

The "effective" fire danger period was used in this study, which was defined as the time interval in which 90% of burns occurred.

Monthly averaged air temperature and precipitation data ($0.5° \times 0.5°$ cell size) for the whole of Siberia were taken from Climatic Research Unit (http://www.cru.uea.ac.uk; [15,16]). Monthly drought index SPEI (Standardized Precipitation Evapotranspiration Index) data were obtained from SPEI Global Drought Monitor (http://sac.csic.es/spei/map/maps.html; [17]; grid cell size was $0.5° \times 0.5°$). SPEI was calculated as the difference between precipitation (P) and potential evapotranspiration (PET) [18]:

$$SPEI_i = P_i - PET_i \tag{4}$$

Solar radiation values were taken from the Solar Radiation and Climate of the Earth database at (http://www.solar-climate.com; [19]). The data were averaged with 1-degree latitude resolution. Data were corrected with respect to solar zenith angle, daylight length and air mass impact along meridian.

Along with satellite data, fire return intervals (FRI) were analyzed within the northern portion of the transect (62°–71°+ N). FRI were calculated based on the dendrochronology analysis of samples taken from trees with visual evidence burn on the bole. FRI was defined as the time interval between consecutive stand-replacing fires. In spite of periodic wildfires, some old trees (>300 year) were present with several fire-scars. Trees were sampled until at least 12 samples were collected. In this study we used earlier obtained data on FRI in northern larch forests [6,20,21] which were analyzed against insolation along the meridian. Test sites where FRI data were obtained are shown on Figure 1.

ESRI ArcGIS software was used for GIS analysis. Statsoft Statistica was used for statistical analysis.

3. Results and Discussion

3.1. Long-Term Wildfire Statistics

Long-term statistics of annual wildfires area and the number of fires in Siberia showed a positive trend ($R^2 = 0.69$ and 0.47, respectively; $p < 0.05$) (Figure 2). The correlation of annual burned area with air temperature anomalies was the highest during the June–July period ($r = 0.67$); correlation with temperature anomalies during the whole fire season (April–September) was lower ($r = 0.56$). Similarly, correlations between wildfires numbers and air temperature anomalies were higher for June–July ($r = 0.60$ vs. $r = 0.55$ for April–September) (Figure 3).

Correlation with long-term precipitation anomalies within all Siberia was non-significant, also these correlations should be significant at a smaller scale.

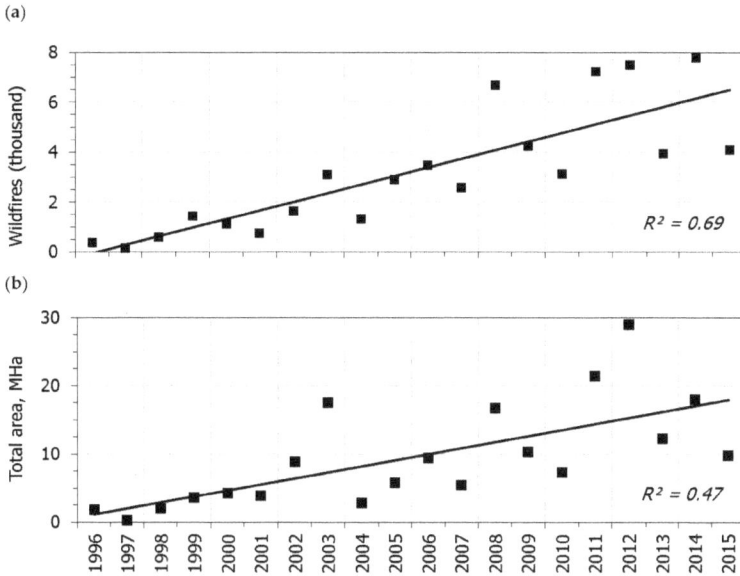

Figure 2. Temporal trends in number of wildfires (**a**); (R^2 = 0.69) and burned areas (**b**); (R^2 = 0.47) in Siberia ($p < 0.05$). Linear trends are shown by a solid line.

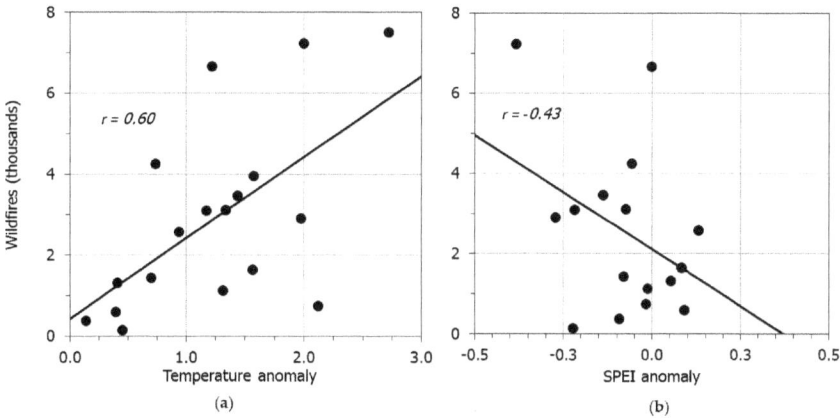

Figure 3. Correlation between wildfire numbers and June–July air temperature (**a**) (r = 0.60) and April–September Standardized Precipitation Evapotranspiration Index (SPEI) (**b**) (r = −0.43) anomalies for the forested area in Siberia.

3.2. Wildfires vs Latitude along Transect

The wildfires distribution along latitude was a quasi-normal type (Figure 4). The fire location maximum was observed at about 52° N with an exponential decrease as latitude increased (Figure 4a). The distribution of wildfires number along the south-north transect (52°–71° N range) was strongly correlated with incoming solar radiation along the latitudinal gradient (r = 0.81; Figure 4a). Wildfires number (as well as relative fire frequency and relative burned area; Figure 4b) showed an exponential decrease southward (latitude range <52°), which is attributed to the extreme topography of the high southern mountains, which is atypical of the northward area. However, along with dependence on solar radiation, fire frequency was also linked to the level of anthropogenic impact [22]. The relative burned area (RBA) and relative fire frequency (RFF) were correlated with solar radiation ($r = 0.87$ and $r = 0.89$, respectively) and were strongly decreased from south to north along the transect (Figure 4b). Mean RBA for the transect is 1.19%. In western Canada, for comparison, RBA was reported to be 0.56% [19]. RFF at high (60°+ N) latitudes (0.065–0.22 per 10^5 ha fires/year) were similar to the fire frequency value for western Canada (about 0.09 per 10^5 ha/year) [23], and considerably higher for the southern part of the transect (0.98–2.67) (Figure 4b).

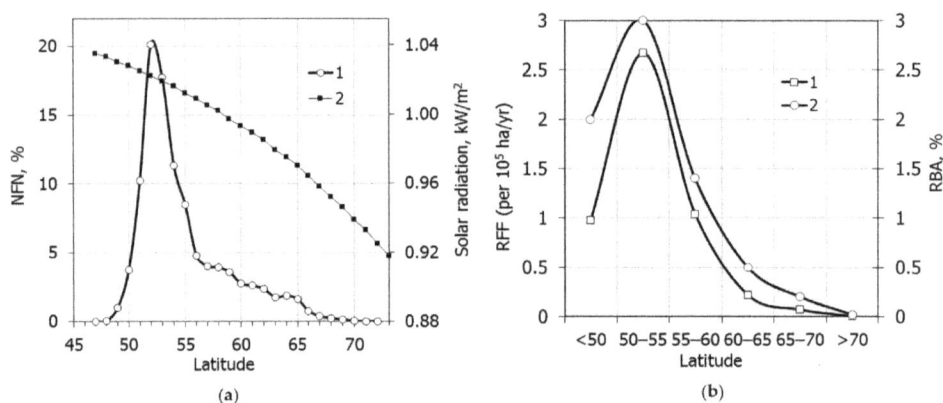

Figure 4. (a) Wildfires number (normalized) distribution along transect (1) and incoming solar radiation (2). Correlation between these two datasets is $r = 0.81$. (b) Distribution of relative fire frequency (RFF; 1) and relative burned area (RBA; 2). Wildfires parameters correlations with incoming solar radiation are $r = 0.81$ (for wildfires number), $r = 0.87$ (for RBA) and $r = 0.89$ (for RFF) (within 52°–71° N range). Analyzed period was 1996–2015 year.

The spatial and temporal variations of relative fire frequency and relative burned area along the transect is shown on Figure 5. The seasonal histogram of relative fire frequency and burned area had two maximums within the range 45°–55° N (corresponding to spring-early summer and to late summer-early fall; both maxima are statistically significant based on Fisher's criteria). Northward of 55° N, seasonal fire frequency and burned area distributions become unimodal (Figure 5a,b).

3.3. Fire Danger Period and Fire Return Intervals

Along the south-north transect the fire danger period decreased from 130 (±32) days in the south to 29 (±10) days in the north (Figure 6a). That decrease is strongly correlated with incoming solar radiation ($r = 0.97$), as well as the number of wildfires, RBA and RFF (Figure 4). Meanwhile, we did not find significant temporal trend in fire danger period duration which was characterized by high variability. For example, the date of the first fire varied by up to 30 days.

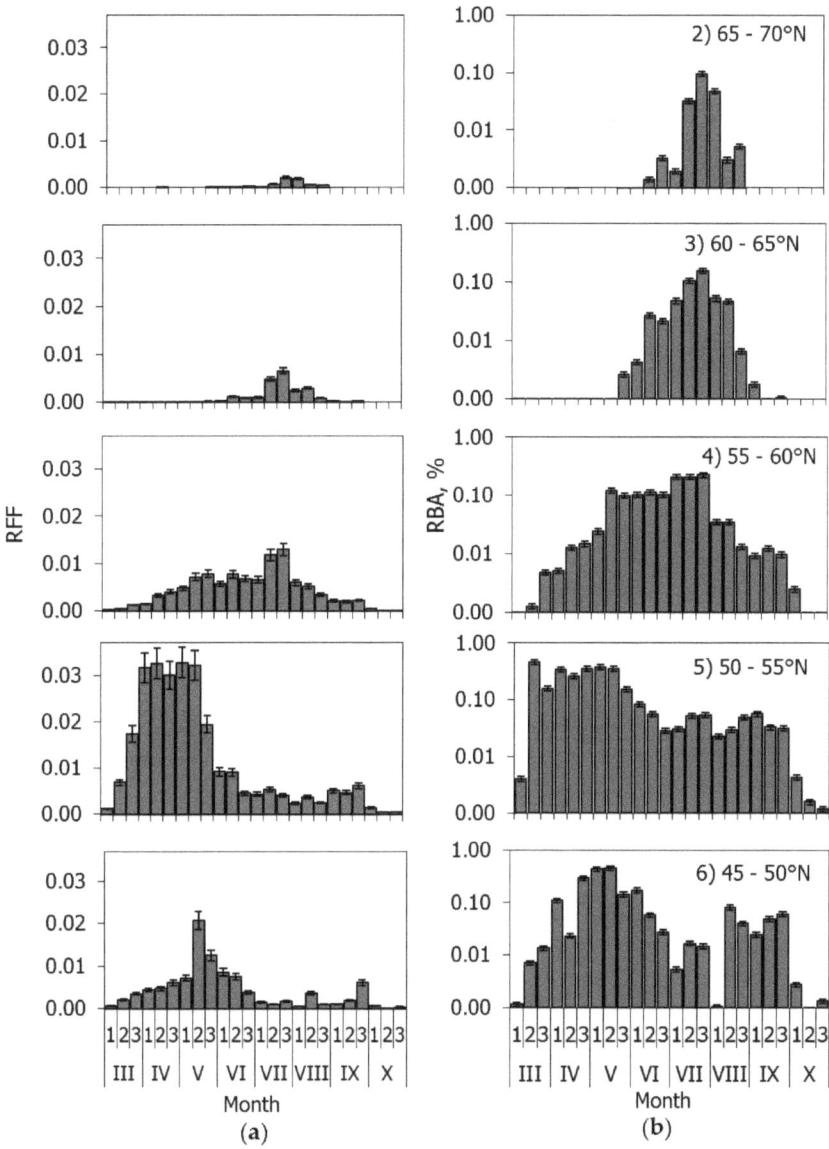

Figure 5. Seasonal distribution of the relative fire frequency (**a**) and relative burned area (**b**) along the transect (see Figure 1). Bars were calculated for mean 10-day period (numbered 1 to 3) from March (III) to October (X) (mean for period 1996–2015). Note: within transect zone 1 (70° + N) only one wildfire (in 3rd decade of July) was detected (not shown on the graph).

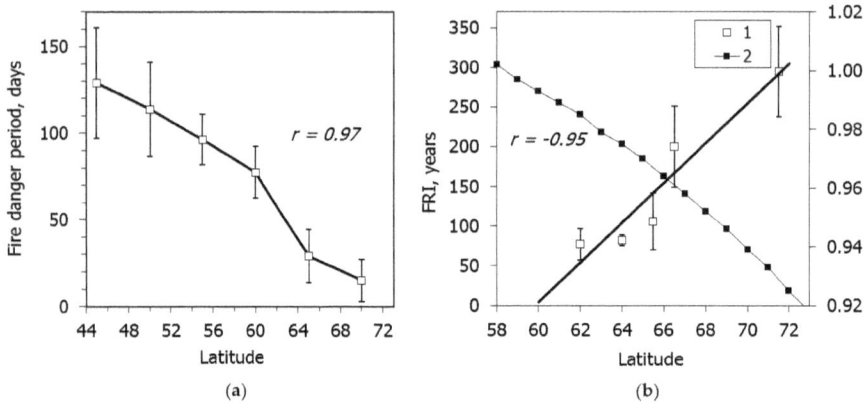

Figure 6. (a) Fire danger period dependence on latitude (correlation with solar radiation is $r = 0.97$); (b) fire return interval (FRI) dependence on (1) latitude and (2) incoming solar radiation ($r = -0.95$; $p < 0.05$). Bars show 95% confidence level.

The fire return interval (we obtained earlier [6,20,21]) increased along the latitudinal gradient (Figure 6b). It varied from 80 years at 64° N to about 200 years at the latitude of the Arctic Circle. The maximum values (about 300 years) were detected for stands at the northern limit of closed larch stands (i.e., 71°+ N). That increase was negatively correlated with incoming solar radiation ($r = -0.95$). For a south-north transect in western Canadian forests (about 60°–70° N 90°–130° W), de Groot et al., 2013 found an FRI value of about 180 years [23], which was similar to FRI in Siberia at the latitude of the Arctic Circle. However, no FRI dependence on the latitude was studied in that paper.

4. Conclusions

The main conclusion is that the number of fires, fire frequency, burned area and the fire danger period, as well as fire return intervals along a south-north transect in Siberia were strongly correlated with incoming solar radiation ($r = 0.81$–0.97). The fire frequency and burned area distributions along the transect were bimodal within 45°–55° N and unimodal at higher latitudes. The fire frequency exponentially decreased northward, whereas fire return intervals increased from 80 years at 62° N to 200 years at the Arctic Circle, and to about 300 years near the northern limit of closed stands (about 71°+ N).

The second main conclusion is that climate-induced fire frequency and burned area are increasing within Siberian forests. During recent decades, positive trends were observed for both number of wildfires ($R^2 = 0.69$) and size of burned areas ($R^2 = 0.47$). Wildfire frequency was also correlated with air temperature anomalies and drought index, SPEI. This result is similar to observations within the North American portion of boreal forests [24,25], and supports the hypothesis of climate-driven increase of fire frequency in boreal forests with the possible turning of boreal forests from carbon sink to a carbon source.

Acknowledgments: The work was supported by Russian Science Foundation (project #14-24-00112). Field measurements and K.J. Ranson's activities were supported by NASA's Earth Science Division.

Author Contributions: E.I. Ponomarev processed satellite database on wildfires, V.I. Kharuk suggested the idea of studies and processed data on fire return interval; E.I. Ponomarev, V.I. Kharuk and K.J. Ranson analyzed data and wrote the paper.

Conflicts of Interest: The authors declare no conflict of interest.

Abbreviations

The following abbreviations are used in this manuscript:

NOAA	National Oceanic and Atmospheric Administration
AVHRR	Advanced Very High Resolution Radiometer
MODIS	Moderate Resolution Imaging Spectroradiometer
TM	The Landsat Thematic Mapper
ETM	Enhanced Thematic Mapper
GIS	Geographic Information System
RFF	relative fire frequency
RBA	relative burned area
SPEI	The Standardized Precipitation Evapotranspiration Index
PET	potential evapotranspiration
NIR	near-infrared
MIR	mid-infrared
TIR	thermal infrared
FRI	fire return interval

References

1. IPCC. Climate Change 2014: Impacts, Adaptation, and Vulnerability. Summaries, Frequently Asked Questions, and Cross-Chapter Boxes. In *A Contribution of Working Group II to the Fifth Assessment Report of the Intergovernmental Panel on Climate Change*; Field, C.B., Barros, V.R., Eds.; World Meteorological Organization: Geneva, Switzerland, 2014; p. 190.

2. Conard, S.G.; Sukhinin, A.I.; Stocks, B.J.; Cahoon, D.R.; Davidenko, E.P.; Ivanova, G.A. Determining effects of area burned and fire severity on carbon cycling and emissions in Siberia. *Clim. Chang.* **2002**, *55*, 197–211. [CrossRef]

3. Shvidenko, A.Z.; Schepaschenko, D.G. Climate Change and Wildfires in Russia. *Contemp. Probl. Ecol.* **2013**, *6*, 50–61. [CrossRef]

4. Federal State Statistics Service. Available online: http://www.gks.ru (accessed on 29 February 2016).

5. Forkel, M.; Thonicke, K.; Beer, C.; Cramer, W.; Bartalev, S.; Schmullius, C. Extreme fire events are related to previous-year surface moisture conditions in permafrost-underlain larch forests of Siberia. *Environ. Res. Lett.* **2012**, *7*, 1–9. [CrossRef]

6. Kharuk, V.I.; Ranson, K.J.; Dvinskaya, M.L.; Im, S.T. Wildfires in northern Siberian larch dominated communities. *Environ. Res. Lett.* **2011**, *6*, 1–7. [CrossRef]

7. Krylov, A.; McCarty, J.L.; Potapov, P.; Loboda, T.; Tyukavina, A.; Turubanova, S.; Hansen, M.C. Remote sensing estimates of stand-replacement fires in Russia, 2002–2011. *Environ. Res. Lett.* **2014**, *9*, 1–8. [CrossRef]

8. Gauthier, S.; Bernier, P.; Kuuluvainen, T.; Shvidenko, A.Z.; Schepaschenko, D.G. Boreal Forest Health and Global Change. *Science* **2015**, *349*, 819–822. [CrossRef] [PubMed]

9. Goldammer, J.G.; Stocks, B.J.; Sukhinin, A.I.; Ponomarev, E.I. Current Fire Regimes, Impacts and the Likely Changes-II: Forest Fires in Russia–Past and Current Trends. In *Vegetation Fires and Global Change: Challenges for Concerted International Action: A White Paper Directed to the United Nations and International Organizations*; Goldammer, J.G., Ed.; Global Fire Monitoring Center (GFMC)/Kessel Publishing House: Eifelweg, Germany, 2013; pp. 51–79.

10. Bartsch, A.; Balzter, H.; George, C. The influence of regional surface soil moisture anomalies on forest fires in Siberia observed from satellites. *Environ. Res. Lett.* **2009**, *4*, 1–9. [CrossRef]

11. Ponomarev, E.I.; Kharuk, V.I. Wildfire Occurrence in Forests of the Altai–Sayan Region under Current Climate Changes. *Contemp. Probl. Ecol.* **2016**, *9*, 29–36. [CrossRef]

12. Bartalev, S.A.; Egorov, V.A.; Ershov, D.V.; Isaev, A.S.; Loupian, E.A.; Plotnikov, D.E.; Uvarov, I.A. Satellite mapping of vegetation in Russia using MODIS spectroradiometer data. *Mod. Problems Remote Sens. Earth Sp.* **2011**, *8*, 285–302. (In Russian)

13. Loboda, T.V.; Csiszar, I.A. Estimating Burned Area from AVHRR and MODIS: Validation Results and Sources of Error. In Proceedings of the 2nd Open All-Russia Conference: Current Aspects of Remote Sensing of Earth from Space, Moscow, Russia, 16–18 November 2004; Space Research Institute: Moscow, Russia, 2004; pp. 415–421.

14. Ponomarev, E.I.; Shvetsov, E.G. Satellite detection of forest fires and geoinformation methods of results calibrating. *Stud. Earth from Sp.* **2015**, *1*, 84–91. (In Russian) [CrossRef]

15. Harris, I.; Jones, P.D.; Osborn, T.J.; Lister, D.H. Updated high-resolution grids of monthly climatic observations—The CRU TS3.10 Dataset. *Int. J. Climatol.* **2014**, *34*, 623–642. [CrossRef]

16. Climatic Research Unit. Available online: http://www.cru.uea.ac.uk (accessed on 29 February 2016).

17. SPEI Global Drought Monitor. Available online: http://sac.csic.es/spei/map/maps.html (accessed on 29 February 2016).

18. Vicente-Serrano, S.M.; Beguería, S.; López-Moreno, J.I. A Multi-scalar drought index sensitive to global warming: The Standardized Precipitation Evapotranspiration Index–SPEI. *J. Clim.* **2010**, *23*, 1696–1718. [CrossRef]

19. The Solar Radiation and Climate of the Earth. Available online: http://www.solar-climate.com (accessed on 29 February 2016).

20. Kharuk, V.I.; Ranson, K.J.; Dvinskaya, M.L. Wildfires dynamic in the larch dominance zone. *Geophys. Res. Lett.* **2008**, *35*, 1–5. [CrossRef]

21. Kharuk, V.I.; Ranson, K.J.; Dvinskaya, M.L. Fire return intervals within the northern boundary of the larch forest in Central Siberia. *Int. J. Wildland Fire* **2013**, *22*, 207–211. [CrossRef]

22. Kovacs, K.; Ranson, K.J.; Sun, G.; Kharuk, V.I. The relationship of the Terra MODIS fire product and anthropogenic features in the Central Siberian landscape. *Earth Interactions* **2004**, *8*, 1–25. [CrossRef]

23. De Groot, W.J.; Cantin, A.S.; Flannigan, M.D.; Soja, A.J.; Gowman, L.M.; Newbery, A. A comparison of Canadian and Russian boreal forest fire regimes. *For. Ecol. Manag.* **2013**, *294*, 23–34. [CrossRef]

24. Girardin, M.P.; Ali, A.A.; Carcaillet, C.; Mudelsee, M.; Drobyshev, I.; Hely, C.; Bergeron, Y. Heterogeneous response of circumboreal wildfire risk to climate change since the early 1900s. *Glob. Chang. Biol.* **2009**, *15*, 2751–2769. [CrossRef]

25. Flannigan, M.; Cantin, A.S.; de Groot, W.J.; Wotton, M.; Newbery, A.; Gowman, L.M. Global wildland fire season severity in the 21st century. *For. Ecol. Manag.* **2013**, *294*, 54–61. [CrossRef]

© 2016 by the authors. Licensee MDPI, Basel, Switzerland. This article is an open access article distributed under the terms and conditions of the Creative Commons Attribution (CC BY) license (http://creativecommons.org/licenses/by/4.0/).

forests

MDPI

Article

Fire Regime in Marginal Jack Pine Populations at Their Southern Limit of Distribution, Riding Mountain National Park, Central Canada

Jacques C. Tardif [1,*], Stephen Cornelsen [2], France Conciatori [1], Eben Blake Hodgin [3] and Marlow G. Pellatt [4,5]

1   Centre for Forest Interdisciplinary Research (C-FIR), University of Winnipeg, 515 Portage Avenue, Winnipeg, MB R3B 29E, Canada; f.conciatori@uwinnipeg.ca
2   Resource Management Specialist, Fire Management Officer, Parks Canada Agency, Riding Mountain National Park, Wasagaming, MB R0J 2H0, Canada; stephen.cornelsen@pc.gc.ca
3   Department of Earth and Planetary Sciences, Harvard University, 20 Oxford Street, Cambridge, MA 02138, USA; hodgin@fas.harvard.edu
4   Parks Canada, Natural Resource Conservation, Protected Areas Establishment and Conservation Directorate, 300-300 West Georgia Street, Vancouver, BC V6B 6B4, Canada; marlow.pellatt@pc.gc.ca
5   School of Resource and Environmental Management, Simon Fraser University, 8888 University Drive, Burnaby, BC V5A 1S6, Canada
*   Correspondence: j.tardif@uwinnipeg.ca; Tel.: +1-204-786-9475

Academic Editors: Yves Bergeron and Sylvie Gauthier
Received: 12 July 2016; Accepted: 23 September 2016; Published: 30 September 2016

Abstract: In central Canada, long fire history reconstructions are rare. In a context where both anthropogenic and climate influences on fire regime have changed, Parks Canada has a mandate to maintain ecological integrity. Here we present a fire history derived from fire-scarred jack pine (*Pinus banksiana* Lamb.) trees growing at their southern distribution limit in Riding Mountain National Park (RMNP). In Lake Katherine Fire Management Unit (LKFMU), a subregion within the park, fire history was reconstructed from archival records, tree-ring records, and charcoal in lake sediment. From about 1450 to 1850 common era (CE) the fire return intervals varied from 37 to 125 years, according to models. During the period 1864–1930 the study area burned frequently (Weibull Mean Fire Intervals between 2.66 and 5.62 years); this period coincided with the end of First Nations occupation and the start of European settlement. Major recruitment pulses were associated with the stand-replacing 1864 and 1894 fires. This period nevertheless corresponded to a reduction in charcoal accumulation. The current fire-free period in LKFMU (1930–today) coincides with RMNP establishment, exclusion of First Nations land use and increased fire suppression. Charcoal accumulation further decreased during this period. In the absence of fire, jack pine exclusion in LKFMU is foreseeable and the use of prescribed burning is advocated to conserve this protected jack pine ecosystem, at the southern margins of its range, and in the face of potential climate change.

Keywords: fire history; boreal mixedwood; *Pinus banksiana*; dendrochronology; fire scars; lake sediment charcoal; First Nations; European settlement; fire exclusion; paleoecology

1. Introduction

1.1. Fire and Prescribed Fire in Central Canada

Wildland fire across boreal Canada remains a primary ecological process, despite fire suppression being the dominant management paradigm [1,2]. In Manitoba, prescribed fire (hazard reduction, silvicultural site preparation, enhancement of wildlife habitat, range burning, and insect/disease control or ecosystem conservation) has not been as widely used as in other Canadian provincial and federal legislations [3]. Prescribed fire has been primarily used to reduce forest encroachment in native prairie ecosystems, to maintain tall-grass prairie [4,5], and to study its impact as a site preparation tool following clear-cutting of jack pine (*Pinus banksiana* Lamb.) stands [6]. In 1983, the use of prescribed fire in National Parks started in Banff and Jasper with the objective of maintaining the natural age-classes distribution of lodgepole pine (*Pinus contorta* Dougl. ex. Loud.) and Douglas-fir (*Pseudotsuga menziesii* (Mirb.) Franco) populations [1,7]. In 2007, the Riding Mountain National Park (RMNP, Manitoba) management plan identified fire regime alteration as one of its ecological integrity concerns [8]. Maintaining or improving ecological integrity by approximating historical long-term fire cycles and minimizing ecological risk has also been identified as a goal in the National Fire Management Plan [9]. Restoration of long-term fire regimes is thus becoming a critical part of land management, particularly in parks and protected areas [1]. Achieving this objective requires an understanding of historical fire regimes and of the extent and degree to which they have been altered. In many landscapes, fire regime characteristics were also influenced by cultural values that dictate human practices such as: Land use, impacts of settlement, forestry practices, and fire suppression policies [10–12]. In addition, uncertainty and controversy still remain about the importance of fire usage by First Nations, Metis, and European settlers [1].

1.2. Jack Pine Distribution and Fire Regime

In North America, jack pine has a wide distribution and the species is typical of fire-prone habitats [13,14]. In RMNP, jack pine forms marginal populations reaching their southern limit of distribution. South of this natural limit in Manitoba, jack pine plantations can be however found [15]. In the boreal forest, the species distribution is regulated by fire [16–18] and populations are usually referred to as even-aged originating from stand-replacing fires. At the northern distribution limit of the species in northern Québec, fire intervals have been short enough to prevent jack pine exclusion [19]. A key requirement for the long-term maintenance of jack pine populations is a fire return interval (FRI) shorter than average life span of individual trees [19,20]. In the absence of fire, jack pine would disappear as a natural component of the boreal landscape [7,16]. Lethal fires that are too frequent will also affect long-term maintenance by preventing regeneration. Less frequently, uneven-aged jack pine populations have been associated with lack of fires [21,22] or non-lethal surface fire regime [17,23]. In northern Québec, jack pine trees growing in contrasting fire regime (mainland versus lake islands) were found to express different serotiny levels, this character being less expressed in non-lethal surface fire regime [17]. In north-central Manitoba, old open uneven-aged mixed jack pine/northern white-cedar (*Thuja occidentalis* L.) stands associated with a non-lethal surface fire regime can be found [24]. In southwestern Manitoba (north of RMNP), open jack pine stands bearing fire scars were also observed [25] in upland meadows in the Duck Mountain Provincial Forest (DMPF).

1.3. Fire Regime in the Boreal Plains of Western Manitoba

In central Canada, little research has been conducted with regards to disturbance dynamics in the boreal and mixedwood forests [25,26]. A study conducted in the boreal plains (e.g., DMPF) indicated the prevalence of stand-replacing fires associated with major drought periods and a lengthening of the fire cycle since pre-European Settlement [25]. Large fires in 1885 were reported in the DMPF [27] and large fires were also reported north in Porcupine Mountain at the end of the 19th century and in 1919 [28]. Some of the most notable fires occurred during the period 1885–1895, which burned almost half of the forested area of the uplands region [25,27,29].

In RMNP, limited specific information exists on fire history [30]. Some authors (i.e., [31,32]) attributed the major fires of 1822, 1853–1855, 1889–1891, and 1918–1919 to RMNP, but these were actually reported for the "B18 mixedwood section of the boreal forest," with study sites located from Manitoba to Saskatchewan [33]. The only fire specifically associated with RMNP was the 1915 Whirlpool fire [33]. In RMNP, various reports have identified fires as having been most prevalent during European settlement (1885–1895) as land was cleared for farming [34]. Two large fires in the early 1890s were reported to have burned over 70% of the western portion of the park and fires in circa 1830 and 1895 burning in the eastern portion [35]. Jack pine stands in the southeastern portion of RMNP were also reported to have burned repeatedly at the turn of the 20th century [34,36]. Even-aged pine stands failing to regenerate due to repeated fire causing open prairie lands were also reported [35,37]. Since the creation of RMNP in 1930, numerous fires have burned into the park from surrounding farmland areas [30,34]. Nonetheless, a lengthening of the fire cycle was reported [20,36]. Fire prevention/suppression policies were implemented from the time of Forest Reserve establishment, up to 1979. In 1979, the Parks Canada policy changed from protection to management and permitted under certain conditions "active management or manipulation of the ecosystems" [38].

1.4. Objectives

In a context pertaining to protected area management, the main objectives of this study were (i) to document the historical variability in fire regime in marginal jack pine populations located at their southern limit of distribution and (ii) to translate this knowledge into ecosystem management strategies. An understanding of the past fire regime and historical legacies could lead to restoration and/or to the identification of factors that would warrant it. First, the characteristics of the recent fire regime (interval, seasonality, and spatial distribution) were reconstructed using exact fire scar dates coupled with establishment/mortality records. Second, indices regarding the anthropogenic or climatic nature of the fire regime were analyzed. Third, macroscopic charcoal particles (>125 μm) recovered from Lake Katherine sediment were quantified to determine changes in the fire regime at a time frame beyond that provided by the archival and tree-ring records. The presence of jack pine stands in RMNP with trees bearing multiple fire scars offered a unique opportunity to reconstruct the fire regime of a portion of the park and to address the anthropogenic or climatic nature of these fires.

2. Materials and Methods

2.1. Study Area

The study area lies within RMNP located in southwestern Manitoba about 250 km northwest of Winnipeg (Figure 1). The park covers 2969 km^2 and constitutes the southeastern extent of the Mixed Wood Section of the Boreal Forest Region [39]. Riding Mountain is part of the Manitoba Escarpment, which rises approximately 300 m from the eastern Manitoba lowlands. The park is primarily on the plateau, transitioning from the first prairie level (the Manitoba Plain) to the second prairie level (Saskatchewan Plain). Riding Mountain also forms the southern limit of the Mid Boreal Uplands ecoregion, which also includes Duck Mountain Provincial Forest (DMPF) and Porcupine Provincial Forest (PPF) to the north [40]. Outside RMNP boundaries, the landscape is dominated by agricultural development.

The vegetation of RMNP is characteristic of the mixed boreal forest [20,31]. The present-day boreal forest existed since about 2500 BP with a parkland phase of grassland and deciduous species dominating from 6500 to 2500 BP [41]. In the well-drained upland portion of RMNP the characteristic boreal forest association predominates with trembling aspen (*Populus tremuloides* Michx.), balsam poplar (*Populus balsamifera* L.), white spruce (*Picea glauca* (Moench) Voss), and balsam fir (*Abies balsamea* (L.) Mill.) dominating. On the sandier, drier, nutrient poor sites, jack pine is found; on moister sites, black spruce (*Picea mariana* (Mill.) B.S.P.) and tamarack (*Larix laricina* (DuRoi) K. Koch) increase in dominance. On the exposed edges of the region, bur oak (*Quercus macrocarpa* Michx.) dominates and, where black chernozemic soils are present, grasslands are embedded in the mixedwood forest. Trembling aspen is the prevalent species across the transition zone, taking the form of small scattered clumps in the grasslands to large stands in the boreal forest [20,31,37,39]. In RMNP, jack pine is confined to an area of approximately 250 km^2 in the SE portion (Figure 1D) and is separated from the contiguous boreal forest by about 80 km [20,42].

The RMNP lies within a mid-boreal climate with short, cool summers and cold winters. At Wasagaming (50°39'18" N, 99°56'31" W, elevation 627.4 m.a.s.l.) for the period 1981–2010, average annual precipitation was 488 mm, with 372 mm falling as rain. June is the wettest month with a mean rainfall of 80 mm [43]. The temperature ranges from −17.5 °C (mean January) to 17.0 °C (mean July), with a mean annual daily temperature of 0.7 °C with extremes ranging from −47.8 °C to 36.5 °C.

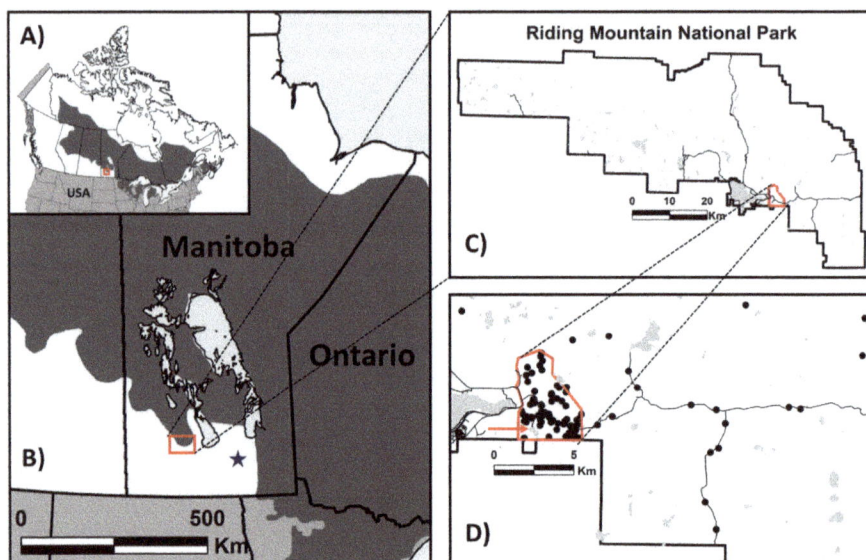

Figure 1. Location of the study area in Canada (**A**) and of Riding Mountain National Park (RMNP) in Manitoba (**B**). The star indicates the location of the city of Winnipeg and the dark grey shaded area represents *Pinus banksiana* range of distribution. The upper right inset (**C**) shows RMNP and the location of the Lake Katherine Fire Management Unit (LKFMU). Clear Lake can be seen left of LKFMU. The lower right inset (**D**) indicates the *P. banksiana* sites (black circles) that were sampled within and outside LKFMU. Lake Katherine can be seen in the lower left corner of LKFMU (arrow), with Clear Lake to the west.

2.2. RMNP History

The history of the Riding Mountain region is not fully documented but human occupation started with the retreat of the Wisconsin ice sheet and the formation of Glacial Lake Agassiz. Clovis projectile points were found on Rolling River [44], about 15 km from our study area. The highlands of RMNP, being transitional parkland with both forest and grassland ecosystems, provided optimum resource and habitat availability on a seasonal basis [45]. During the historical period, the resource-rich base was exploited for seasonal rounds and year-round use by First Nations, with Assiniboine, Cree, and Ojibwa having traditionally occupied portions of the park [30,46]. The Ojibwa moved into the area in the 1700s as a response to the fur trade and by the early 1800s were established as the dominant society of the Riding Mountain region [30,32].

In the Riding Mountain region, European settlement, lumber milling, and land clearing occurred from 1870 to 1930. Timber was sought after for building materials, fuelwood, and railway ties [44]. Intensive logging of the southeastern portion of RMNP was triggered by the construction of the railway to Dauphin in the early 1890s [20], with fire being a continuous issue because of its careless use by settlers clearing homesteads [30]. In 1895, the Riding Mountain Forest Reserve was established to protect the remaining forest and timber supply from exploitation and destruction by fire. In 1906, the Dominion Forest Reserve Act was passed marking the beginning of regulated harvesting and organized fire protection. National Park establishment was in 1930, with timber harvest, grazing, and haying continuing up to the mid-1960s. Strict fire prevention/suppression policies were implemented from the time of Forest Reserve establishment, up to 1979. Prior to its regulation, numerous fires from careless slash and disposal operations burned a large portion of the reserve [30]. In the early days, Forest Rangers also frequently created fire guards along the park boundary by burning meadows early in the spring [1].

In Riding Mountain, European settlement, the establishment of Indian reserves in the 1870s and the creation of the Riding Mountain Forest Reserve were all instrumental in restricting First Nations' use of the area [30]. One reserve at Clear Lake, belonging to the Keeseekoowenin Ojibwa First Nation, fell within the boundaries of the National Park, and in 1936, the people were evicted by Park staff and their homes burned [32,47]. Currently six First Nation communities are located adjacent to or within RMNP on Indian Reserve lands, with the Keeseekoowenin and Rolling River [46] located closest to the study area.

2.3. Field Procedures

Sampling occurred in the southeastern portion of RMNP where jack pine stands predominate. More precisely, sampling mainly took place in Lake Katherine Fire Management Unit (LKFMU; 14U 437000, 5613000), a 30 km^2 zone adjacent to Clear Lake on the southeastern portion of RMNP (Figure 1D). The landform is mainly hummocky to undulating stagnation moraine, at an elevation of 620–695 m.a.s.l., and covered by orthic gray luvisol soils. The LKFMU is located on the boundary of the park at the southwestern edge of the jack pine range which was mapped by Zoltai [42].

Data collection for the tree-ring portion of the study came from two independent studies. First a fire study was initiated by Parks Canada in 2008–2009 following standard methods [48–50]. Jack pine stands were identified and located from 1928, 1959, 1979, and 2004 aerial photos (Figure 2A,B) and 1937 forest inventory maps and systematically surveyed for evidence of fire disturbance. Cross-sections were taken from the fire-scarred trees and remnants (snags, logs, and stumps) with the most numerous scars (Figure 2C). A total of 89 trees with apparent fire scars were cut. Sampled trees were not evenly distributed across the study area. For each sampled tree, the number of visible scars, number of cross-sections taken, number of pieces per cross-section, height of cross-sections above ground, azimuth in degrees of fire-damaged cat faces, live or dead specimens, and Universal Transverse Mercator (UTM) location were recorded. Fire scars were assigned a compass bearing for the direction of the middle of the arc of the killed cambium. Direction of fire spread was estimated by converting the compass bearing for the fire scar orientation to a cardinal direction. The scars were assumed to have formed on the leeward side of the tree during a fire event [23,51]. Only the first scar was used for the direction analysis because subsequent scarring is often more likely to occur after the first burn due to exposed cambium and thinner bark in the region of the first scar [23]. It must be noted that local fuel conditions, weather, and topography also need to be taken into consideration when analyzing the direction of spread [52].

Second, a study initiated in 2009 by the University of Winnipeg DendroEcology Laboratory (UWDEL) aimed at developing long tree-ring chronologies for multi-species including jack pine [53]. Collected samples included both living and dead jack pine trees distributed within RMNP including LKFMU. Given that the objective of this study was to locate old living and dead trees, five to 10 trees (two cores per tree) were usually sampled per site except in young stands and/or where jack pine trees were in low abundance. Cross-sections were collected from dead material when available. This sampling added 110 trees from sites distributed in LKFMU. Another 100 jack pine trees located outside LKFMU were also used in the development of tree-ring chronologies and their establishment/mortality dates were considered for comparison (Figure 1D).

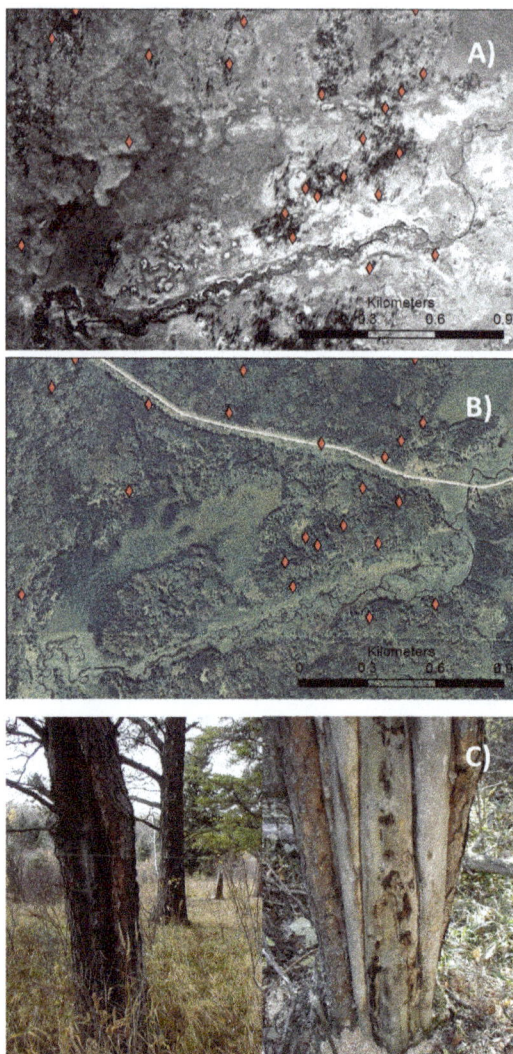

Figure 2. Aerial photographs from a portion of the LKFMU taken in 1929 (**A**) and in 2007 (**B**) showing some of the sampling points and vegetation densification. *Pinus banksiana* trees showing fire scars (**C**).

2.4. Fire Scar and Chronology Development

All cores and disks were prepared following standard procedures including drying, gluing, sanding, and crossdating [25,50,52]. Samples were first visually crossdated using pointer years [53]. For each tree, all fire scar dates as well as tree establishment/mortality dates were recorded (Figure 3A). Determining fire years from tree scars followed standard techniques [54], with visual crossdating performed both before and after the events. Injury scars that were discovered a posteriori and did not correspond to a typical fire scar (Figure 3A) were not included. In addition, the relative position of the fire scar within the annual rings was determined (EE: Early earlywood, ME: Middle earlywood, LE: Late earlywood, LW: Latewood, D: Dormant, or U: Undetermined) to assess the season of fire occurrence [55]. Tree diameter (four radii) and tree age at the first fire scar were also determined in the laboratory for each tree bearing the pith to determine the mean minimum tree size (age) at time of injury. This information allowed for determining the minimum tree size for fire survival [51]. For samples that did not intersect the pith, the distance to it was estimated using a circle template that best fit the curvature of the innermost ring [22]. The number of years to the circle center was then estimated using an age–radius curve derived from samples for which the pith was intercepted. All age structures are thus presented to the year despite presenting a slight bias.

Given that the absence of fire scars does not necessarily imply that an area did not burn at a given time [52], white earlywood rings (WER; [56]) were also systematically compiled. White earlywood rings were observed in both jack pine trees bearing fire scars and in unscarred trees within LKFMU (Figure 3B,C). The presence of WER could provide an indication of crown scorch as they have been associated with crown damages during the dormant season, presumably leading to a carbohydrate deficit responsible for the production of earlywood tracheids with thin secondary walls [56]. The absence of WER in jack pine trees growing outside the LKFMU would support this hypothesis.

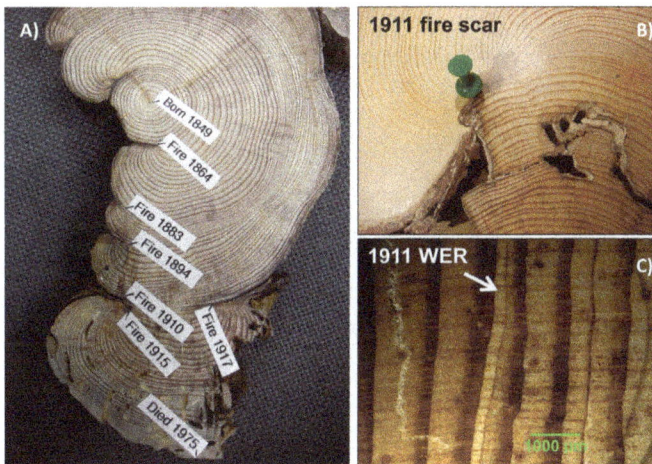

Figure 3. *Pinus banksiana* cross-section presenting six fire scars (**A**); *P. banksiana* cross-section showing a fire scar in 1911 (**B**), and another one showing a white earlywood rings (white arrow) in 1911 (**C**).

A tree-ring reference chronology was also developed from jack pine trees collected within and outside LKFMU. For each sample, annual growth increments were measured along two radii at a precision of 0.001 mm using a Velmex measuring stage coupled with a computer. Both visual crossdating and measurements were statistically validated using the program COFECHA [57], which calculates cross-correlations to a lag of 10 years between each individual standardized measurement series and a reference chronology. All measurement series were then standardized using a 53-year (or more flexible for a few series) cubic smoothing spline using program ARSTAN (Ver 4.4, [58]).

2.5. Analyses of Tree-Ring and Climate Data

The software FHX2 and FHAES Ver. 2.0.1 [59,60] were used to develop the fire chronology and summarize fire intervals statistics. Mean fire return interval (mFRI), Weibull mean fire return interval (WmFRI) and descriptive statistics were computed for both the composite and individual tree fire intervals. The mFRI is the average number of years between fire dates in the composite chronology [48,61]. Results are presented for individual tree intervals and for composite chronologies at multiple scales. Fire statistics for LKFMU were computed for three sets of data: (i) entire fire years (no filter); (ii) fire recorded by 10% or more of the sample trees; and (iii) fires recorded by 25% or more of the sample trees. This approach aimed at highlighting spatial variation among locations, assuming that fires present in more than 10% and 25% of the samples were more widespread [62]. The FHAES program was also used to analyze the seasonality of fire and analyses.

To determine the potential impact of climate on LKFMU fire regime, the association between the jack pine tree-ring chronologies and the reconstructed Canadian Drought Code (CDC) was assessed using correlation analyses. This analysis was done using the CDC derived from a network of chronologies developed for DMPF located about 125 km north of RMNP [63]. Running correlations were also used to assess the association between LKFMU jack pine chronology and the one developed outside LKFMU.

2.6. Lake Katherine Charcoal Sediment Analyses

In addition to the tree-ring studies, a paleoecological study was conducted in the winter of 2010 to quantify macroscopic charcoal particles in Lake Katherine sediment. Lake Katherine is located in the southwest section of LKFMU (Figure 1D). It has an area of 26.7 ha and a depth of 8.5 m [64]. Lake Katherine is also a small closed basin, presumably ground water fed, with no significant water courses in or out. Unconsolidated sediment was extracted with a Glew gravity corer [65] down to 40 cm depth and dated with 13 ^{210}Pb samples using a constant rate of supply model [66,67]. The procedure followed for ^{210}Pb dating was that outlined by MyCore Scientific Inc., (Dunrobin, ON, Canada). (http://mycore.ca). Underlying consolidated lake sediments were extracted with a Livingston Piston corer [68] down to 673 cm and dated with four ^{14}C samples. Dating of the deeper sediments was undertaken on organic material using the ^{14}C Accelerator Mass Spectrometry (AMS) method [69] following the procedure outlined by Beta Analytic (Miami, FL, USA) (http://www.radiocarbon.com). In this study, the upper 1.4 m of sediment was analyzed and the age-depth model was thus derived using the first two radiocarbon dates and the first eleven ^{210}Pb dates. The two ^{14}C dates were derived from (i) charred spruce needles located at a depth of 259 cm and dated to 1710 BP $\pm$ 30 years and (ii) from a piece of wood at a depth of 476 cm and dated to 5035 BP $\pm$ 30 years. The dating of the sediment (^{210}Pb and ^{14}C) allowed the development of an age-depth model providing a calendar date for each centimeter of the sediment core. The age-depth model first required ^{14}C dates to be calibrated using program CALIB 7.0 [70] based on IntCal13 [71]. An age–depth model was developed using MCAgeDepth program [72] following 1000 Monte Carlo simulations to generate confidence intervals (Figure S1).

The Lake Katherine charcoal record was generated by sub-sampling the upper 140 cm of sediment (first 105 cm) at ~1 cm intervals and the remaining sediment at ~4 cm) with a calibrated (1 cm^3) brass subsampler. Samples were soaked in a 10% Sodium Pyrophosphate solution for two days before being wet-sieved at 125 microns. Next, the sieved residue was soaked in a 3% Hydrogen Peroxide solution for six hours before again being wet sieved at 125 microns [73]. Residues were then transferred to a channeled Bogorov tray, where charcoal pieces >125 microns were identified and counted using a stereomicroscope at 10–50× magnification. Charcoal particles were compiled using three size classes (125–250, 250–500, >500 microns), with the total count representing their sum. Among the particles, the number pertaining to the grass cuticle morphotype [74] was also determined. Based on the age–depth model, total charcoal counts were reported as charcoal accumulation rate (CHAR, particles/cm^2/cal year).

The CHAR series was analyzed using the CharAnalysis program [75] to identify fire events and calculate FRIs. In this study, the period ~1450–2010 CE was analyzed and CHAR series was interpolated ($C_{interpolated}$) to equal time steps (four years, i.e., median sample resolution), containing a low frequency ($C_{background}$) and a high frequency (C_{peak}) component [76,77]. The $C_{background}$ was determined using the Lowess smoother robust to outliers using a window width of 300 years. C_{peak} series were identified by subtracting $C_{background}$ from $C_{interpolated}$. The C_{peak} series were further decomposed into a C_{noise} and C_{fire}, with the latter representing in theory significant peaks associated with local fires [66,76]. The Gaussian mixture model was used to determine C_{noise} distribution and a locally defined threshold was used to identify significant C_{peak}. The 90th, 95th, and 99th percentiles of the C_{noise} distribution were considered as a possible threshold separating C_{peak} into 'fire' and 'non-fire' events. The minCountP value was set to 0.99 in all runs given the short FRIs characterizing the recent RMNP fire history. The final determination of the peak analysis parameters were based on maximization of the signal to noise (SNI) index (typically >3) and goodness of fit (GOF; $p < 0.05$) [66,77].

3. Results

3.1. Recent Fire History in LKFMU

The tree-ring reconstructed fire history covers the period 1812 to 2009 (Figure 4). Two hundred and seventy-nine crossdated fire scars were recorded. A total of 28 fire years were recorded, with the first event recorded by two trees or more being observed in 1864 and the last one in 1930. During this period, 17 fires were recorded by more than two and up to 44 scarred trees: 1864, 1882–1883, 1887–1888, 1891, 1893–1894, 1903–1904, 1910–1911, 1915, 1917, 1923, 1925, and 1930. More than 48% of the 91 recording trees displayed a fire scar in 1915 (Figure 4). Prior to 1859 (one scarred tree), no fires were recorded despite sample depth reaching about 20 trees. The 1864 fire was recorded by few trees and this event was followed by an important recruitment pulse lasting from about 1865 to 1875 (Figure 4). The 1880 decade also marked an increase in recorder trees, with 14 trees recording the 1883 fire. The 1894 and perhaps 1887–1888 fires were also followed by important jack pine recruitment, occurring within a decade of few or no fires. Of the recorded fires, numerous ones also occurred consecutively (e.g., 1882–1883, 1887–1888, 1893–1894, 1903–1904, and 1910–1911; see Figure 4).

A total of 181 fire intervals originating from fire scars were recorded in LKFMU (30 km^2) ranging from 1 to 38 years with a mean fire return interval (mFRI) of 10.98 years and a Weibull mean fire return interval (WmFRI) of 11.03 years (Table 1). During the period 1850–1930 the mFRI derived from composite chronologies (no filter, $\geq$10% and $\geq$25% of the tree scarred) was 2.63, 4.18, and 5.60 years, respectively. The WmFRI values were 2.66, 4.19, and 5.62 years, respectively (Table 1). No fire has been recorded in LKFMU over the last 85 years (a fire-free period) corresponding to the period of First Nations land use exclusion, the National Park establishment, and active fire exclusion.

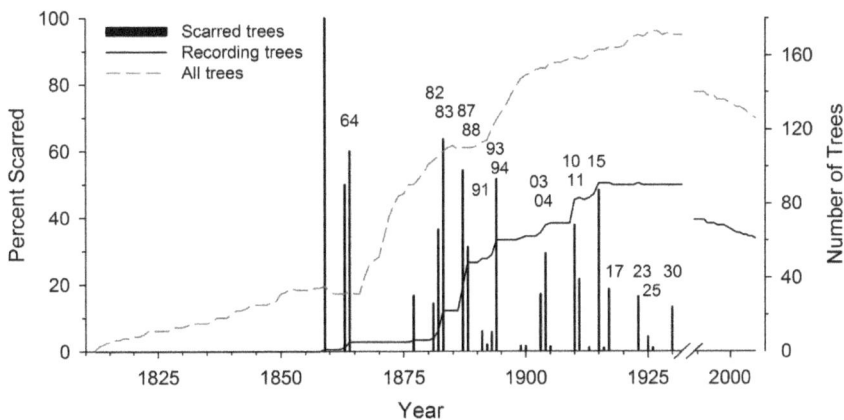

Figure 4. Percent of *Pinus banksiana* trees scarred per fire year. Sample depth (number of trees) is indicated. Percentage of scarred trees (black bars) is calculated from recorder trees (black line), with all sampled trees in the LKFMU indicated by the dashed grey line. No fire scars recorded by at least two trees were observed prior to 1864 or after 1930. The break in the *x*-axis from 1932 to 1989 is simply intended to better capture the figure's details.

Table 1. Descriptive statistics and fire intervals for Lake Katherine Fire Management Unit in Riding Mountain National Park.

	LKFMU	LKFMU	LKFMU	LKFMU
	All Sampled Trees	Composite: 1% Recording Trees Scarred	Composite: 10% Recording Trees Scarred	Composite: 25% Recording Trees Scarred
Year of coverage	1850–1930	1850–1930	1850–1930	1850–1930
Minimum number of sample	1	1	1	1
Total intervals	181	27	17	10
Mean fire interval	10.98	2.63	4.18	5.60
Median fire interval	10.00	1.00	4.00	4.50
Fire frequency	0.09	0.38	0.24	0.18
Weibull mean interval	11.03	2.66	4.19	5.62
Weibull median interval	10.19	2.10	3.42	4.36
Weibull fire frequency	0.10	0.48	0.29	0.23
Minimum fire interval	1	1.00	1.00	1.00
Maximum fire interval	38	13.00	13.00	18.00
Lower exceedance interval	4.19	0.55	0.96	1.08
Upper exceedance interval	18.44	5.17	7.98	11.05
Including WER				
Total intervals	189	No Change	No Change	No Change
Mean fire interval	10.75	No Change	No Change	No Change
Median fire interval	10.00	No Change	No Change	No Change
Fire frequency	0.09	No Change	No Change	No Change
Weibull mean interval	10.79	No Change	No Change	No Change
Weibull median interval	9.96	No Change	No Change	No Change
Weibull fire frequency	0.10	No Change	No Change	No Change
Minimum fire interval	1	No Change	No Change	No Change
Maximum fire interval	38	No Change	No Change	No Change
Lower exceedance interval	4.08	No Change	No Change	No Change
Upper exceedance interval	18.05	No Change	No Change	No Change

In the bottom portion of the table white earlywood rings (WER) were added to the fire scar data.

The jack pine age structure (establishment/mortality) within LKFMU indicated continuous recruitment since the early 1800, with important pulses being observed in the late 1860s and the late 1890s (Figure 5A). These establishment peaks corresponded to periods with longer fire intervals that followed the 1864 and the 1894 fires (Figure 5A). Establishment was also observed around 1810, 1850, and 1920. While our study did not include the quantification of recruitment after 1930, qualitative field observations indicate little to no jack pine regeneration in LKFMU. Regeneration and infill of forest gaps is mainly associated with white spruce and trembling aspen. Jack pine mortality has also been continuous during the reference period and increasing mortality was observed since the late 1970s (Figure 5A). The jack pine age structure outside LKFMU revealed distinct features (Figure 5B). Little indications of past fires were observed in these sites with the exception that some did burn in the spring of 1980 (field observations). Outside LKFMU, recruitment pulses may have occurred in the 1810s, 1830s, and 1890s, as indicated by pith dates. Similarly to LKFMU, jack pine mortality also increased around the late 1970s (Figure 5B), with that observed in 1980 being a consequence of a documented forest fire burning outside LKFMU.

In LKFMU, jack pine trees recorded their first scar at a mean diameter of 4.3 cm at cross-section height (standard deviation: 2.33, range 1.07–15.00, n = 93) and at a mean age of 20.61 years (standard deviation: 12.07, range 6–66, n = 93). These numbers also reached lower values during the two periods of clustered fire scars starting in the early 1860s and late 1880s (Figure 3A; Figure 6). The diameter (age) distribution at which trees recorded their first fire scar also indicated a significant decrease with time in those parameters, suggesting an intensification of fire frequency in the late 1800s (Figure 6).

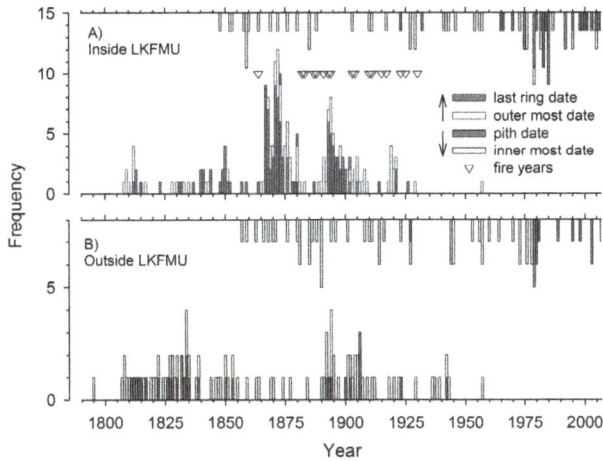

Figure 5. Age structure distribution of *Pinus banksiana* trees sampled within (**A**) and outside (**B**) Lake Katherine Fire Management Unit (LKFMU). The location of sampling sites located inside and outside of LKFMU can be seen in Figure 1D. For the age structure (**A** and **B**), the bottom bars are associated with recruitment whereas the top bars are associated with mortality. The frequency of trees with the pith date (bottom bars in **A** and **B**) and with the last ring complete (upper bars in **A** and **B**) are gray, whereas the white ones indicate the innermost and outermost dates.

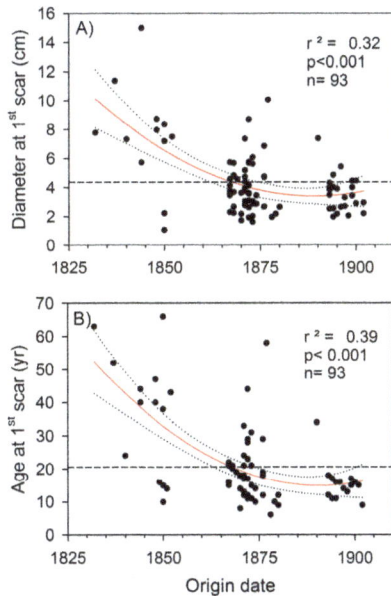

Figure 6. Diameter (**A**) and age (**B**) at first recorded scar as a function of origin date for *Pinus banksiana* trees. The r^2 value associated with a second-order polynomial regression is indicated. The regression lines (with 95% CI) suggest that the first fire scar was recorded at a larger diameter (older age) in trees recruited prior to the 1860s. The horizontal dashed lines indicate the mean value.

3.2. Seasonality and Direction of Fires

The seasonal analysis of fire years indicated that the great majority of the fires were recorded in the dormant season (79.6%), followed by earlywood fires (14.5%; Table 2). Overall, 92.1% of the scars indicate dormant season or spring fires. The majority of the scars identified to the dormant season also had a few trees recording them in the earlywood (Figure 7A). Intriguingly, the fire years 1894 and 1915 stand out with scars recorded from the dormant period to the mid-earlywood position (Figure 7A). Numerous fire scar years were also associated with the production of WER (Figure 7B). Agreement between both proxies was observed in 1883, 1887–1888, 1894, 1910–1911, 1923, and 1930, whereas the WER years 1943 and 1954 did not correspond to any fire scars (Figure 7B).

Table 2. Seasonality of fire scars based on relative intra-annual position of fire scar within the annual growth rings.

	Number	%
Total fire scars	279	100
Scars with season determined	269	96.4
Scars undetermined	10	3.6
Dormant (D)	214	79.6
First third earlywood (E)	39	14.5
Second third earlywood (M)	16	5.9
Last third earlywood (L)	0	0.0
Latewood (A)	0	0.0
DEM	269	100.0
LA	0	0.0

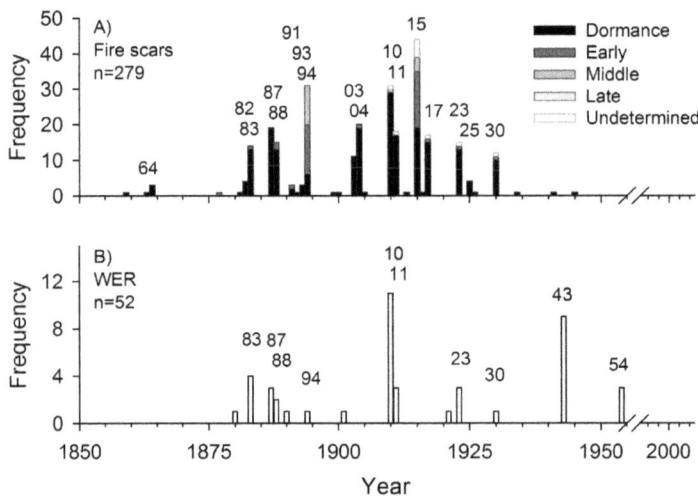

Figure 7. Frequency of fire scars with location within tree rings (**A**) and of white earlywood rings (**B**). The break in *x*-axis from 1955–1989 is simply intended to better capture the figure's details.

The distribution of fire-scarred trees in LKFMU revealed large spatial variability from one fire to another (Figure 8). For example, the 1888, 1903, and 1911 fire scars were mainly restricted to the southeastern portion of LKFMU. Scarred trees were also observed close to Lake Katherine in 1864, 1883, 1887, 1894, 1904, 1910, 1915, and 1917. The distribution of WER also matched that of fire scars very well. While few scars dating from 1864 were found, the 10-year period following this event was characterized by intensive jack pine recruitment. The 1894 (and possibly 1887–1888) fire was also followed by abundant recruitment. Some of the larger fires in LKFMU were observed in 1864, 1887, 1894, 1910, and 1915. The frequency of WER was highest in 1910, a year corresponding to widespread fire occurring mainly during the dormancy period. The 1915 fire was widespread and apparently burned for a long period in the growing season, having been recorded from the dormant season to the early and mid-earlywood (Figure 7). Interestingly, the spatial distribution of fire scarred trees also suggests that LKFMU may have been burned over a two-year period. For example, the position of the scarred trees in 1887–1888, 1903–1904, and 1910–1911 suggested that different portions (northwest versus southeast) of the area were burned each year (Figure 8). In LKFMU, the wind rose with the frequency of spread directions suggested the prevailing fire spread was from the southwest (Figure 8).

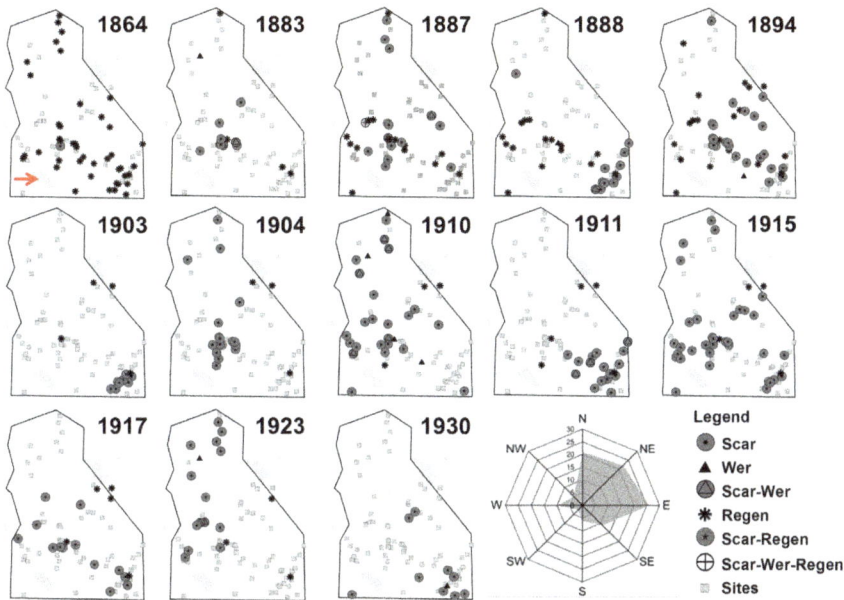

Figure 8. Spatial distribution of sampling sites and fire-scarred *Pinus banksiana* trees for 13 major fire years in the LKFMU. Fire scars, white earlywood rings (WER), and regeneration as well as combinations of them are indicated in the legend. A wind rose depicting the frequency of fire spread directions is also presented. The arrow indicates the location of Lake Katherine.

3.3. Drought and Fire

The tree-ring chronology developed from jack pine trees from inside and outside LKFMU covers the period 1812–2008 and 1804–2008, respectively (Figure 9A). These chronologies were significantly correlated ($r = 0.798$, $p < 0.001$, $n = 189$) for the period 1820–2008 and they were both also significantly correlated with the reconstructed CDC index ($r = -0.513$ and $r = -0.504$, $p < 0.001$, $n = 177$) for the boreal plains of Manitoba (Figure 9). The running correlation between the inside and outside LKFMU chronologies also presented a strong degradation of the common signal from about 1840 to 1910; the correlation reached a minimum during the interval 1855–1904 (Figure 9B). In LKFMU, fire years occurred when the CDC values ranged from 194 to 312, indicating that fires occurred in years when the drought code was high to extreme [78]. Among all fires, the 1915 fire occurred in a very dry summer (Figure 9C).

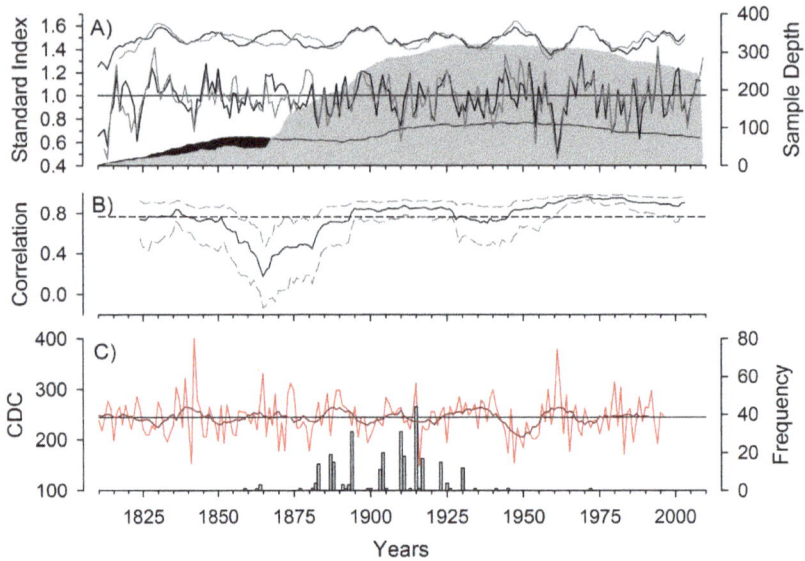

Figure 9. *Pinus banksiana* standard tree-ring chronology and sample depth (**A**) developed from trees inside (gray line) and outside (black line) the LKFMU. The upper lines represent their respective 11-year moving average. Thirty-year running correlations between the two chronologies are plotted with 95% CI (**B**). The dashed line indicates the mean correlation value. Reconstructed Canadian Drought Code (with 11-year moving average) for the Manitoba boreal plains [63] and recorded fire scar within LKFMU (**C**).

3.4. Lake Katherine Charcoal Chronology

The CHAR (Charcoal Accumulation Rate) profile for Lake Katherine indicated a stepwise and substantial reduction in charcoal concentration starting around 1850 (Figure 10A). CHAR from the early record until about 1850 was relatively stable, with a mean count of 100.5 particles (cm^{-2}·year^{-1}). From 1850 to 1930, CHAR decreased to a mean of 32.8 particles and after RMNP establishment (1930), it decreased to a mean of 7.9 particles. After about 1850, both the contribution of grass cuticles and of large charcoal particles (>500 microns) to total CHAR also decreased abruptly (Figure 10B). Over the period analyzed, the number of fire events detected varied according to the threshold used (90th, 95th, and 99th percentiles). The number of fires identified was 9, 7, and 4, respectively (Figure 10A). With five fire intervals detected prior to 1850, the mFRI (90th percentile) was estimated to be 58 years, ranging from 13 to 113 years. Numerous fire events were also detected using the grass morphotype CHAR and the 90th percentile threshold (Figure 10B). The CHAR peak analysis in this study failed to identify the numerous fires that occurred between 1850 and the RMNP establishment (Figure 10A). Between 1850 and 1930, fire events at the 90th percentile were detected in 1858, 1990, and 2002, whereas at the 95th only fires dated to 1858 and 1990 were detected. A fire in 1982 was identified from the grass particle CHAR (Figure 10B). At the 99th percentile, two fire events were detected in the early 1800s. The smooth FRI (600 years smoothing) associated with each percentile indicated long-term FRIs ranging from about 50 to 133 years (Figure 10C).

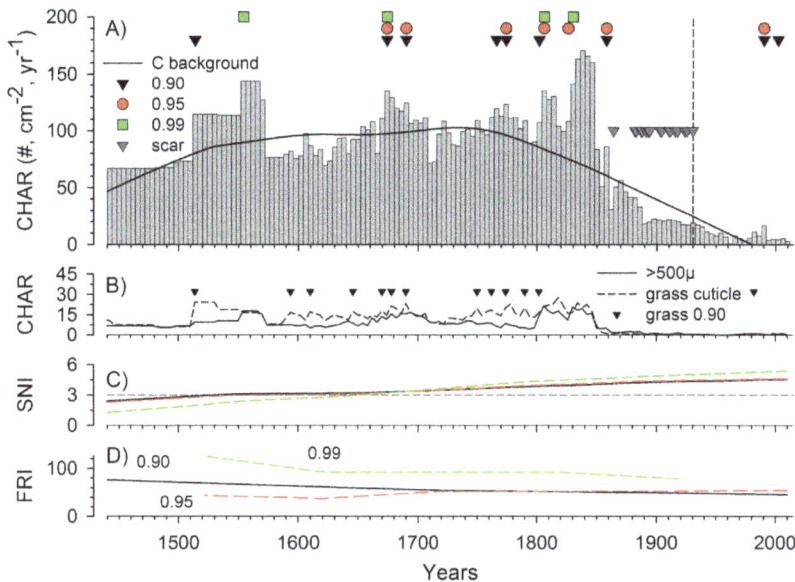

Figure 10. C$_{interpolated}$ (four years) with C$_{background}$ curve using a 300-year window (**A**). The fire events identified using the 0.90, 0.95, and 0.99 thresholds are identified. The gray triangles indicate fire years determined from fire scars occurring in two or more trees. The vertical dashed line indicates the establishment of Riding Mountain National Park. The contributions of grass cuticle morphotype and that of charcoal particles above 500 microns are also indicated, as well as the fire events identified in the former at the 90th percentile (**B**). The signal to noise ratio with SNI = 3 threshold value (short dashed grey line); (**C**) and the fire return intervals (600 years smoothing) associated to each threshold (**D**) are also indicated.

4. Discussion

4.1. LKFMU Fires: European Exploitation/Settlement Period (1850–1930)

The tree-ring evidence suggests that a mixed fire regime may have dominated LKFMU, with dominance of frequent low intensity surface fires during the period 1850–1930. Major recruitment pulses followed two of the largest, most severe fires (1864 and 1894) and periodically followed a number of less severe fires. The absence of repeated and/or widespread fires in the decade following these two fires may have been determinant of successful jack pine establishment. The fire regime during this period had characteristics of both grassland and boreal forest—a transitional regime where ecosystems with frequent light surface fires meet ecosystems with less frequent, high intensity, stand-replacing fires. The LKFMU, due to its position between grasslands and boreal mixedwood, may present characteristics of both regimes [79]. Some of the fire years (e.g., 1864 and 1894) also corresponded to prairie fire years documented by Rannie [80], but it remains unclear whether these would have reached Riding Mountain. During this period, the mFRI was among the shortest documented for jack pine forests (e.g., [18,81–85]).

This period of intense fire activity, however, is not typical of boreal mixedwood and coincided with exploration, the demise of the fur trade, First Nations displacement, European settlement, and forest reserve establishment. The causes for these frequent fires can be traced in the local history and are similar to findings in other areas and parks along the southern edge of the boreal forest [10,11,18,25,26,86]. Conducting forest surveys from 1906 to 1908, Dickson [35] described the fires and resulting forests of the Riding Mountain Forest Reserve. He wrote that "*The damage done to this reserve by fire has been enormous. Large areas have been crossed and recrossed by the most destructive fires. For miles and miles alongside the old Indian trails stretch open prairies and desolated wastes of blackened stumps".* He specifically makes reference to the study area, *"the work of recurrent fires is apparent over large areas, shown particularly in the denuded semi-prairie conditions all about Clear Lake"* and further added that " . . . *fire in the previous 20 years (were) due to careless lumber men, settlers, half-breeds and Indians"* adding that *"ground fire around the outskirts of the reserve, especially those bordering settlement, are annual inflictions"* [35]. Forest rangers also created fire guards along the park boundary, with mention of burning 90 miles (~144 km) of meadows around Riding Mountain in 1911 [87].

These descriptions are sustained by the 1929 aerial photos, indicating that LKFMU was characterized by grassland, aspen parkland, and scattered jack pine stands (Figure 2). Interestingly, despite CHAR peak analysis being unsuitable to capture fires that occurred during this period of very short fire intervals, it corresponded to a drastic decrease in CHAR, suggesting a reduction in biomass burned and/or fire intensity in close proximity to the lake. The few CHAR peaks observed (e.g., 1858) could reflect some of the largest fires that occurred and this within the built-in confidence intervals associated with sediment dating. Given the robust dating of the upper sediment, the low production of charcoal by low-intensity fires (low CHAR values) may have limited the ability to detect fire peaks. Fire event detection from CHAR was reported to work best when large, low-frequency, and high-intensity fires dominate a landscape and also when FRI is at least five times that of the sediment sample resolution [76]. In our case, this would represent fire intervals of about 20 years. Taphonomic processes related to transport and deposition of charcoal may also be involved. Factors like minimal relief and small watershed may have played a role in delaying the delivery of charcoal to Lake Katherine, thereby smoothing the signal associated with low-intensity fires.

Outside LKFMU, the jack pine age structure displayed little similarity with that of the LKFMU. The scarce occurrence of fire scars in these sites suggests the predominance of a crown fire regime. This area was, however, presumably also affected by anthropogenic activities and the logging impacts on these jack pine forests remains unknown. Maps of logging occurrences in the area between 1900 and 1920 exist [20] but little detail was given on species, exact locations, or volumes removed. The high demand for jack pine railway ties in the early 1890s was associated with intensive logging in the southwestern portion of Riding Mountain [20]. No evidence of logging was, however, observed in the field. According to historical accounts, these jack pine stands could have burned in ca. 1830 and 1895 (supported by recruitment centered around these dates) during fires that affected the eastern portion of RMNP [35], and post-fire logging could have removed this evidence. The differences in fire dynamics within and outside LKFMU further suggest that the period 1850–1930 was likely a time of First Nations fire use that transitioned into a time of European fire use. The open forests/grasslands were easier to burn than closed forests and Europeans may have intensified their imprint on an already existing landscape.

4.2. LKFMU Fires: Fire-Exclusion Period (1930 to the Present)

Since the creation of RMNP in 1930, no fire has occurred in the LKFMU [30,36]. Generally the fire-exclusion period corresponds to park establishment, changing land use, end of resource extraction, and park management. This lack of fires also corresponds to a drastic reduction in CHAR. About 200 km southwest of RMNP, a quasi-total elimination of grass fires was also documented following Euro-Canadian settlement [88]. There are a number of possible explanations for the length of the current fire-free interval in LKFMU: (i) due to land use change, fires no longer start in the south and spread northward into the park; (ii) fire prevention and suppression policy within the park has eliminated fire; and (iii) the 20th-century climate may have been less conducive to fire than in the past. While no fires were observed in LKFMU, a number of smaller fires were observed elsewhere in RMNP during the 1930s and 1940s and these were thought to have been deliberately set by farmers [20]. The recent fire record of RMNP (1937–2015) has 382 fires burning a total of 973 km^2 (Cornelsen, unpublished data). Only 7.3% of the fires and 1.4% of the area burned was attributed to lightning fires, with the remainder being human caused. Lightning fire ignitions have been uncommon despite the region receiving an abundance of thunderstorm activity (30 thunderstorm days per year, or 1.5–2.0 lightning strikes/km^2/year) [89]. Despite active fire suppression, large fires have occurred in RMNP in 1940, 1961, and 1980, corresponding to major drought years (Figure 9C). The last significant wildfire occurred in May 1980, when a human-caused fire burned 200 km^2, including most of the parks' mature jack pine stands located outside LKFMU. The CHAR analyses identified a fire in the late 20th century (1982, 1990, or 2002, according to the model) and could reflect a small input from the 1980 fire. This 1980 fire came within 3 km of Lake Katherine and outside its drainage basin, with the lake being downwind for the three days of major fire spreads. Extremely low CHAR background during this period of fire suppression could have enhanced the ability to record fire events, although the signal may be influenced by the addition of even a few dozen charcoal particles that could also be attributed to other factors such as erosion, heavy rain, and sediment delivery.

4.3. LKFMU Fires: Early Period (1450–1850)

The jack pine trees that were established in LKFMU prior to the 1860s did not record any fire scars (or datable ones), but the tree samples for which the pith was present suggest continuous recruitment, with most of these trees dying in the late 19th century. The sparse indications left by jack pine trees that were established in the early 1800s also suggest that fire intervals may have become shorter in the mid-1850s, with trees registering scars at a younger age and a smaller diameter. Supporting this interpretation is the sudden decrease in CHAR from the mid-1800s. The high and relatively stable CHAR values observed during the period 1450–1850 suggests increased fire intensity and/or biomass burned compared to the period 1850–1930. The best estimate of the FRI during this period varies from about 50 to 133 years. Because of the stepwise change in sedimentation rate at the interphase of the two sediment cores, we exercise caution in over-interpreting the increase in CHAR that occurs abruptly at this transition. Interestingly, however, large CHAR peaks occurred in the late 1700s, early 1800s, overlapping the period when Anishinaabe people moved into the Riding Mountain region in the 1700s and were established as the dominant society by the early 1800s [30,32]. We speculate that both the high CHAR values observed during this period as well as the high grass morphotype values may be indicative of recurrent burning to the lake margins, increasing charcoal inputs and background. The practice of burning around lakes, streams, and rivers increases access to numerous plant species and other resources [90]. In Alberta, systematic burning of meadows, river edges, lake shores, and other areas by indigenous peoples was also reported to increase productivity and biodiversity at the landscape level by creating small patchworks of early successional plant communities within later seral stage communities [91].

Little information could be gathered regarding the magnitude of the eco-cultural use of fire by First Nations, Metis, and early white settlers prior to 1850; both uncertainty and controversy exist regarding their impact [1,30,87]. Numerous historical accounts of eco-cultural fires burning in the southeastern Canadian prairies from the latter 18th to the latter 19th century exist [80,88]. Indigenous burnings to promote berry patches may have involved burning on a cycle of 3–5 years, whereas regeneration of aspen and willows around beaver ponds may have been maximized with a 10–12-year fire interval [90,91]. In addition, a substantial number of forest fires in the boreal must have originated from campfires, signal fires, or fires set for the gumming of canoes, hunting, in warfare, or combating insect pests [92]. Numerous examples of the use of prescribed burning by indigenous and white people during the fur trade period in Alberta were also documented [93].

Little information is available describing the use of fire by First Nations in LKFMU. Historical accounts indicate that the occupation of the area by Anishinaabe people started around 1800–1850. In a study of the traditional land and resource use of Riding Mountain, Lake Katherine was often referred to as being a highly valued and visited place [46]. Berries were picked and roots harvested, and people stayed and camped by the lake. In this resource-rich environment, man-made fire was not referred to as a management tool [46]. The author speculated that traditional ecological knowledge of fire usage has been lost given evidence of fire usage by Anishinaabe First Nations [46]. Miller [94] documented the use of controlled fire by the Pikangikum First Nation (Anishinaabe) in northwestern Ontario and the fact that this practice was largely stopped by provincial forest managers in the 1950s, leading to young people having little knowledge of these former fire practices. This pattern appears to be similar to that observed by Tardif (unpublished data), who observed a drastic reduction in fire scars in a disjunct red pine (*Pinus resinosa* Ait.) stand located on Black Island, Manitoba. This reduction in fires coincided with the settlement of the mainland in the late 1870s, the exploitation of Black Island resources, the opening of a pulp mill in Pine Fall in 1927, and the creation of Hecla Provincial Park in 1969. Despite the Hollow Water Anishinaabe referring to the area as a berry picking ground [95], no recollection of past fire usage was communicated.

4.4. Seasonality and Directional Spread of Fire

In LKFMU, the majority of fire scars occurred during the dormant season and the earlywood portion of tree rings. Local observations about cambial growth initiation in jack pine trees from RMNP (Cornelson, unpublished data) and elsewhere [96,97] indicate that production of earlywood tracheids began in late May and that of latewood in mid-July. Fire scars formed in the dormant season may thus be from late April or early May and those in the earlywood from late May or early June. Dormant season fires were assumed to be spring fires due to the relative absence of fires or area burned in the fall in the recent written fire record (Cornelsen, unpublished data). This interpretation is supported by the occurrence of WER, which was often synchronized with fire years, and this anomaly has been associated with crown and/or foliage damage during the dormant season prior to the onset of radial growth [56]. Their high frequency in fire years suggests crown damages associated with low intensity fires prior to the onset of growth. This interpretation is also supported by the absence of WER in jack pine trees growing outside the LKFMU. This is in contrast with the WER of 1943 and 1954, which were also observed outside the LKFMU (and in DMPF and PPF), and were most probably associated with winter and/or spring regional frost damages [56]. Fire data indicate that RMNP has a spring-dominated fire regime, with 6% of fires occurring in April, 90% in May, and 3% in June [36]. In RMNP, the dormant season fires and early spring fires (May) are typically outside of the main lightning season.

The direction associated with the fire scars indicated that in the LKFMU fires mainly spread into the park from the S/SW. This direction of spread is consistent with historical accounts of fires coming off the prairies and up into mountains [34] and the fact that LKFMU is proximate to grasslands, travel corridors, areas of occupation, and hunting/gathering locations of the Anishinaabe First Nations [46].

4.5. LKFMU Fires, Regional Climatic Influences, and Implications for Management

The seasonality results, the recent fire record, and the historical fire record support the suggestion that the majority of the LKFMU fires were human caused. No distinction was made as to exact cause and whether the fire was accidental or purposeful. The period 1850–1930 was characterized by frequent fires and coincided with First Nations fire use transitioning into European settlement and fire use; it is not necessarily typical of boreal mixedwood fire regimes. Interestingly, none of the major fire years corresponded to major light ring years observed in RMNP (1866, 1868, 1873, 1875, 1880, 1885, 1890, 1892, 1907, 1935, and subsequent; [53]), which suggests that cooler-than-average summer temperatures (late spring and late summer) may not have been conductive to natural or anthropic fire ignition. None of the fires were also associated with major flood years of the Red River (1747, 1762, 1826, 1852, 1862, 1950, and 1997; [98]) indicating that heavy snow load in winter and/or a wet spring may also not have been conductive to spring fire ignition. Some of the major fire years (1864, 1894, and/or 1915) were, however, associated with dry conditions [99–101]. Some of these fire years were also observed among others in Saskatchewan [102], Manitoba [25,27], Minnesota [81,103], and Ontario [104–106]. Within central North America, the Hudson Bay Company's archival records for northwestern Ontario indicated that the 1860s marked a transition towards increasing fires up to the end of the 19th century [104].

In LKFMU, the potential implications of the long current fire-free interval (fire exclusion) are greater stand density, accumulation of fuel on the forest floor, encroachment of fire-susceptible species, loss of jack pine and associated species, and increased vulnerability to a stand-replacing fire of high severity and a post-fire vegetation community outside the natural range of variation in reference conditions [7,20,107,108]. Our results indicated that jack pine mortality has been increasing, with the fire-free interval nearing the end of the normal life expectancy of individual trees. In LKFMU, small (young) trees rarely had scars because their cambium may be killed by a passing fire or their foliage killed by crown scorch [51]. Results support this view, with few trees <4.3 cm in basal diameter (<20.6 years) bearing a first fire scar. These findings have implications for the application of prescribed fire treatments. Fire intervals should be short enough to promote regeneration, but sufficiently long to allow survival and continued recruitment [35,79]. Jack pine is known as a prolific cone producer, with trees bearing cones as early as 6–8 years [13,37] and viable seeds being reported in 3–5-year-old stems, with 50% seed germination in 20-year-old stands [16]. Jack pine growing in the open also produced more seeds than in closed stands, where cones may be initiated between 10 and 25 years on average, with the greatest production occurring in 70- to 80-year-old stands [14]. Young jack pine trees also mainly produce non-serotinous cones, with serotinous ones being produced when trees reach a diameter at breast height of about 10 cm [109].

A proposed management prescription for fire restoration in LKFMU would need to adopt a diverse approach with a mixture of intervals, severities, and sizes that create a patchy mosaic. This could be produced with an FRI of 5–10 years, burning only portions of the 30 km^2 study area in any one year, with low-intensity surface fire. This is supported by the presence of scars and of WER, suggesting that many jack pine trees recorded and survived crown scorching without recording scars. Once every 25–30 years (or longer), the application of higher intensity surface fire burning larger areas would create preferential regeneration conditions, assuming an adequate seed bank is present. More frequent surface fire of any intensity could be used to manage fuel loads and increase crown fire resistance, subsequently reducing the risk to adjacent property. Such a prescribed burn program would emulate a dynamic fire regime (historical and anthropogenic) with regards to fire frequency, severity, seasonality, patch size, and extent that can adequately regenerate the jack pine stands and maintain functional ecosystems. Ecological and social objectives could be achieved by restoration of historic fire regimes, even if they are anthropogenic in nature. Such a fire regime would also be in line with current park management objectives. Given the proximity of LKFMU to significant values at risk (Wasagaming and Onanole communities), a frequent, low-intensity, surface fire regime would be safer to manage than an infrequent, high-intensity, crown fire regime (typical of boreal jack pine). From an ecological perspective, the number of small diameter and/or young jack pine trees that survived frequent surface fires until recently indicated successful recruitment.

5. Conclusions

The LKFMU was characterized by short FRIs during the end of First Nations occupation and European settlement period and a long fire exclusion period following the establishment of RMNP. The fire regime in jack pine stands located within LKFMU and at the southern extent of their range indicated that during the period 1850–1930 a surface fire regime predominated with short FRIs. Fire severity was presumably mixed, with two fires (1864 and 1894) accounting for most of the recruitment. Part of LKFMU did burn almost every other year, with the two periods with longer interval corresponding to most of the recruitment. While acknowledging that some of the fires detected may have been of climatic origin, we speculate that the bi-annual nature of many fires may have corresponded with the needs of First Nations or European settlers. Given the age of the jack pine trees, little information is available prior to European settlement, rendering it difficult to assess the use of fire by First Nations.

Despite its limitation in temporal range, the tree-ring data suggested that longer FRIs may have existed in LKFMU prior to the 1850s. In this study, an imperfect calibration existed between fire years derived from tree-ring fire scars and charcoal records. Nonetheless, changes observed in background CHAR have been more revealing than those associated with fire frequency. Prior to 1850, more charcoal biomass was produced per year, with a higher fraction of grass morphotype and of large charcoal particles potentially indicative of more frequent/less distant fires to the lake. Frequent grass/shrub burning at the lake margins intermixed with high-severity fires may provide a different CHAR signature than that associated with more distant burning in jack pine stands. From the tree-ring record, the short FRI and presumably low-intensity fires associated with European settlement corresponded to a major CHAR reduction starting around 1850, which was followed by another CHAR reduction after RMNP establishment. More work will be needed with regards to the identification of fire signatures in a context where total CHAR and/or fire regime have changed. In this study, adopting a sub-sampling strategy of the upper sediment with a finer resolution (less than 1 cm intervals) may have provided a better calibration between the datasets. The addition of a radiocarbon date close to the interphase between the two cores (Glew and Livingston) may also have provided a better estimation of the sedimentation rate. Future pollen quantification from Lake Katherine may also provide needed information regarding the changes in fire regime, allowing us to better decipher the signatures associated with climate and anthropogenic influences. The results of this study provide information about the fire history that has shaped the fire regime of the narrow vegetation community of LKFMU, which transitions from aspen parkland into boreal mixedwood forest. The presence of an island population of jack pine at its southern extent presents an interesting management challenge in that too frequent fires and fire exclusion will both eventually eliminate the species from the region. While historical knowledge and an understanding of the fire regime are of central importance for fire regime restoration, an important question remains: should an era of more frequent cultural fire be included in restoration discussions, either from an ecological or a management perspective [1]? Concurrently, predictions about a warmer climate leading to shorter fire cycles and increasing forest disturbances have been made [110,111], and concerns about climate change and its potential impact on the boreal landscape are increasing worldwide.

Supplementary Materials: The following are available online at http://www.mdpi.com/1999-4907/7/10/219/s1, Figure S1: Age-depth model for Lake Katherine based on 11 ^{210}Pb and two radiocarbon dates. The two sub-figures on the right show the sediment accumulation rate and the sample resolution. The period of analysis presented in this study corresponds to the first 140 cm of sediment.

Acknowledgments: Many thanks go to our field and laboratory assistants: Justin Waito, Elsa Gauthier, Pierre-Antoine Grout, Thomas Amodei, and Brenden Clarke. This research was undertaken, in part, with funding from the Canada Research Chairs Program and Parks Canada. The Natural Sciences and Engineering Research Council of Canada, and the University of Winnipeg also supported this research. We also thank Parks Canada for issuing the permits required to complete the sampling. Thank you to Cliff White and Mark Heathcott for providing inspiration to study fire history, as well as to Reade Tereck and the Riding Mountain Fire Crew members who operated chainsaws and belt sanders. We would also like to thank Reade Tereck, Ross Robinson, and Bob Reside for help with collecting the sediment cores and Terri Lacourse for technical advice. Lastly, we thank the three anonymous reviewers and the associate editor for providing constructive comments on previous drafts of the manuscript.

Author Contributions: This paper amalgamates data from three independent studies: one conducted by Stephen Cornelsen on fire ecology in marginal jack pine populations, another by Jacques C. Tardif and France Conciatori on the dendroclimatology of jack pine, and the last by Eben Blake Hodgin and Marlow G. Pellatt on charcoal accumulation in Lake Katherine sediment. The current study was mainly conceived and designed by both Stephen Cornelsen and Jacques C. Tardif. All authors were involved in sampling. France Conciatori performed most of the dendrochronological laboratory work and was also involved in preparing the manuscript. Eben Blake Hodgin and Marlow G. Pellatt were involved in sampling and laboratory work associated with determination of charcoal accumulation. Jacques C. Tardif was also involved in charcoal data analysis. All authors participated in the writing and reviewing of the manuscript.

Conflicts of Interest: The authors declare no conflict of interest. The founding sponsors had no role in the design of the study; in the collection, analyses, or interpretation of data; in the writing of the manuscript, and in the decision to publish the results.

References

1. White, C.A.; Perrakis, D.D.B.; Kafka, V.G.; Ennis, T. Burning at the edge: Integrating biophysical and eco-cultural fire processes in Canada's parks and protected areas. *Fire Ecol.* **2011**, *7*, 74–106. [CrossRef]
2. Ward, P.C.; Mawdsley, W. Fire management in the boreal forests of Canada. In *Fire, Climate Change, and Carbon Cycling in the Boreal Forest*; Kasischke, E.S., Stocks, B.J., Eds.; Springer: New York, NY, USA, 2000; pp. 274–288.
3. Weber, M.G.; Taylor, S.W. The use of prescribed fire in the management of Canada's forested lands. *For. Chron.* **1992**, *68*, 324–334. [CrossRef]
4. Shay, J.; Kunec, D.; Dyck, B. Short-term effects of fire frequency on vegetation composition and biomass in mixed prairie in south-western Manitoba. *Plant Ecol.* **2001**, *155*, 157–167. [CrossRef]
5. Bleho, B.I.; Koper, N.L.; Borkowsky, C.; Hamel, C.D. Effects of weather and land management on the western prairie fringed-orchid (Platanthera praeclara) at the northern limit of its range in Manitoba, Canada. *Am. Midl. Nat.* **2015**, *174*, 191–203. [CrossRef]
6. Sims, H.P. The effect of prescribed burning on some physical soil properties of jack pine sites in southeastern Manitoba. *Can. J. For. Res.* **1976**, *6*, 58–68. [CrossRef]
7. Weber, M.G.; Stocks, B.J. Forest fires and sustainability in the boreal forests of Canada. *Ambio* **1998**, *27*, 545–550.
8. Parks Canada. *Riding Mountain National Park of Canada: Management Plan*; Parks Canada: Wasagaming, MB, Canada, 2007.
9. Parks Canada. *National Fire Management Plan*; Parks Canada Agency: Ottawa, ON, Canada, 2008.
10. Weir, J.M.H.; Johnson, E.A. Effects of escaped settlement fires and logging on forest composition in the mixedwood boreal forest. *Can. J. For. Res.* **1998**, *28*, 459–467. [CrossRef]
11. Grenier, D.J.; Bergeron, Y.; Kneeshaw, D.; Gauthier, S. Fire frequency for the transitional mixedwood forest of Timiskaming, Québec, Canada. *Can. J. For. Res.* **2005**, *35*, 656–666. [CrossRef]
12. Pellatt, M.G.; Gedalof, Z. Environmental change in Garry oak (*Quercus garryana*) ecosystems: The evolution of an eco-cultural landscape. *Biodivers. Conserv.* **2014**, *23*, 2053–2067. [CrossRef]
13. Rudolph, T.D.; Laidly, P.R. *Pinus banksiana* Lamb. jack pine. *Silvics of North America: Volume 1. Conifers*; Burns, R.M., Honkala, B.H., Eds.; Forest Service: Washington, DC, USA, 1990; pp. 280–293.
14. Sims, R.A.; Kershaw, H.M.; Wickware, G.M. *The Autecology of Major Tree Species in the North Central Region of Ontario*; COFRDA Report 3302; Forestry Canada: Sault Ste. Marie, Ottawa, ON, Canada, 1990.
15. Robson, J.R.M.; Conciatori, F.; Tardif, J.C.; Knowles, K. Tree-ring response of jack pine and scots pine to budworm defoliation in central Canada. *For. Ecol. Manag.* **2015**, *347*, 83–95. [CrossRef]
16. Cayford, J.H.; McRae, D.J. The role of fire in jack pine forests. In *The Role of Fire in Northern Circumpolar Ecosystems*; Wein, R.W., MacLean, D.A., Eds.; John Wiley & Sons: New York, NY, USA, 1983; pp. 183–199.
17. Gauthier, S.; Bergeron, Y.; Simon, J.P. Effects of fire regime on the serotiny level of jack pine. *J. Ecol.* **1996**, *84*, 539–548. [CrossRef]
18. Simard, A.J.; Blank, R.W. Fire history of a Michigan jack pine forest. *Mich. Acad.* **1982**, *15*, 59–71.
19. Desponts, M.; Payette, S. Recent dynamics of jack pine and its northern distribution limit in northern Québec. *Can. J. Bot.* **1992**, *70*, 1157–1167. [CrossRef]
20. Bailey, R.H. *Notes on the Vegetation in Riding Mountain National Park Manitoba*; National Parks Forest Survey No. 2.; Department of Forestry and Rural Development, Forest Management Institute: Ottawa, ON, Canada, 1968.
21. Abrams, M.D. Uneven-aged jack pine in Michigan. *J. For.* **1984**, *82*, 306–307.
22. Conkey, L.E.; Keifer, M.; Lloyd, A.H. Disjunct jack pine (*Pinus banksiana* Lamb.) structure and dynamics, Acadia National Park, Maine. *Écoscience* **1995**, *2*, 168–176.
23. Bergeron, Y.; Brisson, J. Fire regime in red pine stands at the northern limit of the species' range. *Ecology* **1990**, *71*, 1352–1364. [CrossRef]
24. Grotte, K.L.; Heinricks, D.K.; Tardif, J.C. Old-growth characteristics of disjunct Thuja occidentalis stands at their northwestern distribution limit, central Canada. *Nat. Areas J.* **2012**, *32*, 270–282. [CrossRef]
25. Tardif, J. *Fire History in the Duck Mountain Provincial Forest, Western Manitoba. Sustainable Forest Management Network*; University of Alberta: Edmonton, AB, Canada, 2004.

26. Weir, J.M.H.; Johnson, E.A.; Miyanishi, K. Fire frequency and the spatial age mosaic of the mixed-wood boreal forest in western Canada. *Ecol. Appl.* **2000**, *10*, 1162–1177. [CrossRef]

27. Harrison, J.D.B. *The Forests of Manitoba*; Forest Service Bulletin 85; Department of the Interior: Ottawa, ON, Canada, 1934.

28. Stevenson, H.I. *The Forests of Manitoba*; Report No. 9; Manitoba Economic Survey Board: Winnipeg, MB, Canada, 1938.

29. Gill, C.B. Cyclic forest phenomena. *For. Chron.* **1930**, *6*, 42–56. [CrossRef]

30. Barlow, J.L. *Riding Mountain National Park Resource Description and Analysis*; Natural Resource Conservation Prairie Region, Parks Canada: Ottawa, ON, Canada, 1979.

31. Caners, R.T.; Kenkel, N.C. Forest stand structure and dynamics at Riding Mountain National Park, Manitoba, Canada. *Community Ecol.* **2003**, *4*, 185–204. [CrossRef]

32. Canadian Parks and Wilderness Society (CPAWS). *Riding Mountain Ecosystem Community Atlas*; Canadian Parks and Wilderness Society (CPAWS): Winnipeg, MB, Canada, 2004.

33. Rowe, J.S. *Factors Influencing White Spruce Reproduction in Manitoba and Saskatchewan*; Forestry Branch Tech. Note No. 3; Department of Northern Affairs and National Resources: Ottawa, ON, Canada, 1955.

34. Sentar Consultants Ltd. *Riding Mountain National Park: A Literature Review of Historic Timber Harvesting and Forest Fire Activity*; Parks Canada: Wasagaming, MB, Canada, 1992.

35. Dickson, J.R. *The Riding Mountain Forest Reserve*; Forestry Branch Bull. No. 6; Department of the Interior: Ottawa, ON, Canada, 1909.

36. Cornelsen, S.; Vanderschuit, W. *Riding Mountain National Park Fire Management Plan*; Unpublished Report; Parks Canada: Wasagaming, MB, Canada, 2002.

37. Tunstell, G.; Gill, C.B.; Kuhring, G.F. *Silvical Report on Riding and Duck Mountain Forest Reserves*; Unpublished Department of Interior Report; Canada Forest Service: Ottawa, ON, Canada, 1922.

38. Parks Canada. *Parks Canada Policy*; Publication QS-7079-000-EE-A1; Parks Canada: Ottawa, ON, Canada, 1979.

39. Rowe, J.S. *Forest Regions of Canada*; Publication No. 1300; Department of Environment, Canadian Forest Service: Ottawa, ON, Canada, 1972.

40. RSmith, E.; Veldhuis, H.; Mills, G.F.; Eilers, R.G.; Fraser, W.R.; Lelyk, G.W. *Terrestrial Ecozones, Ecoregions, and Ecodistricts, an Ecological Stratification of Manitoba's Landscapes*; Technical Bulletin 98-9E; Land Resource Unit, Brandon Research Centre, Research Branch, Agriculture and Agri-Food Canada: Winnipeg, MB, Canada, 1998.

41. Ritchie, J.C. Absolute pollen frequencies and carbon-14 age of a section of Holocene lake sediment from the Riding Mountain area of Manitoba. *Can. J. Bot.* **1969**, *47*, 1345–1349. [CrossRef]

42. Zoltai, S. *Southern Limit of Coniferous Trees on the Canadian Prairies*; Information Report NOR-X-128; Environment Canada, Forestry Service, Northern Forest Research Center: Edmonton, AB, Canada, 1975.

43. Environment Canada. *Canadian Climate Normals and Averages 1981–2010*; National Climate Archives; Meteorological Service of Canada, Environment Canada: Toronto, ON, Canada, 2016. Available online: http://climate.weatheroffice.ec.gc.ca/climate_normals/index_e.html (accessed on 24 June 2016).

44. Tabulenas, D.T. *A Narrative Human History of Riding Mountain National Park and Area: Prehistory to 1980*; Parks Canada: Wasagaming, MB, Canada, 1983.

45. Nicholson, B.A. Human Ecology and Prehistory of the Forest/Grassland Transition Zone of Western Manitoba. Ph.D. Thesis, Department of Archaeology, Simon Fraser University, Vancouver, BC, Canada, 1987.

46. Peckett, M.K. Anishnabe Homeland History: Traditional Land and Resources Use of Riding Mountain, Manitoba. Master's Thesis, University of Manitoba, Winnipeg, MB, Canada, 1999.

47. Sandlos, J. Not wanted in the boundary: The expulsion of the Keeseekoowenin Ojibway band from Riding Mountain National Park. *Can. Hist. Rev.* **2008**, *89*, 189–221. [CrossRef]

48. Arno, S.F.; Sneck, K.M. *A method for Determining Fire History in Coniferous Forests in the Mountain West*; General Technical Report INT-42; U.S.D.A. Forest Service, Inter Mountain Forest and Range Experimental Station: Fort Collins, CO, USA, 1977.

49. McBride, J.R. Analysis of tree rings and fire scars to establish fire history. *Tree-Ring Bull.* **1983**, *43*, 51–67.

50. Johnson, E.A.; Gutsell, S.L. Fire frequency models, methods, and interpretations. *Adv. Ecol. Res.* **1994**, *25*, 239–287.

51. Gutsell, S.L.; Johnson, E.A. How fire scars are formed: Coupling a disturbance process to its ecological effect. *Can. J. For. Res.* **1996**, *26*, 166–174. [CrossRef]

52. Falk, D.A.; Heyerdahl, E.K.; Brown, P.M.; Farris, C.; Fulé, P.Z.; McKenzie, D.; Swetnam, T.W.; Taylor, A.H.; van Horne, M.L. Multi-scale controls of historical forest-fire regimes: New insights from fire-scar networks. *Front. Ecol. Environ.* **2011**, *9*, 446–454. [CrossRef]

53. Tardif, J.C.; Girardin, M.P.; Conciatori, F. Light rings as bioindicators of climate change in Interior North America. *Glob. Planet. Chang.* **2011**, *79*, 134–144. [CrossRef]

54. Madany, M.H.; Swetnam, T.W.; West, N.E. Comparison of two approaches for determining fire dates from tree scars. *For. Sci.* **1982**, *28*, 856–861.

55. Baisan, C.H.; Swetnam, T.W. Fire history of a desert mountain range: Rincon Mountain wilderness, Arizona, U.S.A. *Can. J. For. Res.* **1990**, *20*, 1559–1569. [CrossRef]

56. Waito, J.; Conciatori, F.; Tardif, J.C. Frost rings and white earlywood rings in Picea mariana trees from the boreal plains, central Canada. *IAWA J.* **2013**, *34*, 71–87. [CrossRef]

57. Holmes, R.L. Computer-assisted quality control in tree-ring dating and measurement. *Tree-Ring Bull.* **1983**, *43*, 69–78.

58. Cook, E.R. A Time Series Approach to Tree-Ring Standardization. Ph.D. Thesis, University of Arizona, Tucson, AZ, USA, 1985.

59. Brewer, P.W.; Velásquez, M.E.; Sutherland, E.K.; Falk, D.A. *Fire History Analysis and Exploration System (FHAES)*, version 2.0.1. Available online: http://www.fhaes.org (accessed on 2 May 2016). [CrossRef]

60. Grissino-Mayer, H.D. FHX2- Software for analyzing temporal and spatial patterns in fire regimes from tree rings. *Tree-Ring Res.* **2001**, *57*, 115–124.

61. Grissino-Mayer, H.D. Modeling fire interval data from the American southwest with the Weibull distribution. *Int. J. Willdland Fire* **1999**, *9*, 37–50. [CrossRef]

62. Lombardo, K.J.; Swetnam, T.W.; Baisan, C.H.; Borchert, M.I. Using bigcone Douglas-fir fire scars and tree rings to reconstruct interior chaparral fire history. *Fire Ecol.* **2009**, *5*, 35–56. [CrossRef]

63. Girardin, M.P.; Tardif, J.C.; Flannigan, M.; Bergeron, Y. Synoptic-scale atmospheric circulation and boreal Canada summer drought variability of the past three centuries. *J. Clim.* **2006**, *19*, 1922–1947. [CrossRef]

64. Kooyman, A.H. *The Aquatic Resources of Riding Mountain National Park. Volume III: Data on Lakes*; Canadian Wildlife Services: Winnipeg, MB, Canada, 1980.

65. Glew, J. A portable extruding device for close interval sectioning of unconsolidated core samples. *J. Paleolimnol.* **1988**, *1*, 235–239. [CrossRef]

66. Brossier, B.; Oris, F.; Finsinger, W.; Asselin, H.; Bergeron, Y.; Ali, A.A. Using tree-ring records to calibrate peak detection in fire reconstructions based on sedimentary charcoal records. *Holocene* **2014**, *24*, 635–645. [CrossRef]

67. PHiguera, E.; Whitlock, C.; Gage, J.A. Linking tree-ring and sediment-charcoal records to reconstruct fire occurrence and area burned in subalpine forests of Yellowstone National Park, USA. *Holocene* **2010**, *21*, 327–341. [CrossRef]

68. Wright, H.E.; Mann, D.H.; Glaser, P.H. Piston corers for peat and lake sediments. *Ecology* **1984**, *65*, 657–659. [CrossRef]

69. Björk, S.; Wohlfarth, B. [14]C Chronostratigraphic techniques in paleolimnology. In *Tracking Environmental Change Using Lake Sediments Volume 1: Basin Analysis, Coring and Chronological Techniques*; Last, W.M., Smol, J.P., Eds.; Kluwer Academic Publishers: Dordrecht, The Netherlands, 2001; pp. 205–245.

70. Stuiver, M.; Reimer, P.J. Extended (super 14) C data base and revised CALIB 3.0 (super 14) C age calibration program. *Radiocarbon* **1993**, *35*, 215–230. [CrossRef]

71. Reimer, P.J.; Bard, E.; Bayliss, A.; Beck, J.W.; Blackwell, P.G.; Ramsey, C.B.; Buck, C.E.; Cheng, H.; Edwards, R.L.; Friedrich, M.; et al. Intcal13 and marine13 radiocarbon age calibration curves 0–50,000 years cal BP. *Radiocarbon* **2013**, *55*, 1869–1887. [CrossRef]

72. Higuera, P.E. *MCAgeDepth 0.1: Probabilistic Age-Depth Models for Continuous Sediment Records. User's Guide*; Montana State University: Bozeman, MT, USA, 2008.

73. Schlachter, K.J.; Horn, S.P. Sample preparation methods and replicability in macroscopic charcoal analysis. *J. Paleolimnol.* **2009**, *44*, 701–708. [CrossRef]

74. Jensen, K.; Lynch, E.A.; Calcote, R.; Hotchkiss, S.C. Interpretation of charcoal morphotypes in sediments from Ferry Lake, Wisconsin, USA: Do different plant fuel sources produce distinctive charcoal morphotypes? *Holocene* **2007**, *17*, 907–915. [CrossRef]
75. Higuera, P.E. *CharAnalysis 0.9: Diagnostic and Analytical Tools for Sediment Charcoal Analysis: User's Guide;* Montana State University: Bozeman, MT, USA, 2009.
76. Higuera, P.E.; Gavin, D.G.; Bartlein, P.J.; Hallett, D.J. Peak detection in sediment-charcoal records: Impacts of alternative data analysis methods on fire-history interpretations. *Int. J. Wildland Fire* **2010**, *19*, 996–1014. [CrossRef]
77. Kelly, R.F.; Higuera, P.E.; Barrett, C.M.; Hu, F.S. A signal-to-noise index to quantify the potential for peak detection in sedimentcharcoal records. *Quat. Res.* **2011**, *75*, 11–17. [CrossRef]
78. Van Wagner, C.E. *Development and Structure of the Canadian Forest Fire Weather Index System;* Forestry Technical Report 35; Canadian Forestry Servive: Ottawa, ON, Canada, 1987.
79. Stambaugh, M.C.; Guyette, R.P.; Godfrey, R.; McMurry, E.R.; Marschall, J.M. Fire, drought, and human history near the western terminus of the Cross Timbers, Wichita Mountains, Oklahoma, USA. *Fire Ecol.* **2009**, *52*, 51–65. [CrossRef]
80. Rannie, W.F. Awful splendour: Historical accounts of prairie fire in southern Manitoba prior to 1870. *Prairie Forum* **2001**, *26*, 17–45.
81. Heinselman, M.L. Fire in the virgin forests of the Boundary Waters Canoe Area, Minnesota. *Quat. Res.* **1973**, *3*, 329–382. [CrossRef]
82. Heinselman, M.L. Fire intensity and frequency as factors in the distribution and structure of northern ecosystems. In Proceedings of the Conference Fire Regimes and Ecosystem Properties, Honolulu, HI, USA, 11–15 December 1981.
83. Lynham, T.J.; Stocks, B.J. The natural fire regime of an unprotected section of the boreal forest in Canada. In *High Intensity Fire in Wildlands: Management Challenges and Options*, Proceedings of the 17th Tall Timbers Fire Ecology Conference, Tallahassee, FL, USA, 18–21 May 1989.
84. Larsen, C.P.S.; MacDonald, G.M. Fire and vegetation dynamics in a jack pine and black spruce forest reconstructed using fossil pollen and charcoal. *J. Ecol.* **1998**, *86*, 815–828. [CrossRef]
85. Cleland, D.T.; Crow, T.R.; Saunders, S.C.; Dickmann, D.I.; Maclean, A.L.; Jordan, J.K.; Watson, R.L.; Sloan, A.M.; Brosofske, K.D. Characterizing historical and modern fire regimes in Michigan (USA): A landscape ecosystem approach. *Landsc. Ecol.* **2004**, *19*, 311–325. [CrossRef]
86. Lefort, P.; Gauthier, S.; Bergeron, Y. The influence of fire weather and land use on the fire activity of the Lake Abitibi area, eastern Canada. *For. Sci.* **2003**, *49*, 509–521.
87. Pyne, S.J. *Awful Splendour: A Fire History of Canada;* UBC Press: Vancouver, BC, Canada, 2008.
88. Boyd, M. Identification of anthropogenic burning in the paleoecological record of the northern prairies: A new approach. *Ann. Assoc. Am. Geogr.* **2002**, *92*, 471–487. [CrossRef]
89. Burrows, W.R.; King, P.; Lewis, P.J.; Kochtubajda, B.; Snyder, B.; Turcotte, V. Lightning occurrence patterns over Canada and adjacent United States from lighning detection network observations. *Atmos. Ocean* **2002**, *40*, 59–81. [CrossRef]
90. Roy-Denis, C. Fire for well-being: Use of prescribed burning in the northern boreal forest. *Earth Common J.* **2015**, *5*, 40–50.
91. Lewis, H.T. *A Time for Burning, Edmonton: Boreal Institute for Northern Studies;* University of Alberta: Edmonton, AB, Canada, 1982.
92. Lutz, H.J. *Aboriginal Man and White Man as Historical Causes of Fires in the Boreal Forest, with Particular Reference to Alaska;* School of Forestry Bulletin No 65; Yale University: New Haven, CT, USA, 1959.
93. Ferguson, T.A. "Careless fires" and "smoaky weather": The documentation of prescribed burning in the Peace–Athabasca trading post journals 1818–1899. *For. Chron.* **2011**, *87*, 414–419. [CrossRef]
94. Miller, A.M. Living with Boreal Forest Fires: Anishinaabe Perspectives on Disturbance and Collaborative Forestry Planning, Pikangikum First Nation, Northwestern Ontario. Ph.D. Thesis, Natural Resource and Environmental Management, University of Manitoba, Winnipeg, MB, Canada, 2010.
95. Raven, G.; Hollow Water First Nation, Lake Winnipeg, MB, Canada. Personal communication, 2004.
96. Johnson, E.A.; Miyanishi, K.; O'Brien, N. Long-term reconstruction of the fire season in the mixedwood boreal forest of Western Canada. *Can. J. Bot.* **1999**, *77*, 1185–1188. [CrossRef]

97. Heinrichs, D.K.; Tardif, J.C.; Bergeron, Y. Xylem production in six tree species growing on an island in the boreal forest region of western Québec, Canada. *Can. J. Bot.* **2007**, *85*, 518–525. [CrossRef]
98. George, S.S.; Nielsen, E. Palaeoflood records for the Red River, Manitoba, Canada derived from anatomical tree-ring signatures. *Holocene* **2003**, *13*, 547–555. [CrossRef]
99. Hill, R.B. *Manitoba: History of Its Early Settlement, Development and Resources*; William Briggs: Toronto, ON, Canada, 1890.
100. Sauchyn, D.J.; Skinner, W.R. A proxy record of drought severity for the southwestern Canadian plains. *Can. Water Res.* **2001**, *26*, 253–272. [CrossRef]
101. George, S.S.; Meko, D.M.; Girardin, M.P.; Macdonald, G.M.; Nielsen, E.; Pederson, G.T.; Sauchyn, D.J.; Tardif, J.C.; Watson, E. The tree-ring record of drought on the Canadian prairies. *J. Clim.* **2009**, *22*, 689–710. [CrossRef]
102. Johnson, E.A.; Miyanishi, K.; Weir, J.M.H. Wildfires in the western Canadian boreal forest: Landscape patterns and ecosystem management. *J. Veg. Sci.* **1998**, *9*, 603–610. [CrossRef]
103. Corson, C.W.; Allison, J.H.; Cheyney, E.G. Factors controlling forest types on the Cloquet forest, Minnesota. *Ecology* **1929**, *10*, 112–125. [CrossRef]
104. Fritz, R.; Suffling, R.; Younger, T.A. Influence of fur trade, famine and forest fires on moose and woodland caribou populations in northwestern Ontario from 1786 to 1911. *Environ. Manag.* **1993**, *17*, 477–489. [CrossRef]
105. Cwynar, L.C. The recent fire history of Barron Township, Algonquin Park. *Can. J. Bot.* **1977**, *55*, 1524–1538. [CrossRef]
106. Burgess, D.M.; Methven, I.R. *The Historical Interaction of Fire, Logging and Pine: A Case Study at Chalk River, Ontario*; Information Report PS-X-66; Canadian Forestry Service, Petawawa Forest Experiment Station: Chalk River, ON, Canada, 1977.
107. Keane, R.E.; Ryan, K.C.; Veblen, T.T.; Allen, C.D.; Logan, J.; Hawkes, B. *Cascading Effects of Fire Exclusion in the Rocky Mountain Ecosystems: A Literature Review*; General Technical Report RMRSGTR-91; U.S. Department of Agriculture: Fort Collins, CO, USA, 2002.
108. Fulé, P.Z. Does it make sense to restore wildland fire in changing climate? *Restor. Ecol.* **2008**, *16*, 526–531. [CrossRef]
109. Gaulhier, S.; Bergeron, Y.; Simon, J.P. Cone serotiny in jack pine: Ontogenetic, positional, and environmental effects. *Can. J. For. Res.* **1993**, *23*, 394–401.
110. Girardin, M.P.; Ali, A.A.; Carcaillet, C.; Gauthier, S.; Hély, C.; le Goff, H.; Terrier, A.; Bergeron, Y. Fire in managed forests of eastern Canada: Risks and options. *For. Ecol. Manag.* **2012**, *294*, 238–249. [CrossRef]
111. Wotton, B.M.; Nock, C.A.; Flannigan, M.D. Forest fire occurrence and climate change in Canada. *Int. J. Wildland Fire* **2010**, *19*, 253–271. [CrossRef]

© 2016 by the authors. Licensee MDPI, Basel, Switzerland. This article is an open access article distributed under the terms and conditions of the Creative Commons Attribution (CC BY) license (http://creativecommons.org/licenses/by/4.0/).

Article

Fire Regimes of Remnant Pitch Pine Communities in the Ridge and Valley Region of Central Pennsylvania, USA

Joseph M. Marschall [1,*], Michael C. Stambaugh [1], Benjamin C. Jones [2], Richard P. Guyette [1], Patrick H. Brose [3] and Daniel C. Dey [3]

[1] Missouri Tree-Ring Laboratory, School of Natural Resources, University of Missouri, 203 ABNR Building, Columbia, MO 65211, USA; stambaughm@missouri.edu (M.C.S.); guyetter@missouri.edu (R.P.G.)
[2] Habitat Planning and Development Division, Pennsylvania Game Commission, Harrisburg, PA 17110, USA; benjjones@pa.gov
[3] U.S. Forest Service, Northern Research Station, Newtown Square, PA 19073, USA; pbrose@fs.fed.us (P.H.B.); ddey@fs.fed.us (D.C.D.)
* Correspondence: marschallj@missouri.edu; Tel.: +1-573-884-9262

Academic Editors: Sylvie Gauthier and Yves Bergeron
Received: 29 July 2016; Accepted: 26 September 2016; Published: 2 October 2016

Abstract: Many fire-adapted ecosystems in the northeastern U.S. are converting to fire-intolerant vegetation communities due to fire suppression in the 20th century. Prescribed fire and other vegetation management activities that increase resilience and resistance to global changes are increasingly being implemented, particularly on public lands. For many fire-dependent communities, there is little quantitative data describing historical fire regime attributes such as frequency, severity, and seasonality, or how these varied through time. Where available, fire-scarred live and remnant trees, including stumps and snags, offer valuable insights into historical fire regimes through tree-ring and fire-scar analyses. In this study, we dated fire scars from 66 trees at two sites in the Ridge and Valley Province of the Appalachian Mountains in central Pennsylvania, and described fire frequency, severity, and seasonality from the mid-17th century to 2013. Fires were historically frequent, of low to moderate severity, occurred mostly during the dormant season, and were influenced by aspect and topography. The current extended fire-free interval is unprecedented in the previous 250–300 years at both sites.

Keywords: Pennsylvania; dendrochronology; fire scars; fire severity; humans

1. Introduction

Natural community restoration is of increasing interest to land managers and scientists throughout the U.S. [1]. Many of these communities are fire dependent and have greatly declined in area due to decades of fire suppression. Prescribed fire and other vegetation management activities (e.g., commercial and non-commercial forest cuttings, mowing/mulching, chemical treatments) are increasingly being applied to increase resilience and resistance to global changes, particularly on public lands [2]. In forests, these activities influence succession and are often applied with the goal of reducing tree density and promoting early successional species and native seedbanks [3]. An understanding of historical ecology, past environmental changes, and long-term ecosystem dynamics is fundamental to such management, and thus provides a reasoned scientific foundation for developing a restoration context [4–6].

Fire disturbances are integral to creating and sustaining diverse ecosystems throughout the eastern U.S. [7–10]. Many fire-dependent ecosystems in the eastern U.S. have remained unburned for nearly a century due to fire-suppression policies initiated in the early 20th century [11–14]. As a result of this current era of fire suppression, fire-dependent plant species and communities are declining in abundance and failing to regenerate throughout the eastern U.S. [15–19], leading to regional losses in biodiversity and habitat quality [10,20–24].

Restoring fire regimes can create unique communities and species assemblages across taxonomic levels of plants and animals in eastern U.S. ecosystems [3,7,10,25,26]. Restoration of historically fire-dependent ecosystems is essential for creating or improving wildlife habitats [27–29]. There is little quantitative data describing historic fire regime attributes such as frequency, severity, and seasonality, or how these varied through time and across regions and habitats in these ecosystems [30]. Defining such fire regime characteristics necessitates regionally-based information because fire regimes vary due to factors such as topography, climate, vegetation, and ignition sources [31,32].

Recurring fires were historically important for maintaining fire-dependent communities in the eastern U.S., particularly pine and/or oak forests, woodlands, and scrublands [33–36]. For the region of interest in this study, the Ridge and Valley Province of central Pennsylvania, the widespread exclusion of fire is deemed as a "major threat" to key habitats for species of greatest conservation need in Pennsylvania's State Wildlife Action Plan [28]. Under its mandate to provide hunting opportunity and conserve wildlife habitat, the Pennsylvania Game Commission (PGC) is embarking on landscape-scale habitat restoration using prescribed fire as a primary tool [37]. Success of these ecological restoration efforts will be enhanced by a greater understanding of the fire regime conditions that maintained functionality of these ecosystems [38].

In this study, we used dated fire scars on cross-sections of dead and live pitch pine (*Pinus rigida*) trees to describe the historical and current fire regimes of two sites in the Ridge and Valley Province of central Pennsylvania. Dating fire scars on old or preserved trees is recognized as one of the best methods to reconstruct fire regimes. Despite little use in previous studies, pitch pine is ideal for reconstructing historical fire events because of its longevity, ability to survive and record multiple fires, and potential for preservation after death due to high resin content. Multiple characteristics of pitch pine indicate that it is a fire-adapted tree species, including basal and epicormic sprouting ability, thick bark, and cone serotiny [39–42].

The objectives of this research were to: (1) quantify historical fire regime characteristics; (2) identify associations among fire frequency/occurrences and major fire environment factors (human populations, climate conditions, drought, topography); and (3) discuss the implications of these findings to restoration efforts and fire and vegetation management.

2. Materials and Methods

2.1. Study Site Descriptions

From field reconnaissance conducted during 2010 and 2011, two study sites were identified based on the presence of fire-scarred pitch pine remnants (stumps and dead trees). These sites are located in the Appalachian Mountain Section of the Ridge and Valley Province [43] of Juniata and Perry Counties, Pennsylvania, USA (Figure 1). In this region, annual mean precipitation is between 107 and 112 cm, and mean annual temperature is 12 °C (source: NOAA). Mean annual snowfall ranges from 79 to 102 cm and occurs from mid-October to late April (source: NOAA). This region is characterized by long paralleling ridges and adjacent valleys (trending west-southwest, approximately 250°) and is currently comprised of a mixture of agricultural lands, forested ridges, and rural communities.

Study sites are separated 25 km north to south by the Juniata River valley on lands owned and managed by PGC (State Game Lands 088 and 107, hereafter SGL 088 and SGL 107). The SGL 088 study site (40°27'17.8" N, 77°25'26.4" W) is located on the south side of the Juniata River on ridgetop, opposing shoulder, and mid-slope positions of Tuscarora Mountain, straddling the boundary of Perry and Juniata Counties (Figure 1). The topography is characterized as a relatively flat ridgetop, approximately 320 m wide, running SW/NE and providing slopes with NW and SE aspects. The SGL

107 study site (40°40′21.8″ N, 77°19′52.0″ W) is located on the opposite (northern) side of the Juniata River in Juniata County. The site is positioned on a sub-apex bench on the south-facing slope of Shade Mountain, at the top of a minor drainage (Laurel Run). According to PGC records, the lands on which the study sites occur were acquired by PGC in the early 1930s, and both sites are comprised of closed-canopy forests that initiated circa 1900.

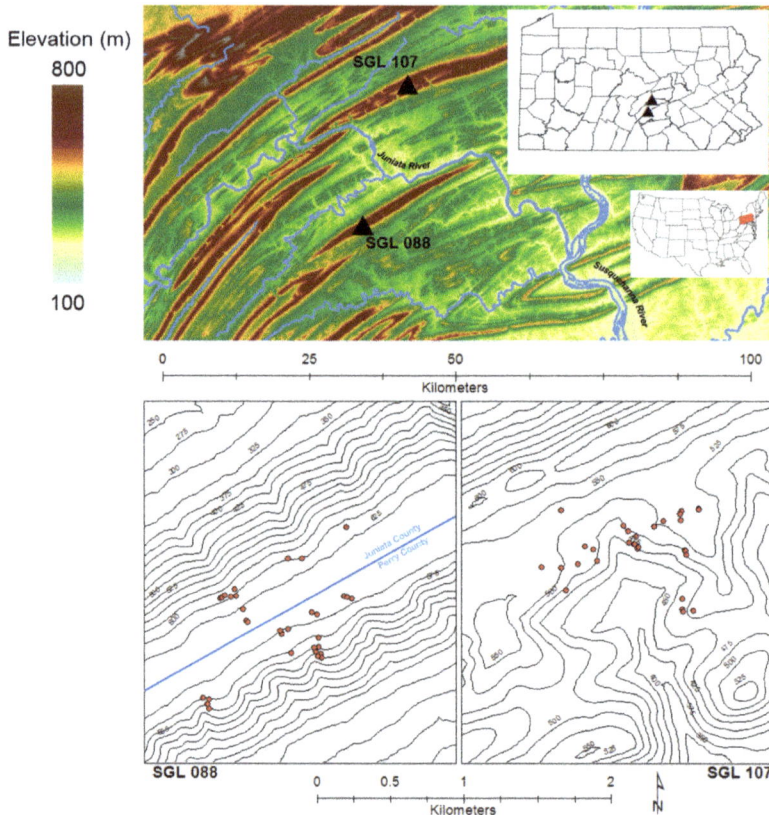

Figure 1. Top panel: Maps showing the two fire history study sites (black triangles) in relation to Pennsylvania (top inset map showing counties), the contiguous United States (bottom inset showing states, Pennsylvania in red), and the regional/local topography. Bottom panels: Topographic maps (25 m contour interval) of the two fire history sites; red dots indicate the locations of fire-scarred pitch pine remnants included in the fire-scar analysis. For SGL 088 topographic map, the blue line marks the Juniata/Perry County boundary, separating samples for analysis by aspect.

Both mountains are capped by resistant Tuscarora Formation quartzite and rise approximately 500 m above the Juniata River valley [44]. Both sites were classified by PGC's ecological classification system as Dry Oak—Heath Forests, noted to occur on xeric, acidic soils with the forest overstories typically dominated by chestnut (*Quercus montana*), black (*Q. velutina*), scarlet (*Q. coccinea*), and white (*Q. alba*) oak, along with other species, including black gum (*Nyssa sylvatica*) and sweet birch (*Betula lenta*) [45]. Based on site inventory data, the forest overstories of both sites are currently fully stocked and dominated by chestnut oak and black gum; 67% and 68% combined for SGL 088 and SGL 107, respectively. At SGL 088, the remaining composition is sweet birch (12%), red maple (*Acer rubrum*, 10%), black oak (7%), and 2% or less of eastern hemlock (*Tsuga canadensis*), serviceberry

(*Amelanchier arborea*), and sassafras (*Sassafras albidum*). The remaining composition at SGL 107 is occupied by northern red oak (*Quercus rubra*, 11%), red maple (10%), white oak (10%), and eastern hemlock (2%). Very minor components of pitch and white pine (*Pinus strobus*), and American chestnut (*Castanea dentata*) were observed at both sites. In PGC's classified Dry Oak—Heath Forests, shrub layers are commonly dominated by ericaceous species, including mountain laurel (*Kalmia latifolia*) and other genera (e.g., *Gaylussacia, Vaccinium*) at such densities that the resulting herbaceous layer is sparse and the primary ground cover is leaf litter [45].

The paucity of living pitch pine, and abundance of pitch pine remnants and fire-intolerant tree species (i.e., red maple, birch) at these sites are indicative of recent prolonged fire intervals [16,46]. The apparent 'mesophication', sensu Nowacki and Abrams (2008) [10], of these ecosystems is of concern to PGC habitat managers because wildlife habitat quality, hunting opportunity, and ecosystem resiliency are decreased. This concern applies to management areas across the Ridge and Valley Province, not just the two sites studied here.

2.2. Data Collection

Fire-Scar Data

Study sites were surveyed for living and dead (stumps and snags) fire-scarred pitch pine trees in May 2014. Full and partial basal cross-sections (~20 cm thick) were removed from trees using a chainsaw (Figure 2). Cross-sections were collected from 39 trees at SGL 088; seven of which were excluded from the study due to rot and/or too few annual growth rings for dendrochronological dating. Cross-sections were collected from 40 trees at SGL 107, six were excluded due to rot and/or too few annual growth rings. The sampling area encompassed approximately 0.70 km^2 for SGL 088, and 0.61 km^2 for SGL 107, measured using the Minimum Bounding Geometry Tool (Convex Hull type) in ArcGIS (v. 10.3). On some trees, multiple cross-sections were collected at different heights above ground to capture the most complete fire record possible. Cross-sections were assigned a sample number, aspect, orientation, and geographic location. Cross-sections were transported to the Missouri Tree-Ring Laboratory at the University of Missouri (Columbia, MO, USA) for surface preparation and analysis.

Figure 2. (**A**) Fire-scarred pitch pine (*Pinus rigida*) stump (sample no. 107032) with charcoal, inset shows side view. (**B,C**) Cross-sectional views of rings and fire scars from this sample, arrows denote fire-scar years. This tree had 27 fire scars during the time period of 1734–1898, with most scars occurring in the dormant season tree-ring position.

2.3. Fire-Scar Data Analysis

Top and bottom surfaces of cross-sections were prepared using an electric orbital sander with progressively finer sandpaper (80 to 1200 grit) to reveal cellular detail of annual rings and fire scars. A radius (pith-to-bark) of each cross-section with the least amount of ring-width variability due to callus was chosen for tree-ring width measurement. All rings were measured to 0.01 mm precision using a binocular microscope and a Velmex measuring stage (Velmex, Bloomfield, New York, NY, USA). Tree ring-width series from each sample were visually cross-dated using ring-width plots [47]. Cross-dating was statistically verified with the COFECHA computer program [48,49]. Fire scars were dated to the year of cambial response to injury and, if possible, to a within-ring location following Kaye and Swetnam (1999) [50]. The number of years with a growing-season fire scar were tallied. Fire scars were identified by the presence of callus tissue, traumatic resin canals, liquefaction of resin, and cambial injury [51].

We used FHX2 software [52] to construct the fire event chronology, analyze fire-scar years, and graph individual tree and composite fire intervals (years between fire events). Fire-scar statistical analysis was restricted to only include time periods which included at least 3 trees in the tree-ring record, resulting in a 349-year record (1663–2013) for SGL 088, and 369-year record (1644–2013) for SGL 107. Mean fire intervals (MFIs), standard deviations, and lower/upper exceedance intervals were computed. The exceedance intervals indicate if a fire interval is significantly longer or shorter than the mean (per time period). Kolmogorov-Smirnov (K-S) Goodness-of-fit tests conducted on the frequency distribution of fire intervals were used to determine whether a Weibull distribution modeled the interval data better than a normal distribution [53]. Weibull median fire intervals were recorded when appropriate. The percentage of trees scarred in fire years and the average per time period (mean percentage of trees scarred) were calculated. K-S tests were conducted using SAS statistical software version 9.4 [54] to determine if MFIs or mean percentage of trees scarred were statistically different ($\alpha < 0.05$) between sites and time periods (intra-site).

The locations of fire-scarred remnant pitch pine wood at SGL 088 on Tuscarora Mountain were well-suited to be divided to test for fire regime differences by aspect. The center of the broad plateau top of Tuscarora Mountain (~320 m wide) marks the boundary between Juniata (to the north) and Perry Counties, and served as the line to divide the fire-scar data for comparison, resulting in 12 samples from the north-facing slope and 20 from the south-facing slope. There is interest among research and land management communities regarding how fire regimes differed by aspect and landscape position [55].

The time period recorded by samples was divided into three sub-periods based on cultural and land-use changes: pre-1754 (pre-European settlement), 1755–1914 (European settlement), and 1915–2013 (fire suppression). We identified 1754 as a division year for the pre-European settlement time period based on the Albany Purchase whereby the region was sold to the United States by the Iroquois [56]. The Albany Purchase led to increased colonial settlement and conflict with Native Americans in the study region [56,57]. The pre-European settlement period included in fire interval analyses (SGL088: 1663–1754; SGL107: 1644–1754) does not reflect a Native American fire regime entirely free of European influence. Colonial settlement along the eastern seaboard of the United States and contact through fur trade activities had already led to displaced Native American populations and increased intra-native conflicts [58,59]. In addition, Native American populations were already significantly reduced across the eastern U.S. due to European diseases [59–62]. We determined 1915 to be the first year of the fire suppression era based on the associated establishment of the Bureau of Forest Protection under the Pennsylvania Department of Forestry [12] which instigated a new period of forest protection policies that included fire suppression [11].

2.4. Fire and Climate Analysis

Associations between historical fire events and drought conditions were tested using superposed epoch and correlation analysis. Superposed epoch analysis (SEA) was conducted separately for fire events at each study site then again for shared fire event years between the sites and between aspects at SGL 088. SEA was conducted for the full period of record and 50-year sub-periods from 1650 to present (e.g., 1650 to 1700, 1700 to 1750). SEA was performed within the Fire History Analysis and Exploration System (FHAES v. 2.0.1) [63]. Drought data consisted of reconstructed summer season Palmer Drought Severity Indices (PDSI) [64]. PDSI data were bootstrapped for 1,000 simulated events to derive confidence limits. Fire event years were paired with PDSI to determine if conditions were significantly wet or dry from six years preceding to four years succeeding fire events. Conditions prior to, during, and following fire events were considered significantly wet or dry when average PDSI values exceeded confidence limits. Separately, Pearson correlations were used to test for relationships between PDSI and the percentages of trees scarred during fire years. Reconstructed PDSI data were obtained for the two nearest gridpoints to the study sites (gridpoints 254, 255) and SEA analyses were conducted separately for each gridpoint.

3. Results

3.1. Tree-Ring and Fire-Scar Data

3.1.1. All-Time Period

The time periods spanned by tree-rings were 1548 to 2013 CE for SGL 088, and 1620 to 2013 CE for SGL 107 (Table 1, Figure 3). Fire-scar statistical analyses were only conducted for the time periods during which at least 3 trees were present in the tree-ring record, 1663–2013 at SGL 088 and 1644–2013 at SGL 107. During this period (all-time period), samples from SGL 088 (*n* = 32) revealed 201 fire scars from 56 different fire years. In comparison, samples from SGL 107 (*n* = 33) recorded more fire scars (*n* = 387) from fewer fire years (*n* = 44). Considering the two sites together, 87 unique fire years were identified; 13 (15%) of which were common to both sites. At SGL 088, composite fire intervals ranged from 1 to 21 years, with an MFI of 5.1 years. At SGL 107, composite fire intervals had a wider range (1–37 years), with a similar MFI of 5.7 years. MFIs did not differ significantly between sites, though the mean percentage of trees scarred did (*p* < 0.0001). At SGL 088, the mean percentage of trees scarred was less than half that of SGL 107 (18.7% vs. 46.4%, respectively; Table 1, Figure 4).

Table 1. Fire-scar history data for SGL 088 and SGL 107 (Ridge and Valley Province, central Pennsylvania, USA).

	All Time		Pre-European Settlement		European Settlement		Fire Suppression	
	1663–2013	1644–2013	≤1754		1755–1914		1915–2013	
Site	SGL088	SGL107	SGL088	SGL107	SGL088	SGL107	SGL088	SGL107
No. scars	201	387	40	39	156	348	5	0
No. fire years	56	44	15	10	38	34	3	0
MFI (years)	5.1	5.7	5.6	12.0 [a]	4.2	4.0 [a]	na	na
Standard deviation	4.2	7.0	4.9	12.1	2.9	3.5	na	na
Range (years)	1–21	1–37	1–21	2–37	1–15	1–21	na	na
WMI (years)	4.3	4.3	4.8	9.2	3.8	3.4	na	na
LEI (no. exceedence)	1.3 (5)	1.0 (0)	1.5 (1)	2.2 (1)	1.3 (4)	1.1 (3)	na	na
UEI (no. exceedence)	9.5 (5)	11.6 (3)	10.6 (1)	24.0 (2)	7.5 (7)	7.4 (2)	na	na
Mean percentage scarred	18.7 [a]	46.4 [a]	14.6 [b]	51.2 [b]	19.7 [c]	44.9 [c]	26.2	na

MFI = mean fire interval, WMI = Weibull median interval, LEI/UEI = Lower and upper exceedance intervals (number of exceedance instances in parentheses). Superscripts of the same letter in a row indicate significant differences (α = 0.05) between sites or time periods (intra-site), na = not applicable due to insufficient number of observations for calculation.

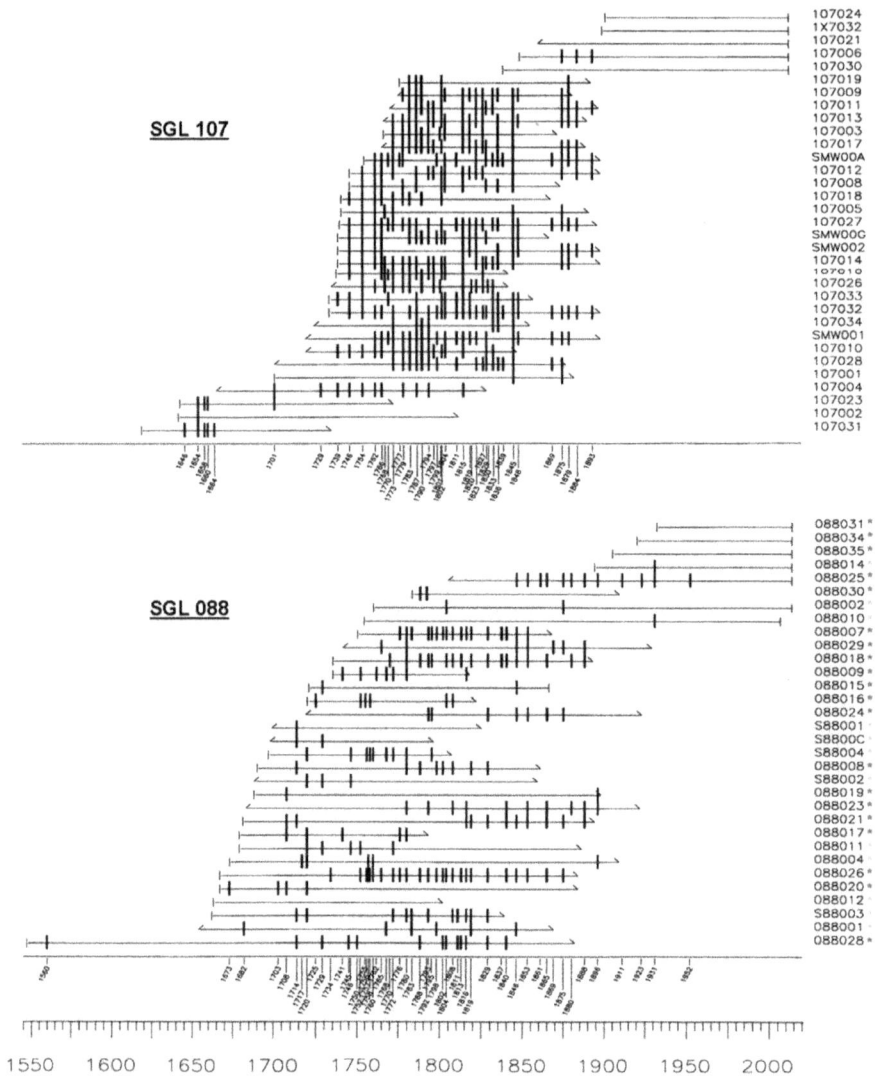

Figure 3. Fire-scar diagrams for SGL 107 and SGL 088. Each horizontal line represents the lifespan of an individual tree (sample number on right). Bold vertical lines are fire-scar years. Slanted or vertical lines at the earliest year shown for each sample indicate either the inner-most ring or pith date, respectively. Similarly, for the last year recorded, a slanted line indicates the outer-most ring present, a vertical line indicates bark year. For the SGL 088 diagram, red and blue asterisks denote trees from the south- and north-slopes, respectively.

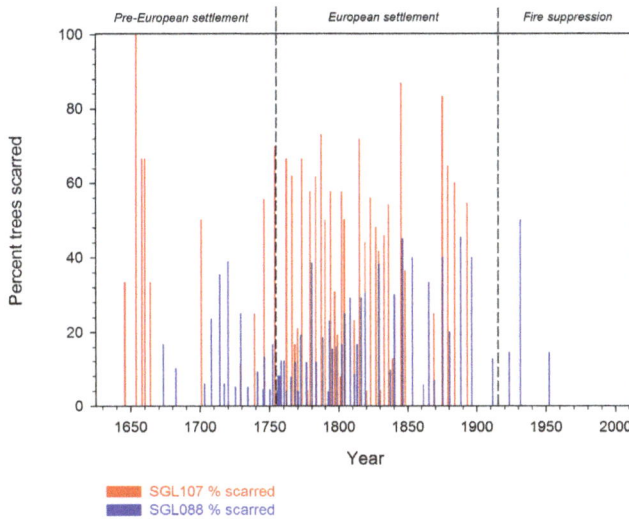

Figure 4. Percentage of trees scarred during fire years with at least 3 trees present in the tree-ring record at the two fire history study sites. Time periods separated by vertical dotted lines.

Season of injury was identifiable on over half of the fire scars at both sites (Table 2). Approximately 40% of the fire scars at both sites were not identified to season of injury due to missing wood (caused by rot) or obscured wood anatomy at the union of injured and uninjured cells. At both sites, a large proportion of fire scars (98.3% and 95.1% for SGL 088 and SGL 107, respectively) with season of injury identified were in the dormant season tree-ring position (Table 2).

Table 2. Fire-scar position and seasonality of fire scars at SGL 088 and SGL 107 by sub-periods.

	All Time		Pre-European Settlement		European Settlement	
	1663–2013	1644–2013	≤1754		1755–1914	
Site	SGL 088	SGL 107	SGL 088	SGL 107	SGL 088	SGL 107
Dormant	57.2%	54.5%	52.5%	33.3%	58.3%	56.9%
Early earlywood	1.0%	2.1%	5.0%	2.6%	0.0%	2.0%
Middle earlywood	0.0%	0.8%	0.0%	2.6%	0.0%	0.6%
Unidentified	41.8%	42.6%	42.5%	61.5%	41.7%	40.5%
Years with growing season fire scar (no.)	1.8% (1)	9.1% (4)	6.7% (1)	20.0% (2)	0.0%	5.9% (2)

Data are not shown for the 1915–2013 period. Five fire scars were recorded after 1914 at SGL 088, 3 of which were in the dormant position/season and 2 with unidentifiable position/season. No fires occurred in the 1915–2013 period at SGL 107.

3.1.2. Pre-European Settlement Period (Pre-1755)

During the pre-European settlement era, samples from SGL 088 recorded 40 fire scars from 15 different fire years, while SGL 107 had a similar number of fire scars ($n = 39$), but from fewer ($n = 10$) fire years. For SGL 088, composite fire intervals ranged from 1 to 21 years, with an MFI of 5.6 years. Composite fire intervals at SGL 107 ranged from 2 to 37 years, with an MFI of 12.0 years. Fire intervals were not significantly different between the sites. The mean percentage of trees scarred at SGL 088 was significantly less than that at SGL 107 ($p = 0.006$, Table 1). During this time period, the majority (61.5%) of fire scars at SGL 107 had unidentified seasonality, more than any other time period record at either site. Most fire scars with identifiable seasonality were in the dormant tree-ring position, though slightly more scars occurred in the growing season relative to the other periods considered (Table 2).

3.1.3. European Settlement Period (1755 to 1914)

During this period, fire occurrence was higher and more similar between sites than during any other time period (Table 1). MFIs were statistically similar at both sites (4.2 and 4.0 years at SGL 088 and SGL 107, respectively), though SGL 107 continued to have a significantly higher mean percentage of trees scarred ($p < 0.0001$). At both sites, most of the fire scars and years assigned seasonality were in the dormant season (Table 2).

3.1.4. Fire-Suppression Period (1915 to 2013)

There were 3 fire years at SGL 088 during this time period, and none at SGL 107. MFIs were not calculated due to an insufficient number of observations; however, if the open interval at the end of each fire chronology is accounted for by considering the bark year to be a fire year, MFIs are greater than 30 and 90 years for SGL 088 and SGL 107, respectively. At SGL 088, 3 of the 5 fire scars were in the dormant position and two were not able to be assigned season of injury.

3.1.5. Intra-Site Fire Regime Characteristics across Time Periods

At SGL 088, no significant differences in mean percentage of trees scarred or MFI were detected between the pre-European and European settlement time periods (Table 1). At SGL 107, no significant difference in the mean percentage of trees scarred between time periods was detected; however, MFI was significantly longer ($p = 0.02$) during the pre-European period than the European settlement period.

3.1.6. Aspect Influence at SGL 088

Twenty trees were sampled on the south-facing slope compared to 12 on the north-facing slope at SGL 088. Years with fire were nearly twice as frequent on south-facing versus north-facing slopes (Table 3, Figure 5), though MFIs were not statistically different. Between aspects, the mean percentage of trees scarred was similar during the all-time and pre-European settlement periods, though more trees on the south-facing slope were scarred on average ($p = 0.039$) during the European settlement period (Table 3). North-facing and south-facing slopes shared 44.6% ($n = 25$ years) of all fire years.

Table 3. Fire-scarring characteristics for samples on the north-facing versus south-facing slopes of Tuscarora Mountain at SGL 088.

	All Time		Pre-European Settlement		European Settlement		Fire Suppression	
	1664–2013	1668–2013	≤1754		1755–1914		1915–2013	
Unit (north vs. south)	north	south	north	south	north	south	north	south
No. scars	47	154	17	22	28	129	2	2
No. fire years	28	53	7	12	20	38	1	2
MFI (years)	9.2	5.4	11.7	7.2	7.4	4.2	na	na
Mean percentage scarred	21.0%	24.4%	28.0%	20.1%	16.0% [a]	25.0% [a]	66.7%	29.2%

Superscript letter designates significantly different means between slopes ($\alpha = 0.05$), na = not applicable due to insufficient number of observations for calculation.

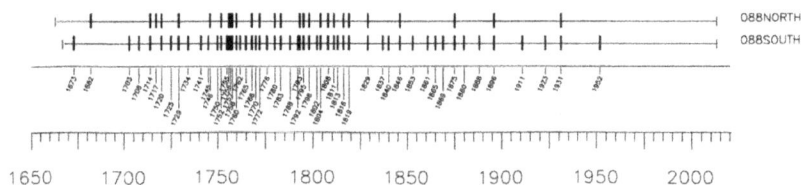

Figure 5. Composite fire-scar history at SGL 088 separated by north and south aspects. For time periods with at least 3 trees present in the tree-ring record, there were 28 fire years recorded by 12 trees on the north-facing slope at SGL 088, compared to 53 fire years recorded by 20 trees on the south-facing slope.

3.2. Fire and Drought

From 1550 to 2003, PDSI values ranged from −5.0 (extreme drought) to 3.8 (very wet) with a mean of −0.28 (near normal). Over the fire-scar record, the climatic conditions before, during, and following fire years were not significantly wet or dry at either study site. When considered in 50-year sub-periods, SGL 088 site showed drought conditions were significantly dry one year prior and in the year of fire for the 1850 to 1900 and 1900 to 1950 sub-periods, respectively. This result was consistent regardless of considering drought data from either drought data gridpoint. For SGL 107, drought-fire associations through SEA were inconsistent when considering sub-periods and drought data gridpoint. Fire years shared between SGL 088 and SGL 107 showed significantly drier conditions 5 years prior and 2 years following fires, whereas 2 years prior to fire had significantly wetter conditions. No significant drought-fire associations were detected for shared fire years between aspects at SGL 088. PDSI was weakly related to percentages of trees scarred at SGL088 ($r = 0.28$, $p = 0.04$). No other correlations between drought and percentages of trees scarred were found.

4. Discussion

4.1. Characterizing Historical Fire Regimes

4.1.1. Fire Frequency

Over the last four centuries, fire regimes of the sites studied here can be characterized as frequent, yet variable through time. In general, MFIs were short (5.1 and 5.7 years; Table 1), despite fire-suppression effects during the 20th century. Fire intervals were positively skewed prior to the fire-suppression era with the majority of intervals being 1 to 5 years in length (Figure 6). At both sites, multiple incidences of annual burning were documented. Similarly, at both sites, long fire-free intervals were documented (Table 1) that likely had important effects on vegetation communities and successional pathways [6].

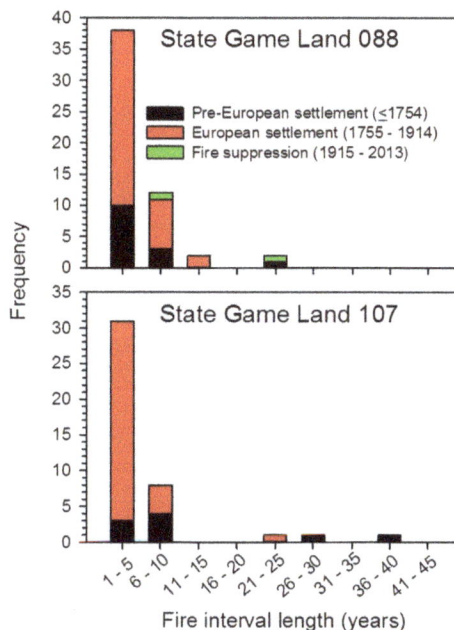

Figure 6. Frequency distributions of fire intervals at SGL 088 and SGL 107 color coded by time period.

The fire frequencies reported here are comparable to those of two previous studies conducted in yellow pine forests of the Ridge and Valley Province [65,66], located approximately 750 and 420 km to the south, respectively. Another study, also to the south, but closer (approximately 180 km) on the Appalachian Plateau, also reported similar fire frequencies from white oaks [67]. Conversely, the MFIs reported here are over 3 times shorter than those reported for red pine (*Pinus resinosa*) stands [35,68] approximately 100 km to the north in the highly dissected Deep Valleys physiographic region of the Appalachian Plateau. We expect that differences in regional topography, climate, and human influences were major contributing factors for these regional fire frequency differences. In Missouri, USA, historical fire frequency was found to be driven by humans, yet mitigated by topographic roughness [31]. At broader regional extents, historical fire frequency is strongly correlated with regional climate variability, particularly across Pennsylvania [32].

MFIs did not differ statistically between sites or time periods, except for the pre-European and European settlement time periods at SGL 107 (Table 1). During the pre-European time period, SGL 107 had two significantly long fire intervals (1664–1701, 1701–1729; Figure 3); interestingly, both were preceded by a brief episode of frequent fires (1646–1664; Figure 3). Though there are relatively few trees recording during this time of infrequent fires (1664–1729), the 4 trees recording were relatively small and likely sufficiently describe fire activity at the site. This change from high to low fire frequency does not appear to be associated with exceptionally wet or dry conditions and may indicate influence by humans. During the early and mid-17th century, large-scale depopulation and emigration events of Native Americans occurred across the eastern seaboard of northeastern North America [59,62,69,70]. If humans were a primary ignition source, then decreased populations would be expected to result in decreased fire frequency—a relationship observed in other locations [68,71–73]. Effects of decreased populations may also account for the conspicuous fire-free period at SGL 107 from 1848 to 1869. During this time, Brose et al. [68] also identified a longer than expected fire interval and suggested that it related to the American Civil War (1861–1865), specifically a decrease in human ignitions caused by men leaving the region to enlist in the Union Army.

Many fire history studies in the eastern U.S. have identified human population, settlement patterns, and commerce activities to be closely associated with changes in fire frequency [16,35,71–76]. Compared to human ignitions, lightning ignitions are rare in the northeastern U.S. Most lightning events are accompanied by rain and high levels of humidity, further implicating humans as an important fire ignition source for maintaining frequent fire regimes [77]. Fire records from 1979 to 2013 show 1.3% of fires reported in Pennsylvania were ignited by lightning, and resulted in 1.5% of the total area burned [78]. Based on the lack of association between drought and fire occurrence, as well as the paucity of lightning-caused ignitions, we suggest that the fire frequency findings in this study are attributable to human factors such as population, occupancy, and migration patterns.

4.1.2. Fire Severity and Extent

Historical fire severity is difficult to document using dendrochronological methods. In non-stand replacement fire regimes, fire severity may be approximated by the percentages of trees scarred [79]. Based on the long-term presence of trees (Figure 3), evidence for stand-replacing fires did not exist; therefore, we expect that historical fires were primarily low- to moderate-severity surface events. A fire scar only indicates that a tree was injured and survived; other metrics of past fire severity (e.g., substrate changes, vegetation mortality, scorch height) are less easily measured, especially over long time periods and following multiple fires. In this study, SGL 107 exhibited significantly higher levels of percentages of trees scarred than SGL 088 (Table 1, Figure 4). Causes for this difference are unknown, but we expect it is due to topographic and landscape position differences as opposed to differences in ignitions (e.g., timing, locations). SGL 088 spans a convex ridge top while SGL 107 is contained within a concave slope at the upper-most section of a single minor drainage (Figure 1). Fires burning upslope at SGL 088 would have to cross north-south aspects, while at SGL 107, a fire burning upslope would have a high probability of burning the entire extent of the study area due to continuous slopes, pre-heating effects, and no aspect changes or other major fire barriers or fire intensity moderating landscape features.

4.1.3. Fire Seasonality

As indicated by the fire scars for which season of injury could be determined, the majority of historical fires occurred in the dormant season. It is important to note that historical fire seasonality is determined by the growth period of the recorder trees and that fires occurring during this dormancy period cannot be further separated into fall or spring occurrences. Tree (i.e., cambial growth) dormancy occurs from approximately September to May in central Pennsylvania, although timing varies from year to year due to climate conditions. More precise determination of historical fire timing within the dormant season could be extrapolated based on Ryan et al. (2013) [3], who speculated that historical, like modern era fire regimes, were separated into spring and fall seasons. Modern era (1940–2015) monthly wildfire records indicate most wildfires occur during March, April, and May, with a much smaller portion of occurrences in October and November [78]. In recent decades, prescribed fire activity has been greatest in the late winter/early spring months. Taken together, these sources suggest that dormant season fires may have predominantly occurred in the late winter/early spring time periods, but may also have included late fall or early winter.

Growing season fires historically had a minor presence at our study sites, and the majority of these events occurred during the pre-European settlement period (Table 2). That fires occurred overwhelmingly during the dormant season is consistent with previous fire history studies in the eastern U.S. [35,65–68]. Distinction of historical fire seasonality has important implications for both fire and vegetation interpretations. For example, Sparks et al. [80] found growing and dormant season fires had differing effects on herbaceous community composition and structure in the shortleaf pine (*P. echinata*) woodlands in the Ouachita Mountains, Arkansas. Albeit infrequent, our findings indicate that growing season fire did occur historically in these pitch pine communities, and may be required to achieve certain desired fire effects.

4.2. Management Implications

The historically high frequency of fire reported at both sites, along with the high density of remnant pitch pine trees (stumps and snags) relative to the very few living pitch pine present, is evidence that these sites have undergone significant ecological changes (e.g., vegetation type conversion, open to closed canopy structure, herbaceous/grass to tree leaf litter fuel type transition, altered carbon and nutrient cycles) coincident with the onset of the fire-suppression era. The resulting extended fire-free period is outside of the historic range of variability reported in this study, and is reflected by the minor presence of pitch pine in the current vegetation community. Overall, the findings of this study support the current understanding of pitch pine ecology, specifically its association with frequent fire occurrence. Based on this history, fire is ecologically appropriate to restore, manage, and perpetuate pitch pine communities into the future.

The differences in historical fire severity between SGL 088 and SGL 107 suggest that aspect and landscape position should be considered when planning management activities. For some historical fires, contrasting fire effects due to aspect existed, and therefore can be expected within a single prescribed fire compartment. Exceptions likely exist in very dry conditions when fuel moisture is not differentiated by aspect. At SGL 088 there were nearly double the number of fire years for all time periods on south- versus north-slopes, likely indicating a different historical vegetation and fuels matrix and/or fuel moisture, i.e., oak and other hardwoods refugia on north-facing slopes, pine and grasses dominating on those facing south [81–83]. These findings, as well as the fact that pine remnants were found in landscape positions below the ridgetop (Figure 2, both sites), demonstrate that fires and fire-adapted species were not relegated only to supposed pyrogenic microsites (e.g., ridgetops), as suggested by Matlack (2013) [55]. Additional studies describing variability in fire regimes across a range of spatial extents would further inform management of larger landscapes.

5. Conclusions

This paper presents evidence that fire regimes of pitch pine communities in the Ridge and Valley Province of central Pennsylvania were historically frequent, of low to moderate severity, and dominated by dormant season fire events. This fire regime information can be used to guide modern fire management and restoration of fire-dependent ecosystems. Additional studies characterizing fire regimes in Pennsylvania would help to refine description of fire regimes and aid in understanding their variability and driving influences.

Acknowledgments: Identification of study sites was supported by the U.S. Forest Service, Northern Research Station. This research was funded by the Pennsylvania Game Commission. We thank David Matthew Bourscheidt for data collection assistance in the field, Erin Abadir for help identifying cultural and land-use periods, and Chris Bobryk for cartography assistance. We are grateful for contributions from two anonymous reviewers, which improved the manuscript.

Author Contributions: J.M.M. conceived the study, conducted fieldwork, analyzed data, wrote the manuscript; M.C.S. conceived the study, conducted fieldwork, analyzed data, and wrote the manuscript, B.C.J. conducted fieldwork and wrote the manuscript, R.P.G. conceived the study and wrote the manuscript, P.H.B. aided in locating study sites and wrote the manuscript, D.C.D. wrote the manuscript.

Conflicts of Interest: The authors declare no conflict of interest.

References

1. Cohesive National Wildland Fire Management Strategy. The Final Phase in the Development of the National Cohesive Wildland Fire Management Strategy. 2014. Available online: https://www.forestsandrangelands.gov/strategy/documents/strategy/CSPhaseIIINationalStrategyApr2014.pdf (accessed on 15 May 2016).
2. Brandt, L.; He, H.; Iverson, L.; Thompson, F.R., III; Butler, P.; Handler, S.; Janowiak, M.; Danielle, S.P.; Swanston, C.; Albrecht, M.; et al. *Central Hardwoods Ecosystem Vulnerability Assessment and Synthesis: A Report from the Central Hardwoods Climate Change Response Framework Project*; USDA Forest Service General Technical Report NRS-124; Northern Research Station: Newtown Square, PA, USA, 2014.
3. Ryan, K.C.; Knapp, E.E.; Varner, J.M. Prescribed fire in North American forests and woodlands: History, current practice, and challenges. *Front. Ecol. Environ.* **2013**, *11*, e15–e24. [CrossRef]
4. Engstrom, R.T.; Gilbert, S.; Hunter, M.L., Jr.; Merriwether, D.; Nowacki, G.J.; Spencer, P. Practical applications of disturbance ecology to natural resource management. In *Ecological Stewardship: A Common Reference for Ecosystem Management*; Johnson, N.C., Malik, A.J., Sexton, W.T., Szabo, R.C., Eds.; Elsevier: Oxford, UK, 1999; Volume 2.
5. Swetnam, T.W.; Allen, C.D.; Betancourt, J.L. Applied historical ecology: Using the past to manage for the future. *Ecol. Appl.* **1999**, *9*, 1189–1206. [CrossRef]
6. Stambaugh, M.C.; Marschall, J.M.; Guyette, R.P. Linking fire history to successional changes of xeric oak woodlands. *For. Ecol. Manag.* **2014**, *320*, 83–95. [CrossRef]
7. Abrams, M.D. Fire and the development of oak forests. *BioScience* **1992**, *42*, 346–353. [CrossRef]
8. Lorimer, C.G. Historical and ecological roles of disturbance in eastern North American forests: 9000 Years of change. *Wildl. Soc. Bull.* **2001**, *29*, 425–439.
9. Ruffner, C.M. Understanding the evidence for historical fire across eastern forests. In *Fire in Eastern Oak Forests: Delivering Science to Land Managers*, Proceedings of a Conference, Columbus, OH, USA, 15–17 November 2005; Dickinson, M.B., Ed.; Gen. Tech. Rep. NRS-P-1; Northern Research Station: Newtown Square, PA, USA, 2006; pp. 40–48.
10. Nowacki, G.J.; Abrams, M.D. The demise of fire and "Mesophication" of forests in the eastern United States. *BioScience* **2008**, *58*, 123–138. [CrossRef]
11. Taber, T.T. Logging Railroad Era of Lumbering in Pennsylvania. In *Sunset along the Susquehanna*; Lycoming Printing Company: Williamsport, PA, USA, 1972; Volume 4.
12. Decoster, L.A. *The Legacy of Penn's Woods: A History of the Pennsylvania Bureau of Forestry*; Pennsylvania Historical and Museum Commission: Harrisburg, PA, USA, 1995.

13. Johnson, A.S.; Hale, P.E. The historical foundations of prescribed burning for wildlife: A Southeastern perspective. In *The Role of Fire in Nongame Wildlife Management and Community Restoration*; Ford, W.M., Russell, K.R., Moorman, C.E., Eds.; U.S.D.A. Forest Service Northeastern Research Station: Newtown Square, PA, USA, 2002; pp. 11–23.

14. Donovan, G.H.; Brown, T.C. Be careful what you wish for: The legacy of Smokey Bear. *Front. Ecol. Environ.* **2007**, *5*, 73–79. [CrossRef]

15. Lorimer, C.G. Development oak forests. *For. Sci.* **1984**, *30*, 3–22.

16. Nowacki, G.J.; Abrams, M.D. Community, edaphic, and historical analysis of mixed oak forests of the Ridge and Valley Province in central Pennsylvania. *Can. J. For. Res.* **1992**, *22*, 790–800. [CrossRef]

17. Harrod, J.C.; Harmon, M.E.; White, P.S. Post-fire succession and 20th century reduction in fire frequency on xeric southern Appalachian sites. *J. Veg. Sci.* **2000**, *11*, 465–472. [CrossRef]

18. McEwan, R.W.; Muller, R.N. Spatial and temporal dynamics in canopy dominance of an old-growth central Appalachian forest. *Can. J. For. Res.* **2006**, *36*, 1536–1550. [CrossRef]

19. Fei, S.L.; Kong, N.N.; Steiner, K.C.; Moser, W.K.; Steiner, E.B. Change in oak abundance in the eastern United States from 1980 to 2008. *For. Ecol. Manag.* **2011**, *262*, 1370–1377. [CrossRef]

20. Rodewald, A.D.; Abrams, M.D. Floristics and avian community structure: Implications for regional changes in eastern forest composition. *For. Sci.* **2002**, *48*, 267–272.

21. McShea, W.J.; Healy, W.M.; Devers, P.; Fearer, T.; Koch, F.H.; Stauffer, D.; Waldon, J. Forestry matters: Decline of oaks will impact wildlife in hardwood forests. *J. Wildl. Manag.* **2007**, *71*, 1717–1728. [CrossRef]

22. Hutchinson, T.F.; Long, R.P.; Ford, R.D.; Sutherland, E.K. Fire history and the establishment of oaks and maples in second-growth forests. *Can. J. For. Res.* **2008**, *38*, 1184–1198. [CrossRef]

23. Ratajczak, Z.; Nippert, J.B.; Collins, S.L. Woody encroachment decreases diversity across North American grasslands and savannas. *Ecology* **2012**, *93*, 697–703. [CrossRef] [PubMed]

24. McCord, J.M.; Harper, C.A.; Greenberg, C.H. Brood cover and food resources for wild turkeys following silvicultural treatments in mature upland hardwoods. *Wildl. Soc. Bull.* **2014**, *38*, 265–272. [CrossRef]

25. Pyne, S.J. *Fire in America*; University of Washington Press: Princeton, NJ, USA, 1982; p. 655.

26. Pausas, J.G.; Keeley, J.E. A burning story: The role of fire in the history of life. *BioScience* **2009**, *59*, 593–601. [CrossRef]

27. McShea, W.J.; Healy, W.M. (Eds.) *Oak Forest Ecosystems: Ecology and Management for Wildlife*; John Hopkins University Press: Baltimore, MD, USA, 2002; p. 432.

28. Pennsylvania Fish and Boat Commission. Pennsylvania State Wildlife Action Plan. 2015. Available online: http://fishandboat.com/swap.htm (accessed on 15 May 2016).

29. Harper, C.A.; Ford, W.M.; Lashley, M.A.; Moorman, C.E.; Stambaugh, M.C. Fire effects on wildlife in the Central Hardwoods and Appalachian Regions. *Fire Ecol.* **2016**, *12*, 127–159.

30. Varner, M.J.; Arthur, M.A.; Clark, S.L.; Dey, D.C.; Hart, J.L.; Schweitzer, C.J. Fire in Eastern North American oak ecosystems: Filling the gaps. *Fire Ecol.* **2016**, *12*, 1–6.

31. Stambaugh, M.C.; Guyette, R.P. Predicting spatio-temporal variability in fire return intervals using a topographic roughness index. *For. Ecol. Manag.* **2008**, *254*, 463–473. [CrossRef]

32. Guyette, R.P.; Stambaugh, M.C.; Dey, D.C.; Muzika, R.M. Predicting fire frequency with chemistry and climate. *Ecosystems* **2012**, *15*, 322–335. [CrossRef]

33. Cronon, W. *Changes in the Land: Indians, Colonists and the Ecology of New England*; Hill and Wang: New York, NY, USA, 1983.

34. Crow, T.R. Reproductive mode and mechanisms for self-replacement of northern red oak (Quercus rubra)—A review. *For. Sci.* **1988**, *34*, 19–40.

35. Brose, P.H.; Dey, D.C.; Guyette, R.P.; Marschall, J.M.; Stambaugh, M.C. The influence of drought and humans on the fire regimes of northern Pennsylvania, USA. *Can. J. For. Res.* **2013**, *43*, 757–767. [CrossRef]

36. Johnson, L.B.; Kipfmueller, K.F. A fire history derived from *Pinus resinosa* Ait. for the Islands of Eastern Lac La Croix, Minnesota, USA. *Ecol. Appl.* **2016**, *26*, 1030–1046. [CrossRef] [PubMed]

37. Pennsylvania Game Commission. PA Game Commission Strategic Plan 2015–2020. Available online: http://www.pgc.pa.gov/InformationResources/AboutUs/Documents/PGC%20Strategic%20Plan%202015-2020.pdf (accessed on 1 June 2016).

38. Bowen, B.W. Preserving genes, species, or ecosystems? Healing the fractured foundations of conservation policy. *Mol. Ecol.* **1999**, *8*, S5–S10. [CrossRef] [PubMed]

39. Little, S. Prescribed burning as a tool of forest management in the northeastern states. *J. For.* **1953**, *51*, 496–500.

40. Vogl, R.J. Fire: A destructive menace or a natural process? In *Recovery and Restoration of Damaged Ecosystems*, Proceedings of the International Symposium, Blacksburg, VA, USA, 23–25 March 1975; Cairns, J., Jr., Dickson, K.L., Herricks, E.E., Eds.; University Press of Virginia: Charlottesville, VA, USA, 1977; pp. 261–289.

41. Richardson, D.M. *Ecology and Biogeography of Pinus*; Cambridge University Press: Cambridge, UK, 1998.

42. Keeley, J.E. Ecology and evolution of pine life histories. *Ann. For. Sci.* **2012**, *69*, 445–453. [CrossRef]

43. Sevon, W.D. *Physiographic Provinces of Pennsylvania, Map 13*, 4th ed.; Pennsylvania Geologic Survey: Harrisburg, PA, USA, 2000.

44. Schultz, C.H. (Ed.) *The Geology of Pennsylvania*; Geological Survey, Department of Conservation of Natural Resources: Harrisburg, PA, USA, 1999.

45. Stone, B.; Gustafson, D.; Jones, B. *Manual of procedure for State Game Land Cover Typing*; Commonwealth of Pennsylvania Game Commission, Bureau of Wildlife Habitat Management, Forest Inventory and Analysis Section, Forestry Division: Harrisburg, PA, USA, 2006, revised 2007; p. 79.

46. Ruffner, C.M.; Abrams, M.D. Relating land-use history and climate to the dendroecology of a 326-year-old Quercus prinus talus slope forest. *Can. J. For. Res.* **1998**, *28*, 347–358. [CrossRef]

47. Stokes, M.A.; Smiley, T.L. *Introduction to Tree-Ring Dating*; University of Chicago Press: Chicago, IL, USA, 1968.

48. Holmes, R.L. Computer-assisted quality control in tree-ring dating and measurement. *Tree-Ring Bull.* **1983**, *43*, 69–78.

49. Grissino-Mayer, H.D. Evaluating crossdating accuracy: A manual and tutorial for the computer program COFECHA. *Tree-Ring Res.* **2001**, *57*, 205–221.

50. Kaye, M.W.; Swetnam, T.W. An assessment of fire, climate, and Apache history in the Sacramento Mountains, New Mexico. *Phys. Geogr.* **1999**, *20*, 305–330.

51. Smith, K.T.; Sutherland, E.K. Fire-scar formation and compartmentalization in oak. *Can. J. For. Res.* **1999**, *29*, 166–171. [CrossRef]

52. Grissino-Mayer, H.D. FHX2-Software for analyzing temporal and spatial patterns in fire regimes from tree rings. *Tree-Ring Res.* **2001**, *57*, 115–124.

53. Grissino-Mayer, H.D. Modeling fire interval data from the American Southwest with the Weibull distribution. *Int. J. Wildl. Fire* **1999**, *9*, 37–50. [CrossRef]

54. SAS Institute Inc. *SAS/STAT® 14.1 User's Guide*; SAS Institute Inc.: Cary, NC, USA, 2015.

55. Matlack, G.R. Reassessment of the use of fire as a management tool in deciduous forests of eastern North America. *Conserv. Biol.* **2013**, *27*, 916–926. [CrossRef] [PubMed]

56. Muller, E.K. *A Concise Historical Atlas of Pennsylvania*; Temple University Press: Philadelphia, PA, USA, 1989; p. 44.

57. Jordan, J.W. *A History of the Juniata Valley and Its People*; Lewis Historical Publishing Co.: New York, NY, USA, 1913; Volume 1, p. 556.

58. Richter, D.K. *Facing East from Indian Country: A Native History of Early America*; Harvard University Press: Cambridge, MA, USA, 2001.

59. Mann, C.C. *1491: New Revelations of the Americas before Columbus*; Knopf Publishing: New York, NY, USA, 2005.

60. Hulbert, A.B. *David Zeisberger's History of the Northern American Indians in 18th Century Ohio, New York, and Pennsylvania*; Wennawoods Publishing: Lewisburg, PA, USA, 1910.

61. Wallace, P.A. *Thirty Thousand Miles with John Heckewelder*; Wennawoods Publishing: Lewisburg, PA, USA, 1958.

62. Williams, M. *Americans and Their Forests: A Historical Geography*; Cambridge University Press: New York, NY, USA, 1989; p. 599.

63. *Fire History Analysis and Exploration System (FHAES)*. 2015. Available online: http://www.fhaes.org (accessed on 15 June 2016).

64. Cook, E.R.; Meko, D.M.; Stahle, D.W.; Cleaveland, M.K. North American summer PDSI reconstructions. In *IGBP PAGES/World Data Center for Paleoclimatology Data Contribution Series #2004–045*; NOAA/NGDC Paleoclimatology Program: Boulder, CO, USA, 2004.

65. Flatley, W.T.; Lafon, C.W.; Grissino-Mayer, H.D.; LaForest, L.B. Fire history, related to climate and land use in three southern Appalachian landscapes in the eastern United States. *Ecol. Appl.* **2013**, *23*, 1250–1266. [CrossRef] [PubMed]

66. Aldrich, S.R.; Lafon, C.W.; Grissino-Mayer, H.D.; DeWeese, G.G. Fire history and its relations with land use and climate over three centuries in the central Appalachian Mountains, USA. *J. Biogeogr.* **2014**, *41*, 2093–2104. [CrossRef]

67. Shumway, D.L.; Abrams, M.D.; Ruffner, C.M. A 400-year history of fire and oak recruitment in an old-growth oak forest in western Maryland, USA. *Can. J. For. Res.* **2001**, *31*, 1437–1443. [CrossRef]

68. Brose, P.H.; Guyette, R.P.; Marschall, J.M.; Stambaugh, M.C. Fire history reflects human history in the Pine Creek Gorge of north-central Pennsylvania. *Nat. Areas J.* **2015**, *35*, 214–223. [CrossRef]

69. Williams, G.W. *References on the American Indian Use of Fire in Ecosystems*; US Department of Agriculture, Forest Service: Washington, DC, USA, 2003.

70. Nevle, R.J.; Bird, D.K.; Ruddiman, W.F.; Dull, R.A. Neotropical human-landscape interactions, fire, and atmospheric CO_2 during European conquest. *The Holocene* **2011**, *21*, 853–864. [CrossRef]

71. Guyette, R.P.; Muzika, R.M.; Dey, D.C. Dynamics of an anthropogenic fire regime. *Ecosystems* **2002**, *5*, 472–486.

72. Guyette, R.P.; Spetich, M.A.; Stambaugh, M.C. Historic fire regime dynamics and forcing factors in the Boston Mountains, Arkansas, USA. *For. Ecol. Manag.* **2006**, *234*, 293–304. [CrossRef]

73. Stambaugh, M.C.; Guyette, R.P.; Marschall, J.M. Fire history in the Cherokee Nation of Oklahoma. *Hum. Ecol.* **2013**, *41*, 749–758. [CrossRef]

74. Guyette, R.P.; Dey, D.C.; Stambaugh, M.C.; Muzika, R.M. *Fire Scars Reveal Variability and Dynamics of Eastern Fire Regimes*; GTR NRS-P-1; USDA Forest Service: Newtown Square, PA, USA, 2006; p. 304.

75. Muzika, R.M.; Guyette, R.P.; Stambaugh, M.C.; Marschall, J.M. Fire, drought, and humans in a heterogeneous Lake Superior landscape. *J. Sustain. For.* **2015**, *34*, 49–70. [CrossRef]

76. Stambaugh, M.C.; Guyette, R.P.; Marschall, J.M.; Dey, D.C. Scale dependence of oak woodland historical fire intervals: Contrasting the Barrens of Tennessee and Cross Timbers of Oklahoma, USA. *Fire Ecol.* **2016**, *12*, 65–84.

77. Abrams, M.D.; Nowacki, G.J. Native Americans as active and passive promoters of mast and fruit trees in the eastern USA. *Holocene* **2008**, *18*, 1123–1137. [CrossRef]

78. Pennsylvania Department of Conservation and Natural Resources. Wildfire Statistics. 2016. Available online: http://www.dcnr.state.pa.us/forestry/wildlandfire/firestatistics/index.htm (accessed on 25 May 2016).

79. Farris, C.A.; Baisan, C.H.; Falk, D.A.; Yool, S.R.; Swetnam, T.W. Spatial and temporal corroboration of a fire-scar-based fire history in a frequently burned ponderosa pine forest. *Ecol. Appl.* **2010**, *20*, 1598–1614. [CrossRef] [PubMed]

80. Sparks, J.C.; Masters, R.E.; Engle, D.M.; Palmer, W.; Bukenhofer, G.A. Effects of late growing-season and late dormant-season prescribed fire on herbaceous vegetation in restored pine-grassland communities. *J. Veg. Sci.* **1998**, *9*, 133–142. [CrossRef]

81. Foti, T.L.; Glenn, S.M. The Ouachita Mountain landscape at the time of settlement. In *Restoration of Old-Growth Forests in the Interior Highlands of Arkansas and Oklahoma*; Henderson, D., Hedrick, L.D., Eds.; Ouachita National Forest and Winrock International Institute for Agricultural Development: Morillton, AR, USA, 1991.

82. Abrams, M.D.; Ruffner, C.M. Physiographic analysis of witness-tree distribution (1765–1798) and present forest cover through north central Pennsylvania. *Can. J. For. Res.* **1995**, *25*, 659–668. [CrossRef]

83. Guldin, J.M. Restoration and management of shortleaf pine in pure and mixed stands—Science, empirical observation, and the wishful application of generalities. In *Shortleaf Pine Restoration and Ecology in the Ozarks*, Proceedings of a Symposium, Springfield, MO, USA, 7–9 November 2006; Kabrick, J.M., Dey, D.C., Gwaze, D., Eds.; Gen. Tech. Rep. NRS-P-15; Department of Agriculture, Forest Service, Northern Research Station: Newtown Square, PA, USA, 2007; pp. 47–58.

© 2016 by the authors. Licensee MDPI, Basel, Switzerland. This article is an open access article distributed under the terms and conditions of the Creative Commons Attribution (CC BY) license (http://creativecommons.org/licenses/by/4.0/).

Article

Fire Regime Characteristics along Environmental Gradients in Spain

María Vanesa Moreno * and Emilio Chuvieco

Environmental Remote Sensing Research Group, Department of Geology, Geography and Environment, University of Alcala, Colegios 2, 28801 Alcalá de Henares, Spain; emilio.chuvieco@uah.es
* Correspondence: vanesa.moreno@uah.es; Tel.: +34-918-854-438

Academic Editors: Yves Bergeron and Sylvie Gauthier
Received: 26 June 2016; Accepted: 1 November 2016; Published: 4 November 2016

Abstract: Concern regarding global change has increased the need to understand the relationship between fire regime characteristics and the environment. Pyrogeographical theory suggests that fire regimes are constrained by climate, vegetation and fire ignition processes, but it is not obvious how fire regime characteristics are related to those factors. We used a three-matrix approach with a multivariate statistical methodology that combined an ordination method and fourth-corner analysis for hypothesis testing to investigate the relationship between fire regime characteristics and environmental gradients across Spain. Our results suggest that fire regime characteristics (i.e., density and seasonality of fire activity) are constrained primarily by direct gradients based on climate, population, and resource gradients based on forest potential productivity. Our results can be used to establish a predictive model for how fire regimes emerge in order to support fire management, particularly as global environmental changes impact fire regime characteristics.

Keywords: fire density; fire ecology; fire management; fire seasonality; fourth-corner; RLQ; pyrogeography

1. Introduction

Fire regimes play an important role in many terrestrial ecosystems [1–3], and changes in fire regime characteristics affect the structure and composition of vegetation, which in turn might affect different ecosystem characteristics, such as biodiversity [1,2]. Several authors have reported changes in fire regime characteristics, such as fire density (i.e., the number of fires per unit area) and seasonality (the period of the year during which fires occur) as a result of climate warming [4] and demographic factors [5–8]. Others have shown that humans now influence fire activity in areas where climate has been the main historical driving force such as in some tropical [9] or boreal regions [10,11]. Regional studies have emphasised the role of humans in future fire regimes, such as in Africa [7], but climate has a stronger role at the global scale [12]. To predict the potential effect of global environmental change on future fire regimes, we must therefore understand the relationship between fire regime characteristics and the environment, particularly how both climate [12,13] and socio-economic factors [8,14–16] interact with fire ignition and fuel accumulation processes.

Pyrogeographical theory suggests that fire regimes are structured by spatial gradients of climate, vegetation and ignitions, which constrain fuel flammability, fuel loading and ignition sources [13,17–19]. However, such relationships are complex because of their feedback and interactions, and some of them remains unclear [17,20,21]. For example, climate is a major driver of fire activity, while fire activity influences climate by emitting carbon dioxide (CO_2) and by causing changes to land albedo.

Meyn et al. [22] proposed a pyrogeographical model based on the premise that fire activity varies across fuel loading and fuel moisture gradients. This theory was later supported by Batllori et al. [23], who studied Mediterranean ecosystems. Similarly, Pausas et al. [24] emphasised the role of a productivity gradient in the Mediterranean ecosystem in south-east Australia and Spain. Krawchuk and Moritz [21] and Pausas and Ribeiro [25] identified a resource gradient at the global scale. At a global scale, Chuvieco et al. [26] observed a diversity of fire regimes, and Archibald et al. [27] emphasised the presence of different fire regimes in various biomes.

In Spain, the current understanding of the relationships between fire regime and environmental characteristics is based on studies that have been carried out at local or regional levels, and there is a limited understanding of how those relationships vary regionally. La Page et al. [5] related fire seasonality to human activities in northern Spain. Pausas and Paula [28] emphasised the role of fuel in fire-climate relationships and suggested that the vegetation structure controls the fire and climate relationship using data for the whole country of Spain except Basque Country and Navarra. However, Pausas and Fernández-Muñoz [29] suggested that climate currently has a greater effect on fire regimes than fuel in the province of Valencia on the Mediterranean Coast of Spain. Moreover, Moreno et al. [30] demonstrated that changes in the number of fires and burned areas in three Spanish regions over two seasons coincided with changes in weather, land use and fire management, thus indicating different regional and seasonal roles.

To gain a better understanding of environmental gradients that constraint fire regime characteristics from a regional perspective, we examined the spatial structure of fire regimes and environmental gradients and tested for significant relationships between them. In addition, we propose a methodological approach that could be used for other regional or global studies. This paper presents the main structure of fire regime characteristics, density and seasonality of fire activity, and environmental gradients in Spain as well as the statistical significance of the associations.

2. Materials and Methods

2.1. Study Area

The study area includes the Spanish territory of the Iberian peninsula, divided in a 10 km $\times$ 10 km Universal Transverse Mercator (UTM) grid, which covers a surface area of 466,614 km^2 (Figure 1).

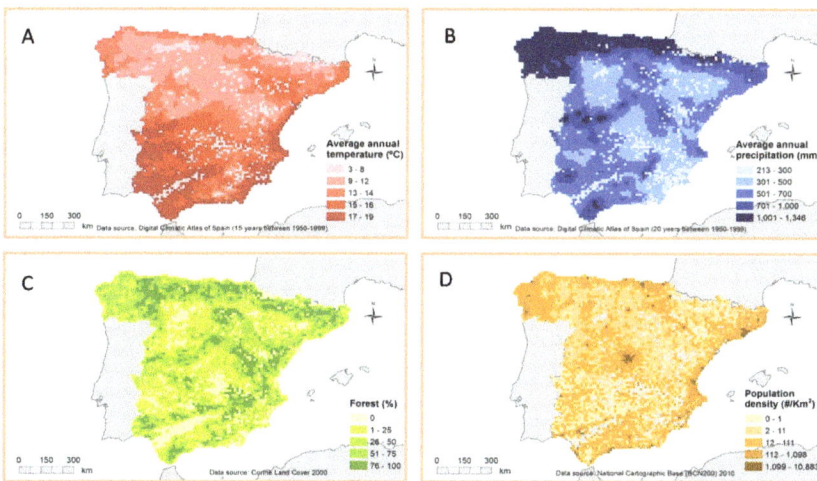

Figure 1. Spatial distribution over the study area of the average annual temperature (°C) (**A**) and the average annual precipitation (mm) (**B**) from the Digital Climatic Atlas of Spain [31] (20 and 15 years between 1950–1999, respectively), the population density (inhabitants/km^2) (**D**) from the National Cartography Base (BCN200; 2010) and the forest land cover (%) (**C**) from the Corine Land Cover 2000 map.

The average annual temperature ranges between 3–19 °C, showing a general decrease from west/northwest to east/southeast, and an average annual precipitation between 213–1348 mm, showing a general increase from east/southeast to west/northwest. The northwest is characterized by mild winters and warm summers and abundant precipitations over the year. The west is characterized by cool and wet winters, and warm and dry summers. The east is characterized by mild winters and warm and dry summers. The southeast is characterized by mild winters and very warm and dry summers.

The crisis in traditional agriculture in the 1960s caused a massive immigration from rural areas to medium and large cities, resulting in an increase of population densities in the main industrial cores and along the coast, as a result of the touristic attraction. Furthermore, the abandonment of marginal areas and traditional grazing practices led to natural afforestation of previously cultivated areas, and consequently an increase in the accumulation of fuel [30].

2.2. Fire Regime Groups and Fire Regime Characteristics

We used a fire regime classification system developed by Moreno and Chuvieco [32] using the data of all fires that burned ≥ 1 ha, regardless of the ignition source, from the Spanish General Statistics of Forest Fires (EGIF) database from 1988–2007 and georeferenced to a 10 km $\times$ 10 km UTM grid.

Using a statistical approach, Moreno and Chuvieco [32] computed six fire regime metrics, density (number of fires or ha per km^2 per year), interannual variability (standard deviation of density) and seasonality (the number of months with significant density) as a function of both the number of fires and the burned area. These six fire regime metrics were analysed using Principal Component Analysis (PCA). The first PCA axis was related to both the density and interannual variability of the number of fires and burned areas (loadings ≥ 0.75), whereas the second PCA axis was related to the seasonality of the number of fires and burned areas (loadings ≥ 0.69), where both axes were more strongly related to the burned area than to the number of fires. For further analysis, the first PCA axis was considered as the density of fire activity and the second PCA axis as the seasonality of fire activity. Finally, the two PCA axes were used in a cluster analysis to generate fire regime groups.

The resulting fire regime classification system [32] included the presence of four fire regime groups defined by two fire regime characteristics, density, and seasonality of fire activity (Table 1), which were partitioned into two respective states, high and low density, and long and short seasonality. For example, a fire regime group characterised by a high density and long seasonality indicated that both the density of fires (number of fires per km^2 per year) and the burned areas (ha per km^2 per year) were high and also that the number of months with significant density in terms of the number of fires and burned areas was large.

Table 1. Fire regime groups and fire regime characteristics data [32] used in this study.

Fire Regime Groups	Number of 10 km $\times$ 10 km Cells	Fire Regime Characteristics	
Fire regime 1	703	High density	Long seasonality
Fire regime 2	1083	High density	Short seasonality
Fire regime 3	1582	Low density	Short seasonality
Fire regime 4	1284	Low density	Long seasonality

Density refers to the number of fires or ha per km^2 per year. Seasonality refers to the number of months with significant density.

2.3. Environmental Gradients

Based on pyrogeographical theory [13,17–19] and considering fires as a 'herbivore' according to some authors [2,33], we expected that fire regime characteristics would change along three types of environmental gradients (Table 2): (i) direct gradients based on climate and humans because both affect fuel flammability and fire ignition processes; (ii) resource gradients based on forest potential productivity and land cover because fuel is consumed by fires; and (iii) indirect gradients based on livestock grazing and land cover changes from afforestation, forest degradation and the abandonment of traditional activities, which indirectly affect ignition, structure and the amount of fuel [34].

Our direct gradients were based on climate and humans as measured by three indicators: (i) a measure of humidity, which was calculated as the ratio of the average annual precipitation and the average annual temperature [35]; (ii) a seasonal temperature fluctuation, which was calculated as the difference between the average temperature of the coldest and warmest months [35]; and (iii) human population pressure, as measured by population density (inhabitants/km^2). Meteorological data were obtained from the Digital Climatic Atlas of Spain [31], and population data were obtained from the National Cartographic Base (BCN200) established by the National Geographic Institute.

Table 2. Environmental gradients and characteristics data used in this study.

Environmental Gradients		Environmental Characteristics	Data Source
Direct gradients	Climate	Humidity	Digital Climatic Atlas of Spain [31] (20 and 15 years between 1950–1999 respectively)
		Seasonal temperature fluctuation	
	Human	Population density	National Cartographic Base (BCN200; 2010), National Geographic Institute
Resource gradients	Forest potential productivity	Forest potential productivity 1	Forest Potential Productivity map of Spain, 2000 [36]
		Forest potential productivity 2	
		Forest potential productivity 3	
		Forest potential productivity 4	
		Forest potential productivity 5	
		Forest potential productivity 6	
		Forest potential productivity 7	
	Land cover	Needle-leaf trees	Corine Land Cover 2000 map, National Geographic Institute
		Broadleaf trees	
		Mixed trees	
		Shrubs	
		Herbaceous vegetation	
		Cultivated and managed vegetation/agriculture (including mixtures)	
Indirect gradients	Livestock grazing	Livestock density	Rural census 2001, National Statistics Institute
	Land cover changes	To forest	Corine Land Cover Change 1990–2000 map, National Geographic Institute
		To shrubs	
		From managed	

Our resource gradients were based on forest potential productivity, which is defined on a scale ranging from 1–7 as the original source does, in which 1 was most productive and 7 was least productive, and land cover, which was defined by six categories: needle-leaf trees, broadleaf trees, mixed trees, shrubs, herbaceous vegetation and cultivated and managed vegetation/agriculture (including mixtures). The forest potential productivity data were obtained from the Forest Potential Productivity map of Spain defined using climate and soil data [36]. The land cover data were obtained from the 2000 Corine Land Cover map compiled by the National Geographic Institute, which were reclassified into broader categories using a simplified legend [37,38].

Our indirect gradients were based on livestock grazing, which was defined by livestock density (cattle, sheep and goats per km^2), and land cover changes defined by the following three changes: (i) land cover changes to forests: from cultivated and managed vegetation/agriculture (including mixtures), herbaceous vegetation and shrubs or barren to needle-leaf trees, broadleaf trees and mixed/other trees; (ii) land cover changes to shrubs: from needle-leaf trees, broadleaf trees and mixed/other trees to shrubs or barren; and (iii) land cover changes from managed: from cultivated and managed vegetation/agriculture (including mixtures) to needle-leaf trees, broadleaf trees, mixed/other trees, shrubs or barren and herbaceous vegetation. The livestock data were obtained from the 2001 rural census compiled by the National Statistics Institute. The land cover change data were obtained from the Corine Land Cover Change map from 1990–2000 compiled by the National Geographic Institute, which was reclassified into categories using the identical generalized legend used for land cover.

All of the data layers were projected as a UTM Zone 30 and the European Datum 1950 (ED50) using a 10 km × 10 km grid, which is the standard grid used to compile fire statistics by the Spanish forest service. The climate data were resampled from 1 km × 1 km by assigning the average of the 1 km × 1 km values to each 10 km × 10 km cell. For the forest potential productivity, land cover and land cover change characteristics, we computed the proportion of each class within a 10 km × 10 km cell. The population and livestock data were resampled from a municipal level to grid size as a function of the proportion of area of each municipality into each cell.

2.4. RLQ Analysis

An RLQ analysis is a three-matrix ordination approach proposed by Dolédec et al. [39] based on a fourth-corner matrix D ($m \times q$), which describes the fire regime characteristics-environment relationship (Figure 2). Matrix L ($p \times n$) describes the presence or absence of the p = 4 fire regime groups at n = 4652 sites. Each n site corresponds with each cell from the 10 km × 10 km UTM grid. Matrix Q ($p \times q$) describes the q = 2 fire regime characteristics for each p fire regime groups. Matrix R ($m \times n$) describes the m = 20 environment characteristics for each n site.

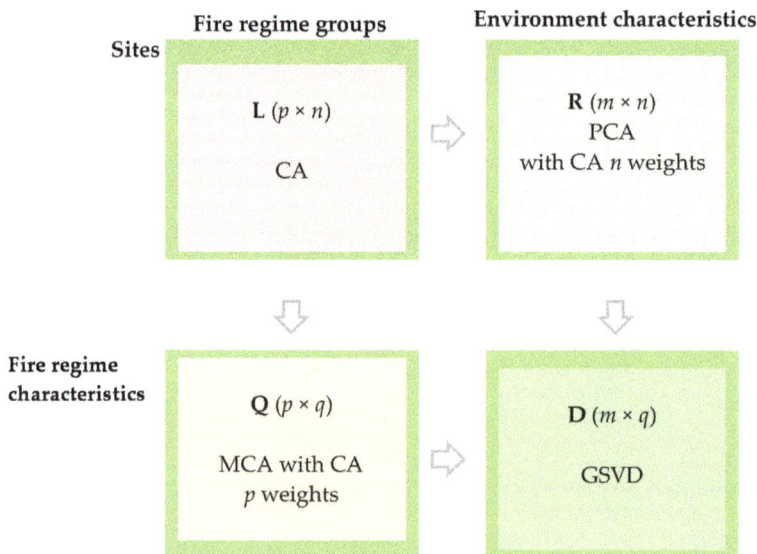

Figure 2. Graphical representation of the RLQ analysis. Matrix L represents the p = 4 fire regime groups at n = 4652 sites and was analysed using correspondence analysis (CA). Matrix Q represents the q = 2 fire regime characteristics for each p fire regime group and was analysed using multiple correspondence analysis (MCA) with CA p weights. Matrix R represents the m = 20 environment characteristics for each n site and was analysed by principal component analysis (PCA) with CA n weights. Matrix D represents the fire regime characteristics-environment relationship and was analysed by a generalized singular value decomposition (GSVD).

An ordination method was applied to the RLQ matrices to identify the main structure of variation. To understand how fire regime groups are organized, correspondence analysis (CA) was applied to the fire regime groups (Matrix L) because it is suitable for presence-absence data and dimensionally homogeneous and positive values [40]. Multiple correspondence analysis (MCA) was applied to the fire regime characteristics (Matrix Q) with CA fire regime group weights to study the fire regime characteristics because that ordination method is suitable for qualitative data [40]. Principal Component Analysis (PCA) was applied to the environmental characteristics (Matrix R), previously transformed using a logarithmic transformation to ensure a multi-normal distribution, with CA site weights because PCA is required to study quantitative data [40].

2.5. Fourth-Corner Analysis

A fourth-corner analysis is also a three-matrix approach, proposed by Legendre et al. [41], and was used to test the fire regime characteristics-environment relationship, corresponding to matrix D.

The measurement of resemblance between the standardized quantitative characteristics and the standardized qualitative characteristics recoded into binary variables was based on Pearson's r [41].

Statistical significance was tested by 9999 permutations using model 2, proposed by Legendre et al. [41]. Model 2 tested for significant differences under the null hypothesis (H_0), a random distribution of the fire regime groups among the sites, vs. an alternative hypothesis (H_1), a non-random distribution of the fire regime groups among the sites. This comparison was based on the premise that fire regime groups are observed at sites where their characteristics are constrained by environmental gradients. The model permuted entire rows of matrix L, which was equivalent to permuting the rows of matrix R and maintaining co-inertia among the fire regime groups and the link to matrix Q. The significance level was adjusted using the false discovery rate method proposed by Benjamini and Hochberg [42] for multiple testing. We used model 2 because the link between fire regime groups (Matrix L) and fire regime characteristics (Matrix Q) was fixed in our data. Therefore, the combination of the two permutation models, which better controls the Type I error, as proposed by Dray and Legendre [43] and improved by ter Braak et al. [44], was not suitable for our data.

2.6. Combining RLQ and Fourth-CornerAnalyses

The RLQ and four-corner analyses were combined by adapting the framework proposed by Dray et al. [45]. First, a global multivariate test of statistical significance between the fire regime (Matrix Q) and environmental characteristics (Matrix R) and the fire regime groups (Matrix L) based on the total co-inertia of the RLQ analysis was tested using 9999 permutations of entire rows of matrix L. This analysis tested for significant differences under the null hypothesis (H_0) of independence of fire regime characteristics among environmental gradients vs. an alternative hypothesis (H_1) of non-independence. This comparison was based on the premise that fire regime characteristics are constrained by environmental gradients. The Monte Carlo method was used, which is equivalent to SRLQ proposed by Dray and Legendre [43].

RLQ and fourth-corner analyses were then combined by performing a fourth-corner analysis on the main axis that resulted from the RLQ analysis.

All analyses were computed with the statistical RStudio software using the ade4 package version 1.5-2 [46].

3. Results

3.1. Global Multivariate Statistic

The null hypothesis of the global multivariate statistic based on the total co-inertia of the RLQ analysis is that fire regime and environmental characteristics are independent, and because the result was highly significant (*p*-value = 0.0001), we can conclude that there is a global relationship between fire regime characteristics (Matrix Q) and environmental characteristics (Matrix R) based on the fire regime groups (Matrix L).

3.2. Decomposition of Inertia

The results of the RLQ analysis, summarized in Table 3, show that the first RLQ axis represented 96% and the second RLQ axis 4% of the total co-inertia between the fire regime characteristics and the environment. The inertia of the environmental scores was well preserved on the first two RLQ axes, 86%, and the total inertia of the fire regime characteristics was 100%. However, the correlation was low for the second RLQ axis, 0.13.

Table 3. Eigenvalue decomposition summary of the RLQ analysis results.

Fire Regime Characteristics-Environment (RLQ Axes)	eig	covar	sdR	sdQ	corr
First RQL axis	0.45 (96%)	0.67	1.88	0.69	0.52
Second RLQ axis	0.02 (4%)	0.14	1.51	0.73	0.13
Environmental characteristics (Matrix R)	**inertia**	**max**	**ratio**		
First RLQ axis	3.54	4.10	0.87		
First and second RLQ axes	5.84	6.78	0.86		
Fire regime characteristics (Matrix Q)	**inertia**	**max**	**ratio**		
First RLQ axis	0.47	0.53	0.90		
First and second RLQ axes	1.00	1.00	1.00		
Fire regime groups (Matrix L)	**corr**	**max**	**ratio**		
First RLQ axis	0.52	1.00	0.52		
Second RLQ axis	0.13	1.00	0.13		

Fire regime characteristics-environment (RLQ axes) compare the eigenvalues (eig), covariance (covar), standard deviations (sdR and sdQ) and correlation (corr) of the two sets of scores projected on the RLQ axes. Environmental characteristics (Matrix R) and fire regime characteristics (Matrix Q), respectively, compare the inertia projected on the RLQ axes (inertia) to the maximum inertia on the axes of the PCA (max) and MCA ordination (max) in addition to a measure of concordance between the two projections (ratio). Correlation fire regime groups (Matrix L) compares the correlation projected on the RLQ axes (corr) to the maximum correlation on the axes of the CA ordination (max) in addition to a measure of concordance between the two projections (ratio).

3.3. Fire Regime Characteristics-Environmental Structures and Relationships

The null hypothesis of model 2 states that fire regime groups are randomly distributed across the study area, and because we can reject the null hypothesis, we can conclude that the fire regime groups were not randomly distributed across the sites. This indicates that fire regime groups were observed at sites where fire regime characteristics were constrained by environmental gradients.

The structure of fire regime characteristics, shown in Figure 3, changed along a complex environmental gradient represented by the first and second RLQ axes.

The first RLQ axis identifies and is significantly positively correlated with high density and long seasonality of fire activity (fire regime 1) at sites defined by humidity, density of population, high forest potential productivity (categories 1–2), broadleaf trees, mixed trees, shrubs and herbaceous land covers, density of livestock and land cover changes and negatively correlated with low density and short seasonality of fire activity (fire regime 2) at sites defined by seasonal temperature fluctuation, low forest potential productivity (categories 3–7), as well as by needle-leaf trees and managed land covers.

The second RLQ axis identifies and is significantly positively correlated with high density and short seasonality of fire activity (fire regime 2) at sites defined by medium forest potential productivity (category 3) and shrubs and negatively correlated with low density and long seasonality of fire activity (fire regime 4) at sites defined by density of population.

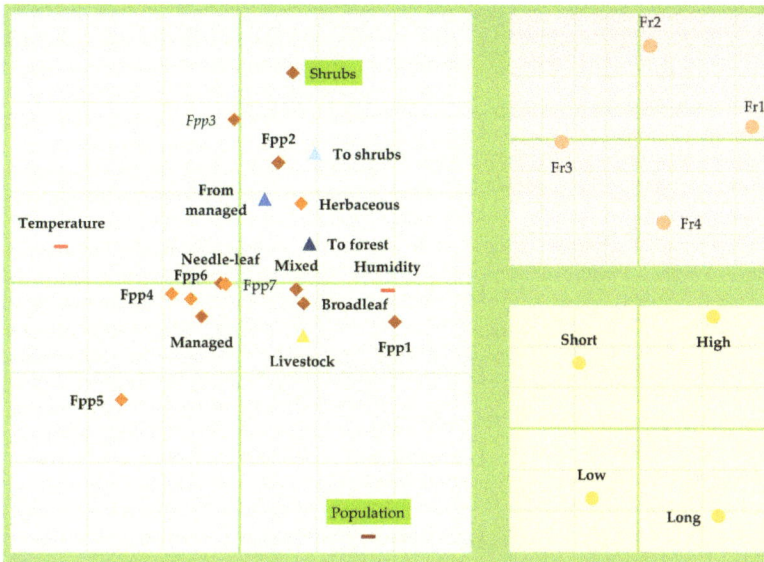

Figure 3. The RLQ scores of fire regime groups, fire regime characteristics and environmental characteristics on the first and second RLQ axes and significance of both characteristics with both axes. Fire regime groups (**upper right**; grid size: 0.5 × 0.5), fire regime characteristics (**lower right**; grid size: 0.5 × 0.5) and environmental characteristics (**left**; grid size: 0.2 × 0.2). Characteristics significantly correlated with the first RLQ axis are emphasised in bold, with the second RLQ axis in italics, with both RLQ axes in green and not significant in regular. Fire regime (Fr) groups, from 1–4. High and low density of fire activity (High and Low). Long and short seasonality of fire activity (Long and short). Environmental characteristics symbolized by a dash represent indirect gradients based on climate and humans, the rhombi represent resource gradients based on forest potential productivity and land cover and triangles represent indirect gradients based on livestock grazing and land cover changes. Population density (Population), Seasonal temperature fluctuation (Temperature), Forest potential productivity (Fpp) categories, from 1–7. Cultivated and managed vegetation/agriculture (incl. mixtures) (Managed). Livestock density (Livestock).

4. Discussion

Our results indicate that the selected fire regime characteristics, namely density and seasonality of fire activity, responded to environmental gradients, thus suggesting that fire regimes are spatially structured. Density and seasonality of fire activity changed along environmental gradients, as expected, and were structured primarily by direct and resource gradients and secondarily by indirect gradients, suggesting that climate, humans and forest potential productivity have an important role in the fire regime characteristics, constraining flammability and the amount of fuel and ignitions.

The first RLQ axis positively identified fire regime 1 with high density and long seasonality of fire activity at sites defined by humidity, but negatively identified fire regime 3 with low density and short seasonality of fire activity at sites defined by seasonal temperature fluctuation. The relationship between humidity and high density and long seasonality of fire activity is related to the fact that in more humid environments, the amount of fuel is higher. For example, these fire regime characteristics are found in the Atlantic region of Spain, one of the most fire-prone areas, where the fires tend to be smaller (medium to small) at the end of the winter and the beginning of the spring season, and larger during the summer and the beginning of autumn [32]. Besides, at the end of the winter and the beginning of the spring season, this region is affected by the Foehn winds that originate from southwest of the Iberian Peninsula. These results are consistent with the results reported by some authors who investigated the effects of precipitation and summer drought on fire seasonality [47] and fire activity in Spain [48] and other Mediterranean countries [49,50].

Furthermore, fire regime 1, characterized by high density and long seasonality of fire activity, was also found at sites defined by population. Lloret and Mari [51] suggested that humans affect fire ignition and seasonality in north-eastern Spain because of the absence of winter burning to improve pastures, thus causing the currently higher fire activity in the summer than during the medieval fire regimes when winter burning was practiced. This result is consistent with La Page et al. [5], who argued the effects of human activity on fire ignition and seasonality in northern Spain and Portugal. For example, Syphard et al. [52] also reported that humans are now a major determinant of fire seasonality and the location of fires in California and Archibald [20] in Africa. Moreover, the second RLQ axis negatively identified fire regime 4 with low density and long seasonality of fire activity at sites defined by population density. The difference in density of fire activity between fire regime 4 and 1 may be explained by the fact that density of fire activity is higher under intermediate population densities, as in other Mediterranean areas [52], and therefore the fire regime 4 is constrained by ignitions and potentially by other indirect gradients such as fire management.

Density and seasonality of fire activity change along a resource gradient defined by forest potential productivity and land covers, which is consistent with other studies performed in Spain [28,53] and other Mediterranean areas [24] or at the global scale [25]. Moreover, Vazquez de la Cueva et al. [53] observed stronger associations of forest potential productivity with fire activity than population density, thus suggesting that fuel amount is the main control on fire activity. This result was consistent with our study because the population density present in the RLQ score and the correlation were lower, although both were significant. One reason for the persistence of fires in northern Spain is because of its high forest potential productivity which results in the traditional agricultural burning to maintain pastures and clear land [47]. Shrubs were the only land cover that were positively correlated to high fire density and long seasonality (fire regime 1), as identified by the first RLQ axis, as well as by high density and short seasonality of fire activity (fire regime 2), identified by the second RLQ axis. The middle forest potential productivity (category 3) was also correlated with high density and short seasonality of fire activity (fire regime 2), as identified by the second RLQ axis. The difference in seasonality of fire activity may be related to the fact that seasonality is longer under high forest productivity [28], suggesting that fire regime 2 is constrained more by fuel flammability, determined by the seasonal temperature fluctuation, than by fuel amounts determined by humidity. Needle-leaf trees were the only tree cover that were negatively correlated to high density and to long fire seasonality of fire activity (fire regime 1), as identified by the first RLQ axis. This result may be related to the fact that these trees used to be located at depopulated higher elevations and were affected by lightning-caused fires, which currently account for less than 5% of fires in Spain and are more seasonal [32]. Furthermore, managed land cover and low forest potential productivity (categories 4–6) defined the sites where fire regime 3 with low density and short seasonality of fire activity were found.

Our results indicate that fire activity is preferably selective for high and middle environments and avoids low environments within Spain. High forest potential productivity was correlated with high density and long seasonality of fire activity, middle forest potential productivity was correlated with high density and short seasonality of fire activity, and low forest productivity was correlated with low density and short seasonality of fire activity. These results contrast with other studies arguing that fire is preferably selective for middle environments and avoids extreme environments [19,21,28].

Indirect gradients defined by livestock density and land cover changes exhibited lower scores and correlations, although they were significant. Livestock density was positively correlated with high density and long seasonality of fire activity (fire regime 1). These results are consistent with other studies, in which changes to fire regimes have been related to changes in livestock grazing in areas with a high forest potential productivity where fire has been traditionally used as a tool to regenerate pastures for livestock grazing, such as in Spain [30], or as a proxy for potential ignitions in alpine ecosystems [54]. By contrast to other regions, such as southern Africa, where an increase in livestock grazing implied a decrease in burned areas [55]. Additionally, land cover changes were positively correlated with high density and long seasonality. Land cover changes to shrubs resulting from forest degradation presented a higher correlation followed by land cover changes to forest resulting from afforestation. Land cover changes from managed areas resulting from the abandonment of traditional activities exhibited lower correlation. Our results are consistent with other authors, who observed a strong correlation between fire activity and land abandonment and forest extension in Spain [56], which were related as major driving forces of fire regime changes [30].

5. Conclusions

Our results indicate that environmental gradients constrain fire regime characteristics across Spain. A strong positive correlation indicated that seasonal temperature fluctuations, humidity, population density and forest potential productivity are the main constraints of density and seasonality of fire activity, therefore suggesting that environmental global change may have a strong effect on future fire regimes. For example, in Spain, an increase in humidity might lead to an increase in density and seasonality of fire activity by indirectly increasing the forest potential productivity; however, it will also depend primarily on the evolution of the seasonal temperature and population density, and secondarily on the resources and indirect environmental gradients.

The combination of the three-matrix approaches is a robust method for identifying relationships between fire regime characteristics and environmental gradients. Additional fire regime characteristics, such as the distinction between human and natural fires, and indirect environmental gradients, such as fire management, remain for future studies in order to establish a more complete understanding of the environmental constraints of fire regimes. This methodology can be used to test the hierarchical environmental control of fire at multiple scales, and the results can be used to model fire regime responses to environmental changes to refine management to preserve our ecosystems.

Acknowledgments: The contribution of the first author was partially supported by the International Fellowship Program of Bancaja and the University of Alcala. Comments from David L. Martell and anonymous reviewers are appreciated.

Author Contributions: María Vanesa Moreno designed the study, performed the analyses, interpreted the data, and wrote and reviewed the manuscript. Emilio Chuvieco contributed to the writing and reviewing of the manuscript.

Conflicts of Interest: The authors declare no conflict of interest.

Abbreviations

The following abbreviations are used in this manuscript:

UTM	Universal Transverse Mercator
PCA	Principal Component Analysis
BCN200	National Cartography Base
ED50	European Datum 1950
CA	Correspondence Analysis
MCA	Multiple Correspondence Analysis
FPP	Forest Potential Productivity

References

1. Bowman, D.M.J.S.; Balch, J.K.; Artaxo, P.; Bond, W.J.; Carlson, J.M.; Cochrane, M.A.; D'Antonio, C.M.; DeFries, R.S.; Doyle, J.C.; Harrison, S.P.; et al. Fire in the earth system. *Science* **2009**, *324*, 481–484. [CrossRef] [PubMed]
2. Bond, W.J.; Keeley, J.E. Fire as a global "herbivore": The ecology and evolution of flammable ecosystems. *Trends Ecol. Evol.* **2005**, *20*, 387–394. [CrossRef] [PubMed]
3. Gill, A.M. Fire and the Australian flora: A review. *Aust. For.* **1975**, *38*, 4–25. [CrossRef]

4. Westerling, A.L.; Hidalgo, H.G.; Cayan, D.R.; Swetnam, T.W. Warming and earlier spring increase western U.S. Forest wildfire activity. *Science* **2006**, *313*, 940–943. [CrossRef] [PubMed]
5. Le Page, Y.; Oom, D.; Silva, J.M.N.; Jönsson, P.; Pereira, J.M.C. Seasonality of vegetation fires as modified by human action: Observing the deviation from eco-climatic fire regimes. *Glob. Ecol. Biogeogr.* **2010**, *19*, 575–588. [CrossRef]
6. Magi, B.I.; Rabin, S.; Shevliakova, E.; Pacala, S. Separating agricultural and non-agricultural fire seasonality at regional scales. *Biogeosciences* **2012**, *9*, 3003–3012. [CrossRef]
7. Archibald, S.; Staver, A.C.; Levin, S.A. Evolution of human-driven fire regimes in Africa. *Proc. Natl. Acad. Sci. USA* **2011**, *109*, 847–852. [CrossRef] [PubMed]
8. Bowman, D.M.J.S.; Balch, J.; Artaxo, P.; Bond, W.J.; Cochrane, M.A.; D'Antonio, C.M.D.; DeFries, R.; Johnston, F.H.; Keeley, J.E.; Krawchuk, M.A.; et al. The human dimension of fire regimes on earth. *J. Biogeogr.* **2011**, *38*, 2223–2236. [CrossRef] [PubMed]
9. Cochrane, M.A.; Alenca, A.; Schulze, M.D.; Souza, C.M.; Nepstad, D.C.; Lefebvre, P.; Davidson, E.A. Positive feedbacks in the fire dynamic of closed canopy tropical forests. *Science* **1999**, *284*, 1832–1835. [CrossRef] [PubMed]
10. Mollicone, D.; Eva, H.D.; Achard, F. Ecology: Human role in Russian wild fires. *Nature* **2006**, *440*, 436–437. [CrossRef] [PubMed]
11. Martell, D.L.; Sun, H. The impact of fire suppression, vegetation, and weather on the area burned by lightning-caused forest fires in Ontario. *Can. J. For. Res.* **2008**, *38*, 1547–1563. [CrossRef]
12. Pechony, O.; Shindell, D. Driving forces of global wildfires over the past millennium and the forthcoming century. *Proc. Natl. Acad. Sci. USA* **2010**, *107*, 19167–19170. [CrossRef] [PubMed]
13. Krawchuk, M.A.; Moritz, M.A.; Parisien, M.-A.; van Dorn, J.; Hayhoe, K. Global pyrogeography: The current and future distribution of wildfire. *PLoS ONE* **2009**, *4*, e5102. [CrossRef] [PubMed]
14. Vannière, B.; Colombaroli, D.; Chapron, E.; Leroux, A.; Tinner, W.; Magny, M. Climate versus human-driven fire regimes in Mediterranean landscapes: The Holocene record of Lago dell'Accesa (Tuscany, Italy). *Quat. Sci. Rev.* **2008**, *27*, 1181–1196. [CrossRef]
15. Chuvieco, E. Global impacts of fire. In *Earth Observation of Wildland Fires in Mediterrranean Ecosystems*; Chuvieco, E., Ed.; Springer: Berlin, Germany, 2009; Volume 2, p. 257.
16. Harrison, S.P.; Marlon, J.R.; Bartlein, P.J. Fire in the earth system. In *Changing Climates, Earth Systems and Society*; Dodson, J., Ed.; Springer: Dordrecht, The Netherlands, 2010; pp. 21–48.
17. O'Connor, C.D.; Garfin, G.M.; Falk, D.A.; Swetnam, T.W. Human pyrogeography: A new synergy of fire, climate and people is reshaping ecosystems across the globe. *Geogr. Compass* **2011**, *5*, 329–350. [CrossRef]
18. Moritz, M.A.; Morais, M.E.; Summerell, L.A.; Carlson, J.M.; Doyle, J. Wildfires, complexity, and highly optimized tolerance. *Proc. Natl. Acad. Sci. USA* **2005**, *102*, 17912–17917. [CrossRef] [PubMed]
19. Parisien, M.-A.; Moritz, M.A. Environmental controls on the distribution of wildfire at multiple spatial scales. *Ecol. Monogr.* **2009**, *79*, 127–154. [CrossRef]
20. Archibald, S. Managing the human component of fire regimes: Lessons from Africa. *Phil. Trans. R. Soc. B* **2016**, *371*, 20150346. [CrossRef] [PubMed]
21. Krawchuk, M.A.; Moritz, M.A. Constraints on global fire activity vary across a resource gradient. *Ecology* **2011**, *92*, 121–132. [CrossRef] [PubMed]
22. Meyn, A.; White, P.S.; Buhk, C.; Jentsch, A. Environmental drivers of large, infrequent wildfires: The emerging conceptual model. *Prog. Phys. Geogr.* **2007**, *31*, 287–312. [CrossRef]
23. Batllori, E.; Parisien, M.-A.; Krawchuk, M.A.; Moritz, M.A. Climate change-induced shifts in fire for Mediterranean ecosystems. *Glob. Ecol. Biogeogr.* **2013**, *22*, 1118–1129. [CrossRef]
24. Pausas, J.G.; Bradstock, R.A. Fire persistence traits of plants along a productivity and disturbance gradient in Mediterranean shrublands of south-east Australia. *Glob. Ecol. Biogeogr.* **2007**, *16*, 330–340. [CrossRef]
25. Pausas, J.G.; Ribeiro, E. The global fire-productivity relationship. *Glob. Ecol. Biogeogr.* **2013**, *22*, 728–736. [CrossRef]
26. Chuvieco, E.; Giglio, L.; Justice, C. Global characterization of fire activity: Toward defining fire regimes from earth observation data. *Glob. Chang. Biol.* **2008**, *14*, 1488–1502. [CrossRef]
27. Archibald, S.; Lehmann, C.E.R.; Gómez-Dans, J.L.; Bradstock, R.A. Defining pyromes and global syndromes of fire regimes. *Proc. Natl. Acad. Sci. USA* **2013**, *110*, 6442–6447. [CrossRef] [PubMed]

28. Pausas, J.G.; Paula, S. Fuel shapes the fire-climate relationship: Evidence from Mediterranean ecosystems. *Glob. Ecol. Biogeogr.* **2012**, *21*, 1074–1082. [CrossRef]

29. Pausas, J.G.; Fernández-Muñoz, S. Fire regime changes in the western Mediterranean Basin: From fuel-limited to drought-driven fire regime. *Clim. Chang.* **2012**, *110*, 215–226. [CrossRef]

30. Moreno, M.V.; Conedera, M.; Chuvieco, E.; Pezzatti, G.B. Fire regime changes and major driving forces in Spain from 1968 to 2010. *Environ. Sci. Policy* **2014**, *37*, 11–22. [CrossRef]

31. Ninyerola, M.; Pons, X.; Roure, J. *Atlas Climático Digital de la Península Ibérica. Metodología y Aplicaciones en Bioclimatología y Geobotánica*; Universitat Autònoma de Barcelona: Bellaterra, Spain, 2005.

32. Moreno, M.V.; Chuvieco, E. Characterising fire regimes in Spain from fire statistics. *Int. J. Wildland Fire* **2013**, *22*, 296–305. [CrossRef]

33. Moreira, F.; Rego, F.C.; Ferreira, P.G. Temporal (1958–1995) pattern of change in a cultural landscape of northwestern Portugal: Implications for fire occurrence. *Landsc. Ecol.* **2001**, *16*, 557–567. [CrossRef]

34. Martínez, J.; Vega-García, C.; Chuvieco, E. Human-caused wildfire risk rating for prevention planning in Spain. *J. Environ. Manag.* **2009**, *90*, 1241–1252. [CrossRef] [PubMed]

35. Kohnke, H.; Stuff, R.G.; Miller, P.A. Quantitative relations between climate and soil formation. *Z. Pflanzenernähr. Bodenkd.* **1968**, *119*, 24–33. [CrossRef]

36. Sánchez Palomares, O.; Sánchez Serrano, F. *Mapa de la Productividad Potencial Forestal de España*; Ministerio de Agricultura, Alimentación y Medio Ambiente: Madrid, Spain, 2000.

37. Herold, M.; Mayaux, P.; Woodcock, C.E.; Baccini, A.; Schmullius, C. Some challenges in global land cover mapping: An assessment of agreement and accuracy in existing 1 km datasets. *Remote Sens. Environ.* **2008**, *112*, 2538–2556. [CrossRef]

38. Moreno, M.V.; Chuvieco, E. Validation of global land cover products for the Spanish peninsular area. *Rev. Teledetec.* **2009**, *31*, 5–22.

39. Dolédec, S.; Chessel, D.; Braak, C.J.F.; Champely, S. Matching species traits to environmental variables: A new three-table ordination method. *Environ. Ecol. Stat.* **1996**, *3*, 143–166. [CrossRef]

40. Legendre, P.; Legendre, L. Ordination in reduced space. In *Developments in Environmental Modelling*; Pierre, L., Louis, L., Eds.; Elsevier: Amsterdam, The Netherlands, 1998; Volume 20, Chapter 9; pp. 387–480.

41. Legendre, P.; Galzin, R.; Harmelin-Vivien, M.L. Relating behavior to habitat: Solutions to the fourth-corner problem. *Ecology* **1997**, *78*, 547–562. [CrossRef]

42. Benjamini, Y.; Hochberg, Y. Controlling the false discovery rate: A practical and powerful approach to multiple testing. *J. R. Stat. Soc. Ser. B* **1995**, *57*, 289–300.

43. Dray, S.; Legendre, P. Testing the species traits-environment relationships: The fourth-corner problem revisited. *Ecology* **2008**, *89*, 3400–3412. [CrossRef] [PubMed]

44. Ter Braak, C.J.F.; Cormont, A.; Dray, S. Improved testing of species traits—Environment relationships in the fourth-corner problem. *Ecology* **2012**, *93*, 1525–1526. [CrossRef] [PubMed]

45. Dray, S.; Choler, P.; Dolédec, S.; Peres-Neto, P.R.; Thuiller, W.; Pavoine, S.; ter Braak, C.J.F. Combining the fourth-corner and the RLQ methods for assessing trait responses to environmental variation. *Ecology* **2013**, *95*, 14–21. [CrossRef]

46. Dray, S.; Dufour, A.-B. The ade4 package: Implementing the duality diagram for ecologists. *J. Stat. Softw.* **2007**, *22*, 1–20. [CrossRef]

47. Vélez, R. Incenidos forestales y su relacion con el medio rural. *Rev. Estud. Agrosoc.* **1986**, *136*, 195–224.

48. Pausas, J.G. Change in fire and climate in the eastern Iberian Peninsula (Mediterranea Basin). *Clim. Chang.* **2004**, *63*, 337–350. [CrossRef]

49. Viegas, D.; Viegas, M. A relationship between rainfall and burned area for Portugal. *Int. J. Wildland Fire* **1994**, *4*, 11–16. [CrossRef]

50. Pereira, M.G.; Trigo, R.M.; da Camara, C.C. Synoptic patterns associated with large summer forest fires in Portugal. *Agric. For. Meteorol.* **2005**, *129*, 11–25. [CrossRef]

51. Lloret, F.; Marí, G. A comparison of the medieval and the current fire regimes in managed pine forests of Catalonia (NE Spain). *For. Ecol. Manag.* **2001**, *141*, 155–163. [CrossRef]

52. Syphard, A.D.; Radeloff, V.C.; Keeley, J.E.; Hawbaker, T.J.; Clayton, M.K.; Stewart, S.I.; Hammer, R.B. Human influence on California fire regimes. *Ecol. Appl.* **2007**, *17*, 1388–1402. [CrossRef] [PubMed]

53. Vázquez de la Cueva, A.; García del Barrio, J.M.; Ortega, M.; Sánchez, O. Recent fire regimes in peninsular Spain in relation to forest potential productivity and population density. *Int. J. Wildland Fire* **2006**, *15*, 397–405. [CrossRef]

54. Pezzatti, G.B.; Zumbrunnen, T.; Bürgi, M.; Ambrosetti, P.; Conedera, M. Fire regime shifts as a consequence of fire polity and socio-economic development: An analysis based on the change point approach. *For. Policy Econ.* **2013**, *29*, 7–18. [CrossRef]

55. Archibald, S.; Roy, D.P.; Brian, W.; Wilgen, V.; Scholes, R.J. What limits fire? An examination of drivers of burn area in southern Africa. *Glob. Chang. Biol.* **2009**, *15*, 613–630. [CrossRef]

56. Vázquez, A.; Moreno, J.M. Spatial distribution of forest fires in Sierra de Gredos (central Spain). *For. Ecol. Manag.* **2001**, *147*, 55–65. [CrossRef]

© 2016 by the authors. Licensee MDPI, Basel, Switzerland. This article is an open access article distributed under the terms and conditions of the Creative Commons Attribution (CC BY) license (http://creativecommons.org/licenses/by/4.0/).

forests

MDPI

Article

Effects of Lakes on Wildfire Activity in the Boreal Forests of Saskatchewan, Canada

Scott E. Nielsen [1,*], Evan R. DeLancey [2], Krista Reinhardt [1] and Marc-André Parisien [2]

[1] Department of Renewable Resources, University of Alberta, 751 General Services Building, Edmonton, AB T6G 2H1, Canada; kfrench@ualberta.ca

[2] Northern Forestry Centre, Canadian Forest Service, Natural Resources Canada, 5320 122nd Street, Edmonton, AB T5H 3S5, Canada; edelance@ualberta.ca (E.R.D.); marc-andre.parisien@canada.ca (M.-A.P.)

* Correspondence: scottn@ualberta.ca; Tel.: +1-780-492-1656

Academic Editors: Yves Bergeron and Sylvie Gauthier
Received: 21 August 2016; Accepted: 31 October 2016; Published: 5 November 2016

Abstract: Large lakes can act as firebreaks resulting in distinct patterns in the forest mosaic. Although this is well acknowledged, much less is known about how wildfire is affected by different landscape measures of water and their interactions. Here we examine how these factors relate to historic patterns of wildfire over a 35-year period (1980–2014) for the boreal forest of Saskatchewan, Canada. This includes the amount of water in different-sized neighborhoods, the presence of islands, and the direction, distance, and shape of nearest lake of different sizes. All individual factors affected wildfire presence, with lake sizes ≥5000 ha and amount of water within a 1000-ha surrounding area the most supported spatial scales. Overall, wildfires were two-times less likely on islands, more likely further from lakes that were circular in shape, and in areas with less surrounding water. Interactive effects were common, including the effect of direction to lake as a function of distance from lakeshore and amount of surrounding water. Our results point to a strong, but complex, bottom-up control of local wildfire activity based on the configuration of natural firebreaks. In fact, fire rotation periods predicted for one area varied more than 15-fold (<47 to >700 years) depending on local patterns in lakes. Old-growth forests within this fire-prone ecosystem are therefore likely to depend on the surrounding configuration of larger lakes.

Keywords: boreal forest; firebreaks; fire frequency; fire refugia; lake shadows; landscape patterns; Saskatchewan; spatial scale; wildfire

1. Introduction

Boreal forests are shaped by wildfires that are affected by spatial and temporal factors that influence both individual fire characteristics (e.g., size, shape, severity) and more generally the local landscape's fire cycle [1,2]. Although temporal factors affect when a fire occurs, its intensity (energy release), and its severity (ecological impacts), the landscape structure and physiography influence where fires are more likely to burn, and thus the spatial pattern of the forest mosaic [3–6]. More complex landscapes have more heterogeneous patterns in vegetation, which results in more complicated fire patterns [4,7]. One element of a landscape's physiography that is particularly important in the boreal forest is the amount and location of water bodies, since these features can act as natural firebreaks and thus barriers to fire spread [8–10].

There are a number of different landscape characteristics of water bodies that may affect the adjacent upland's fire patterns and local fire cycles, including the size, shape, and orientation of water bodies [11]. Large lakes are especially important in stopping the spread of fires, whereas smaller water bodies (or other natural firebreaks) may stop the spread of smaller fires and those fires perpendicular to the dominant wind direction [12]. In the boreal and hemi-boreal forest of central Canada where winds are dominantly from the west, forest stands on the eastern sides of larger lakes often have longer fire cycles [11]. Indeed, in some places, old-growth forests may be much more common on the eastern shore of large lakes and on islands within the lake [8,11,13]. Likewise, fire-intolerant tree species, such as white cedar (*Thuja occidentalis* L.) that are uncommon in upland forest locations away from lakes, are often associated with the downwind stretches of lake shorelines [14]. Although white cedar is most often considered a wet-adapted species given where it most often grows, it is also known to occur in the most extreme dry conditions along bedrock ridges and cliffs where fires are limited [15], thus suggesting that some combination of fire and moisture limits its distribution, and not solely water availability [16]. Landscape patterns of lakes may therefore not only influence distribution of where fires occur and thus associated patterns in forest age, but also forest composition.

Although it is well acknowledged that natural firebreaks such as lakes, a common feature of the boreal biome, exert a strong (if localized) bottom-up control on boreal wildfires, the mechanisms by which this occurs requires further examination. For instance, the "edge effect" of lakes, whereby wildfires are less likely closer to the edge of large lakes, has been described previously by others [9,14,17], but it is still unclear how this is affected by lake size, lake shape, and orientation (bearing) around lakes. Given that both landscape patterns and prevailing winds are non-random [12], the effect of lake proximity on wildfires should be dependent on and interacting with other factors. Similar to individual lake size, the overall proportion of natural firebreaks (including lakes) has been shown to affect wildfire activity in the boreal forest of Sweden [18]. This concept of vegetation fragmentation is in fact one of the principles guiding landscape fuel reduction [19]. However, simulation studies suggest that there is more to the "more nonfuel equates to fewer fires" concept, as the effectiveness of firebreaks is also a function of landscape configuration and fire regime characteristics (e.g., fire size, ignition patterns) [20]. A better understanding of the effects of large lakes on boreal wildfire activity thus implies the need to capture both primary effects of adjacent lake characteristics (e.g., proximity, proportion) and the synergistic effects of factors describing configuration and orientation.

The objective of this paper is to examine the landscape effects of lake size, lake shape, amount of surrounding water, presence of islands, and the distance and direction to lakes on wildfire patterns in the boreal forest of Saskatchewan, Canada. Specifically, we examine how the distance to different sized lakes interacts with direction to lakes, shape of lakes, and amount of surrounding water to influence wildfire patterns, thus helping to identify the conditions where old-growth forests are most likely within a fire-prone forested ecosystem. We did this by examining a 35-year history (1980–2014) of wildfires in northern Saskatchewan's boreal forest where fire perimeters have been mapped using aerial imagery and compared this information with landscape measures of water.

2. Materials and Methods

2.1. Study Area

The study area was defined as the Boreal Shield and Boreal Plain ecozones of northern Saskatchewan, Canada (Figure 1). This 361,924 km^2 area is bounded in the north by Lake Athabasca, Black Lake, and Wollaston Lake, which delimits the northern limit of boreal forests and the start of taiga forests. Taiga forests were excluded from this study given the smaller size of this ecozone in Saskatchewan and lower amounts of forest cover. The Boreal Plain ecozone represents the southern part of the study area (178,225 km^2) where boreal forest transitions to the Prairie ecozone. The area's largest human imprint is, by far, at this transition (i.e., the Boreal Transition ecoregion) (Figure 1c). Large tracts of land in this area have been converted to agriculture and, given its higher population density than areas to the north, are subject to more intensive fire suppression effort. The Boreal Plain is underlain by sedimentary rock with thick glacial deposits with water bodies being common (8.4% water). Forest exploitation occurs in the Boreal Plain but is relatively minor and is localized. The Boreal Shield ecozone covers the northern parts of the study area (183,699 km^2) and is characterized by rolling terrain of Precambrian bedrock (Canadian Shield) with shallow soils derived from thin glacial deposits and numerous water bodies (19.9% of the ecozones is water). There is no forestry in the Boreal Shield. North of the Boreal Transition ecoregion, the anthropogenic impact on fire activity in Saskatchewan is generally considered to be relatively minor, given the extreme nature of fire-conducive weather (i.e., thereby limiting fire suppression effectiveness) and the lack of fire management activities beyond the immediate vicinity of communities [21].

The boreal ecozones of Saskatchewan have a continental climate with cold winters and warm summers with an average annual (1981–2010) temperature of 0.2 °C and an average annual precipitation of 486 mm with 31% falling as snow [22]. Common tree species to both ecozones include black and white spruce (*Picea mariana* (Mill.) BSP and *P. glauca* (Moench) Voss), jack pine (*Pinus banksiana* Lamb.), eastern larch (*Larix larciana* (Du Roi) K. Koch), balsam fir (*Abies balsamea* (L.)), trembling aspen (*Populus tremuloides* Michx.), balsam poplar (*Populus balsamifera* L.), and white birch (*Betula papyrifera* Marsh.). Fire is common to the region with some of the highest fire frequencies recorded in the Canadian boreal forests with annual area burned ranging from a 0.1% to >1.5% [1,22]. The area north of the Churchill River, which approximately divides the Boreal Plain in the south from the Boreal Shield in the north, maintains a relatively natural fire regime as there is little forest protection (i.e., "let it burn" policy). Active fire suppression in this area only occurs around communities and infrastructure (no active timber leases/management). Although there is fire suppression south of the Churchill River, wildfires are still common, including large intense wildfires such as those that occurred around La Ronge during the summer of 2015.

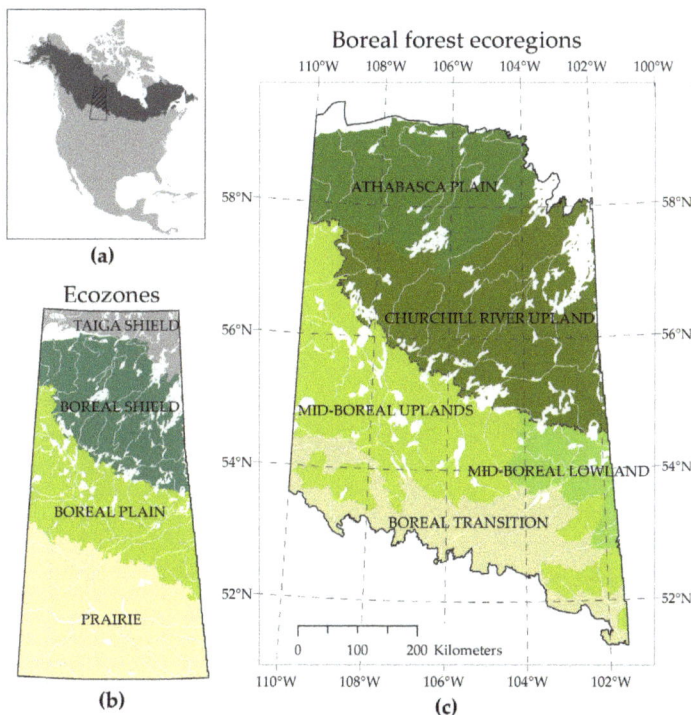

Figure 1. Location of study area in Saskatchewan, Canada. (**a**) Boreal and taiga forests of North America (dark gray), Saskatchewan boundary (hollow polygon), and boreal ecozone (crosshatch); (**b**) Ecozones of Saskatchewan with major lakes and rivers in white; (**c**) Ecoregions within the boreal ecozone of Saskatchewan where fire patterns were studied (note boundary line delineating plains from shield).

2.2. Historical Mapping of Wildlife Patterns

We used mapped fire perimeters from 1980 to 2014 (35 years) for Saskatchewan's Boreal ecozones and a 50-km buffer around the study area to ensure moving window landscape analyses were not biased by edge effects. Fire data were obtained from the Canadian Forest Service National Fire Database [2]. Only fires greater than 200 ha were retained as those smaller than that size are inconsistently reported (fires <200 ha account for less than 3%–4% of the area burned, [1]). Fire perimeters were then rasterized into a binary burned/unburned (0/1) grid using a cell size of 25 m (0.0625 ha). Estimates of fire frequency were derived for each ecoregion to compare with local predicted variation in fire frequency from our models. Fire frequency was defined as the fire rotation period based on the proportion of landscape burned over the study time period [16]. For this study, we defined the study period as 35 years (1980–2014). Rather than estimate, as standard practice, the proportion of the landscape burned based on total area, we first removed water pixels to represent the area that was truly "burnable". This resulted in shorter fire rotation periods, especially for the Boreal Shield where the amount of water approached 20%.

2.3. Landscape Measures of Water Bodies

Water bodies and island boundaries were defined by National Hydrographic Network (NHN) data [23]. Most water bodies in the boreal ecozone were lakes (81.5% for all water features ≥500 ha and 96.4% for all water features ≥5000 ha). Five landscape predictors relating to landscape measures of water were derived from NHN data and used to relate to local patterns in wildfire occurrence (Table 1). This included: (1) amount (proportion) of water in surrounding area; (2) distance (in 100 m units) to nearest water body; (3) direction to nearest water body ("eastness" index); (4) shape index (round to more irregular) of nearest water body; and (5) presence of islands. Table 1 includes a description of expected (predicted) linear responses of wildfire likelihood for each factor.

Table 1. Landscape predictors of wildfire presence based on landscape measures of water bodies, including scale (ha) at which they were measured. All predictor variables were assumed to linearly relate to the probability of wildfire at a site.

Landscape Measure of Water	Scales (ha)	Prediction (Probability of a Site (25 m Pixel) Burning)
Amount (proportion)	1000, 10,000, 100,000	*Negative*—Wildfire decreases as amount of water surrounding a site increases
Distance to (100 m units)	Water body: 500, 1000, 5000, and 10,000	*Positive*—Wildfire increases as the distance from water increases
Direction to (eastness index)	Water body: 500, 1000, 5000, and 10,000	*Positive*—Wildfire increases as the direction to water becomes more eastern (i.e., west side of lakes)
Shape index (irregularity)	Water body: 500, 1000, 5000, and 10,000	*Negative*—Wildfire decreases as shape of water bodies becomes more irregular
Islands	N.A. (binary)	*Negative*—Wildfire decreases on islands after accounting for other landscape measures of water

Amount (proportion) of water in the surrounding area was measured in moving circular windows of three different sizes (1000, 10,000, and 100,000 ha) using the focal statistics tool in ArcGIS. Distance to water was measured using the Euclidean distance tool in ArcGIS and transformed to log10 scale (distance +1) since distance values where highly left skewed. Direction to nearest water body was calculated using the formula: COS ("bearing to lake"−90), with the bearing calculated in a geographic information system (GIS) using the ArcGIS raster calculator tool. Using this index a maximum value of 1 represented locations that were on the west side of water bodies with the bearing to water bodies being east. Conversely, a minimum value of −1 represented a location that was on the east side of water bodies with the bearing to water bodies being west. Finally, north and south directions had an equivalent and intermediate value of 0. Shape of water bodies was calculated with the spatial statistics tool in the Patch Analyst extension [24] where more round-shaped water bodies approached a minimum possible value of 1, while more irregular-shaped water bodies were much greater than 1 (our data ranged from ~2 to 44).

We evaluated a range of spatial scales for each of the water variables, except islands, which were simply noted as a binary variable based on whether the location was on a mainland (0) or island (1). The island variable was included in addition to the other landscape measures of water to test whether such a well-defined firebreak further reduced the probability of a wildfire beyond that of which can be explained by the other landscape measures of water. For amount of surrounding water, we examined three moving window sizes of 1000, 10,000, and 100,000 ha (Figure 2).

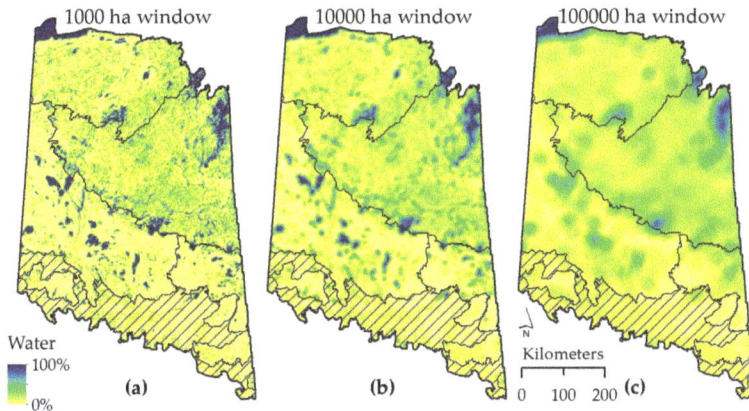

Figure 2. Amount of water in different moving window sizes in the boreal forest of Saskatchewan, Canada. Window sizes include: (**a**) 1000 ha; (**b**) 10,000 ha; (**c**) 100,000 ha. Areas of dark blue represent areas dominated by water, while yellow areas represent areas with very low amounts of surrounding water.

In comparison to surrounding amount of water, distance to nearest water body (Figure 3), direction (eastness) to nearest water body (Figure A1), and shape of nearest water body (Figure A2) were assessed for four different water body size thresholds of 500, 1000, 5000, and 10,000 ha. For instance, the 500-ha threshold required that water bodies be ≥500 ha in size before measuring their distance, direction, and shape. All variables were mapped in a GIS at a 25-m resolution.

2.4. Regression Modeling

We randomly sampled 50,000 non-water sites (25-m pixels) in the Boreal ecozone that were at least (minimum distance) 100 m apart. This large sample size was chosen given high local variability in landscape factors (e.g., distance from water, amount of surrounding water, etc.) and due to the large size of our study area. Because the Boreal Transition ecoregion along the southern border of the boreal forest was dominated by agricultural conversion (65%) with wildfire activity altered [22], we subsequently removed this ecoregion from all analyses. This resulted in a total of 41,713 sample locations with an overall sample intensity of 0.175 locations per km^2 and an average nearest neighbor distance between sample locations of 1325 m (SD = 716 m). For each sample pixel, we queried the presence of recent (1980–2014) wildfires (binary response variable), the ecoregion identity, and all landscape measures of water. Although other factors, such as vegetation configuration, fire-conducive weather, the spatial-temporal patterns of fire ignitions, and anthropogenic features (e.g., roads, cutblocks) also affect fire spread, these were not considered in this study in order to focus on the effect of large water bodies.

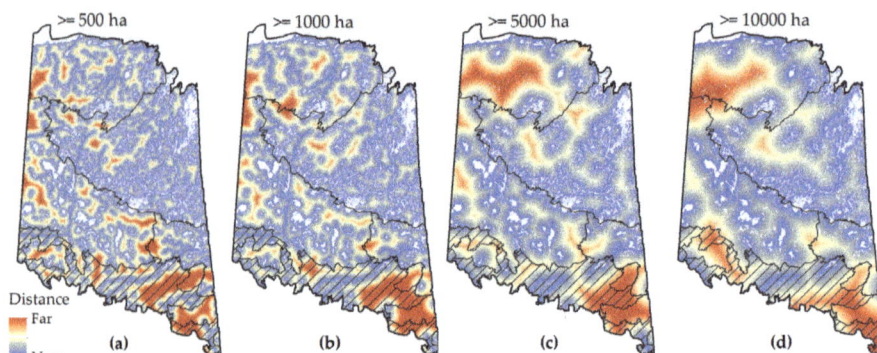

Figure 3. Distance to nearest water body of defined sizes in the boreal forest of Saskatchewan, Canada. Maps represent distance to water at (**a**) ≥500 ha; (**b**) ≥1000 ha; (**c**) ≥5000 ha; (**d**) ≥10,000 ha. Areas of red are further from water; while areas of blue are close to water (water bodies are white).

We used a mixed effects logistic regression model (xtmelogit command in STATA; [25]) with ecoregion set as a random effect (intercept) to estimate the probability that a site (25-m pixel) would burn from wildfires over the 35-year fire history based on the amount of surrounding water, distance to water, direction to water body (eastness index), shape irregularity of water feature, and presence of islands (Table 1). Correlations among linear terms were assessed for each model to identify possible collinearity issues as defined by Pearson correlations greater than |0.7|. The random effect accounted for non-independence of samples (random pixels) within ecoregions, as well as accounting for ecoregion differences in fire frequency (i.e., a random intercept for each ecoregion).

Because we expected *a priori* that the presence of wildfires around water bodies depended on a combination of landscape factors and their interactions (particularly with distance), we fit a number of two-way and three-way interactions (Table 2). Six two-way interactions were tested that included: (1) distance to water and amount of surrounding water; (2) distance and direction to water; (3) distance and shape of nearby water; (4) shape of nearby water and amount of surrounding water; (5) shape and direction of nearby water; and (6) direction to nearby water and amount of surrounding water (Table 2). Finally, we considered three three-way interactions that included: (1) amount of surrounding water and direction and distance to nearest water body; (2) amount of surrounding water and distance and shape of nearest water bodies; and (3) amount of surrounding water and direction (eastness) and shape of nearest water body. See Table 2 for a list of hypothesized predictions.

2.5. Model Building

Model building followed a hierarchical approach given the large number of possible scales and variable combinations. First, each of the four water body size thresholds (500, 1000, 5000, and 10,000 ha; Table 1) were evaluated to determine which water body size threshold most affected the likelihood of a site burning from wildfires. All models included the island variable and the distance to nearest water body, direction (eastness) to nearest water body, and shape of nearest water body. Second, we evaluated support among the three different window sizes measuring amount of surrounding water (1000, 10,000, 100,000 ha; Table 1) after holding the water body size threshold identified in the first analysis constant. Third, we assessed support of interactions for the most supported "linear" model (scale) that included both the selected water body size threshold variable and window size measuring the amount of surrounding water. Only linear responses among predictors and wildfire presence were tested as we did not expect *a priori* non-linear responses between water variables and wildfire probability.

Table 2. Hypothesized interactions among landscape predictors of wildfire presence based on landscape measures of water bodies. Scale of variables chosen reflects the best fit scale from linear terms.

Interactions	Prediction (Probability of a Site (25-m Pixel) Burning)
Distance × Amount	Positive—Wildfire increases more with distance from lakes when surrounded by more water
Distance × Direction	Positive—Wildfire increases more with distance from lakes with an east bearing (west shores that represent the source of prevailing winds)
Distance × Shape	Negative—Wildfire decreases more with distance from lakes when nearby lakes are irregularly shaped
Direction × Amount	Negative—Wildfire decreases more on west shores when surrounded by more water
Direction × Shape	Negative—Wildfire decreases more on west shores if nearby lake is irregularly shaped
Shape × Amount	Negative—Wildfire decreases more when nearby lake is irregular shaped and surrounded by more water
Amount × Distance × Direction	Distance depends on amount of surrounding water and direction (can be further from east shores when surrounding amount of water is high)
Amount × Distance × Shape	Distance depends on both amount of surrounding water and shape of lakes
Amount × Direction × Shape	Direction depends on amount of surrounding water and shape of lakes

2.6. Model Selection and Assessment

Model selection (support) was evaluated using Akaike's Information Criteria (AIC), which assesses the trade-off between model fit and complexity [26], thus representing a measure of model parsimony [27]. For the most supported final model, we examined model predictions (responses) relative to our hypothesized predictions (Tables 1 and 2) based on the model parameters and predicted responses (graphs and maps). Basic descriptive statistics of model predictions by ecoregion were also used to help interpret local landscape variation in wildfire likelihood as compared to ecoregion-scale averages. Finally, we assessed model predictive accuracy of the most supported model based on a non-parametric Receiver Operating Characteristic (ROC) Area Under the Curve (AUC) estimate using the "roctab" command in STATA [25] with the 95% confidence interval reported. Models with AUC values ranging from 0.5 to 0.7 were considered to have "low" model accuracy, values between 0.7 and 0.9 were considered to have "good" model accuracy, and finally those above 0.9 were considered to have "high" model accuracy [28,29]. A model that is no different than random would have an AUC of 0.5.

2.7. Predicting Local Variation in Fire Frequency

Local fire frequency was estimated from model predictions based on the probability of wildfire over the 35-year study period. For the purpose of comparison with ecoregion estimates, we modified the definition of the fire rotation period [e.g., 16] for model-based local predictions of wildfire by swapping the proportion of the landscape burned with the probability (p) of the site being burned (e.g., fire rotation period = $35/p$). When site (pixel) probabilities are summed across the landscape they estimate the total predicted area burned and thus when divided by landscape area (in this case "burnable" area) they represent the proportion of landscape burned. However, unique in this situation is that local landscape conditions can be assessed for how they change the fire rotation period. Local variation in the fire rotation period due to the landscape effects of water was then compared to regional estimates to illustrate the influence of water bodies on local patterns in the fire regime.

3. Results

3.1. Fire History (1980–2014)

A total of 38.9% of Saskatchewan's boreal forest burned between 1980 and 2014 (Figure 4), resulting in an average fire rotation period of 90 years. However, when considering individual ecozones the

Boreal Shield burned substantially more area at 62.6%, resulting in an average fire rotation period of 56 years. In contrast, the Boreal Plain burned 17.5% of the area resulting in a fire rotation period of 200 years; however, when removing the Boreal Transition ecoregion, 23.9% of the Boreal Plain ecozone burned resulting in an average 146 year fire rotation period.

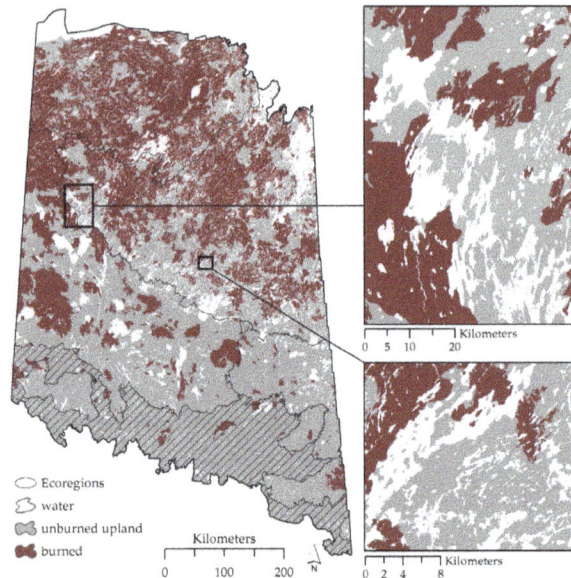

Figure 4. Wildfire patterns in the boreal ecoregions of Saskatchewan, Canada between 1980 and 2014. Mapped water bodies are in white. Stippled area represents the Boreal Transition that was removed from analyses due to significant agricultural conversion and few wildfires.

3.2. Wildfire Patterns as It Relates to Landscape Characteristics of Water Bodies

3.2.1. Water Body (Lake) Size Thresholds

The most supported water body size threshold for measuring distance, direction, and shape to nearby water bodies (all models included the presence of islands) was 5000 ha (Table 3). The second most supported scale was 10,000 ha, thereby indicating that moderately-large to large water bodies were more influential in acting as firebreaks than smaller (<5000 ha) water bodies. Given that only 3.6% of water features ≥5000 ha were defined as rivers, we simplify hereto our terminology of water features from "water bodies" to "lakes".

3.2.2. Scale of Window Size for Amount of Surrounding Water

After holding lake size threshold for distance, direction, and shape of nearest lake constant at the 5000 ha size (Table 3), the window size (scale) measuring the amount (proportion) of surrounding water was most supported at a 1000 ha window size (Table 4). This indicated a more local effect of surrounding water on wildfire probability, particularly given the lack of support for the largest window size examined of 100,000 ha (ΔAIC = 93.0).

Table 3. Evaluation of lake threshold sizes (scales) on probability of a site being burned in the boreal forest ecozone of Saskatchewan, Canada between 1980 and 2014. Null model represents mean probability of burning across all sites. Models are ranked from most to least supported using Akaike's Information Criteria (AIC). Number of parameters in the model (K), change in AIC from the top-model (ΔAIC), and Akaike weights (w_i) are listed.

Model, Scale (ha)	K	AIC	ΔAIC	w_i
5000	6	48,466.2	0.0	1.00
10,000	6	48,675.0	208.8	<0.01
500	6	48,833.2	367.0	<0.01
1000	6	48,857.6	391.4	<0.01
Null (mean) model	1	57,281.4	8815.2	<0.01

Table 4. Evaluation of different window size thresholds of the amount of surrounding water on wildfire presence (1980–2014) in the boreal forest ecozone of Saskatchewan, Canada. Null model represents the best fitting model using island, distance to lake, direction to lake, and shape of lake with a lake size threshold of 5000 ha (see Table 3). Models are ranked from most to least supported using Akaike's Information Criteria (AIC). See Table 3 for definition of terms.

Model, Scale (ha)	K	AIC	ΔAIC	w_i
1000	7	48,256.0	0.0	0.99
10,000	7	48,265.6	9.7	0.01
100,000	7	48,349.0	93.0	<0.01
Null water size model	6	48,466.2	210.2	<0.01

3.2.3. Support for Interacting Landscape Effects on Wildfires

When considering two-way interactions between landscape measures of water (Table 2), the most supported (lowest AIC) interaction term was between amount of surrounding water (1000 ha) and direction to nearest water body $\geq$5000 ha, followed by direction and distance to nearest water body (Table A1). Three other two-way interactions were marginally more supported than the null base model of linear factors without interactions. Finally, the interaction between distance and shape of nearest water body had no model support (higher AIC values than that of the null model; Table A1). Models considering different combinations of two-way interactions supported a model with four of the five two-way interactions with only the interaction of distance and shape to nearest lake less supported than the base null model (Table A2). Finally, two of the three tested three-way interactions were supported (Amount $\times$ Distance $\times$ Direction and Amount $\times$ Distance $\times$ Shape) after holding the most supported two-way interactions constant (Table A3).

3.2.4. Model Parameters, Predictions, and Accuracy

The most supported model had 11 landscape variables (Table 5) and good overall model predictive accuracy with a ROC AUC of 0.759 (95% CI = 0.754–0.763). Main hypothesized predictions for linear variables (see Table 1 for list of predictions) were supported for four of the five linear terms in the models (predictions in Table 1; Table 5 with results) and four of the five two-ways interactions (see Table 2 for predictions). Correlation among linear terms in the final model (i.e., model collinearity) was minimal (Pearson | r | < 0.442). Supported hypothesized predictions included island, distance to lake, amount of surrounding water, and shape of lake, while the unsupported prediction was related to direction to lake. All of these factors, however, depended on interactions with each other making interpretations complex. Finally, three 3-way interactions were evident for distance, amount, and direction of lakes (negatively related) and distance, amount, and shape of lakes (positively related) (Table 5).

Table 5. Model parameters for the most supported model describing the probability of wildfire between 1980–2014 as a function of landscape characteristics of water in the boreal forest of Saskatchewan, Canada. Note parameter coefficient (β) is also reported as an odds ratio. Random intercepts for ecoregions of Mid-boreal Uplands, Mid-boreal Lowlands, Churchill River Upland, and Athabasca Plain were −0.330, −2.569, 1.267, and 1.635, respectively.

Variable	β	S.E.	95% C.I. Lower	Upper	Odds ratio
Island	−0.684	0.142	−0.962	−0.406	0.504
Amount of water (1000 ha)	−2.179	0.190	−2.551	−1.806	0.133
Distance to water (5000 ha)	0.304	0.027	0.252	0.356	1.356
Direction to water (5000 ha)	−0.420	0.100	−0.615	−0.224	0.657
Shape of nearby water (5000 ha)	−0.023	0.002	−0.026	−0.020	0.977
Amount × Direction	1.313	0.357	0.614	2.012	3.718
Distance × Direction	0.337	0.043	0.252	0.422	1.401
Amount × Shape	0.011	0.011	−0.011	0.033	1.011
Direction × Shape	0.005	0.002	0.002	0.009	1.005
Distance × Amount × Direction	−1.397	0.193	−1.776	−1.019	0.247
Distance × Amount × Shape	0.012	0.006	<0.001	0.025	1.012
Intercept	−1.075	0.833	−2.706	0.558	0.341

When considering linear terms, the probability of wildfire was approximately two-times less likely on islands (odds ratio = 0.504) than mainland sites after controlling for other landscape factors, 36% more likely (odds ratio = 1.36) per 10-fold increase in distance from lakes, 7.7 times (odds ratio = 0.13) more likely for areas with no surrounding water compared with areas completely surrounded by water, and 1.5 times less likely (odds ratio = 0.66) on the west side (eastern direction) of lakes compared to north or south orientations, although the orientation-effect depended on the distance from the lake, amount of surrounding water, and shape of nearby lake (Table 5). Finally, wildfire presence decreased marginally (odds ratio = 0.98) as the lake shape became more irregular. Figure 5 illustrates the predicted responses for each landscape variable in the Churchill River Uplands (excluding islands) with variability in responses representing the variation in predicted wildfire probability as influenced by other landscape factors of water. Major differences along landscape gradients in water were most apparent for distance to lake (Figure 5a) and amount of surrounding water (Figure 5b). The effect of distance to lakes was most pronounced within 2.5 km where wildfire likelihood was dramatically reduced and also much lower when amount of surrounding water was greater than 20% (Figure 5).

Due to the complexity of interpreting interactions from coefficients, model predictions were also graphed based on different combinations of factors (Figure 6), as well as mapped for one example area in the Churchill Uplands Ecoregion (Figure 7). Predictions in Figure 6 illustrate changes in probability of wildfire for a 35-year period as a function of amount (proportion) of surrounding water (1000-ha window) and distance to nearest lake ≥5000 ha for different distance classes of 10, 100, 1000, and 10,000 m. This was done for both the west-side (Figure 6a) and east-side of lakes (Figure 6b) since the amount of water, distance from lake, and direction to lake represented the strongest interaction. The interaction between the distance to large lakes and the proportion of water is typical of the hypothesized predictions of wildfire patterns on the west side of lakes that represent in this region the direction of prevailing winds, with the likelihood of wildfire being greater further from lakes when amount of surrounding water was low (Figure 6a). However, there was less variation in wildfire likelihood by distance to lake for the east side of lakes when there was little surrounding water (Figure 6b). This variation is likely the result of complex interactions among the different landscape factors, given that this effect is not observed in the singular relationships illustrated in Figure 5. Regardless, a strong "edge effect" of lakes can be observed on both sides of lakes as illustrated by the consistent low likelihood of burning when adjacent to lakes (10 m). Specifically, the risk ratio of wildfire probability on the west side of lakes was 3.5 times more likely for sites 100 m distant from lakes than 10 m distant and 6.6 times more likely for sites 10,000 m distant from lakes than 10 m distant (Figure 6).

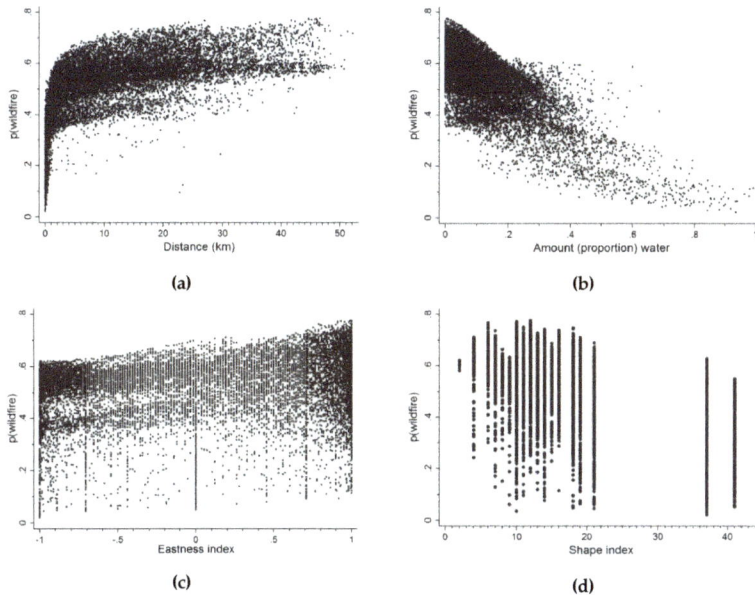

Figure 5. Model predictions of wildfire probability, p(wildfire), at sample locations (25-m pixel) in the Churchill River Upland over a 35-year period (1980–2014) as a function of: (**a**) distance from lake ≥5000 ha; (**b**) amount (proportion) of water within 1000 ha; (**c**) eastness to nearest lake ≥5000 ha (−1 is east side/west bearing to water; +1 is west side/east bearing); and (**d**) shape of nearest lake ≥5000 ha (larger values are more irregular shaped). Variation in predictions for any one variable represents the effects of other linear and interacting factors.

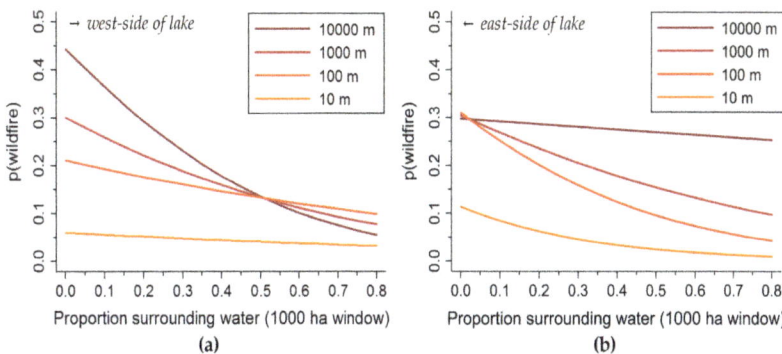

Figure 6. Model predictions of wildfire probability, p(wildfire), for sites (25-m pixels) within the Churchill River Upland ecoregion over a 35-year period (1980–2014) as a function of the amount (proportion) of water in 1000 ha surrounding windows and distance classes (10, 100, 1000, 10,000 m) to lakes ≥5000 ha for either the west-side of lakes (**a**) or east-side of lakes (**b**). All sites were assumed to be mainland sites with lake shape held at its mean value.

Figure 7 illustrates local spatial patterns in predicted wildfire probability (assuming a 35-year period) for the southern part of Reindeer Lake in the Churchill River Upland ecoregion where fire is common (i.e., 62.6% of Boreal Shield uplands burned in the 35-year period). Differences in predicted wildfire activity were evident when comparing wildfire probability around Reindeer Lake (large lake in center and top of map) with smaller, circular-shaped lakes (but still >5000 ha) in the southeast corner of the map (Figure 7). Wildfire probabilities on some islands (especially eastern sides of islands) and some shorelines along Reindeer Lake were predicted to be less than 0.05 (per 35-year period) resulting in a fire rotation period that exceeded 700 years. Many of the larger islands and immediate shorelines of larger lakes had wildfire probabilities between 0.05 and 0.25 or a fire rotation period between 140 and 700 years. In contrast to these longer fire rotation periods, some areas were estimated to have wildfire probabilities exceeding 0.75 over the 35-year period with a corresponding fire rotation period of less than 47 years. Overall, landscape variability in fire rotation periods within this small area was predicted to vary more than 15-fold depending on local landscape characteristics of the lakes. This local variability in probability of wildfire and thus variability in fire rotation cycle was also pronounced in the Mid-Boreal Upland and Athabasca Plain ecoregions, but less so for the Mid-Boreal Lowland ecoregion (Table A4, Figure A3).

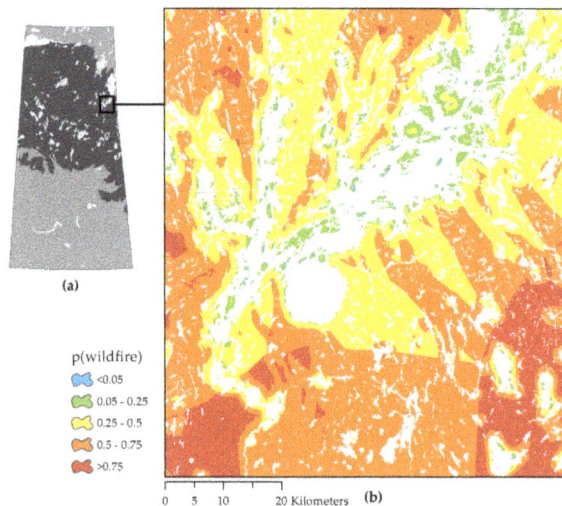

Figure 7. Model predictions of the likelihood of a wildfire, p(wildfire), over a 35 year period (1980–2014) as a function of the presence of islands and the amount of water, distance, direction, and shape of nearest large lake. (**a**) Map of Saskatchewan indicating location of boreal forests (dark gray) and the location of mapped predictions (small black polygon with leader line); (**b**) predicted probability of wildfire, p(wildfire), per 35-year period for the area around the southern parts of Reindeer Lake in northeast Saskatchewan, Canada.

4. Discussion

The results of this study support the idea that spatial patterns, and more specifically the local likelihood, of wildfires are not uniformly distributed within the boreal forest. Natural firebreaks, particularly large lakes, represent a strong bottom-up control on wildfire activity. Despite the stochastic nature of wildfire, all of the hypothesized landscape measures of water affected wildfire patterns. This includes the amount of water, the distance to large lakes, the direction to large lakes, the shape of the lake and, finally, insularity. Although the primary effect of these individual factors has been documented in different biomes of the world, we show here that they interact amongst themselves to yield highly heterogeneous patterns of wildfire likelihood.

Perhaps the most common measure of assessing the effect of natural firebreaks on wildfire occurrence is the distance of wildfire from those breaks. When fires burn less frequently near lakes, "fire shadow" patterns develop [14]. Our results are in agreement with those reported elsewhere in the Canadian boreal forest that the effect of large lakes may extend several kilometers beyond the edge of the lakes and wetlands [9,10]. The most pronounced effects of lakes are, however, within the first 100 m of the lakeshore [17]. Our results suggest a 3.5-fold increase in risk of wildfire between 10 and 100 m lake distances. The increase in wildfire risk with distance from lakes stabilized at ~2.5 km, which is similar to that reported elsewhere [30].

Amount of open water in the surrounding landscape also influenced wildfire patterns. In northern Saskatchewan's boreal forest, the amount of surrounding water appears to be as important to wildfire distribution as proximity to lakes. These results are similar to those in boreal Sweden, where mean fire intervals were correlated to wetland density, provided that these wetlands were moist enough to limit fire ignition and spread [18]. Given the strength of this control, fragmented landscapes simply do not burn as well as those with highly continuous fuels. Indeed, the proportion of natural firebreaks is often identified as a key variable explaining broad-scale wildfire patterns in the boreal forests of Canada [31,32]. The specific mechanism by which reductions in wildfire frequency occurs is, however, complex. Not only are lakes (and other non-fuels) limiting the potential spread of large wildfires, but they also eliminate possible sources of ignition. Fire shadows therefore develop from both a lack of ignition in lakes and the impossibility of a fire growing out of the nonfuel [33].

To provide a more comprehensive assessment of the effects of large lakes on boreal wildfire activity, this study incorporated a number of known or suspected factors that affect fire ignition and spread. For instance, our results support others who demonstrate that orientation of landscape features may impede or promote (e.g., river valleys parallel to dominant winds) the spread of large wildfires [12]. Although this remains to be fully investigated in the boreal forest, the orientation of natural firebreaks may not just affect the frequency of wildfires, but also the type. For instance, crown fires may wrap around firebreaks and burn as less-intense surface fires [34]. The shape of firebreaks, including lakes, is also important [35]. We found that more irregular-shaped lakes reduce wildfire activity in Saskatchewan, but that the effect of this factor is highly dependent on other factors, notably size and orientation to nearest large lake. For example, elongated features perpendicular to an incoming fire may provide a more effective fuel break than a round-shaped lake [12]. However, if the lake is too narrow wildfires may simply breach (i.e., through fire spotting) the firebreak.

Interactions among factors further highlight the complexity of the relationship between wildfire patterns and natural firebreaks. Interestingly, the top-two most supported interactions in our analysis (amount × direction and distance × direction) include the orientation relative to the lake, which emphasizes the importance of the direction of incoming fires in identifying and predicting potential fire refugia. Whereas our results point to important multiplicative effects between variables, the interpretation of these interactions is not straightforward. For instance, predicted patterns of wildfire probability on the east side of large lakes were less related to distance to lake when amount of surrounding water was low. This could be interpreted as a higher-level interaction among factors. Given the high density of large lakes in the region, the effect of a given lake on wildfire patterns is assuredly influenced by that of nearby lakes. Likewise, we found evidence for three-way interactions that support the idea that there is a high degree of complexity in the fire–environment relationship, which in turn leads to complex landscape patterns [36].

Whereas the likelihood of wildfire occurrence is highly dependent of transient factors, such as forest type [37] and daily fire weather [38], our results highlight the effects of quasi-permanent landscape features that can reduce wildfire likelihood for decades or centuries. Areas close to large or numerous lakes are simply more likely to lead to long-term fire refugia, which can be defined as parts of the landscape where intense crown fires are rare. These areas are therefore likely to support old-growth components that are not common elsewhere in the landscape [39]. Fire refugia have a potentially important—though still poorly understood—role in the maintenance of biodiversity and ecological processes in the boreal forest (but see [40]). In a matrix of high fire frequency, areas of the boreal forest that rarely burn may support isolated populations of organisms not found elsewhere in the landscape. For instance, species ill-adapted to fire, such as balsam fir, have survived on islands of large lakes in northern Québec [41], while common fire-adapted species, such as jack pine, have been absent from fire-sheltered sites for millennia in northern Wisconsin [42]. Spatial variability in wildfire occurrence also affects fundamental ecosystem properties [13] that, in turn, further affect community composition [43].

Results from this modeling study are contingent on their assumptions and data quality. For instance, some wildfire perimeters do not include unburned islands which would attenuate the strength of the effect of islands on wildfire probability. The somewhat coarse resolution of the fire perimeter mapping will invariably affect the strength of the other relationships, although we do not expect these to be directionally biased in a way that adversely affects our inferences. Likewise, we did not consider non-water related variables that are known to affect wildfires (e.g., land cover, daily fire weather, lightning, topography [44]). Lack of inclusion of these factors limits the predictive ability of our models, although the five landscape water variables considered here were predictive and largely supported our hypothesized relationships. Of note is the effect of humans, which is pervasive (if not intense) in the boreal plains portion of our study area [45]. Although large boreal wildfires are virtually uncontrollable and burn more or less "freely" once they escape initial attack, humans may have a subtle yet considerable influence on wildfire activity through direct (igniting or extinguishing fires) or indirect (land-use change) means [46].

5. Conclusions

Natural firebreaks, particularly large lakes, represent a strong bottom-up control on wildfire activity in the boreal forest of Saskatchewan, Canada. Landscape measures of water including presence of islands, amount of surrounding water, and distance and direction from lakeshore interacted to yield highly heterogeneous patterns of wildfire likelihood. These patterns were strongest for lake sizes ≥5000 ha and in the immediate 1000 ha surrounding area. Overall, we found that long-term fire refugia were more likely in places near lakeshores of irregularly-shaped larger (≥5000 ha) lakes and in areas (1000-ha window) surrounded by high amounts of water. This has implications for forest management and conservation of sites most likely to contain old-growth elements.

Acknowledgments: We thank Natural Sciences and Engineering Research Council (NSERC) of Canada for provided funding to support Scott E. Nielsen.

Author Contributions: Scott E. Nielsen, Krista Reinhardt, and Marc-André Parisien conceived and designed the study; Scott E. Nielsen, Evan R. DeLancey, and Krista Reinhardt prepared spatial data; Scott E. Nielsen analyzed the data; Scott E. Nielsen and Marc-André Parisien wrote the paper with contributions from Evan R. DeLancey and Krista Reinhardt.

Conflicts of Interest: The authors declare no conflict of interest.

Appendix A

Tables A1–A4, and Figures A1–A3 contain additional supporting information on landscape variables, model selection, and summary statistics.

Table A1. Evaluation of support for individual two-way interaction terms among landscape water variables that are predicted to affect local occurrence of fires in the boreal forest ecozone of Saskatchewan, Canada between 1980 and 2014. Null model here represents the best fitting model from Table 3 (island presence, amount of surrounding water [1,000 ha], and distance to lake, direction to lake, and shape of nearest lake ≥5000 ha). Models are ranked from most to least supported using Akaike's Information Criteria (AIC). "Response" represents the direction of response with bold, italicized text supporting our initial hypotheses from Table 3. See Table 3 in the text for definition of terms used in the table. Note that "N.A." is "not applicable".

Model	Response	K	AIC	ΔAIC	w_i
Amount × Direction	*Negative*	8	48,198.9	0.00	1.00
Distance × Direction	*Positive*	8	48,218.3	19.3	<0.01
Shape × Amount	Positive	8	48,252.6	53.6	<0.01
Direction × Shape	*Negative*	8	48,255.4	56.4	<0.01
Distance × Amount	*Negative*	8	48,255.8	56.8	<0.01
Null model (water size & amount)	N.A.	7	48,256.0	57.0	<0.01
Distance × Shape	N.A.	8	48,258.0	59.0	<0.01

Table A2. Evaluation of support for individual two-way interaction terms among landscape water variables that are predicted to affect local patterns in location of wildfires in the boreal forest ecozone of Saskatchewan, Canada between 1980 and 2014. Null model here represents the best fitting model from Table 4 and null model used in Table A1. Models are ranked from most to least supported using Akaike's Information Criteria (AIC). See Table 3 in the text for definition of other terms used in the table.

Model	K	AIC	ΔAIC	w_i
Top four two-way interactions	11	48,184.3	0.0	0.68
Top three two-way interactions	10	48,186.3	2.0	0.25
Top two two-way interactions	9	48,188.7	4.4	0.07
Top single two-way interaction	8	48,198.9	14.6	<0.01
Top five two-way interactions	12	48,255.8	71.5	<0.01
Null model (water size and amount)	7	48,256.0	71.7	<0.01

Table A3. Evaluation of support for models with three-way interaction terms among landscape water variables that are predicted to affect local patterns in location of wildfires in the boreal forest ecozone of Saskatchewan, Canada between 1980 and 2014. Null model here represents the best fitting two-way interaction model (Table A2). Models are ranked from most to least supported using Akaike's Information Criteria (AIC). See Table 3 in the text for definition of other terms used in the table.

Model	K	AIC	ΔAIC	w_i
4a Amount × Distance × Direction; Amount × Distance × Shape	13	48,131.5	0.0	0.43
5 (all three three-way interactions)	14	48,132.3	0.8	0.29
1 Amount × Distance × Direction	12	48,133.4	1.9	0.16
4b Amount × Distance × Direction; Amount × Direction × Shape	13	48,134.1	2.6	0.12
4c Amount × Direction × Shape; Amount Dist × Shape	13	48,181.4	49.9	<0.01
2 Amount × Shape × Distance	12	48,182.0	50.5	<0.01
3 Amount × Direction × Shape	12	48,183.6	52.2	<0.01
Null model (top four two-way interactions and linear terms)	11	48,184.3	52.8	<0.01

Table A4. Summarized model predictions by Ecoregion depicting the likelihood of wildfire over a 35-year period and its associated fire rotation period. Statistics reported include the 1st, 50th, and 99th centiles.

Ecoregion	p(wildfire)			Fire Rotation Period (Years)		
	1%	50%	99%	1%	50%	99%
Mid-boreal Upland	0.094	0.268	0.450	78	131	372
Mid-boreal Lowland	0.011	0.036	0.082	429	975	3253
Churchill River Upland	0.230	0.548	0.730	48	64	152
Athabasca Plain	0.442	0.714	0.841	42	49	79
All ecoregions	0.018	0.441	0.822	43	79	1909

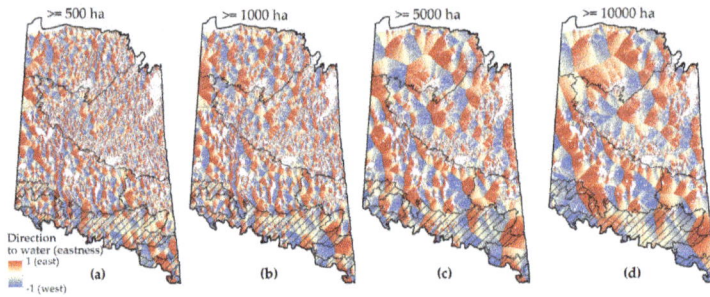

Figure A1. Direction to nearest water body of defined sizes in the boreal ecozones of Saskatchewan, Canada using an "eastness" index where an east direction is scaled to 1 and a west direction scaled to −1. Threshold water body sizes of: (**a**) ≥500 ha; (**b**) ≥1000 ha; (**c**) ≥5000 ha; (**d**) ≥10,000 ha. Areas of red color have an east bearing to water, while areas of blue have a west bearing to water (water bodies are shown in white).

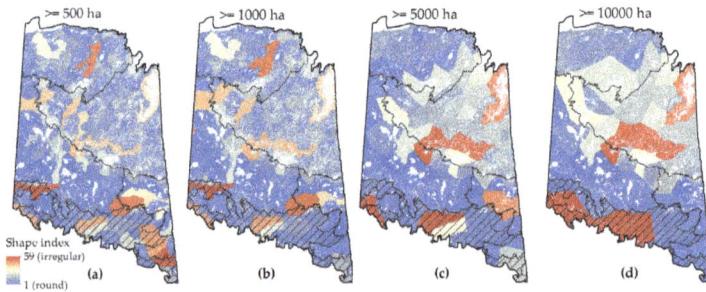

Figure A2. Shape index of nearest water body of defined threshold sizes in the boreal ecozones of Saskatchewan, Canada. Shape of water body size threshold of: (**a**) ≥500 ha; (**b**) ≥1000 ha; (**c**) ≥5000 ha; (**d**) ≥10,000 ha. Areas of red color have more irregular shaped nearby water bodies, while areas of blue have more round-shaped nearby water bodies (water bodies are shown in white).

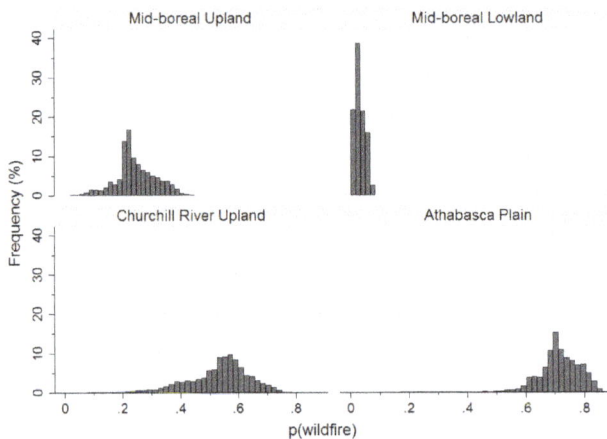

Figure A3. Distribution (variation) in predicted wildfire likelihood (probability of burning within 35 years, 1980–2014) by ecoregion in Saskatchewan, Canada's boreal forest based on histograms of model predictions.

References

1. Stocks, B.J.; Mason, J.A.; Todd, J.B.; Bosch, E.M.; Wotton, B.M.; Amiro, B.D.; Flannigan, M.D.; Hirsch, K.G.; Logan, K.A.; Martell, D.L.; et al. Large forest fires in Canada, 1959–1997. *J. Geophys. Res.* **2002**, *108*, FFR 5-1–FFR 5-12. [CrossRef]

2. Parisien, M.-A.; Peters, V.S.; Wang, Y.; Little, J.M.; Bosch, E.M.; Stocks, B.J. Spatial patterns of forest fires in Canada 1980–1999. *Int. J. Wildland Fire* **2006**, *15*, 361–374. [CrossRef]

3. Dansereau, P.-R.; Bergeron, Y. Fire history in the southern boreal forest of northwestern Quebec. *Can. J. For. Res.* **1993**, *23*, 25–32. [CrossRef]

4. Turner, M.G.; Romme, W.H. Landscape dynamics in crown fire ecosystems. *Landsc. Ecol.* **1994**, *9*, 59–77. [CrossRef]

5. Niklasson, M.; Granström, A. Numbers and sizes of fires: Long-term spatially explicit fire history in a Swedish boreal landscape. *Ecology* **2000**, *81*, 1484–1499. [CrossRef]

6. Schulte, L.A.; Mladenoff, D.J.; Burrows, S.N.; Sickley, T.A.; Nordheim, E.V. Spatial Controls of Pre–Euro-American Wind and Fire Disturbance in Northern Wisconsin (USA) Forest Landscapes. *Ecosystems* **2005**, *8*, 73–94. [CrossRef]

7. Heinselman, M.L. Fire and Succession in the Conifer Forests of Northern North America. In *Forest Succession: Concept and Application*; West, D.C., Shugart, H.H., Botkin, D.B., Eds.; Springer-Verlag: New York, NY, USA, 1981; pp. 374–405.

8. Bergeron, Y. The influence of island and mainland lakeshore landscapes on Boreal forest fire regimes. *Ecology* **1991**, *72*, 1980–1992. [CrossRef]

9. Larsen, C.P.S. Spatial and temporal variations in boreal forest fire frequency in northern Alberta. *J. Biogeogr.* **1997**, *24*, 663–673. [CrossRef]

10. Cyr, D.; Bergeron, Y.; Gauthier, S.; Larouche, A.C. Are the old-growth forests of the Clay Belt part of a fire-regulated mosaic? *Can. J. For. Res.* **2005**, *35*, 65–73. [CrossRef]

11. Heinselman, M.L. *The Boundary Waters Wilderness Ecosystem*; University of Minnesota Press: Minneapolis, MN, USA, 1996; pp. 72–73.

12. Mansuy, N.; Boulanger, Y.; Terrier, A.; Gauthier, S.; Robitaille, A.; Bergeron, Y. Spatial attributes of fire regime in eastern Canada: influences of regional landscape physiography and climate. *Landsc. Ecol.* **2014**, *29*, 1157–1170. [CrossRef]

13. Wardle, D.A.; Hörnberg, G.; Zackrisson, O.; Kalela-Brundin, M.; Coomes, D.A. Long-term effects of wildfire on ecosystem properties across an island area gradient. *Science* **2003**, *300*, 972–975. [CrossRef] [PubMed]

14. Heinselman, M.L. Fire in the Virgin Forests of the Boundary Waters Canoe Area, Minnesota. *Quat. Res.* **1973**, *3*, 329–382. [CrossRef]

15. Larson, D.W.; Kelly, P.E. The extent of old-growth *Thuja occidentalis* on cliffs of the Niagara Escarpment. *Can. J. Bot.* **1991**, *69*, 1628–1636. [CrossRef]

16. Frelich, L.E. *Forest Dynamics and Disturbance Regimes*, 1st ed.; Cambridge University Press: Cambridge, UK, 2002; pp. 1–266.

17. Araya, Y.H.; Remmel, T.K.; Perera, A.H. What governs the presence of residual vegetation in boreal wildfires? *J. Georg. Syst.* **2016**, *18*, 159–181. [CrossRef]

18. Hellberg, E.; Niklasson, M.; Granström, A. Influence of landscape structure on patterns of forest fires in boreal forest landscapes in Sweden. *Can. J. For. Res.* **2004**, *34*, 332–338. [CrossRef]

19. Loehle, C. Applying landscape principles to fire hazard reduction. *For. Ecol. Manag.* **2004**, *198*, 261–267. [CrossRef]

20. Parisien, M.A.; Parks, S.A.; Miller, C.; Krawchuk, M.A.; Heathcott, M.; Moritz, M.A. Contributions of ignitions, fuels, and weather to the spatial patterns of burn probability of a boreal landscape. *Ecosystems* **2011**, *14*, 1141–1155. [CrossRef]

21. Parisien, M.-A.; Hirsch, K.G.; Lavoie, S.G.; Todd, J.B.; Kafka, V. *Saskatchewan Fire Regime Analysis*; Information Report NOR-X-394; Natural Resources Canada, Canadian Forest Service, Northern Forestry Centre: Edmonton, AB, Canada, 2004; pp. 1–49.

22. Saskatchewan, L.R. Canadian Climate Normals 1981–2010 Station Data. Available online: http://climate.weather.gc.ca/climate_normals/index_e.html (accessed on 16 March 2016).

23. Government of Canada. Atlas of Canada National Hydro Network. Natural Resources Canada: Edmonton, AB, Canada, 2015. Available online: http://geogratis.cgdi.gc.ca/geogratis (accessed on 1 May 2016).

24. Rempel, R.S.; Kaukinen, D.; Carr, A.P. *Patch Analyst and Patch Grid*; Ontario Ministry of Natural Resources, Centre for Northern Forest Ecosystem Research: Thunder Bay, ON, Canada, 2012.

25. StataCorp. *Stata: Release 13*; Statistical Software; StataCorp LP: College Station, TX, USA, 2013.

26. Akaike, H. Likelihood of a model and information criteria. *J. Econom.* **1981**, *16*, 3–14. [CrossRef]

27. Burnham, K.P.; Anderson, D.R. *Model Selection and Inference*; Springer: New York, NY, USA, 1998; pp. 1–31.

28. Swets, J.A. Measuring the accuracy of diagnostic systems. *Science* **1988**, *240*, 1285–1293. [CrossRef] [PubMed]

29. Manel, S.; Williams, H.C.; Ormerod, S.J. Evaluating presence–absence models in ecology: the need to account for prevalence. *J. Appl. Ecol.* **2001**, *38*, 921–931. [CrossRef]

30. Erni, S.; Arseneault, D.; Parisien, M.-A.; Bégin, Y. Spatial and temporal dimensions of fire activity in the fire-prone eastern Canadian taiga. *Global Change Biol.* **2016**. [CrossRef] [PubMed]

31. Boulanger, Y.; Gauthier, S.; Gray, D.R.; Le Goff, H.; Lefort, P.; Morissette, J. Fire regime zonation under current and future climate over eastern Canada. *Ecol. Appl.* **2013**, *23*, 904–923. [CrossRef] [PubMed]

32. Parisien, M.-A.; Parks, S.A.; Krawchuk, M.A.; Little, J.M.; Flannigan, M.D.; Gowman, L.M.; Moritz, M.A. An analysis of controls on fire activity in boreal Canada: comparing models built with different temporal resolutions. *Ecol. Appl.* **2014**, *24*, 1341–1356. [CrossRef]

33. Parks, S.A.; Miller, C.; Holsinger, L.M.; Baggett, S.; Bird, B.J. Wildland fire limits subsequent fire occurrence. *Int. J Wildland Fire* **2016**, *25*, 182–190. [CrossRef]

34. Frelich, L.E.; Lorimer, C.G. Natural disturbance regimes in hemlock-hardwood forests of the upper Great Lakes region. *Ecol. Monograph* **1991**, *61*, 145–164. [CrossRef]

35. Andison, D.W. The influence of wildfire boundary delineation on our understanding of burning patterns in the Alberta foothills. *Can. J. For. Res.* **2012**, *42*, 1253–1263. [CrossRef]

36. Parisien, M.-A.; Miller, C.; Parks, S.A.; DeLancey, E.R.; Robinne, F.-N.; Flannigan, M.D. The spatially varying influence of humans on fire probability in North America. *Environ. Res. Lett.* **2016**, *11*, 075005. [CrossRef]

37. Cumming, S.G. A parametric model of the fire-size distribution. *Can. J. For. Res.* **2001**, *31*, 1297–1303. [CrossRef]

38. Flannigan, M.D.; Harrington, J.B. A study of the relation of meteorological variables to monthly provincial area burned by wildfire in Canada (1953–80). *Bull. Am. Meteorol. Soc.* **1988**, *27*, 441–452. [CrossRef]

39. Robinson, N.M.; Leonard, S.W.J.; Ritchie, E.G.; Bassett, M.; Chia, E.K.; Buckingham, S.; Gibb, H.; Bennett, A.F.; Clarke, M.F. Refuges for fauna in fire-prone landscapes: Their ecological function and importance. *J. Appl. Ecol.* **2013**, *50*, 1321–1329. [CrossRef]

40. Ouarmim, S.; Ali, A.A.; Asselin, H.; Hély, C.; Bergeron, Y. Evaluating the persistence of post-fire residual patches in the eastern Canadian boreal mixedwood forest. *Boreas* **2015**, *44*, 230–239. [CrossRef]

41. Sirois, L. Distribution and dynamics of balsam fir (*Abies balsamea* [L.] Mill.) at its northern limit in the James Bay area. *Ecoscience* **1997**, *4*, 340–352. [CrossRef]

42. Lynch, E.A.; Calcote, R.; Hotchkiss, S.C.; Tweiten, M. Presence of lakes and wetlands decreases resilience of jack pine ecosystems to late-Holocene climatic changes. *Can. J. For. Res.* **2014**, *44*, 1331–1343. [CrossRef]

43. Clarke, P.J. Habitat islands in fire-prone vegetation: do landscape features influence community composition? *J. Biogeogr.* **2002**, *29*, 677–684. [CrossRef]

44. Birch, D.S.; Morgan, P.; Kolden, C.A.; Abatzoglou, J.T.; Dillon, G.K.; Hudak, A.T.; Smith, A.M.S. Vegetation, topography and daily weather influenced burn severity in central Idaho and western Montana forests. *Ecosphere* **2015**, *6*, 1–23. [CrossRef]

45. Lehsten, V.; de Groot, W.; Sallaba, F. Fuel fragmentation and fire size distributions in managed and unmanaged boreal forests in the province of Saskatchewan, Canada. *For. Ecol. Manag.* **2016**, *376*, 148–157. [CrossRef]

46. Peterson, G.D. Contagious disturbance, ecological memory, and the emergence of landscape pattern. *Ecosystems* **2002**, *5*, 329–338. [CrossRef]

© 2016 by the authors. Licensee MDPI, Basel, Switzerland. This article is an open access article distributed under the terms and conditions of the Creative Commons Attribution (CC BY) license (http://creativecommons.org/licenses/by/4.0/).

Section 2:
Impact on Vegetation

forests

MDPI

Article

Fires of the Last Millennium Led to Landscapes Dominated by Early Successional Species in Québec's Clay Belt Boreal Forest, Canada

Maxime Asselin [1], Pierre Grondin [2,*], Martin Lavoie [1] and Bianca Fréchette [3]

[1] Département de Géographie and Centre D'études Nordiques, Université Laval,
Québec, QC G1V 0A6, Canada; maxime.asselin.1@ulaval.ca (M.A.); martin.lavoie@cen.ulaval.ca (M.L.)
[2] Ministère des Forêts, de la Faune et des Parcs du Québec, Direction de la Recherche Forestière,
Québec, QC G1P 3W8, Canada; pierre.grondin@mffp.gouv.qc.ca
[3] Centre de Recherche Geotop, Université du Québec à Montréal, C.P. 8888, Succursale Centre-ville,
Montréal, QC H3C 3P8, Canada; frechette.bianca@uqam.ca
* Correspondence: pierre.grondin@mffp.gouv.qc.ca; Tel.: +1-418-643-7994 (ext. 6653)

Academic Editors: Yves Bergeron and Sylvie Gauthier
Received: 3 August 2016; Accepted: 7 September 2016; Published: 15 September 2016

Abstract: This study presents the long-term (over the last 8000 years) natural variability of a portion of the *Picea mariana*-moss bioclimatic domain belonging to Québec's Clay Belt. The landscapes are dominated by mesic-subhydric clay and early successional forests composed of *Populus tremuloides*, *Pinus banksiana* and *Picea mariana*. The natural variability (fires and vegetation) of one of these landscapes was reconstructed by means of pollen and macroscopic charcoal analysis of sedimentary archives from two peatlands in order to assess when and how such landscapes were formed. Following an initial afforestation period dominated by *Picea* (8000–6800 cal. Years BP), small and low-severity fires favored the development and maintenance of landscapes dominated by *Picea* and *Abies balsamea* during a long period (6800–1000 BP). Over the last 1000 years, fires have become more severe and covered a larger area. These fires initiated a recurrence dynamic of early successional stands maintained until today. A decline of *Abies balsamea* has occurred over the last centuries, while the pollen representation of *Pinus banksiana* has recently reached its highest abundance. We hypothesize that the fire regime of the last millennium could characterize Québec's Clay Belt belonging to the western *Picea mariana*-moss and *Abies balsamea-Betula papyrifera* domains.

Keywords: boreal forest; forest fires; Holocene; pollen analysis; sedimentary charcoal; Québec; vegetation history

1. Introduction

Boreal forest landscapes develop under the combined influence of climate, natural (fires, insect outbreaks) and anthropogenic (logging, fires of human origin) disturbances, as well as physical environment [1,2]. This combination of factors generated the contemporary landscape diversity defined, at different spatial scales, through hierarchies of ecological classification [3,4]. These classifications consider the physical features, such as the abundance of lakes, the area covered by peatlands or sandy soils and the relief. All these physical features have a strong influence on the long-term history of fire and vegetation [5–15]. Knowing the long-term natural variability of landscapes, it will become easier to define forest strategies in regard to ecosystem management and climatic changes [16–20].

In each boreal landscape, the species are distributed along a toposequence of surficial deposits, slope and drainage conditions, from bedrock and well-drained soils (till, clay or sand) to poorly drained organic soils. In northeastern North America, jack pine (*Pinus banksiana* Lambert) and black spruce (*Picea mariana* (Miller) Britton, Sterns & Poggenburgh) are well adapted to sites characterized by bedrock or sandy soil [21,22]. Thick, moderately well-drained soils support a vegetation more demanding with respect to the nutrient regime, such as trembling aspen (*Populus tremuloides* Michaux), white birch (*Betula papyrifera* Marshall) and balsam fir (*Abies balsamea* (Linnaeus) Miller), whereas poorly-drained soils are mainly colonized by *P. mariana* and *Larix laricina* (Du Roi) K. Koch. Under the influence of natural disturbances, mainly fires, the vegetation of each portion of the toposequence changes with time. For example, after fires on well-drained rich soils, the light-demanding early successional species (*P. tremuloides*, *B. papyrifera*) give way to late successional ones (*P. mariana*, *A. balsamea*), thereby defining a successional forest dynamic [23]. However, in other parts of the landscape, as on low altitude and undulated or flat relief where *P. mariana* and *P. banksiana* are dominant, fires can occur so frequently spaced in time that cohorts of early successional species succeed one another, creating a recurrence dynamic of stands dominated by these species [24,25]. Under such circumstances, stands dominated by early successional species can be considered in equilibrium with climate and disturbance regimes. These examples show that, in the context of the eastern Canadian boreal forest, *P. mariana* can be considered as both an early and a late successional species.

In an integrative study of paleoecological data covering the major biomes of Québec (Canada) initiated in order to demonstrate the specificity of each biome, Blarquez et al. [26] suggest that in the boreal coniferous forest dominated by *Picea mariana*, biomass burning was higher during the mid-Holocene period (~6000 to 4000 cal. Years BP) than during the late Holocene (over ~4000 years). Some sites show a brief increase of fire frequency around 1000 cal. Years BP, possibly related to the Medieval climatic optimum [10,27,28]. The majority of the study lakes considered by Blarquez et al. [26] are located in the western part of the *Picea mariana* domains (moss and lichen). If fires were less frequent during the late Holocene, they were, however, larger in extent [27]. The widespread general decline of fire frequency during the late Holocene in the *P. mariana* domains would be explained by climatic factors, i.e., an increase in annual precipitation, as well as a decrease of temperature (mainly during July), a shorter growing season and lower summer insolation [10,26,27,29–35]. Blarquez et al. [26] also suggest that fire frequency increased throughout the Holocene in the boreal mixedwood. Numerous sites studied in this biome are, however, located in eastern Québec, where hydroclimatic conditions differ greatly from those prevailing in the western part of the province. Indeed, annual precipitation is more than 200 mm higher in eastern Québec [36]. Moreover, the mixedwood includes three bioclimatic domains according to Québec classification; these are, from north to south, *Abies balsamea-Betula papyrifera*, *Abies balsamea-Betula alleghaniensis* and *Acer saccharum-Betula alleghaniensis* [4]. Specific fire reconstruction studies in the *Abies balsamea-Betula papyrifera* domain show a gradual increase of fire events during the late Holocene, and some of these studies were undertaken in the Clay Belt of Québec [37,38] and Ontario [12] (Figure 1).

The aim of this study is to explain the long-term natural variability of the vegetation and fire in a Québec Clay Belt landscape dominated by early successional species (*Populus tremuloides*, *Pinus banksiana* and *Picea mariana*) (Figure 1). These stands are abundant throughout the Clay Belt. Considering the type of landscape (vegetation, surficial deposit) and the increase of fires during the last millennium in some areas dominated by clay, we hypothesize that the mesic-subhydric clay of our studied landscape was also affected by increased fire activity during the late Holocene, which promoted the maintenance until now of early successional species through a recurrent forest dynamic. To test this hypothesis, we first analyzed the contemporary vegetation of a 4000 km^2 study area with the objective of understanding the current forest dynamics. We then reconstructed the long-term vegetation and fire history through pollen and macroscopic charcoal analyses of two selected peatlands in order to obtain insight into the forest dynamics at the margin of the peatlands.

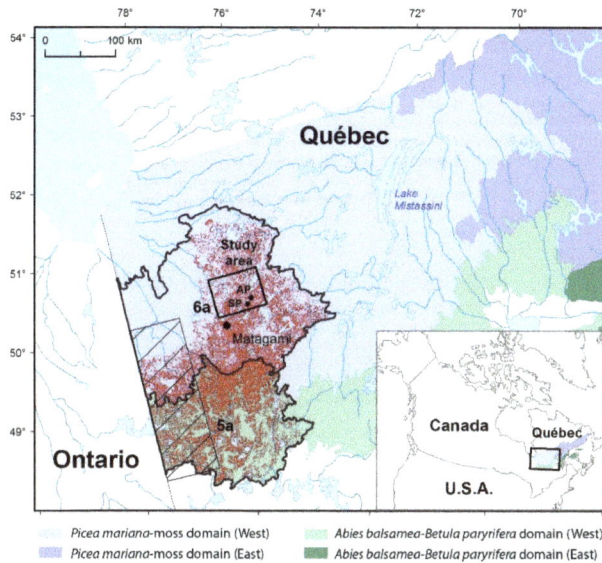

Figure 1. Location of the study area (4000 km²) in Québec's *Picea mariana*-moss bioclimatic domain, Canada [4]. The study area is located in ecological region 6a (Plaine du lac Matagami). The two peatlands studied are codified as: AP: Aspen peatland and SP: Shadow peatland. Ecological region 5a (Plaine du lac Abitibi) is part of the western *Abies balsamea-Betula papyrifera* bioclimatic domain. The sector with oblique lines located close to the border with Ontario corresponds to the fire origin map created by Bergeron et al. [7]. The red dots indicate the clay deposit distribution (QMFFP forest maps) corresponding to Québec's Clay Belt.

2. Materials and Methods

2.1. Study Area

The study area covers 4000 km² within the *Picea mariana*-moss bioclimatic domain, more specifically within the 'Plaine du lac Matagami' ecological region [4] (Figure 1). Annual mean temperature at the Matagami weather station [36] is between −1 and −2 °C. July is the warmest month of the year (mean: 16.8 °C) and January the coldest (mean: −20 °C). Total annual precipitation varies between 800 and 1000 mm, of which 25%–30% falls as snow. This region is characterized by a flat to gently undulated relief, low to mid altitudes (200–400 m) and the presence of some large lakes. An important portion of the ecological region belongs to the Clay Belt, a vast half-moon shaped area stretching from eastern Ontario to western Québec. The Clay Belt is characterized by thick clayey glaciolacustrine deposits left by postglacial lakes Barlow (southern area) and Ojibway (northern area), which reached their maximum water level 8000 years ago when the retreating Laurentide ice sheet was further north [39]. Soon after and for a short period, a readvancing ice sheet surged southward into glacial Lake Ojibway, incorporating glacial material into glaciolacustrine sediments and forming the compact Cochrane Till. This till, the low altitude and the flat relief all played an important role in the development of large peatlands in the northeastern part of the ecological region.

A reconstruction of the contemporary fire history (1700–2000 AD) in a 15,000 km² area located along the Québec-Ontario border and belonging mainly to the Clay Belt showed that fires burned large areas especially during the 1820 and 1910 periods [7,32,40] (Figure 1). Since then, only a few fires have occurred, most of them in the *Abies balsamea-Betula papyrifera* domain, and mainly of anthropogenic origin. The post-fire forest dynamics are characterized by two main successional pathways. The first is mostly observed in the northwestern part dominated by Cochrane Till. In this sector, numerous

Picea mariana forests have been paludified and this process of peat accumulation, controlled by flat relief, lower altitude and fire activities, is still active at the edge of the peatlands [41]. The second successional pathway is typical of mesic and subhydric clay, mostly undulated relief and higher altitude (Clay Belt). *Populus tremuloides*, *Pinus banksiana* and *P. mariana* early successional forests evolve towards *P. mariana-Abies balsamea* late successional forests if the time elapsed after fire is long enough [22]. Due to the vast area covered by fires of the 1910 period [7,40], this late successional forest is currently rare.

2.2. Contemporary Forest Landscapes

During the first phase of this study, the current forest composition and dynamics of the 4000 km^2 study area, belonging to the Clay Belt, were defined by using forest maps produced by the Québec Ministry of Forests, Fauna and Parks (QMFFP). These maps (scale 1:20,000) were produced by photointerpretation of aerial photographs taken in the 2005–2010 period. In total, 38,887 forest stands were delineated and defined on the basis of their vegetation composition and structure (density, age) as well as physical characteristics (surficial deposit, drainage, slope and altitude). The age of forest stands identified on maps was studied with the aim of establishing links with Bergeron et al. [7,40]. However, these links were difficult to make because many stands on the QMFFP maps were identified as pertaining to the 70-year-old age-class, suggesting fires in the 1950 period. This age was estimated by photointerpretation mainly on the basis of tree height and density, and without information on the contemporary fire history. Moreover, older forests were classified into the 120-year maximum age class. In order to acquire knowledge on the fire contemporary history of the studied area, we created a fire origin map. Fieldwork (2013 and 2014) was conducted at 140 sites. For each site, we collected fire scars or, in their absence, a cross section of the collar from one to five trees per site. In total, 313 trees were sampled from the dominant individuals. In the laboratory, samples were dried and sanded, and the number of rings was counted using a binocular magnifier. Years of stand origin show a dominance of trees belonging to the ca. 1820 and ca. 1910 periods. As expected, our results are comparable to those of Bergeron et al. [7,40] and not to those of QMFFP forest maps. We also identified the 1970 period, which was characterized by some small fire events. Based on this information, the majority of forest stands referred to on the QMFFP maps as being 90 years old and, erroneously, 70 years old were assigned to the 1910 period of origin, whereas the majority of the 120 year-old stands were considered as having an origin close to 1820.

The 38,887 stands were subjected to a redundancy analysis (RDA). The RDA was conducted simultaneously on matrices characterizing forest composition (Y matrix, 38,887 records, five species: *Picea mariana*, *Abies balsamea*, *Pinus banksiana*, *Populus tremuloides*, *Betula papyrifera*) and site characteristics (X matrix, 38,887 records, five variables: altitude, slope, surficial deposit, drainage, period of origin). The RDA characterized each forest stand according to five ordination axes. All axes were submitted to *K*-means partitioning in order to group forest stands with comparable vegetation and site characteristics. Each group of stands composes a landscape [2,42].

2.3. Study Sites and Sampling

During the second phase of this study, two ombrotrophic peatlands located in a landscape belonging to the Clay Belt and dominated by early successional stands (*Populus tremuloides*, *Pinus banksiana*, *Picea mariana*) were selected for the reconstruction of the Holocene fire and vegetation history. The two peatlands are situated 10.5 km apart (Figures 1 and 2). Aspen peatland (hereafter "Aspen", 50°00.46′ N; 77°00.14′ W) covers an area of 30 ha and lies at an altitude of 283 m. Shadow peatland (hereafter "Shadow", 49°56.57′ N; 77°06.43′ W) covers an area of 15 ha at an altitude of 280 m. Both are dominated by Ericaceae on a carpet of *Sphagnum* moss. Shrubby *P. mariana* (height 14 m) form a low density stratum (20%–40%). In Aspen, the coring site for organic sediments lies 100 m from both sides of the forest, whereas in Shadow the coring site is located 25 m from the forest border. These two locations were selected to obtain a long temporal sequence (Aspen) and a shorter one but with a stronger signal of the local forest dynamics (Shadow). Sampling was conducted by excavating a trench with a shovel, and then cutting peat monoliths (30 cm × 30 cm × 30 cm) from top to bottom. The absence of water from the trench during excavation made it possible to collect a complete section of peat right down to the underlying marine clay. Monoliths were stored in the laboratory at 4 °C until analysis.

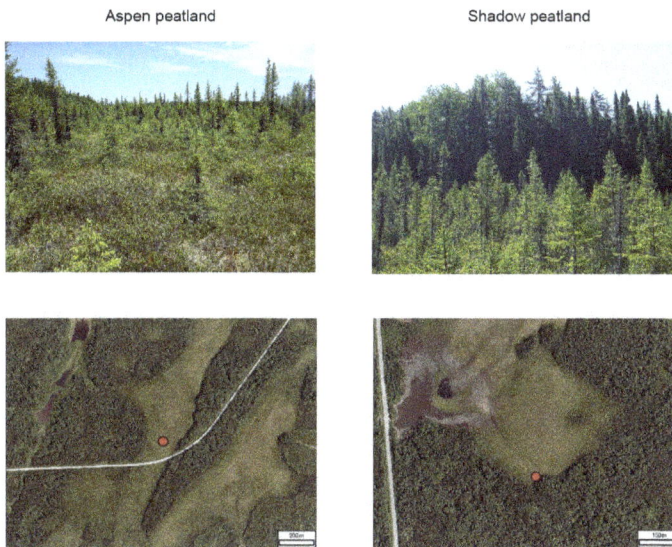

Aspen peatland   Shadow peatland

Figure 2. Study sites (Aspen and Shadow peatlands) where paleoecological reconstructions were conducted. **Upper** portion: overview of the peatlands and their forest environment. **Lower** portion: aerial photographs and location of sedimentary cores (**red** dots).

2.4. Characterization of the Contemporary Forest Surrounding the Peatlands

To characterize the forest surrounding each peatland, three parallel transects (50 m × 4 m; 200 m^2) 10 m apart were delineated. Live tree stems (>2 cm in diameter at breast height) were identified and counted. The stems representing forest regeneration (<2 cm in diameter) were identified and numbered in plots (1 m^2) distributed every 5 m along each transect (10 plots per transect). A cross section of the collar was cut from five of the largest stems in order to determine their age. Finally, a moss sample was collected from the soil surface in the forest surrounding Aspen, as well as from a stand dominated by *Picea mariana* and *Abies balsamea* located 10 km north of Aspen, to obtain a picture of the pollen representation of current vegetation. In the laboratory, the stem counts of individuals exceeding 2 cm per species and per diameter class were transformed into relative basal area per species. Regenerating tree stems were analyzed to define the frequency of species per sampling site (stocking of regeneration). The basal samples were sanded and growth rings counted to determine the stand year of origin.

2.5. Stratigraphy and Chronology

Prior to specific analyses, sediments were cleaned and cut into continuous 1-cm thick slices. The general composition of the organic matrix (*Sphagnum*, herbaceous, brown mosses, wood remains) was determined qualitatively from analysis of subsamples [43,44]. A total of 20 samples were subjected to radiocarbon dating by accelerator mass spectrometry (AMS). They were first prepared in the radiochronology laboratory of Université Laval's Centre for Northern Studies, then dated at Kerk Laboratory, University of California (Irvine). Dated samples consist of charcoal fragments, with the exception of one sample from the base of the Aspen sedimentary core (seeds). CALIB 7.0.4 software and the IntCal13 database [45,46] were used to determine the probability distributions in calibrated years for each of the ^{14}C datings ($\pm 2\sigma$). The median of the probability distribution was selected as the calibrated date. Dated charcoal fragments with overlapping calendar age distributions were considered as originating from the same fire [25]. Net vertical peat accumulation rates (mm·year^{-1}) were calculated between calibrated dates. The models were developed using CLAM 2.2 software [47], applying a linear interpolation between each level dated (1000 iterations). All results are expressed as calibrated years BP.

2.6. Macrocharcoal Analysis

The fire history of forests surrounding the two peatlands was reconstructed by charcoal analysis following a modified version of the procedure used by Hörnberg et al. [48]. Subsamples (2 cm^3) were taken at 2 cm intervals and soaked in a potassium hydroxide (KOH) solution (5%) for 24 h, then sieved through a 0.425 mm mesh screen. The remaining particles were bleached in a sodium hypochlorite (NaClO) solution (10%) to distinguish charcoal from dark organic matter. Charcoal particles were counted in a petri dish using a stereomicroscope (20× magnification). They were picked from the petri dish, placed in weighing boats and oven-dried overnight (± 50 °C). The mass of charcoal (anthracomass) was determined for each sample using an electronic balance, and data are expressed as anthracomass (mg·2 cm^{-3}) [49,50].

2.7. Reconstruction of Vegetation History

To reconstruct vegetation history, 1 cm^3 subsamples (3 cm^3 in the upper portion of the cores corresponding to the acrotelm of the peatlands) were collected for pollen analyses. Analyses were conducted at 2 cm intervals for sections with high charcoal content (as determined by prior charcoal analysis) and at 4 cm intervals for sections with low charcoal content. Subsamples, as well as two samples of surface moss, were processed using chemical treatments with KOH (10%), HCl (10%) and an acetolysis solution [51]. *Lycopodium* tablets with a known concentration were added to each subsample prior to preparation, in order to calculate pollen concentration (grains cm^{-3}). Pollen counts were made at 400× magnification. At least 300 grains of terrestrial vascular plant pollen (excluding Ericaceae and Cyperaceae) were counted for each level, using an optical microscope (400× magnification). Results are expressed in percentages. The curves of only the most abundant pollen taxa are presented in the diagrams. The diagrams were subdivided into pollen assemblage zones (PAZ) using the stratigraphically constrained cluster analysis CONISS program of Tilia software [52]. Percentages were first transformed by a square root, in order to increase the importance of rare or poorly represented taxa (e.g., *Populus*) and to reduce the importance of those strongly represented (e.g., *Betula*).

To synthesize changes in vegetation through time, principal component analysis (PCA) was used on the combined Aspen and Shadow pollen datasets. Because the variables are dimensionally homogeneous, a dispersion (variance/covariance) matrix was used [42,53]. The number of pollen types was reduced by selecting taxa with a value greater than or equal to 1% in at least one sample, and creating a collective category among the pollen types (herbs). Cyperaceae and Ericaceae pollen were excluded. This selection resulted in 11 pollen types. For PCA, the relative frequencies (in percent) of the pollen types were square-root transformed. This transformation was performed in order to optimize the signal-to-noise ratio and stabilize the variances. PCA calculations were performed using R software version 3.2.2 (http://cran.r-project.org/).

3. Results

3.1. Contemporary Forest Landscapes of the Study Area

In order to understand the distribution and the dynamic of the vegetation and to select the most appropriate landscape for the paleoecological reconstruction, the study area (4000 km^2) was divided into eight landscapes (Figure 3; Table S1). Each shows specific characteristics in regard to forest composition, altitude, surficial deposit, slope, drainage and fire period of origin. The landscapes are mainly distributed along an eastward altitudinal gradient (225–250 to 300–375 m) corresponding to an increase of relatively well-drained forests on clay. Along the gradient, the proportion of slopes higher than 4% increases continuously, whereas the area covered by open peatlands decreases. Landscape P4b (553 km²) was selected because it contains the highest proportion of stands dominated or sub-dominated by early successional species (mainly *Populus tremuloides*). Stands composed mainly of *Picea mariana* and growing on mineral soils are abundant. Clay covers a large part of it, and drainage is mainly mesic or subhydric. Altitude varies from 275 to 300 m. The majority of forest stands originated from fires of the 1910 period. Late successional *Picea mariana*-moss and *Abies balsamea* stands originating mainly from fires of the 1820 period are very rare (less than 2%), and landscape P3 contains the highest proportion of this last community (close to 10%; Table S1).

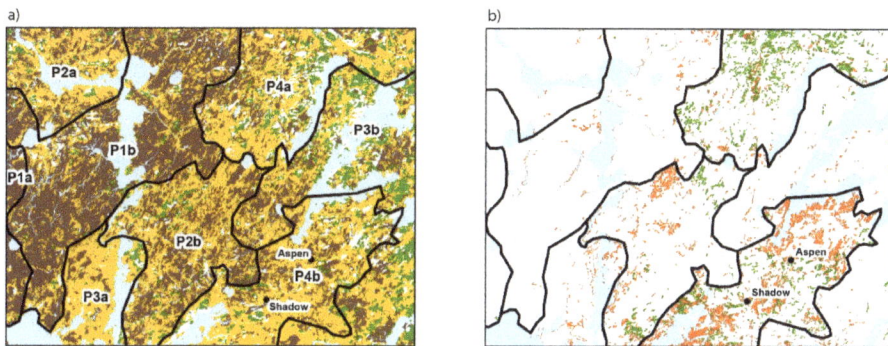

Figure 3. Delineation of the boundaries of the landscapes (P1–P4) in the study area. (**a**) Distribution of the surface deposits: organic deposits (**brown**), clay (**orange**) and till (**green**); (**b**) Distribution of the forest stands dominated by *Populus tremuloides* (**orange**) and *Pinus banksiana* (**green**). Aspen and Shadow peatlands are located in landscape P4b. The information on surficial deposits and forest cover was compiled from QMFFP forest maps. The description of the landscapes is presented in Table S1.

3.2. Contemporary Forest Surrounding the Two Peatlands Studied

The contemporary forests surrounding the Aspen and Shadow peatlands developed following a fire that occurred in 1917. The forest bordering Aspen is deciduous, dominated by *Populus tremuloides* and subdominated by *Picea mariana*, whereas *Pinus banksiana* is sparse (Figure 4). The number of *P. tremuloides* stems greater than 10 cm in diameter is close to 700 ha^{-1} whereas the basal area approaches 30 m²·ha^{-1}. Regeneration essentially consists of *P. mariana* (stocking of 70%). The shrubs *Alnus alnobetula* subsp. *crispa* (Aiton) Raus and *Alnus incana* subsp. *rugosa* (Du Roi) R.T. Clausen) are abundant (90%). The pollen assemblage of the surface moss sample is mainly dominated by *Picea* (43%), *P. banksiana* (36%), *Alnus* (10%) and *Betula* (5%) (Table 1). Although *P. tremuloides* is dominant in the forest, it is entirely absent from the pollen assemblage. The forest surrounding Shadow is mixed. The number of stems belonging to the 10 cm diameter class or greater is on the order of 300 ha^{-1} for *P. mariana*, 200 for *P. tremuloides* and 100 for *P. banksiana*. The basal area of each of these three species is 10 m²·ha^{-1}. Undergrowth vegetation is dominated by *Alnus* (stocking of 80%). Forest regeneration is sparse, with a few *P. tremuloides* (10%) and *P. mariana* (5%); the *P. mariana* regeneration is far less abundant than at Aspen. The pollen assemblage of the moss sample collected in the *Picea mariana-Abies balsamea* stand situated about 10 km north of Aspen (landscape P3b; Figure 3) primarily consists of *Picea* (49%), *P. banksiana* (23%), *Alnus* (12%), *Betula* (10%) and *Abies balsamea* (2%) (Table 1). In this landscape, *P. mariana-A. balsamea* stands cover close to 5% of the surface area (Table S1).

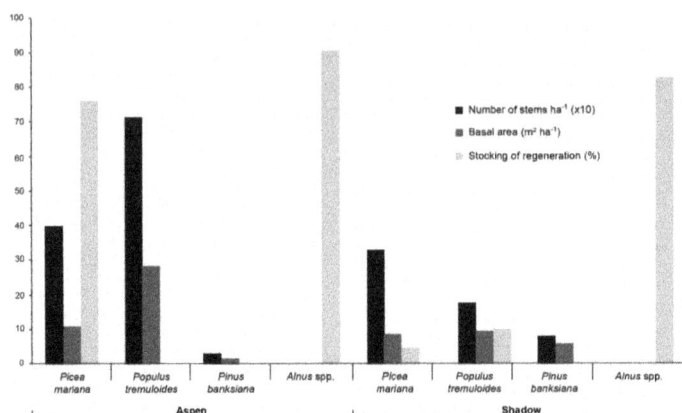

Figure 4. Characteristics of the contemporary forests surrounding Aspen and Shadow peatlands.

Table 1. Pollen representation (selected species) of current vegetation of two stands in the study area. The first stand is located in the forest close to Aspen peatland. The second is located in landscape P3b (Figure 3).

Site and Stand Dominant Species	*Picea mariana*	*Pinus banksiana*	*Abies balsamea*	*Betula*	*Populus tremuloides*	*Alnus*
Aspen peatland P. tremuloides, P. banksiana, P. mariana	43.3%	36.2%	0.9%	5.4%	0%	10.6%
10 km north of Aspen peatland P.mariana, A. balsamea	49.4%	23.1%	2.1%	9.9%	0%	12%

3.3. Stratigraphy and Chronology of Sedimentary Cores

At Aspen, the peat lies on clay and is 134 cm thick at the sampling point (Figure 5). At the base, a peat composed of brown mosses (134–110 cm) is overlain by peat that is a mixture of *Sphagnum* and herbaceous remains (110–40 cm) and poorly decomposed *Sphagnum* (40–0 cm). A date of 7990 BP was obtained for the onset of peat accumulation (Table 2, Figure 5). After a first period (7990–5370 BP) during which the sedimentary accumulation rate averaged 0.18 mm·year^{-1}, the rate remained low until very recently (5370–290 BP; 0.03–0.14 mm·year^{-1}). At Shadow, the peat is 96 cm thick at the sampling point. The matrix consists essentially of *Sphagnum* with numerous wood remains. A charcoal fragment from the point of contact with clay was dated to 4110 BP. The numerous pieces of wood found at the organo-mineral contact during excavation of the trench suggest that the late initiation of peat inception at the sampling point is linked to a process of paludification of an ancient forest originally present locally. The sedimentary accumulation rate was variable over time (0.07 to 0.37 mm·year^{-1}). The very high rate of accumulation in the upper portion of samples from both peatlands is the result of less compaction and decomposition of the peat compared to that at lower levels. No charred layer resulting from a fire in situ that could have caused a hiatus in sediment accumulation was observed.

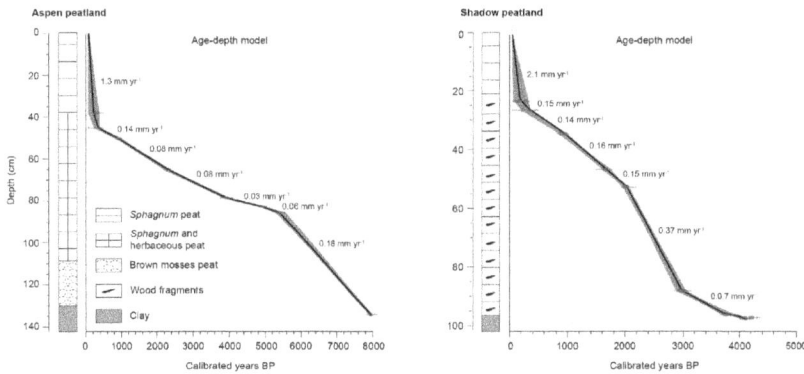

Figure 5. Stratigraphy and age-depth model of the sedimentary cores collected at the Aspen and Shadow peatlands.

Table 2. Radiocarbon ages from Aspen and Shadow peatlands, calibrated with an IntCal13 calibration curve [46] using Calib software version 7.0.4 [45] or Clam [47].

Site and Depth (cm)	Laboratory Number [a]	Material Dated	[14]C Date (Years BP)	Calibrated Age Range [b] (Years BP)	Median [c] (cal. Years BP)
Aspen					
37–38	145,391–4797	Charcoal	180 ± 30	136–224	180
38–39	150,957–5063	Charcoal	75 ± 15	33–73	*50*
44–45	145,611–4828	Charcoal	235 ± 20	281–307	290
49–50	145,608–4825	Charcoal	955 ± 20	796–874	840
52–53	150,956–5062	Charcoal	915 ± 15	840–910	*880*
64–65	150,951–5057	Charcoal	2275 ± 15	2306–2346	2330
77–78	150,952–5058	Charcoal	3590 ± 20	3839–3929	3880
81–82	150,955–5061	Charcoal	4345 ± 20	4856–4963	4910
84–85	145,607–4824	Charcoal	4705 ± 20	5325–5409	5370
132–133	145,602–4819	Seeds	7185 ± 25	7956–8026	7990
Shadow					
21–22	145,390–4796	Charcoal	120 ± 30	51–149	100
24–25	145,609–4826	Charcoal	240 ± 20	281–308	300
32–33	145,603–4820	Charcoal	945 ± 20	796–874	835
41–42	145,612–4829	Charcoal	1760 ± 20	1608–1724	**1670**
44–45	145,604–4821	Charcoal	1680 ± 20	1539–1619	1580
45–46	150,954–5060	Charcoal	1655 ± 15	1524–1573	**1550**
50–51	145,605–4822	Charcoal	2010 ± 20	1920–1998	1960
85–86	145,606–4823	Charcoal	2790 ± 25	2842–2956	2900
93–94	145,610–4827	Charcoal	3425 ± 20	3613–3720	3680
94–95	145,582–4783	Charcoal	3745 ± 25	4068–4157	4110

[a] First number: UCIAMS (University of California); second number: ULA (Laval University); [b] Calibrated age range at 95% confidence intervals from CLAM 2.2.; [c] Calibrated age from the CALIB 7.0.4 program and the IntCal13 database. Dates excluded in the Clam age-depth model are shown in bold.

3.4. Reconstruction of the Fire History

Based on data from the anthracomasses and radiocarbon dates of the two sedimentary cores, at least eight fires were identified at Aspen and Shadow. For both peatlands, some successive layers are characterized by high charcoal concentrations. At Aspen (Figure 6), four of the eight dated fires occurred before 2000 BP (5370, 4910, 3880 and 2330 BP), and four in the last 1000 years (880–840, 290, 180 and 50 BP). Some charcoals were found at a depth of about 60 and 92 cm, but were too small and too few to make dating possible. At Shadow (Figure 7), three of the eight fires occurred before

2900 BP (4110, 3680 and 2900 BP), and the five others were dated from 1960, 1670, 1550–1580, 835, 300 and 100 BP. Only the last millennium registered relatively synchronous fires between the two sites, specifically those dating from 880–840 (Aspen) and 835 BP (Shadow), 290 BP (Aspen) and 300 BP (Shadow), as well as 50 BP (Aspen) and 100 BP (Shadow). The most recent ^{14}C dates are related to dendrochronological data: the date of 180 BP (Aspen) is associated to fires of the 1820 period, and those of 100 and 50 BP to fires of the 1920 period.

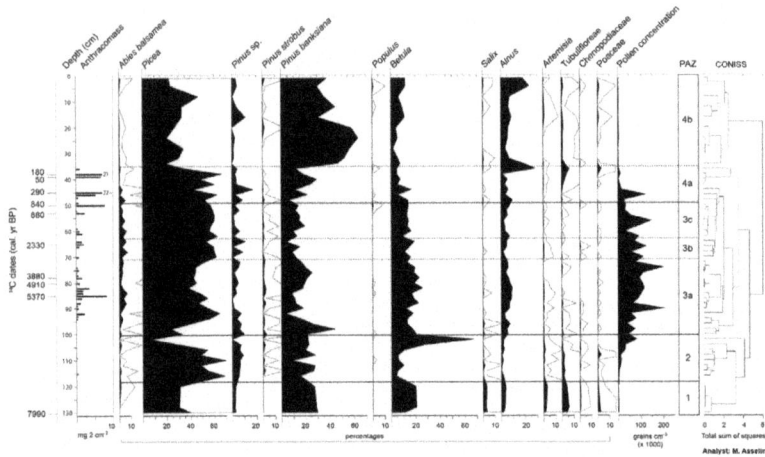

Figure 6. Pollen diagram (percentages; selected taxa) and charcoal content (mg·2 cm^{-3}) of the sedimentary core collected at Aspen peatland. Open curves show a 10× exaggeration.

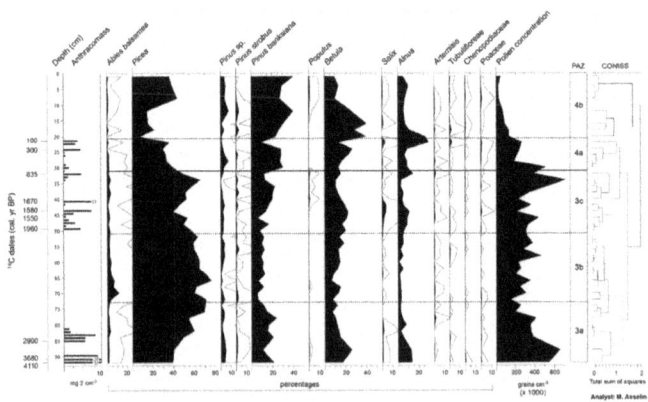

Figure 7. Pollen diagram (percentages; selected taxa) and charcoal content (mg·2 cm^{-3}) of the sedimentary core collected at Shadow peatland. Open curves show a 10× exaggeration.

3.5. Vegetation History

Four pollen assemblage zones (PAZs 1 to 4) can be defined in the pollen diagrams of the two sites (Figures 6 and 7). The sedimentary core collected at Aspen covers a longer period (8000 years) than that from Shadow (4100 years); this is why PAZs 1 and 2 (the oldest zones) are observable only in the diagram of the former. In PAZ 1 (ca. 8000 to ca. 7600 BP), at the base of the Aspen diagram, the pollen assemblages are relatively rich in herbaceous taxa (*Artemisia*, Tubuliflorae, Chenopodiaceae, Poaceae) and shrubs (*Salix*, *Alnus*). Arboreal taxa are primarily represented by *Picea* (probably *Picea mariana*), *Pinus banksiana* and *Betula*. The *Betula* pollen in this zone could be a mixture of *Betula papyrifera* and the shrubby species *Betula glandulosa* Michaux. The low pollen concentrations suggest that this first stage in plant colonization after the retreat of postglacial Lake Ojibway was characterized by open forests (afforestation period). PAZ 2 (ca. 7600 to ca. 6800 BP) shows a gradual increase of pollen concentrations and a decline in the representation of herbaceous taxa, interpreted as a signal of a densification of the forest cover. Pollen assemblages are essentially dominated by *Picea* (40%–50%) and *P. banksiana* (20%–30%). Maximum values are recorded for *Betula* (50%–60%) at the summit of the zone.

PAZ 3 (ca. 6800 to ca. 780–735 BP) of both diagrams covers a long period during which *Abies balsamea* is well represented (maximum: 10%), as is *Picea* (45%–65%). This zone is divided into three subzones. Subzone 3a extends over more than 3000 years (ca. 6800 to ca. 3100 BP at Aspen; >ca. 2600 BP at Shadow). It shows relatively high percentages of *Pinus banksiana*, *Betula* and *Alnus*. In subzone 3b (ca. 3100–2200 BP at Aspen; ca. 2600–2000 BP at Shadow), a drop in the pollen representation of *P. banksiana* and *Alnus* can be noted. Finally, subzone 3c (ca. 2200–2000 to ca. 780–735 BP) is characterized by a slight increase of *P. banksiana* (Aspen) and *P. banksiana* and *Alnus* (Shadow).

On both sites, PAZ 4 corresponds to the last ca. 800 years, and can be divided into two subzones: 4a (ca. 780–745 to ca. 160–115 BP) and 4b (<ca. 160–115 BP). Pollen concentrations are very low because *Sphagnum* peat in the top section is less decomposed and less compact, which dilutes the pollen in the peat. The pollen representation of *Abies balsamea* declines from the base to the summit of subzone 4a, then drops in subzone 4b, particularly in the diagram of Aspen. In both diagrams, the beginning of subzone 4b is marked by a notable increase of *Pinus banksiana* (>40% at Aspen; >30% at Shadow), and the species reaches its maximum percentage in this subzone.

The major trend of the PAZs for both sites was compared objectively by applying principal component analysis (PCA) to the combined Aspen and Shadow pollen datasets. The first axis of PCA accounts for 42.0% of the variance within data and represents the major vegetation gradient (Figure 8a). High (positive) PCA axis 1 scores are driven by *Pinus banksiana*, *Alnus* and *Betula*, three early successional species in these landscapes, whereas low (negative) PCA axis 1 scores are driven by *Picea* and *Abies balsamea* (Figure 8b), two late successional species. The trend in PCA axis 1 scores for the Shadow site is similar to that for the Aspen site. The four PAZ are also highlighted by axis 1. At Shadow and Aspen, there is a pronounced change at about 100–150 BP (PAZ 4b/4a). The change is more gradual at Shadow but more abrupt at Aspen. From ca. 2500 BP to today, a linear trend can be observed in PCA axis 1 scores. This trend is best evidenced at Shadow, which is situated closer to the forest. There is a gradual transition from negative to positive scores, showing that *P. banksiana* was increasing to the detriment of *A. balsamea* and *Picea*.

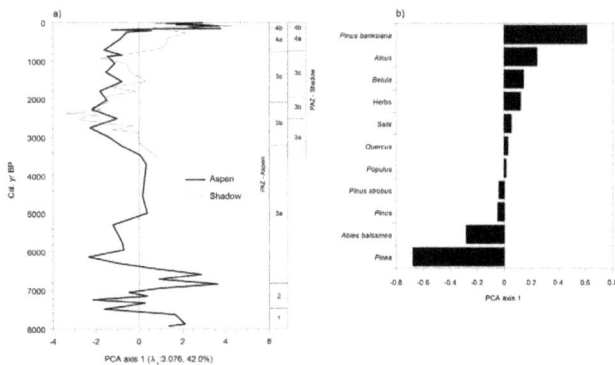

Figure 8. Principal component analysis (PCA) results for the combined pollen assemblages of Aspen and Shadow sites. (**a**) Stratigraphic plot of PCA axis 1 sample scores for the Aspen and Shadow sites over time. Horizontal lines separate pollen assemblage zones (PAZs). The first eigenvalue and percentage of variance are given; (**b**) PCA axis 1 variable loadings.

4. Discussion

This study presents the long-term natural variability (fire and vegetation) of a portion of the *Picea mariana*-moss bioclimatic domain belonging to Québec's Clay Belt. This variability has been described through paleoecological analysis of sediment cores from two bogs 10.5 km apart, located in a forested landscape dominated by early successional species (*Populus tremuloides*, *Pinus banksiana*, *Picea mariana*). The sediment core from Aspen (100 m from the forest) provides a continuous history (vegetation and fire events) of 8000 years, whereas that from Shadow (25 m from the forest) covers a period of 4000 years. The results increase our knowledge about the Holocene history (fire and vegetation) of western Québec boreal forests, mainly based on paleoecological studies of lake sediments. The results from the two peatlands support our initial hypothesis specifying that increased fire activity during the late Holocene on mesic-subhydric clay promoted the development and maintenance of early successional species through a fire recurrence dynamic. Four main periods were recognized in the long-term vegetation and fire history of the forests surrounding our study sites. For each, we compare our results with those of lakes studied in western Québec *Picea mariana*-moss and *Abies balsamea-Betula papyrifera* domains (Figure S1; Table S2). These lakes are located in low-altitude landscapes dominated by peatlands (Pessière, Geais, Profond, Raynald, Loutre, Garot), in mid-altitude landscapes with undulated clay (Cèdres, Schön, Francis, Pas de Fond) and relatively high-altitude landscapes with undulated till (Twin, Richard, Aurélie, Nans). The latter are located to the east of our study area, near Lake Mistassini.

4.1. The Afforestation Phase (ca. 8000 to ca. 6800 BP)

At Aspen, only a few isolated charcoal pieces were found in the sediments that represent the period between the beginning of vegetation colonization following the draining of the waters of proglacial Lake Ojibway (ca. 8000 BP) and the densification of the forest cover (forest phase) around 6800 BP. The rarity of charcoal during this first period corresponding to afforestation (PAZs 1 and 2) can be linked to the low initial density of the forest, which may have been too sparse to support recurrent fire activity. At Aspen, the presence of brown mosses and seeds of *Menyanthes trifoliata* Linnaeus indicates minerotrophic and water-saturated soil at the time and supports the hypothesis that the very humid local conditions present initially could have prevented fires from propagating toward the center of the peatland. The reconstructions of fire history from lake sediments in western Québec *Picea mariana*-moss domain also suggest a low fire frequency during the afforestation phase [10].

4.2. A Landscape Dominated by Late and Early Successional Forest Stands (ca. 6800 to ca. 2500 BP)

The beginning of the forest phase at Aspen (ca. 6800 BP; PAZ 3a) is marked by an increase in the pollen representation of *Abies balsamea*, a late successional species that is generally under-represented in pollen assemblages [54]. Until 2500 BP, at least four fires occurred in the surrounding forests (5370, 4910, 3880 and 2330 BP). The presence of smaller quantities of charcoal fragments at other levels of the core indicates that it may be possible that other fires occurred during this period. In the forests bordering Shadow, fires occurred at dates different from those at Aspen (4110, 3680, 2900 BP). Since the two sites are only 10.5 km apart, this asynchronism suggests that the fires covered small areas and were not severe. We hypothesize that this type of fire regime promoted the development of *Picea mariana-Abies balsamea* stands. Early successional species (*Betula papyrifera*, *Pinus banksiana*, *Alnus* spp.) were also well represented in the pollen diagrams, suggesting that the landscapes were composed of stands belonging to late and early successional stages. Our results concerning the fires differ from those based on lake sediments which show a relatively high fire frequency (expressed by a number of fires per millenium). However, there is a wide variation in the maximum frequency of fires and the period during which this rate was maintained [10,15,27,31,34].

4.3. Landscapes Dominated by Picea mariana and Abies balsamea (ca. 2500 to ca. 2100 BP)

Anthracomass values for the period corresponding to ca. 2500 to ca. 2100 BP (PAZ 3b) are low at Aspen, whereas no charcoal particles were found at Shadow. The pollen representation of *Abies balsamea* is still relatively high, whereas that of *Pinus banksiana*, *Betula* and *Alnus* spp. drops. It is estimated that this period corresponds to a dominance of *Picea mariana* and *A. balsamea* forests in the studied landscape, and perhaps in the majority of landscapes belonging to Québec's Clay Belt. The reconstructions of fire history from lake sediments in the western *Picea mariana*-moss domain also show a general decrease in fire frequency during this period, but with a certain variability among sites [10,28,31,34].

4.4. Increase of Early Successional Species (ca. 2100 to ca. 800 BP)

Anthracomass values for the period corresponding to ca. 2100 to ca. 800 BP are relatively low at Aspen (PAZ 3c). They are higher at Shadow, however, where three fires occurred in a short lapse of time (1960, 1670 and 1580–1550 BP). The pollen representation of *Abies balsamea* remains constant, while an increase of *Pinus banksiana*, *Betula* and *Alnus* is observed, especially at Shadow. An increase of *P. tremuloides* is not evident in the two pollen diagrams because this species is strongly under-represented in pollen assemblages [55]. The variability of fires and vegetation observed between the two peatlands also characterizes lake sites. Fire frequency is relatively high around ca. 1000 BP for some lakes (Profond, Raynald, Richard and Nans; Figure S1) but low for others (Geais, Loutre, Twin, Aurélie). These differences could be associated with the short duration of the Medieval Warm Period, which caused a varied fire pattern in the western portion of *Picea mariana*-moss domain [28]. Differences in physical features probably do not explain this variability because lakes with high and low fire frequency are located both in peatlands (Profond, Raynald, Geais, Loutre) and till environments (Richard, Nans, Twin, Aurélie).

4.5. More Severe Fires Favored the Development of Landscapes Dominated by Early Successional Species (from 800 BP to the Present-Day)

In the last 1000 years, three of the four dated fires recorded at Aspen and Shadow appear to have affected the two sites simultaneously (840/880–835, 290–300, 50–100 BP). Another fire (180 BP) was only recorded at Aspen. The fact that the forests surrounding the two peatlands were affected by the same fires on three occasions strongly suggests that they once covered wider areas and were more severe than those of the previous periods. In this region, two fire periods that were both severe and extended over a wide area occurred recently, in the 19th (ca. 1820) and 20th (ca. 1910) centuries [7,40]. The fire dating from 180 BP (Aspen) would correspond to ca. 1820, and the fire dating from 50–100 BP to ca. 1910. The link between the recent fires that occurred in the forest bordering the two peatlands and the deposition of charcoal in organic sediments is therefore well-established, suggesting that the same process characterized all of the Holocene period [56], at least in the case of large and severe fires. The fires occurring in our study area (4000 km^2) and in the 'Plaine du lac Matagami' during the last two centuries are considered natural because this region was sparsely inhabited until recently. The village of Matagami (Figure 1) was founded only in 1963 to open access to the James Bay hydroelectric reservoirs. The fires of the 1910 period are so widely distributed in the western *Picea mariana* moss domain that any possible role played by Aboriginal peoples living in the area during this fire period would have been minor [2].

The fires of the 19th and 20th centuries may have played an important role in the decrease of *A. balsamea* as shown on the two pollen diagrams of Aspen and Shadow (Figures 6 and 7). The relative abundance of *Abies balsamea* until recently is different from the proposed fragmentation of *Abies balsamea* forests initiated as early as 3500 BP [57]. During the last millennium, the pollen assemblages of Aspen and Shadow are dominated by *Pinus banksiana* and are at their highest level on the scale of the last 8000 years. An increase of *Pinus banksiana* in the two last centuries is also recorded elsewhere, notably at Schön Lake located close to the studied peatlands [15]. This recent dominance of early successional species is summarized in Figure 8, where positive scores of the first ordination axis are reported.

Most of the lake sediments studied by previous authors, and located in a peatland environment (Pessière, Geais, Profond, Raynald, Loutre, Figure S1), show a decrease in fire frequency during the last or two last millennia [10,31,34]. However, some lakes indicate an increase in fire frequency during this period in the western *Picea mariana*-moss domain. This is the case of Schön Lake, located close to the studied peatlands, to the north, in an environment dominated by clay [15,34] (Figure S1). Remy et al. [15] suggest that fire frequency increased since 2000 BP in the western *Picea mariana* moss (lakes Garot, Schön, Figure S1) and *Picea mariana* lichen (lakes Loup, Nano, Trèfle, Marie-Ève) domains. Most of these lakes are located in an environment with well-drained surficial deposits; Garot Lake, with an environment dominated by peatlands, is the only exception. Recent methodological advances in fire event reconstruction may lead to new results, and mainly at the extremities of the sediment core [58]. For example, while Ali et al. [10] show a decrease in fire frequency during the last 2000 years BP at Lake Geais, Oris et al. [34] indicate an increase since 1000 years BP for the same lake. An increase of fire frequency has been observed at lakes Francis and Pas de Fond, located in the Clay Belt portion of the western *Abies balsamea-Betula papyrifera* bioclimatic domain [37,38] (Figure S1). Both lakes are associated with a very marked cut ca. 2000 BP, characterized by the transition from a fire frequency higher than 200 years to less than 200 years. Thus, a higher fire incidence in the last millennia could characterize the entire Québec Clay Belt regardless of bioclimatic domain. All these fire activities and climate conditions led to landscapes dominated by early successional species, mainly *Pinus banksiana* and *Populus tremuloides* (Figure 9) [59,60].

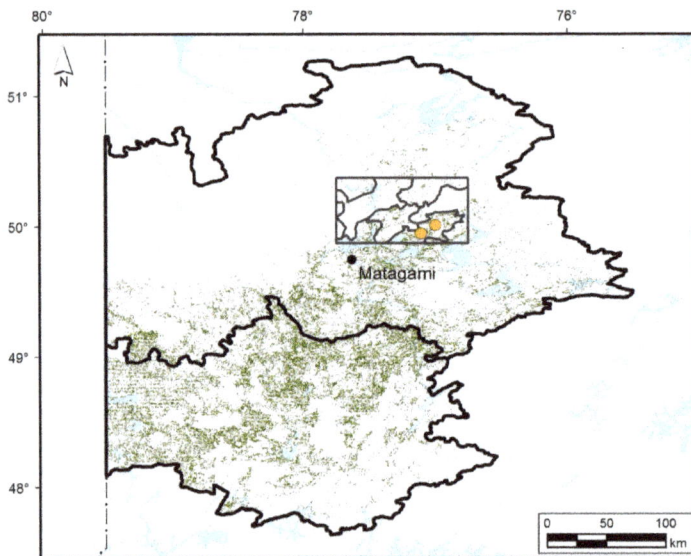

Figure 9. Forest stands dominated by *Populus tremuloides* on clay (**green**) in ecological region 6a-Plaine du lac Matagami (*Picea mariana*-moss domain, upper part) and 5a-Plaine du lac Abitibi (*Abies balsamea-Betula papyrifera* domain, lower part). The study area (rectangle), the landscapes delineated in this area (*n* = 8) and the two studied peatlands (**orange** points) are illustrated.

Finally, does the last millennium, particularly the last two centuries, correspond to a Holocene period characterized by a decrease in fire occurrence and severity, or does this period constitute the most important period of the entire Holocene in regard to large and severe fires? It may be that, during the recent millennia, landscapes dominated by peatlands show a fire decrease while those composed mainly of clay or other mineral soils an increase. The answer could lie in the type of surficial deposit (peatland vs. clay and other mineral soils) rather than the bioclimatic domain (*Picea mariana* vs. *Abies* domains) [8,11].

4.6. Contemporary Forest Dynamics

The forests bordering the two peatlands and the entire landscape studied (P4b, Figure 3) were strongly affected by contemporary fires. We estimate that the fires of the last millennia favoured a recurrence dynamic of *Populus tremuloides*, *Pinus banksiana* and *Picea mariana* stands. Today, forest regeneration is mainly composed of *P. mariana* accompanied by some *P. tremuloides*. Based on this type of regeneration, the forests should evolve towards mature *P. mariana-P. tremuloides* stands [22]. *Abies balsamea* is presently rare in the landscapes dominated by early successional forests, and this is probably a response to the two close and severe fire periods (1820–1910) that occurred in less than 200 years. If sufficient time elapses before the next fire occurs (approximately 200–250 years) [18], *Abies* could reach maturity and regenerate abundantly. Late successional forests dominated by *Picea mariana* and *Abies balsamea*, such as those that characterized the landscape during the Holocene (ca. 6800 to ca. 800 BP), could thus form once again. However, this possibility remains unlikely in the context of climate change because an increase in frequency and severity of fires is predicted for the western Québec *Picea mariana*-moss domain [32,33,60].

5. Conclusions

This study carried out in Québec's Clay Belt (Canada) provides new knowledge about the role of forest fires in landscapes dominated by early successional species (*Populus tremuloides*, *Pinus banksiana*, *Picea mariana*) during the last millennia. More severe and extensive fires stimulated the development and the continuity of these forests through a recurrence dynamic. Such a fire regime differs from that which prevailed for thousands of years during the Holocene, and allowed the growth of the late successional *P. mariana* and *Abies balsamea* forests in the western *Picea-mariana*-moss domain. This information on long-term natural variability provides insights that can guide the development and implementation of sound, ecosystem forest and fire management plans in the context of climate change.

Our study opens the door to the possibility that two types of fire regimes occurred during the Holocene in Québec's Clay Belt: the first, defined by a decrease in fire frequency during the late Holocene, characterized landscapes dominated by peatlands. The second shows an increase of fires over the last millennium and is related to landscapes mainly formed by mesic-subhydric clays. This second regime could characterize the entire Québec Clay Belt, regardless of bioclimatic domain (*Picea mariana*-moss and *Abies balsamea-Betula papyrifera*). In order to demonstrate the presence of these two regimes, studies coupling analysis of lake sediments, peatlands and mineral soils should be conducted jointly in the future.

Supplementary Materials: The following are available online at www.mdpi.com/1999-4907/7/9/205/s1. Figure S1. Delimitation of the study area in western Québec's *Picea mariana*-moss bioclimatic domain, Canada, and in ecological region 6a (Plaine du lac Matagami). Ecological region 5a (Plaine du lac Abitibi) is part of the western *Abies balsamea-Betula papyrifera* bioclimatic domain. The long transect with oblique lines located close to the border with Ontario corresponds to the fire origin map created by Bergeron et al. [7]. The red dots indicate the clay distribution according to QMFFP forest maps. This clay corresponds to the Québec Clay Belt. Codes with a P (peatland) or an L (lakes) correspond to paleoecological studies considered in this study. AP: Aspen peatland, SP: Shadow peatland, LA: Lake Aurélie, LCE: Lake Cèdres, LF: Lake Francis, LGA: Lake Garot, LGE: Lake Geais, LL: Lake Loutre, LN: Lake Nans, LPE: Lake Pessière, LPF: Lake Pas de Fond, LPR: Lake Profond, LRA: Lake Raynald, LRI: Lake Richard, LS: Lake Shön, LT: Lake Twin. Table S1. Description of the eight forest landscapes delineated in the study area (Figure 3). The two peatlands studied are located in landscape P4b. Results are expressed in relative importance of surface area. Table S2. Description of the contemporary vegetation surrounding the lakes studied in the western part of Québec's boreal forest. The description is based on QMFFP maps and considers a 15 km^2 radius around each lake or peatland.

Acknowledgments: This research received financial support through grants from the Fonds de recherche du Québec—Nature et technologies (FRQNT). Scholarships to M. Asselin were provided by Mitacs-Accélération Québec. The authors would like to thank A. Delwaide, M. Bourgon-Desroches, V. Poirier and J. Noël for their technical contributions. A. Ali contributed advice regarding the analysis and interpretation of fire history from lacustrine sediments. He also authorized us to use the pollen and charcoal data of the lakes he studied in western Québec. The English translation of the text was revised by K. Grislis and G. Mercier. We extend our thanks to two anonymous reviewers and the two Academic Editors (S. Gauthier, Y. Bergeron) whose suggestions helped to greatly improve this manuscript.

Author Contributions: All authors contributed to all aspects of the work, including designing the experiments, analyzing the data, and writing the paper.

Conflicts of Interest: The authors declare no conflict of interest.

References

1. White, P.S. Pattern, process, and natural disturbance in vegetation. *Bot. Rev.* **1979**, *45*, 229–299. [CrossRef]
2. Grondin, P.; Gauthier, S.; Borcard, D.; Bergeron, Y.; Noël, J. A new approach to ecological land classification for the Canadian boreal forest that integrates disturbances. *Landsc. Ecol.* **2014**, *29*, 1–16. [CrossRef]
3. Pojar, J.; Klinka, J.; Meidinger, D.V. Biogeoclimatic ecosystem classification in British-Columbia. *For. Ecol. Manag.* **1987**, *22*, 119–154. [CrossRef]
4. Saucier, J.P.; Grondin, P.; Robitaille, A.; Gosselin, J.; Morneau, C.; Richard, P.J.H.; Brisson, J.; Sirois, L.; Leduc, A.; Morin, H.; et al. Écologie Forestière. In *Manuel de Foresterie*, 2nd ed.; Éditions MultiMondes: Québec, QC, Canada, 2009; pp. 165–316.
5. Rowe, J.S.; Scotter, G.W. Fire in the boreal forest. *Quat. Res.* **1973**, *3*, 444–464. [CrossRef]
6. Zackrisson, O. Influence of forest fires on the North Swedish boreal forest. *Oikos* **1977**, *29*, 22–32. [CrossRef]
7. Bergeron, Y.; Gauthier, S.; Flannigan, M.; Kafka, V. Fire regimes at the transition between mixedwood and coniferous boreal forest in northwestern Québec. *Ecology* **2004**, *85*, 1916–1932. [CrossRef]
8. Lynch, J.A.; Hollis, J.L.; Hu, F.S. Climatic and landscape control of the boreal forest fire regime: Holocene records from Alaska. *J. Ecol.* **2004**, *92*, 477–489. [CrossRef]
9. Cyr, D.; Bergeron, Y.; Gauthier, S.; Larouche, A.C. Are the old growth forests of the Clay Belt part of a fire-regulated mosaic? *Can. J. For. Res.* **2005**, *35*, 65–73. [CrossRef]
10. Ali, A.A.; Carcaillet, C.; Bergeron, Y. Long-term fire frequency variability in the eastern Canadian boreal forest: The influences of climate vs. local factors. *Glob. Chang. Biol.* **2009**, *15*, 1230–1241. [CrossRef]
11. Mansuy, N.; Gauthier, S.; Robitaille, A.; Bergeron, Y. The effects of surficial deposit–drainage combinations on spatial variations of fire cycles in the boreal forest of eastern Canada. *Int. J. Wildland Fire* **2010**, *19*, 1083–1098. [CrossRef]
12. Genries, A.; Finsinger, W.; Asnong, H.; Bergeron, Y.; Carcaillet, C.; Garneau, M.; Hély, C.; Ali, A.A. Local versus regional processes: Can soil characteristics overcome climate and fire regimes by modifying vegetation trajectories? *J. Quat. Sci.* **2012**, *27*, 745–756. [CrossRef]
13. Kuosmanen, N.; Fang, K.; Bradshaw, R.H.W.; Clear, J.L.; Seppa, H. Role of forest fires in Holocene stand-scale dynamics in the unmanaged taiga forest of northwestern Russia. *Holocene* **2014**, *24*, 503–514. [CrossRef]
14. Senici, D.; Chen, H.Y.H.; Bergeron, Y.; Ali, A.A. The effects of forest fuel connectivity on spatiotemporal dynamics of Holocene fire regimes in the central boreal forest of North America. *J. Quat. Sci.* **2015**, *30*, 365–375. [CrossRef]
15. Remy, C.C.; Lavoie, M.; Girardin, M.; Hély, C.; Bergeron, Y.; Grondin, P.; Oris, F.; Asselin, H.; Ali, A.A. Wildfire size alters long-term vegetation trajectories in boreal forests of eastern North America. *J. Biogeogr.* In press.
16. Landres, P.B.; Morgan, P.; Swanson, F.J. Overview of the use of natural variability concepts in managing ecological systems. *Ecol. Appl.* **1999**, *9*, 1179–1188.
17. Kuuluvainen, T. Natural variability of forests as a reference for restoring and managing biological diversity in boreal Fennoscandia. *Silva Fenn.* **2002**, *36*, 97–125. [CrossRef]
18. Gauthier, S.; Vaillancourt, M.A.; Leduc, A.; de Grandpré, L.; Kneeshaw, D.; Morin, H.; Drapeau, P.; Bergeron, Y. *Aménagement Ecosystémique en Forêt Boréale*; Presses de l'Université du Québec: Montréal, QC, Canada, 2008.
19. Conedera, M.; Tinner, W.; Neff, C.; Meurer, M.; Dickens, A.F.; Krebs, P. Reconstructing past fire regimes: Methods, applications, and relevance to fire management and conservation. *Quat. Sci. Rev.* **2009**, *28*, 555–576. [CrossRef]

20. Cyr, D.; Gauthier, S.; Bergeron, Y.; Carcaillet, C. Forest management is driving the eastern North American boreal forest outside its natural range of variability. *Front. Ecol. Environ.* **2009**, *7*, 519–524. [CrossRef]

21. Cogbill, C.V. Dynamics of the boreal forests of the Laurentian Highlands, Canada. *Can. J. For. Res.* **1985**, *15*, 252–261. [CrossRef]

22. Blouin, J.; Berger, J.-P. *Guide de reconnaissance des types écologiques de la région écologique 6a-Plaine du lac Matagami et 6b-Plaine de la baie de Rupert*; ministère des Ressources naturelles et de la Faune, Direction des inventaires forestiers, Division de la classification écologique et productivité des stations: Québec, QC, Canada, 2005.

23. Frelich, L.E.; Reich, P.B. Spatial patterns and succession in a Minnesota southern-boreal forest. *Ecol. Monogr.* **1995**, *65*, 325–346. [CrossRef]

24. Swain, A.M. Environmental changes during the past 2000 years in north-central Wisconsin: Analysis of pollen charcoal and seeds from varved lake sediments. *Quat. Res.* **1978**, *10*, 55–68. [CrossRef]

25. Frégeau, M.; Payette, S.; Grondin, P. Fire history of the central boreal forest in eastern North America reveals stability since the mid-Holocene. *Holocene* **2015**, *25*, 1912–1922. [CrossRef]

26. Blarquez, O.; Ali, A.A.; Girardin, M.P.; Grondin, P.; Fréchette, B.; Bergeron, Y.; Hély, C. Regional paleofire regimes affected by non-uniform climate, vegetation and human drivers. *Sci. Rep.* **2015**, *5*, 13356. [CrossRef] [PubMed]

27. Ali, A.A.; Blarquez, O.; Girardin, M.P.; Hély, C.; Tinquaut, F.; El-Guellab, A.; Valsecchi, V.; Terrier, A.; Bremond, L.; Genries, A.; et al. Control of the multimillenial wildfire size in boreal North America by spring climatic conditions. *Proc. Natl. Acad. Sci. USA* **2012**, *109*, 20966–20970. [CrossRef] [PubMed]

28. El-Guellab, A.; Asselin, H.; Gauthier, S.; Bergeron, Y.; Ali, A.A. Holocene variations of wildfire occurrence as a guide for sustainable management of the northeastern Canadian boreal forest. *For. Ecosyst.* **2015**, *2*, 1–7. [CrossRef]

29. Viereck, L.A. Wildfire in the taiga of Alaska. *Quat. Res.* **1973**, *3*, 465–495. [CrossRef]

30. Viau, A.E.; Gajewski, K. Reconstructing millennial-scale, regional paleoclimates of boreal Canada during the Holocene. *J. Clim.* **2009**, *22*, 316–330. [CrossRef]

31. Hély, C.; Girardin, M.P.; Ali, A.A.; Carcaillet, C.; Brewer, S.; Bergeron, Y. Eastern boreal North American wildfire risk of the past 7000 years: A model-data comparison. *Geophys. Res. Lett.* **2010**, *37*, L14709. [CrossRef]

32. Girardin, M.P.; Ali, A.A.; Carcaillet, C.; Gauthier, S.; Hély, C.; Le Goff, H.; Terrier, A.; Bergeron, Y. Fire in managed forests of eastern Canada: Risks and options. *For. Ecol. Manag.* **2013**, *294*, 238–249. [CrossRef]

33. Girardin, M.P.; Ali, A.A.; Carcaillet, C.; Blarquez, O.; Hély, C.; Terrier, A.; Genries, A.; Bergeron, Y. Vegetation limits the impact of a warm climate on boreal wildfires. *New Phytol.* **2013**, *199*, 1001–1011. [CrossRef] [PubMed]

34. Oris, F.; Asselin, H.; Finsinger, W.; Hély, C.; Blarquez, O.; Ferland, M.E.; Bergeron, Y.; Ali, A.A. Long-term fire history in northern Québec: Implications for the northern limit of commercial forests. *J. Appl. Ecol.* **2014**, *51*, 675–683. [CrossRef]

35. Fréchette, B.; Richard, P.J.H.; Grondin, P.; Lavoie, M.; Larouche, A.C. *Histoire Postglaciaire de la Végétation et du Climat des Pessières et des Sapinières de l'Ouest du Québec*; Direction des Inventaires Forestiers, Gouvernement du Québec: Québec, QC, Canada, 2016.

36. Environment Canada. National Climate Data and Information Archive. Canadian Climate Normals 1971–2000. 2013. Available online: http://www.climate.weatheroffice.ec.gc.ca (accessed on 17 September 2015).

37. Carcaillet, C.; Bergeron, Y.; Richard, P.J.H.; Fréchette, B.; Gauthier, S.; Prairie, Y.T. Change of fire frequency in the eastern Canadian boreal forests during the Holocene: Does vegetation composition or climate trigger the fire regime? *J. Ecol.* **2001**, *89*, 930–946. [CrossRef]

38. Carcaillet, C.; Richard, P.J.H.; Bergeron, Y.; Fréchette, B.; Ali, A.A. Resilience of the boreal forest in response to Holocene fire-frequency changes assessed by pollen diversity and population dynamics. *Int. J. Wildland Fire* **2010**, *19*, 1026–1039. [CrossRef]

39. Vincent, J.S.; Hardy, L. L'évolution et l'extension des lacs glaciaires Barlow et Ojibway en territoire québécois. *Geogr. Phys. Quat.* **1977**, *31*, 357–372. [CrossRef]

40. Bergeron, Y.; Gauthier, S.; Kafka, V.; Lefort, P.; Lesieur, D. Natural fire frequency for the Eastern Canadian boreal forest: Consequences for sustainable forestry. *Can. J. For. Res.* **2001**, *31*, 384–391. [CrossRef]

41. Lecomte, N.; Bergeron, Y. Successionnel pathways on different surficial deposits in the coniferous boreal forest of the Québec Clay Belt. *Can. J. For. Res.* **2005**, *35*, 1984–1995. [CrossRef]

42. Legendre, P.; Legendre, L. *Numerical Ecology*, 3rd ed.; Elsevier: Amsterdam, The Netherlands, 2012.
43. Bhiry, N.; Robert, É.C. Reconstruction of changes in vegetation and trophic conditions of a palsa in a permafrost peatland, subarctic Québec, Canada. *Écoscience* **2006**, *13*, 55–65. [CrossRef]
44. Lavoie, M.; Filion, L.; Robert, É.C. Boreal peatland margins as repository sites of long-term natural disturbances of balsam fir/spruce forests. *Quat. Res.* **2009**, *71*, 295–306. [CrossRef]
45. Stuiver, M.; Reimer, P.J. Extended ^{14}C database and revised CALIB radiocarbon calibration program. *Radiocarbon* **1993**, *35*, 315–330. [CrossRef]
46. Reimer, P.J.; Bard, E.; Bayliss, A.; Beck, J.W.; Blackwell, P.G.; Ramsey, C.B.; Buck, C.E.; Cheng, H.; Edwards, R.L.; Friedrich, M.; et al. INTCAL 13 and MARINE13 radiocarbon age calibration curves 0–50,000 years cal BP. *Radiocarbon* **2013**, *55*, 1869–1887. [CrossRef]
47. Blaauw, M. Methods and code for "classical" age-modelling of radiocarbon sequences. *Quat. Geochron.* **2010**, *5*, 512–518. [CrossRef]
48. Hörnberg, G.; Ohlson, M.; Zackrisson, O. Stand dynamics, regeneration patterns and long-term continuity in boreal old-growth *Picea abies* swamp-forests. *J. Veg. Sci.* **1995**, *6*, 291–298. [CrossRef]
49. Magnan, G.; Lavoie, M.; Payette, S. Impact of fire on long-term vegetation dynamics of ombrotrophic peatlands in northwestern Québec, Canada. *Quat. Res.* **2012**, *77*, 110–121. [CrossRef]
50. Kasin, I.; Blanck, Y.; Storaunet, O.; Rolstad, J.; Ohlson, M. The charcoal record in peat and mineral soil across a boreal landscape and possible linkages to climate change and recent fire history. *Holocene* **2013**, *23*, 1052–1065. [CrossRef]
51. Faegri, K.; Kaland, P.E.; Krzywinski, K. *Textbook of Pollen Analysis*, 4th ed.; John Wiley & Sons: New York, NY, USA, 1989.
52. Grimm, E.C. CONISS: A FORTRAN 77 program for stratigraphically constrained cluster analysis by the method of incremental sum of squares. *Comput. Geosci.* **1987**, *13*, 13–35. [CrossRef]
53. Birks, H.J.B. Quantitative Palaeoenvironmental Reconstructions. In *Statistical Modelling of Quaternary Science Data*; Maddy, D., Brew, J.S., Eds.; Quaternary Research Association: Cambridge, UK, 1995; pp. 161–254.
54. Richard, P.J.H. Origine et dynamique postglaciaire de la forêt mixte au Québec. *Rev. Paleobot. Palynol.* **1993**, *79*, 31–68. [CrossRef]
55. Comtois, P.; Payette, S. Représentation pollinique actuelle et subactuelle des peupleraies boréales au Nouveau-Québec. *Geogr. Phys. Quat.* **1984**, *38*, 123–133.
56. Brossier, B.; Oris, F.; Finsinger, W.; Asselin, H.; Bergeron, Y.; Ali, A. Using tree-ring to calibrate peak detection in fire reconstruction based on sedimentary charcoal records. *Holocene* **2014**, *24*, 635–645. [CrossRef]
57. Ali, A.; Asselin, H.; Larouche, A.C.; Bergeron, Y.; Carcaillet, C.; Richard, P.J.H. Changes in fire regime explain the Holocene rise and fall of *Abies balsamea* in the coniferous forests of western Québec, Canada. *Holocene* **2008**, *18*, 693–703. [CrossRef]
58. Blarquez, O.; Girardin, M.P.; Leys, B.; Ali, A.A.; Aleman, J.C.; Bergeron, Y.; Carcaillet, C. Paleofire reconstruction based on an ensemble-member strategy applied to sedimentary charcoal. *Geophysic. Res. Lett.* **2013**, *40*, 2667–2672. [CrossRef]
59. Le Goff, H.; Flannigan, M.D.; Bergeron, Y.; Girardin, M.P. Historical fire regime shift related to climate teleconnections in the Waswanipi area, central Quebec, Canada. *Int. J. Wildland Fire* **2007**, *16*, 607–618. [CrossRef]
60. Bergeron, Y.; Cyr, D.; Girardin, M.; Carcaillet, C. Will climate change drive 21st century burn rates in Canadian boreal forest outside its natural variability: Collating global climate model experiment with sedimentary charcoal data. *Int. J. Wildland Fire* **2010**, *19*, 1127–1139. [CrossRef]

© 2016 by the authors. Licensee MDPI, Basel, Switzerland. This article is an open access article distributed under the terms and conditions of the Creative Commons Attribution (CC BY) license (http://creativecommons.org/licenses/by/4.0/).

forests

MDPI

Article

Vegetation Mortality within Natural Wildfire Events in the Western Canadian Boreal Forest: What Burns and Why?

Colin J. Ferster [1], Bianca N. I. Eskelson [1,*], David W. Andison [1,2] and Valerie M. LeMay [1]

[1] Department of Forest Resources Management, University of British Columbia, Vancouver, BC V6T 1Z4, Canada; Colin.Ferster@ubc.ca (C.J.F.); Andison@bandaloop.ca (D.W.A.); Valerie.LeMay@ubc.ca (V.M.L.)
[2] Bandaloop Landscape-Ecosystem Services, 1011 Hendecourt Road, North Vancouver, BC V7K 2X3, Canada
* Correspondence: Bianca.Eskelson@ubc.ca; Tel.: +1-604-827-0629

Academic Editors: Yves Bergeron and Sylvie Gauthier
Received: 17 June 2016; Accepted: 19 August 2016; Published: 26 August 2016

Abstract: Wildfires are a common disturbance event in the Canadian boreal forest. Within event boundaries, the level of vegetation mortality varies greatly. Understanding where surviving vegetation occurs within fire events and how this relates to pre-fire vegetation, topography, and fire weather can inform forest management decisions. We used pre-fire forest inventory data, digital elevation maps, and records of fire weather for 37 naturally-occurring wildfires (1961 to 1982; 30 to 5500 ha) covering a wide range of conditions in the western Canadian boreal forest to investigate these relationships using multinomial logistic models. Overall, vegetation mortality related to a combination of factors representing different spatial scales. Lower vegetation mortality occurred where there was lower fuel continuity and when fires occurred under non-drought conditions. Higher classification accuracy occurred for class extremes of no mortality (i.e., unburned areas within the burn event) and high mortality; partial vegetation mortality classes were harder to distinguish. This research contributes to the knowledge required for natural pattern emulation strategies, and developing responses to climate change.

Keywords: historic natural wildfire; multinomial logistic model; ecosystem-based management

1. Introduction

The boreal forest provides a vast array of goods and services, including timber products, wildlife habitat, water quality and quantity, recreation, carbon sequestering, hunting and fishing opportunities, minerals, and gas and oil [1]. Concerns about the sustainability of the boreal forest and its ecosystem services [2] has inspired a management approach where knowledge of historical disturbance patterns provides the primary foundation for planning activities [3]. The natural pattern (NP) concept proposes that landscape compositions and structures that are similar to historic conditions are more likely to maintain historical levels of biodiversity [4]. Further, the NP concept is the cornerstone of ecosystem-based management (EBM) [5] and has been embraced by provincial regulators in Canada [6] and forest certification agencies [7].

In the Canadian boreal forest (hereafter termed "boreal"), wildfires are responsible for the majority of the historical disturbance patterns [8,9]. For many years, the boreal was referred to as a stand-replacing ecosystem, meaning that most fires are highly severe, killing a large percentage of the trees [8]. However, more recent evidence suggests that historical boreal wildfires left a large amount of unburned and partially burned vegetation within the fire boundary [10,11]. Moreover, the amount and spatial patterns of surviving remnant vegetation within the area of a wildfire event (hereafter referred to as "remnants") can vary greatly both between and within fires. For example, remnants in wildfires in the Boreal Plains Ecoregion tend to create stepping stones, while those of the Foothills Ecoregion tend to form spatially contiguous corridors [12]. The amount, but also the spatial arrangement of remnants is important for forest succession, habitat, refugia, and revegetation patterns [13,14].

In general, our understanding of how much vegetation survives in boreal wildfires far exceeds that of the spatial patterns of remnants and the associated underlying factors. Survival levels within boreal fires have been linked to fire weather, size and duration, pre-fire forest structure, vegetation types, topography, the locations and density of water bodies, and fuel types and arrangement [15–19]. Despite the breadth of research, the natural range of variation (NRV) of both the amount and spatial locations of remnants within boreal fires remains largely unexplained.

One possible explanation for this limited understanding is under-estimating the underlying complexity of the relationships between vegetation survival and causal factors. Across large landscapes and over several decades, individual fires respond to local differences in fire weather, topography, ecological zone, and fuel-type conditions [12,20,21]. This means that capturing the NRV of wildfire patterns requires extensive information across very large areas. To address this, a number of studies have used forest inventory data (e.g., [22]), the national wildfire database (e.g., [23]), and satellite imagery to capture NRV of historical fire patterns (e.g., [15,24]). While all of these data are readily available, none of these data capture smaller remnants and/or partial mortality. Recent studies have shown that this fine-scale complexity accounts for a significant part of wildfire remnants in the western boreal [12]. To capture the high complexity and variability of remnant patterns among fire events, large sample sizes are needed. While detailed analyses of burning conditions on individual or a small number of fires can provide valuable local insights (e.g., [17,25]), the findings of such studies are only a subset of NRV. Similarly, studies that include only a few explanatory variables (e.g., [26]) could lead to misleading conclusions about which factors are most important with regards to vegetation survival.

Further contributing to the challenges of understanding wildfire burning patterns is the inconsistent terminology. The meanings of "fire severity", "burn severity", "vegetation survival" and "remnants" not only differ but may also vary from one study to another. Many studies using remotely sensed data use the Normalized Difference Vegetation Index (NDVI) and Normalized Burn Ratio (NBR) to represent fire severity and burn severity, respectively. These indices capture vegetation mortality, but also reflect losses of dead wood and organic matter, ash deposition, and changes to soil structure and chemistry [18,27]. The Composite Burn Index (CBI) also captures burn severity, but does not capture vegetation mortality levels consistently. Jain and Graham [28] found that burn severity can vary greatly for canopy versus ground levels. Definitions of wildfire remnants have ranged from large, entirely unburned islands of trees within a burned polygon (e.g., [22]) to all types and levels of surviving vegetation within the general vicinity of a wildfire event (e.g., [23]). Andison [29] found that a three-fold difference in remnants can occur with even very subtle definition differences. As an example, Eberhart and Woodard [30] found that remnants comprised 5%–15% of the area of historical wildfire events, compared to 20%–60% by Andison and McCleary [12] for fires from a similar study area. These inconsistencies can lead to pattern artefacts [31], can reduce the ability to compare and compile NRV knowledge across studies, and create unnecessary confusion.

In this paper, we examined relationships between physical and environmental variables linked to fire behaviour at a range of spatial scales and levels of vegetation mortality within historic fire events to inform NRV management strategies in the western Canadian boreal. For this, we used an existing, spatially extensive, high resolution photo-based historical wildfire dataset from western boreal Canada that also includes pre-fire vegetation data. We then obtained information on a number of possible biotic and abiotic factors related to fire-caused vegetation mortality levels including topography metrics and local fire weather data. Using these variables at the fire event and within-event spatial scales, we built multinomial logistic (i.e., multinomial logit) models to predict vegetation mortality classes as a mechanism for examining the roles of these factors in determining the amounts and locations of post-fire remnants.

2. Materials and Methods

2.1. Study Area

The historic wildfire data used for this study was obtained from an existing database of 129 fires distributed across the Boreal Plains, Canadian Shield, Foothills, and Rocky Mountains Ecoregions of Canada (Figure 1).

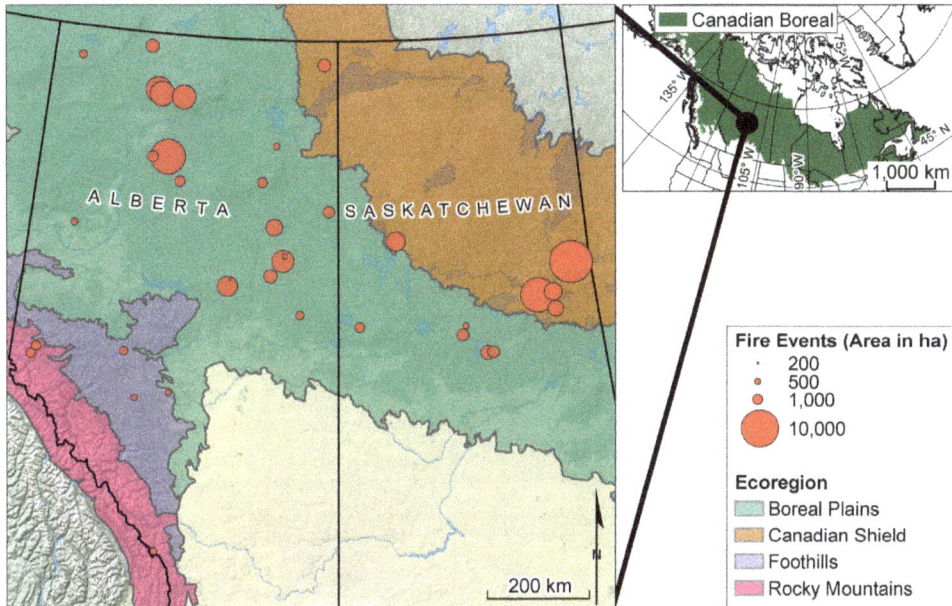

Figure 1. Location of fire events. The Ecoregions are from Acton and Natural Regions Committee [32,33], the full extent of the Canadian boreal forest in the overview map is from Brandt [34], and base map is from Kelso and Patterson [35].

The Boreal Plains Ecoregion is generally a flat or gently rolling landscape with thick glacially shaped soils and subtle upland areas interspersed with lower-lying wetlands. In the south, pure aspen (*Populus tremuloides* (Michx.) and other *Populus* spp.) stands are common, with patches of white spruce (*Picea glauca* (Moench) Voss) associated with moist soils. Wetlands are often dominated by shrubs. Further north in the Boreal Plains Ecoregion, upland vegetation is composed of mixed stands of aspen, balsam poplar (*Populus balsamifera* (L.)), and white spruce, with extensive lowland wetland areas with black spruce (*Picea mariana* (Mill.) B.S.P.), larch (*Larix laricina* (Du Roi) K. Koch), shrubs, and sedges. Throughout the region, stands of jack pine (*Pinus banksiana* (Lamb.)), lodgepole pine (*Pinus contorta* (Douglas) var. *latifolia* (Engel.) Critchfield), and hybrids of these two species can occur on sandy well-drained soils [32,33]. The Canadian Shield is characterized by glacially scoured granite bedrock, often barren and exposed at the surface or covered by thinner soils, resulting in rolling hills, coniferous or mixed forest stands where the soil is sufficiently thick, and numerous lakes and wet areas. Black spruce dominates the area, mixing with jack pine on uplands, and larch on lowland areas. Hardwoods and mixed stands are found on certain sites with favourable soil and aspect [32,33]. The Foothills Ecoregion is distinguished from the Boreal Plains Ecoregion by steeper slopes, higher altitudes, and more common lodgepole pine stands. Upland stands are typically deciduous or mixed-wood, with wetlands dominated by stunted black spruce, larch, or shrubby vegetation. In the Rocky Mountains, there is mountainous terrain with a larger range of elevation, steeper slopes, and orthographic precipitation. At low elevations, closed conifer stands are typical, while at higher elevations, more open conifer stands and herbaceous meadows, grasslands, or low shrubs typically occur. At high elevations, the growing season may be too short to support trees, resulting in alpine and subalpine environments. Given the steep slopes and large terrain features, vegetation cover can be strongly influenced by slope location and microsite (in particular, the availability of moisture and solar energy); for example, open coniferous stands with grasslands may be more commonly found on south facing slopes and closed conifer stands on north facing slopes [33].

2.2. Data

Thirty-seven fire events had available data on pre-fire vegetation, topography, and fire weather. Fire events in our sample were distributed across four Ecoregions: Boreal Plains (24); Canadian Shield (seven); Foothills (three); and the Rocky Mountains (three) (Figure 1).

2.2.1. Mortality Maps

The wildfire database used for this study had strict eligibility criteria, specifically:

1. No evidence of pre-fire anthropogenic activity;
2. Minimal or no fire suppression;
3. No post-fire salvage logging; and
4. High quality, high resolution aerial photos available within five years post-fire (see [29] for details).

The photo negatives required to cover the area of each fire were obtained and digitally scanned at 10 μm and imported into Softcopy [36]. The scale of the aerial photographs was either (a) better than 1:20,000 or (b) large format plates greater than 1:31,860 of sufficient quality to resolve individual trees (see [12,29] for details). In the Softcopy environment, photo-pairs were used to define the outer boundaries of each fire (i.e., the "shell"), and map all remnants using six mortality classes: class 0 = no mortality, 0%; class 1 = 1% to 25%; class 2 = 26% to 50%; class 3 = 51% to 75%; class 4 = 76% to 94%; and class 5 = ≥95%. Vegetation mortality was measured as:

1. The percent dead tree crowns attributed to the fire event for forested areas;
2. The percent cover of dead shrubs and bushes for non-forested areas with other woody plants; or
3. The percent area scorched for non-forested areas with only grass or bryophytes.

For consistency, the same person interpreted all fire events. A minimum mapping unit of 10 m × 10 m was used, representing an area occupied by no more than three tree crowns. The resulting mortality maps were scanned, digitized, and spatially registered to 1:50,000 topographic maps. Thus the remnants tested in this study are the equivalent of "island remnants" sensu Andison [29].

2.2.2. Pre-Fire Vegetation

Pre-fire data were collected using high resolution photos acquired within five years of the fire event [29]. The photo negatives covering each fire were digitally scanned at 10 μm. The extent of vegetation interpretation included a 100 m buffer on the outside of the shell of each fire to capture the boundary area. Interpretation protocols followed the standardized vegetation inventory methods as defined by both Alberta and Saskatchewan [37,38]. The parameters used to define polygons included percent crown cover, species composition, height class, and age class for forested areas, vegetation type for non-forested areas, and soil moisture class. Since the specific criteria for photo-interpretation varied by province, variables chosen for this study were harmonized (Table 1). Additionally, midpoints of classes were used to represent percent height class, age class, and crown cover instead of including these as class variables in all analyses.

Table 1. Biotic and abiotic attributes.

Type	Variable	Description
Site	Ecoregion	Ecological land classification [32,33]
Vegetation polygon	Event area (m²)	Event area within boundary defined by vegetation mortality including unburned islands.
	Age (years)	Year of the fire—estimated year of stand origin (photo-interpreted). Zero for non-forested polygons.
	Fuel category	See Table 2.
	Height (m)	Average height of the dominant and codominant trees of the leading species. Zero for non-forest.
	Overstory crown closure (%)	Percent ground area covered by crowns of the dominant and codominant trees. Zero for non-forest.
	Soil moisture class	Interpreted classification of dry, mesic, or wet.
	Understory crown closure (%)	Cover of trees forming a distinct understory layer. Zero for non-forest or no understory layer.
	Understory height (m)	Average height of trees forming a distinct understory layer. Zero for non-forest or no understory layer.
	Ladder fuel index	Ratio of overstory and understory height multiplied by understory crown closure (following Alexander et al. [39]).
Topography: All variables calculated using the 30 m Digital Elevation Model (DEM).	Elevation (m)	
	Slope (degrees)	
	Curvature	Surface curvature. Positive is convex (ridges), zero is flat, and negative is concave (valleys) [40].
	Solar aspect index (TSRAI)	Circular aspect: 0 on northeast slopes, 0.5 for flat ground to 1 on southwest slopes [41].
	Compound topographic index (CTI)	Potential soil moisture, based on slope and catchment area. High values have potentially high soil moisture [42].
	SCOSA + SSINA	Slope * cosine (aspect) and slope * sine (aspect). Represents the "northness" and "eastness" of a pixel, respectively [43].
	Slope position	Relative elevation. Areas with values higher than 1 are higher than the surroundings and vise-versa [44].
Fire weather	Fine fuel moisture code (FFMC)	Moisture content of litter and fine materials indicating the ease of ignition and flammability of fine fuels.
	Duff moisture code (DMC)	Related to the average moisture content of organic soil layers to a moderate depth.
	Drought Code (DC)	Indicates seasonal drought and the expected smouldering in deep organic soil layers and large logs.
	Initial Spread Index (ISI)	Expected rate of fire spread given wind and FFMC. ISI threshold is the proportion of fire days with ISI > 8.7 [45].
	Build Up Index (BUI)	A rating of the total amount of fuel available for combustion by combining DMC and DC.
	Fire Weather Index (FWI)	Index of potential fire intensity given ISI and BUI. FWI threshold is the proportion of fire days FWI > 19 [45].

Using these pre-fire vegetation and soil attributes, a fuel category was assigned to each vegetation polygon within the buffered area of influence of each fire based on the Canadian Fire Behavior Prediction (FBP) fuels types [46]. For forested areas, fuel categories were assigned based on overstory species and stand age; several new categories were assigned to represent local conditions (i.e., FX for balsam-fir (*Abies balsamea* (L.) Mill.) and LX for larch (Table 2). Additionally, codes for non-forest vegetated areas were added.

Table 2. Classification of fuel category based on FBP fuel type [46].

Type	Variable	Description
Forest	C2	Boreal spruce: Spruce (*Picea* spp.) as the leading species.
	C3	Mature pine: Pine (*Pinus* spp.) as the leading species, age > 40 years.
	C4	Immature Pine (*Pinus* spp.) as the leading species, age ≤ 40 years.
	D1	Leafless aspen: Trembling aspen (*Populus tremuloides* (Michx.)) and other types of aspen as the leading species, fire event started before June 1st (i.e., commonly before leaves flushed).
	D2	Leaf-on aspen: Trembling aspen (*Populus tremuloides* (Michx.)) and other types of aspen (*Populus* spp.) as the leading species, fire event started after June 1st (i.e., commonly after leaves flushed).
	FX	Fir: Balsam fir (*Abies balsamea* (L.) Mill.) as the leading species.
	LX	Larch: Larch (*Larix laricina* (Du Roi) K. Kosh) as the leading species.
	M1	Boreal mixedwoods, leafless: Conifer and deciduous species mixed in the overstory, fire event started before June 1st.
	M2	Boreal mixedwoods, leaf-on: Mixed conifer and deciduous species in the overstory, fire event started after June 1st.
Non-forest		Seven classes: Open shrub, closed shrub, open muskeg, treed muskeg, brush and alder, bryophytes, and grassland.

2.2.3. Topography

Digital elevation data were downloaded to cover the buffered area of influence for each fire event from the Canadian Digital Elevation Dataset (CDED) which has a spatial resolution of 30 m (available from http://geobase.ca, accessed 27 October 2014). A suite of topographic indices were calculated that relate to elevation, slope, aspect, and landscape position (Table 1) using the Toolbox for Surface Gradient and Geomorphometric Modeling extension software version 2.0-0 for ArcGIS version 10.3.1 [40].

2.2.4. Fire Weather

Historic fire weather data from Amiro et al. [47] were used for this research. These data represent daily fire weather for Canadian wildfires >200 ha from 1959 to 1999, created by spatially interpolating weather station measures for 21 days following the recorded fire ignition date. Using these data, six daily fire weather indices were calculated based on the moisture availability of fuels and fire behaviour (Table 1). For this study, only three of the six available fire weather indices were selected, representing different aspects of fire behaviour [45,48]: ISI, to represent short-term changes in the availability of fine fuels; DC, to represent longer-term seasonal weather conditions and the availability of deep organic soil and large logs; and FWI as an overall index of potential fire energy output. For ISI and FWI, Podur and Wotton [45] calculated threshold values that distinguish fire spread events, where fire intensity is high, tree crowns are consumed, and the rate of spread is rapid, and non-spread events, where fire intensity is low, most burning occurs on the ground surface, and the rate of fire spread is low. In this study additional summary statistics were calculated for ISI and FWI as the proportion of days where each index exceeded the respective threshold for spread conditions (ISI = 8.7 and FWI = 19;

following Podur and Wotton [45]). For each of these indices, daily values were summarized (i.e., median, maximum, minimum, and range) for either the duration of the fire minus one day, or for the full 21-days from the beginning of the fire event if the event was longer than 21 days.

2.2.5. Other Data

Other data collected and used in the analyses included Ecoregion, as defined by a blending of the Alberta and Canadian Ecoregions (see [12] for details). The event area was calculated (including unburned islands) and assigned to each pixel within each respective fire event.

2.3. Multinomial Logistic Models for Vegetation Mortality Class

For each fire event, the vegetation mortality class and all described biotic and abiotic attributes were tabulated by intersecting all layers with the center point of 30 m pixels, which was the lowest common spatial resolution of the vegetation polygons, vegetation mortality, and topography data. Although originally included in the data, pixels from the 100 m buffer around each fire boundary were not used for model training and testing because large portions of these boundary regions were subject to fire event-ending weather conditions that were different than the summary fire weather. From the remaining 433,475 pixels, 60% were randomly selected for model fitting to provide a computationally manageable balance between the data used for training and testing, with the majority used to train the model. Multinomial logistic (also known as multinomial logit) models were then fit to predict vegetation mortality class from subsets of biotic and abiotic attributes. These models provide cumulative probabilities of each class (i.e., class 1 versus others, classes 1 and 2 versus others, etc.) which can then be used to obtain the probability of each class via subtraction [49]. Because the focus of this study was to investigate which biotic and abiotic attributes affect vegetation mortality levels within fire events, equal weights were assigned to each mortality class in model fitting. Since there were a large number of attributes that might affect vegetation mortality and these attributes are likely related, particularly with the groups defined in Table 1 (i.e., vegetation, topography, and fire weather), the process used to select among predictor variables to obtain a final model was:

1. Vectors of means (for continuous variables in Table 1) and proportions (for ecoregion plus other class variables in Table 1) were calculated by vegetation mortality level using all pixels to investigate univariate relationships with vegetation mortality and to indicate importance of each possible predictor variable. The results from this were used to provide an overview that described the general relationships.
2. A correlation matrix was used to identify pairs of variables with moderately high multicollinearity ($r > 0.7$) as an aide to pre-selection among highly related variables. Using this as a guide along with interpretability of relationships as reported in other studies, a subset of the fire weather variables was retained in further analyses, namely: DC (median) to represent the drought conditions for the fire event; FWI (min) to represent the minimum potential energy output during the burning period; and ISI (proportion of days above threshold) to represent the proportion of days during the burning period that had high intensity fire-spread conditions [45]. Similarly, from the topography variables, slope location and surface curvature were strongly correlated; slope location was selected since it related to biomass consumption in fire events in previous studies [50]. From the vegetation variables, the understory index was strongly correlated with understory cover, so the simpler and more interpretable understory cover variable was retained.
3. Following Hosmer et al. [51], multinomial models were initially fitted using each predictor variable (i.e., univariate models) and ranked based on the Akaike Information Criterion (AIC) [52] along with the percent of correctly classified pixels using the model fitting data.
4. For each variable group (i.e., site, vegetation, topography, and weather), a stepwise selection process was followed by including all variables of the variable group and then removing variables one at a time. Variables previously dropped were then considered for entry back into the model

at later steps. Again, AIC and the percent of correctly classified pixels were used to evaluate variable importance (i.e., whether to retain or drop a variable). However, supporting literature regarding relationships between mortality and biotic and abiotic variables was also used in deciding whether to retain or drop a variable.

5. Since class variables can affect the coefficient associated with each continuous variable, interaction terms were then added and evaluated as in Step 4. Some interactions resulted in singular matrices and were removed. For example, some fuel categories only occurred within given ecoregions (e.g., muskeg did not occur in the Rocky Mountains or Foothills), and others only occurred within certain terrain (e.g., muskeg only occurred on flat ground) resulting in no differences in some attributes (i.e., all grasslands had an overstory height of 0).

6. Once a subset of predictor variables was selected from each group of variables, these were merged together to obtain a model using all variable groups. Interactions between class variables and continuous variables across variable groups were then evaluated with regards to model improvements.

7. Since predictor variables eliminated in previous steps might become important in a later step (for example, in combination with ecoregion or other variables), these were added again to the overall combined model and evaluated.

The vegetation mortality class with the highest predicted probability using the selected multinomial logistic model was assigned for the classification. The model that was finally selected was further assessed using confusion matrices for the reserved test data (40% of pixels within the fire event). The kappa coefficient was calculated [53] to indicate classification performance relative to a random assignment of pixels (-1 to $+1$; $+1$ indicates perfect agreement; <0 indicate performance is no better than random). For the final mode, the weighted kappa coefficient [54] was calculated to assess the degrees of agreement between cells in the confusion matrix.

3. Results

3.1. General Relationships

The highest vegetation mortality class accounted for 68% of pixels overall, although that ranged from 64% in the Boreal Plains to 82% in the Foothills (Table 3). Fires in the Boreal Plains and Canadian Shield had the highest levels of partially burned areas (33% and 27% respectively, compared to 16% and 18%) (Table 3). No clear trend was evident for fire event area.

Table 3. Numbers of fire events and pixels by Ecoregion, and percentages by vegetation mortality class.

	Number of		Vegetation Mortality					
Ecoregion	Fire Events	Pixels	0%	1%–25%	26%–50%	51%–75%	76%–94%	≥95%
Boreal plains	24	246,951	3%	5%	7%	8%	13%	64%
Canadian Shield	7	166,425	1%	5%	7%	7%	8%	72%
Foothills	3	8378	2%	7%	7%	2%	0%	82%
Rocky Mountains	3	11,721	3%	5%	6%	2%	5%	78%
All	37	433,475	2%	5%	7%	7%	10%	68%

For the variables from the vegetation polygons, the only variable that showed a consistent trend was overstory crown closure, which increased with higher mortality classes ranging from 42.5% in mortality class 0 to 49.6% in class 5 (See Table A1). For age, height, understory height, and soil moisture, no obvious trend was evident. For ladder fuel index and understory crown closure, partially burned pixels were common. Partially burned pixels had lower ladder fuel index (6.1–6.6) than pixels with no (8.2) or high (8.5) mortality. Partially burned pixels had also lower understory crown closure (11.9%–13.1%) than either high (16.8%) or low mortality (14.3%) pixels (Table A1).

Sixty-eight percent of all forested pixels were in the highest mortality class, although the numbers varied by fuel-type. Immature pine (C4) had the highest proportion of pixels with high vegetation mortality at 79%. Boreal spruce (C2), mature pine (C3), and boreal mixedwoods (M1 and M2) pixels had between 67%–71% of their pixels in the high mortality class. The lowest levels of high vegetation mortality occurred in fir (FX) and larch (LX) (49% and 51%, respectively) although the number of pixels in FX was very small (Table A2).

Non-forested fuel types within the highest mortality class accounted for 63% of the pixels, which is 5% lower than the overall percentage for the forested class (Table A2). Although the FBP fuel-types do not differentiate non-forested fuel types, the percentage of high mortality pixels in closed shrub is only 42%, compared to 62% for open muskeg, and 72% for treed muskeg (Table A2).

Areas with no vegetation mortality were observed at higher elevations (>700 m) than areas with partial and high mortality (<615 m) (Table A3). Areas with steeper slopes were associated with both unburned areas and areas with very high vegetation mortality (2.8–2.9 degrees), while slopes ranged between 2.1 and 2.5 degrees in areas mortality classes 1, 2, and 3 (Table A3). More prominent slope positions (i.e., + slope position values) and convex slopes (i.e., + curvature values) had higher vegetation mortality (classes 3, 4, and 5) than less prominent positions and concave slopes (e.g., gullies and valleys) (Table A3). For SCOSA and SSINA, representing "eastness" and "northness", respectively (see Table 1. for definitions), southeast slopes (negative SCOSA and positive SSINA) were associated with unburned areas, and south slopes associated with higher vegetation mortalities (negative SCOSA, SSINA averaging near zero). There was no obvious pattern between vegetation mortality and CTI or TSRAI (Table A3).

In terms of fire weather variables, high median, minimum, and maximum values of ISI, DC, and FWI were associated with low to moderate levels of vegetation mortality (Table 4). Wider ranges of ISI, DC, and FWI were associated with both unburned areas and increasing levels of vegetation mortality. For both the ISI and FWI threshold values, a greater number of days above the threshold for spread conditions was associated with low and moderate levels of vegetation mortality (Table 4).

Table 4. Means (and standard deviations) of daily fire weather variables (see Table 1 for definitions) by vegetation mortality class. For all variables, higher values indicate more favourable burning conditions; for the range variables, positive numbers indicate changes toward more favourable burning conditions and negative numbers indicate changes to less favourable burning conditions.

Variable	Vegetation Mortality					
	0%	1%–25%	26%–50%	51%–75%	76%–94%	≥95%
ISI Median	4.8 (2.6)	6.2 (4.2)	5.4 (3.4)	5.2 (3)	4.5 (2.2)	4.3 (2.1)
ISI Min	1.6 (2.3)	2.7 (4.7)	1.6 (4)	1.3 (3.3)	0.8 (1.9)	0.7 (1.7)
ISI Max	15.5 (3.2)	13 (4.2)	12.7 (3.5)	13.5 (3.8)	13.4 (4.4)	12.2 (4.1)
ISI Range	−13.1 (5.7)	−7.7 (7.8)	−9.6 (6.7)	−10.7 (7)	−10.7 (8)	−8.7 (8.5)
ISI Threshold	0.2 (0.2)	0.3 (0.3)	0.2 (0.2)	0.2 (0.2)	0.1 (0.1)	0.1 (0.1)
DC Median	207.2 (61.9)	254.2 (118.9)	290.1 (110.6)	282.9 (107.2)	275.1 (105.1)	276.2 (104.8)
DC Min	166.1 (56.4)	217.3 (109.4)	247 (105.7)	238.5 (103.9)	225.6 (103.4)	229.6 (102.2)
DC Max	15.5 (3.2)	13.0 (4.2)	12.7 (3.5)	13.5 (3.8)	13.4 (4.4)	12.2 (4.1)
DC Range	62.9 (51.5)	52.6 (65.4)	74.9 (58)	80.4 (53)	79.6 (62.6)	70.4 (76.2)
FWI Median	14.3 (6.9)	17.5 (8.4)	16.5 (6.2)	16.3 (6.1)	14.2 (6.1)	13.8 (6.3)
FWI Min	5.1 (7.4)	7.8 (10.1)	4.7 (8.1)	4.1 (7.2)	2.8 (4.8)	2.7 (5.1)
FWI Max	35.8 (5.3)	31.5 (8.6)	32.5 (6.4)	34.1 (6.4)	33.1 (8.2)	31.2 (8)
FWI Range	−29.2 (11.9)	−18.1 (18.2)	−24.3 (16.7)	−26.1 (17.1)	−25.2 (18.9)	−20.8 (21.2)
FWI Threshold	0.3 (0.3)	0.4 (0.4)	0.4 (0.3)	0.4 (0.3)	0.3 (0.2)	0.3 (0.2)

3.2. Model Fitting and Classification

For the site variables, preliminary analyses indicated a nonlinear trend between vegetation mortality classes and the fire event area. As a result, the logarithm of fire event area was used in all models as in previous studies (e.g., [15]). The model using fire event area, Ecoregion, and the interaction term between them resulted in the best model fit (i.e., lowest AIC relative to the null model) and classification performance (Table B1, Model 5). The interaction term indicates that the effects of fire event area vary with Ecoregion. Based on these results, both Ecoregion and fire event area were considered for inclusion in multivariable models using variables across variable groups.

Of the models using a single vegetation variable, fuel category had the best fit (Table B2, Model 2); however, the classification performance was poor. Despite the poor classification, fuel category was included for consideration in multivariable models, given this impacts fire behaviour [46]. Variables related to fuel continuity (overstory and understory crown closure) had poorer fits, but better classification results (Table B2, Models 3 and 5). As a result, the fuel continuity variables were also selected for possible inclusion in the multivariable models. Of the two variables representing soil moisture, CTI was selected over soil moisture classes since it is a continuous variable at a smaller spatial scale (Table B2, Model 4 vs. Table B3, Model 3). Understory height and age followed with regards to model fit. Both were considered for the multivariate models, given the impacts of ladder fuels on fire behaviour [55] and of stand age on fire impacts [19,56]. Finally, overstory height was selected for multivariable models, given the importance of crown structure for fire behaviour [55]. Using all of these selected variables resulted in a better fit than using individual variables (Table B2, Model 9). The classification performance of the model with all selected variables (Table B2, Model 9) was not improved by adding interaction terms between the canopy variables for the two respective layers (Table B2, Model 10). It was not possible to add further interaction terms for the fuel category variable, due to rank deficiency and linear combinations. Removing the overstory height variable slightly improved the model classification and resulted in fewer terms in the model (Table B2, Model 11).

The topographic variables with the best fit and classification accuracy were elevation and CTI (Table B3, Models 2 and 3). Using the two highest ranked individual variables together had nearly as good of a fit and classification performance as using all topographic variables (Table B3, Model 11). Adding an interaction term between the variables further improved fit and classification performance (Table B3, Model 12). Adding the previously removed variables to the model did not improve fit or classification accuracy.

Within the fire weather variable group, DC had the best model fit, while ISI had the best classification accuracy (Table B4). DC was the best variable for classifying unburned areas (Table B4, Model 2; kappa = 0.05, 0, 0, 0, 0, and 0.07 for classes 0 through 5, respectively), FWI was the best variable for classifying areas with intermediate levels of vegetation mortality (Table B4, Model 3; kappa = 0.03, 0, 0.03, 0.05, −0.04, and 0.10 for classes 0 through 5, respectively), and ISI was the best variable for classifying areas with high vegetation mortality (Table B4, Model 4; kappa = 0.01, 0, −0.02, 0, 0.02, and 0.15 for classes 0 through 5, respectively). All three variables were selected for the final multivariable model for this group of variables (Table B4, Model 5). Adding interactions between all selected variables improved model fit (Table B4, Model 6).

Combining selected variables from each variable group improved both fit and classification performance particularly when interaction terms between variables and Ecoregion were included (Table 5, Model 3). No further improvements were obtained by including variables eliminated in previous steps. In particular, using soil moisture (classes) instead of CTI did not improve the model (AIC 1352 higher and equal overall kappa), nor did using ISI alone without the other fire weather variables (AIC 6649 higher and overall kappa 0.01 lower). Adding overstory height improved the fit (AIC 1744 lower), but not the classification performance (equal kappa). Adding additional topography variables made relatively small changes to the fit (changes in AIC < 1750) and did not improve the classification performance (equal kappa).

Using the final model (Table 5, Model 3) resulted in a classification accuracy of 39%. The percent of pixels correctly classified was best for the unburned and highest vegetation mortality categories (the kappa coefficients were 0.18, 0.06, 0.03, 0.03, 0.00 and 0.19 for class 0 through class 5, respectively). For intermediate levels of vegetation mortality, the classification accuracy was only slightly better than random, and class 4 had the worst classification accuracy (Table 6). Many of the misclassifications were in adjacent categories. For example, the weighted kappa, which takes into consideration misclassifications in neighbouring classes by assigning a weighted penalty proportional to the squared distance to the correct class, was considerably better at 0.26. Finally, by counting classifications within ±1 category as correct, which is justifiable from a management standpoint, 64% of pixels were classified in the correct classes with kappa coefficients of 0.26, 0.29, 0.21, 0.37, 0.31, and 0.32 for mortality classes 0 through 5, respectively.

Table 5. Fit statistics for models using combinations of variables across variable groups.

Model	Variables [a]	AIC	Δ AIC [b]	% Correct	Kappa
1 (null)		932,037	0	17%	0.00
2	Ecoregion * area, fuel category, overstory crown closure, understory crown closure, understory height, age, elevation * CTI, DC * FWI * ISI	888,482	43,555	40%	0.07
3	Ecoregion * (overstory crown closure, understory crown closure, understory height, age, elevation * CTI), DC * FWI * ISI	870,911	61,126	39%	0.10

[a] Interaction terms indicated by *. For example x * z means that the model includes variables x and z, as well as the interaction between them; [b] Δ AIC is relative to the null model with a larger value indicating a better model.

Table 6. Confusion matrix for the final model using a combination of site, vegetation, topography, and fire weather variables.

Predicted Vegetation Mortality Class	Actual Vegetation Mortality Class					
	0%	1%–25%	26%–50%	51%–75%	76%–94%	≥95%
0%	66%	31%	14%	16%	16%	7%
1%–25%	19%	19%	15%	13%	15%	9%
26%–50%	2%	7%	12%	13%	9%	8%
51%–75%	1%	5%	8%	8%	5%	5%
76%–94%	7%	19%	27%	24%	22%	23%
≥95%	4%	19%	25%	25%	33%	48%

Application of the selected model to visually compare predicted to actual vegetation mortality classes for a randomly selected fire from each Ecoregion provided further insights (Figure 2). Overall, the actual and predicted vegetation mortality class maps are similar. However, the fire event in the Boreal Plains Ecoregion (Figure 2a) had long and narrow patches with high vegetation mortality (high length to breadth ratio, see [46]) oriented from southwest to northeast, suggesting that it may have been driven by high-velocity winds. Further, assuming fire propagation followed this wind direction, remnants formed on the lee-side of non-flammable areas, while high vegetation mortality occurred in windward locations. In this case, the direction of fire propagation may have also been important, given that it appears that live vegetation remnants formed while the fire was spreading in a downhill direction. However, without fire propagation maps and detailed information on wind direction, it is impossible to know how wind velocity, topography, and fire propagation affected vegetation mortality. The fire event in the Canadian Shield had large areas of very high vegetation mortality (Figure 2b). Within the classification, several low lying wet areas, indicated by predicted low-levels of mortality in typical meandering patterns at the bottom of drainages, were misclassified as having lower vegetation mortality than was observed (Figure 2f). Within the Foothills, the classification had very similar patterns of vegetation mortality; in particular, remnants appeared as linear shapes in both the observed and predicted maps. These remnants could serve as wildlife corridors and were present in both the observed and predicted maps (Figure 2c,g). Finally, for sampled fire events in the Rocky Mountains, the model appeared to make misclassifications within fuel types that may have been related to differences in aspect, which was not accounted for in the final model (Figure 2d,h).

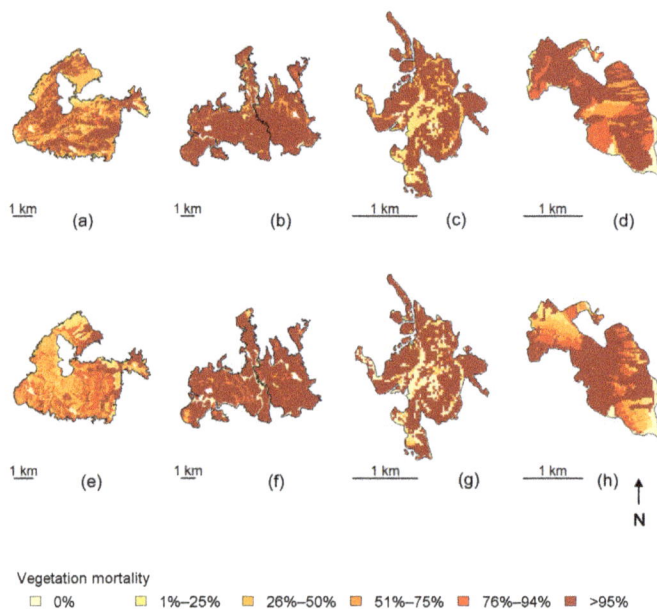

Figure 2. Map of observed vegetation mortality classes (top row: (**a–d**)) versus predicted classes (bottom row: (**e–h**)) for randomly selected fire events from the Boreal Plains ((**a**) and (**e**)), Canadian Shield ((**b**) and (**f**)), Foothills ((**c**) and (**g**)), and Rocky Mountains ((**d**) and (**h**)).

4. Discussion

4.1. Factors Contributing to Vegetation Survival

The results suggest that a combination of factors influenced fire-caused vegetation mortality at two distinct scales. At the event scale, overall vegetation mortality was moderately influenced by Ecoregion reflecting the broad ecological patterns first identified by [12], and found by others [15,16]. Fires within both the Foothills and Rocky Mountains had lower proportions of remnants in partial mortality classes and higher proportions of high mortality than other Ecoregions. In terms of fire weather, the influence of DC likely reflects a logical relationship between fewer remnants and higher levels of seasonal drought conditions associated with lower moisture content of large woody debris and deep organic soils. However, the direct relationship between FWI minimum values and the amount of low to moderate vegetation mortality was unexpected. One possible explanation for this is the nature of diurnal burning patterns. For example, during weather conditions favourable to even high-energy output wildfires, patches with low to moderate vegetation mortality may be created at night under less favourable burning conditions (i.e., lower temperatures and higher relative humidity). Another possible explanation is that fuel-types that normally have a low risk of flammability (e.g., wetlands) will burn under highly favourable fire weather conditions, but even then, only partially.

Within events, the most influential factor affecting vegetation mortality was fuel type, which is not surprising given its prominence in fire prediction models (e.g., [46]). Combining fuel-type with forest structure variables such as crown closure further improved the ability to predict vegetation mortality levels, which is also consistent with what we know about the strong relationship between crown closure and crown fires [57]. Elevation and soil moisture were the next most important within-event scale variables, although one could argue that elevation also acts at the event scale. The relationship between vegetation mortality and elevation also varied greatly between Ecoregions: in the Canadian Shield, higher vegetation mortality was associated with higher elevations, while in the Foothills, higher elevations were associated with lower mortality. In the Canadian Shield, high elevation sites tend to be height-of-land ridges, peaks and knolls, which tend to be hot and dry. In the Foothills, "high-elevation-sites" is a relative term, in that all fire events lie within the eastern slopes of the Rocky Mountains, which tend to be colder and wetter.

4.2. Complexity of Relationships

Many of the relationships with remnants found in this study have been noted by others, and are intuitively consistent. For example, Ecoregion was previously associated with the proportion of vegetation mortality within fire events [12,26], certain fuel types and percent cover variables have been linked with vegetation mortality and burn severity [28,46,56], topography has been related to burn severity [15,50], and fire weather variables exhibited weak relationships to burn severity [21]. While this study confirms many of these previously found relationships, it reveals a new level of complexity in terms of the processes involved in the formation of remnants. The relative influences of, and interactions between, both large-scale and within-fire attributes were examined in this study. Rarely is there a single physical or environmental factor influencing the amount or location of remnants acting alone. Acknowledgment of this complexity helps guide both research and management in several ways. First, studying within-fire vegetation survival improves our knowledge of factors affecting fire-related mortality and thereby improves our ability to manage forests using the NP concept. To facilitate this, access to high resolution spatial data is an advantage in that it allows a more precise evaluation of these relationships. The use of high quality spatial data sets should become the rule, not the exception. Second, given the large variability of remnant patterns noted here both within and between fire events, extensive and detailed data sets that cover the range of conditions are needed across large areas in order to capture the NRV of remnant patterns at local and landscape scales. Finally, both factors that change slowly in time and those varying within and between years affect vegetation mortality within wildfires. Ecological type and topography variables change only over a very long time-scale. In contrast, fire weather attributes are highly variable in shorter time periods. Although improving our knowledge of within-fire vegetation mortality is critical for altering factors that impact vegetation mortality, some factors will always be beyond the control of forest managers (e.g., fire weather). However, a decision support tool that helps predict the impacts of fire weather and other factors on vegetation mortality would be very valuable given expected impacts of climate changes on future wildfires.

From a management perspective, decision support tools can and should differentiate between variation that can be accounted for (i.e., deterministic) and that which cannot (i.e., random). The deterministic part of the natural variability attributable to regional-scale factors such as ecological zone, and local-scale factors such as vegetation type and topography can be modelled, predicted and emulated by careful planning. The random portion of remnant NRV attributable to fire weather can be applied more broadly across a region or landscape as intrinsic variability. Integrating both types of variation into forest planning is consistent with an NRV approach. The division of remnant NRV may also have value as a tool for understanding and potentially mitigating the impacts of climate change. The relationship between remnants and fire weather may help us better predict how wildfire remnant patterns might shift in the future, and how forest management activities might offset those changes as desired.

4.3. Classification Accuracy and Model Complexity

While the overall classification accuracy found in this study is comparable to those from other studies, the comparison is misleading for several reasons. First, this study included fire events covering an extensive land area with diverse ecological, topographical, and climatic conditions. Other broad regional studies tended to report lower accuracy with more complex models [21,31,58] than studies using individual or a very limited numbers of fire events [25,50]. Second, the increased variation represented in this more extensive area coupled with finer-scale variables enabled a more thorough investigation of factors and highlights the importance of considering local and regional conditions. Lastly, we included six mortality classes, compared to two or three classes for most other comparable studies. If we ignore all of the partial mortality classes, our classification accuracy was much higher than any of those reported in other studies.

Our models struggled most with identifying partial mortality. Partial mortality is a well-documented weakness of imagery-based interpretations [24,59]. We had hoped to overcome this challenge by using more accurate and precise mortality data based on high resolution photographic negatives, the results suggest that there are other factors involved in predicting vegetation mortality levels. One possibility is that mortality continues over several years post-fire [60] which means that all post-fire "snapshots" (including those used in our study) represent just one of several remnant vegetation patterns over time. Another possibility is that moderate mortality is associated with areas that experienced less severe fire conditions, such as overnight fires and/or the back or flanks of a fire where the relative influence of fire weather, topography and fuel-type may change, or even reverse [25]. Finally, the factors affecting vegetation mortality may differ among levels. Fire weather conditions may be critical for the formation of moderate levels of vegetation mortality, while fuel-type and topography differences may over-ride fire weather and play a more dominant role for determining areas that are either unburned or burned with high mortality. The higher proportion of fires that burned with high levels of island remnants found in the Foothills relative to the Boreal Plains reported by Andison and McCleary [12] support this concept. In any case, given the prevalence of partial mortality noted recently in boreal wildfires, this would seem an obvious target for future research.

We also learned some important lessons from a methodological perspective. One of the issues that arose in this and other related studies is how to examine the many variables that may be related to fire-caused vegetation mortality. Random Forests [61] is a commonly utilized machine learning approach mainly used in conjunction with remote sensing-based response variables [16,21,25,50], while very few studies have used logistic regression and vegetation mortality-based response variables [57]. In contrast, the multinomial logit model approach used in this research required a deliberate and thoughtful process for the selection of dependent variables. While documenting this process can be tedious, it highlights the complexity of relationships for vegetation mortality and importance of interpreting the relationships. The process we used included separating variables into logical groups which also improved our ability to interpret relationships.

5. Conclusions

Understanding the processes involved in the formation of surviving remnant vegetation within boreal forest wildfires has been more challenging than we have collectively assumed. Many prior fire remnant vegetation studies relied on available spatial data (of varying quality), a small number of fire events, and only a few predictor variables, all of which could impact the results and interpretation, ultimately resulting in an incorrect representation of the true nature of historical, natural patterns. In this study, we used high resolution and high quality spatial data for a large number of fire events over an extensive area to improve post-fire remnant vegetation mortality prediction models over results previously reported for regional studies. Moreover, we demonstrated that vegetation survival occurs as a result of a complex interaction between several biotic and abiotic factors at different spatial and temporal scales. More specifically, we found that the most influential general variables such as Ecoregion and drought code (DC) interacted with fuel-type, canopy closure, and soil moisture in very specific ways. While the influences of some variables (such as soil moisture, fuel-type, and drought code) were universally applicable, their relative influence in our predictive model varied by fire event and ecological region.

The findings from this study suggest that adapting fine-scale forest management patterns that are truly ecosystem-based involves a more holistic interpretation of fire patterns knowledge. However well-intentioned, traditional regulatory models, that mandate the amount and location of remnants based on deterministic rules are unlikely to capture the full historic range. Historic bounds are defined very generally by smaller-scale, in situ attributes such as fuel-type, soil moisture and canopy closure, many of which can be manipulated by forest managers. However, our results suggest that a substantial portion of the natural variation may not be directly related to in-situ variables. Fire weather, acting in concert with in-situ factors, cannot generally be manipulated by forest managers. Managing towards achieving natural variation objectives will require a more results-based approach that is challenging to integrate into practice from a regulatory perspective, but at the same time, this will offer the flexibility needed for management of a broader range of ecosystem values.

Acknowledgments: Thank you to the fRI Research Healthy Landscapes Program, Saskatchewan Ministry of Environment, and Mistik Management. Thanks to Doug Turner, Chris Dallyn, Kathleen Gazey, Al Balisky, Cliff McLauchlan, Roger Nesdoly, Kim Rymer, and David Stevenson for sharing their insight and guidance during a field visit. This work was funded by the Healthy Landscapes Program of fRI Research and an NSERC Engage to LeMay, Andison, and Ferster with Mistik Management.

Author Contributions: C.J.F. wrote the manuscript and conducted the analyses. B.N.I.E. wrote parts of the manuscript, particularly on interpretations of models, and provided guidance for the analyses and presentation of results. D.W.A. formulated research questions, advised the analyses, provided data, and wrote sections of the manuscript in particular, relating the work to previous research, current forest management, and conclusions. V.M.L. advised on the statistical analyses and presentation of results, and wrote sections of the manuscript.

Conflicts of Interest: The authors declare no conflict of interest.

Appendix A. Data Summaries

Table A1. Means (standard deviations) of vegetation variables (see Table 1 for definitions) by vegetation mortality class.

Variable	Vegetation Mortality					
	0%	1%–25%	2%–50%	51%–75%	76%–94%	≥95%
Age (years)	39.9 (20.4)	42.6 (24.4)	40 (21.9)	41.4 (20.4)	41.7 (19.8)	38.8 (21)
Height (m)	11.5 (5.4)	9.9 (4.4)	9.5 (4.3)	10.3 (4.6)	11.0 (4.7)	10.4 (3.7)
Ladder fuel index (no units)	8.2 (13.8)	6.3 (10.6)	6.6 (10.1)	6.4 (9.6)	6.1 (9.3)	8.5 (11.6)
Overstory crown closure (%)	42.5 (23.6)	43.6 (23.9)	43.2 (21.9)	46.4 (22.1)	49.3 (22.2)	49.6 (23.8)
Understory crown closure (%)	14.3 (20.4)	11.9 (18.8)	13.1 (18.7)	12.4 (17.1)	12.2 (17.3)	16.8 (21.1)
Understory height (m)	3.3 (3.7)	2.5 (3.2)	2.6 (3.0)	2.9 (3.2)	2.8 (3.1)	2.9 (3.0)

Table A2. Relative frequency (as a %) and percentage of pixels by vegetation mortality class for each fuel category (see Table 2 for definitions).

Fuel Category	Vegetation Mortality						Frequency
	0%	1%–25%	2%–50%	51%–75%	76%–94%	≥95%	
Forest							
C2—Boreal spruce	2%	6%	6%	6%	9%	71%	27.62%
C3—Mature pine	2%	5%	6%	8%	11%	67%	17.73%
C4—Immature pine	1%	4%	4%	5%	7%	79%	8.1%
D1—Leafless aspen	5%	11%	4%	4%	17%	59%	0.83%
D2—Leaf-on aspen	1%	7%	15%	4%	6%	67%	0.41%
FX—Fir	16%	16%	6%	13%	0%	49%	0.03%
LX—Larch	1%	4%	14%	16%	14%	51%	6.95%
M1—Boreal mixedwoods, leafless	1%	3%	8%	7%	11%	70%	2.81%
M2—Boreal mixedwoods leaf-on	3%	4%	6%	8%	12%	67%	26.23%
All forest	2%	5%	7%	8%	11%	68%	90.71%

Table A2. *Cont.*

Fuel Category	Vegetation Mortality						Frequency
	0%	1%–25%	2%–50%	51%–75%	76%–94%	≥95%	
	Non-forest						
Open shrub	1%	2%	0%	1%	0%	96%	0.06%
Closed shrub	6%	15%	14%	10%	13%	42%	2.27%
Open muskeg	10%	10%	3%	9%	5%	62%	1.21%
Treed muskeg	2%	6%	7%	4%	10%	72%	5.49%
Brush and alder	0%	15%	0%	25%	0%	60%	0.06%
Bryophytes	0%	0%	0%	0%	100%	0%	0%
Grassland	1%	4%	21%	28%	3%	43%	0.2%
All non-forest	4%	9%	8%	7%	10%	63%	9.29%
All forest and non-forest	2%	5%	7%	7%	10%	68%	100%

Table A3. Means (and standard deviations) for topography variables (see Table 1 for definitions) by vegetation mortality class. Slope position index, surface curvature, and SCOSA + SSINA were scaled by a factor of 1000; TSRAI was scaled by a factor of 10.

Variable	Vegetation Mortality					
	0%	1%–25%	2%–50%	51%–75%	76%–94%	≥95%
Elevation	711.8 (234.5)	561.4 (249.3)	606.1 (262)	614.1 (252.1)	611.5 (260.4)	598.5 (268.3)
CTI	9.0 (2.1)	9.5 (2.3)	9.1 (2)	9.1 (2)	9.0 (2)	8.9 (2)
Slope (degrees)	2.9 (2.6)	2.1 (3.2)	2.5 (3.3)	2.5 (3.2)	2.8 (3.7)	2.9 (3.6)
Slope position	−7.2 (145.7)	−7.0 (161.8)	−3.0 (188)	1.4 (174.8)	2.7 (183.8)	11.0 (196.7)
Curvature	−4.1 (84.2)	−3.9 (93.8)	−1.7 (108.5)	0.8 (101.1)	1.5 (106.5)	6.2 (113.4)
TSRAI	5.1 (3.6)	5.5 (3.2)	4.9 (3.3)	5.1 (3.4)	5.1 (3.3)	5.3 (3.4)
SCOSA	−7.1 (46.4)	−7.4 (49.2)	−4.7 (49)	−2.1 (39.4)	−5.8 (57.4)	−7.1 (56.8)
SSINA	5.7 (48.8)	−1.3 (46.7)	3.0 (55.8)	−6.1 (62.2)	−0.8 (62.7)	−1.5 (61.8)

Appendix B. Model Comparisons

When using each of the site variables individually in the model, Ecoregion had a lower AIC and higher kappa than fire event area (Table B1, Model 2 versus 3). However, previous studies have found that fire event area is related to the proportions of different levels of vegetation mortality [15,16]. Using the two variables together in a model resulted in an improved model fit and higher classification accuracy than using either variable alone (Table B1, Model 4).

Table B1. Fit statistics for models using Ecoregion and fire event area (area).

Model	Variables [a]	AIC	Δ AIC [b]	% Correct	Kappa
1 (Null)		932,037	0	17%	0.00
2	Ecoregion	929,929	2107	30%	0.03
3	Area	932,030	7	32%	0.01
4	Ecoregion, area	929,230	2807	38%	0.04
5	Ecoregion * area	920,267	11,769	37%	0.07

[a] Interaction terms indicated by *. For example x * z means that the model includes variables x and z, as well as the interaction between them; [b] Δ AIC is relative to the null model with a larger value indicating a better model.

Table B2. Fit statistics for models using vegetation polygon variables.

Model	Variables [a]	AIC	Δ AIC [b]	% Correct	Kappa
1 (Null)		932,037	0	17%	0.00
2	Fuel category	925,179	6858	20%	−0.01
3	Overstory crown closure	928,397	3640	36%	0.02
4	Soil moisture class	931,447	590	44%	0.02

Table B2. *Cont.*

Model	Variables [a]	AIC	Δ AIC [b]	% Correct	Kappa
5	Understory crown closure	931,840	197	34%	0.02
6	Understory height	931,912	125	36%	−0.02
7	Age	931,918	119	26%	0.01
8	Overstory height	931,948	88	32%	0.00
9	Fuel category, overstory crown closure, understory crown closure, understory height, age, overstory height	917,106	14,930	43%	0.07
10	Fuel category, age, overstory crown closure * overstory height, understory crown closure * understory height	915,557	16,479	42%	0.07
11	Fuel category, overstory crown closure, understory crown closure, understory height, age	917,529	14,507	43%	0.07

[a] Interaction terms indicated by *. For example x * z means that the model includes variables x and z, as well as the interaction between them; [b] Δ AIC is relative to the null model with a larger value indicating a better model.

Using TSRAI resulted in an AIC that was worse than the null model (Table B3, Model 7) and this variable was not further considered. Removing variables from the model with all selected variables in reverse order of the individual fits resulted in small decreases in model fit and classification accuracy (Table B3, Models 8 through 10).

Table B3. Fit statistics for models using topographic variables.

Model	Variables [a]	AIC	Δ AIC [b]	% Correct	Kappa
1 (Null)		932,037	0	17%	0.00
2	Elevation	930,250	1787	36%	0.02
3	CTI	931,265	771	41%	0.02
4	SCOSA, SSINA	931,726	310	29%	0.02
5	Slope	931,741	296	23%	0.00
6	Slope position	931,760	277	33%	0.01
7	TSRAI	932,038	−1	39%	0.00
8	Elevation, CTI, SCOSA, SSINA, Slope, Slope position	928,148	3889	33%	0.02
9	Elevation, CTI, SCOSA, SSINA, Slope	928,282	3755	34%	0.02
10	Elevation, CTI, SCOSA, SSINA	928,454	3583	34%	0.02
11	Elevation, CTI	929,482	2555	34%	0.02
12	Elevation * CTI	929,186	2851	32%	0.03

[a] Interaction terms indicated by *. For example x * z means that the model includes variables x and z, as well as the interaction between them; [b] Δ AIC is relative to the null model with a larger value indicating a better model.

Table B4. Fit statistics for models using fire weather variables.

Model	Variables [a]	AIC	Δ AIC [b]	% Correct	Kappa
1 (Null)		932,037	0	17%	0.00
2	DC	923,964	8072	43%	0.04
3	FWI	924,210	7827	51%	0.05
4	ISI	926,688	5349	53%	0.06
5	DC, FWI, ISI	920,298	11,739	34%	0.02
6	DC * FWI * ISI	914,502	17,535	47%	0.03

[a] Interaction terms indicated by *. For example x * z means that the model includes variables x and z, as well as the interaction between them; [b] Δ AIC is relative to the null model with a larger value indicating a better model.

References

1. Laestadius, L.; Nogueron, R.; Lee, P.; Askenov, D.; Smith, W. *Canada's Large Intact Forest Landscapes (2006 Update)*; Global Forest Watch Canada: Edmonton, AB, Canada, 2006.
2. The Canadian Boreal Forest Agreement Canadian Boreal Forests Agreement. Available online: http://cbfa-efbc.ca/wp-content/uploads/2014/12/CBFAAgreement_Full_NewLook.pdf (accessed on 10 March 2016).
3. Landres, P.B.; Morgan, P.; Swanson, F.J. Overview of the Use of Natural Variability Concepts in Managing Ecological Systems. *Ecol. Appl.* **1999**, *9*, 1179–1188.
4. Grumbine, R.E. What Is Ecosystem Management? *Conserv. Biol.* **1994**, *8*, 27–38. [CrossRef]
5. Franklin, J.F. Preserving Biodiversity: Species, Ecosystems, or Landscapes? *Ecol. Appl.* **1993**, *3*, 202–205. [CrossRef]
6. OMNR. *Forest Management Guide for Natural Disturbance Forest Management Guide for Natural Disturbance*; Ontario Ministry of Natural Resources: Toronto, ON, Canada, 2001.
7. Forest Stewardship Council Canada Working Group. *National Boreal Standard*; Forest Stewardship Council Canada Working Group: Toronto, ON, Canada, 2004.
8. Johnson, E.A. *Fire and Vegetation Dynamics*; Cambridge University Press: Cambridge, UK, 1992.
9. Payette, S. Fire as a controlling process in the North American boreal forest. In *A Systems Analysis of the Global Boreal Forest*; Shugart, H.H., Leemans, R., Bonan, G.B., Eds.; Cambridge University Press: Cambridge, UK, 1992; pp. 144–169.
10. Delong, S.C.; Kessler, W.B. Ecological characteristics of mature forest remnants left by wildfire. *For. Ecol. Manag.* **2000**, *131*, 93–106. [CrossRef]
11. Perera, A.H.; Buse, L.J. *Ecology of Wildfire Residuals in Boreal Forests*; John Wiley & Sons, Ltd.: Chichester, UK, 2014.
12. Andison, D.W.; McCleary, K. Detecting regional differences in within-wildfire burn patterns in western boreal Canada. *For. Chron.* **2014**, *90*, 59–69. [CrossRef]
13. Rowe, J.; Scotter, G.W. Fire in the boreal forest. *Quat. Res.* **1973**, *3*, 444–464. [CrossRef]
14. Banks, S.C.; Dujardin, M.; McBurney, L.; Blair, D.; Barker, M.; Lindenmayer, D.B. Starting points for small mammal population recovery after wildfire: Recolonisation or residual populations? *Oikos* **2011**, *120*, 26–37. [CrossRef]
15. Duffy, P.A.; Epting, J.; Graham, J.M.; Rupp, T.S.; McGuire, A.D. Analysis of Alaskan burn severity patterns using remotely sensed data. *Int. J. Wildl. Fire* **2007**, *16*, 277–284. [CrossRef]
16. Wu, Z.; He, H.S.; Liang, Y.; Cai, L.; Lewis, B.J. Determining relative contributions of vegetation and topography to burn severity from LANDSAT imagery. *Environ. Manag.* **2013**, *52*, 821–836. [CrossRef] [PubMed]
17. Lentile, L.B.; Smith, F.W.; Shepperd, W.D. Influence of topography and forest structure on patterns of mixed severity fire in ponderosa pine forests of the South Dakota Black Hills, USA. *Int. J. Wildl. Fire* **2006**, *15*, 557–566. [CrossRef]
18. Keeley, J.E. Fire intensity, fire severity and burn severity: A brief review and suggested usage. *Int. J. Wildl. Fire* **2009**, *18*, 116–126. [CrossRef]
19. Thompson, J.R.; Spies, T.A.; Olsen, K.A. Canopy damage to conifer plantations within a large mixed-severity wildfire varies with stand age. *For. Ecol. Manag.* **2011**, *262*, 355–360. [CrossRef]

20. Mansuy, N.; Boulanger, Y.; Terrier, A.; Gauthier, S.; Robitaille, A.; Bergeron, Y. Spatial attributes of fire regime in eastern Canada: Influences of regional landscape physiography and climate. *Landsc. Ecol.* **2014**, *29*, 1157–1170. [CrossRef]

21. Birch, D.S.; Morgan, P.; Kolden, C.A.; Abatzoglou, J.T.; Dillon, G.K.; Hudak, A.T.; Smith, A.M.S. Vegetation, topography and daily weather influenced burn severity in central Idaho and western Montana forests. *Ecosphere* **2015**, *6*, 1–23. [CrossRef]

22. Delong, C.S.; Tanner, D. Managing the pattern of forest harvest: Lessons from wildfire. *Biodivers. Conserv.* **1996**, *5*, 1191–1205. [CrossRef]

23. Burton, P.J.; Parisien, M.-A.; Hicke, J.A.; Hall, R.J.; Freeburn, J.T. Large fires as agents of ecological diversity in the North American boreal forest. *Int. J. Wildl. Fire* **2008**, *17*, 754–767. [CrossRef]

24. San-Miguel, I.; Andison, D.W.; Coops, N.C.; Rickbeil, G.J.M. Predicting post-fire canopy mortality in the boreal forest from dNBR derived from time series of Landsat data. *Int. J. Wildl. Fire* **2016**, *25*, 762–774. [CrossRef]

25. Viedma, O.; Quesada, J.; Torres, I.; De Santis, A.; Moreno, J.M. Fire Severity in a Large Fire in a Pinus pinaster Forest is Highly Predictable from Burning Conditions, Stand Structure, and Topography. *Ecosystems* **2014**, *18*, 237–250. [CrossRef]

26. Madoui, A.; Leduc, A.; Gauthier, S.; Bergeron, Y. Spatial pattern analyses of post-fire residual stands in the black spruce boreal forest of western Quebec. *Int. J. Wildl. Fire* **2010**, *19*, 1110–1126. [CrossRef]

27. Kasischke, E.S.; Turetsky, M.R.; Ottmar, R.D.; French, N.H.F.; Hoy, E.E.; Kane, E.S. Evaluation of the composite burn index for assessing fire severity in Alaskan black spruce forests. *Int. J. Wildl. Fire* **2008**, *17*, 515–526. [CrossRef]

28. Jain, T.B.; Graham, R.T. The Relation Between Tree Burn Severity and Forest Structure in the Rocky Mountains. *Restoring Fire-Adapted Ecosystems: Proceedings of the 2005 National Silviculture Workshop*; Powers, R.F., Ed.; USDA Forest Service, Pacific Southwest Research Station: Albany, CA, USA, 2007; p. 306.

29. Andison, D.W. The influence of wildfire boundary delineation on our understanding of burning patterns in the Alberta foothills. *Can. J. For. Res.* **2012**, *42*, 1253–1263. [CrossRef]

30. Eberhart, K.E.; Woodard, P.M. Distribution of residual vegetation associated with large fires in Alberta. *Can. J. For. Res.* **1987**, *17*, 1207–1212. [CrossRef]

31. Parisien, M.; Peters, V.S.; Wang, Y.; Little, J.M.; Bosch, E.M.; Stocks, B.J. Spatial patterns of forest fires in Canada, 1980–1999. *Int. J. Wildl. Fire* **2006**, *15*, 361–374. [CrossRef]

32. Acton, D.; Padbury, G.; Stushnoff, C. *The Ecoregions of Saskatchewan*; University of Regina Press: Regina, SK, Canada, 1998.

33. Natural Regions Committee. ; Downing, D.J.; Pettapiece, W.W. (Compiler) *Natural Regions and Subregions of Alberta*; Government of Alberta: Edmonton, AB, Canada, 2006.

34. Brandt, J.P. The extent of the North American boreal zone. *Environ. Rev.* **2009**, *17*, 101–161. [CrossRef]

35. Kelso, N.V.; Patterson, T. Natural Earth 2015. North American Cartographic Association, University of Wisconsin-Madison, and Florida State University. Available online: http://www.naturalearthdata.com/ (accessed on 15 December 2015).

36. Jordan, T. Desktop Mapping System Softcopy Photo Mapper: DMS Softcopy Version 5.1. 2004. Available online: http://www.tommyjordan.com/DMS_Specs.htm (accessed on 24 August 2016).

37. SFVI. *Saskatchewan Forest Vegetation Inventory*, 4th ed.; Saskatchewan Environment—Forest Service: Prince Albert, SK, Canada, 2004.

38. AVI. Alberta vegetation inventory interpretation standards. *Vegetation Inventory Standards and Data Model Documents*; Resource Information Management Branch, Alberta Sustainable Resource Development: Edmonton, AB, Canada, 2005.

39. Alexander, J.D.; Seavy, N.E.; Ralph, C.J.; Hogoboom, B. Vegetation and topographical correlates of fire severity from two fires in the Klamath-Siskiyou region of Oregon and California. *Int. J. Wildl. Fire* **2006**, *15*, 237–245. [CrossRef]

40. Evans, J.; Oakleaf, J.; Cushman, S.; Theobald, D. An ArcGIS Toolbox for Surface Gradient and Geomorphometric Modeling. Available online: http://evansmurphy.wix.com/evansspatial (accessed on 10 March 2016).

41. Roberts, D.; Cooper, S. Concepts and techniques of vegetation mapping. *Land Classifications Based on Vegetation: Applications for Resource Management*; USDA Forest Service Gen. Tech. Rep. INT-257; Ferguson, D.,

Morgan, P.; Johnson, F.D., Eds.; USDA Forest Service Intermountain Forest and Range Experiment Station: Ogden, UT, USA, 1989; pp. 90–96.

42. Gessler, P.E.; Moore, I.D.; McKenzie, N.J.; Ryan, P.J. Soil-landscape modelling and spatial prediction of soil attributes. *Int. J. Geogr. Inf. Syst.* **1995**, *9*, 421–432. [CrossRef]

43. Stage, A. Notes: An expression for the effect of aspect, slope, and habitat type on tree growth. *For. Sci.* **1976**, *22*, 457–460.

44. Berry, J. Beyond Mapping Use Surface Area for Realistic Calculations. *Geo World* **2002**, *15*, 20–21.

45. Podur, J.; Wotton, B.M. Defining fire spread event days for fire-growth modelling. *Int. J. Wildl. Fire* **2011**, *20*, 497–507. [CrossRef]

46. Taylor, S.; Pike, R.; Alexander, M. *Field Guide to the Canadian Forest Fire Behavior Prediction (FBP) System*; Victoria, B.C., Edmonton, A.B., Eds.; Joint publication of the Canadian Forest Service and the British Columbia Ministry of Forests: Victoria, BC, Canada, 1997.

47. Amiro, B.D.; Todd, J.B.; Wotton, B.M.; Logan, K.A.; Flannigan, M.D.; Stocks, B.J.; Mason, J.A.; Martell, D.L.; Hirsch, K.G. Direct carbon emissions from Canadian forest fires, 1959–1999. *Can. J. For. Res.* **2001**, *31*, 512–525. [CrossRef]

48. Wang, Z.; Grant, R.F.; Arain, M.A.; Chen, B.N.; Coops, N.; Hember, R.; Kurz, W.A.; Price, D.T.; Stinson, G.; Trofymow, J.A.; et al. Evaluating weather effects on interannual variation in net ecosystem productivity of a coastal temperate forest landscape: A model intercomparison. *Ecol. Model.* **2011**, *222*, 3236–3249. [CrossRef]

49. Agresti, A. *Categorical Data Analysis*, 3rd ed.; John Wiley & Sons: Hoboken, NJ, USA, 2013.

50. Kane, V.R.; Cansler, C.A.; Povak, N.A.; Kane, J.T.; Mcgaughey, R.J.; Lutz, J.A.; Churchill, D.J.; North, M.P. Mixed severity fire effects within the Rim fire: Relative importance of local climate, fire weather, topography, and forest structure. *For. Ecol. Manag.* **2015**, *358*, 62–79. [CrossRef]

51. Hosmer, D.W.; Lemeshow, S.; Sturdivant, R.X. *Applied Logistic Regression*, 3rd ed.; John Wiley & Sons, Inc.: Hoboken, NJ, USA, 2013.

52. Akaike, H. Information theory and an extension of the maximum likelihood principle. In *2nd International Symposium on Information Theory*; Csáki, F., Petrov, B.N., Eds.; Akadémiai Kiado: Budapest, Hungary, 1973; pp. 267–281.

53. Cohen, J. A Coefficient of Agreement for Nominal Scales. *Educ. Psychol. Meas.* **1960**, *20*, 37–46. [CrossRef]

54. Cohen, J. Weighted kappa: Nominal scale agreement with provision for scaled disagreement or partial credit. *Psychol. Bull.* **1968**, *70*, 213–220. [CrossRef] [PubMed]

55. Morrow, B.; Johnston, K.; Davies, J. *Rating Interface Wildfire Threats in British Columbia*; Ministry of Forests and Range Protection Branch: Victoria, BC, Canada, 2008.

56. Hanel, C. *Effects of Fire on Vegetation in the Interior Douglas-Fir Zone*; University of British Columbia: Vancouver, BC, Canada, 2000; Volume 1991.

57. Kafka, V.; Gauthier, S.; Bergeron, Y. Fire impacts and crowning in the boreal forest: Study of a large wildfire in western Quebec. *Int. J. Wildl. Fire* **2001**, *10*, 119–127. [CrossRef]

58. Dillon, G.K.; Holden, Z.A.; Morgan, P.; Crimmins, M.A.; Heyerdahl, E.K.; Luce, C.H. Both topography and climate affected forest and woodland burn severity in two regions of the western US, 1984 to 2006. *Ecosphere* **2011**, *2*, 1–33. [CrossRef]

59. Keeley, J.E.; Brennan, T.; Pfaff, A.H. Fire Severity and Ecosytem Responses Following Crown Fires in California Shrublands. *Ecol. Appl.* **2008**, *18*, 1530–1546. [CrossRef] [PubMed]

60. Angers, V.A.; Gauthier, S.; Drapeau, P.; Jayen, K.; Bergeron, Y. Tree mortality and snag dynamics in North American boreal tree species after a wildfire: A long-term study. *Int. J. Wildl. Fire* **2011**, *20*, 751–763. [CrossRef]

61. Breiman, L. Random forests. *Mach. Learn.* **2001**, *45*, 5–32. [CrossRef]

© 2016 by the authors. Licensee MDPI, Basel, Switzerland. This article is an open access article distributed under the terms and conditions of the Creative Commons Attribution (CC BY) license (http://creativecommons.org/licenses/by/4.0/).

forests

MDPI

Article

Burn Severity Dominates Understory Plant Community Response to Fire in Xeric Jack Pine Forests

Bradley D. Pinno * and Ruth C. Errington

Natural Resources Canada, Canadian Forest Service, Northern Forestry Centre Edmonton, Edmonton, AB T6H 3S5, Canada; ruth.errington@canada.ca
* Correspondence: brad.pinno@canada.ca; Tel.: +1-780-430-3829

Academic Editors: Yves Bergeron and Sylvie Gauthier
Received: 29 February 2016; Accepted: 9 April 2016; Published: 15 April 2016

Abstract: Fire is the most common disturbance in northern boreal forests, and large fires are often associated with highly variable burn severities across the burnt area. We studied the understory plant community response to a range of burn severities and pre-fire stand age four growing seasons after the 2011 Richardson Fire in xeric jack pine forests of northern Alberta, Canada. Burn severity had the greatest impact on post-fire plant communities, while pre-fire stand age did not have a significant impact. Total plant species richness and cover decreased with disturbance severity, such that the greatest richness was in low severity burns (average 28 species per 1-m^2 quadrat) and plant cover was lowest in the high severity burns (average 16%). However, the response of individual plant groups differed. Lichens and bryophytes were most common in low severity burns and were effectively eliminated from the regenerating plant community at higher burn severities. In contrast, graminoid cover and richness were positively related to burn severity, while forbs did not respond significantly to burn severity, but were impacted by changes in soil chemistry with increased cover at pH >4.9. Our results indicate the importance of non-vascular plants to the overall plant community in this harsh environment and that the plant community is environmentally limited rather than recruitment or competition limited, as is often the case in more mesic forest types. If fire frequency and severity increase as predicted, we may see a shift in plant communities from stress-tolerant species, such as lichens and ericaceous shrubs, to more colonizing species, such as certain graminoids.

Keywords: *Pinus banksiana*; burn severity; composite burn index; revegetation; forest regeneration; lichen

1. Introduction

Large fires are expected to become more common in northern boreal forests in the future with a changing climate [1], and these fire events are often characterized by highly variable burn severity [2–4]. Many studies have examined tree regeneration after variable severity burns in the boreal forest [5,6], but in terms of overall plant diversity, understory species are most important, especially in the tree species-poor, xeric jack pine (*Pinus banksiana*) forests of the northern boreal [7]. Understory plant communities also play critical roles in maintaining key ecosystem processes, such as nutrient cycling, habitat for wildlife and overstory succession [8,9].

The regeneration mechanism of understory plants differs from that of the jack pine canopy trees, *i.e.*, an aerial seedbank of serotinous cones that releases seeds immediately after burning [10,11]. In contrast, after fire in the northern boreal forest, understory plants may resprout from roots or rhizomes [7,12,13], germinate from seeds in the soil seed bank [14,15], germinate from seeds carried in from off-site [16] or encroach from surrounding areas [17]. The relative importance of these different regeneration mechanisms differs by species and will likely be impacted by disturbance severity, stand age and soil properties. Previous work on jack pine regeneration after the same fire as this current study showed that stand age and burn severity were the main drivers of tree regeneration [2], but it is not clear if these same drivers are controlling understory plant community re-establishment.

The regeneration of understory plants may be impacted by high burn severities by altering seed bed availability, with most species requiring exposed mineral soil to germinate. High burn severities can also change the composition of surviving vegetation from which resprouting is possible by reducing or eliminating stored seedbanks in the organic material where most seeds are found and by eliminating competition [18]. Stand age has a clear impact on species composition related to the successional stage of forest development with more shade-tolerant species being found in older forests and species that are more adapted to the specific soil and site conditions rather than to disturbance [19]. This influences the species present on site and capable of vegetative reproduction, but may not influence regeneration from seed. Soil properties may also impact understory plant regeneration after disturbance to a greater degree than tree regeneration, as understory plant species are more sensitive to changes in soil chemical and physical properties than are tree species growing on the same sites. However, how these factors interact with each other and with environmental conditions to influence plant community establishment in boreal jack pine forests in the years following fire is not clear.

Most studies on forest understory plant community response to disturbance focus on vascular plants [20–22], which are the dominant plant groups in most mesic boreal forests and are more responsive to disturbance than non-vascular species [8]. However, in harsh environments within the boreal forest, such as the xeric jack pine-dominated forests that are the focus of this study, non-vascular lichens and bryophytes are a more significant component [7]. This also holds true for other harsh environments, such as bogs at the other end of the moisture gradient within the boreal forest, where the understory community is also dominated by non-vascular plant species, such as *Sphagnum* mosses. Less is known about the regeneration ecology of these non-vascular species when compared to herbaceous and shrub vegetation, and the response of the non-vascular community to varying levels of disturbance severity is not well understood [23,24].

We studied understory plant communities in response to a range of natural burn severities and pre-fire stand ages in the xeric jack pine boreal forest of north-eastern Alberta, Canada. This forest area is not currently being developed for timber, but will likely be impacted, at least in part, in the future by expanding oil sands developments. Therefore, it is important to gain a better understanding of how the plant communities in this area respond to varying levels of disturbance severity and identify any potential risks to ecosystem sustainability in a post-disturbance landscape. The specific questions we asked are: (1) How do burn severity and pre-fire stand age impact understory plant community development post-fire in these xeric, pure jack pine boreal forests? (2) Do all plant groups respond in a similar manner to these drivers?

2. Methods

2.1. Study Area and Fire Description

We examined plant community development four growing seasons after the Richardson Fire, a 576,000-ha fire in north-eastern Alberta, Canada (Figure 1). This fire resulted from human ignition, burned from May until August 2011 and was the second largest documented fire in western Canadian history. During periods of extreme fire weather, there were spread rates of over 30 km per day with modelled head fire intensities in excess of 10,000 kW· m^{-1}.

The dominant forest type in the region is jack pine growing on sandy dystric Brunisol soils on aeolian parent material with common understory vegetation consisting of *Arctostaphylos uva-ursi*, *Vaccinium myrtilloides* and *Cladonia arbusculata* subsp. *mitis*. The local site type is referred to as the xeric "a" eco-site in the Canadian Shield and Boreal Mixedwood Ecological Areas [25]. The southern and western portions of the Richardson Burn also contain stands of more mesic trembling aspen

(*Populus tremuloides*) and white spruce (*Picea glauca*) forest types, but this work focused on the jack pine-dominated portion of the burn, which accounts for 65% of the total burned area. The climate in the area is continental with long, cold winters (average January temperature $-19\,°C$) and short, cool summers (average July temperature $17\,°C$) (based on Fort McMurray climate normal from Environment Canada). Median annual precipitation is 455 mm, which is less than the average annual potential evaporation of 480 mm [26]. The xeric site type and moisture-limiting climate indicate that these forests are very challenging environments for many types of plants to grow, and this is supported by the low basal area (average 16.3 $m^2 \cdot ha^{-1}$) and short canopy height (average 11.8 m) for mature jack pine stands [2].

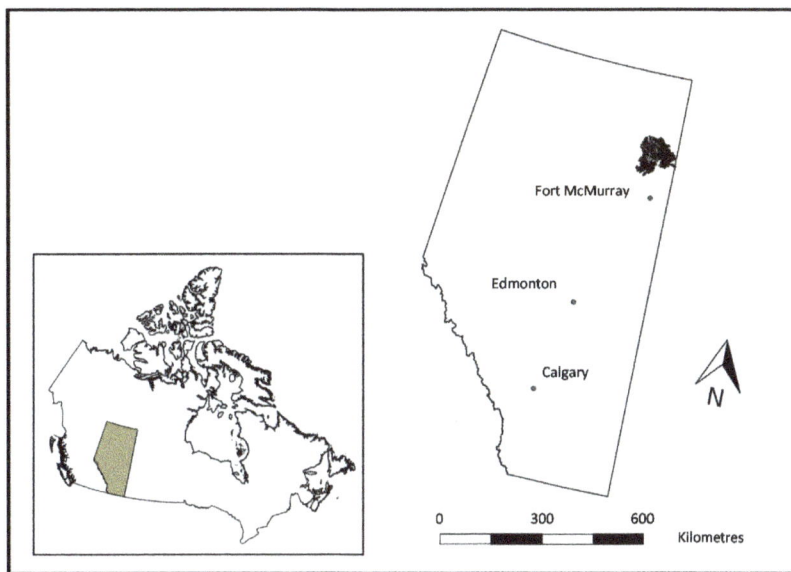

Figure 1. Location of the 2011 Richardson Fire in northeastern Alberta, Canada.

2.2. Field Sampling

The sampling design was based on a previous project [2], which examined jack pine regeneration after the Richardson Burn in stands of different pre-fire age and burn severity. For the current study, we used this matrix of three burn severities (low, moderate and high) and two pre-fire stand ages (young and old) and measured understory plant community composition in six or seven stands within each of the six categories for a total of 38 stands. Selected stands had pure jack pine overstories with the pre-fire stand age and burn severity determined in the year immediately following the fire as part of the pine regeneration study. Burn severity was determined using the Composite Burn Index (CBI), which incorporates measures of forest floor, understory and overstory burn severity [27], and was then categorized as low, moderate or high. High burn severity stands had complete overstory mortality and greater than 50% forest floor consumption; moderate burn severity stands had on average 75% overstory mortality and 25%–50% forest floor consumption; and low burn severity stands had <25% overstory mortality and only light charring of the forest floor. Pre-fire stand ages were determined from cores taken at breast height of representative canopy trees. Stands were sampled according to pre-fire stand age, with stands less than 30 years old considered "young" and stands greater than 60 years old considered "old". Soil samples were taken from the upper 15 cm of mineral soil and were analyzed for texture, percent carbon, electrical conductivity (EC) and pH, none of which were significantly different among burn severity or pre-fire stand age classes.

Plot centres were established at a random point within each selected stand, and then, four 1 m × 1 m quadrats were located 10 m from the plot centre along cardinal bearings. This size of quadrat was used, as it is the standard in the region and will allow for comparison with other studies. At each quadrat, all vascular and non-vascular plants, including lichens, were identified to the species level, and cover was visually estimated to the nearest percent with cover values of 0.5% and + (present but at lower than 0.5% cover) also recorded. A complete stand level species list was also developed for the area enclosed by the four quadrats, an area of approximately 300 m^2, by completing a walkaround survey. For quantitative analyses, the average cover at the stand level was used with + assigned a value of 0.05% and presence in the area enclosed by the quadrats, but not in the quadrats themselves, assigned a cover value of 0.005%. Total species richness (the total number of understory species per stand) and average percent cover of vegetation (the average of the four quadrats) were determined for each stand. Richness and cover by species group, *i.e.*, lichen, bryophyte, graminoid, forb, shrub, were also determined. Species nomenclature was standardized for vascular plants [28], bryophytes [29] and lichens [30].

2.3. Statistical Analyses

Non-metric multidimensional scaling (NMDS) ordinations were used to identify patterns of understory species composition with the Sorensen (Bray–Curtis) distance used as a measure of ecological dissimilarity in the NMDS ordinations conducted using PC-ORD [31]. To quantify the multi-variate differences between groups, we used the multi-response permutation procedure (MRPP) in PC-ORD with Bonferroni correction of *post hoc* pairwise comparisons conducted in R [32]. These ordinations were completed for all plant species together and then for the vascular plants and non-vascular plants (lichens and bryophytes) separately. Indicator species analysis [33], conducted in PC-ORD, was used to identify characteristic species for each burn severity class. This method combines the relative abundance of a species with its relative frequency of occurrence within each burn severity type, producing a maximum value (100%) when all individuals of a species are found in a single burn severity type and when the species occurs in all samples within that type. Significance of each species as an indicator was tested for the burn severity type in which it reached its maximum value using a randomization procedure.

Two-way ANOVA followed by Tukey *post hoc* tests were used to compare species richness and cover among burn severity and pre-fire stand age groups. Regression analysis was used to determine the specific relationship between species groups' richness and cover and continuous site and environmental variables, including burn severity and soil pH. Regression tree analysis was used to quantify distinct thresholds in the response variables. These analyses were conducted using Systat13 (Systat Software Inc., Chicago, IL, USA).

3. Results

There was a total of 95 plant (including lichen) species (Appendix A) found across all stands with the most numerous species group being lichens with 32 different species found followed by forbs with 31 species found. Stand-level species richness decreased with burn severity and pre-fire stand age (Figure 2a), such that the greatest richness was found in young stands with low burn severity, and the lowest richness was found in old stands with high burn severity. Vascular plant richness, when considered separately, showed no significant differences among burn severity ($p = 0.885$) or age classes ($p = 0.537$). Average stand level total plant cover ranged from 4%–63% and also decreased with burn severity, but was not affected by pre-fire stand age (Figure 2b).

The ordination including all vegetation (Figure 3) and supported by MRPP clearly shows differing plant communities among burn severity classes (A = 0.072, $p < 0.001$), but there were no significant differences among pre-fire age classes (A = 0.010, $p = 0.083$). Axis 1 is positively correlated with burn severity and graminoid cover, while lichen richness and cover were negatively correlated (Figure 3a). Axes 2 and 3 function mostly as a soil chemistry gradient with pH and EC positively related to forb and shrub richness (Figure 3a,b).

Including only vascular vegetation in the ordination clearly separates the stand replacing moderate and high severity burns from the low severity burns and results in a main gradient defined by burn severity (A = 0.068, $p < 0.001$) and a secondary gradient associated with soil variables, but there was no difference among pre-fire stand age classes (A = -0.004, $p = 0.645$) (Figure 4a). When only non-vascular species are included in the ordination, high severity burns are clearly separated with burn severity again being the primary factor along Axis 1 (A = 0.045, $p = 0.003$), but there is no difference among pre-fire stand age classes (A = -0.003, $p = 0.302$) (Figure 4b). In this ordination, the secondary gradient is negatively correlated with greater lichen richness and positively with forb cover.

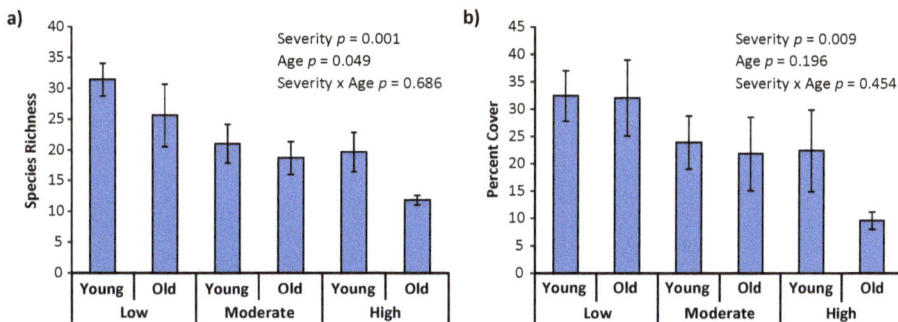

Figure 2. (**a**) Species richness (total number of plant species found in each stand) and (**b**) percent cover (total cover of all species averaged from four 1 m × 1 m quadrats per stand), in relation to burn severity (as measured by the Composite Burn Index) and pre-fire stand age.

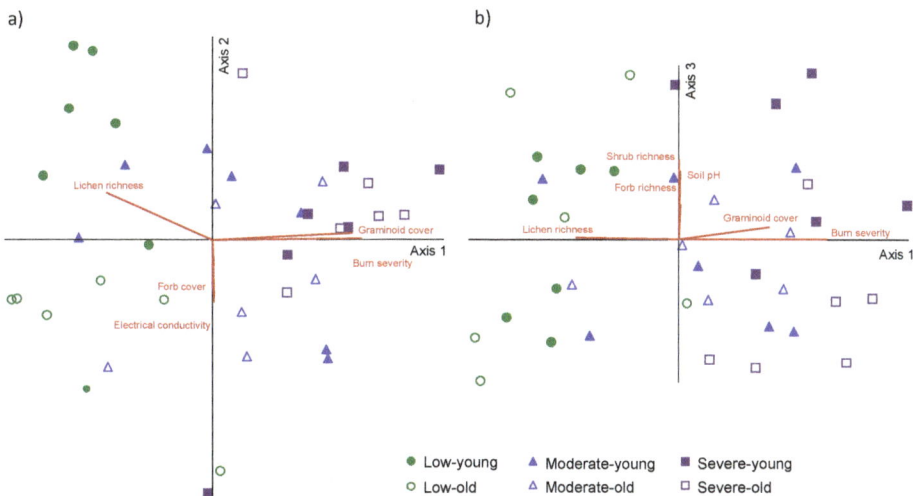

Figure 3. Non-metric multidimensional scaling ordination of all plant community data: (**a**) Axes 1 and 2; (**b**) Axes 1 and 3. The environmental overlays present the most significant relationships for each species group, burn severity and soil chemistry with $r > 0.5$.

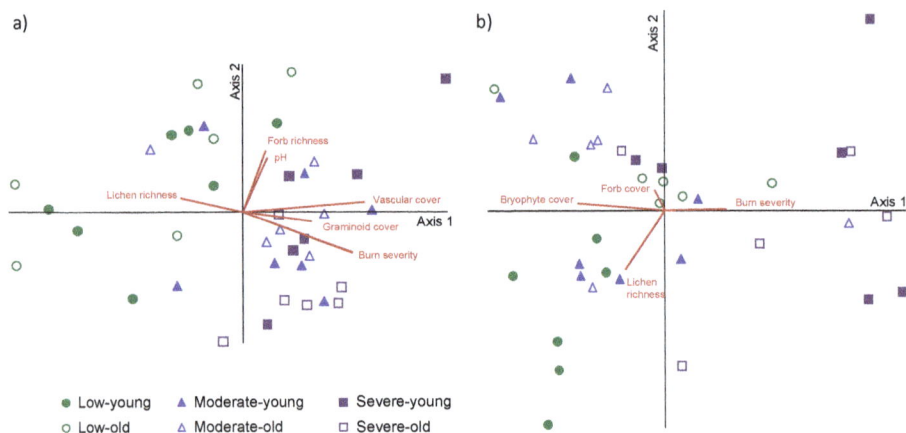

Figure 4. Non-metric multidimensional scaling ordination including only (**a**) vascular plants or (**b**) non-vascular plants. The environmental overlays present the most significant relationships for each species group, burn severity and soil chemistry with $r > 0.5$.

Lichen species, mainly from the genus *Cladonia* with the most common species being *Cladonia arbusculata* subsp. *mitis* and *Cladonia gracilis* subsp. *turbinata*, which were found in 16 out of 38 stands each, accounted for 70% of total cover and 38% of the total species richness in low severity burns where they were most common. Lichen cover showed a negative relationship to burn severity with a significant threshold at CBI 1.9, with lichens being effectively eliminated from the plant community beyond this point (Figure 5a). Bryophyte cover showed a similar pattern to lichens with a negative relationship to burn severity, but with an elimination threshold at a higher burn severity of 2.5 CBI (Figure 5b). The most common bryophytes were *Ceratodon purpureus* and *Polytrichum piliferum*, which were found in 37 and 30 of 38 stands, respectively. *P. piliferum* along with the two lichen species *C. arbusculata* subsp. *mitis* and *C. gracilis* subsp. *turbinata* were the strongest indicators of low severity burns (Table 1), but *C. purpureus*, a ubiquitous species associated with disturbance and open sandy soils, was not a significant indicator of burn severity (Table 1).

Graminoid cover, on the other hand, was quite low (average 2.1% across all burn severities), but was positively related to burn severity (Figure 5c). Graminoid species richness was also positively correlated to burn severity ($r = 0.468$, $p = 0.003$) with the rhizomatous, quickly colonizing *Carex siccata* indicative of moderate and high severity burns (Table 1).

Forb cover was low overall (average 1.2%) and was not related to burn severity ($p = 0.802$) or pre-fire stand age ($p = 0.409$), but it was related to soil pH with significantly higher ($p < 0.05$) forb cover on soils with pH > 4.9 (average 4.1 %) compared to soils with pH < 4.9 (average 0.8%). The most common forbs were *Solidago simplex*, *Campanula rotundifolia* and *Apocynum androsaemifolia*, which were found in 24, 20 and 15 of 38 plots, respectively, across the range of soil pH. On the higher pH soils, however, *Maianthemum canadense* and *Cornus canadensis* became dominant, with the highest forb cover values found in any stands associated with these two species. Colonizing species with small windblown seeds, such as *Chamerion angustifolium* and *Symphyotrichum laeve*, were each only found in nine of 38 stands. Although not as ubiquitous, the annual forbs *Erigeron canadensis* and *Leucophysalis grandiflora* were both significant indicators of high severity burns (Table 1).

Shrub cover averaged 3.8% and was not related to burn severity ($p = 0.520$), pre-fire stand age ($p = 0.189$) nor measured soil properties (pH, $p = 0.272$). The most common species were all low shrubs, mainly ericaceous, with *Arctostaphylos uva-ursi*, *Hudsonia tomentosa* and *Vaccinium myrtilloides* being found in 38, 24 and 22 of 38 stands, respectively, and there were no significant shrub indicators of burn severity.

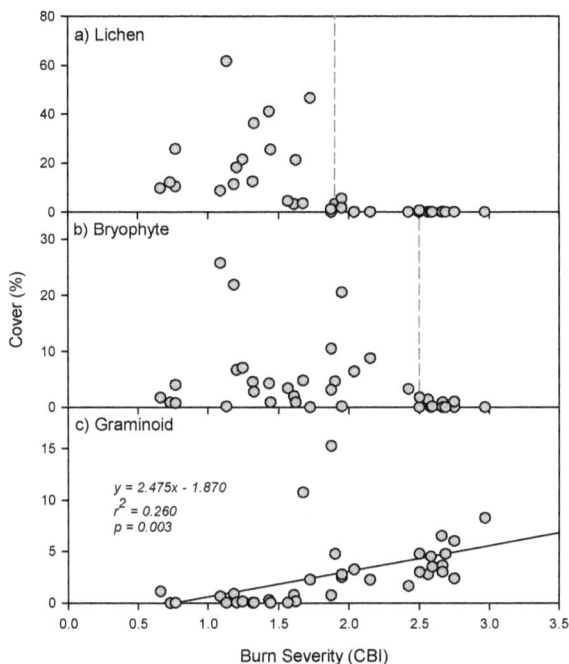

Figure 5. Relationships between the cover of plant groups and burn severity as measured by the Composite Burn Index (CBI) for (**a**) lichens, (**b**) bryophytes and (**c**) graminoids. The thresholds for lichen cover at 1.9 CBI and bryophyte cover at 2.5 CBI were determined from regression tree analysis at a significance level of $p < 0.05$.

Table 1. Significant indicator species with an indicator value >25 for each burn severity class as determined by the method of Dufrêne and Legendre [33]. Each species was only tested for significance in the group for which it had its maximum indicator value.

Severity	Indicator Species	Indicator Value	p	Growth Form	Life Strategy
Low	*Cladonia arbusculata* subsp. *mitis*	82.1	0.0002	lichen	perennial
	Cladonia gracilis subsp. *turbinata*	76	0.0008	lichen	perennial
	Polytrichum piliferum	75.4	0.0034	moss	perennial
	Pyrola chlorantha	64.7	0.001	forb	perennial
	Cladonia pyxidata	59.1	0.0002	lichen	perennial
	Cladonia deformis	54.3	0.0008	lichen	perennial
	Trapeliopsis granulosa	50.1	0.011	lichen	perennial
	Cladonia sulphurina	47.7	0.0012	lichen	perennial
	Cladonia subulata	45.7	0.003	lichen	perennial
	Cetraria ericetorum	44.7	0.0042	lichen	perennial
	Cladonia uncialis	43.6	0.0432	lichen	perennial
	Cladonia botrytes	41.7	0.0056	lichen	perennial
	Parmeliopsis ambigua	38.5	0.0074	lichen	perennial
	Cladonia cristatella	38	0.0124	lichen	perennial
	Peltigera malacea	37	0.0184	lichen	perennial
	Cladonia cornuta	36.6	0.0334	lichen	perennial
	Dicranum polysetum	35.3	0.0238	moss	perennial
	Vulpicida pinastri	30.8	0.0284	lichen	perennial
	Cladonia borealis	28	0.0498	lichen	perennial
Moderate	*Carex siccata*	29	-	graminoid	perennial
High	*Carex siccata*	69.6	0.0252	graminoid	perennial
	Erigeron canadensis	31.9	0.0338	forb	annual
	Leucophysalis grandiflora	25	0.0272	forb	annual

4. Discussion

There are distinct plant communities developing four years after burns of differing severity, and three out of five main species groups (lichens, bryophytes and graminoids) responded significantly to burn severity. Low severity burns favour the persistence of lichens and bryophytes and have higher overall species richness than high severity burns. In contrast, the vegetation community after high severity burns is comprised mainly of annual forbs and colonizer-type graminoids seeding in from off site along with resprouting forbs and shrubs, resulting in lower overall species diversity and cover. Species richness did not increase with moderate levels of disturbance, as has been suggested by other studies [34,35], but instead, species richness declined with burn severity in our study. While peaks in species richness at intermediate levels of disturbance are actually the exception rather than the rule in published studies [36], the intermediate disturbance hypothesis and underlying mechanisms are fundamental concepts in community ecology, and it is important to understand under what situations these mechanisms function. In our study, in a climatically marginal forest area and on a xeric site type, the number of species capable of surviving there is much less than are available to a more mesic site type, indicating that the plant community developing here may be environmentally limited rather than recruitment or competition limited. Certainly, it is the trade-off between early successional species adapted for colonization (colonizers) *versus* later successional species better able to compete for resources (competitors) that is thought to be one of the primary mechanisms responsible for the peak in species richness at intermediate levels of disturbance [35,37]. While no peak in richness at moderate burn severity was observed in our study, differential species responses to disturbance were observed.

High severity burns, which result in a loss of more than 50% of the total forest floor, result in greatly reduced species richness, as much of the seedbank and resprouting organs are found in the forest floor [12,15]. The new forest plant community must therefore rely mainly on colonizing species from outside, of which there appear to be few other than select graminoids and forbs, capable of surviving in this harsh environment. It also appears that competition between plants is not a major component in determining community composition in this forest given the low overall vegetation cover (16% ± 4% (mean ± SE) in high severity burns) and the corresponding high level of bare ground, again indicating that environmental and not internal controls (*i.e.*, competition) are mainly responsible for controlling plant community development. In contrast, after the same fire event, mesic boreal mixedwood sites had much greater vegetation cover and species richness [22] indicative of higher competition levels among plants and suggesting that this xeric jack pine forest responds differently to fire than do nearby mesic forests.

Lichens and bryophytes were completely eliminated from the plant community at high burn severity, a response frequently evident in non-vascular plants following high severity burns [38]. Although generally considered to be dispersal limited [39], lichens are expected to begin initial recolonization one to three years following the fire, with pioneer lichen species, such as *Cladonia coccifera* (likely equivalent to our *C. borealis*) and *C. gracilis* [7]. Both of these pioneer species were found in our study, although they were confined to low and moderate severity burns and likely represent relict populations from pre-fire communities. Four years post-fire, lichens are only just beginning to recolonize the most severely-burnt sites, with *Cladonia* species present in seven of the twelve stands, but in early stages of regeneration (basal scales only). This is consistent with a generalized successional sequence found throughout lichen-dominated forests of the northern taiga and tundra regions [40]. The bryophytes *Polytrichum piliferum* and *P. juniperinum* were both detected in all age and burn severity classes, along with the ubiquitous *Ceratodon purpureus*. The presence of these species is typical of post-fire succession in northern boreal forests and indicates that these moss species are not dispersal limited, as is hypothesized for lichens. Increased bryophyte diversity in the low and moderate severity burns is also typical of later successional sequences and indicates persistence after low severity burns rather than recolonization. While this general successional sequence of non-vascular plants has been documented in black spruce (*Picea mariana*) [41], lodgepole pine (*Pinus contorta* var. *latifolia*) [42] and jack pine/lichen forests [7], xeric jack pine lichen forests in northern Alberta may follow an alternate successional trajectory leading to a final park-like pine/lichen stage, instead of the more closed canopy pine/feathermoss stage [7].

Unlike bryophytes and lichens, graminoid cover increased with burn severity. Most of these species are thought to seed in from outside areas, but they may have been a component of the seedbank. Other studies from more southern, climatically-favourable jack pine forests have described the rapid expansion of graminoids, particularly *Carex*, after disturbance with possible negative implications for tree regeneration [43], but the low cover values for graminoids in our study (average 4.4% in high severity) indicate that is not a concern in this forest area. The lack of graminoids in low severity burns with little disturbance to the forest floor also indicates that these species are transient members of the plant community and are likely not a substantial part of the mature forest understory.

Forb cover and richness were not related to burn severity, but were related to soil chemistry with increased cover above pH 4.9. This is comparable to more forb-rich aspen stands of the region, which were found to have pH ($\pm$ SE) values of 5.28 ± 0.33 and 4.77 ± 0.19 for mature and post-fire stands, respectively [22]. The lack of response to burn severity may reflect the vegetative regeneration strategy of many common perennial forb species, including the species most abundant on high pH soils, *Maianthemum canadense* and *Cornus canadensis* [44,45]. Even on sites with very low competition levels and correspondingly high suitable seedbeds in the form of exposed mineral soil, forb richness did not increase, suggesting that colonizing forb species with widely-dispersed seed rain, such as *Chamerion angustifolium* and *Symphyotrichum laeve* [46,47], were not able to thrive in this harsh environment. Instead, the sites with the most abundant forb cover relied on resprouting of existing forbs.

Shrub cover and richness were not related to burn severity, pre-fire stand age or any measured environmental variable. Most shrubs can regenerate vegetatively, and this appears to be the case here with similar shrub species found across the range of site types and burn severities. The low shrub cover (average 3.8% across all sites) may therefore be more indicative of the slow growth rates of the dominant ericaceous shrubs, rather than a limitation in recruitment to the site given that *Arctostaphylos uva-ursi* was found in every sampled stand.

5. Conclusions

In conclusion, we demonstrate the relatively large range of potential plant communities developing following a large, spring fire in xeric, boreal jack pine forests. Although there are indications that soil chemical factors play a secondary role in structuring the forb community, four years post-fire, the variation in plant communities is primarily due to burn severity and is evident in both the vascular and non-vascular components of the flora. If fire frequency and severity continue to increase as predicted [1], these results indicate that we may see a shift in plant communities from stress-tolerant, slower growing species, such as lichens and ericaceous shrubs, to more colonizing species, such as select graminoids and forbs, which appear to be able to tolerate the harsh environmental conditions of this forest. The implications of this potential shift in understory plant communities on wildlife habitat, particularly for the woodland caribou, which rely on lichen for much of their diet, is not known. As this study was based on only a single, albeit massive, fire event, caution is needed in extrapolating these results to other areas. The clear message from this study, however, is that the potential range of post-disturbance plant communities is dependent on burn severity at the site level.

Acknowledgments: Thanks to Edith Li for field assistance and Brent Frey and Andre Arsenault for reviewing an earlier draft of the manuscript. Funding for this project was provided by Natural Resources Canada, Canadian Forest Service.

Author Contributions: Bradley D. Pinno and Ruth C. Errington both contributed to all aspects of the work, including designing the experiment, analyzing the data and writing the paper.

Conflicts of Interest: The authors declare no conflict of interest.

Appendix A

Table A1. Complete species list with the number of stands the species was found in grouped according to burn severity and pre-fire age class.

Burn Severity		Low		Moderate		High		
Age Class		Old	Young	Old	Young	Old	Young	
Number of plots per stand type		N = 6	N = 7	N = 6	N = 7	N = 6	N = 6	
Species	Layer							Life Strategy
Achillea millefolium Linnaeus	herb	0	2	0	0	0	0	perennial
Agrostis scabra Willdenow	herb	5	6	5	7	6	5	perennial
Alnus viridis (Chaix) de Candolle subsp. crispa (Aiton) Turrill	shrub	0	0	0	1	0	1	perennial
Amelanchier alnifolia (Nuttall) Nuttall ex M. Roemer	shrub	1	1	1	0	0	2	perennial
Anemone multifida Poiret	herb	2	1	1	0	0	2	perennial
Apocynum androsaemifolium Linnaeus	herb	3	3	2	3	1	4	perennial
Aralia nudicaulis Linnaeus	herb	3	0	0	0	0	1	perennial
Arctostaphylos uva-ursi (Linnaeus) Sprengel	shrub	6	7	6	7	6	6	perennial
Bryoria simplicior (Vainio) Brodo & D. Hawksw.	lichen	0	1	0	0	0	0	perennial
Calamagrostis stricta subsp. inexpansa (A. Gray) Greene	herb	1	1	1	0	0	2	perennial
Campanula rotundifolia Linnaeus	herb	3	5	3	3	1	4	perennial
Capnoides sempervirens (Linnaeus) Borkhaussen	herb	0	0	0	0	0	0	biennial
Carex foenea Willdenow	herb	0	1	2	4	2	4	perennial
Carex praticola Rydberg	herb	0	0	1	0	0	0	perennial
Carex c.f. richardsonii R. Brown	herb	1	0	1	0	0	1	perennial
Carex siccata Dewey	herb	4	5	6	7	6	6	perennial
Carex tonsa (Fernald) E.P. Bicknell	herb	5	7	5	7	5	6	perennial
Carex umbellata Schkuhr ex Willdenow	herb	0	0	1	0	1	0	perennial
Ceratodon purpureus (Hedwig) Bridel	bryophyte	6	7	6	7	6	5	perennial
Cetraria ericetorum Opiz	lichen	2	5	0	2	0	0	perennial
Chamerion angustifolium (Linnaeus) Holub subsp. angustifolium	herb	0	1	3	1	1	3	perennial
Cladonia amaurocraea (Flörke) Schaerer	lichen	0	1	0	0	0	0	perennial
Cladonia borealis S. Stenroos	lichen	0	5	0	2	0	0	perennial
Cladonia botrytes (K.G. Hagen) Willd.	lichen	1	5	0	2	0	0	perennial
Cladonia cariosa (Ach.) Sprengel	lichen	0	1	0	0	0	0	perennial
Cladonia cornuta (L.) Hoffm.	lichen	0	5	0	3	0	0	perennial
Cladonia crispata (Ach.) Flotow	lichen	0	4	1	2	0	0	perennial

Table A1. *Cont.*

Burn Severity			Low	Moderate	High
Cladonia cristatella Tuck.	lichen	perennial	5	2	0
Cladonia deformis (L.) Hoffm.	lichen	perennial	7	3	0
Cladonia c.f. fimbriata (L.) Fr.	lichen	perennial	1	0	0
Cladonia gracilis (L.) Willd. subsp. *turbinata* (Ach.) Ahti	lichen	perennial	7	3	0
Cladonia macilenta Hoffm.	lichen	perennial	0	0	0
Cladonia macrophylla (Schaerer) Stenh.	lichen	perennial	0	1	0
Cladonia arbusculata (Wallr.) Flotow subsp. *mitis* (Sandst.) Ruoss	lichen	perennial	7	3	0
Cladonia multiformis G. Merr.	lichen	perennial	1	0	0
Cladonia pyxidata (L.) Hoffm.	lichen	perennial	6	1	0
Cladonia rangiferina (L.) F.H. Wigg.	lichen	perennial	6	5	3
Cladonia sp. P. Browne	lichen	perennial	6	5	0
Cladonia stygia (Fr.) Ruoss	lichen	perennial	5	0	0
Cladonia subulata (L.) F.H. Wigg.	lichen	perennial	5	1	0
Cladonia sulphurina (Michaux) Fr.	lichen	perennial	5	2	0
Cladonia uncialis (L.) Weber ex F.H. Wigg.	lichen	perennial	4	3	0
Cladonia verticillata (Hoffm) Schaerer	lichen	perennial	0	2	0
Collomia linearis Nuttall	herb	annual	2	0	0
Comandra umbellata (Linnaeus) Nuttall	herb	perennial	1	1	0
Cornus canadensis Linnaeus	herb	perennial	1	0	1
Crepis tectorum Linnaeus	herb	annual	0	0	1
Dichanthelium acuminatum (Swartz) Gould & C.A. Clarke subsp. *fasciculatum* (Torrey) Freckmann & Lelong	herb	perennial	0	1	0
Dicranum polysetum Swartz	bryophyte	perennial	4	1	0
Diphasiastrum complanatum (Linnaeus) Holub	herb	perennial	1	0	1
Erigeron canadensis Linnaeus	herb	annual	0	0	3
Evernia mesomorpha Nyl.	lichen	perennial	1	0	0
Festuca saximontana Rydberg	herb	perennial	0	0	0
Flavocetraria nivalis (L.) Kärnefelt & A. Thell	lichen	perennial	2	0	1
Fragaria virginiana Miller	herb	perennial	1	0	1
Galium boreale Linnaeus	herb	perennial	2	0	1
Geocaulon lividum (Richardson) Fernald	herb	perennial	0	0	1
Geranium bicknellii Britton	herb	annual or biennial	0	1	2
Hieracium umbellatum Linnaeus	herb	perennial	2	0	1
Hudsonia tomentosa Nuttall	herb	perennial	3	6	4

Table A1. Cont.

Burn Severity			Low	Moderate	High
Hylocomium splendens (Hedwig) Shimper in P. Bruch and W.P. Shimper	bryophyte	perennial	1	0	0
Leucophysalis grandiflora (Hooker) Rydberg	herb	annual	0	0	3
Leymus innovatus (Beal) Pilger subsp. innovatus	herb	perennial	0	0	1
Linnaea borealis Linnaeus	herb	perennial	3	1	1
Maianthemum canadense Desfontaines	herb	perennial	3	2	3
Melampyrum lineare Desrousseaux	herb	annual	1	0	0
Oryzopsis asperifolia Michaux	herb	perennial	1	1	1
Packera paupercula (Michaux) Á. Löve & D. Löve	herb	perennial	1	0	0
Parmeliopsis ambigua (Wulfen) Nyl.	lichen	perennial	2	1	0
Peltigera malacea (Ach.) Funck	lichen	perennial	2	1	0
Peltigera rufescens (Weiss) Humb.	lichen	perennial	0	0	0
Pinus banksiana Lambert	tree	perennial	4	7	6
Piptatheropsis pungens (Torrey ex Sprengel) Romaschenko, P.M. Peterson & Soreng	herb	perennial	4	5	5
Pleurozium schreberi (Wildenow ex Bridel) Mitten	bryophyte	perennial	2	0	0
Polytrichum juniperinum Hedwig	bryophyte	perennial	5	3	2
Polytrichum piliferum Hedwig	bryophyte	perennial	6	6	2
Populus tremuloides Michaux	tree	perennial	1	2	1
Prunus pensylvanica Linnaeus f.	tree, shrub	perennial	3	2	3
Ptilidium ciliare (L.) Hampe	bryophyte	perennial	2	1	0
Ptilium crista-castrensis (Hedwig) De Notaris	bryophyte	perennial	1	0	0
Pyrola chlorantha Swartz	herb	perennial	4	2	1
Rosa acicularis Lindley	shrub	perennial	1	0	1
Salix bebbiana Sargent	shrub	perennial	0	0	0
Selaginella densa Rydb.	herb	perennial	0	0	1
Sibbaldia tridentata (Aiton) Paule & Soják	herb	perennial	1	0	2
Solidago simplex Kunth var. simplex	herb	perennial	3	4	3
Stereocaulon alpinum Laurer ex Funck	lichen	perennial	0	0	0
Symphyotrichum ciliolatum (Lindley) Á. Löve & D. Löve	herb	perennial	3	1	0
Symphyotrichum laeve (Linnaeus) Á. Löve & D. Löve var. laeve	herb	perennial	2	0	3
Trapeliopsis granulosa (Hoffm.) Lumbsch	lichen	perennial	2	3	0
unknown seedling		#N/A	0	1	0
Vaccinium myrtilloides Michaux	shrub	perennial	4	4	4
Vaccinium vitis-idaea Linnaeus	shrub	perennial	3	2	2
Viola adunca Smith	herb	perennial	1	1	2
Vulpicida pinastri (Scop.) J.-E. Mattsson & M.J. Lai	lichen	perennial	1	0	0

References

1. deGroot, W.J.; Flannigan, M.D.; Cantin, A.S. Climate change impacts on future boreal fire regimes. *For. Ecol. Manag.* **2012**, *294*, 35–44.
2. Pinno, B.D.; Errington, R.C.; Thompson, D.K. Young jack pine and high severity fire combine to create potentially expansive areas of understocked forest. *For. Ecol. Manag.* **2013**, *310*, 517–522. [CrossRef]
3. Hollingsworth, T.N.; Johnstone, J.F.; Bernhardt, E.L.; Chapin, F.S., III. Fire severity filters regeneration traits to shape community assembly in Alaska's boreal forest. *PLoS ONE* **2013**, *8*, e56033. [CrossRef] [PubMed]
4. Lentile, L.B.; Smith, F.W.; Shepperd, W.D. Patch structure, fire-scar formation, and tree regeneration in a large mixed-severity fire in the South Dakota Black Hills, USA. *Can. J. For. Res.* **2005**, *35*, 2875–2885. [CrossRef]
5. Johnstone, J.F.; Chapin, F.S., III. Effects of soil burn severity on post-fire tree recruitment in boreal forest. *Ecosystems* **2006**, *9*, 14–31. [CrossRef]
6. Arsenault, D. Impact of fire behaviour on postfire forest development in a homogeneous boreal landscape. *Can. J. For. Res.* **2001**, *31*, 1367–1374.
7. Carroll, S.B.; Bliss, L.C. Jack pine–lichen woodland on sandy soils in northern Saskatchewan and northeastern Alberta. *Can. J. Bot.* **1982**, *60*, 2270–2282. [CrossRef]
8. Hart, S.A.; Chen, H.Y.Y. Vegetation dynamics of North American boreal forests. *Crit. Rev. Plant Sci.* **2006**, *25*, 381–397. [CrossRef]
9. Venier, L.A.; Pearce, J.L. Boreal forest landbirds in relation to forest composition, structure, and landscape: implications for forest management. *Can. J. For. Res.* **2007**, *37*, 1214–1226. [CrossRef]
10. Cayford, J.H.; McRae, D.J. The ecological role of fire in jack pine forests. In *The Role of Fire in Northern Circumpolar Ecosystems*; Wein, R.W., MacLean, D.A., Eds.; John Wiley & Sons: Hoboken, NJ, USA, 1983; pp. 183–199.
11. Greene, D.F.; Johnson, E.A. Modelling recruitment of *Populus tremuloides*, *Pinus banksiana*, and *Picea mariana* following fire in the mixedwood boreal forest. *Can. J. For. Res.* **1999**, *29*, 462–473.
12. Archibold, O.W. Buried viable propagules as a factor in postfire regeneration in northern Saskatachewan. *Can. J. Bot.* **1979**, *57*, 54–58. [CrossRef]
13. Whittle, C.A.; Duchesne, L.C.; Needham, T. Soil seed bank of a jack pine (*Pinus banksiana*) ecosystem. *Int. J. Wildland Fire.* **1998**, *8*, 67–71. [CrossRef]
14. Morgan, P.; Neuenschwander, L.F. Seed-bank contributions to regeneration of shrub species after clear-cutting and burning. *Can. J. Bot.* **1988**, *66*, 169–172. [CrossRef]
15. Moore, J.M.; Wein, R.W. Viable seed populations by soil depth and potential site recolonization after disturbance. *Can. J. Bot.* **1977**, *55*, 2408–2412. [CrossRef]
16. Qi, M.; Scarratt, J.B. Effects of harvesting method on seed bank dynamics in a boreal mixedwood forest in northwestern Ontario. *Can. J. Bot.* **1998**, *76*, 872–883. [CrossRef]
17. Hautala, H.; Tolvanen, A.; Nuortila, C. Regeneration strategies of dominant boreal forest dwarf shrubs in response to selective removal of understorey layers. *J. Veg. Sci.* **2001**, *12*, 503–510. [CrossRef]
18. Roberts, M.R. Response of the herbaceous layer to natural disturbance in North American forests. *Can. J. Bot.* **2004**, *82*, 1273–1283. [CrossRef]
19. Hunt, S.L.; Gordon, A.M.; Morris, D.M.; Marek, G.T. Understory vegetation in northern Ontario jack pine and black spruce plantations: 20-year successional changes. *Can. J. For. Res.* **2003**, *33*, 1791–1803. [CrossRef]
20. Strong, W.L. Secondary vegetation and floristic succession within a boreal aspen (*Populus tremuloides* Michx.) clearcut. *Can. J. Bot.* **2004**, *82*, 1576–1585. [CrossRef]
21. Macdonald, S.E.; Fenniak, T.E. Understory plant communities of boreal mixedwood forests in western Canada: Natural patterns and responses to variable-retention harvesting. *For. Ecol. Manag.* **2007**, *242*, 34–48. [CrossRef]
22. Errington, R.C.; Pinno, B.D. Early successional plant community dynamics on a reclaimed oil sands mine in comparison with natural boreal forest communities. *Ecoscience* **2016**, in press.
23. Frego, K.A. Regeneration of four boreal bryophytes: colonization of experimental gaps by naturally occurring propagules. *Can. J. Bot.* **1996**, *74*, 1937–1942. [CrossRef]
24. Haeussler, S.; Bergeron, Y. Range of variability in boreal aspen plant communities after wildfire and clear-cutting. *Can. J. For. Res.* **2004**, *34*, 274–288. [CrossRef]

25. Beckingham, J.D.; Archibald, J.H. *Field Guide to Ecosites of Northern Alberta*; Canadian Forest Service, Northwest Region, Northern Forestry Centre: Edmonton, AB, Canada, 1996.

26. Ozoray, G.; Hackbarth, D.; Lytviak, A.T. Earth Sciences Report 78-6. In *Hydrogeology of the Bitumount-Namur Lake Area, Alberta*; Alberta Research Council: Edmonton, AB, Canada, 1980.

27. Key, C.H.; Benson, N.C. Landscape assessment: Ground measure of severity, the composite burn index: And remote sensing of severity, the Normalized Burn Ratio. In *FIREMON: Fire Effects Monitoring and Inventory System*; Lutes, RMRS-GTR-164; Lutes, D.C., Keane, R.E., Caratti, J.F., Key, C.H., Benson, N.C., Sutherland, S., Gangi, L.J., Eds.; USDA Forest Service, Rocky Mountain Research Station: Ogden, UT, USA, 2006; pp. 1–51.

28. Brouillet, L.; Coursol, F.; Meades, S.J.; Favreau, M.; Anions, M.; Bélisle, P.; Desmet, P. VASCAN, the Database of Vascular Plants of Canada. Available online: http://data.canadensys.net/vascan/ (accessed on 4 December 2015).

29. Flora of North America Editorial Committee. *Flora of North America North of Mexico*; Oxford University Press: New York, NY, USA; Oxford, UK, 1993.

30. Esslinger, T.L. *A Cumulative Checklist for the Lichen Forming, Lichenicolous and Allied Fungi of the Continental United States and Canada (Version 20)*; North Dakota State University: Fargo, ND, USA, 2015.

31. McCune, B.; Mefford, M.J. *PC-ORD. Multivariate Analysis of Ecological Data*; Version 6 MjM Software: Gleneden Beach, OR, USA, 2011.

32. R Core Team. *R: A Language and Environment for Statistical Computing*; R Foundation for Statistical Computing: Vienna, Austria, 2013.

33. Dufrêne, M.; Legendre, P. Species assemblages and indicator species: The need for a flexible asymmetrical approach. *Ecol. Monogr.* **1997**, *67*, 345–366. [CrossRef]

34. Mayor, S.J.; Cahill, J.F., Jr.; He, F.; Sólymos, P.; Boutin, S. Regional boreal biodiversity peaks at intermediate human disturbance. *Nature Comm.* **2012**, *3*, 1142. [CrossRef] [PubMed]

35. Grime, J.P. Competitive exclusion in herbaceous communities. *Nature* **1973**, *242*, 344–347.

36. Mackey, R.L.; Currie, D.J. The diversity-disturbance relationship: Is it generally strong and peaked? *Ecology* **2001**, *82*, 3479–3492.

37. Shea, K.; Roxburgh, S.H.; Rauschert, E.S. Moving from pattern to process: coexistence mechanisms under intermediate disturbance regimes. *Ecol. Letters* **2004**, *7*, 491–508. [CrossRef]

38. Webb, E.T. Survival, persistence, and regeneration of the reindeer lichens, *Cladina stellaris*, *C. rangiferina*, and *C. mitis* following clearcut logging and forest fire in northwestern Ontario. *Rangifer* **1998**, *10*, 41–47. [CrossRef]

39. Zouaoui, S.; Boudreault, C.; Drapeau, P.; Bergeron, Y. Influence of time since fire and micro-habitat availability on terricolous lichen communities in black spruce (*Picea mariana*) boreal forests. *Forests* **2014**, *5*, 2793–2809. [CrossRef]

40. Ahti, T.; Oksanen, J. Epigeic lichen communities of taiga and tundra regions. *Vegetatio* **1990**, *86*, 39–70. [CrossRef]

41. Maikawa, E.; Kershaw, K.A. Studies on lichen-dominated systems. XIX. The postfire recovery sequence of black spruce-lichen woodland in the Abitau Lake Region, NWT. *Can. J. Bot.* **1976**, *54*, 2679–2687. [CrossRef]

42. Coxson, D.S.; Marsh, J. Lichen chronosequences (postfire and postharvest) in lodgepole pine (*Pinus contorta*) forests of northern interior British Columbia. *Can. J. Bot.* **2001**, *79*, 1449–1464.

43. Abrams, M.D.; Dickmann, D.I. Early revegetation of clear-cut and burned jack pine sites in northern lower Michigan. *Can. J. Bot.* **1982**, *60*, 946–954. [CrossRef]

44. Pavek, D.S. Maianthemum canadense. In *Fire Effects Information System*; USDA Forest Service, Rocky Mountain Research Station, Fire Sciences Laboratory: Fort Collins, CO, USA, 1993.

45. Gucker, C.L. Cornus canadensis. In *Fire Effects Information System*; USDA Forest Service, Rocky Mountain Research Station, Fire Sciences Laboratory: Fort Collins, CO, USA, 2012.

46. Pavek, D.S. Chamerion angustifolium. In *Fire Effects Information System*; USDA Forest Service, Rocky Mountain Research Station, Fire Sciences Laboratory: Fort Collins, CO, USA, 1992.

47. Sullivan, J. Symphyotrichum leave. In *Fire Effects Information System*; USDA Forest Service, Rocky Mountain Research Station, Fire Sciences Laboratory: Fort Collins, CO, USA, 1992.

© 2016 by the authors. Licensee MDPI, Basel, Switzerland. This article is an open access article distributed under the terms and conditions of the Creative Commons Attribution (CC BY) license (http://creativecommons.org/licenses/by/4.0/).

forests

MDPI

Article

Climate Change Refugia, Fire Ecology and Management

Kate M. Wilkin [1,*], David D. Ackerly [2] and Scott L. Stephens [1]

1   Department of Environmental Science, Policy, and Management, University of California, Berkeley, CA 94720, USA; sstephens@berkeley.edu
2   Department of Integrative Biology, University of California, Berkeley, CA 94720, USA; dackerly@berkeley.edu
*   Correspondence: Kate.Wilkin@berkeley.edu; Tel.: 510-642-4934

Academic Editors: Yves Bergeron and Sylvie Gauthier
Received: 29 January 2016; Accepted: 23 March 2016; Published: 30 March 2016

Abstract: Early climate change ideas warned of widespread species extinctions. As scientists have probed more deeply into species responses, a more nuanced perspective emerged indicating that some species may persist in microrefugia (refugia), including in mountainous terrain. Refugia are habitats that buffer climate changes and allow species to persist in—and to potentially expand under—changing environmental conditions. While climate and species interactions in refugia have been noted as sources of uncertainty, land management practices and disturbances, such as wildland fire, should also be considered when assessing any given refugium. Our landscape scale study suggests that cold-air pools, an important type of small-scale refugia, have unique fire occurrence, frequency, and severity patterns in frequent-fire mixed conifer forests of California's Sierra Nevada: cold-air pool refugia have less fire and if it occurs, it is lower severity. Therefore, individuals and small populations are less likely to be extirpated by fire. Active management, such as restoration and fuels treatments for climate change adaptation, may be required to maintain these distinctive and potentially important refugia.

Keywords: mixed conifer forest; arid forest; fire ecology; fire management; refugia; climate change; vulnerability

1. Introduction

Early models of climate change impacts predicted widespread species extinctions as the rate of climate change outpaced the ability of plants and animals to migrate and track suitable climate [1–3]. Subsequent investigations suggest a more nuanced perspective, indicating that while species extinctions may still be dramatic, some species are likely to persist in microrefugia (referred to hereafter as refugia), including in mountainous terrain [4–10]. The heterogeneous conditions created by complex montane topography create local areas, refugia, where aspects of current climate may be maintained within the region for more than 100 years during climate change [6,11]. As the ice sheets retreated following the last ice age, these refugia are believed to have played an integral role in the rapid expansion of many species by providing source propagules for rapid species migration [8–10,12–14].

Many studies have sought to define refugia based on biological [7] or climatic [5] evidence. Keppel *et al.*'s (2011) biological definition of refugia is "habitats that components of biodiversity retreat to, persist in and can potentially expand from under changing environmental conditions." Dobrowski *et al.*, (2011) describe refugia as locations where extant climates (temperature and available water) are maintained during periods of climate change. Together they form a holistic definition, a habitat that buffers climate and allows species to persist in and to potentially expand from in response to changing environmental conditions. Refugia with relictual species, such as a disjunct southern population persisting from past large-scale populations during a past ice age, may or may not continue to function as refugia with climate change and changing disturbance regimes. Here, we focus on the importance of fire regimes as a component of potential refugia during periods of climate change. For fire sensitive species, landscape locations with reduced fire frequency or severity will serve as refugia with distinct disturbance regimes. Here we address the question of how some refugia in montane ecosystems may buffer both climate change and fire disturbance, with the potential to maintain species through episodes of climate change.

Refugia have attracted attention as important conservation areas [4,15], however the full range of conservation threats in these areas, including wildfire, have not been fully explored. Keppel and Wardell-Johnson (2012) highlight how refugia play a potentially important role in climate buffering and may offer *in situ* conservation benefits in the face of climate change and its biological effects [4]. We expand 'climate change and its biological effects' to explicitly include ecological processes such as fire. In addition, we consider the importance of land management practices because of how policies, such as fire suppression, have drastically changed fire frequency and severity patterns in many forests [16]. Climate, species interactions, ecological processes, and land management combine to create conservation challenges and opportunities for individual refugia (Figure 1).

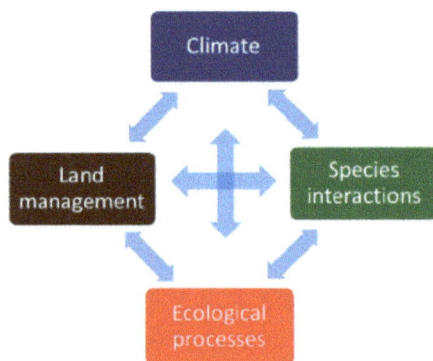

Figure 1. Refugia ecology is complex and affected by climate, land management, species interactions, ecological processes, and their interactions.

Fire is one example of a changing ecological process. Fire frequency and severity are changing world-wide due to land management [16–20] and climate change [19,21–23]. Refugia populations are at a greater risk from changing fire patterns [21] due to their predisposition to local extinction because of their small, isolated nature, especially if the plants are fire avoiders [24]. Fire could alter species occurrence directly by killing vegetation or indirectly by altering vegetation's moderating effect on climate [25,26] and the selection pressures on vegetation [27]. In more extreme cases fire can locally extirpate plants if fires are more frequent [28], less frequent [29], or more severe than historical fire regimes [30]. Some define refugia as areas with complete absence of fire [31]; however in arid regions with frequent fire, a reduction in frequency or severity may also create distinct conditions that serve as biological refugia for some taxa.

In this paper we focus on cold-air pool refugia (CAPs), a particular example of climatic refugia occurring in mountainous regions including our study region in the Sierra Nevada of California [5]. CAPs have lower temperatures and more frequent minimum temperatures, below any chosen threshold. And maximum temperatures are also lower, with fewer days exceeding certain thresholds. CAPS often have greater moisture availability than their surrounding landscape, due to reduced evaporative demand and water accumulation into low-lying areas [32,33]. While overall temperatures are warming, the weather patterns that produce cold-air pooling at a landscape scale are projected to increase, so the frequency and duration of cold-pools may increase in the Sierra Nevada with climate change [34,35]. Alternatively, if the weather patterns that produce cold-air pooling decrease, CAPs may warm at a greater rate than the landscape, yet still have lower average temperatures than the landscape. Overall CAPs have and will likely continue to have cooler and moister climates that their surrounding landscape although the magnitude of these differences is uncertain.

Plants respond to environmental conditions within CAPs and these sites often have plants characteristically found at higher elevations or latitudes. Similar climatic refugia forests are known as frost hollows in Quebec's boreal temperate forests [36] and cove forests in southern Appalachia especially in Great Smoky Mountain National Park [37]. Another well-known example is subalpine fir (*Abies lasiocarpa*), which descend more than 300 m into river valleys with cold-air pools in Idaho [38,39]. In a more visually extreme case, the subalpine tree line is inverted due to cold-air pooling where mixed conifer forests grow above subalpine treeless regions on Mt. Hotham in Australia [40]. In the White Mountains of California, bristlecone pine are expanding downhill into local cold-air pools as temperatures warm [41]. Lesser known examples in the CAPs of Yosemite National Park in California include the Merced Grove, with disjunct distributions of species more characteristic of the Pacific Northwest, such as mountain lady's slipper (*Cypripedium montanum*) and Old-man's-beard Lichen (*Alectoria sarmentosa*), growing together with Sierran species. These areas are dominated by ponderosa pine (*Pinus ponderosa*) intermingled with sugar pine (*Pinus lambertiana*), white fir (*Abies concolor*), and mountain hemlock (*Tsuga mertensiana*) with western hazelnut (*Corylus cornuta*) and mountain dogwood (*Cornus nuttallii*) in the understory [42] (Figure 2). As the climate changes and species ranges shift, we may see different species become restricted to CAPs, and the fate of the examples noted here are unknown.

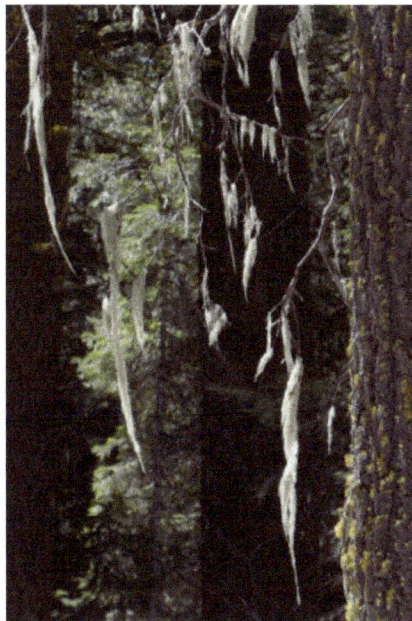

Figure 2. Refugia where species retreated to after the last ice age, such as regions of Yosemite National Park's Merced Giant Sequoia Grove, have cooler climates with disjunct species characteristic of the Pacific Northwest such as *Alectoria sarmentosa* (Witch's hair lichen) pictured here. Yosemite National Park has used prescribed fires to protect areas from large, nearly unstoppable fires such as the catastrophic Rim Fire in 2013. Photo by Martin Hutten.

Fire refugia and climate change refugia are similar, but they are also distinct. These refugia share some topographical similarities, and are both often associated with cooler, wetter places on the landscape [5,43,44]. Fire refugia are areas where fire has been excluded [45] whereas climate change refugia are areas that buffer climate change and the fire regime may be dramatically or subtlety different than the surrounding landscape [5]. Both likely have distinct fire regimes from their surrounding areas, although the magnitude may be different.

CAPs have unique climates and plants and they may serve as *in situ* conservation opportunities in the face of climate change. However, little is known about CAPs' fire ecology and risk. Therefore, we investigate the fire ecology of CAPs. Specifically we ask:

1. Do cold-air pools have similar fire occurrence as their surrounding landscape?
2. Do cold-air pools have similar fire frequency as their surrounding landscape?
3. If fires occur, do cold-air pools have similar fire severity patterns as their surrounding landscape?

2. Materials and Methods

2.1. Study Site

The study was conducted in mixed conifer forests in central California's Sierra Nevada Range within Yosemite National Park (37.8499° N, 119.5677° W), approximately 240 km inland from the Pacific Ocean. Historical fires were frequent and burnt every 6 to 15 years [46,47]. Fire suppression occurred from the late 1800s to the mid-1900s. Beginning in the 1970s, fire patterns were allowed to return to historical patterns, as lightning strike fires were allowed to burn freely through nearly one-third of the study area and prescribed fires occurred over a large area as well [23,46,47]. A large fire deficit persists for more than two-thirds of our study area. Vegetation is mixed conifer forest co-dominated by fire resilient species including ponderosa pine (*Pinus ponderosa*), Jeffrey pine (*Pinus jeffreyi*), white fir (*Abies concolor*), and red fir (*Abies magnifica*), and scattered meadows and shrublands. The study region is from 1000 to 2600 m elevation and has hot, dry summers followed by cold, snowy winters.

2.2. Study Design

Spatial data was prepared in ArcGIS 10.2 (ESRI, Redlands CA, USA) and projected using bilinear interpolation into NAD 83 UTM 11N. Relevant spatial data was compiled from private and public sources, including predicted location of CAPs (based on methodology of [32,48]), fire history polygons from 1930 until 2012 [49], and Relative differenced Normalized Burn Ratio (RdNBR) fire severity categories based on changes in tree cover from 1984 to 2012 [50,51] (Figure 3). Predicted CAP values included absent (no cold-air pooling occurs), marginal (areas with no clear signal for cold-air pooling due to topography and weather patterns and present (areas with potential for cold-air pooling to occur). The predicted CAPs are areas with potential to pool cold-air and there is variation in the frequency and duration of cold-air pooling. CAPs include areas with short and infrequent cold-air pooling that have similar climate and fire history to the surrounding area that are false-positives for climate change refugia. Conversely, areas predicted to lack CAPs may have CAPs present and thus represent false-negatives from the modeling procedure. Fire severity classes correlate to the change in canopy cover and tree basal area whereby very low and low severity fires have between 0 and 25% change, moderate severity fires have between 26 and 75% change, and high severity fires have more than 76% change in canopy cover and tree basal area [52].

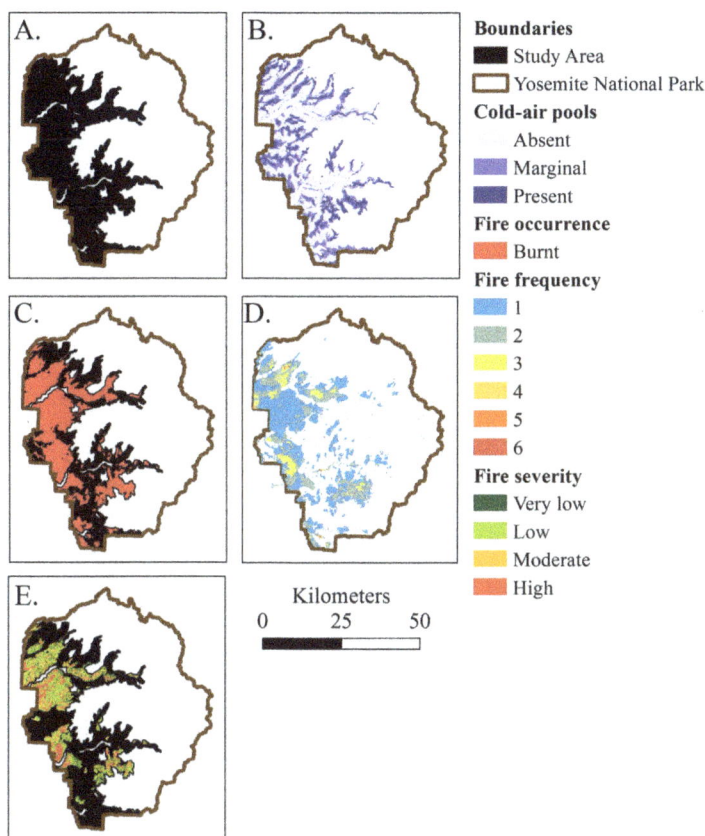

Figure 3. (**A**) The study area included mixed conifer forests of Yosemite National Park from 1000 to 3600 m in elevation, which encompassed about 170,000 ha; (**B**) Predicted cold-air pools (CAPs) followed drainages and had a semi-regular pattern throughout the study area. The study area was dominated by areas without CAPs (100,000 ha) followed by areas with marginal CAPs (40,000 ha), and lastly CAPs (30,000 ha); (**C**) Fires have occurred through about 60% study area from 1930 to 2012; (**D**) Areas within the study area have burnt up to 6 times; (**E**) Fire severity distribution from 1984 to 2012 was dominated by low severity followed by moderate severity fires. Fire severity area included: very low severity 16,000 ha, low severity 44,000 ha, moderate severity 28,000 ha, and high severity 12,000 ha.

Spatial data was clipped with a USGS 10 meter Digital Elevation Model to restrict it to the mixed conifer zone [53]. Spatial autocorrelation of cold-air pools, fire frequency, and fire severity were tested with ArcGIS's Spatial Autocorrelation Global Moran's I test from 100–300 m and statistical significance was assessed with the z-score. The data were not autocorrelated at 100 m, but they were autocorrelated for other distances tested (Figures A1–3). Therefore, a 100 m point grid was created in ArcGIS with the Create Fishnet Geoprocessing Tool. Equal sample sizes from all levels of CAPs (absent, marginal, present) were randomly subsampled without replacement for both analyses, including fire occurrence, and if fire occurs, its severity (using the sample function in R 3.1.2 [54]).

2.3. Analysis

All analyses were conducted in R 3.1.2 [54]. We constructed statistical models to test if cold-air pools are related to fire Equation (1).

$$\gamma \sim \text{Cold air pool} \tag{1}$$

where γ is the response variable, either fire occurrence (absent: 0, present: 1), fire frequency (0 to 6), or fire severity (very low, low, moderate, and high). Cold-air pool categories include absent, marginal, and present.

We used linear models to test the significance of fire occurrence and frequency. We used binomial distribution for fire occurrence and a Gaussian distribution for fire severity, based on the four ordered fire severity classes. All model dispersion and residuals were reviewed. We used chi-squared test to test the significance of fire severity.

3. Results

CAPs were significantly less likely to have experienced a fire than their surrounding landscape (Figure 4A) ($P < 0.0001$). Areas without CAPs burnt 60% of the time on average,, marginal CAPs burnt 52% of the time on average, and CAPs burnt 44% of the time on average. CAPs also had significantly fewer fires than their surrounding landscape as well (Figure 4B) ($P < 0.0001$). Areas without CAPS had a fire frequency of 0.95 on average, areas with marginal CAPs had a fire frequency of 0.78 on average, and areas with CAPs had a fire frequency of 0.60 on average.

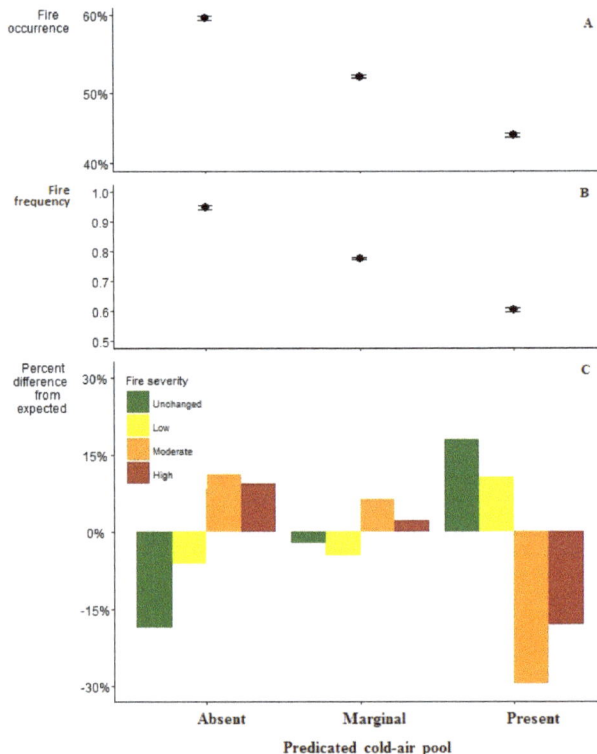

Figure 4. (**A**) Regression estimates for area burnt grouped by their predicted CAP level where dots represent the regression coefficients and whiskers are the 95% confidence intervals; (**B**) Regression estimates for number of times burnt grouped by their predicted CAP level where dots represent the regression coefficients and whiskers are the 95% confidence intervals; (**C**) Chi-square estimates for the percent differences in the actual fire severity from expected by their predicted CAP level.

Additionally, if CAPs burned, they had significantly less severe fires than their surrounding landscapes ($P < 0.0001$). Areas with CAPs had 16% more very low severity fire, 10% more low severity fire, 29% less moderate severity fire, and 16% less high severity fire than expected. Areas with marginal CAPs had 3% less very low severity fire, 4% less low severity fire, 7% more moderate severity fire, and 3% more high severity fire than expected. Areas without CAPs had 16% less very low severity fire, 6% less low severity fire, 11% more moderate severity fire, and 9% more high severity fire.

4. Discussion

CAP refugia had significantly different fire patterns from the surrounding landscape likely due to a combination of their vegetation, topography, or microclimate. Surprisingly, these trends persisted despite our landscape-scale study using predicted CAPs which may have reduced the strength of our results. The fire regime differences between CAP refugia and their surrounding landscape are likely a conservative estimate because prediction of a CAP does not include its frequency, duration or if biological species respond to its effects. The actual magnitude of the effects might be greater if we had directly field mapped CAPs.

Refugia in rugged terrain separated by fire barriers such as rock or water are more likely to have decreased fire occurrence than their surrounding terrain through these bottom-up controls [55]. Other refugia may lack a physical barrier to fire, but their distinct microclimates can influence fire behavior [44,45]. CAPs may have direct effects on fire behavior (temperature, wind, and fuel moisture) as well as indirect effects mediated through vegetation and fuel characteristics (amount and size distribution of fuel, fuel continuity, fuel moisture, forest structure, and relative humidity) [56]. These effects are realized as reductions in fire energy (commonly called intensity) and changes in dominant vegetation (commonly called severity), and in some cases even reducing fire extent if the fire self-extinguishes (Figure 5).

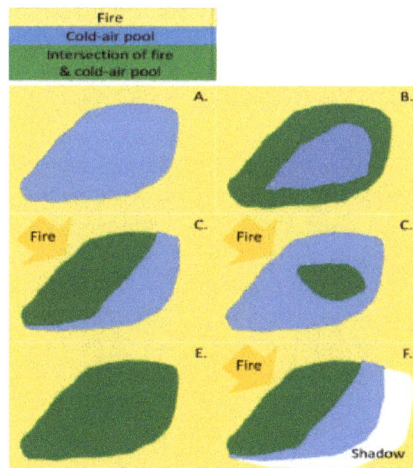

Figure 5. The interaction of fire and CAPs may be dependent upon fire behavior including the fire's direction, magnitude, and intensity. The yellow background indicates where fire occurred and the golden arrows indicate the fire's direction of movement. (**A**) Fires which move slowly (low magnitude) and release little energy (low intensity) may respond quickly to a refugium's microenvironment and not penetrate the CAP, whereas (**B**) fires with high magnitude and intensity may respond slowly to a CAP (burn a buffer around the perimeter); (**C**) and/or a larger region near the flame front; (**D**) create a spot fire ignites within the CAP which grows especially during dry windy conditions; or (**E**) can burn the entire CAP. (**F**) There also may be a CAP fire shadow where a reduction in fire extent or severity occurs because of its proximity to a CAP and the subsequent reduction in fire presence or severity.

The interaction between CAPs and fire may be influenced by diurnal patterns because drivers of fire behavior vary diurnally including temperature, wind, and humidity. CAPs may be cooler in the evening and morning but reach similar maximum daytime temperatures as surrounding areas, but fire effects may be lessened by higher humidity and residual fuel moisture [32,56,57]. Overall, our analysis demonstrates that the unique climate and topographic placement of CAPs reduce fire frequency and intensity and thus, quite possibly, the impact of fire on these ecosystems.

The interaction between CAPs and fire may be moderated by effects of season and weather on fire behavior. Historically fuel moisture would have been lower in fall due to seasonal drought and thus there may have been an increase in fire occurrence and severity during this time. Climate change may magnify this trend because extreme fire weather (hot, dry and windy weather) is becoming more common, which allows fires to grow rapidly and reach unprecedented fire intensity and size [58]. As fire behavior becomes more extreme, bottom-up controls such as microclimate, topography, and fire barriers are weaker [55]. If fire becomes driven by extreme events such as drought, high winds, or in extreme cases where the combination of weather, topography and fuels catalyze fire-generated weather phenomena such as plume-dominated fires, then historic fire barriers of refugia may no longer function [59]. Large, severe fire was evident in the 2013 Rim Fire, which burned over 100,000 ha, impacting a noticeable portion of our study area [60]. Refugia that historically have had physical and climatic barriers to fire may be more susceptible to high-severity mega-fires becaue they produce effects well outside of desired ranges [20].

Refugia are commonly embedded within riparian areas in our study region. Other studies focused on riparian forest fire ecology suggest that these areas did burn historically, that these areas have demonstrated the greatest rate of fuel accumulation during fire suppression, and are at greatest risk for uncharacteristic high severity fire now due to fuel accumulations [61,62]. While historic fires only influenced aquatic communities in the short-term if at all [63], uncharacteristic large, high severity fires may have a larger influence. Uncharacteristic high severity fires reduce canopy cover, change the peak water temperatures in creeks, and may sometimes negatively influence sensitive aquatic species [64,65].

Implications for Management

More than 100 years of fire suppression in the Sierra Nevada have fostered dense tree stands with ladder fuels including small trees reaching from the forest floor to its' canopy. Today's mixed conifer forests are more susceptible to high severity fire than ever before [66]. There is a consensus that we need to reduce tree density and fuels to make forests resilient to climate-change induced disturbances [67–69]. While prescribed fire and mechanical fuel reduction treatments may have short-term effects on the inhabitants of refugia, the long-term lack of fire exacerbates climate change's increasing disturbance threats to biodiversity, including increased fire frequency and severity [22,70].

Refugia, especially in arid regions like Yosemite National Park have greater moisture, fuel production, and historically lower fire occurrence than the surrounding landscape [5,32,62]. Refugia fire occurrence will likely increase relative to historical levels due to the combination of fire suppression and climate change induced extreme fires or mega-fires [20,71]. Even as fire risk/frequency increases, refugia are likely to maintain distinct fire regimes relative to surrounding areas. Refugia commonly occur in riparian areas, which were heavily altered by fire suppression resulting in extraordinary amounts of fuel (more than five times greater than historic levels), leaving them uncharacteristically susceptible to high severity fire that might be quite detrimental to biodiversity [62]. As climate becomes drier, these fuels have reduced moisture and are therefore available to burn for a larger portion of the year [72]. These additional fuels can contribute to more severe fires.

High-severity fire risk could be mitigated with prescribed fire, mechanical treatments, or managed wildfire within the CAP and at the landscape level [73]. Fuel reduction could have unintended consequences, since trees moderate climate. Trees filter solar radiation, providing a buffer between CAPs and the atmosphere; removing trees can increase maximum surface temperatures 10 to

40 °C [26]. Oddly, fire suppression which increases tree density may have also made CAPs climates cooler. To reduce the potential to increase temperatures dominant trees should be preserved in fuel treatments [74]. Nonetheless, managers must understand the trade-offs for refugia management and thresholds between reducing fire risk and altering the very environment that they wish to protect.

Stand-replacing fires surrounding CAPs are also of concern because large-scale high severity fires can kill many trees within the CAP air shed and in doing so may also change the local climate. Killing trees that moderate the temperature and soil moisture in the greater air shed may reduce the cold-air source and thus the frequency and duration of cold-air pooling. Trees likely have strong biological feedback on climate in gentle sloping environments because there are not strong topographical drivers of cold-air movement. Steeper slopes have faster cold-air run-off and trees likely play a smaller role here. Fuel and restoration treatments could be completed in areas surrounding CAPs to protect cold-air sources from severe fire. As before, caution is needed when applying fuel treatments because trees moderate temperature and minimizing these impacts will be a higher priority with climate change.

5. Conclusions

Conservation planners are advised that their "highest priority (is) to reduce negative edge effects and improve *in situ* management of existing habitat patches" [75]. Refugia do exactly this, allowing *in situ* management of habitat patches [75]. Refugia are complex habitats influenced by species interactions, climate, and fire that interact with one another. Therefore, protecting the land associated with refugia is not sufficient to protect the biological and physical properties of refugia; *additional management actions are necessary*. Many of these actions, like strategic fuel treatments or managed wildfire, are already recommended for both forest restoration and climate change adaptation [70,73]. Refugia are also susceptible to disturbances, therefore redundancy on the landscape is necessary to utilize this conservation strategy. Actions to manage refugia are similar to manager's current tool kits for conservation, but the need for these actions in refugia may be an additional incentive to complete them. Managers will be asked to make decisions about refugia without understanding their full ecological complexity and they must understand that refugia are not static.

Acknowledgments: The George Melendez Wright Youth Climate Change Initiative supported Wilkin for this project that catalyzed National Park Service collaborations that enriched the final product. Special thanks to Yosemite National Park employees (Alison Colwell, Martin Hutten, Kent Van Wagtendonk, Mitzi Thornley, and Linda Mazzu) and Devils Postpile National Monument employees (Monica Buhler and Deanna Dulen) for their support with project development, grant and manuscript reviews, and their insights into managing climate change refugia. This paper would not have been possible without the Climate Refugia Workshop in Eugene, Oregon in August of 2012 sponsored by the Ecological Society of America. K. W. is grateful to those who coordinated this meeting (Dan Gavin and Erin Herring) and the participants who inspired analysis of refugia management for the future with climate change, especially Zack Holden and Arndt Hampe. Special thanks to Jessica Lundquist for fostering a sound understanding of cold-air pools and to Axel Kuhn for producing and sharing a Sierra Nevada data layer based on Lundquist (2008).

Author Contributions: Kate Wilkin, David Ackerly and Scott Stephens conceived and designed the analysis; Kate Wilkin performed the experiments; Kate Wilkin analyzed the data; Kate Wilkin wrote the paper.

Conflicts of Interest: The authors declare no conflict of interest. The founding sponsors had no role in the design of the study; in the collection, analyses, or interpretation of data; in the writing of the manuscript, and in the decision to publish the results.

Abbreviations

The following abbreviations are used in this manuscript:

MDPI:	Multidisciplinary Digital Publishing Institute
DOAJ:	Directory of open access journals
CAP:	Cold-air pool
RdNBR:	Relative differenced Normalized Burn Ratio
Refugia:	microrefugia

Appendix A. Spatial autocorrelation

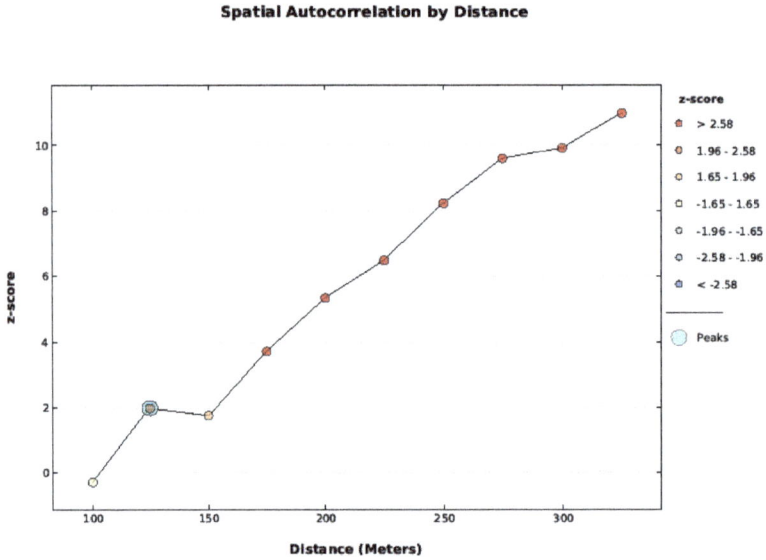

Figure A1. Cold-air pool data was tested with the Global Moran's I applied to Incremental Spatial Autocorrelation starting at 100 m and increasing at 50 m intervals. The data was not spatially auto correlated at 100 m, but became correlated by 125 m and beyond.

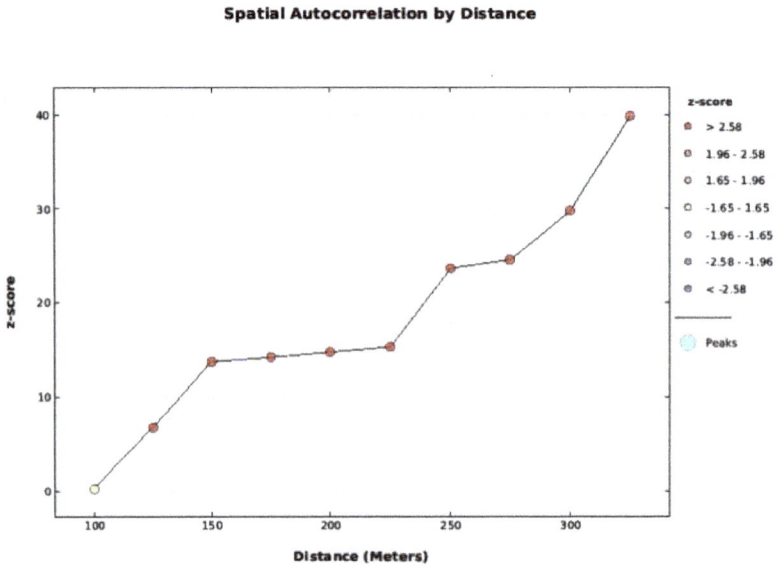

Figure A2. Fire frequency data was tested with the Global Moran's I applied to Incremental Spatial Autocorrelation starting at 100 m and increasing at 50 m intervals. The data was not spatially auto correlated at 100 m, but become increasingly correlated as distance increased.

Spatial Autocorrelation by Distance

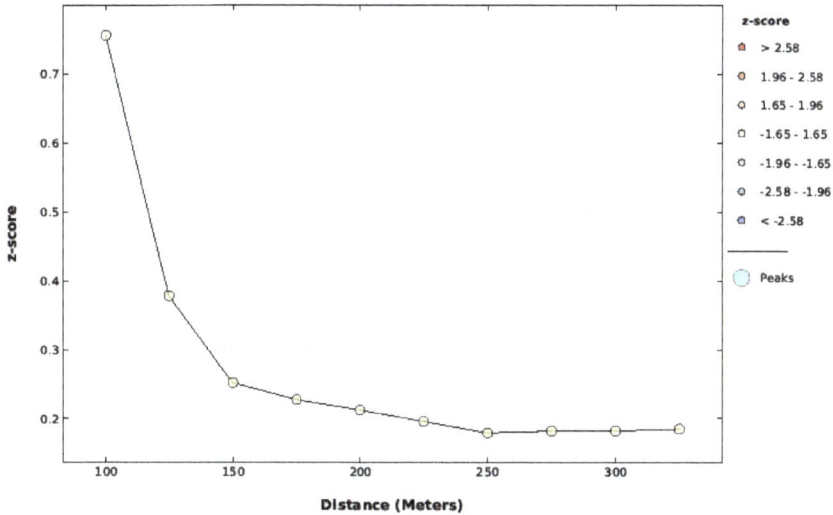

Figure A3. Fire severity data was tested with the Global Moran's I applied to Incremental Spatial Autocorrelation starting at 100 m and increasing at 50 m intervals. The data was not spatially auto correlated between 100 and 300 m.

References

1. Walther, G.R.; Post, E.; Convey, P.; Menzel, A.; Parmesank, C.; Beebee, T.J.C.; Fromentin, J.; Hoegh-GuldbergI, O.; Bairlein, F. Ecological responses to recent climate change. *Nature* **2002**, *416*, 389–395. [CrossRef] [PubMed]
2. Root, T.L.; Price, J.T.; Hall, K.R.; Schneider, S.H.; Rosenzweig, C.; Pounds, J.A. Fingerprints of global warming on wild animals and plants. *Nature* **2003**, *421*, 57–60. [CrossRef] [PubMed]
3. Thomas, C.D.; Cameron, A.; Green, R.E.; Bakkenes, M.; Beaumont, L.J.; Collingham, Y.C.; Erasmus, B.F.; De Siqueira, M.F.; Grainger, A.; Hannah, L. Extinction risk from climate change. *Nature* **2004**, *427*, 145–148. [CrossRef] [PubMed]
4. Keppel, G.; Wardell-johnson, G.W. Refugia: Keys to climate change management. *Glob. Chan. Biol.* **2012**, 2389–2391. [CrossRef]
5. Dobrowski, S.Z. A climatic basis for microrefugia: The influence of terrain on climate. *Glob. Chan. Biol.* **2011**, *17*, 1022–1035. [CrossRef]
6. Loarie, S.R.; Duffy, P.B.; Hamilton, H.; Asner, G.P.; Field, C.B.; Ackerly, D.D. The velocity of climate change. *Nature* **2009**, *462*, 1052–1055. [CrossRef] [PubMed]
7. Keppel, G.; van Niel, K.P.; Wardell-Johnson, G.W.; Yates, C.J.; Byrne, M.; Mucina, L.; Schut, A.G.T.; Hopper, S.D.; Franklin, S.E. Refugia: Identifying and understanding safe havens for biodiversity under climate change. *Glob. Ecol. Biogeogr.* **2011**, 393–404. [CrossRef]
8. Taberlet, P.; Cheddadi, R. Quaternary refugia and persistence of biodiversity. *Science* **2002**, *297*, 2009–2010. [CrossRef] [PubMed]
9. Petit, R.J.; Aguinagalde, I.; de Beaulieu, J.-L.; Bittkau, C.; Brewer, S.; Cheddadi, R.; Ennos, R.; Fineschi, S.; Grivet, D.; Lascoux, M.; *et al.* Glacial refugia: Hotspots but not melting pots of genetic diversity. *Science* **2003**, *300*, 1563–1565. [CrossRef] [PubMed]

10. Petit, R.J.; Csaikl, U.M.; Bordács, S.; Burg, K.; Coart, E.; Cottrell, J.; van Dam, B.; Deans, J.D.; Dumolin-Lapègue, S.; Fineschi, S.; *et al.* Chloroplast DNA variation in european white oaks: Phylogeography and patterns of diversity based on data from over 2600 populations. *For. Ecol. Manag.* **2002**, *156*, 5–26. [CrossRef]

11. Dobrowski, S.Z.; Abatzoglou, J.; Swanson, A.K.; Greenberg, J.A.; Mynsberge, A.R.; Holden, Z.A.; Schwartz, M.K. The climate velocity of the contiguous united states during the 20th century. *Glob. Chan. Biol.* **2012**, 241–251. [CrossRef] [PubMed]

12. Taberlet, P.; Fumagalli, L.; Wust-Saucy, A.-G.; Cosson, J.-F. Comparative phylogeography and postglacial colonization routes in europe. *Mol. Ecol.* **1998**, *7*, 453–464. [CrossRef] [PubMed]

13. Willis, K.J.; Whittaker, R.J. The refugial debate. *Science* **2000**, *287*, 1406–1407. [CrossRef] [PubMed]

14. Pearson, R.G. Climate change and the migration capacity of species. *Trends Ecol. Evol.* **2006**, *21*, 111–113. [CrossRef] [PubMed]

15. Groves, C.; Game, E.; Anderson, M.; Cross, M.; Enquist, C.; Ferdaña, Z.; Girvetz, E.; Gondor, A.; Hall, K.; Higgins, J.; *et al.* Incorporating climate change into systematic conservation planning. *Biodivers. Conserv.* **2012**, *21*, 1651–1671. [CrossRef]

16. Miller, J.D.; Collins, B.M.; Lutz, J.A.; Stephens, S.L.; van Wagtendonk, J.W.; Yasuda, D.A. Differences in wildfires among ecoregions and land management agencies in the Sierra Nevada region, California, USA. *Ecosphere* **2012**, *3*, 1–20. [CrossRef]

17. Anderson, K. *Tending the wild: Native American Knowledge and the Management of California's Natural Resources*; University of California Press: Berkeley, CA, USA, 2005; p. 558.

18. Falk, D.A.; Heyerdahl, E.K.; Brown, P.M.; Farris, C.; Fulé, P.Z.; McKenzie, D.; Swetnam, T.W.; Taylor, A.H.; van Horne, M.L. Multi-scale controls of historical forest-fire regimes: New insights from fire-scar networks. *Front. Ecol. Environ.* **2011**, *9*, 446–454. [CrossRef]

19. Trouet, V.; Taylor, A.H.; Wahl, E.R.; Skinner, C.N.; Stephens, S.L. Fire-climate interactions in the american west since 1400 ce. *Geophys. Res. Lett.* **2010**, *37*, L04702. [CrossRef]

20. Stephens, S.L.; Burrows, N.; Buyantuyev, A.; Gray, R.W.; Keane, R.E.; Kubian, R.; Liu, S.; Seijo, F.; Shu, L.; Tolhurst, K.G. Temperate and boreal forest mega-fires: Characteristics and challenges. *Front. Ecol. Environ.* **2014**, *12*, 115–122. [CrossRef]

21. Moritz, M.A.; Parisien, M.-A.; Batllori, E.; Krawchuk, M.A.; van Dorn, J.; Ganz, D.J.; Hayhoe, K. Climate change and disruptions to global fire activity. *Ecosphere* **2012**, *3*, 1–22. [CrossRef]

22. Westerling, A.; Bryant, B. Climate change and wildfire in california. *Clim. Chan.* **2008**, *87*, 231–249. [CrossRef]

23. Swetnam, T.W. Fire history and climate change in giant sequoia groves. *Science* **1993**, *262*, 885–889. [CrossRef] [PubMed]

24. Rowe, J.S. Concepts of fire effects on plant individuals and species. *The Role of Fire in Northern Circumpolar Ecosystems* **1983**, *18*.

25. Rambo, T.R.; North, M.P. Canopy microclimate response to pattern and density of thinning in a sierra nevada forest. *For. Ecol. Manag.* **2009**, *257*, 435–442. [CrossRef]

26. Ford, K.R.; Ettinger, A.K.; Lundquist, J.D.; Raleigh, M.S.; Lambers, J.H.R. Spatial Heterogeneity in Ecologically Important Climate Variables at Coarse and Fine Scales in a High-Snow Mountain Landscape. *PLoS ONE* **2013**, *8*, e65008. [CrossRef] [PubMed]

27. Bond, W.J.; Keeley, J.E. Fire as a global 'herbivore': The ecology and evolution of flammable ecosystems. *Trends Ecol. Evol.* **2005**, *20*, 387–394. [CrossRef] [PubMed]

28. Keeley, J.E.; Ne'eman, G.; Fotheringham, C. Immaturity risk in a fire-dependent pine. *J. Mediterr. Ecol.* **1999**, *1*, 41–48.

29. Menges, E.S.; Hawkes, C.V. Interactive effects of fire and microhabitat on plants of florida scrub. *Ecol. Appl.* **1998**, *8*, 935–946. [CrossRef]

30. Collins, B.M.; Roller, G.B. Early forest dynamics in stand-replacing fire patches in the northern sierra nevada, california, USA. *Landsc. Ecol.* **2013**, *28*, 1801–1813. [CrossRef]

31. Nordén, B.; Dahlberg, A.; Brandrud, T.E.; Fritz, Ö.; Ejrnaes, R.; Ovaskainen, O. Effects of ecological continuity on species richness and composition in forests and woodlands: A review. *Ecoscience* **2014**, *21*, 34–45. [CrossRef]

32. Lundquist, J.D.; Pepin, N.; Rochford, C. Automated algorithm for mapping regions of cold-air pooling in complex terrain. *J. Geophys. Res.* **2008**, *113*, D22. [CrossRef]

33. Dobrowski, S.Z.; Abatzoglou, J.T.; Greenberg, J.A.; Schladow, S. How much influence does landscape-scale physiography have on air temperature in a mountain environment? *Agric. For. Meteorol.* **2009**, *149*, 1751–1758. [CrossRef]

34. Daly, C.; Conklin, D.R.; Unsworth, M.H. Local atmospheric decoupling in complex topography alters climate change impacts. *Int. J. Clim.* **2010**, *30*, 1857–1864. [CrossRef]

35. Pepin, N.C.; Daly, C.; Lundquist, J. The influence of surface versus free-air decoupling on temperature trend patterns in the western United States. *J. Geophys. Res.* **2011**, *116*, 16. [CrossRef]

36. Dy, G.; Payette, S. Frost hollows of the boreal forest as extreme environments for black spruce tree growth. *Can. J. For. Res.* **2007**, *37*, 492–504. [CrossRef]

37. Shanks, R.E. Climates of the great smoky mountains. *Ecology* **1954**, *35*, 354–361. [CrossRef]

38. Axelrod, D.I. *The Eocene Thunder Mountain flora of Central Idaho*; University of California Press: Berkeley, CA, USA, 1998; Vol. 142.

39. Daubenmire, R. Mountain topography and vegetation patterns. *Northwest Sci.* **1980**, *54*, 146–152.

40. Wearne, L.J.; Morgan, J.W. Floristic composition and variability of subalpine grasslands in the mt hotham region, north-eastern victoria. *Aust. J. Bot.* **2001**, *49*, 721–734. [CrossRef]

41. Millar, C.I.; Westfall, R.D.; Delany, D.L.; Flint, A.L.; Flint, L.E. Recruitment patterns and growth of high-elevation pines in response to climatic variability (1883–2013), in the western great basin, USA. *Can. J. For. Res.* **2015**, *45*, 1299–1312. [CrossRef]

42. Colwell, A. *Yosemite National Park Special Status Plants (Dataset)*; Yosemite National Park: Berkeley, CA, USA, 2012.

43. Eberhart, K.E.; Woodard, P.M. Distribution of residual vegetation associated with large fires in alberta. *Can. J. For. Res.* **1987**, *17*, 1207–1212. [CrossRef]

44. Camp, A.; Oliver, C.; Hessburg, P.; Everett, R. Predicting late-successional fire refugia pre-dating european settlement in the wenatchee mountains. *For. Ecol. Manag.* **1997**, *95*, 63–77. [CrossRef]

45. Ouarmim, S.; Asselin, H.; Hély, C.; Bergeron, Y.; Ali, A.A. Long-term dynamics of fire refuges in boreal mixedwood forests. *J. Quat. Sci.* **2014**, *29*, 123–129. [CrossRef]

46. Scholl, A.E.; Taylor, A.H. Fire regimes, forest change, and self-organization in an old-growth mixed-conifer forest, yosemite national park, USA. *Ecol. Appl.* **2010**, *20*, 362–380. [CrossRef] [PubMed]

47. Collins, B.M.; Stephens, S.L. Managing natural wildfires in sierra nevada wilderness areas. *Front. Ecol. Environ.* **2007**, *5*, 523–527. [CrossRef]

48. Lundquist, J.D.; Cayan, D.R. Surface temperature patterns in complex terrain: Daily variations and long-term change in the central Sierra Nevada, California. *J. Geophys. Res.* **2007**, *112*, D11124. [CrossRef]

49. Yosemite National Park. *Yosemite National Park Fire History Polygons from 1930 to 2011 (Dataset)*; Yosemite National Park: Berkeley, CA, USA, 2012.

50. Miller, J.D. *Yosemite National Park Wildfire Fire Severity from 1984 to 2010*; USDA: McClellan, CA, USA, 2012.

51. Miller, J.D.; Thode, A.E. Quantifying burn severity in a heterogeneous landscape with a relative version of the delta normalized burn ratio (dnbr). *Remote. Sens. Environ.* **2007**, *109*, 66–80. [CrossRef]

52. Miller, J.D.; Knapp, E.E.; Key, C.H.; Skinner, C.N.; Isbell, C.J.; Creasy, R.M.; Sherlock, J.W. Calibration and validation of the relative differenced normalized burn ratio (rdnbr) to three measures of fire severity in the sierra nevada and klamath mountains, California, USA. *Remote. Sens. Environ.* **2009**, *13*, 645–656. [CrossRef]

53. Gesch, D.B. *The National Elevation Dataset. American Society for Photogrammetry and Remote Sensing*; American Society for Photogrammetry and Remote Sensing: Bethesda, MD, USA, 2007.

54. R Development Core Team. *R: A Language and Environment for Statistical Computing*; R Foundation for Statistical Computing: Vienna, Austria, 2008.

55. Heyerdahl, E.K.; Brubaker, L.B.; Agee, J.K. Spatial controls of historical fire regimes: A multiscale example from the interior west, USA. *Ecology* **2001**, *82*, 660–678.

56. Brown, J.K. *Handbook for inventorying downed woody material*; USDA, Ed.; Intermountain Forest and Range Experiment Station: Ogden, UT, USA, 1974.

57. Fosberg, M.A.; Rothermel, R.C.; Andrews, P.L. Moisture content calculations for 1000-hour timelag fuels. *For. Sci.* **1981**, *27*, 19–26.

58. Collins, B.M. Fire weather and large fire potential in the northern sierra nevada. *Agric. For. Meteorol.* **2014**, *189–190*, 30–35. [CrossRef]

59. Turner, M.G.; Romme, W.H.; Gardner, R.H. Prefire heterogeneity, fire severity, and early postfire plant reestablishment in subalpine forests of yellowstone national park, wyoming. *Int. J. Wildland Fire* **1999**, *9*, 21–36. [CrossRef]
60. Lydersen, J.M.; North, M.P.; Collins, B.M. Severity of an uncharacteristically large wildfire, the rim fire, in forests with relatively restored frequent fire regimes. *For. Ecol. Manag.* **2014**, *328*, 326–334. [CrossRef]
61. Van de Water, K.; North, M. Fire history of coniferous riparian forests in the sierra nevada. *For. Ecol. Manag.* **2010**, *260*, 384–395. [CrossRef]
62. Van de Water, K.; North, M. Stand structure, fuel loads, and fire behavior in riparian and upland forests, sierra nevada mountains, USA; a comparison of current and reconstructed conditions. *For. Ecol. Manag.* **2011**, *262*, 215–228. [CrossRef]
63. Beche, L.A.; Stephens, S.L.; Resh, V.H. Effects of prescribed fire on a sierra nevada (california, USA) stream and its riparian zone. *For. Ecol. Manag.* **2005**, *218*, 37–59. [CrossRef]
64. Rieman, B.; Clayton, J. Wildfire and native fish: Issues of forest health and conservation of sensitive species. *Fisheries* **1997**, *22*, 6–15. [CrossRef]
65. Hitt, N.P. Immediate effects of wildfire on stream temperature. *J. Freshw. Ecol.* **2003**, *18*, 171–173. [CrossRef]
66. Stephens, S.; Agee, J.K.; Fulé, P.; North, M.; Romme, W.; Swetnam, T.; Turner, M.G. Managing forests and fire in changing climates. *Science* **2013**, *342*, 41–42. [CrossRef] [PubMed]
67. Millar, C.I.; Stephenson, N.L.; Stephens, S.L. Climate change and forests of the future: Managing in the face of uncertainty. *Ecol. Appl.* **2007**, *17*, 2145–2151. [CrossRef] [PubMed]
68. Collins, B.M.; Miller, J.D.; Thode, A.E.; Kelly, M.; van Wagtendonk, J.W.; Stephens, S. Interactions among wildland fires in a long-established sierra nevada natural fire area. *Ecosystems* **2009**, *12*, 114–128. [CrossRef]
69. Schwilk, D.W.; Keeley, J.E.; Knapp, E.E.; McIver, J.; Bailey, J.D.; Fettig, C.J.; Fiedler, C.E.; Harrod, R.J.; Moghaddas, J.J.; Outcalt, K.W.; *et al.* The national fire and fire surrogate study: Effects of fuel reduction methods on forest vegetation structure and fuels. *Ecol. Appl.* **2009**, *19*, 285–304. [CrossRef] [PubMed]
70. Westerling, A.L.; Gershunov, A.; Brown, T.J.; Cayan, D.R.; Dettinger, M.D. Climate and wildfire in the western united states. *Bull. Am. Meteorol. Soc.* **2003**, *84*, 595–604. [CrossRef]
71. Williams, J. Exploring the onset of high-impact mega-fires through a forest land management prism. *For. Ecol. Manag.* **2013**, *294*, 4–10. [CrossRef]
72. Batllori, E.; Parisien, M.; Krawchuk, M.A.; Moritz, M. Climate change-induced shifts in fire for mediterranean ecosystems. *Glob. Ecol. Biogeogr.* **2013**, *22*, 1118–1129. [CrossRef]
73. Stephens, S.L.; Millar, C.I.; Collins, B.M. Operational approaches to managing forests of the future in Mediterranean regions within a context of changing climates. *Environ. Res. Lett.* **2010**, *5*, 024003. [CrossRef]
74. Agee, J.K.; Skinner, C.N. Basic principles of forest fuel reduction treatments. *For. Ecol. Manag.* **2005**, *211*, 83–96. [CrossRef]
75. Oliver, T.H.; Smithers, R.J.; Bailey, S.; Walmsley, C.A.; Watts, K. A decision framework for considering climate change adaptation in biodiversity conservation planning. *J. Appl. Ecol.* **2012**, *49*, 1247–1255. [CrossRef]

© 2016 by the authors. Licensee MDPI, Basel, Switzerland. This article is an open access article distributed under the terms and conditions of the Creative Commons Attribution (CC BY) license (http://creativecommons.org/licenses/by/4.0/).

forests

MDPI

Article

Burning Potential of Fire Refuges in the Boreal Mixedwood Forest

Samira Ouarmim [1,*], Laure Paradis [2], Hugo Asselin [1], Yves Bergeron [1,4], Adam A. Ali [1,2] and Christelle Hély [1,2,3]

[1]   Chaire Industrielle CRSNG-UQAT-UQÀM en Aménagement Forestier Durable, Université du Québec en Abitibi-Témiscamingue, 445 Boulevard de l'Université, Rouyn-Noranda, QC J9X 5E4, Canada; hugo.asselin@uqat.ca (H.A.); Yves.Bergeron@uqat.ca (Y.B.); ali@univ-montp2.fr (A.A.A.); christelle.hely-alleaume@univ-montp2.fr (C.H.)
[2]   Institut des Sciences de l'Evolution de Montpellier, ISEM UMR 5554 CNRS-IRD-Université Montpellier-EPHE, Avenue Eugène Bataillon, Montpellier 34095, Cedex 5, France; laure.paradis@univ-montp2.fr
[3]   Ecole Pratique des Hautes Etudes, EPHE, PSL Research University, 4-14 rue Ferrus, Paris 75014, France
[4]   Centre D'étude de la Forêt, Université du Québec à Montréal, CP 8888, succ. Centre-ville, Montréal, QC H3C 3P8, Canada
*   Correspondence: samira.ouarmim@gmail.com; Tel.: +33-467-144-288

Academic Editor: Timothy A. Martin
Received: 8 July 2016; Accepted: 8 October 2016; Published: 21 October 2016

Abstract: In boreal ecosystems, wildfire severity (i.e., the extent of fire-related tree mortality) is affected by environmental conditions and fire intensity. A burned area usually includes tree patches that partially or entirely escaped fire. There are two types of post-fire residual patches: (1) patches that only escaped the last fire; and (2) patches with lower fire susceptibility, also called fire refuges, that escaped several consecutive fires, likely due to particular site characteristics. The main objective of this study was to test if particular environmental conditions and stand characteristics could explain the presence of fire refuges in the mixedwood boreal forest. The FlamMap3 fire behavior model running at the landscape scale was used on the present-day Lake Duparquet forest mosaic and on four other experimental scenarios. FlamMap3 was first calibrated using BehavePlus and realistic rates of fire spread obtained from the Canadian Fire Behavior Prediction system. The results, based on thousands of runs, exclude the effects of firebreaks, topography, fuel type, and microtopography to explain the presence of fire refuges, but rather highlight the important role of moisture conditions in the fuel beds. Moist conditions are likely attributed to former small depressions having been filled with organic matter rather than present-day variations in ground surface topography.

Keywords: fire refuges; fire modeling; fire susceptibility; FlamMap3; fuel moisture

1. Introduction

Fire is one of the dominant ecological drivers affecting vegetation patterns and dynamics in the circumboreal region [1,2]. Fire effects vary spatially depending on fire behavior [3,4] and the fire return interval. A burned area usually includes residual patches that partially or entirely escaped fire [5–8]. Two types of post-fire residual patches have been distinguished in eastern North American boreal mixedwood forests [9]: (1) "transient residual patches" that only escaped the last fire, probably due to peculiar but temporary unsuitable conditions for fire propagation; and (2) "fire refuges" that escaped several consecutive fires, likely due to specific site conditions. Although fire refuges represent a small proportion of the total area burned, they could provide unique habitats in post-fire successional landscapes [8]. Indeed, the ecological continuity recorded in fire refuges (unlike in transient residual patches, which only escaped the last fire) [9] could provide refuges for species with specific biodiversity signatures associated with old successional stages [10,11] that could be taken into account in biological conservation strategies for boreal ecosystems [12].

Fire behavior varies spatially depending on fuel features (composition, load, moisture, and spatial arrangement), landscape structure and composition, soil type, topographic constraints, and weather conditions [3,4,13]. Numerous studies have documented the spatial distribution of post-fire residual patches at the landscape scale (e.g., [14–19]). They showed that the occurrence of residual patches within burned areas could be related to topography, soil moisture, or wind dynamics during fire, fuel load, or presence of firebreaks. However, the respective roles of these factors likely vary among regions and residual patch types (transient or refuge), but this has never been tested before. Here we propose to evaluate the role of different ecological factors in the occurrence and long-term persistence of fire refuges in the mixedwood boreal forest of northeastern North America. We hypothesized that environmental conditions in fire refuges are less prone to fire activity as compared to the surrounding forest matrix. Therefore, fire refuges were compared with the surrounding forest matrix under different environmental conditions (mainly weather and fuel moisture) to determine if fire refuges have a lower propensity to burn than the other forest cover types present in the landscape.

A non-destructive approach, based on modeling, was selected for estimating the fire susceptibility of fire refuges. Fire refuges are old, conifer-dominated stands (time since the last fire is at least 250 years) within a surrounding matrix (less than 250 years old) of broadleaved, mixed, or coniferous stands. Therefore, a qualitative assessment of stands based only on composition might not allow for the differentiation of refuges from the matrix. Stands were thus quantitatively characterized based on fuel type, fuel load, and tree species composition and structure. We developed an original three-step methodology, based on the combination of three fire behavior models, often used in the North American context. First, we qualitatively characterized stand fuels using the Canadian Fire Behavior Prediction system (hereafter FBP, [20,21]), which allowed us to simulate realistic values of fire behavior within stands. Second, we quantitatively calibrated the characterization of stand fuel types and loads using the BehavePlus system [22,23]. BehavePlus allowed us to discriminate the different stand types while still predicting realistic fire behavior at the stand level. Finally, we simulated fires and their respective areas burned at the landscape level using the FlamMap3 model [24,25]. FlamMap3 is based on the BehavePlus system for fire propagation across pixels and allows computation of burned area probability maps in order to test stand susceptibility to fire.

2. Materials and Methods

2.1. Study Area

The reference forest mosaic used in this study is an 11,000-ha natural forest mosaic encompassing the Lake Duparquet Research and Teaching Forest (Figure 1). The study area is located in the eastern Canadian boreal mixedwood forest [26], and was previously used to test the effect of landscape composition on fire size distribution [27]. The studied landscape is characterized by balsam fir (*Abies balsamea* (L.) Mill.), paper birch (*Betula papyrifera* Marsh.), white spruce (*Picea glauca* (Moench)), trembling aspen (*Populus tremuloides* Michx.), and eastern white cedar (*Thuja occidentalis* L.) as the main tree species [28]. The geomorphology is characterized by the presence of a massive clay deposit left by pro-glacial lakes Barlow and Ojibway [29]. The climate is cold temperate with a mean annual temperature of 0.7 °C and mean annual precipitation of 890 mm [30]. The closest meteorological station is located at La Sarre, 42 km north of the study area.

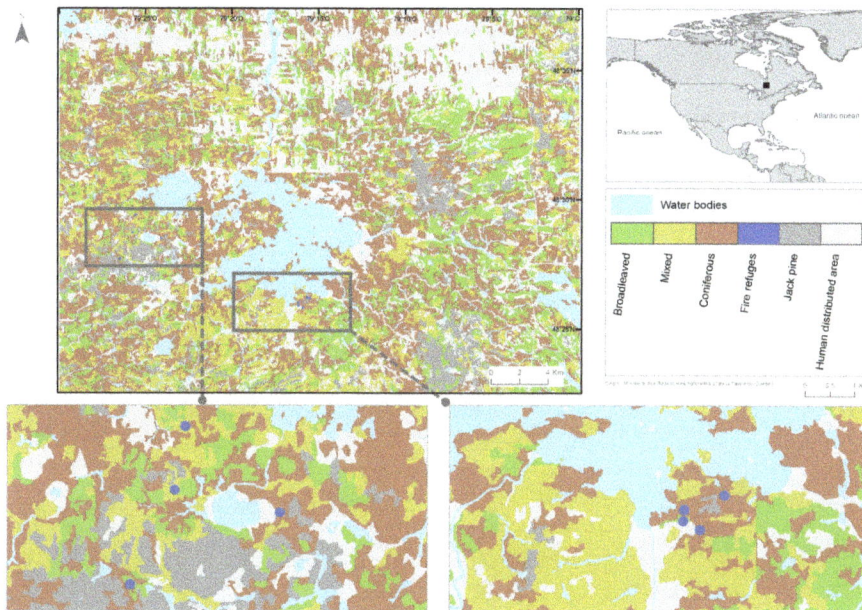

Figure 1. Forest stand types (fuel models) and location of the studied fire refuges in the Lake Duparquet Research and Teaching Forest.

2.2. Site Selection

Typical post-fire succession on mesic clay deposits in the eastern Canadian boreal mixedwood forest involves a gradual change from post-fire pioneering stands dominated by shade-intolerant deciduous tree species (trembling aspen or paper birch) during the first ca. 75 years, to mixed stands with an important white spruce component in the next ca. 75 years, to coniferous stands dominated by balsam fir and eastern white cedar after ca. 150 years [28]. It was thus possible, using ecoforestry maps produced by the Quebec Ministry of Forests, Wildlife and Parks [31], to perform an exhaustive fire refuge census by distinguishing, within an otherwise relatively homogeneous forest matrix, stands considered as post-fire residual patches due to contrasting composition and structure representative of late-successional (older) stands. From this preliminary selection [9], thirteen post-fire residual patches were found in areas where the last known fire, reconstructed from dendrochronology, occurred in 1944 or 1923 (depending on site location) [28,32], and the second-to-last fire in 1717 or 1760 [28,32]. These stands have been previously identified [9] as coniferous old-growth forest patches (with balsam fir and eastern white cedar) embedded in a matrix of younger deciduous forests (with trembling aspen or white birch).

From the palaeoecological reconstruction of fire activity based on radiocarbon dating of macroscopic soil charcoal peaks in stratigraphic sections sampled from these post-fire residual patches [9], eight of the 13 patches were identified as fire refuges that had escaped two or more consecutive fires (1923 or 1944 and 1760 or 1717). The five other patches only escaped the most recent fire (1923 or 1944). These were therefore recorded as transient residual patches (Table 1), not used in this study focusing on refuges, and therefore considered as regular coniferous stands thereafter.

Table 1. Characteristics of the eight sampled fire refuges.

Stand Type	Organic Matter Thickness (cm)	Slope (°)
Refuge	49	0
Refuge	50.5	0
Refuge	98	0
Refuge	9	3
Refuge	59	1
Refuge	149	2
Refuge	25	0
Refuge	11	4

2.3. Field Sampling

Stand composition, structure, and fuel loads were measured in each fire refuge according to the sampling design set by Hély et al. [33], along and in the vicinity of a single 30-m sided equilateral triangle [34]. The same stand sampling methodologies were used to avoid false significant differences in fuel or structures between fire refuges and other matrix stands (previously sampled by Hély et al. [33]) due to changes in sampling method. In each stand, forest structure and canopy characteristics (tree species and diameter at breast height, total height, and canopy base height) were therefore estimated from 24 trees (12 dominant and 12 suppressed) that were selected in the triangle vicinity based on the point-centered quadrant method [34].

Loads of all fuel types defined in the BehavePlus system [22] were measured within each stand along or apart from the triangle sides depending on fuel type [33]. Woody debris from the three American time-lag classes, representative of desiccation times (1 h, 10 h, and 100 h time lags) and corresponding approximately to diameters <0.6 cm, 0.6–2.5, cm and 2.7–7.6 cm, respectively [22], were measured using the line intersect method [35,36]. The same species coefficients and equations as in Hély et al. [37] were used to estimate fuel loads from twig and branch numbers as they had been adapted to the boreal mixedwood forest. Shrub and herbaceous loads were measured in six 1-m² quadrats [38], evenly spaced along the 90-m triangle transect. Shrub loads were estimated from basal stem diameter measurements using species dry weight-basal diameter relationships set from shrub samples previously collected in the Duparquet area [39–41]. Herbs were collected to obtain oven-dried weight. Litter and duff layer depths were measured in six quadrats (25 cm × 25 cm) and total litter and duff material was separately collected to obtain oven-dried weights.

Stand characteristics of young (deciduous), intermediate (mixed), and old (coniferous) stands representative of the mixedwood boreal forest matrix of the studied landscape were obtained from Hély [42]. They were merged to the stand characteristics of fire refuges (Table 2) sampled in the present study to create a stand fuel load dataset to be applied to the ecoforestry map polygons (Figure 1). The few *Pinus banksiana* stands present in the landscape were classified separately, as these stands represent young but not deciduous stands and were not considered in the analyses.

Table 2. Mean fuel bed depth and loads of each fuel type composing mixedwood boreal forest stand types for BehavePlus and FlamMap3 models.

Fuel Type/Stand Type	Fuel Bed Depth (cm)	Litter Load	1 h Fuel Load (t/ha)	10 h Fuel Load (t/ha)	100 h Fuel Load (t/ha)
Broadleaved	3	4	0.06	1.2	0.8
Mixed	4.7	4.6	0.1	1.2	1
Coniferous	10.2	4.5	0.3	2.1	2.9
Fire refuges	10.2	6.9	0.26	0.24	3.1

2.4. Weather Data and Associated Fuel Moisture Scenarios

To simulate fire ignition and early fire behavior under different weather conditions, two fire weather indices from the Canadian Fire Weather system [43] were selected. They represent moderate and high fire danger (Fire Weather Index (FWI) = 5 and 15, respectively), as used by the SOPFEU (Quebec Society for the protection of forests against fire) in the studied region. Two days representative of each fire danger were used in order to take into account wind or dry air effects and to partially capture intrinsic weather variability (Table 3).

Table 3. Details of the fire weather indices [33].

FWI	Scenario	Wind Speed (km/h)	FFMC	ISI	BUI
Moderate (5)	1	9	87.4	4.6	11.5
	2	9	72.4	1.1	76.7
High (15)	3	22	89	11.5	15.1
	4	5	86.8	3.4	92.5

Note: FFMC, Fine Fuel Moisture Content, which is a numerical rating of the moisture content of litter and other cured fine fuels. This code is an indicator of the relative ease of ignition and flammability of fine fuel. ISI, Initial Spread Index, which is a rating of the expected rate of fire spread. It combines the effects of wind and FFMC on rate of spread without the influence of fuel quantity. BUI, Buildup index, which is a numerical rating of the total amount of fuel available for combustion. FWI, Fire Weather Index, which is a rating of fire intensity that combines ISI and BUI. It is suitable as a general index of fire danger throughout the forested areas of Canada [44].

These weather characteristics were transformed into fuel moisture content, based on the range of values from scenarios provided by BehavePlus (Table 4). The fine fuel moisture content (FFMC), duff moisture code (DMC), and drought code (DC) of the Canadian Fire Weather Index [43] were assumed to match with 1 h, 10 h, and 100 h moisture contents, respectively, used in both BehavePlus and FlamMap3 fire models, knowing that dead fuel moisture of extinction (maximum fuel moisture content, which limits fire propagation) is usually set at 30% in BehavePlus. For the calibration process, all stands (broadleaved, mixed, coniferous, and fire refuge stands) were attributed the same water content (1 h, 10 h, and 100 h moistures). Wind direction used in the FlamMap3 model corresponded to the main wind direction (from the south-southwest) over the studied area and is representative of the fire-season wind [45].

Table 4. Fuel moisture scenarios for BehavePlus and FlamMap3 models.

FWI	Scenario	Wind Speed (km/h)	1-h	10-h	100-h
Moderate (5)	1	9	8	13	14
	2	9	10	10	11
High (15)	3	22	8	13	14
	4	5	6	7	8

2.5. Parameterization of Fuel Models at the Stand Level

First, the FBP model was used under the four different weather conditions (Table 3) in order to determine fire behavior variability for each stand type in the landscape mosaic. The FBP System is indeed regarded as producing good predictions of fire behavior when compared to natural or experimental fire records [46–48].

Slope and elevation were maintained constant (0° and 300 m above sea-level, respectively) in FBP and BehavePlus runs to ease comparison and because the study area has an overall flat topography. In the FBP model, stand types are characterized by tree species composition and density, and fuels are qualitatively described [20]. Fire behavior variables (e.g., rate of spread (ROS) and head fire intensity (HFI)) are predicted from empirical relationships computed from many fire measurements recorded during both wildfires and experimental fires, and covering a large range of weather conditions [49]. Moreover, different fire behavior relationships exist for spring (without leaves) and summer (with leaves) stand types that include broadleaved species. Spring fire behavior relationships were selected, because springs sustain faster and more intense fires than summers [33,50]. Spring broadleaved and mixed fuel types (i.e., Dl and Ml, respectively) were therefore chosen, as they were deemed representative of the present boreal mixedwood forest mosaic. The coniferous percentage was increased in M1 stands to age stands toward late successional states dominated by coniferous trees.

The second methodological step involved the BehavePlus model. BehavePlus is a non-spatially explicit deterministic model based on the physical properties and combustion properties of fuel types (see below). The generic fuel model "Moderate load broadleaf litter" available in the BehavePlus system [23] was used for all stand types, replacing fuel load values (for dead (1 h, 10 h, 100 h time lag) and living (herbaceous and woody shrubs) fuels, respectively), fuel bed depth, and canopy structure by those measured in situ (Table 2). Based on the same topographical (slope and elevation) and weather conditions as for the FBP model, BehavePlus fuel models were calibrated by comparing BehavePlus simulated rates of spread (ROS) with those obtained from the FBP. Hély et al. [48] have concluded that systematically slower ROS were simulated by Behave (earlier version of BehavePlus) as compared to the ROS simulated by FBP, and that these differences were likely due to the exclusion of the duff layer in Behave simulations. We addressed and solved this problem by adding the duff load to the 1-h fuel load. Consequently, the fuel bed depth was also increased in the calibration process until BehavePlus ROS predictions were in the same range as those from the FBP predictions with still realistic fuel bed depths in the range of observed values (Table 1).

2.6. Settings for Simulations at the Landscape Level

Spatially explicit fire simulations at the landscape mosaic level were performed using the FlamMap3 model [24,25]. FlamMap3 is based on BehavePlus fuel models and fire behavior at the pixel level. FlamMap3 propagates fire from an ignition location (randomly selected or not) through neighboring pixels over a given simulation time (see below). Potential fire behavior calculations include surface ROS [51], crown fire initiation [52], and crown ROS [53]. The ignition-propagation scheme is repeated a given number of times set by the user (with random ignition locations). FlamMap3 provides an output map for each fire behavior variable (e.g., ROS and Head Fire Intensity (HFI)). It also provides a burn probability map reporting for each pixel the number of times fire went through it compared to the total number of ignitions.

Slope (°), elevation (m above sea-level), aspect (°N), and composition maps for the Lake Duparquet landscape, including fire refuges, were rasterized (225 m^2 spatial resolution, i.e., 15 × 15 m) using the *ArcGIS* software to produce Ascii grids required as input data in FlamMap3. A landscape fuel map was created by applying to the stand composition ecoforestry map the four fuel models representative of mean broadleaved, mixed, coniferous, and *Pinus banksiana* stands, respectively. A fifth fuel model, representative of mean fuel conditions measured in fire refuges, was specifically defined and projected on the fire refuge locations. Water bodies and recently logged stands were represented as a generic non-fuel model and considered as firebreaks. Moreover, as for BehavePlus at the stand level, for each fuel model (stand type), fuel load (kg/ha) (Table 2), canopy cover (%), tree height (m), crown bulk density (kg/m^3), and height-to-live crown base (m) were used to create the FlamMap3 required maps based on stand attributes.

In order to simulate fires, whose size distribution would express the natural variability recorded in the boreal mixedwood forest [27], four simulation times in FlamMap3 (63, 105, 408, and 4835 min) were selected, corresponding to the four quartiles (25%, 50%, 75%, 100%) of the fire size distribution (Figure 2) computed from Quebec public archives (1994–2007) provided by the SOPFEU [54]. Simulation times were calculated using the mean pixel-based ROS computed from preliminary FlamMap3 output run over the entire landscape. However, as the smallest simulated fires (63-min fires) could not cross several pixels or reach fire refuge stands (except when ignition occurred within), results were only reported for simulations with fires whose size was at least equal to the median archived fire size in Quebec. Hereafter, the median size fires were called small fires, the 75% size fires were called medium fires, and the 100% size fires were called large fires.

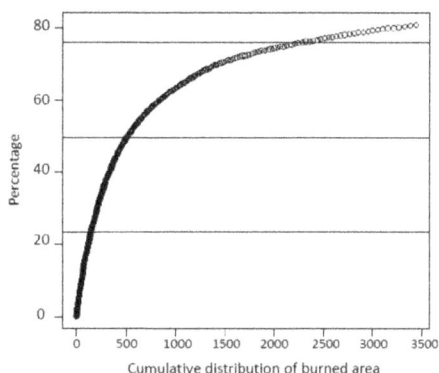

Figure 2. Cumulative distribution of area burned (ha) recorded in the boreal mixedwood forest.

2.7. Modeling Experiments

Five modeling experiments were performed to test the effect of different stand characteristics and their combinations on the occurrence of fire refuges. Each experimental design produced a burn probability map.

The first experiment (Table A1) tested the effect of observed stand attributes (fuel type proportions, loads, and stand structure) combined with the indirect effect of stand location (the occurrence of fire breaks in the vicinity, topography). A given fuel moisture scenario was applied to all stands throughout the landscape. This experiment included a total 120,000 simulations representing 12 runs of 10,000 ignitions, each run being a combination of one simulation time (over three) and one weather/moisture scenario (over four) with 10,000 random ignition locations.

The second experiment (Table A1) tested the effect of low occurrence of fire refuges in the landscape. To proceed, the number of fire refuges was artificially increased to 800 stands and randomly placed in the landscape to represent 2% of the forest mosaic area. To save time, computer space, as well as to keep high computer performance, only 40,000 randomly ignited fires were simulated, based on the four weather-moisture scenarios and the 408-min simulation time (75th percentile) only. The potential differences in burn probabilities computed with this second experiment and those extracted from the previous simulations with fire refuges in their real location but only with the corresponding 408-min simulation time could be attributed to their low occurrence in the landscape.

The third experiment (Table A1) tested the combined effect of firebreaks and topography on the occurrence and persistence of fire refuges. Fire refuges were manually placed in the vicinity of water bodies and in particular topographical situations (depressions) to test their effect on the probability of burning. In this experiment, as for the previous one, only 40,000 fire ignitions were performed using only the 408 min simulation time.

In the fourth experiment (Table A1), the fire refuge microtopography was tested (stand scale), as the flat macrotopography (landscape scale) was not expected to explain the location of fire refuges. Indeed, the thick organic matter layer previously measured in most of the fire refuges [9] could reflect the presence of shallow topographical depressions created just after the drainage of Holocene proglacial lake Obijway-Barlow. These depressions would have been filled with organic matter during the Holocene. This assumption comes from the fact that topography at ground surface in fire refuges did not differ from the surrounding forest matrix while fire refuges had significantly thicker organic matter accumulation [55]. Therefore, in order to reproduce the initial conditions in fire refuges (without organic matter accumulation), the presence of small depressions was tested by artificially changing the altitudinal conditions of fire refuges. As for the two last experiments, only 40,000 new randomly ignited fires were simulated (with 10,000 per moisture scenario based on the 408 min simulation time).

Finally, as the thick organic matter in fire refuges has been assimilated to peat [55] and therefore very moist ground conditions, the fifth experiment (Table A1) tested the effect of fuel moisture on the probability of burning. To proceed, the moisture content was artificially increased for fire refuges only, without exceeding the fuel moisture of extinction, and without changing fuel moisture in the forest matrix (kept constant for a given weather/moisture scenario as in the other experiments). We ended with fire refuge fuel moistures of 21%, 24%, and 26% (for 1 h, 10 h, and 100 h dead fuel, respectively). Once again, only 40,000 randomly ignited fires were simulated based on the median 408 min simulation time and the four general weather-moisture scenarios used in the other experiments.

2.8. Statistical Analyses

All analyses were performed using the R software [56]. The analyses were performed not directly on the burn probability output from FlamMap3, but rather on stand propensity to burn as defined as the ratio of cumulative area burned probability for each stand type (burned proportion) on its representativeness (stand proportion in the forest mosaic matrix). For a given stand type, a ratio higher (lower) than 1 highlights a relative propensity to burn or to propagate fire (to escape fire or to slow down fire spread). This was necessary due to the fact that fire refuges represent only a few stands in the overall studied landscape and likely are even less abundant in the simulated area burned.

For each modeling experiment, stand type propensities to burn were tested using Chi-square and contrast tests in order to compare their cumulated burned proportion (observed values) to their cover representativeness in the forest matrix (theoretical percentage values). Second, we looked for significant differences in burn probabilities among forest stand types (i.e., among fuel model types) using one-way non-parametric analyses of variance on rank scores (Kruskal-Wallis non-parametric test) followed by a Tukey multiple comparison test.

Using the simulation dataset from the first experiment, we tested the effect of stand attributes (fuel type proportions, loads, and stand structure) including as well the combined effects of macro-topography and firebreak locations (experiment 1). The similar overall response through weather scenarios and through simulation times helped us in restricting the last four simulation experiments to the 75 percentile simulation time of 408 min.

The number of fire refuges (experiment 2), their location as compared to firebreaks (experiment 3), their maintenance over micro-topographic depression buildup (experiment 4), as well as their fuel moisture content (experiment 5) were tested using the same statistical approach, but only applied to the series of 408 min runs with the four fuel moisture scenarios.

3. Results

At the end of the parameterization procedure, ROS predicted by the FBP and BehavePlus systems were in good agreement (R^2 = 0.8), ranging from 0.2 to 9 m/min. A slight overestimation from BehavePlus in conifer stands was noted, with the heaviest fuel loads under highest fire risk with windy conditions. This satisfactory agreement between both fire behavior models running at the stand level confirmed that the BehavePlus model could be adapted to the boreal mixedwood forest if the fermentation layer (deep litter) was taken into account in fuel load characterization. This adjustment of fuel load yielded more realistic values for all stand types.

While the large number of iterations (10,000 fire ignitions) for each scenario and experiment almost always resulted in rejection of the null hypothesis, we chose to interpret only the most significant signals in stand propensity to burn with a special focus on fire refuge type. We also chose to interpret only the general pattern among stand types in order to evaluate the ecological relevance of the landscape fire model, discarding minor differences which were difficult to interpret from an ecological standpoint.

Simulation comparisons from the first modeling experiment based on actual forest mosaic stand characteristics (location, number, and fuel types) but with set weather/fuel moisture scenarios, showed, for each fire size class, that fire refuges seem to have a high propensity to burn (up to 3.46), except for scenario 4 (Figure 3). Moreover, the relative propensity to burn of fire refuges seems to be higher than those of any of the other stand types (Table 5). Between-stand differences decreased with increased fire danger (scenarios 3 and 4) and fire duration.

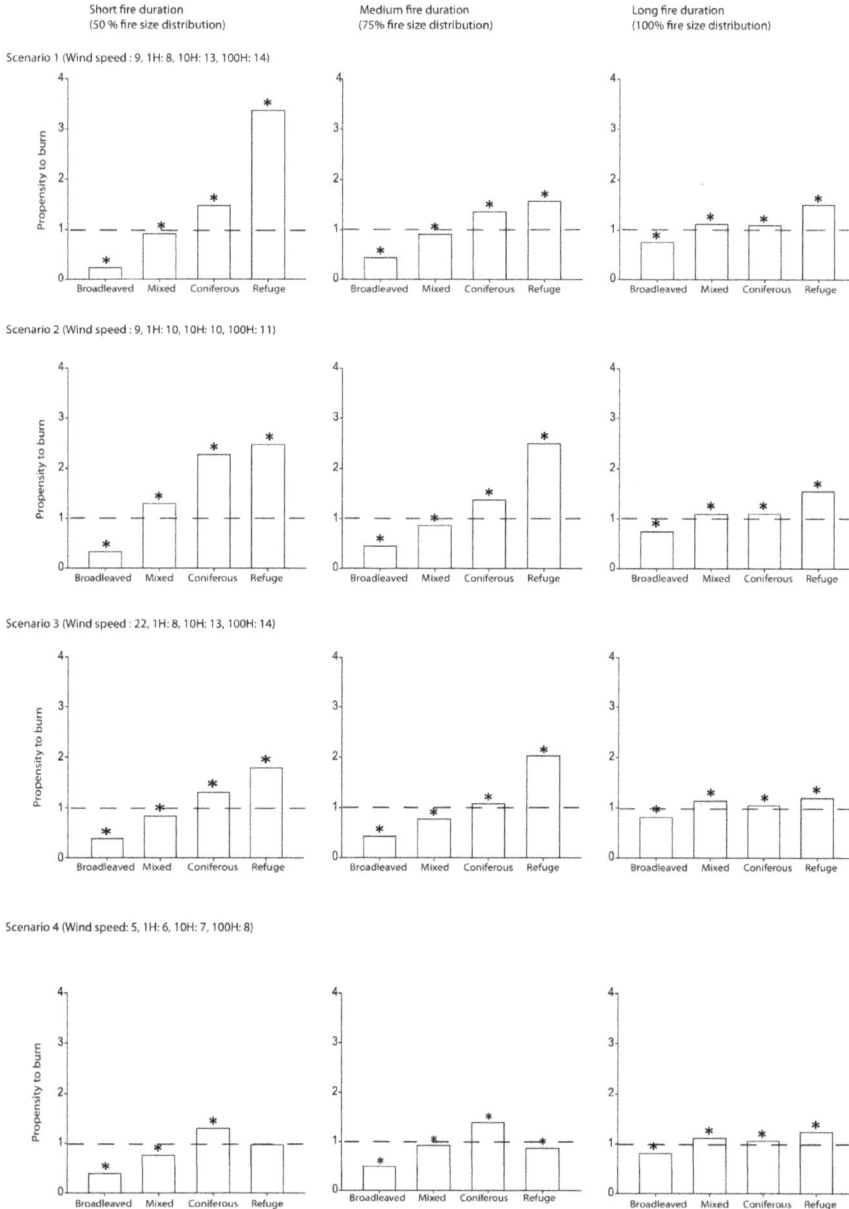

Figure 3. Propensity of stand types to burn as a function of fire danger (scenarios) and fire duration (a proxy for small, medium, and large fires based on the 50%, 75%, and 100% quartiles of the burned area distribution from the 1994–2004 SOPFEU fire database [54]). In each of the twelve panels, we tested the stand type propensity to burn as compared to its landscape representativeness using Chi-square and contrast tests. * for significant departures from 1 (burning equal to representativeness, dashed line).

Table 5. Comparisons of burn probabilities among stand types, based on actual mosaic composition and spatial arrangement, as a function of fire danger (weather/fuel moisture scenarios) and fire duration (a proxy for small, medium, and large fires based on the 50%, 75%, and 100% quartiles of the burned area distribution from the 1994–2004 SOPFEU fire database [54]). Significant differences from Kruskal-Wallis and Tukey tests (with $p < 0.05$) are represented by different letters.

Fire Duration (Quartile of the Quebec Fire Size Distribution)	Scenario (Weather)	Broadleaved Stands	Mixedwood Stands	Coniferous Stands	Refuge Stands
105 min (50%)	1	d	c	b	a
	2	c	b	a	a
	3	d	c	b	a
	4	d	c	a	b
408 min (75%)	1	d	c	b	a
	2	d	c	b	a
	3	d	c	b	a
	4	c	b	a	b
4835 min (100%)	1	b	a	a	a
	2	c	b	b	a
	3	d	c	b	a
	4	d	b	c	a

The fifth modeling experiment (Figure 4, Table 6), testing the effect of very moist fuels in fire refuges compared to other stands, was the only experiment that systematically showed significant reduction in fire refuge propensity to burn through all fire danger scenarios (Figure 4, right panel). The reduction was so strong that the fire refuge propensity to burn shifted to a propensity not to burn (ratio < 1) and fire refuge stands were systematically ranked in third position, just above deciduous stands but below mixed and coniferous stands. For all other modeling experiments, propensity to burn generally increased from broadleaved to coniferous stands, and those from fire refuges were among the highest (Figure 4, first three columns and Table 6). Differences in propensity to burn of fire refuges compared to those from their actual locations (second experiment in Figure 4 versus Figure 3) were likely due to the random ignition effect.

Table 6. Comparisons of burn probabilities among stand types as a function of experiments using only the medium fire duration (408 min runs) representing the 75% quartile of the burned area distribution from the 1994–2004 SOPFEU fire database [54]. Significant differences from Kruskal-Wallis and Tukey tests (with $p < 0.05$) among stand types are represented by different letters.

Experiment	Scenario (Weather)	Broadleaved Stands	Mixedwood Stands	Coniferous Stands	Refuge Stands
Number of refuges	1	d	c	b	a
	2	d	c	b	a
	3	d	c	a	b
	4	d	c	a	b
Topography and firebreaks	1	d	c	b	a
	2	d	c	a	b
	3	b	c	d	a
	4	d	c	a	b
Small depression	1	d	c	b	a
	2	d	c	b	a
	3	d	c	b	a
	4	c	b	a	b
High fuel moisture	1	d	b	a	c
	2	c	b	a	c
	3	d	b	a	c
	4	d	b	a	c

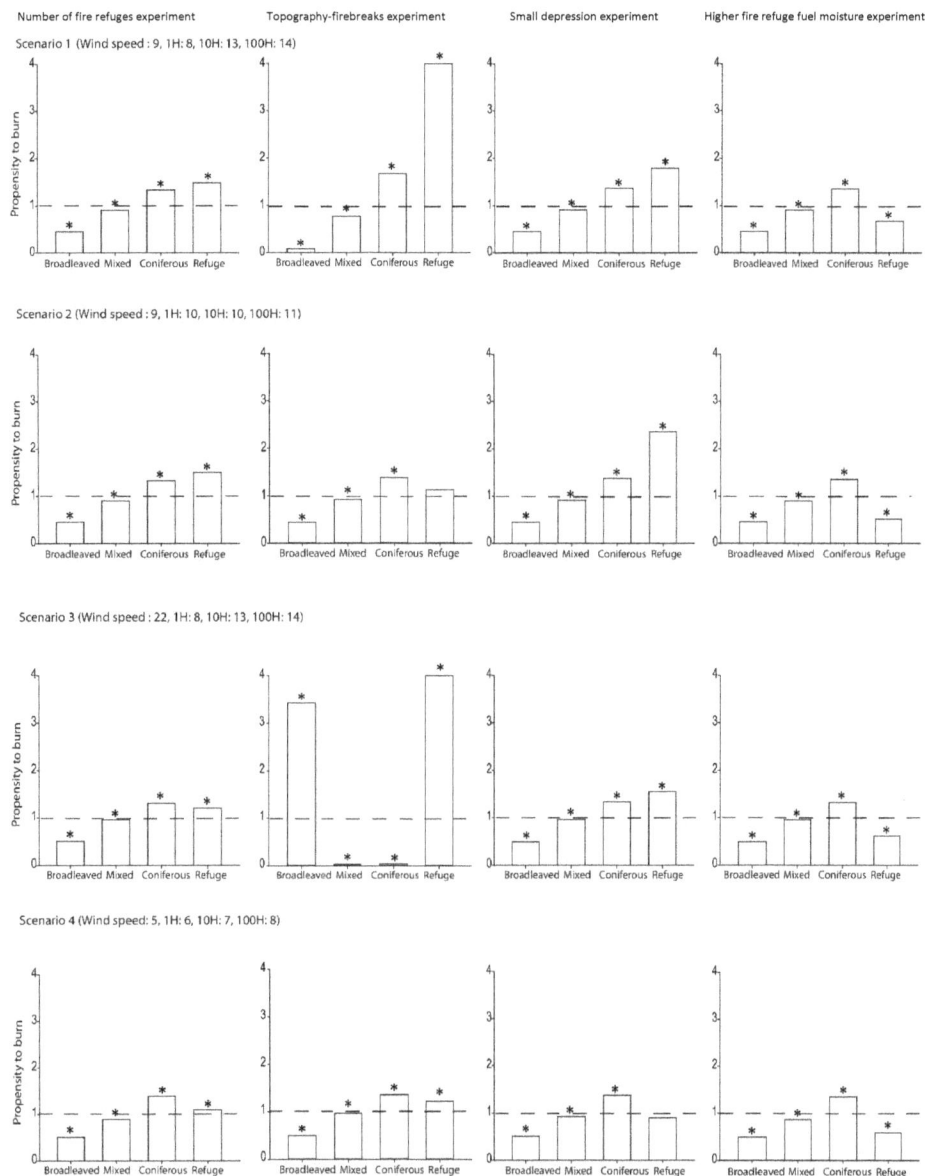

Figure 4. Propensity of stand types to burn as a function of experiments using only the medium fire duration (408 min runs) representing the 75% quartile of the burned area distribution from the 1994–2004 SOPFEU fire database [54]. In each of the twelve panels, we tested the stand type propensity to burn as compared to its landscape representativeness using Chi-square and contrast tests. * for significant departures from 1 (burning equal to representativeness dashed line). The propensity to burn of fire refuges in scenarios 1 and 3 of the topography-firebreak experiment reached the value of 8 (although we topped the y-axis at 4 to ease readability).

4. Discussion

Results from the present study are based on a large number of fire behavior simulations at the landscape level, combining different fire sizes and weather/moisture scenarios, and testing different environmental factors possibly explaining the occurrence of fire refuges, a peculiar type of post-fire residual patches [18,33,57,58]. Moreover, due to the influence of environmental features such as fire severity, location in the landscape, and fuel characteristics, fire refuges, unlike other transient residual patches, are assumed to not be randomly distributed [59].

Although the large number of fire ignitions almost systematically pushed the statistical interpretation toward rejection of the null hypothesis, we clearly and logically found an increase in propensity to burn from broadleaved to coniferous stands, suggesting that FlamMap3 is an efficient model to study the relationship between vegetation and fire in the mixedwood boreal forest. Our analysis also suggests that stand fire hazard increases through the successional sequence in canopy tree species replacement [33]. This increase in propensity to burn only results here from differential fuel load accumulation and fuel spatial arrangement (input in FlamMap3) [27,33], but not from changes in surface fuel quality (composition) since it has not been directly considered as input in the FlamMap system. The late successional coniferous forest stands accumulate fuels, leading to increased forest combustibility known as build-up [4,60,61]. This trend was confirmed in most simulation results except for three simulations whose patterns could only be explained by the randomness of ignition locations (Figure 4), which were likely overrepresented in some types (broadleaved, refuges, or even in water bodies and human disturbed areas).

A second logical result from the real forest mosaic composition (i.e., stand proportions) was the effect of weather conditions, which induced a decrease in stand propensity to burn with increasing fire weather risk, as well as with increasing fire duration (i.e., fire size). Indeed, both changes allowed fires to spread more easily and therefore increased the probability to reach many stand types.

Regarding the main objective of the present study, we showed that fire refuges in the eastern Canadian mixedwood boreal forest burned more than the surrounding forest matrix when fuel moisture per fuel type (1 h, 10 h, 100 h, litter, and live fuels, respectively) was held constant throughout the landscape. From this first analysis, it is clear that in their current location, the association of fuel types and loads, topography at ground level, and firebreaks in their vicinity cannot explain the presence of fire refuges nor their spatial distribution in the forest mosaic. This was confirmed by the results of the second experiment with more numerous fire refuges randomly located in the landscape, as fire refuges maintained a higher propensity to burn than the surrounding forest matrix. Hence, fire refuges in the Lake Duparquet area actually present fuel characteristics favorable to fire ignition and fire spread in terms of shape, size, density, loading, chemical properties, and spatial configuration [62], but not moisture.

Similarly, our results excluded the predominant effects of firebreaks and topography in the occurrence of refuges based on the first, third, and the fourth modeling experiments. This result contrasts with those from previous studies explaining that the spatial occurrence of post-fire residual stands was influenced by firebreaks, such as rock outcrops and water bodies, which disrupt horizontal fuel continuity, thus preventing fire propagation [3,17,18,55]. Despite the relatively high proportion of water bodies in the study area, their role as firebreaks does not seem to explain the lower propensity to burn of fire refuges. However, this study cannot exclude the effect of firebreaks on other residual patches (transient residual patches) in the landscape [63]. Indeed, the role of lakes intercepting and stopping fire spread more likely depends on wind direction during fire and on ignition location in the landscape in comparison with residual patches location. The importance of protected topographical positions was also reported as an important factor in interior forests of eastern Washington [64] and south central Wyoming [65], where old forests are found. The generally flat topography in our study area as compared to the abovementioned regions likely explains why no link between the occurrence of fire refuges and landforms was found here. Moreover, while we tested for the presence of Holocene small depressions as initial conditions favorable to fire refuge creation, they appeared to have no negative effect on fire behavior and therefore cannot fully explain the occurrence of these persistent unburned stands in the landscape. However, fire refuges burned *less* than other stand types (except pure broadleaved stands) when they were assigned higher fuel moisture content in the simulations as observed and measured in the field. Because fire refuges escaped multiple stand-replacing fires that have occurred in the surrounding forest matrix, high fuel moisture appears to be the critical factor reducing fire intensity, as also reported in previous studies [63,66]. Moreover, it is worth noting that under past extreme fire weather conditions, even fire refuges sometimes have burned [9]. Beyond weather condition, fuel moisture is usually related to topography and aspect [67]. The aspect determines the solar-flux (cool, wet north and east facing aspects) and can impact soil moisture, which has important influences on fire behavior [48,68,69]. In the present study, due to the flat topography, the high moisture content of the litter in fire refuges seems to be rather related to the indirect effect of shallow depressions that have been filled with organic matter though time (centuries to millennia [9]), in which water tends to accumulate and peat to develop.

While the results of this study seem to be in accordance with palaeoecological analyses performed in situ, comparisons with other studies or regions are difficult. First, differences between fire refuges and other transient residual patches have never been tested before. Second, the role of firebreaks and topography can vary regionally based on the macro- and micro-topography of the study area. Therefore, to understand the general pattern of fire refuge occurrences in a given landscape, more investigations are needed locally to take into account the present day mosaic specificities and potential past changes as revealed by palaeoecological analyses. Hence, given the potential importance of fire refuges in the landscape (biodiversity hot spots [10,11]), they should be subjected to special conservation efforts.

5. Conclusions

To conclude, fuel moisture content appears to be the most important factor influencing the distribution of fire refuges at the landscape scale in the Eastern Canadian mixedwood boreal forest where topography on clay belt is relatively flat. This result is in good agreement with palaeoecological analyses performed in the same stands, which showed the occurrence of aquatic taxa and moisture tolerant tree species in fire refuges [70]. Hence, fire refuges could be considered as "soaked powder kegs", having well enough fuel to burn, but too much humidity. As dryer climate conditions are expected in the northeastern North American boreal forest over the next decades [71], fire conditions leading to the burning of fire refuges could become more frequent. Simulation studies using various climate change scenarios are necessary to evaluate potential effects on the persistence of fire refuges.

Acknowledgments: This study was part of S. Ouarmim's PhD project (NSERC studentship). It was also part of the "Cold Forests" International Research Group on boreal and mountain forests (GDRI Forêts froides) funded by CNRS, the University of Montpellier, and Ecole Pratique des Hautes Etudes. It was supported by Tembec, the Natural Sciences and Engineering Research Council of Canada, and the Fonds de Recherche du Québec-Nature et Technologies (FQRNT). Financial support for ^{14}C dating was provided by the PALEOFEUX program supported by the Institut National des Sciences de l'Univers (INSU, CNRS, France), and the French National program ARTEMIS. We thank Claude-Michel Bouchard, Philippe Duval, and Mélanie Desrochers for providing maps of the study area. We thank Danielle Charron for helping us organize fieldwork, and Jennifer Bergeron, Carine Côté-Germain, Lauriane Mietton, Mickael Paut, Mathilde Robert-Girard, and Julie Magnier for helping us in the field. We also thank the Programme Samuel-de-Champlain coopération recherche scientifique et technologique, commission permanente de coopération Franco-Québécoise and BoreAlp project supported by NSERC, OURANOS, and le Programme de soutien aux initiatives internationales du gouvernement du Québec.

Author Contributions: Samira Ouarmim, Christelle Hély, Yves Bergeron, Adam Ali, and Hugo Asselin conceived the design sampling; Samira Ouarmim performed the sampling; Samira Ouarmim, Laure Paradis, and Christelle Hély analyzed the data; Samira Ouarmim, Christelle Hély, Hugo Asselin, Adam Ali, and Yves Bergeron wrote the paper.

Conflicts of Interest: The authors declare no conflict of interest.

Appendix A. Experiments of the Study

Table A1. Recapitulative table of the different experiments as a function of weather conditions (scenarios).

Experiment	Fire Duration (Fire Size Distribution)	Wind Speed	Fuel Moisture	Refuge Abundance
Fuel and refuge location	50%	Moderate	Moderate	As in the field
		Moderate	High	As in the field
		High	Moderate	As in the field
		Low	Low	As in the field
Fuel and refuge location	75%	Moderate	Moderate	As in the field
		Moderate	High	As in the field
		High	Moderate	As in the field
		Low	Low	As in the field
Fuel and refuge location	100%	Moderate	Moderate	As in the field
		Moderate	High	As in the field
		High	Moderate	As in the field
		Low	Low	As in the field
Number of fire refuges	75%	Moderate	Moderate	800
		Moderate	High	800
		High	Moderate	800
		Low	Low	800
Topography and firebreaks	75%	Moderate	Moderate	As in the field
		Moderate	High	As in the field
		High	Moderate	As in the field
		Low	Low	As in the field
Small depression	75%	Moderate	Moderate	As in the field
		Moderate	High	As in the field
		High	Moderate	As in the field
		Low	Low	As in the field
High fuel moisture	75%	Moderate	Moderate	As in the field
		Moderate	High	As in the field
		High	Moderate	As in the field
		Low	Low	As in the field

References

1. Zackrisson, O. Influence of forest fires on the North Swedish boreal forest. *Oikos* **1977**, *29*, 22–32. [CrossRef]
2. Payette, S. Fire as a controlling process in the North American boreal forest. In *A Systems Analysis of the Global Boreal Forest*; Shugart, H.H., Leemans, R., Bonan, G.B., Eds.; Cambridge University Press: Cambridge, UK, 1992; pp. 144–169.
3. Swanson, F.J.; Kratz, T.K.; Caine, N.; Woodmansee, R.G. Landform effects on ecosystem patterns and processes. *BioScience* **1988**, *38*, 92–98. [CrossRef]
4. Agee, J.K. *Fire Ecology of Pacific Northwest Forests*; Island Press: Washington, DC, USA, 1993; p. 497.
5. Gluck, M.J.; Rempel, R.S. Structural characteristics of post-wildfire and clearcut landscapes. *Environ. Monit. Assess.* **1996**, *39*, 435–450. [CrossRef] [PubMed]
6. Wallenius, T.H.; Kuuluvainen, T.; Vanha-Majamaa, I. Fire history in relation to site type and vegetation in Vienansalo wilderness in eastern Fennoscandia, Russia. *Can. J. For. Res.* **2004**, *34*, 1400–1409. [CrossRef]
7. Wallenius, T.H.; Pitkänen, A.; Kuuluvainen, T.; Pennanen, J.; Karttunen, H. Fire history and forest age distribution of an unmanaged *Picea abies* dominated landscape. *Can. J. For. Res.* **2005**, *35*, 1540–1552. [CrossRef]
8. Burton, P.J.; Parisien, M.A.; Hicke, J.A.; Hall, R.J.; Freeburn, J.T. Large fires as agents of ecological diversity in the North American boreal forest. *Int. J. Wildland Fire* **2008**, *17*, 754–767. [CrossRef]

9. Ouarmim, S.; Ali, A.A.; Asselin, H.; Hély, C.; Bergeron, Y. Evaluating the persistence of post-fire residual patches in the eastern Canadian boreal mixedwood forest. *Boreas* **2015**, *44*, 230–239. [CrossRef]
10. Selva, S.B. Using calicioid lichens and fungi to assess ecological continuity in the Acadian forest ecoregion of the Canadian Maritimes. *For. Chron.* **2003**, *79*, 550–558. [CrossRef]
11. Rivas Plata, E.; Lücking, R.; Lumbsch, H.T. When family matters: An analysis of Thelotremataceae (*Lichenized Ascomycota*: Ostropales) as bioindicators of ecological continuity in tropical forests. *Biodivers. Conserv.* **2008**, *17*, 1319–1351. [CrossRef]
12. Angelstam, P.K. Maintaining and restoring biodiversity in European boreal forests by developing natural disturbance regimes. *J. Veg. Sci.* **1998**, *9*, 593–602. [CrossRef]
13. Whelan, R.J. *The Ecology of Fire*; Cambridge University Press: Cambridge, UK, 1995; p. 346.
14. Foster, D.R. The history and pattern of fire in the boreal forest of southeastern Labrador. *Can. J. Bot.* **1983**, *61*, 2459–2470. [CrossRef]
15. Eberhart, K.E.; Woodard, P.M. Distribution of residual vegetation associated with large fires in Alberta. *Can. J. For. Res.* **1987**, *17*, 1207–1212. [CrossRef]
16. Cyr, D.; Gauthier, S.; Bergeron, Y. Scale-dependent determinants of heterogeneity in fire frequency in a coniferous boreal forest of eastern Canada. *Landsc. Ecol.* **2007**, *22*, 1325–1339. [CrossRef]
17. Madoui, A.; Leduc, A.; Gauthier, S.; Bergeron, Y. Spatial pattern analyses of post-fire residual stands in the black spruce boreal forest of western Quebec. *Int. J. Wildland Fire* **2011**, *19*, 1110–1126. [CrossRef]
18. Keeton, W.S.; Franklin, J.F. Fire-related landform associations of remnant old-growth trees in the southern Washington Cascade Range. *Can. J. For. Res.* **2004**, *34*, 2371–2381. [CrossRef]
19. Dragotescu, I.; Kneeshaw, D.D. A comparison of residual forest following fires and harvesting in boreal forests in Quebec, Canada. *Silva Fenn.* **2012**, *46*, 365–376. [CrossRef]
20. Forestry Canada Fire Danger Group. *Development and Structure of the Canadian Forest Fire Behavior Prediction System*; Information Report ST-X-2; Forestry Canada: Science and Sustainable Development Directorate: Ottawa, ON, Canada, 1992; pp. 8–46.
21. Wotton, B.M.; Alexander, M.E.; Taylor, S.W. *Updates and Revisions to the 1992 Canadian Forest Fire Behavior Prediction System*; Information Report GLC-X-10; Canadian Forest Service: Sault Ste. Marie, ON, Canada, 2009; pp. 1–30.
22. Burgan, R.E.; Rothermel, R.C. *BEHAVE: Fire Behavior Prediction and Fuel Modeling System-Fuel Subsystem*; General Technical Report INT-167; U.S. Department of Agriculture, Forest Service: Rocky Mountain Research Station: Ogden, UT, USA, 1984; pp. 1–53.
23. Andrews, P.L. *BehavePlus Fire Modeling System: Past, Present, and Future*; U.S. Department of Agriculture, Forestry Service: Rocky Mountain Research Station: Missoula, MT, Canada, 2008; pp. 1–13.
24. Stratton, R.D. Assessing the effectiveness of landscape fuel treatments on fire growth and behavior. *J. For.* **2004**, *102*, 32–40.
25. Finney, M.A. An overview of FlamMap fire modeling capabilities. In *Fuel Management–How to Measure Success*; Andrews, P.L., Butler, B.W.C., Eds.; U.S. Department of Agriculture, Forest Service: Rocky Mountain Research Station: Portland, OR, USA, 2006; pp. 213–220.
26. Harvey, B. The Lake Duparquet research and teaching forest: Building a foundation for ecosystem management. *For. Chron.* **1999**, *75*, 389–393. [CrossRef]
27. Hély, C.; Fortin, M.J.; Anderson, K.H.; Bergeron, Y. Landscape composition influences local pattern of fire size in the eastern Canadian boreal forest: Role of weather and landscape mosaic on fire size distribution in mixedwood boreal forest using the Prescribed Fire Analysis System. *Int. J. Wildland Fire* **2010**, *19*, 1099–1109. [CrossRef]
28. Bergeron, Y. Species and stand dynamics in the mixed woods of Quebec's southern boreal forest. *Ecology* **2000**, *81*, 1500–1516. [CrossRef]
29. Vincent, J.S.; Hardy, L. L'évolution et l'extinction des lacs glaciaire Barlow et Ojibway en territoire québécois. *Geogr. Phys. Quat.* **1977**, *31*, 357–372.
30. Environment Canada. Canadian Climate Normals 1971–2000. Canadian Climate program; Atmospheric Environment Service: Downsview, ON, Canada, 2011. Available online: http://climate.weather.gc.ca/climate_normals/index_e.html (accessed on 10 October 2016).
31. Forests, Wildlife and Parks. Available online: http://www.mrn.gouv.qc.ca/forets/inventaire/fiches/couches-peuplements-ecoforestiers.jsp (accessed on 10 October 2016).

32. Dansereau, P.-R.; Bergeron, Y. Fire history in the southern boreal forest of northwestern Quebec. *Can. J. For. Res.* **1993**, *23*, 25–32. [CrossRef]

33. Hély, C.; Bergeron, Y.; Flannigan, M.D. Effects of stand composition on fire hazard in mixed-wood Canadian boreal forest. *J. Veg. Sci.* **2000**, *11*, 813–824. [CrossRef]

34. McRae, D.J.; Alexander, M.E.; Stocks, B.J. *Measurement and Description of Fuels and Fire Behavior on Prescribed Burns: A Handbook*; Report O-X-287; Canadian Forest Service, Great Lakes Forest Research Center: Sault Ste. Marie, ON, Canada, 1979; pp. 1–37.

35. Van Wagner, C.E. The line intersect method in forest fuel sampling. *For. Sci.* **1968**, *14*, 20–26.

36. Van Wagner, C.E. *Practical Aspects of the Line Intersect Method*; Report PI-X-12; Canadian Forest Service, Petawawa National Forestry Institute: Ottawa, ON, Canada, 1980; pp. 1–10.

37. Hély, C.; Bergeron, Y.; Flannigan, M.D. Coarse woody debris in the southeastern Canadian boreal forest: Composition and load variations in relation to stand replacement. *Can. J. For. Res.* **2000**, *30*, 674–687. [CrossRef]

38. Brown, J.K.; Oberheu, R.D.; Johnston, C.M. *Handbook for Inventorying Surface Fuels and Biomass in Interior West*; General Technical Report Int-129; U.S. Department of Agriculture, Forest Service, Rocky Mountain Research Station: Ogden, UT, USA, 1982; pp. 1–18.

39. Aubin, I. Végétation de Sous-Bois et Disponibilité de la Lumière Dans la Forêt Boréale du Sud-Ouest Québécois. Master's Thesis, Université du Québec à Montréal, Montréal, QC, Canada, 1999.

40. Paquette, M. Effets de la composition forestière initiale et du temps depuis le dernier feu sur la dynamique des combustibles et du comportement du feu dans la pessière à mousse de la ceinture d'argile du Québec. Master's thesis, Université du Québec à Montréal, Montréal, QC, Canada, 2011.

41. Terrier, A.; Paquette, M.; Gauthier, S.; Girardin, M.P.; Pelletier-Bergeron, S.; Bergeron, Y. Influence of fuel load dynamics on carbon emission by wildfire: old-growth forests in the Clay Belt boreal landscape generate less carbon. *Forests* **2016**. under review.

42. Hély, C. Influence de la Végétation et du Climat Dans le Comportement des Incendies en Forêt Boréale Mixte Canadienne. Ph.D. Thesis, Université du Québec à Montréal, Montréal, QC, Canada, 2000.

43. Van Wagner, C.E. *Development and Structure of the Canadian Forest Fire Weather Index System*; Forestry Technical Report 35; Canadian Forestry Service, Petawawa National Forestry Institute: Ottawa, ON, Canada, 1987; pp. 1–30.

44. Canadian Forestry Service. *Canadian Forest Fire Danger Rating System*; User's Guide; Canadian Forestry Service Fire Danger Group, Three-Ring Binder: Ottawa, ON, Canada, 1987; unnumbered publication.

45. Bergeron, Y.; Brisson, J. Fire regime in red pine stands at the Northern limit of species range. *Ecology* **1990**, *71*, 1352–1364. [CrossRef]

46. Stocks, B.J. Fire behavior in immature jack pine. *Can. J. For. Res.* **1987**, *17*, 80–86. [CrossRef]

47. Stocks, B.J. Fire potential in the spruce budworm-damaged forests of Ontario. *For. Chron.* **1987**, *63*, 8–14. [CrossRef]

48. Hély, C.; Flannigan, M.D.; Bergeron, Y.; McRae, D. Role of vegetation and weather on fire behavior in the Canadian mixedwood boreal forest using two fire behavior prediction systems. *Can. J. For. Res.* **2001**, *31*, 430–441. [CrossRef]

49. Stocks, B.J. Fire behavior in mature jack pine. *Can. J. For. Res.* **1989**, *19*, 783–790. [CrossRef]

50. Hirsch, K.G. *Canadian Forest Fire Behavior Prediction (FBP) System: User's Guide*; Canadian Forestry Service, Northern Forestry Centre: Edmonton, AB, Canada, 1996; pp. 1–122.

51. Rothermel, R. *A Mathematical Model for Predicting Fire Spread in Wildland Fuels*; Research Parper INT-116; U.S. Department of Agriculture, Forest Service, Rocky Mountain Research Station: Ogden, UT, USA, 1972; pp. 1–40.

52. Van Wagner, C.E. Conditions for the start and spread of crown fire. *Can. J. For. Res.* **1977**, *7*, 24–34. [CrossRef]

53. Rothermel, R. *Predicting Behavior and Size of Crown Fires in the Northern Rocky Moutains*; Research Parper INT-438; U.S. Department of Agriculture, Forest Service, Rocky Mountain Research Station: Ogden, UT, USA, 1991; pp. 1–44.

54. Societé de protection des forêts contre le feu (SOPFEU). Available online: http://www.sopfeu.qc.ca/fr/sopfeu/organisation/organisation (accessed on 10 October 2016).

55. Ouarmim, S.; Asselin, H.; Bergeron, Y.; Ali, A.A.; Hély, C. Stand structure of fire refuges in the eastern Canadian boreal mixedwood forest. *For. Ecol. Manag.* **2014**, *324*, 1–7. [CrossRef]

56. R Development Core Team. *R: A Language and Environment for Statistical Computing*; R Foundation for Statistical Computing: Vienna, Austria, 2007. Available online: http://www.R-project.org (accessed on 10 October 2016).

57. Cyr, D.; Bergeron, Y.; Gauthier, S.; Larouche, A. Are the old-growth forests of the Clay Belt part of a fire-regulated mosaic? *Can. J. For. Res.* **2005**, *35*, 65–73. [CrossRef]

58. Román-Cuesta, R.M.; Gracia, M.; Retana, J. Factors influencing the formation of unburned forest islands within the perimeter of a large forest fire. *For. Ecol. Manag.* **2009**, *258*, 71–80. [CrossRef]

59. Bradstock, R.A.; Bedward, M.; Gill, A.M.; Cohn, J. Which mosaic? A landscape ecological approach for evaluating interactions between fire regimes, habitat and animals. *Wildl. Res.* **2005**, *32*, 409–423. [CrossRef]

60. Barrett, S.W.; Arno, S.F.; Key, C.H. Fire regimes of western larch-lodgepole pine forests in Glacier National Park, Montana. *Can. J. For. Res.* **1991**, *21*, 1711–1720. [CrossRef]

61. Arno, S.F.; Reinhardt, E.D.; Scott, J.H. *Forest Structure and Landscape Patterns in the Subalpine Lodgepole Pine Type: A Procedure for Quantifying Past and Present Conditions*; General Technical Report INT-294; U.S. Department of Agriculture, Forest Service, Rocky Mountain Research Station: Ogden, UT, USA, 1993; pp. 1–17.

62. Burgan, R.E. *Concepts and Interpreted Examples in Advanced Fuel Modeling*; General Technical Report INT-238; U.S. Department of Agriculture, Forest Service, Rocky Mountain Research Station: Ogden, UT, USA, 1987; pp. 1–39.

63. Denneler, B.; Asselin, H.; Bergeron, Y.; Begin, Y. Decreased fire frequency and increased water levels affect riparian forest dynamics in southwestern boreal Quebec, Canada. *Can. J. For. Res.* **2008**, *38*, 1083–1094. [CrossRef]

64. Camp, A.E.; Oliver, C.D.; Hessburg, P.; Everett, R. Predicting late-successional fire refugia pre-dating European settlement in the Wenatchee Mountains. *For. Ecol. Manag.* **1997**, *95*, 63–77. [CrossRef]

65. Romme, W.H.; Knight, D.H. Fire frequency and subalpine forest succession along a topographic gradient in Wyoming. *Ecology* **1981**, *62*, 319–326. [CrossRef]

66. Camp, A.E. Predicting Late-Successional Fire Refugia from Physiography and Topography. Ph.D. Dissertation, University of Washington, Seattle, WA, USA, 1995.

67. Miyanishi, K.; Johnson, E.A. Process and patterns of duff consumption in the mixedwood boreal forest. *Can. J. For. Res.* **2002**, *32*, 1285–1295. [CrossRef]

68. Agee, J.K. The landscape ecology of western forest fire regimes. *Northwest Sci.* **1998**, *72*, 24–34.

69. Gardner, R.H.; Romme, W.H.; Turner, M.G. Predicting forest fire effects at landscape scales. In *Spatial Modeling of Forest Landscape Change*; Mladenoff, D.J., Baker, W.L., Eds.; Cambridge University Press: Cambridge, UK, 1999; pp. 163–185.

70. Ouarmim, S.; Asselin, H.; Hély, C.; Bergeron, Y.; Ali, A.A. Long-term dynamics of fire refuges in boreal mixedwood forests. *J. Quat. Sci.* **2014**, *29*, 123–129. [CrossRef]

71. Flannigan, M.D.; Logan, K.A.; Amiro, B.D.; Skinner, W.R.; Stocks, B.J. Future area burned in Canada. *Clim. Chang.* **2005**, *72*, 1–16. [CrossRef]

© 2016 by the authors. Licensee MDPI, Basel, Switzerland. This article is an open access article distributed under the terms and conditions of the Creative Commons Attribution (CC BY) license (http://creativecommons.org/licenses/by/4.0/).

Section 3:
Impact on Ecosystems

![forests logo] *forests*

MDPI

Review

Disturbance Agents and Their Associated Effects on the Health of Interior Douglas-Fir Forests in the Central Rocky Mountains

Andrew D. Giunta [1],*, Michael J. Jenkins [1], Elizabeth G. Hebertson [2] and Allen S. Munson [2]

[1] Department of Wildland Resources, Utah State University, Logan, UT 84322, USA; mike.jenkins@usu.edu
[2] U.S. Department of Agriculture, Forest Service, Forest Health Protection, Ogden, UT 84403, USA; lghebertson@fs.fed.us (E.G.H.); smunson@fs.fed.us (A.S.M.)
* Correspondence: adgiunta@gmail.com; Tel.: +1-801-560-8414

Academic Editors: Yves Bergeron and Sylvie Gauthier
Received: 18 January 2016; Accepted: 30 March 2016; Published: 6 April 2016

Abstract: Interior Douglas-fir is a prevalent forest type throughout the central Rocky Mountains. Past management actions, specifically fire suppression, have led to an expansion of this forest type. Although Douglas-fir forests cover a broad geographic range, few studies have described the interactive effects of various disturbance agents on forest health conditions. In this paper, we review pertinent literature describing the roles, linkages, and mechanisms by which disturbances, including insect outbreaks, pathogens, fire, and other abiotic factors, affect the development, structure, and distribution of interior montane forests primarily comprised of Douglas-fir. We also discuss how these effects may influence important resource values such as water, biodiversity, wildlife habitat, timber, and recreation. Finally, we identify gaps where further research may increase our understanding of these disturbance agents, their interacting roles, and how they influence long-term forest health.

Keywords: interior Douglas-fir forest; Douglas-fir beetle; western spruce budworm; disturbance; forest health; mixed-severity fire

1. Introduction

Disturbances exert strong influences over forest development and are expressed on a wide range of temporal and spatial scales [1–3]. Over the past century land management practices including timber harvesting, livestock grazing, and fire suppression have greatly altered disturbance regimes across the western USA. In mixed conifer forests the consequences have been increased tree densities, unnatural fuel accumulations and the expansion of fire-intolerant species (Figure 1) [4,5]. Particularly, this is the case where stand conditions are dissimilar to fire-adapted forests that historically had short-interval, low-severity surface fire regimes (e.g., ponderosa pine *Pinus ponderosa* Laws.) [6,7].

For example, the absence of frequent surface fires in some locations allowed for the expansion of shade-tolerant white-fir (*Abies concolor* Lindl.) and interior Douglas-fir (*Pseudotsuga menziesii* var. *glauca* Mirb. Franco) into open ponderosa pine stands. These species now form dense understories, effectively lowering canopy base heights, increasing ladder fuels, and elevating the hazard of high-severity fires [6,8].

Logging activities, including widespread clear-cutting during the 19th and early 20th century, throughout the interior west also created landscapes comprised of forests similar in size and age [9]. These forests are now reaching maturity, resulting in stands which are now suitable habitats for bark beetles [10]. In addition, warmer climate conditions favoring bark beetle success have led to an expansion of recent outbreaks, which have increased in severity and hectares infested [11,12]. A rise in bark beetle activity since the early 1990s has occurred across a range of forest types from low-elevation pinyon pine (*Pinus edulis* Engelm.) [13,14] to upper-elevation lodgepole pine (*Pinus contorta* Dougl. var. *latifolia* Engelm.) [15,16] and Engelmann spruce (*Picea engelmannii* Parry ex Engelm.) forests [17,18]. In interior Douglas-fir forests, the primary insect pest is the Douglas-fir beetle (DFB; *Dendroctonus pseudotsugae* Hopkins, Curculionidae: Scolytinae), which utilizes Douglas-fir exclusively [19,20].

Figure 1. Increased dead and down fuel following windthrow and Douglas-fir beetle colonization in a mixed conifer forest in the Dixie National Forest, Utah, USA (Photo: A. Giunta).

Since the implementation of the Forest Ecosystems and Atmospheric Research Act of 1988 [21] and the Healthy Forest Restoration Act of 2003 [22], studies investigating the roles of both natural and anthropogenic disturbances on forest health degradation and associated impacts on wildlife habitat, timber production, water quality, recreation, aesthetics, grazing, and biodiversity have increased [23,24]. More recently, interest has grown in understanding how multiple and different disturbances will interact and affect a landscape [25]. In the Rocky Mountains, multiple studies on the interactive effects of fire and insects have been conducted with an emphasis in subalpine spruce-fir forests [26–30].

The complexity of the interactions between multiple disturbance agents in interior Douglas-fir forests and subsequent forest health effects has not readily been quantified or assessed. Thus, from a management perspective, it is important to understand how the potential interactions of multiple disturbances affect ecosystem patterns and processes, and how these in turn affect the vulnerability and susceptibility of forests within ecosystems to subsequent disturbances [31].

In this paper, we used the published literature to construct a synthesis of disturbance agents that primarily regulate vegetative dynamics within interior Douglas-fir forests in the central Rocky Mountains. These disturbances include DFB, western spruce budworm (WSBW; *Choristoneura freeman* Freeman, Lepidoptera: Tortricidae) and Douglas-fir dwarf mistletoe (*Arceuthobium douglasii* Engelm.). We focus on how the interactions of these disturbance agents influence the distribution, development, structure and health of interior Douglas-fir forests within the central Rocky Mountains. We start with a discussion of interior Douglas-fir forest ecology and the effects of abiotic disturbance agents: fire, wind, snow avalanches, and their effects on these forests. We then discuss the role of biotic agents including DFB, WSBW, Douglas-fir dwarf mistletoe, root diseases, and anthropogenic influences (e.g., fire management, logging) and how each affects the health of these forests and their role as inciting agents to other disturbances. Finally, we identify gaps in our understanding of these agents, their interactions, and their relationship to managing forest health. This information is designed to assist land managers with making ecologically-based decisions and devising appropriate strategies for long-term management of interior Douglas-fir forests.

2. Interior Douglas-Fir Forests

The composition and structure of interior Douglas-fir forests are rich and diverse due to the influences of a unique suite of biogeoclimatic, genetic, and disturbance factors [32]. The complex plant community assemblages in these forests can also be attributed in part to the broad ecological amplitude of the dominant overstory species Douglas-fir, which is one of the most widely distributed conifers in western North America [33–35] (Figure 2).

Figure 2. Geographical distribution of coastal Douglas-fir (*Pseudotsuga menziesii* var. *menziesii*) outlined in green, and interior Douglas-fir outlined in blue. Digital representation from, [36].

Douglas-fir is highly adaptive to an array of site conditions that range across xeric to mesic gradients [35,37]. The geographic extent of the interior variety of this species extends from north-central British Columbia (55° N) to central Mexico (19° N) and is well established across an elevation range between 580 and 3500 m [38,39]. Throughout this range, climate and soil largely influence the site conditions of where this species will grow [40]. At southern latitudes, interior Douglas-fir distribution is limited by moisture availability, and is often restricted to north slopes at middle to high elevations in predominantly mesic sites [41,42]. For example, in the Santa Catalina Mountains of Arizona, Douglas-fir is the dominant conifer species above 2450 m [43].

In the northern portion of its range, the majority of the precipitation falls as snow, while in its southern distribution within the US (southern Utah, Arizona, New Mexico), precipitation is most abundant during the growing season, due to the influence of monsoonal moisture [44]. At northern latitudes, its growth is influenced by the length of the growing season and limited by cold temperatures [45]. The overall climate experienced by interior Douglas-fir throughout the central Rocky Mountains is characterized as a continental climate consisting of long, cold winters and hot, dry summers.

Unlike coastal Douglas-fir which is considered moderately shade-intolerant and is succeeded by more shade-tolerant western hemlock (*Tsuga heterophylla* Raf. Sarg.) and western red cedar (*Thuja plicata* Donn ex. D. Don), interior Douglas-fir is considered fairly shade-tolerant and is generally considered a climax species [46].

In the central Rocky Mountains, interior Douglas-fir is largely distributed within the mid-elevation montane zone which ranges between 900 and 1500 m [45]. In the overlapping montane and subalpine zones, Douglas-fir intermixes with spruce-fir forests dominated by subalpine-fir (*Abies lasiocarpa* Hook. Nutt.) and Engelmann spruce, with scattered pockets of limber pine (*Pinus flexilis* James) and Great Basin bristlecone pine (*Pinus longaeva* Bailey). At mid-elevations, Douglas-fir occurs with lodgepole pine and white fir [47,48]. At the lower end of its elevation range, Douglas-fir is often dispersed with ponderosa pine and woodlands comprised of piñon pine (*Pinus edulis* Engelm.), juniper (*Juniperus* spp.), bigtooth maple (*Acer grandidentatum* Nutt.), and Gambel oak (*Quercus gambelii* Nutt.) [49]. Common understory associates include ninebark (*Physocarpus malvaceus* Greene, Kuntze), mountain snowberry (*Symphoricarpos oreophilus* A. Gray) [50,51], chokecherry (*Prunus virginiana* L.), big sagebrush (*Artemisia tridentata* Nutt.), serviceberry (*Amelanchier alnifolia* Medik.), and currants (*Ribes* spp.) [47,51,52].

Commercially, interior Douglas-fir forests are an important resource for the forest products industry, providing lumber, plywood, house logs, and fuel wood [40,53]. Interior Douglas-fir forest communities also provide a critical wildlife habitat for a variety of bird species. These include Ruby Crowned Kinglets (*Regulus calendula*), Evening Grosbeaks (*Coccothraustes vespertinus*), Western Flycatchers (*Empidonax occidentalis*), and Northern Goshawks (*Accipiter gentilis*), which require habitats associated with mature forests [53,54].

3. Abiotic Disturbance Agents

3.1. Fire

Wildfires are one of the most important disturbance agents strongly influencing vegetative patterns across North America [55–57]. The effects of fire over a landscape are measured using a multitude of parameters including frequency, intensity, severity, and the spatial and temporal extent of a burn [58–61]. Collectively, these measures constitute the basis for describing an environments' fire regime [62]. The most common method for classifying a fire regime is through a severity index which qualitatively describes how fire intensity affects an ecosystem, and is often related to the amount of biomass lost above and below ground [63]. High-severity fire regimes are characterized as those where fire transitions from surface fuels into the crowns of trees, consuming a majority of overstory vegetation [59]. Fire of this type is termed crowning when the fire is actively spreading from tree crown to tree crown [64,65]. In contrast, low-severity fire regimes are typified by frequent (4–30 year) low-intensity fires where surface fuels, including litter, moss, and herbaceous material, are charred or consumed while overstory canopy is minimally damaged or killed [63].

Interior Douglas-fir forests including those mixed with ponderosa pine throughout the central Rocky Mountains are characterized by a mixed-severity fire regime [37,66], one of the most complex and under-studied fire regimes in the western US [67,68]. Under this classification, forest stands experience natural fires across severity levels that range from low to medium to high [35], and with a variable fire return interval between 30 and 100 years [69]. The complexity of the fire regime is driven by the combined influence of both frequent low-severity surface fires and infrequent high-severity stand-replacing fires that create forest stand mosaics across the landscape varying in tree age and density [70–72]. Throughout the central Rocky Mountains, two prominent mixed conifer forest types dominate the landscape. The warm-dry type experiences more frequent non-lethal fires, and the cool-moist type experiences infrequent lethal fires that create even-age patches [73]. Overall, "individual mixed-severity fires typically leave a patchy, erratic pattern of mortality on the landscape, which fosters development of highly diverse communities" [69] (p. 226).

3.1.1. Direct Fire Effects

The direct effect of fire leads to either instantaneous tree mortality during initial fire passage, or delayed mortality resulting from severe injury through damage to foliage, cambium, fine roots, and conductive tissues, affecting physiological processes which are important for tree growth and development [74,75]. In the forest canopy, two types of crown damage determine the likelihood of fire-induced tree mortality. These include crown scorch, where the foliage is killed by hot gases above the flames, and crown consumption, where foliage and occasionally small twigs directly support combustion [76]. Two important parameters for predicting post-fire tree mortality associated with crown scorch include crown scorch volume and crown scorch height. Crown scorch volume is measured as the percent of the crown scorched [77–79] and crown scorch height is the level where heat is lethal to living foliage [76]. Scorch height is dependent upon fireline intensity, wind speed, and air temperature. The physiological effects of crown scorch can lead to a decrease in carbohydrate production, further weakening a tree's response to stress and lowering its resistance to insects, drought, and other disturbances [80].

Heat-induced damage to tree boles can also affect a tree's likelihood of survival after a wildfire [80,81]. Bole charring resulting in cambial death is dependent upon both the amount of heat received by a tree and the insulating capacity of the bark [81]. Older, large-diameter trees tend to have thicker bark with a greater capacity for absorbing heat, thus providing greater resistance to injury [82,83]. Bark of mature Douglas-fir is often comprised of a high percentage of cork, which can aid in the thermal diffusion of heat [81]. Douglas-fir stands with a greater proportion of large-diameter trees are likely to survive low-intensity fires. In the event bole scorch does not produce a fatal response through cambial injury, partial basal girdling and root damage may lead to moisture stress and reduced resistance to insects and diseases [79,84].

Sustained smoldering combustion of litter, duff, or downed woody material within surface and ground fuel layers can lead to root injury and mortality. Soil temperature, soil moisture, root spatial distribution, heat residence time, and fuel loading greatly influence the degree of root damage during a burn [75,85]. Temperatures as low as 48–60 °C have been attributed to root desiccation or death [86,87]. Swezy and Agee [88] found that prescribed surface fires in ponderosa pine stands in Oregon, USA , led to lethal temperatures (greater than 60 °C) that penetrated five centimeters in soil depth, affecting the greatest concentration of fine root mass (1–2 cm diameter). Other studies also concluded that both low- and high-severity burns reduce overall fine root mass [89,90]. Although these studies were conducted in ponderosa pine stands, interior Douglas-fir have shallow lateral roots that are also susceptible to fire damage [91]. Loss of root biomass can have significant implications for decreased essential nutrients and water, and can increase stress and susceptibility of affected stands to insects and diseases [92,93]. Furthermore, root systems anchor soil to prevent erosion, and a reduction or loss of root systems can increase runoff [94].

3.1.2. Indirect Fire Effects

Beyond direct mortality and consumption of forest biomass, fire can have many indirect effects on interior Douglas-fir forests. The disparity in fire severity is related to site influences including topography, aspect, and fuel loading [95], and will create unique fuel complexes in each stand influenced by microsite temperature, precipitation, fuel moisture content, stand densities, and the presence or absence of ladder fuels [96,97]. Variable fire intensities within the mixed-severity regime drive the composition of forests comprised of seral, fire-dependent species and mature fire-resistant species forming multistoried, mixed-aged stands [69,98]. At a landscape scale, even-age forest structures are most common where stand-replacing fires are prevalent, and between these even-age patches are mixed-aged stands where frequent surface fires are dominant [99].

Human activities during Euro-American settlement in the western US altered fire regimes in interior Douglas-fir forests. Throughout the past century, fire suppression actions stemming from fire exclusion policies dating back to the early 20th century have reshaped the landscape. These actions led to extended fire-free periods in montane forests which have allowed understory conifers to develop in formerly open stands [100] (Figure 3).

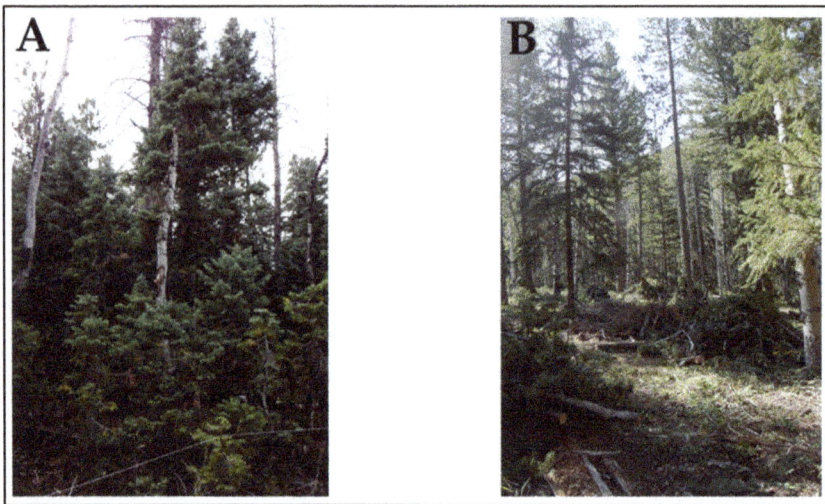

Figure 3. (A) Dense regeneration of white-fir and interior Douglas-fir saplings in the understory of an unmanaged interior Douglas-fir stand in the Dixie National Forest, Utah, USA (Photo A: A. Giunta); **(B)** A managed interior Douglas-fir stand in the Ashley National Forest, Utah, USA (Photo B: D. Malesky, USDA Forest Service, Forest Health Protection, Ogden, UT, USA).

In a fire reconstruction study in southwestern Montana, an increase in Douglas-fir density within grassland-sage communities coincided with the exclusion of surface fires due to anthropogenic influences starting in the mid-1880s [101]. The escalation in understory conifer growth has contributed to an increase in ladder fuels, providing a mechanism for the transition of surface fires into forest canopies, increasing the potential for crowning and the occurrence of high-severity, stand-replacing fires [69]. Fire exclusion–induced changes in forest composition, structure, and fuel loads affect interior Douglas-fir stands and may alter historic low-severity fire regimes [102]. Episodic droughts coupled with dense canopy cover, close intercrown distances and large fuel accumulations create environments that are conducive to extreme fire weather favoring the initiation and spread of crown fires.

3.2. Additional Abiotic Disturbances

Wind and snow avalanches are two additional natural disturbance agents that influence forest composition, structure, and forest soils [103–106]. Primary effects of wind include damage and breakage to the tops of crowns, branch breakage, uprooting, and snapping of trees [107]. Wind events can create canopy gaps that vary in size from a few individual trees to landscape scale (hundreds of hectares) [108]. Gap openings increase available light used by surrounding trees for increased growth, or benefit suppressed trees in the understory [105]. Furthermore, windthrown trees create, "suitable bark beetle habitat, increase fuel loads, and limit mobility of wildlife and forest recreationists" [109] (p. 446). Uprooted trees also expose soil and creates heterogeneity in soil properties [110].

Forest snow avalanches are typically small in size, but large infrequent events can be destructive [111]. Avalanche paths form where there is an abundance of snowfall through natural storm deposition or wind transport in steep terrain (greater than 30° slope angles), allowing for the release and acceleration of a snowslide [112,113]. Much like wind, snow avalanches affect forests through the breakage of stems and branches, uprooting, and the creation of severe wounds on the uphill side of trees [114–116]. Large infrequent avalanches often kill overstory trees with little impact on regeneration and can create a rapid buildup of large, coarse, woody material that gets deposited throughout an avalanche run-out zone [117–119].

Although these two disturbance agents are most closely associated with subalpine and tree line forest zones [120,121], wind events and avalanches do at times affect the interior Douglas-fir zone. McGregor *et al.* [122] reported a strong wind event in November 1981, which blew down thousands of trees in forests in Idaho, USA. Many windthrown Douglas-fir were selected for use in trials of the DFB anti-aggregate pheromone MCH (3-methyl-2-cyclohexen-1-one) following this event. Within the Wasatch Mountains of Utah, USA, snow avalanches that occurred in interior Douglas-fir forests created extensive debris piles that subsequently became infested by DFB. Trapping and MCH application by US Forest Service Forest Health Protection staff were employed to prevent DFB spread into neighboring stands [123–125] (Figure 4).

Figure 4. (**A**) Snow avalanche path through an interior Douglas-fir forest in the Ashley National Forest, UT, USA (Photo A: D. Blackford, USDA Forest Service, Forest Health Protection, Ogden, UT, USA); (**B**) Interior Douglas-fir debris in the avalanche runout zone, Ashley National Forest, UT, USA. (Photo B: D. Blackford).

4. Biotic Disturbances

4.1. Douglas-Fir Beetle

Forest stand development patterns in montane interior Douglas-fir forests are not regulated by fire alone. Bark beetles are also a major disturbance agent that have a large ecological role in reshaping forests [12,24]. Interactions between bark beetles and their hosts have coevolved over the past 200 million years [126]. Insect-induced tree mortality influences the development, senescence, and rebirth of stands, which in turn affects energy flows and nutrient cycles [127]. *Dendroctonus* species (Coleoptera: Curculionidae, Scolytinae) are particularly capable of reshaping stand structure, composition, and function [128–130]. Endemic populations attack old, large, and weakened trees, which removes trees from the overstory and promotes the recruitment of the next generation of trees in a stand [12]. Periodically, epidemic populations occur and are able to kill live, healthy trees in great numbers [131,132].

Within interior Douglas-fir forests, the most prevalent bark beetle species is the DFB [19,20,123,133]. Density-independent factors that influence the population dynamics of this insect include the availability and suitability of host trees, weather conditions, and disturbances (e.g., windthrow, avalanches) that produce downed host material [134,135]. Often, freshly felled or downed trees greater than 20 cm in diameter, contain sufficiently thick phloem with essential nutrients that are required for successful brood production [136,137]. Newly felled trees typically lack effective defense mechanisms including a decrease or cessation of resin production which makes them attractive targets. Resin is a key compound containing monoterpenes and sesquiterpenes that entrap or elevate toxin levels fatal to beetles or pathogens vectored by beetles [138,139]. In previous studies, trees with low oleoresin pressure have been associated with an increased susceptibility to bark beetle attacks [140]. Consequently, high DFB population increases are often related to disturbances including windthrow or snow avalanches that produce an abundance of downed trees [141], or trees physiologically weakened by drought [142], fire [84,143], ice damage [144], defoliation [145], and diseases [146,147]. Evidence of successful host colonization is determined by the presence of entrance holes, emergence holes, egg galleries, and frass accumulations on the bole and near the base of trees [133,148,149].

Stand and bark beetle population dynamics are often highly interrelated [150]. Coulson [150] stated, "Tree age, diameter, and phloem thickness are all correlated and in turn are related to beetle survival, *i.e.*, the large-diameter trees with thick phloem accommodate large beetle populations and have high survival rates" (p. 433). In stands with mixed diameter classes, the percentage of trees killed during outbreaks is related to tree diameter, with the greatest number of trees being killed in the diameter class representing the highest basal area [151].

Unlike lodgepole pine or spruce/fir forests which typically form dense, uniform, even-age forests in the central Rocky Mountains as a result of stand-replacing fire or logging, the intrinsic characteristics of interior Douglas-fir forests combined with the generally less aggressive nature of DFB limit landscape-scale DFB-induced mortality. Throughout the central Rocky Mountains, interior Douglas-fir stands often occur in mixed-species stands, or in small groups surrounded by non-host trees at the edge of their upper and lower elevation limits. Thus, suitable DFB hosts are typically distributed unevenly throughout a forest [152]. Where Douglas-fir is a dominant overstory component, past logging and fire history have created forest mosaics in which Douglas-fir oscillates in age and density, limiting the extent of potential hosts [135]. Typically, small groups of trees are attacked [153]. However, under certain conditions, drought coupled with a supply of recently downed trees can facilitate the development of DFB populations from endemic into epidemic levels. At this stage of an outbreak, DFB are able to overcome host resistance and initiate attacks on standing live trees, where groups of 100 or more can become infested [129,153] (Figure 5).

Figure 5. Interior Douglas-fir stand conditions before (**A**); during (**B**); and following (**C**) Douglas-fir beetle colonization in Utah, USA. (**A**) An uninfested phase—green (G); (**B**) currently infested phase—red (R); and (**C**) older mortality phase—gray (GY). (Photo A: M. Jenkins, B: A. Giunta, C: M. Jenkins).

A stand hazard rating system developed by Weatherby and Thier [154] for southern Idaho, USA, suggests that the highest potential for tree mortality due to a DFB outbreak is in stands with basal areas greater than 23.2 m^2/ha, a proportion of Douglas-fir greater than 50%, an average stand age above 120 years, and an average diameter at breast height greater than 50 cm.

4.2. Western Spruce Budworm

Insect defoliators have an important influence on the condition of interior Douglas-fir forests. Some of the most important impacts of insect defoliation are tree mortality, rotation delays, and increased susceptibility to secondary insects and disease [155]. The WSBW is considered one of the most widespread and destructive defoliators in western coniferous forests, particularly where Douglas-fir and true firs are the primary tree species in a stand [156–158]. Depending on environmental and biological conditions, the timing of WSBW outbreaks is highly variable. The periodicity and duration of outbreaks can range from two to over 35 years [159,160].

The life history requirements for WSBW are highly interdependent upon forest stand structure and conditions. This insect preferentially feeds on the current years' growth where larvae penetrate swelling buds that have the highest food quality and offer the best protection from predators [145,161]. Bud phenology, specifically bud development and the timing of budburst, can greatly influence WSBW population dynamics. Trees that exhibit delayed budburst have been associated with a reduction in WSBW success as new bud formation occurs after second instar larvae emerge from hibernation and initiate feeding [162]. Douglas-fir forests with a large proportion of trees that are genetically predisposed to delayed budburst will likely have a greater resistance to WSBW infestations. Site location can also affect biological processes that influence WSBW survival. Sites with warm soils and warmer microclimates have been linked to earlier budburst timing [163]. Dry sites situated along south and west aspects where earlier bud development coincides with larvae feeding after winter emergence can be associated with higher larvae survival. Often, stands with fewer host trees have been shown to have lower levels of mortality within various size classes [132].

Multi-age, multi-level forest canopies in stands dominated by host trees provide optimal WSBW habitat as second instar larvae are dependent upon a successful canopy descent to reach host resources [164]. Weather factors including wind also exert a large control over the success rate of locating a suitable host. During the past 50 years, singular overlapping and repeated outbreaks of WSBW have greatly altered the structure and composition of montane forests along the Colorado Front Range [132].

The greatest impact of WSBW within infested stands is on subcanopy and understory layers where larvae feed on host regeneration within the understory (Figure 6). Conifer seedlings and saplings have relatively few needles and buds, and new growth can become deformed or killed by only a few larvae [157]. In one study, Hadley and Veblen [165] used dendrochronological analysis to reconstruct past WSBW and DFB attacks throughout the Colorado Front Range. Results from their study indicated WSBW outbreaks were responsible for high (greater than 50%) mortalities of seedlings, saplings, and small-diameter trees. Future regeneration within a stand is further impeded by WSBW feeding on developing cones and seeds [166]. Frank and Jenkins [167] found that a higher percentage of larvae feed on seed cones as opposed to pollen cones. This could affect future regeneration since Douglas-fir is known to have infrequent cone crops every two to seven years at lower elevations [168], and every one in 11 years at higher elevations [169].

Western spruce budworm also negatively affects overstory host trees. Consecutive years of feeding can lead to decreased stem growth, top kill, and, in some cases, tree mortality [170]. A loss in tree volume due to decreased growth rates can lead to an overall decrease in a stand's basal area, which could impact timber harvest projections if merchantable stands were to become infested.

Often, the absence and, more importantly, release patterns of growth rings in mature Douglas-fir and other host species (e.g., white fir), coincide with WSBW outbreaks [171,172]. Using tree-ring reconstructions, Swetnam and Lynch [172] found that overstory trees in Devil's Gulch, located in northern Colorado, USA, experienced 60% mortality from WSBW feeding. In another tree-ring study conducted near Pemberton, British Columbia, Canada, Alfaro *et al.* [173] found a 39% reduction in the number of host trees per hectare within three years after a WSBW outbreak. Most sampled stands experienced host growth reduction evidenced by reduced tree ring widths during WSBW outbreaks. This study also indicated older, less vigorous stands with suppressed Douglas-fir trees were most susceptible to WSBW infestations. In mixed-species stands such as Douglas-fir/ponderosa pine, WSBW outbreaks tend to shift species dominance towards ponderosa pine [120,165].

Figure 6. Western spruce budworm defoliation on interior Douglas-fir saplings in southern Idaho, USA. (Photo: Carl Jorgensen, USDA Forest Service, Forest Health Protection, Boise, ID, USA).

4.3. Douglas-Fir Dwarf Mistletoe

Dwarf mistletoes (*Arceuthobium* spp.) are one of the most important, widespread disease agents in North American conifer forests, and are found throughout montane forest ecosystems [174–176]. All mistletoe species are host specific and slow spreading, making stand composition, tree size, and structure important for their persistence in a forest community [177]. The plants form obligate hemiparisitic relationships with host plants, extracting vital water and minerals through haustorium from their hosts [178,179]. This process depletes essential photosynthetic reserves used for growth and maintenance by host trees [176,180]. Although dwarf mistletoes are capable of complete photosynthesis, upwards of 60% of their carbohydrates can be extracted from their hosts [179]. Tree response to infection results in dense abnormal growth of host twigs that form branch clusters termed witches' brooms [181]. This irregular growth pattern changes branch structure, function, and can eliminate cone production by infected branches [182,183]. Additional degenerative, induced effects on host plants include stem and height growth reductions, top kill, and reduced forest productivity [181,184].

Douglas-fir dwarf mistletoe is the most damaging species that parasitizes Douglas-fir [174,185]. Spread of this disease is initiated through the movement of the parasite to previously uninfected branches of a single tree or between trees [184]. The female plant produces fruit and seed that mature in fall (September, October). Seeds are under high internal water pressure within the fruit which, when abscised from the parent plant, are explosively propelled through the air at upwards of 22 m per s. Seeds contain a sticky viscin coating which allows them to attach to hosts [186]. Spread rates of Douglas-fir dwarf mistletoe are often accelerated in multi-storied Douglas-fir stands where understory trees receive abundant seed rain from infected overstory trees, as reported in southwestern interior Douglas-fir stands [184]. When Douglas-fir is the climax member of the community, as often is the case in interior Douglas-fir stands, there is typically not a shortage of hosts for Douglas-fir dwarf mistletoe, which can persist unless a severe disturbance leads to the loss of its host species [181]. Seedlings and saplings, especially those with main stem infections, readily succumb to this parasite [186]. The presence of non-host species can slow the spread of the disease agent. Stands with open canopies are often more susceptible compared to dense stands, because stands with high densities create shading conditions which retard Douglas-fir dwarf mistletoe growth [186].

4.4. Root Diseases

Pathogens and, in particular, root diseases are an important component in forest ecosystems, and they exert strong influences over forest dynamics including structure, composition, and function [187,188]. Within a stand, fungi spread via spores transported by rhizomorphs in the soil, or through direct root contact between infected and uninfected hosts [189]. Specifically, fungi infect the cambial tissues of roots and root collars where root tissues have evidence of staining and decay. The foliage of root-diseased trees typically appears chlorotic and thin. Trees lose needles from the lower crown upward, and from the inside (near the stem) outward [190]. Trees sometimes respond to infection by producing copious amounts of resin near the base of the stem [191]. Trees may also produce a stress cone crop in response to infection [189]. Eventually, damage to roots leaves trees girdled and host trees die. Connections between infection sites create root disease centers characterized by circular openings in the main canopy that range from approximately one-tenth of a hectare to 400 or more hectares in size [189]. These "mortality centers" are associated with groups of dying and dead conifers. One issue with root diseases is that they are persistent in a stand, and can survive as saprophytes on dead wood material for decades [189]. The severity of infestation is often amplified by disturbances including fire suppression and logging where fungi can colonize stumps and roots of cut trees and eventually spread to healthy trees [187,190,192].

In the Rocky Mountains, *Armillaria* spp., and specifically *Armillaria ostoyae*, is the most important and widespread of all root pathogens [193,194]. It has a broad host range including *Abies* and *Pinus* species. In dry, interior conifer forests, it aggressively infects interior Douglas-fir, colonizing and killing healthy trees in all age classes [190]. The patterns of *Armillaria* spread follow two main pathways: either via distinct mortality patches, circular in nature with mortality mostly confined to the leading edge of the patch, or as dispersed mortality, forming continuous coverage over a site [195,196].

This pathogen greatly affects forest community structure. Once established, a slow progression of the fungi into non-infected portions of a stand creates an initial pulse of mortality. Once trees along the edge die, canopy gaps are created which benefit the regeneration of the next tree cohorts. Seedlings and saplings are also vulnerable to this disease which can retard the development of the stand. Surviving trees, once reaching maturity, become a vital resource that can support further development of the disease which continually cycles its way through a stand, forming a wave pattern of mortality [195].

Species composition within a stand can be affected by the creation of canopy gaps. In the northern Rockies (interior British Colombia, Canada), gaps created by the mortality of interior Douglas-fir associated with *Armillaria* may become filled with more disease-resistant and shade-tolerant western hemlock, western red cedar, or even subalpine fir, though this species is also susceptible to *Armillaria* [187,191]. In the central Rocky Mountains, Douglas-fir which is often the climax overstory species in its community would likely continue to persist in a stand.

Once *Armillaria* is established, infected trees become susceptible to windthrow or fall over on their own from weakened root systems [141,193]. This can lead to an increase in hazardous trees if root disease centers are located in developed recreation areas including campgrounds or trailheads [197]. Furthermore, root disease also affects tree species not infected by creating openings in forests where healthy trees along gap margins can be exposed to high winds and subsequent windthrow [141].

5. Disturbance Interactions

Interactions between disturbance agents including fire, DFB, WSBW, Douglas-fir dwarf mistletoe, and root diseases affect the health of interior Douglas-fir forests, and can predispose forests to subsequent disturbances (Figure 7). Specific disturbance agent interactions are addressed below.

5.1. Fire and Douglas-Fir Beetle

Following non-stand-replacing fire events, wildfire impacts can predispose stands to subsequent bark beetle disturbances. Trees impacted by crown scorch, bole charring, and root damage associated with fire become attractive targets for DFB [198,199]. Furniss [137] reported 70% of fire-injured Douglas-fir were infested by DFB following a fire in southern Idaho, USA. The proportion of attacks was highest in trees that experienced low-to-moderate cambium injury. Trees with high levels of cambium injury experienced fewer attacks as phloem and other essential resources were damaged beyond utilization by beetles. In other studies, DFB mass attacked trees with 25%–50% cambial damage and greater-than-50% crown scorch [200,201]. Colonization patterns following fire progress from fire-damaged trees to healthy live trees over time as suitable phloem resources became scarce in successive fire-damaged classes [143,200]. Trees completely defoliated by crown fires also result in the complete burning or severe scorching of the inner bark, especially in thin-barked trees, and were not suitable for bark beetle use [198,199].

Bark beetle attack dynamics affect forest structure differently compared to fire. With low-intensity fire, often smaller-diameter and younger tree cohorts are killed, while larger-diameter trees survive, because "temperatures in the plume at the height of the canopy are too low" [202] (p. 483). Thus, mature trees within a stand continue to produce seeds contributing to regeneration. With bark beetle attacks, the Douglas-fir beetle seeks old, large-diameter trees, which are also mature, seed-producing trees resulting in stands with a younger age class, and often a reduction in reproductive output [203–205]. Following low-intensity fire, canopy structure is likely to remain intact aside from occasional torched trees. Surviving trees maintain shade cover, which can increase soil moisture retention benefiting regeneration and other established vegetation. Furthermore, periodic low-intensity fire reduces the amount of surface fuels and decreases the overall fuel load in a forest. In contrast, bark beetle colonization contributes to an increase in litter loading when dead needles begin to fall from a tree, a one- to four-year period post-attack [203]. Increased amounts of coarse woody debris can accumulate on the forest floor following overstory tree mortality. The loss of overstory trees creates canopy openings that favor the growth of herbaceous vegetation and grasses [11,129].

5.2. Douglas-Fir Beetle and Forest Fuel Changes

Only recently have the interactive effects between bark beetle–induced changes and fuel complexes been studied thoroughly. Much of the recent research is associated with forest systems that experience infrequent high-severity fire regimes (e.g., lodgepole pine forests infested by mountain pine beetle [206,207], and Engelmann spruce forests infested by spruce beetle [29,208]. Jenkins *et al.* [129] provide an extensive review of fuel complex changes during typical bark beetle rotations in these forest types. It has been hypothesized that DFB-induced alterations to fuel complexes will differ from upper elevation forest types due to drier sites, more open stand conditions, lower biomass loads, and lower tree and canopy base heights [37,57,209]. The spatial pattern of DFB-caused tree mortality also complicates how fuel complexes change within stands.

During a DFB outbreak, the most notable changes begin in the canopy where dead tree foliage begins to desiccate, fading from green to yellow and finally to red. Dead needles drop to the forest floor one to four years post-colonization where in increase of fine surface fuels occurs (litter, woody material less than 7.62 cm in diameter) through branch and canopy breakage. With the loss of overstory cover, a slight increase in herbaceous material follows as increased sunlight reaches the forest floor. Over time (20 years or longer), dead, standing beetle-killed snags begin to fall, increasing the amount of large woody material (greater than 7.62 cm) in the surface fuels layer (Figure 8) [37,129,210].

Donato *et al.* [37] inventoried changes to surface and aerial fuels in interior Douglas-fir forests across four different DFB outbreak stages. This included green (unattacked), red (recent mortality, one to three years post-attack), gray (older mortality, 4–14 years post-attack), and silver (older mortality, 25–30 years post-attack) across a range of Douglas-fir habitats in the Greater Yellowstone Ecosystem (GYE). The results from their study indicated that significant reductions in available canopy fuel load and bulk density occurred as time increased post-outbreak (4–14 years), while significant changes to surface fuels were minor, aside from an increase in 1000 h fuels during the silver stage. In a similar study conducted in northern Utah, Giunta [210] found that in surface fuels, a significant increase in litter depth (cm) and litter loading (kg/m^3) was associated with needle loss during the time bark beetle–killed trees began losing red needles. With canopy loss, an increase in herbaceous biomass followed in stands where the majority of trees were grey as increased levels of light were able to infiltrate the forest floor. In other studies, DFB-induced changes to interior Douglas-fir stands resulted in the basal area of 40%–70%, a reduction in the mean diameter at breast height of 8%–40%, and a three-fold increase in grasses and herbaceous plants in infested stands [21,211].

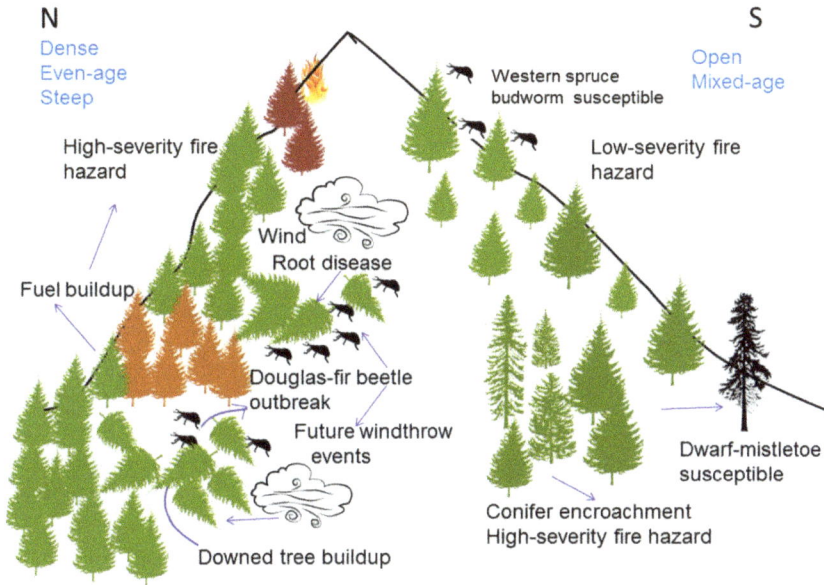

Figure 7. Conceptual schematic of interacting biotic and abiotic disturbance agents and associated forest health issues in interior Douglas-fir forests. (Illustration: A. Giunta).

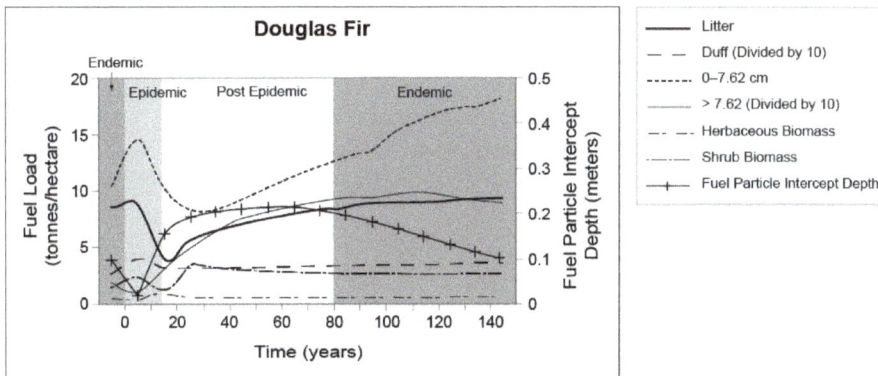

Figure 8. Select fuel characteristic type changes in interior Douglas-fir forests over the course of a Douglas-fir beetle rotation. (Figures and data from Jenkins *et al.* [129]).

Alterations to canopy fuels have been attributed to a series of physiological changes directly resulting from the colonization of host trees by beetles and the introduction of the associated blue-stain fungi, principally the species *Ophiostoma pseudotsugae* (Rumb.) von Arx (*Ceratostomella pseudotsugae* Rumbold) [212]. Blue-stain fungi penetrate the sapwood and phloem tissues adjacent to larval galleries, contributing to the disruption of water transport between root systems and foliage [139]. Needles desiccate over time and foliage fades from green to red [37]. It is during the red needle stage where the probability of torching and crowning can increase as canopy foliar moisture declines to its lowest moisture content level and dead needles are still retained in the canopy [37,129].

Giunta [210] measured foliar moisture content in infested Douglas-fir trees across all four crown condition classes (green, green-infested, yellow, and red) associated with DFB outbreaks. His findings showed that yellow and red crown condition classes had a significantly ($p < 0.0001$) lower foliar moisture content based on percent of oven-dry weight compared to green and green-infested foliage.

5.3. Western Spruce Budworm and Fire

Many open-canopy stands that were once maintained by frequent surface fires have been replaced by densely-stocked, closed canopy stands composed of mixed-age shade-tolerant species across the landscape compared to pre-settlement forested landscapes [171,213].

The effects of WSBW on fuel loads, fire occurrence, and fire behavior are not fully understood, but are certainly different from DFB [214]. Western spruce budworm directly alters aerial fuels by consuming the current year's needles, whereas DFB does not feed on needles, and needle loss occurs once a tree is dead. A reduction in canopy bulk density decreases the likelihood of torching and crown fire initiation and spread [215]. Although no studies on the effect of WSBW on fire and fuel loads have been reported for the central Rocky Mountains, research has been conducted in Douglas-fir stands in the Pacific Northwest, USA. Flower *et al.* [216] examined dendrochronological records of fire and WSBW outbreaks in interior Douglas-fir forests in Oregon, USA, and western Montana, USA. Their research showed no synchronous pattern between WSBW outbreaks and increased fire occurrence. In another study after a WSBW outbreak in Washington, USA, Hummel and Agee [217] measured decreases in canopy closure and a reduction in the density of small-diameter (less than 20 cm) trees over an eight-year period. Incorporating these inputs from stand data, they used the fire behavior model BEHAVE [218] to simulate fire spread in their study site. Crown characteristics that contributed to crown fire initiation (e.g., canopy base height and canopy bulk density) remained stable and did not indicate any significant increase in crown or torch potential. In a similar study (central Oregon to western Montana, USA), Gavin *et al.* [214] used the Wildland-Urban Interface Fire Dynamic Simulator (WFDS) to model fine-scale fuel changes associated with WSBW infestations. They found defoliation consistently reduced both the vertical and horizontal spread of crown fire across a range of surface fire intensities. They also discovered that a row of defoliated trees released substantially less heat compared to a row of non-defoliated trees due to the lack of fuel and associated decrease in flame intensities. Hummel and Agee [217] also found that coarse, woody fuel loads increased by 50% following a WSBW outbreak in the central Cascades, USA. Their plot data inputted into surface fire models predicted a significant increase in surface fire flame lengths. Site-specific fuel models are important to achieve better fire behavior predictions [218].

5.4. Dwarf Mistletoe, Fire, and Other Disturbances

Episodic natural fires help prevent Douglas-fir dwarf mistletoe from spreading by continually removing infected overstory trees from stands and killing infected and uninfected understory tree hosts [209]. Ingrowth and expansion of Douglas-fir due to the lack of fire have increased the abundance of susceptible hosts. Stand density increases have raised fuel accumulations and increased the spread of Douglas-fir mistletoe across the interior west [186].

Douglas-fir dwarf mistletoe impacts on infected trees primarily affect canopy fuels of infected trees. Koonce and Roth [219] reported 73% greater aerial fuels (live and dead witches' brooms) in dwarf mistletoe–infested stands compared to non-infested stands. Witches' brooms typically form in the lower portion of tree crowns. This growth formation of closely spaced small branches traps fallen needles, creating vertically oriented fine fuels and increasing stand flammability. The development of vertical ladder fuels provides a mechanism for surface fire to transition into crowns, increasing the wildfire hazard in interior Douglas-fir stands [176,209,220]. Dwarf mistletoe in Douglas-fir forests on the Bitterroot National Forest in western Montana likely contributed to the high fire intensity observed during the summer of 2000 [221].

The interaction of Douglas-fir dwarf mistletoe with agents of disturbance other than fire can also adversely affect forest health. The combined impacts of Douglas-fir dwarf mistletoe and WSBW, for example, can increase seedling and sapling mortality rates or elevate the susceptibility of mature trees to DFB. An increase in overall tree mortality can increase fuel accumulations contributing to poor stand health [222].

5.5. Western Spruce Budworm and Douglas-Fir Beetle

Previous insect disturbances in a stand can serve as initial stressors to tree vigor, which subsequently diminishes tree defenses and increases the likelihood of successful DFB colonization [147,165]. As an example, in 2011, DFB-caused tree mortality in southern Idaho, USA, Nevada, USA, and Utah, USA more than doubled [223]. This trend continued through 2013, the last year of reported data from the US Forest Service Forest Health Protection Program. This increase was partly attributed to stress induced by several years of WSBW defoliation [224]. In one study it was found that trees in Colorado, USA, infested by DFB had reduced growth attributed to a previous WSBW outbreak [225]. Negrón [132] (p. 82) mentioned, "In the Colorado Front Range, it seems that the primary disturbance agent, although not the only one that triggers DFB outbreaks, is WSBW defoliation". In Logan Canyon, Utah, USA, Fredericks and Jenkins [145] found that trees defoliated by WSBW reduced host tree defenses against DFB. Their observation is consistent with other research [132,165,226]. It also has been found that other defoliators including the Douglas-fir tussock moth (*Orgyia pseudotsugata* McDunnough, Lepidoptera: Lymantriidae) decrease plant vigor, predisposing hosts to subsequent DFB attack [147,227].

5.6. Root Disease, Windthrow, and Douglas-Fir Beetle

The presence of root disease can predispose a stand to additional disturbance agents which can result in tree mortality and decreased stand health. Root pathogens such as *Armillaria* infect healthy trees, which decreases plant vigor and predisposes trees to attack by insects [189]. *Armillaria* primarily weakens root systems, leaving infected trees highly susceptible to windthrow [228]. Wind-felling frequently triggers bark beetle epidemics where they take advantage of abundant breeding material which becomes available following such events [205].

Endemic Douglas-fir beetle populations are often correlated with root diseases [122], which can lead to subsequent build-up of downed trees following wind events. The abundance of fresh slash material enables DFB populations to increase and mass attack surrounding standing live trees, facilitating the development of epidemic populations [229].

6. Anthropogenic Influences

6.1. Historic Human Mediated Disturbance

In addition to natural disturbances, anthropogenic influences cause changes within interior Douglas-fir forests. Human-mediated disturbances have also influenced the present structure and composition of interior Douglas-fir forests. Prior to European settlement, Native Americans indigenous to the central Rocky Mountains used fire as a tool for improving wildlife habitat and hunting grounds [230,231]. These activities likely influenced local fire patterns in interior Douglas-fir forests. Fire type (ground, surface, or canopy), frequency, and extent of Native American fire use and the subsequent effect on altering fire regimes have been debated [231]. Although the scale of their fire practices is likely smaller than originally thought, elucidating the degree to which fire was intentionally used by Native Americans to manipulate interior Douglas-fir forests has proven difficult to assess as conventional fire reconstruction methods are limited in their abilities to differentiate historic natural ignitions from human-caused ignitions [232].

Human-mediated disturbance within these forests increased with the arrival of European settlers. During the mid- to late-19th century, many settlers along the Wasatch Front, Utah, USA, utilized local timber from Douglas-fir forests to construct homesteads, towns, waterwheels, and other goods [233]. Unrestricted logging practices left many slopes devoid of trees, leading to the development of even-age stand structures that currently exist throughout this forest type within the Wasatch Mountains [234].

Throughout the mid-20th century, interior Douglas-fir has remained a viable commercial species [45]. The wood is extremely strong and used as structural timber, railroad ties, plywood, and for pulp [43,235]. The impacts of logging vary depending on the size, intensity, and type of harvesting practices employed (e.g., group tree selection *versus* clear-cutting) [9]. Logging, in some cases, has led to a more homogeneous stand structure, higher tree densities, lack of structural diversity, and loss of old, mature trees [99]. In other cases, previously logged stands have had a number of large trees left uncut, which resulted in a similar volume of large mature trees found in nearby unlogged stands [99].

6.2. Forest Restoration

Early forest management practices were conducted on the premise of a limitless supply of resources. This type of mindset contributed to forest degradation following land use practices including, logging, grazing, and fire suppression [71,236]. Throughout the 21st century, natural resource managers have been shifting management strategies from sole resource extraction to include restoration principles. Often, the goals of forest restoration are to improve the resiliency and ecosystem function of a stand, and return it to a state within the historic range of conditions prior to Euro-American influence [71,237]. Treatments are often considered in cases where anthropogenic activities have greatly altered stand structure, ecosystem function, and composition [236]. In interior Douglas-fir and other forests characterized as having a mixed-severity fire regime, the complex mosaic of forest structures present across the landscape make it difficult to implement effective restoration plans. Even with good intentions, some restoration procedures can create unintentional forest health issues.

6.2.1. Pre-Fire Restoration Treatments

Fire management practices, primarily fuel treatment prescriptions, have greatly influenced forest health. Often, the focus of fire management objectives is to reduce hazardous fuels by decreasing stand density and ladder fuels through thinning [65,238]. These methods are effective at reducing overall canopy bulk density and increasing canopy base height in a stand, which reduces the hazard of crown fire. Fuel reduction treatments, however, can result in unintended consequences including exacerbating the incidence and severity of root diseases [239]. Mechanical damage to tree boles during tree removal operations may also weaken trees, leaving them more susceptible to insect infestation and infection by decay fungi [240]. For example, increasing the proportion of large-diameter trees when stands are thinned from below can increase their susceptibility to DFB [204].

6.2.2. Fire Restoration Treatments

Prescribed fire is another fire management tool widely used to reduce dense accumulations of fuels, remove logging debris, improve wildlife habitat, and manage vegetation [59,241]. Before the era of aggressive fire suppression, periodic fire maintained vegetative diversity on the landscape which helped mitigate the effects of bark beetle attacks. Since then, the lack of fire has created more uniform stands capable of supporting the spread of bark beetles or enhancing the effects of defoliators [220,242].

The reintroduction of fire after years of suppression can have unintended forest health consequences. These fires can burn with greater intensity and severity, especially where tree mortality has increased due to insects and diseases. Widespread tree mortality also causes dead fuels to accumulate for decades, increasing the hazard of high-intensity fire over time [243]. This might particularly be true at mid-elevations (2400–2700 m) where stands that had a naturally mixed-severity fire regime are now altered by fire suppression and transition into higher elevation forests. This may increase the potential for high-intensity fires ignited at lower elevations to burn into higher elevation stands where hazardous fuel accumulations may result in unnatural levels of fire damage and more severe resource effects [240].

6.2.3. Post-Fire Restoration Treatments

A common practice following wildfire is salvage logging, which involves the removal of fire-killed trees to recover economic value before degradation by decay [244]. The impact of these activities can lead to increased sediment erosion [245], loss of snags that provide wildlife habitat, and shifts in wildlife composition toward bird and invertebrate species that prefer more open habitats [245]. In some instances, post-wildfire logging may actually increase future short-term fire risk through the rapid accumulation of coarse, woody debris associated with salvage logging activities. Furthermore, soil disturbance and logging slash can inhibit seedling growth and result in a net reduction in post-fire regeneration [246]. Sanitation harvesting is another technique used during post-fire remediation and involves the removal of both live and dead trees affected by fire. Live fire-damaged trees are often removed to decrease future insect infestations.

7. Future Research Needs

The complex and heterogeneous nature of interior Douglas-fir forests often makes assessing forest health difficult, and devising appropriate management strategies challenging. Maintaining ecosystem function, enhancing biodiversity, and reducing the risk of catastrophic fires in these forests necessitates that management plans consider a holistic, integrated, and adaptive approach. Implementing such an approach requires further research to better understand the effects of multiple disturbance agents on interior Douglas-fir forest communities [247]. We have identified several gaps where further research may increase our understanding of disturbance agents, their interacting roles, and influences on long-term forest health.

(1) Forest conditions including "stand structure, fuel characteristics, and fire regimes" have been greatly altered in part due to management practices that started in the 20th century, where "forest structure, fuel characteristics, and fire regimes of the mixed-conifer forests in the western United States have been dramatically altered" [248,249] (p. 22). One important issue is understanding the historic role and extent of fire in these ecosystems. The present accumulations of live and dead fuels have resulted in a shift toward higher-severity fire behavior. Although high-severity crown fires are not outside the historical range of variability for this fire regime type, the frequency and overall size of these types of fires appear to be increasing [66]. Additional fire reconstruction studies across various geographic scales can help elucidate the natural fire regimes of interior montane forests [231,250–252].

(2) The inherent variability of interior Douglas-fir stands occurring on sites ranging from xeric to mesic and from lower-montane to subalpine zones necessitates that forest managers have a better understanding of how geographic locales influence forest fuel conditions. As stand composition and structure have shifted, so too have changes to fuel complexes.

(3) Research is needed to determine how forest insects, particularly bark beetles and defoliators, affect fuels and fire behavior across a wider range of geographic locales and whether bark beetle alterations of surface and canopy fuels can elevate the potential for fire to spread into upper elevation forests.

(4) Little is known about the combined effects of DFB and WSBW on interior Douglas-fir forests and, in turn, on ecosystem values across various spatial scales.

(5) Information regarding climate change influences on the occurrence, timing, frequency, extent, and duration of disturbances at various temporal and spatial scales for interior Douglas-fir forests is lacking.

(6) A greater understanding of DFB population dynamics in interior Douglas-fir stands is needed.

(7) Douglas-fir dwarf mistletoe and DFB interactions are poorly understood.

(8) Research on how snow avalanches and other abiotic disturbances affect interior Douglas-fir forest health is lacking.

8. Conclusions

Interior Douglas-fir forests are a principal forest type in the central Rocky Mountains. The adaptability of Douglas-fir to a variety of site conditions across a broad latitudinal range results in unique and diverse plant communities that provide for numerous ecosystem and social values. Both natural disturbance processes, including fire, wind, insect outbreaks, pathogens, and human-mediated disturbances will continue to have an important role in shaping these forest communities.

Disturbance-induced alterations to the mixed-severity fire regime characteristic of this forest type will continue to exert a large influence over future stand development. Past forest management policies promoted the advancement of more insect and pathogen outbreaks, modifying the susceptibility to future disturbance events (e.g., windthrow, landslides, snow avalanches), and will continue to affect overall forest health. Current and future forest management decisions should weigh the costs and benefits of implementing certain practices (e.g., harvesting, prescribed fire) so that these activities do not adversely affect the health in certain, already vulnerable interior Douglas-fir forests.

The diversified nature of interior Douglas-fir forest communities discourages a "one size fits all" approach to management. Rather, the management of interior Douglas-fir forests in the future will necessitate developing more holistic, integrated, and adaptive management strategies to maintain forest and ecosystem health to meet multiple management objectives.

Acknowledgments: Funding for this project was provided by the Joint Fire Science Program (Project #11-1-4-16) and supported by the Ecology Center and the Utah Agricultural Experiment Station at Utah State University. We give special thanks to Michaela Teich for assistance with manuscript preparation and review, Wanda Lindquist for graphics assistance, and Laura Dunning for manuscript formatting. The comments from four anonymous reviewers have greatly improved the final manuscript.

Author Contributions: Andrew Giunta was the primary author of this paper and conducted the literature review. Elizabeth Hebertson contributed to the writing, editing, and assisted with literature review. Michael Jenkins provided edits and reviews. Allen Munson provided edits, contributed to the writing, and reviewed all manuscript drafts.

Conflicts of Interest: The authors declare no conflict of interest. The funding sponsors had no role in the design of the study; in the collection, analyses, or interpretation of data; in the writing of the manuscript, and in the decision to publish the results.

References

1. Oliver, C.D. Forest development in North America following major disturbances. *For. Ecol. Manag.* **1981**, *3*, 153–168. [CrossRef]
2. Delcourt, H.R.; Delcourt, P.A.; Webb, T., III. Dynamic plant ecology: The spectrum of vegetational change in space and time. *Quat. Sci. Rev.* **1983**, *1*, 153–175. [CrossRef]
3. White, P.S.; Pickett, S.T.A. Natural disturbance and patch dynamics: An introduction. In *The Ecology of Natural Disturbance and Patch Dynamics*; Pickett, S.T.A., White, P.S., Eds.; Academic Press: Orlando, FL, USA, 1985; pp. 3–13.
4. Parsons, D.J.; DeBenedetti, S.H. Impact of fire suppression on a mixed-conifer forest. *For. Ecol. Manag.* **1979**, *2*, 21–33. [CrossRef]

5.   Westerling, A.L.; Hidalgo, H.G.; Cayan, D.R.; Swetnam, T.W. Warming and earlier spring increase western U.S. forest wildfire activity. *Science* **2006**, *313*, 940–943. [CrossRef] [PubMed]

6.   Fulé, P.Z.; Covington, W.W.; Moore, M.M. Determining reference conditions for ecosystem management of southwestern ponderosa pine forests. *Ecol. Appl.* **1997**, *7*, 894–908. [CrossRef]

7.   Allen, C.D.; Savage, M.; Falk, D.A.; Suckling, K.F.; Swetnam, T.W.; Schulke, T.; Stacey, P.B.; Morgan, P.; Hoffman, M.; Klingel, J.T. Ecological restoration of southwestern ponderosa pine ecosystems: A broad perspective. *Ecol. Appl.* **2002**, *12*, 1418–1433. [CrossRef]

8.   Cocke, A.E.; Fulé, P.Z.; Crouse, J.E. Forest change on a steep mountain gradient after extended fire exclusion: San Francisco Peaks, Arizona, USA. *J. Appl. Ecol.* **2005**, *42*, 814–823. [CrossRef]

9.   Hejl, S.J.; Hutto, R.L.; Preston, C.R.; Finch, D.M. Effects of silvicultuer treatments in the Rocky Mountains. In *Ecology and Management of Neotropical Migratory Birds: A Synthesis and Review of Critical Issues*, 1st ed.; Martin, T.E., Finch, D.M., Eds.; Oxford University Press: New York, NY, USA, 1995; pp. 220–244.

10. Bentz, B. Bark beetle outbreaks in western North America. Causes and consequences. In Proceedings of the Bark Beetle Symposium, Snowbird, UT, USA, 15–17 November 2005.

11. Raffa, K.F.; Aukema, B.H.; Bentz, B.J.; Carroll, A.L.; Hicke, J.A.; Turner, M.G.; Romme, W.H. Cross-scale drivers of natural disturbances prone to anthropogenic amplification: The dynamics of bark beetle eruptions. *BioScience* **2008**, *58*, 501–517. [CrossRef]

12. Bentz, B.J.; Régnière, J.; Hansen, M.; Hayes, J.L.; Hicke, J.A.; Kelsey, R.G.; Negrón, J.F.; Seybold, S.J. Climate change and bark beetles of the western United States and Canada: Direct and indirect effects. *BioScience* **2010**, *60*, 602–613. [CrossRef]

13. Shaw, J.D.; Steed, B.E.; DeBlander, L.T. Forest inventory and analysis (FIA) annual inventory answers the question: What is happening to pinyon-juniper woodlands? *J. For.* **2005**, *103*, 280–285.

14. Kleinman, S.J.; DeGomez, T.E.; Snider, G.B.; Williams, K.E. Large-scale pinyon ips (*Ips confusus*) outbreak in the southwestern United States tied with elevation and land cover. *J. For.* **2012**, *110*, 194–200. [CrossRef]

15. Wulder, M.A.; White, J.C.; Bentz, B.; Alvarez, M.F.; Coops, N.C. Estimating the probability of mountain pine beetle red-attack damage. *Remote Sens. Environ.* **2006**, *101*, 150–166. [CrossRef]

16. Kurz, W.A.; Dymond, C.C.; Stinson, G.; Rampley, G.J.; Neilson, E.T.; Carroll, A.L.; Ebata, T.; Safranyik, L. Mountain pine beetle and carbon forest feeback to climate change. *Nature* **2008**, *452*, 987–990. [CrossRef] [PubMed]

17. Jenkins, M.J.; Hebertson, E.G.; Munson, A.S. Spruce beetle biology, ecology and management in the Rocky Mountains: An addendum to spruce beetle in the Rockies. *Forests* **2014**, *5*, 21–71. [CrossRef]

18. Page, W.G.; Jenkins, M.J.; Runyon, J.B. Spruce beetle-induced changes to Engelmann spruce foliage flammability. *For. Sci.* **2014**, *60*, 691–702. [CrossRef]

19. Schmitz, R.F.; Gibson, K.E. *Douglas-Fir Beetle*; Forest and Insect Disease Leaflet. USDA Forest Service: Washington, DC, USA, 1996; pp. 5–7.

20. Negrón, J.F.; Anhold, J.A.; Munson, S.A. Within-stand spatial distribution of tree mortality caused by the Douglas-fir beetle (Coleoptera: Scolytidae). *Environ. Entomol.* **2001**, *30*, 215–224. [CrossRef]

21. Kolb, T.E.; Wagner, M.R.; Covington, W.W. Forest health from different perspectives, forest health through silviculture. In Proceedings of the 1995 National Silviculture Workshop, Mescalaro, NM, USA, 8–11 May 1995; pp. 5–13.

22. Healthy Forest Restoration Act. "HR 1904". PL108-148. U.S. Department of Agriculture, 2003. Available online: http://www.fs.fed.us/emc/applit/includes/hfr2003.pdf (accessed on 2 March 2016).

23. Coulson, R.N.; Stephens, F.M. Impacts of insects in forest landscapes: Implications for forest health management. In *Invasive Forest Insects, Introduced Forest Trees, and Altered Ecosystems: Ecological Pest Management in Global Forests of a Changing World*; Payne, T.D., Ed.; Springer-Verlag: New York, NY, USA, 2006; pp. 101–125.

24. Fettig, C.J.; Klepzig, K.D.; Billings, R.F.; Munson, S.A.; Nebeker, T.E.; Negrón, J.F.; Nowak, J.T. The effectiveness of vegetation management practices for prevention and control of bark beetle infestations in coniferous forests of the western and southern United States. *For. Ecol. Manag.* **2007**, *238*, 24–53. [CrossRef]

25. Turner, M.G. Disturbance and landscape dynamics in a changing world. *Ecology* **2003**, *84*, 362–371.

26. Kulakowski, D.; Veblen, T.T.; Bebi, P. Effects of fire and spruce beetle outbreak legacies on the disturbance regime of a subalpine forest in Colorado. *J. Biogeogr.* **2003**, *30*, 1445–1456. [CrossRef]

27. Kulakowski, D.; Veblen, T.T. Effects of prior disturbances on the extent and severity of wildfire in Colorado subalpine forests. *Ecology* **2007**, *88*, 759–769. [CrossRef] [PubMed]
28. Bigler, C.; Kulakowski, D.; Veblen, T.T. Multiple disturbance interactions and drought influence fire severity in Rocky Mountain subalpine forests. *Ecology* **2005**, *86*, 3018–3029. [CrossRef]
29. Jorgensen, C.A.; Jenkins, M.J. Fuel complex alterations associated with spruce beetle-induced tree mortality in intermountain spruce-fir forests, USA. *For. Sci.* **2011**, *57*, 232–240.
30. O'Connor, C.D.; Lynch, A.M.; Falk, D.A.; Swetnam, T.W. Post-fire forest dynamics and climate variability affect spatial and temporal properties of spruce beetle outbreaks on a Sky Island mountain range. *For. Ecol. Manag.* **2015**, *336*, 148–162. [CrossRef]
31. Kulakowski, D.; Jarvis, D.; Veblen, T.T.; Smith, J. Stand replacing fires reduce susceptibility of lodgepole pine to mountain pine beetle outbreaks in Colorado. *J. Biogeogr.* **2012**, *39*, 2052–2060. [CrossRef]
32. Meidinger, D.; Pojar, J. *Ecosystems of British Columbia*; British Columbia Ministry of Forests, 1991; p. 330. Available online: http://www.for.gov.bc.ca/hfd/pubs/Docs/Srs/SRseries.htm (accessed on 18 April 2015).
33. Arno, S.F. *Forest Regions of Montana*; Research Paper INT-218; USDA Forest Service, Intermountain Forest and Range Experiment Station: Ogden, UT, USA, 1979; p. 39.
34. Silen, R.R. *Genetics of Douglas-Fir*; Research Paper WO-35; U.S. Department of Agriculture, Forest Service: Washington, DC, USA, 1978; p. 34.
35. Simard, S.W. The foundational role of mycorrhizal networks in self-organization of interior Douglas-fir forests. *For. Ecol. Manag.* **2009**, *258*, 95–107. [CrossRef]
36. Little, E.L., Jr. *Atlas of United States Trees*; U.S. Department of Agriculture: Washington, DC, USA, 1971; Vol. 1, pp. 1–9.
37. Donato, D.C.; Harvey, B.J.; Romme, W.H.; Simard, M.; Turner, M.G. Bark beetle effects on fuel profiles across a range of stand structures in Douglas-fir forests of Greater Yellowstone. *Ecol. Appl.* **2013**, *23*, 3–20. [CrossRef] [PubMed]
38. Fowells, H.A. *Silvics of Forest Trees of the United States*; No. 2171; USDA Forest Service Agriculture Handbook: Washington, DC, USA, 1965; pp. 546–556.
39. Wright, J.W.; Kung, F.H.; Read, R.A.; Lemmien, W.A.; Bright, J.N. Genetic variation in Rocky Mountain Douglas-fir. *Silv. Genet.* **1971**, *20*, 54–60.
40. Van Hooser, D.D.; Wadell, K.L.; Mills, J.R.; Tymcio, R.P. The interior Douglas-fir resource: Current status and projections to the year 2040. In Interior Douglas-Fir, the Species and Its Management, Proceedings of the Interior Douglas-fir: The Species and Its Management, Pullman, WA, USA, 27 February–1 March 1991; Baumgartner, D.M., Lotan, J.E., Eds.; Washington State University: Pullman, WA, USA, 1991; pp. 9–14.
41. Whittaker, R.H.; Niering, W.A. Vegetation of the Santa Catalina mountains Arizona. I. Ecological classification and distribution of species. *J. Ariz. Acad. Sci.* **1964**, *3*, 9–34. [CrossRef]
42. Whittaker, R.H.; Niering, W.A. Vegetation of the Santa Catalina mountains, Arizona. V. biomass, production, and diversity along the elevation gradient. *Ecology* **1975**, *56*, 771–790. [CrossRef]
43. Niering, W.A.; Lowe, C.H. Vegetation of the Santa Catalina Mountains, Arizona, USA: Community types and dynamics. *Vegetatio* **1984**, *58*, 3–28.
44. Lavendar, D.P.; Hermann, R.K. *Douglas-Fir: The Genus Pseudotsuga*, 1st ed.; Oregon State University: Corvallis, OR, USA, 2014; pp. 19–24.
45. Hermann, R.K.; Lavendar, D.P. *Pseudotsuga menziesii* (Mirb.) Franco Douglas-fir. *Silvics of North American Conifers*; Burns, R.M., Honkala, B.H., Eds.; U.S. Department of Agriculture, Forest Service: Washington, DC, USA, 1990; Volume 1, pp. 527–540.
46. Kilgore, B.M. Fire in ecosystem distribution and structure: Western forests and scrublands. *Fire Regimes and Ecosystem Properties*; General Technical Report WO-GTR-26; Mooney, H.A., Bonnickson, T.M., Christensen, N.L., Lotan, J.E., Reiners, W.A., Eds.; U.S. Department of Agriculture, Forest Service: Washington, DC, USA, 1981; pp. 58–89.
47. Bailey, R.G. *Descriptions of the Ecoregions of the United States*; Miscellaneous Publication 1391; U.S. Department of Agriculture, Forest Service: Washington, DC, USA, 1980; p. 77.
48. Eyre, F.H. *Forest Cover Types of the United States and Canada*, 1st ed.; Society of American Foresters: Washington, DC, USA, 1980; p. 148.
49. Harlow, W.M.; Harrrar, E.S.; White, F.M. *Textbook of Dendrology: Covering the Important Forest Trees of the United States and Canada*, 6th ed.; McGraw-Hill: New York, NY, USA, 1979; p. 510.

50. Habeck, J.R.; Mutch, R.W. Fire dependent forests in the northern Rocky Mountains. *Quat. Res.* **1973**, *3*, 408–424. [CrossRef]

51. Mauk, R.L.; Henderson, J.A. *Coniferous Forest Habitat Types of Northern Utah*; General Technical Report INT-170; U.S. Department of Agriculture, Forest Service, Intermountain Range and Experiment Station: Ogden, UT, USA, 1984.

52. Youngblood, A.P.; Mauk, R.L. *Coniferous Forest Habitat Types of Central and Southern Utah*; General Technical Report INT-187; Department of Agriculture, Forest Service, Intermountain Range and Experiment Station: Ogden, UT, USA, 1985.

53. Keegan, C.E.; Swanson, L.D.; Wichman, D.P.; VanHooser, D.D. *Montana's Forest Products Industry: A Descriptive analysis 1969–1988*; Bureau of Business and Economic Research: Missoula, MT, USA, 1990; p. 52.

54. Lanner, R.M. *Trees of the Great Basin—A Natural History*, 1st ed.; University of Nevada Press: Reno, NV, USA, 1984; p. 215.

55. Lilleholm, R.J.; Kessler, W.B.; Merrill, K. Stand density index applied to timber and goshawk habitat objectives in Douglas-fir. *Environ. Manag.* **1993**, *17*, 773–779. [CrossRef]

56. McKenzie, D.; Gedalof, Z.; Peterson, D.L.; Mote, P. Climate change, wildfire, and conservation. *Conserv. Biol.* **2004**, *18*, 890–902. [CrossRef]

57. Hicke, J.A.; Johnson, M.C.; Hayes, J.L.; Preisler, H.K. Effects of bark beetle-caused tree mortality on wildfire. *For. Ecol. Manag.* **2012**, *271*, 81–90. [CrossRef]

58. Jenkins, M.J.; Page, W.G.; Hebertson, E.G.; Alexander, M.E. Fuels and fire behavior dynamics in bark beetle attacked forests in western North America and implications for fire management. *For. Ecol. Manag.* **2012**, *275*, 23–34. [CrossRef]

59. Pyne, S.J.; Andrews, P.L.; Laven, R.D. *Introduction to Wildland Fire*, 2nd ed.; Wiley: New York, NY, USA, 1996; p. 769.

60. Brown, J.K. Introduction and fire regimes. *Wildland Fire in Ecosystems: Effects on Fire on Flora*; General Technical Report RMRS-42; Brown, J.K., Smith, J.K., Eds.; U.S. Department of Agriculture, Forest Service, Rocky Mountain Research Station: Ogden, UT, USA, 2000.

61. Morgan, P.; Hardy, C.C.; Swetnam, T.W.; Rollins, M.G.; Long, D.G. Mapping fire regimes across time and space: Understanding coarse and fine-scale fire patterns. *Int. J. Wildland Fire* **2001**, *10*, 329–324. [CrossRef]

62. Halofsky, J.E.; Donato, D.C.; Hibbs, D.E.; Cambell, J.L.; Cannon, M.D.; Fontaine, J.B.; Thompson, J.R.; Anthony, R.G.; Bormann, B.T.; Kayes, L.J.; *et al.* Mixed-severity fire regimes: Lessons and hypotheses from Klamath-Siskiyou Ecoregion. *Ecosphere* **2011**, *2*, 1–19. [CrossRef]

63. Keeley, J.E. Fire intensity, fire severity and burn severity: A brief review. *Int. J. Wildland Fire* **2009**, *18*, 116–126. [CrossRef]

64. Van Wagner, C.E. Conditions for the start and spread of crown fire. *Can. J. For. Res.* **1977**, *7*, 23–34. [CrossRef]

65. Agee, J.K.; Skinner, C.N. Basic principles of forest fuel reduction treatments. *For. Ecol. Manag.* **2005**, *211*, 83–96. [CrossRef]

66. Schoennagel, T.; Veblen, T.T.; Romme, W. Interaction of fire, fuels, and climate across Rocky Mountain forests. *BioScience* **2004**, *54*, 661–676. [CrossRef]

67. Chappell, C.B.; Agee, J.K. Fire severity and tree seedling establishment in *Abies magnifica* forests, southern Cascades, Oregon. *Ecol. Appl.* **1996**, *6*, 628–640. [CrossRef]

68. Agee, J.K. The complex nature of mixed-severity fire regimes. In *Mixed Severity Fire Regimes: Ecology and Management*; Lagene, L., Zelnik, J., Cadwallader, S., Hughes, B., Eds.; Washington State University: Pullman, WA, USA; The Association for Fire Ecology: Spokane, WA, USA, 2004; pp. 1–10.

69. Arno, S.F.; Parsons, D.J.; Keane, R.E. *Mixed-Severity Fire Regimes in the Northern Rocky Mountains: Consequences of Fire Exclusion and Options for the Future*; Research Paper RMRS-P-15-VOL-5; U.S. Department of Agriculture, Forest Service, Rocky Mountain Research Station: Fort Collins, CO, USA, 2000; pp. 225–232.

70. Brown, J.K. Fire regims and their relevance to ecosystem manaagment. In Proceedings of the Society of American Foresters National Convention, Anchorage, AK, USA, 18–22 September 1994; Society of American Foresters: Bethesda, MD, USA, 1995; pp. 171–178.

71. Baker, W.L.; Veblen, T.T.; Sherriff, R.L. Fire, fuels, and restoration of ponderosa pine-Douglas-fir forests in the Rocky Mountains, USA. *J. Biogeogr.* **2007**, *34*, 251–269. [CrossRef]

72. Jain, T.B.; Battaglia, M.A.; Han, H.S.; Graham, R.T.; Keyes, C.R.; Fried, J.S.; Sandquist, J.E. *A Comprehensive Guide to Fuel Management Practices for Dry Mixed Conifer Forests in the Northwestern United States: Inventory and Model-Based Economic Analysis of Mechanical Fuel Treatments*; Research Note RMRS-RN-64; U.S. Department of Agriculture, Forest Service, Rocky Mountain Research Station: Fort Collins, CO, USA, 2014.

73. Romme, W.H.; Floyd, M.L.; Hanna, D. *Historical Range of Variability and Current Landscape Condition Analysis: South Central Highlands Section, Southwestern Colorado & Northwestern New Mexico*; Colorado Restoration Institute, Colorado State University: Fort Collins, CO, USA; Region 2 U.S. Department of Agriculture, Forest Service: Golden, CO, USA, 2009; Available online: http://www.coloradoforestrestoration.org/CFRIpdfs/2009_HRVSouthCentral-Highlands.pdf (accessed on 1 March 2016).

74. Wyant, J.G.; Omni, P.N.; Laven, R.D. Fire induced tree mortality in a Colorado ponderosa pine/Douglas-fir stand. *For. Sci.* **1986**, *32*, 49–59.

75. Stephens, S.L.; Finney, M.A. Prescribed fire mortality of Sierra Nevada mixed conifer tree species: Effects of crown damage and forest floor combustion. *For. Ecol. Manag.* **2002**, *162*, 261–271. [CrossRef]

76. Van Wagner, C.E. Height of crown scorch in forest fires. *Can. J. For. Res.* **1973**, *3*, 373–378. [CrossRef]

77. Methven, I.R. *Prescribed Fire, Crown Scorch and Mortality: Field and Laboratory Studies on Red and White Pine*; Research Report PS-X-31; Canadian Forest Service, Petawawa Forest Experimental Station: Chalk River, ON, Canada, 1971; p. 10.

78. Peterson, D.L. Crown scorch volume and scorch height: Estimates of post-fire tree condition. *Can. J. For. Res.* **1985**, *15*, 596–598. [CrossRef]

79. Chambers, J.L.; Dougherty, P.M.; Hennessey, T.C. Fire: Its effects on growth and physiological processes in conifer forests. In *Stress Physiology and Forest Productivity*, 1st ed.; Hennessey, T.C., Dougherty, P.M., Kossuth, S.V., Johnson, J.D., Eds.; Martinaus Nijhoff: Hingham, MA, USA, 1986; pp. 171–191.

80. Peterson, D.L.; Ryan, K.C. Modeling postfire conifer mortality for long-range planning. *Environ. Manag.* **1986**, *10*, 797–808. [CrossRef]

81. Ryan, K.C.; Reinhardt, E.D. Predicting post fire severity in seven western conifers. *Can. J. For. Res.* **1988**, *18*, 1291–1297. [CrossRef]

82. Peterson, D.L.; Arbaugh, M.J. Postfire survival in Douglas-fir and lodgepole pine: Comparing the effects of crown and bole damage. *Can. J. For. Res.* **1986**, *16*, 1175–1179. [CrossRef]

83. Costa, J.J.; Oliveria, L.A.; Viegas, D.X.; Neto, L.P. On the temperature distribution inside a tree under fire conditions. *Int. J. Wildland Fire* **1991**, *1*, 87–96. [CrossRef]

84. Hood, S.; Bentz, B. Predicting postfire Douglas-fir beetle attacks and tree mortality in the northern Rocky Mountains. *Can. J. For. Res.* **2007**, *37*, 1058–1069. [CrossRef]

85. Hungerford, R.D.; Harrington, M.G.; Frandsen, W.H.; Ryan, K.C.; Niehoff, G.J. Influence of fire on factors that affect site productivity. In Proceedings of the Management and Productivity of Western-Montane Forest Soils, Boise, ID, USA, 10–12 April 1990; Harvey, A.E., Neuenschwander, L.F., Eds.; General Technical Report INT-280. U.S. Department of Agriculture, Forest Service, Intermountain Research Station: Ogden, UT, USA, 1991; pp. 32–50.

86. Neary, D.G.; Klopatek, C.C.; DeBano, L.F.; Ffolliott, P.F. Fire effects on below ground sustainability: A review and synthesis. *For. Ecol. Manag.* **1999**, *122*, 51–71. [CrossRef]

87. Egan, D. *Protecting Old Trees from Prescribed Burning*; Working Paper No. 24; Ecological Restoration Institute: Flagstaff, AZ, USA, 2011; p. 5.

88. Swezy, D.M.; Agee, J.K. Prescribed fire effects on fine-root and tree mortality in old-growth ponderosa pine. *Can. J. For. Res.* **1991**, *21*, 626–634. [CrossRef]

89. Stendell, E.; Horton, T.R.; Bruns, T.D. Early effects of prescribed fire on the structure of the fungal ectomycorrhizal community in a Sierra Nevada ponderosa pine forest. *Mycol. Res.* **1999**, *103*, 1353–1359. [CrossRef]

90. Hart, S.C.; Classen, A.T.; Wright, R.J. Long-term interval burning alters fine root and mycorrhizal dynamics in a ponderosa pine forest. *J. Appl. Ecol.* **2005**, *42*, 752–761. [CrossRef]

91. Rogstad, A. Recovering From Wildfire a Guide to Arizona's Forest Owners. 2002. Available online: http://ag.arizona.edu/pubs/natresources/az1294/ (accessed on 1 March 2016).

92. Gordon, W.S.; Jackson, R.B. Nutrient concentrations in fine roots. *Ecology* **2000**, *81*, 275–280. [CrossRef]

93. Reubens, B.; Poesen, J.; Danjon, F.; Geudens, G.; Muys, B. The role of fine and coarse roots in shallow slope stability and soil erosion control with a focus on root system architecture: A review. *Trees* **2007**, *21*, 385–402. [CrossRef]

94. Gyssels, G.; Poesen, J.; Bochet, E.; Li, Y. Impact of plant roots on the resistance of soils to erosion by water: A review. *Prog. Phys. Geogr.* **2005**, *29*, 189–217. [CrossRef]

95. Lertzman, K.; Riccius, P.; Fall, J. Does fire isolate elements of the landscape mosaic? *Bull. Ecol. Soc. Am.* **1997**, *78*, 92.

96. Brown, J.K. The unnatural fuel build up issue. In Proceedings of the Symposium and Workshop on Wilderness Fire, Missoula, MT, USA, 15–18 November 1983; Lotan, J.E., Kilgore, B.M., Fischer, W.C., Mutch, R.W., Eds.; General Technical Report INT-182. U.S. Department of Agriculture, Forest Service, Intermountain Forest and Range Experiment Station: Ogden, UT, USA, 1983; pp. 127–128.

97. Turner, M.G.; Romme, W.H. Landscape dynamics in crown fire ecosystems. *Landsc. Ecol.* **1994**, *9*, 59–77. [CrossRef]

98. Agee, J.K. The landscape ecology of western forest fire regimes. *Northwest Sci.* **1998**, *72*, 24–34.

99. Kaufmann, M.R.; Regan, C.M.; Brown, P.M. Heterogeneity in ponderosa pine/Douglas-fir forests: Age and size structure in unlogged and logged landscapes of central Colorado. *Can. J. For. Res.* **2000**, *30*, 698–711. [CrossRef]

100. Covington, W.W.; Moore, M.M. Post settlement changes in natural fire regimes and forest structure: Ecological restoration of old-growth ponderosa pine forests. *J. Sustain. For.* **1994**, *2*, 153–181. [CrossRef]

101. Heyerdahl, E.K.; Miller, R.F.; Parson, R.A. History of fire and Douglas-fir establishment in a savanna and sagebrush-grassland mosaic, southwestern Montana, USA. *For. Ecol. Manag.* **2006**, *230*, 107–118. [CrossRef]

102. Daniels, L.D. Climate and fire: A case study of the Cariboo forest, British Columbia. In Proceedings of the Symposium on the Ecology and Management of Mixed Severity Fire Regimes Conference, Spokane, WA, USA, 17–19 November 2004; Taylor, L., Zelnik, J., Cadwallader, S., Hughes, B., Eds.; pp. 235–246.

103. Schaetzl, R.F.; Burns, S.F.; Johnson, D.L.; Small, T.W. Tree uprotting: Review of impacts on forest ecology. *Vegetatio* **1989**, *79*, 65–176.

104. Patten, R.S.; Knight, D. Snow avalanches and vegetation pattern in Cascade Canyon, Grand Teton National Park. *Arct. Alp. Res.* **1994**, *26*, 35–41. [CrossRef]

105. Mitchel, S.J. A diagnostic framework for windthrow risk estimation. *For. Chron.* **1998**, *74*, 100–105. [CrossRef]

106. Ulanova, N.G. The effects of windthrow on forests at different spatial scales. A review. *For. Ecol. Manag.* **2000**, *135*, 155–167. [CrossRef]

107. Bouget, C.; Duelli, P. The effects of windthrow on forest insect communities: A literature review. *Biol. Conserv.* **2004**, *118*, 281–299. [CrossRef]

108. Everham, E.M., III; Browkaw, N.V.L. Forest damage and recovery from catastrophic wind. *Bot. Rev.* **1996**, *62*, 114–175. [CrossRef]

109. Mitchel, S.J. The windthrow triangle: A relative windthrow hazard assessment procedure for forest managers. *For. Chron.* **1995**, *71*, 446–450. [CrossRef]

110. Beaty, S.W.; Stone, E.L. The variety of microsites created by tree falls. *Can. J. For. Res.* **1986**, *16*, 539–548. [CrossRef]

111. Teich, M.; Bartlet, P.; Gret-Regamey, A.; Bebi, P. Snow avalanches in forested terrain: Influence of forest parameters, topography, and avalanche characteristics on runout distance. *Arct. Antarct. Alp. Res.* **2012**, *44*, 509–519. [CrossRef]

112. Luckman, B.H. The geomorphic activity of snow avalanches. Geografiska Annaler. Series A. *Phys. Geogr.* **1977**, *59*, 31–48.

113. McClung, D.; Schaerer, P. *The Avalanche Handbook*, 3rd ed.; The Mountaineers Books: Seattle, WA, USA, 2006; p. 338.

114. Burrows, C.L.; Burrows, V.L. Procedures for the study of snow avalanche chronology using growth layers of woody plants. *Inst. Arct. Alp. Res.* **1976**, *23*, 54.

115. Hebertson, E.G. Snow Avalanche Disturbance in Intermountain Spruce-Fir Forests and Implications for the Spruce Bark Beetle (Coleoptera: Scolytidae). Ph.D. Thesis, Utah State University, Logan, UT, USA, 2004.

116. Malanson, G.P.; Butler, D.R. Transverse pattern of vegetation on avalanche paths in the northern Rocky Mountains, Montana. *Gt. Basin Nat.* **1984**, *44*, 453–458.

117. Oliver, C.D.; Adams, A.B.; Zasoski, R.J. Disturbance patterns and forest development in a recently deglaciated valley in northwestern Cascade range of Washington USA. *Can. J. For. Res.* **1984**, *15*, 221–232. [CrossRef]
118. Butler, D.R.; Malanson, G.P. Non-equilibrium geomorphic processes and patterns on avalanche paths in the northern Rocky Mountains, U.S.A. *Z. Geomorphol.* **1990**, *34*, 257–270.
119. Walsh, S.J.; Butler, D.R.; Allen, T.R.; Malanson, G.P. Influence of snow patterns and snow avalanches on the alpine treeline ecotone. *J. Veg. Sci.* **1994**, *5*, 657–672. [CrossRef]
120. Veblen, T.T.; Kitzberger, T.; Donnegan, J. Climatic and human influences on fire regimes in ponderosa pine forests in the Colorado Front Range. *Ecol. Appl.* **2000**, *10*, 1178–1195. [CrossRef]
121. Bebi, P.; Kulakowski, D.; Rixen, C. Snow avalanche disturbances in forest ecosystems-State of research and implications for management. *For. Ecol Manag.* **2009**, *257*, 1883–1892. [CrossRef]
122. McGregor, M.D.; Furniss, M.M.; Oaks, R.D.; Gibson, K.E.; Meyer, H.E. MCH Pheromone for preventing Douglas-fir beetle infestation in windthrown trees. *J. For.* **1984**, *82*, 613–616.
123. Blackford, D.C. *Aspen Grove Trailhead Area Pleasant Grove RD, Uinta, NF*; Report No. OFO-TR-05-15; U.S. Department of Agriculture, Forest Service Forest Health Protection: Ogden, UT, USA, 2005.
124. Blackford, D.C.; Guyon, J. *Recreation Sites Insect and Disease Activity Duchesne/Roosevelt RD, Ashley NF, Functional Assistance Trip Report*; Report No. OFO-TR-06-05; U.S. Department of Agriculture, Forest Service, Forest Health Protection: Ogden, UT, USA, 2005.
125. Hebertson, E.G. *BLM Richfield Office-Henry Mountains Field Station Functional Assistance Visit*; Report No. OFO-TR-06-03; U.S. Department of Agriculture, Forest Service, Forest Health Protection: Ogden, UT, USA, 2005.
126. Raffa, K.F.; Berryman, A.A. Interacting selective pressure in conifer-bark beetle systems: A basis for reciprocal adaptations? *Am. Nat.* **1987**, *129*, 234–262. [CrossRef]
127. Samman, S.; Logan, J. *Assessment and Response to Bark Beetle Outbreaks in the Rocky Mountain Area*; Report to Congress from Forest Health Protection, Washington Office, Forest Service, U.S. Department of Agriculture. General Technical Report RMRS-GTR-62; U.S. Department of Agricutlure Forest Service Rocky Mountain Research Station: Fort Collins, CO, USA, 2000; p. 46.
128. Hardwood, W.G.; Rudinsky, J.A. *The Flight and Olfactory Behavior of Checkered Beetles (Coleoptera: Cleridae) Predatory on the Douglas-Fir Beetle*; Technical Bulletin 95; Agriculture Experiment Station, Oregon State University: Corvallis, OR, USA, 1966; p. 36.
129. Jenkins, M.J.; Hebertson, E.; Page, W.G.; Jorgensen, C.A. Bark beetles, fuels, fires and implications for forest management in the Intermountain West. *For. Ecol. Manag.* **2008**, *254*, 16–34. [CrossRef]
130. Anderson, M.N. Mechanisms of odor coding in coniferous bark beetles: From neuron to behavior and application. *Psyche J. Entomol.* **2012**. [CrossRef]
131. Furniss, M.M.; McGregor, M.D.; Foiles, M.W.; Patridge, M.D. *Chronology and Characteristics of a Douglas-Fir Beetle [Dendroctonus pseudotsugae] Outbreak in Northern Idaho [Pseudotsuga menziesii]*; General Technical Report, INT-59; U.S. Department of Agriculture, Forest Service, Service, Intermountain Forest and Range Experiment Station: Ogden, UT, USA, 1979. Available online: http://agris.fao.org/agris-search/search.do?recordID=US7935406 (accessed on 1 March 2016).
132. Negrón, J.F. Probability of infestation and extent of mortality associated with the Dogulas-fir beetle in Colorado Front Range. *For. Ecol. Manag.* **1998**, *107*, 71–85. [CrossRef]
133. Rudinsky, J.A. Scolytid beetles associated with Douglas-fir: Response to terpenes. *Science* **1966**, *152*, 218–219. [CrossRef] [PubMed]
134. Ross, D.; Daterman, G. Using pheromone-baited traps to control the amount and distribution of tree mortality during outbreaks of the Douglas-fir beetle. *For. Sci.* **1997**, *43*, 65–70.
135. Furniss, M.M. The Douglas-fir beetle in western forests a historical perspective—Part 1. *Am. Entomol.* **2014**, *60*, 84–96. [CrossRef]
136. Lejeune, R.; McMullen, L.; Atkins, M. Influence of logging on Douglas fir beetle populations. *For. Chron.* **1961**, *37*, 308–314. [CrossRef]
137. Furniss, M.M. Susceptibility of fire-injured Douglas-fir to bark beetle attack in southern Idaho. *J. For.* **1965**, *63*, 8–11.
138. Lewisohn, E.; Gijzen, M.; Croteau, R. Defense mechanisms of conifers. Differences in constitutive and wound-induced monoterpene biosynthesis among species. *Plant Physiol.* **1991**, *96*, 44–49. [CrossRef]

139. Franceschi, V.R.; Krokene, P.; Christiansen, E.; Krekling, T. Anatomical and chemical defenses of conifer bark against bark beetles and other pests. *New Phytol.* **2005**, *167*, 353–376. [CrossRef] [PubMed]
140. Cates, R.G.; Alexander, H. Host resistance and susceptibility. In *Bark Beetles in North American Conifers, A System for the Study of Evolutionary Biology*, 1st ed.; Mitton, J.B., Sturgeon, K.B., Eds.; University of Texas Press: Austin, TX, USA, 1982; pp. 212–263.
141. Goheen, D.J.; Hansen, E. Effects of pathogens and bark beetles on forests. In *Beetle-Pathogen Interactions in Conifer Forests*, 1st ed.; Schowalter, T.D., Filip, G.M., Eds.; Academic Press: London, UK, 1993; pp. 175–196.
142. Mattson, W.; Haack, R. Role of drought in outbreaks of plant-eating insects. *Bioscience* **1987**, *37*, 110–118. [CrossRef]
143. Cunningham, C.; Jenkins, M.; Roberts, D. Attack and brood production by the Douglas-fir beetle *Dendroctonus pseudotsugae* (Coleoptera: Scolytidae) in Douglas-fir, *Pseudotsuga menziesii* var. *glauca* (Pinaceae), following a wildfire. *West. N. Am. Nat.* **2005**, *65*, 70–79.
144. McGregor, M.; Furniss, M.M.; Bousfield, W.E.; Almas, D.P.; Gravelle, P.J.; Oakes, R.D. *Evaluation of the Douglas-Fir Beetle Infestation North Fork Clearwater River Drainage, Northern Idaho, 1970–73*; Report No. 74-7; U.S. Department of Agriculture, Forest Service, Northern Region State and Private Forestry, Forest Pest Management: Missoula, MT, USA, 1974; p. 17.
145. Fredericks, S.E.; Jenkins, M.J. Douglas-fir beetle {*Dendroctonus pseudotsugae* Hopkins, Coleoptera: Scolytidae} brood production on Douglas-fir defoliated by western spruce budworm (*Choristoneura occidentalis* Freeman, Lepidoptera: Tortricidae) in Logan Canyon, UT. *Gt. Basin Nat.* **1988**, *48*, 348–351.
146. McMullen, L.H.; Atkins, M.D. On the flight and host selection of the Douglas-fir beetle, *Dendroctonus pseudotsugae* Hopk. (Coleoptera: Scolytidae). *Can. Entomol.* **1961**, *94*, 1309–1325. [CrossRef]
147. Wright, L.C.; Berryman, A.A.; Wickman, B.E. Abundance of the fir engraver, *Scolytus ventralis*, and the Douglas-fir beetle, *Dendroctonus pseudotsugae*, following tree defoliation by the Douglas-fir tussock moth, *Orgyia seudotsugata*. *Can. Entomol.* **1984**, *116*, 293–305. [CrossRef]
148. Edmonds, R.L.; Eglitis, A. The role of Douglas-fir and wood borers in the decomposition of and nutrient release from Douglas-fir logs. *Can. J. For. Res.* **1989**, *19*, 853–859. [CrossRef]
149. Kegley, S. *Douglas-Fir Beetle Management*; Forest Health Protection and State Forestry Organizations. Chapter 4. U.S. Department of Agriculture, Forest Service. Availabel online: http://www.fs.usda.gov/Internet/FSE_DOCUMENTS/stelprdb5187396.pdf (archived on 16 May 2015).
150. Coulson, R.N. Population dynamics of bark beetles. *Annu. Rev. Entomol.* **1979**, *24*, 417–447. [CrossRef]
151. Negrón, J.F.; Schaupp, W.C.; Gibson, K.E.; Anhold, J.; Hansen, D.; Their, R.; Mocettini, P. Estimating extent of mortality associated with the Douglas-fir beetle in the central and northern Rockies. *West. J. Appl. For.* **1999**, *14*, 121–127.
152. Atkins, M.D. Behavioral variation among scolytids in relation to their habitat. *Can. Entomol.* **1966**, *98*, 285–288. [CrossRef]
153. Lawrence, R. Early Detection of Douglas-fir beetle infestation with subcanopy resolution hyperspectral imagery. *West. J. Appl. For.* **2003**, *18*, 202–206.
154. Weatherby, J.C.; Thier, R.W. *A preliminary Validation of Douglas-Fir Beetle Hazard Rating System. Mountain Home Ranger District, Boise National Forest, 1992*; Forest Pest Management Report No. R4-93-05; U.S. Department of Agriculture, Forest Service: Boise, ID, USA, 1993; p. 7.
155. Kulman, H.M. Effects of insect defoliation on growth and mortality of trees. *Annu. Rev. Entomol.* **1971**, *16*, 289–324. [CrossRef]
156. Brookes, M.H.; Colbert, J.J.; Mitchell, R.G.; Stark, R.W. *Western Spruce Budworm*; Technical Bulletin 1694; U.S. Department of Agriculture, Forest Service, Cooperative State Research Service: Washington, DC, USA, 1987; p. 108.
157. Fellin, D.G.; Dewey, J.E. *Western Spruce Budworm*; Forest Insect and Disease Leaflet 53; U.S. Department of Agriculture, Forest Service: Washington, DC, USA, 1982; p. 10.
158. Swetnam, T.W.; Lynch, A.M. Multicentury, regional scale patterns of western spruce budworm outbreaks. *Ecol. Monogr.* **1993**, *63*, 399–424. [CrossRef]
159. Myers, J.H. Synchrony in outbreaks of forest Lepidoptera: A possible example of the moran effect. *Ecology* **1998**, *79*, 1111–1117. [CrossRef]
160. Ryerson, D.E.; Swetnam, T.W.; Lynch, A.M. A tree-ring reconstruction of western spruce budworm outbreaks in the San Juan Mountains, Colorado, U.S.A. *Can. J. For. Res.* **2003**, *33*, 1010–1028. [CrossRef]

161. Murdock, T.Q.; Taylor, S.W.; Flower, A.; Mehlenbacher, A.; Montenegro, A.; Zwiers, F.W.; Alfaro, R.; Spittlehouse, D.L. Pest outbreak distribution and forest management impacts in a changing climate in British Columbia. *Environ. Sci. Policy* **2013**, *6*, 75–89. [CrossRef]

162. Cates, R.G.; Redak, R.; Henderson, C.B. Natural product defensive chemistry of Douglas-fir, western spruce budworm success and forest management practices. *J. Appl. Entomol.* **1983**, *96*, 173–182.

163. Wulf, N.; Cates, R. Site and stand characteristics. *Western Spruce Budworm*; Technical Bulletin 1694; Brookes, M.H., Campbell, R.W., Colbert, J.J., Mitchell, R.G., Stark, R.W., Eds.; U.S. Department of Agriculture, Forest Service: Washington, DC, USA, 1987; pp. 90–115.

164. Carlson, C.E.; Fellin, D.G.; Schmidt, W.C.; O'Laughlin, J.; Pfister, R.D. The western spruce budworm in northern Rocky Mountain forests: A review of ecology, insecticidal treatments and silvicultural practices. In Proceedings of the Symposium Management of Second-Growth Forests: The State of Knowledge and Research Needs, Missoula, MT, USA, 14 May 1982; O'Laughlin, J., Pfister, R.D., Eds.; pp. 76–103.

165. Hadley, K.; Veblen, T. Stand response to western spruce budworm and Douglas-fir bark beetle outbreaks, Colorado Front Range. *Can. J. For. Res.* **1993**, *23*, 479–491. [CrossRef]

166. Dewey, J.E. Damage to Douglas-fir cones by *Choristoneura occidentalis*. *J. Econ. Entomol.* **1970**, *63*, 1804–1806. [CrossRef]

167. Frank, C.J.; Jenkins, M.J. Impact of the western spruce budworm on buds, developing cones and seeds of Douglas-fir in west-central Idaho. *Environ. Entomol.* **1987**, *16*, 304–308. [CrossRef]

168. Owens, J.N. *The Reproductive Cycle of Douglas-Fir*; Research Paper BC-P-8; Canadian Forest Service Pacific Forest Research Center: Victoria, BC, Canada, 1976; p. 23.

169. Lowry, W. Apparent meteorological requirements for abundant cone crops in Douglas-fir. *For. Sci.* **1966**, *12*, 185–192.

170. Fellin, D.G.; Shearer, R.C.; Carlson, C.E. Western spruce budworm in the northern Rocky Mountains. *West. Wildlands Nat. Res. J.* **1983**, *9*, 2–7.

171. Swetnam, T.W.; Thompson, M.A.; Sutherland, E.K. *Using Dendrochronology to Measure Radial Growth of Defoliated Trees*; Handbook No. 639; U.S. Department of Agriculture, Forest Service, Cooperative State Research Service: Washington, DC, USA, 1988; p. 39.

172. Swetnam, T.W.; Lynch, A.M. A tree-ring reconstruction of western spruce budworm history in the southern Rocky Mountains. *For. Sci.* **1989**, *35*, 962–986.

173. Alfaro, R.I.; Van Sickle, G.A.; Thomson, A.J.; Wegwitz, E. Tree mortality and radial growth losses caused by the western spruce budworm in a Douglas-fir stand in British Columbia. *Can. J. For. Res.* **1982**, *12*, 780–787. [CrossRef]

174. Hawksworth, F.G.; Wiens, D. *Dwarf Mistletoes: Biology, Pathology, and Systematics*; Agricultural Handbook 709; U.S. Department of Agriculture, Forest Service: Washington, DC, USA, 1996; p. 410.

175. Watson, D.M. Mistletoe—A keystone resource in forests and woodlands worldwide. *Annu. Rev. Ecol. Syst.* **2001**, *32*, 219–249. [CrossRef]

176. Hoffman, J.T. *Management Guide for Dwarf Mistletoe Arceuthobium* spp.; U.S. Department of Agriculture, Forest Service, 2010. Available online: http://www.fs.usda.gov/Internet/FSE_DOCUMENTS/stelprdb 5187427.pdf (accessed on 28 May 2015).

177. Smith, R.H. *Xylem Resin in the Resistance of the Pinaceae to Bark Beetles*; General Technical Report PSW-1; U.S. Department of Agriculture, Forest Service, Pacific Southwest Forest and Range Experiment Station: Berkeley, CA, USA, 1972; p. 7.

178. Pate, J.S. Mineral relationships of parasites and their hosts. In *Parasitic Plants*, 1st ed.; Press, M.C., Graves, J., Eds.; Chapman & Hall: London, UK, 1995; pp. 80–102.

179. Shaw, D.C.; Watson, D.M.; Mathiasen, R.L. Comparison of dwarf mistletoes (*Arceuthobium* spp., Viscaceae) in the western United States with mistletoes (*Amyema* spp., Loranthaceae) in Australia-ecological analogs and reciprocal models for ecosystem management. *Aust. J. Bot.* **2004**, *52*, 481–498. [CrossRef]

180. Lamont, B. Mineral nutrition of mistletoes. In *Biology of Mistletoes*, 1st ed.; Calder, D.M., Bernhardt, P., Eds.; The Academic Press: Sydney, Australia, 1983; p. 348.

181. Tinnin, R.O.; Hawksworth, F.G.; Knutson, D.M. Witches' broom formation in conifers infected by *Arceuthobium* spp.: An example of parasitic impact upon community dynamics. *Am. Midl. Nat.* **1982**, *107*, 351–359. [CrossRef]

182. Kuijt, J. *The Biology of Parasitic Flowering Plants*, 1st ed.; University of California Press: Berkeley, CA, USA, 1969; p. 246.

183. Hawksworth, F.G.; Wiens, D. *Biology and Classification of Dwarf Mistletoes (Arceuthobium)*; Agriculture Handbook Book Number 401; U.S. Department of Agriculture, Forest Service: Washington, DC, USA, 1972; p. 234. Available online: http://naldc.nal.usda.gov/naldc/download.xhtml?id=CAT87208731 (accessed on 23 February 2016).

184. Geils, B.W.; Mathiasen, R.L. Intensification of dwarf mistletoe on Southwestern Douglas-fir. *For. Sci.* **1990**, *36*, 955–969.

185. Sala, A.; Carey, E.V.; Callawy, R.M. Dwarf mistletoe affects whole-tree water relations of Douglas-fir and western larch primarily through changes in leaf to sapwood ratios. *Oecologia* **2001**, *126*, 42–52. [CrossRef]

186. Hadfield, J.S.; Mathiasen, R.L.; Hawksworth, F.G. *Douglas-Fir Dwarf Mistletoe*; Forest Insect and Disease Leaflet 54; U.S. Department of Agriculture, Forest Service: Washington, DC, USA, 2000; p. 9.

187. Castello, J.D.; Leopold, D.J.; Smallidge, P.J. Pathogens, patterns, and processes in forest ecosystems. *BioScience* **1995**, *45*, 16–24. [CrossRef]

188. Ayres, M.P.; Lombardero, M.J. Assessing the consequences of global change for forest disturbance from herbivores and pathogens. *Sci. Total Environ.* **2000**, *262*, 263–286. [CrossRef]

189. Williams, R.E.; Shaw, C.G., III; Wargo, P.M.; Sites, W.H. *Armillaria Root Disease*; Forest Insect and Disease Leaflet 78; U.S. Department of Agriculture, Forest Service, Northern Area State and Private Forestry, 1986. Available online: http://www.na.fs.fed.us/spfo/pubs/fidls/armillaria/armillaria.htm?) (accessed on 23 February 2016). Forest Insect and Disease Leaflet 78.

190. Wargo, P.M.; Shaw, C.G., III. *Armillaria* root rot: The puzzle is being solved. *Plant Dis.* **1985**, *69*, 826–832. [CrossRef]

191. McDonald, G.I.; Martin, N.E.; Harvey, A.E. *Armillaria in the Northern Rockies: Pathogenicity and Host Susceptibility on Pristine and Disturbed Sites*; Research Note INT-371; U.S. Department of Agriculture, Forest Service, Intermountain Research Station: Ogden, UT, USA, 1987. Available online: https://archive.org/details/armillariainnort371mcdo (accessed on 23 February 2016).

192. Shaw, C.G., III, Kile, G.A., Eds.; *Armillaria Root Disease*; Agricultural Handbook No. 691; U.S. Department of Agriculture, Forest Service: Washington, DC, USA, 1991.

193. Wickman, B.E. *Tree Mortality and Top-Kill Related to Defoliation by the Douglass-Fir Tussock Moth in the Blue Ridge Mountains Outbreak*; Forest Research Paper PNW-223; U.S. Department of Agriculture, Forest Service Pacific Northwest Research Station: Portland, OR, USA, 1978; p. 39.

194. Johnson, D.W. *Forest Pest Training Manual*; U.S. Department of Agriculture, Forest Service, Forest Pest Management, Rocky Mountain Region: Lakewood, CO, USA, 1985.

195. Haggle, S.K. *Managmement Guide for Armaliaria Root Disease*; U.S. Department of Agriculture, Forest Service, Forest Health Protection and State Forestry Organizations, 2008. Available online: http://www.fs.usda.gov/Internet/FSE_DOCUMENTS/stelprdb5187208.pdf (accessed on 24 February 2016).

196. Guillaumin, J.J.; Legrand, P. *Armillaria* root rots. In *Infectious Forest Diseases*, 1st ed.; Gonthier, P., Nicolotti, G., Eds.; CABI: Boston, MA, USA, 2013; pp. 159–178.

197. Worral, J.J.; Sullivan, K.F.; Harrington, T.C.; Steimel, J.P. Incidence, host relations and population structure of *Armillaria ostoyae* in Colorado campgrounds. *For. Ecol. Manag.* **2004**, *192*, 191–206. [CrossRef]

198. Amman, G.D. Bark beetle-fire associations in the Greater Yellowstone area. In Proceedings of the Fire and the Environment Symposium, Knoxville, TN, USA, 20–24 March 1991; Nordvin, S.C., Waldrop, T.A., Eds.; pp. 313–320.

199. Amman, G.D.; Ryan, K.C. *Insect Infestation of Fire-Injured Trees in the Greater Yellowstone Area*; Research Note INT-398; U.S. Department of Agriculture, Forest Service, Intermountain Forest and Range Experiment Station: Ogden, UT, USA, 1991; p. 9.

200. Ryan, K.C.; Amman, G.D. Interactions between fire-injured trees and insects in the greater Yellowstone area. In Plants and Their Environment, Proceedings of the First Biennial Scientific Conference on the Greater Yellowstone Ecosystem, Yellowstone National Park, WY, USA, 16–17 September 1991; Despain, D.G., Ed.; National Park Service: Denver, CO, USA, 1995; pp. 259–271.

201. Rasmussen, L.; Amman, G.; Vandygriff, J.; Oakes, R.; Munson, S.; Gibson, K. *Bark Beetle and Wood Borer Infestation in the Greater Yellowstone Area during Four Postfire Years*; Research Paper INT-RP-487; U.S. Department of Agriculture, Forest Service, Intermountain Forest and Range Experiment Station: Ogden, UT, USA, 1996; p. 9.

202. Dickinson, M.B.; Johnson, E.A. Fire effets on trees. In *Forest Fires Behavior and Ecological Effects*, 1st ed.; Johnson, E.A., Miyanishi, K., Eds.; Academic Press: New York, NY, USA, 2001; pp. 477–521.

203. Stevens-Rumman, C.P.; Morgan, C.P.; Hoffman, C. Bark beetles and wildfires: How does forest recovery change with repeated disturbances in mixed conifer forests? *Ecosphere* **2015**, *6*, 1–17. [CrossRef]

204. Furniss, M.M.; Livingston, R.L.; McGregor, M.D. Development of a stand susceptibility classification for Douglas-fir beetle. *Hazard-Rating Systems in Forest Insect Pest Management: Symposium Proceedings, Athens, GA, USA, 31 July–1 August 1980*; Hedden, R.L., Barras, S., Coster, J.E., Eds.; U.S. Department of Agriculture, Forest Service: Washington, D.C., USA, 1981; pp. 115–128.

205. Christiansen, E.; Waring, R.H.; Berryman, A.A. Resistance of conifers to bark beetle attack: Searching for general relationships. *For. Ecol. Manag.* **1987**, *22*, 89–106. [CrossRef]

206. Klutsch, J.G.; Negrón, J.F.; Costello, S.L.; Rhoades, C.C.; West, D.R.; Popp, J.; Caissie, R. Stand characteristics and downed woody debris accumulations associated with a mountain pine beetle (*Dendroctonus ponderosae* Hopkins) outbreak in Colorado. *For. Ecol. Mang.* **2009**, *258*, 641–649. [CrossRef]

207. Collins, B.J.; Rhoades, C.C.; Battaglia, M.A.; Hubbard, R.M. The effects of bark beetle outbreaks on forest development, fuel loads and potential fire behavior in salvage logged and untreated lodgepole pine forests. *For. Ecol. Manag.* **2012**, *284*, 260–268. [CrossRef]

208. DeRose, R.J.; Long, J.N. Wildfire and spruce beetle outbreak: Simulation of interacting disturbances in the central Rocky Mountains. *Ecoscience* **2009**, *16*, 28–38. [CrossRef]

209. Parker, T.J.; Clancy, K.M.; Mathiasen, R.L. Interactions among fire, insects and pathogens in coniferous forests of the interior western United States and Canada. *Agric. For. Entomol.* **2006**, *8*, 167–189. [CrossRef]

210. Giunta, A. Douglas-Fir beetle mediated changes to fuel complexes, foliar moisture content, and terpenes in interior Douglas-fir forests of the central Rocky Mountains. Master's Thesis, Utah State University, Logan, UT, USA, 2016.

211. McMillin, J.D.; Allen., K.K. *Impacts of Douglas-Fir Beetles on Overstory and Understory Conditions of Douglas-Fir Stands*; General Technical Report R2-64; U.S. Department of Agriculture, Forest Service, Rocky Mountain Research Station: Golden, CO, USA, 2000; p. 17.

212. Von Arx, J.A. Ueber die Ascomycetengattungen Ceratostomella Sacc., Ophiostoma Syd. Und Rostrella Zimmerman. *Antonie Leeuwenhoek* **1952**, *18*, 201–213. [CrossRef] [PubMed]

213. Gruell, G.E. *Fire and Vegetative Trends in Northern Rockies: Interpretations from 1871–1982 Photographs*; General Technical Report INT-130; U.S. Department of Agriculture, Forest Service Intermountain Research Station: Missoula, MT, USA, 1983; p. 117.

214. Gavin, D.; Flower, A.; Cohn, G.; Heyerdahl, E.; Parsons, R. *Interactions of Insects, Fire and Climate on Fuel Loads and Fire Behavior in Mixed Conifer Forest*; Final Report for JFSP Project 09-1-06-5; 2013. Available online: http://www.firescience.gov/projects/09-1-06-5/project/09-1-06-5_final_report.pdf (accessed on 27 May 2015).

215. Black, S.H.; Kulakowski, D.; Noon, B.R.; DellaSalla, D.A. Do bark beetle outbreaks increase wildfire risks in the central U.S. Rocky Mountains? Implications from recent research. *Nat. Areas J.* **2013**, *33*, 59–65. [CrossRef]

216. Flower, A.; Gavin, D.G.; Heyerdahl, E.K.; Parson, R.A.; Cohn, G.M. Western spruce budworm outbreaks did not increase fire risk over the last three centuries: A dendrochronological analysis of inter-disturbance synergism. *PLoS ONE* **2014**. [CrossRef] [PubMed]

217. Hummel, S.; Agee, J.K. Western spruce budworm defoliation effects on forest structure and potential fire behavior. *Northwest Sci.* **2003**, *77*, 159–169.

218. Burgan, R.E.; Rothermel, R.C. *BEHAVE: Fire Behavior Prediction and Fuel Modeling System FUEL Subsystem*; General Technical Report INT-167; U.S. Department of Agriculture, Forest Service, Intermountain Forest and Range Experiment Station: Ogden, UT, USA, 1984; p. 126.

219. Koonce, A.L.; Roth, L.F. The effects of dwarf mistletoe on fuel in precommercial ponderosa pine stands. In Proceedings of the Eighth Conference on Fire and Forest Meteorology: Weather—The Drive Train Connecting the Solar Engine to Forest Ecosystems, Detroit, MI, USA, 29 April–2 May 1985; Donoghue, L.R., Martin, R.E., Eds.; Society of American Foresters: Bethesda, MD, USA, 1985; pp. 66–72.

220. Dickman, A.; Cook, S. Fire and fungus in a mountain hemlock forest. *Can. J. Bot.* **1989**, *67*, 2005–2016. [CrossRef]
221. Jain, T.B.; Battaglia, M.A.; Han, H.S.; Graham, R.T.; Keyes, C.R.; Fried, J.S.; Sandquist, J.E. *A Comprehensive Guide to Fuel Management Practices for Dry Mixed Conifer Forests in the Northwestern United States*; General Technical Report RMRS-GTR-292; U.S. Department of Agriculture, Forest Service, Rocky Mountain Research Station: Golden, CO, USA, 2012; p. 331.
222. Hagle, S.K.; Gibson, K.E.; Tunnock, S. *Field Guide to Diseases and Insect Pests of Northern and Central Rocky Mountain Conifers*; General Technical Report R1-03-08; U.S. Department of Agriculture, Forest Service, State and Private Forestry, Northern and Intermountain Regions: Missoula, MT, USA; Ogden, UT, USA, 2003; p. 197.
223. Man, G. *Major Forest and Insect Disease Conditions in the United States: 2011 Update*; Forest Service Report FS-1000; U.S. Department of Agriculture: Washington, DC, USA, 2012; pp. 14–16.
224. Jenkins, M.L. *Major Forest and Insect Disease Conditions in the United States: 2013*; Forest Service Report FS-1054; U.S. Department of Agriculture, Forest Service, Forest Health Protection: Washington, DC, USA, 2015; pp. 18–20.
225. Lessard, E.D.; Schmid, J.M. Emergence, attack densities, and host relationships for the Douglas-fir beetle, (*Dendroctonus pseudotsugae* Hopkins) in northern Colorado. *Gt. Basin Nat.* **1990**, *50*, 333–338.
226. Schmid, J.M.; Mata, S.A. *Natural Variability of Specific Forest Insect Populations and Their Associated Effects in Colorado*; General Technical Report RM-GTR-275; U.S Department of Agriculture, Forest Service: Washington, DC, USA, 1996.
227. Berryman, A.A.; Wright, L.C. Defoliation, tree condition, and bark beetles. *The Douglas-Fir Tussock Moth a Synthesis*; Technical Bulletin 1585; Stark, M.H., Campell, R.W., Eds.; U.S. Department of Agriculture, Forest Service: Washington, DC, USA, 1978; pp. 81–87.
228. Singh, P. Research and management strategies for major tree diseases in Canada: Synthesis part 1. *For. Chron.* **1993**, *69*, 151–162. [CrossRef]
229. Huber, D.P.W.; Borden, J.H. Angiosperm bark volatiles disrupt response of Douglas-fir beetle, *Dendroctonus pseudotsugae*, to attractant-baited traps. *J. Chem. Ecol.* **2001**, *27*, 217–233. [CrossRef] [PubMed]
230. Kay, C.E. Aboriginal overkill and native burning: Implications for modem ecosystem management. *West. J. Appl. For.* **1995**, *10*, 121–126.
231. Keane, R.E.; Ryan, K.C.; Veblen, T.T.; Allen, C.D.; Logan, J.A.; Hawkes, B. The cascading effects of fire exclusion in Rocky Mountain ecosystems. In *Rocky Mountain Futures: An ecological Perspective*, 1st ed.; Baron, J.S., Ed.; Island Press: Washington, DC, USA, 2002; pp. 133–152.
232. Baker, W.L. Indians and fire in the Rocky Mountains: The wilderness hypothesis reviewed. In *Fire Native Peoples, and the Natural Landscape*, 1st ed.; Vale, T.R., Ed.; Island Press: Washington, DC, USA, 2002; pp. 41–77.
233. Peterson, C.S.; Speth, L.E. A History of the Wasatch-Cache National Forest. Utah State University, 1980. Available online: http://www.fs.usda.gov/Internet/FSE_DOCUMENTS/stelprdb5053310.pdf (accessed on 1 March 2016).
234. Alexander, T.G.; Fish, R.J. The Forest Service in Utah. Available online: http://www.uen.org/utah_history_encyclopedia/f/FOREST_SERVICE.html (accessed on 1 March 2016).
235. Owston, P.W.; Stein, W.I. *Pseudotsuga Carr.*, Douglas-fir. *Seeds of Woody Plants in the United States*; Agriculture Handbook 450; Schopmeyer, C.S., Ed.; U.S. Department of Agriculture, Forest Service: Washington, DC, USA, 1974; pp. 674–683.
236. Noss, R.F.; Franklin, J.F.; Baker, W.L.; Schoennagel, T.; Moyle, P.B. Managing fire-prone forests in the western United States. *Front. Ecol. Environ.* **2006**, *4*, 481–487. [CrossRef]
237. Landres, P.; Morgan, P.; Swanson, F. Overview of the use of natural variability concepts in managing ecological systems. *Ecol. Appl.* **1999**, *9*, 1179–1188.
238. Graham, R.T.; Harvey, A.E.; Jain, T.B.; Tonn, J.R. *Effects of Thinning and Similar Stand Treatments on Fire Behavior in Western Forests*; General Technical Report PNW-GTR-463; U.S. Department of Agriculture, Forest Service, Pacific Northwest Research Station: Portland, OR, USA, 1999.
239. Rippy, R.C.; Stewart, J.E.; Zambino, P.J.; Klopfenstein, N.B.; Tirocke, J.M.; Kim, M.S.; Thies, W.G. *Root Diseases in Coniferous Forests of the Inland West: Potential Implications of Fuels Treatments*; General Technical Report, RMRS-GTR-141; U.S. Department of Agriculture, Forest Service, Rocky Mountain Research Station: Fort Collins, CO, USA, 2005.

240. Edmonds, R.L.; Agee, J.K.; Gara, R.I. *Forest Health and Protection*, 1st ed.; Waveland Press, Inc.: Long Grove, IL, USA, 2000; pp. 359–397.

241. Fernandes, P.M.; Hermínio, S.B. A review of prescribed burning effectiveness in fire hazard reduction. *Int. J. Wildland Fire* **2003**, *12*, 117–128. [CrossRef]

242. Hessburg, P.F.; Mitchell, R.G.; Filip, G.M. *Historical and Current Roles of Insects and Pathogens in Eastern Oregon and Washington Forested Landscapes*; General Technical Report, PNW-GTR-327; U.S. Department of Agriculture, Forest Service, Pacific Northwest Research Station: Portland, OR, USA, 1994.

243. Arno, S.F. Forest fire history in the northern Rockies. *J. For.* **1980**, *78*, 460–465.

244. McIver, J.D.; Starr, L. A literature review on the environmental effects of postfire logging. *West. J. Appl. For.* **2001**, *16*, 159–168.

245. Potts, D.F.; Peterson, D.L.; Zuuring, H.R. *Watershed Modeling for Fire Management Planning in the Northern Rocky Mountains*; Research Paper PSW-177; U.S. Department of Agriculture, Forest Service, Pacific Southwest Forest and Range Experiment Station: Berkely, CA, USA, 1985.

246. Donato, D.C.; Fontaine, J.B.; Campbell, J.L.; Robinson, W.D.; Kauffman, J.B.; Law, B.E. Post-wildfire logging hinders regeneration and increases fire risk. *Science* **2006**. [CrossRef] [PubMed]

247. Seidl, R.; Spies, T.A.; Peterson, D.L.; Stephens, S.L.; Hicke, J.A. Searching for resilience: Addressing the impacts of changing disturbance regimes on forest ecosystem services. *J. Appl. Ecol.* **2016**, *53*, 120–129. [CrossRef] [PubMed]

248. Graham, R.T.; McCaffrey, S.; Jain, T.B. *Science Basis for Changing Forest Structure to Modify Wildfire Behavior and Severity*; General Technical Report RMRS-120; U.S. Department of Agriculture, Forest Service: Fort Collins, CO, USA, 2004.

249. Stephens, S.L.; Moghaddas, J.J. Experimental fuel treatment impacts on forest structure, potential fire behavior, and predicted tree mortality in a California mixed conifer forest. *For. Ecol. Manag.* **2005**, *215*, 21–36. [CrossRef]

250. Allen, C.D.; Betancourt, J.L.; Swetnam, T.W. Landscape changes in the southwestern United States: Techniques, long-term data sets, and trends. *Perspectives on the Land-Use History of North America: A Context for Understanding Our Changing Environment*; Sisk, T.D., Ed.; U.S. Geological Survey: Reston, VA, USA; pp. 71–84.

251. Caprio, A.C.; Graber, D.M. Returning fire to the mountains: Can we successfully restore the ecological role of pre-Euroamerican fire regimes to the Sierra Nevada. In Proceedings of the Wilderness Science in a Time of Change Conference, Missoula, MT, USA, 23–27 May 1999; Cole, D.N., McCool, S.F., Borrie, W.T., O'Laughlin, J., Eds.; U.S. Department of Agriculture, Forest Service: Proceedings RMRS-P-15-VOL-5. U.S. Department of Agriculture, Forest Service, Rocky Mountain Research Station: Ogden, UT, USA, 2000; pp. 223–241.

252. Heyerdahl, E.K.; Lertzman, K.; Wong, C.M. Mixed-severity fire regimes in dry forests of southern interior British Columbia, Canada. *Can. J. For. Res.* **2012**, *42*, 88–98. [CrossRef]

© 2016 by the authors. Licensee MDPI, Basel, Switzerland. This article is an open access article distributed under the terms and conditions of the Creative Commons Attribution (CC BY) license (http://creativecommons.org/licenses/by/4.0/).

![forests logo] *forests*

MDPI

Article

Regional Instability in the Abundance of Open Stands in the Boreal Forest of Eastern Canada

Rija Rapanoela [1], Frédéric Raulier [1,*,†] and Sylvie Gauthier [2,†]

1   Centre d'étude de la forêt, Faculté de foresterie, de géographie et de géomatique, Université Laval,
    2405 rue de la terrasse, Québec, QC G1V 0A6, Canada; Rija.Rapanoela.1@ulaval.ca
2   Natural Ressources Canada, Canadian Forest Service, Laurentian Forestry Centre, 1055 du PEPS,
    P.O. Box 10380, Stn. Sainte-Foy, Québec, QC G1V 4C7, Canada; Sylvie.Gauthier@RNCan-NRCan.gc.ca
*   Correspondence: frederic.raulier@sbf.ulaval.ca; Tel.: +1-418-656-2131 (ext. 6742)
†   These authors contributed equally to this work.

Academic Editor: Timothy A. Martin
Received: 15 March 2016; Accepted: 5 May 2016; Published: 12 May 2016

Abstract: Fires are a key disturbance of boreal forests. In fact, they are the main source of renewal and evolution for forest stands. The variability of fire through space and time results in a diversified forest mosaic, altering their species composition, structure and productivity. A resilient forest is assumed to be in a state of dynamic equilibrium with the fire regime, so that the composition, age structure and succession stages of forests should be consistent with the fire regime. Dense spruce-moss stands tend, however, to diminish in favour of more open stands similar to spruce-lichen stands when subjected to more frequent and recurring disturbances. This study therefore focused on the effects of spatial and temporal variations in burn rates on the proportion of open stands over a large geographic area (175,000 km^2) covered by black spruce (*Picea mariana* (Mill.) Britton, Sterns, Poggenb.). The study area was divided into 10 different zones according to burn rates, as measured using fire-related data collected between 1940 and 2006. To test if the abundance of open stands was unstable over time and not in equilibrium with the current fire regime, forest succession was simulated using a landscape dynamics model that showed that the abundance of open stands should increase progressively over time in zones where the average burn rate is high. The proportion of open stands generated during a specific historical period is correlated with the burn rate observed during the same period. Rising annual burn rates over the past two decades have thereby resulted in an immediate increase in the proportion of open stands. There is therefore a difference between the current proportion of open stands and the one expected if vegetation was in equilibrium with the disturbance regime, reflecting an instability that may significantly impact the way forest resources are managed. It is apparent from this study that forestry planning should consider the risks associated with the temporal variability of fire regimes on the forest ecosystem, as the resulting changes can have a significant impact on biodiversity and allowable cut estimates.

Keywords: boreal forest; fire; succession; black spruce; resilience; vulnerability; landscape

1. Introduction

The boreal forest is the largest forest area in Canada. In Eastern Canada, it is dominated mainly by black spruce-moss stands that are composed solely of black spruce (*Picea mariana* (Mill.) Britton, Sterns, Poggenb.) or of a combination of black spruce, jack pine (*Pinus banksiana* Lamb.) or balsam fir (*Abies balsamea* (L.) Mill.). There is a transition area between the forest tundra in the north and the black spruce-moss forest in the south: the black spruce-lichen woodland, which is composed of stands that are less dense as well as less productive [1]. By definition, Spruce-lichen woodlands are open-structure forests with a lichen cover of over 40% [2,3]. A number of studies have shown that the lesser abundance of dense spruce-moss stands in favour of more open stands that are similar to spruce-lichen stands [2,4–6] is related to recurring disturbances [2,4–6], at least in regions where dense and open stands co-occur [7].

According to Turner *et al.* [8], landscapes can be divided into three categories according to the extent and frequency of the disturbances to which they are subjected. Landscapes traditionally considered to be in equilibrium with the disturbance pattern in place are characterized by small (compared to the size of the landscape in question) and locally infrequent disturbances. They also return to a state of equilibrium more rapidly in fact than the length of the disturbance cycle. Stable systems are characterized by medium-sized disturbances occurring on an intermediate basis. These systems return to a stable state in a moderate amount of time, equivalent to the length of the disturbance cycle. Potentially unstable systems are characterized by substantial disturbances (compared to the size of the area in question) occurring more frequently. Furthermore, unstable landscapes take longer than the span of the disturbance cycle to return to their original state.

The variability of fire regimes through space and time results in a diversified mosaic of species, altering their composition, structure and productivity [9]. In fact, forest composition [10,11], structure [12,13] and productivity [5,14] are all related to the fire regime. For boreal forests, forest succession models used for forest planning assume that, without harvest, the vegetation currently found in an area is adapted to its natural disturbance regime [15–19]. For instance, the concept of fire cycle is used in forest ecosystem management to define a minimum target value of old-growth forests to maintain in a landscape [17] or a maximum rate of clear-cut harvesting [16]. The fire cycle corresponds to the time required to burn an area equivalent to the study area [20] and in boreal forests of eastern Canada, this fire cycle varies between one and a few centuries. It is therefore defined at a time scale somewhat larger than that used for forest management planning. Significant temporal and even regional variations in burn rates observed in the boreal forest [16,21–26] however cast doubt on an unquestioned use of this assumption. Namely, an important increase in the burn rates of a number of areas in the North American boreal forest has been observed over the past few decades [25–28] and such an increase should possibly be accounted for when designing sustainable forest management strategies.

The primary objective of this study was to analyze the impact of the variation in fire frequency on the openness of the forest. We wished to assess whether the forest presents a greater abundance of open stands in areas where the current burn rate surpasses rates recorded in the recent past. We therefore tested the hypothesis that the abundance of open forest stands varied according to variation in decadal burn rate. In order to do so, we used a transition matrix model to assess succession. Matrix models are probability models [29–31] used to predict the long-term demographic dynamics of a population about which we have little information [29,32]. Transition matrices represent the probability of different states within systems evolving into other types at a certain moment in time. The succession of the different states depends solely on the current state of the system [30,33]. When undisturbed, these different states always evolve towards a stable long-term equilibrium, independent from a system's initial state [18,30]. Thanks to this characteristic, we were able to assess whether there is in fact an equilibrium between the current proportion of open stands and the current burn rate by simulating the evolution of the abundance of open stands using a landscape dynamics model spanning 150 years and by testing the stability of the proportion of open stands over time.

2. Materials and Methods

2.1. Study Area

The area that was under study (Figure 1) is located in the province of Quebec (Canada), extending from 49° N to 53° N latitude and from 70° W to 76° W longitude around Lake Mistassini. This area covering approximately 175,000 km^2 is predominantly covered in the south by black spruce-moss forest and by black Spruce-lichen woodland in the north. The forest mainly reflects the influence of the physical and climate characteristics of the area, as well as the impact of repeated fires [14]. Essentially, the area is marked by lower burn rates in the south (<0.30% year^{-1}, based on data gathered on fires between 1940 and 2009 [25], and by higher burn rates in the north (between 0.30% and 1.2% year^{-1}) (Figure 1; Table 1). The characteristics of these different fire regimes were established for 10 separate areas (fire zones) by Mansuy *et al.* [25]. Together with the fire regime, the climate, and particularly the temperature (number of growing degree-days above 5 °C) are key factors to the stands' density and productivity [14]. The forest is denser and more productive in the southwest sector (growing degree-days starting at 1200 °C· year^{-1}) than in the northeast sector (800 °C· year^{-1}) (Figure 1; Table 1). The average annual temperature is 1.9 °C in the southwest sector and −6.0 °C than in the northeast sector [25].

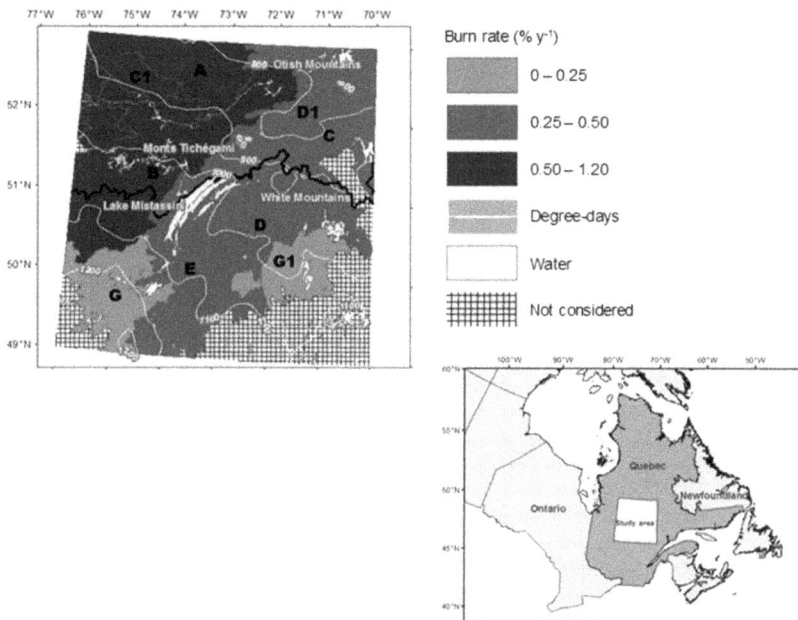

Figure 1. Burn rates [25] and degree-days [34] for the study area. The hatched area lies outside of the spruce-moss bioclimatic domain [9] or a regional burn rate could not be estimated [25]. The northern limit of commercial forests was set in 2002 by the Ministère des Ressources Naturelles et de la Faune [35].

Table 1. General information on the 10 fire zones of the study area.

Region	Area (km^2)	Degree Day (°C· year^{-1})	Burn Rate (%· year^{-1}) [25]	Abundance of Open Stands (Median) in 2006 (%)
A	18,525	900	1.11	89
B	35,157	1000	0.78	88
C1	4265	900	0.67	54
C	42,468	940	0.48	85
D1	6492	800	0.42	81
D	14,291	1010	0.37	93
E	11,522	1130	0.28	44
G1	15,043	1000	0.16	34
G	13,509	1200	0.14	28

2.2. Forest Data

This area straddles the northern limit for timber allocations as established by the Ministère des Forêts, de la Faune et des Parcs (MFFP) [35] in 2002. This northern limit more or less follows the 51st parallel (Figure 1). The forest data used in this study stem from two different inventory programs: the northern forest inventory program for the area above this limit, and the regular inventory program to the south. These two programs were homogenized in 2006 by the MFFP in order to generate forest maps that included surficial deposits, moisture regime, forest cover type (softwood, deciduous, or mixed), understory vegetation, cover density class, development stage, potential vegetation and disturbance of origin over an area of at least 8 ha [25,36]. South of the 51st parallel, forest maps were based on the interpretation of aerial photographs, which were taken between 1990 and 2001, and updated in 2006 by the MFFP to account for recent disturbances. North of the 51st parallel, forest vegetation was classified from satellite images (Landsat 2005), while aerial photographs at the scale of 1:40,000 were used to map surficial deposits and moisture regimes.

In our study, we used 945 permanent and temporary sample plots that predominantly consisted of black spruce and jack pine (with a 75% minimum basal area coverage per species in one plot). A total of 248 temporary and 291 permanent sample plots are located north of the 51st parallel. Temporary sample plots were measured in 2006 and 2007 with the northern forest inventory program and permanent sample plots were remeasured between 1990 and 2001 during the third regular forest inventory program. Plots are evenly distributed across the study area ([14]: their Figure S1). Within each sample plot (of 400 m^2), the species and trunk diameter measured at breast height (DBH) was noted for each merchantable tree (DBH > 9 cm). Three to five dominant or co-dominant trees (taller than two-third of the canopy height [36]) were randomly selected to record their age (core at 1 m height) and total height.

2.3. Data on Forest Fires

The annual burn rate corresponds to the annual area burned, divided by the total terrestrial area (excludes lakes and other water bodies, but includes forested peatlands). The annual areas burned have been used to calculate the burn rate for each region [25]. The history of the areas burned within our study area comes from the spatial database from the MFFP for the period of 1940 to 2006. The fire map has been compiled from various sources: satellite images, aerial photographs, maps and archives. Therefore, the older fires listed are likely to be incomplete and cover about 14% of the territory north of the study area, and instead of having dates of the fires being accurate within one year, they fall within a range of 5 or 10 years [37]. In order to account for the variability of the information sources regarding burned areas, a floating average for a 7-year period (average cycle of repetition of large fires), as described by Gauthier *et al.* [37], was applied to obtain the annual burn rate. This method produced 61 different annual burn rates for each fire zone, from which 10,000 random draws were conducted to estimate the frequency distribution of the annual burn rate for each fire zone. A regional average burn rate was also calculated for three separate twenty-year periods (1947–1966, 1967–1986, and 1987–2006) in order to detect any temporal variations.

2.4. Description of the Landscape Dynamics Model

The structure of the forest landscape is the result of complex interactions between geomorphology, climate, disturbances and natural succession. We used a forest landscape dynamics model, the "Vermillion Landscape Model" or VLM [38] to simulate the landscape dynamics. VLM, which has been described in detail in [38–41], has been implemented in the SELES modelling tool (Spatially Explicit Landscape Event Simulator [42]) by Fall [38] to be compatible with forest map data produced by the MFFP of Québec. Different versions of this model have been used to study the sustainability of forest management strategies in northern temperate or boreal forests of Québec in interaction with fire, insect defoliation, natural succession and road building [38–41,43]. The landscape is described by a set of raster layers (e.g., forest type, stand age, and soil drainage). Processes that influence forest dynamics (fire, natural succession, and logging) are described as landscape agents (submodels) that modify properties of raster cells through time. VLM has been simplified to meet the objective of our study by keeping only succession and fire as active landscape agents (File S1).

2.4.1. Landscape Description in VLM

The VLM model uses raster layers as inputs. Due to its considerable size, the territory was divided into cells or pixels of 16 ha (400×400 m^2) in order to correspond to twice the average size of the forest polygons of the study area. The analyses performed as part of our study focused on the opening of black spruce stands, because shade-intolerant stands that are predominantly composed of softwood conifers represented by jack pine are much less abundant [14]. The territory was therefore stratified according to three criteria: dominant species, degree of cover opening and age class. Because mapping north of the 51st parallel was performed using satellite images, without the identification of coniferous species and estimation of stand age, it was first necessary to estimate the stands' species composition (distinguishing between black spruce and jack pine) and age class.

Species dominance within target populations (and therefore within forest map polygons) was first estimated using two logistic regression models calibrated by Rapanoela *et al.* [14] for the same study area and with the plot dataset described above (Section 2.2). As black spruce largely dominates in the study area, Rapanoela *et al.* [14] first assumed that all stands were composed of black spruce by default. The first regression model was then used to estimate the probability of jack pine occurrence mainly as a function of elevation, drainage, developmental stage and understory cover ([14]: Table 2): jack pine tends to occur more frequently on dry to mesic sites of low elevation, with a developmental stage qualified as "regenerated" or "young" and in sites dominated by lichens in the understory vegetation. Probability of jack pine dominance is more related to the presence of coarse soil deposits, often on hilly areas in the northwestern part of the study area.

Table 2. List of variables selected with logistic regression to explain the variability of abundance of open stands by fire zones and increase in open stands after 150 years of simulations with a forest succession model in interaction with natural disturbances.

Abundance of Open Stands			Probability of Increase of Open Stands		
Variables	Estimate (SE)	Wald χ^2	Variables	Estimate (SE)	Wald χ^2
Degree day	0.0078 (0.0007)	106.7	Last frost day	0.0595 (0.0109)	29.8
Degree day $\times$ burn rate	-1.058 (0.217)	23.9	Total precipitation	-0.00886 (0.0019)	22.0
Burn rate	794 (207)	14.7	Burn rate	1.248 (0.269)	21.6

The age of forest polygons was estimated using two methods. South of the 51st parallel, the age of some stands could be determined by the date of the last disturbance indicated on the forest maps and it has been recalculated by subtracting the year of origin of the disturbance (fire or cutting) from the year of production of the forest map (2006). For other polygons where a date of last disturbance was not available, the age was estimated by a non-parametric method (k-NN) [44]. This calculation method consists of estimating unknown values for forest attributes within an area unit (target polygon) by averaging the values of attributes of similar reference surface units (inventory plots) [45]. Baseline predictors were the cartographic attributes (Section 2.2) and climatic variables. Climatic variables were chosen for their potential impact on succession dynamics and forest productivity in the study area [14,46]. BioSIM 9 [34] was used to estimate these climate variables for each forest polygon within the study area. BioSim adjusts data from the closest weather stations to account for differences in exposure, elevation, latitude and longitude between these stations and the stands targeted. The similarity between the characteristics of the reference polygons and the target polygon (year) was calculated using the Gower distance [47]. We followed the process described by [14] for the variable selection to keep only one variable among correlated variables and to remove variables that do not significantly explain distances between target and reference polygons. The number k of nearest neighbours was chosen by minimizing the root mean squared error of the estimates (RMSE), determined by cross-validation [44,48]. The prediction quality of the age of stands was assessed with the coefficient of determination between predicted and observed ages and the absence of bias. With this procedure, stand age was estimated from the weighted average of 16 nearest neighbours, neighbourhood being assessed with Gower distances estimated with nine variables: six stand cartographic attributes (development stage, cover density class, potential vegetation, surficial deposit, slope and elevation) and three climatic variables (total annual radiation ($MJ \cdot m^{-2} \cdot year^{-1}$), annual snow precipitation ($mm \cdot year^{-1}$) and aridity index ($mm \cdot year^{-1}$, sum of the difference between Thorthwaite's potential evapotranspiration and monthly precipitation [49])). The RMSE (6.7 year) largely exceeded the mean residual bias (−1.4 year) and the fit to observed values was considered acceptable, with a coefficient of determination of 26%.

Two classes were considered when assigning dominant species type, that being shade-tolerant conifers (Rt) and shade-intolerant conifers (Ri), and two classes of canopy openings, that is, open (O) and closed stands (C). The types of canopy opening were determined according to the forest standard mapping of the Nordic Ecoforest Inventory Program: stands where the cover percentage of the canopy of commercial species is greater than 40% ("A", "B" and "C" density classes) were classified as closed, and those where the cover percentage was less than 40% ("D" and "L" density classes) were classified as open. For each fire zone, we simulated the evolution of 4 separate strata: strata closed and open composed of shade-tolerant or shade-intolerant conifers (RtC; RtO; RiC; RiO). Age values were regrouped into six 20-year age classes (0 to 20 (10), 21 to 40 (30), 41 to 60 (50), 61 to 80 (70), 81 to 100 (90), and >100 year old (100^+)).

2.4.2. Succession Submodel

We used the approach developed by Fall *et al.* [38] to design an empirical semi-Markov model of succession. This approach is based on the hypothesis that trends in the distribution of stand patterns by age reflect the current succession process and will continue in the future [50]. First, we assumed that stands within the same fire zone follow the same succession dynamic [46]: a transition matrix was therefore calibrated for each fire zone. The transition probabilities were estimated by age groups based on 20-year periods for each of the four strata (RtC; RtO; RiC; RiO) in a fire zone based on the forest map of the territory. Following a fire, the age of the cell is reinitiated and the composition of a stand after a fire is randomly selected from the abundance of strata in each fire zone for the first age group (0–20 years). A cell can change succession paths randomly over time, with probabilities being derived based on how the strata are represented in the fire zone for its corresponding age group.

2.4.3. Fire Submodel

The empirical distribution of the annual burn rate (Section 2.4) was used to simulate the annual burn rate. The number of fires followed a negative exponential distribution. For each fire, a spark cell was selected at random, and the fire is propagated in all directions until it encounters an obstacle (water body or recent burn) and the number of cells defined by the planned fire area is reached. Recent burns could reburn from the next simulation period.

2.4.4. Simulation Runs

Simulations of the evolution of the abundance of open stands were performed to cover a 150-year time frame by five-year periods. Such a time period corresponds to the beginning of the conversion of even-aged stands to uneven-aged stands [51,52] and to the forest management planning horizon in Quebec. The number of simulations was set at 100. In our case, after 150 years, the coefficient of variation applicable to the abundance of open stands was less than 0.4% after 100 simulations in all fire zones.

2.5. Interpretation of Simulation Results

The primary objective of this study was to explain the abundance and variation over time of areas with open stands, according to fire zones, climate variables, physical variables, and burn rates.

If there is an equilibrium between the vegetation and disturbance rate, the breakdown of the land area into strata should remain approximately constant through time. We therefore measured the absence of equilibrium by subtracting the abundance of open stands after 150 years (as per our simulation) from the numbers observed in 2006 for each fire zone. We then attempted to explain the difference between the initial and final proportions. As we had 100 simulations, we built frequency distributions of these differences by fire zone to express the probability that the abundance of open stands could change.

We also tried to explain the abundance of open stands observed at the simulation start (in 2006). A stand's age is equivalent to the time elapsed since its original disturbance. Variations in the abundance of open stands by age group apparent in the empirical semi-Markovian succession models for each fire zone should explain the absence of a stable equilibrium between the vegetation and fire regime. The abundance of open stands for three age groups (0–20 years, 20–40 years and 40–60 years) as observed in 2006 was therefore calculated to identify and explain any trends in three burn rate temporal periods (1947–1966, 1967–1986, and 1987–2006) (Section 2.4) as a function of biophysical variables (see Sections 2.2 and 2.3). We then applied with this data a backward selection model for variables, using the LOGISTIC procedure from SAS. All independent variables were tested and those that contributed the least to the model were eliminated, according to a 5% threshold. The best model was selected with the Akaike information criterion (AIC) and the adjusted coefficient of determination (adjusted R^2) [53].

3. Results

The study area can be divided into two zones according to the abundance of open stands observed in 2006 for each fire zone (Figure 2). Indeed, open stands are less abundant in areas located in the southern areas covered by the study (average abundance of less than 50% in D, E, G, and G1 areas), while they dominate the northern portions of the territory (A, B, C, C1, and D1 areas), precisely where the burn rate is higher (>0.3% year^{-1}) (Figure 1). This division also coincides with a climatic transition zone with a colder climate zone in the north (A, B, C, C1, and D1 areas) and warmer in the south (D, E, G, and G1 areas) (Figure 1). The division between the two zones straddles the northern limit for timber allocations.

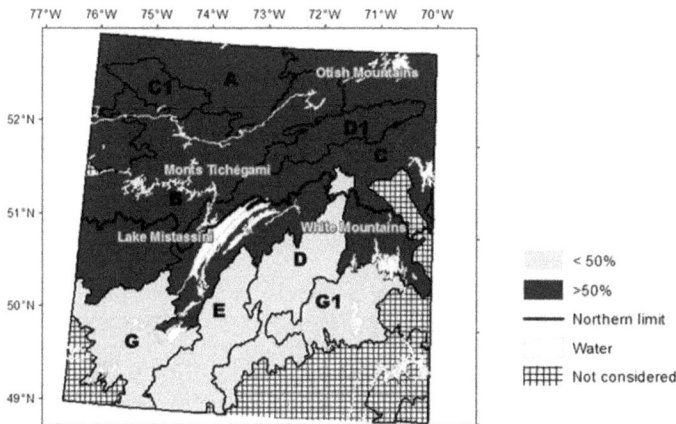

Figure 2. Average proportion by fire zones of the abundance of open stands in 2006.

3.1. Temporal Variation in the Abundance of Open Stands

Simulations using the landscape dynamics model show a sometimes substantial variation in the abundance of open stands over the 150-year period covered for virtually all ten fire zones (Figure 3). In the area dominated by open stands, only the D1 area seemed to show an equilibrium between the abundance of open stands and the disturbance rate. However, the abundance of open stands seems to increase by more than 10% over 150 years in two other fire zones (areas A and C1). Both of these zones are in fact located the furthest north among the areas covered by the study. The increase is not as significant for B and C areas that abut the area dominated by closed stands (Figure 3). In the area dominated by closed stands, only the D area seems to show a substantial increase in the abundance of open stands and this feature is related to the fact that this area is one that has sustained the highest burn rate between 1940 and 2006 in this fire zone (Figure 1).

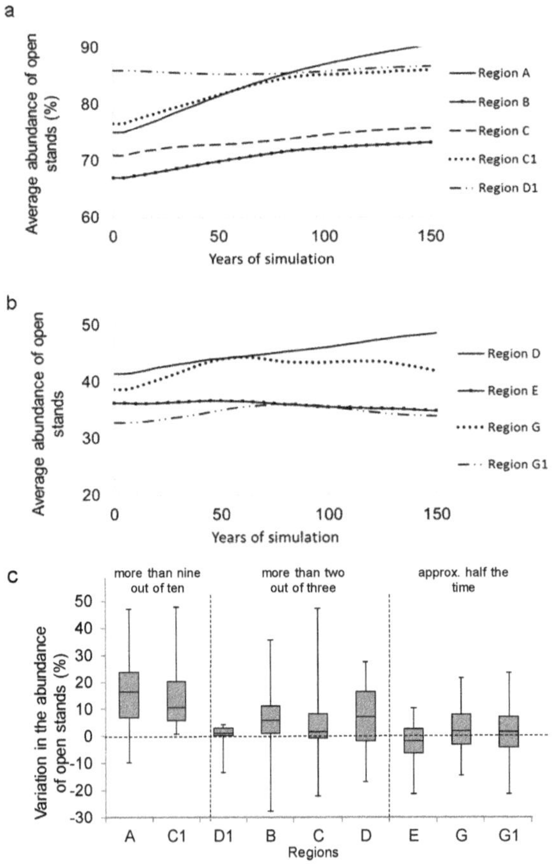

Figure 3. Dynamics of the average abundance of open stands in fire zones dominated by open stands (**a**); and in fire zones dominated by closed stands (**b**) with 100 simulations over 150 years of a forest succession model in interaction with natural disturbances. (**c**) Box and whisker plots representing the frequency distributions of the differences between final and initial abundances of open stands of 100 simulations over 150 years. Cumulative frequencies of positive changes were regrouped into frequency classes (more than nine out of ten: 0.90–0.99; more than two out of three: 0.66–0.90; and approximately half the time: 0.33–0.66).

3.2. Long-Term Change in Abundance of Open Stands

A transition matrix has been calibrated for each fire zone based on the empirical strata's abundance distribution by age groups. The existence of different transition rates for each age group within a fire zone may partly explain the changes observed in the abundance of open stands during the simulations (Figure 3). The classification of open stands by date of origin (1947–1966, 1967–1986, and 1987–2006) shows the actual changes in their relative abundance over long periods (Figure 4a). A general increase in the abundance of open stands was recorded between 1987 and 2006, except for the D1 area. Where burn rates are the highest (in A, B, C1, D, and D1 areas), with the exception of the C fire zone, the increase was higher (up to 70%) between 1987 and 2006 (Figure 4a). This increase in the abundance of open stands observed during the past 20 years is related, with one exception (in the area C), with a recent increase in the regional burn rate (Figure 4b).

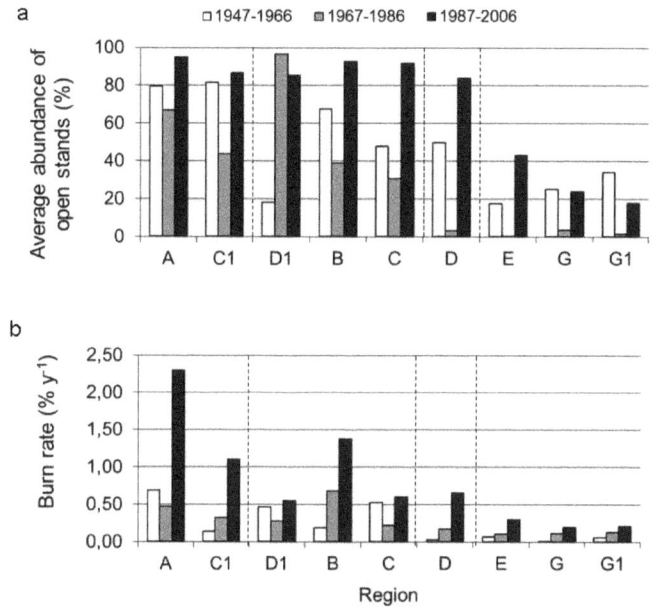

Figure 4. Average abundance of open stands by fire region for three 20-year periods (1947–1966, 1967–1986, and 1987–2006) (**a**); and corresponding regional burn rates by periods (**b**).

3.3. Factors Responsible for the Variation in the Abundance of Open Stands

The abundance of open stands observed at the simulation start (2006) by fire zone is explained mainly by the number of degree-days and the average burn rate (Table 2). The abundance of open stands exceeds 50% when the degree-days of growth are lower than 1000 °C· year^{-1} or when the burn rate is above 0.5% year^{-1} (Figure 5). The logistic model explains 60% of the regional variation in the abundance of open stands (Figure 6).

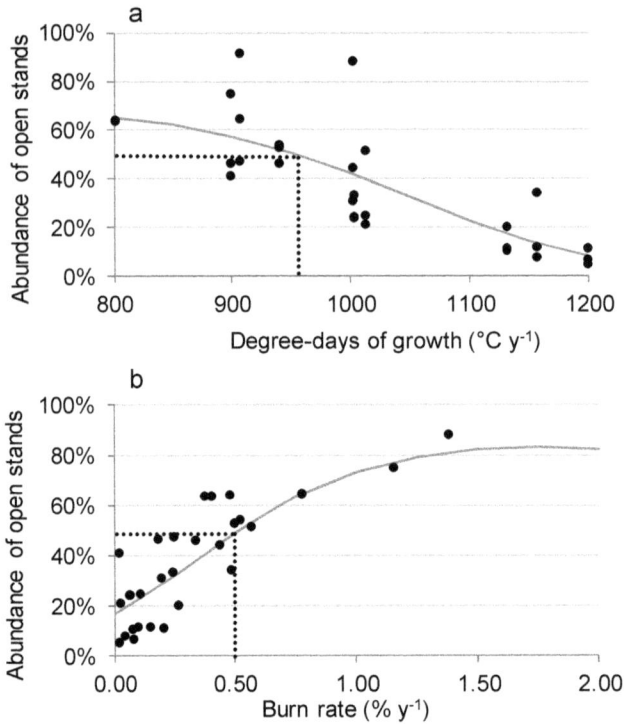

Figure 5. Relationship between the abundance of open stands by fire zone and degree-days of growth (**a**) and burn rate (**b**). Continuous lines refer to the logistic regression presented in Table 2.

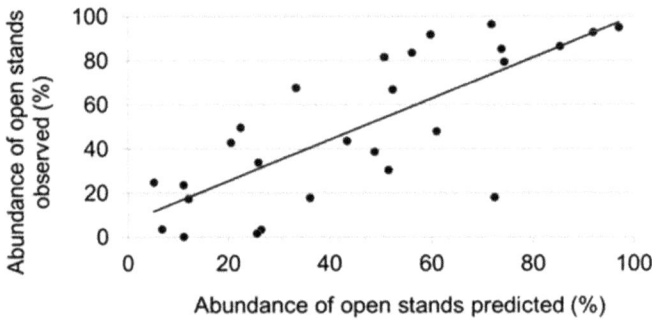

Figure 6. Predicted *vs.* observed abundance of open stands by fire zone. Predicted values stem from the logistic regression presented in Table 2.

The increase in the abundance of open stands after 150 years of simulations is explained mainly by the total precipitation, the last day of frost and the burn rate (Table 1). The increase was more significant in the A and C1 areas, which are drier and/or cooler (total precipitation < 900 mm· year^{-1}; average last day of frost > 170 Julian day), and/or when the burn rate is higher (>0.5% year^{-1}). According to the simulation results (Figure 3c, Figure 7), the abundance of open stands increases in more than 90% of the simulations in the A and C1 areas and in more than two-third of the simulations in B, C, D, and D1 areas.

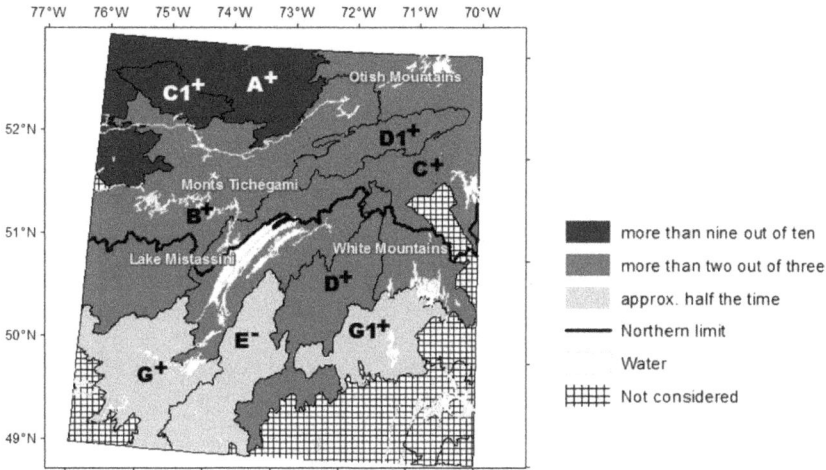

Figure 7. Frequency of change in the abundance of open stands after 150 years of simulation of a forest succession model in interaction with natural disturbances.

4. Discussion

4.1. Correlation of Fire Frequency with the Abundance of Open Stands

The abundance of open stands by fire zone is mainly explained by the burn rate and the regional climate [12] (Figure 2), but also by the periodic variations in the burn rate. Our results show that the vegetation is not in equilibrium with the disturbance regime in all regions and that it may respond immediately to changes in the fire burn rate over a time scale shorter than that of fire cycle (Figure 4). This vegetation/fire regime discrepancy (or lack of resilience) is accentuated when the degree-days of growth are low (less than 1000 °C· year^{-1}) and/or when the burn rate is higher than 0.5% year^{-1}. These results are consistent with the analysis of Chapin *et al.* [54] on the resilience of boreal forests and changes in forest composition: these forests are resilient to disturbances but when biophysical factors become strongly limiting to forest species, these forests start to lack resilience and their composition may change gradually. Some of the fire zones of our study area are indeed subject to a rather cold climate that does not help trees get established after a disturbance [55,56]. Simulations showed that the abundance of open stands should increase over a 150-year period in all fire zones, except for area E, if the burn rate remains equivalent to that observed between 1987 and 2006. As the importance of the simulated changes is related to climate (Figure 4, Table 2), they are predicted to occur over a clear north–south gradient (Figure 7).

The burn rate is defined as the mean annual area burned in a given territory [20,57]. In fact, the burn rate varies according to the area being calculated and the periods of time selected being covered by the calculation [58]. This can lead to considerable variability of burn rate values or to their overestimation [59,60] that obscures the influence of fire regimes on the actual distribution of ecological patterns [61]. Landscape dynamics models often apply the assumption that fire regimes do not change for long periods of time, while significant variations in burn rates are observed from one year to the next, and from one decade to the next [58,62]. Our analysis indicates that periodic variations of the burn rate between 1947 and 2006 (Figure 5) had an immediate impact on the abundance of open stands (Figure 4). Such variations have also been observed throughout the entire study area and are related to the recent increase in the burn rate (1987 to 2006), except for area C. The increase in the abundance of open stands in this fire zone is probably due to a greater amount of precipitation because of the higher altitude that may have mitigated the severity of fires [63,64]. Indeed, when fires are less severe, they do not entirely consume the soil's organic matter, and the absence of mineralized soil limits the regeneration of seeds [1,65–68]. Conversely, severe and more frequent fires promote the regeneration of pioneer species, such as jack pine. This is what is occurring in area A, where a higher proportion of jack pine has been noted by Rapanoela *et al.* [14], even though the stand density there is less [14] due to more frequent deficient postfire recovery [46].

Our results are based on simulations with empirical models: they are not based on an understanding of the underlying processes that drive forest succession. The semi-Markovian transition matrices used in the present study were calibrated from forest maps and therefore suffer from the defects of chronosequences, in which time is substituted by space [69,70]. They reflect past average effects of underlying processes that occurred at specific times and these effects may not exactly repeat themselves in the future. For instance, Markovian transition matrices are aspatial and ignore the importance of local neighbouring effects on forest succession [71]: if the abundance of open forest changes with time, local effects (historical legacies) should also change, which would impact the prediction of successive forest succession events. Forest succession is also conditioned by fire severity [65–68] that has been shown to be related to specific fire events [72]. As a consequence, changes of abundances and their variability, as simulated here (Figure 3c), cannot be used for predictive purposes (for instance to account for climate change effects [73]). Therefore, the present results served to demonstrate that the abundance of open forest stands has indeed increased in response to a recent increase in burn rates and to show the existence of a state of non-equilibrium over a relatively short time scale (in comparison with black spruce longevity), in line with the variability of the burn rate.

4.2. Natural Dynamics of the Spruce-Moss Forest and Lichen Woodland

The boreal forest is a vast ecosystem that was formed about 8000 years ago. From north to south, it encompasses four bioclimatic domains: forest tundra, Spruce-lichen woodland, spruce-moss forest and the balsam fir-white birch domain [9,74,75]. The spruce-moss forest is part of the sub-area of the boreal forest, where stands are relatively dense. Forest cover is essentially dominated by black spruce, and fire is the primary disruption that causes forest renewal [76,77]. Spruce-lichen woodland and spruce-moss forest used to be considered two separate communities [9,75,76]. The existence of these two communities under the same environmental and climatic conditions has led to the conclusion that the Spruce-lichen woodland and spruce-moss forest are stable alternative states [1,78]. However, our findings challenge this notion of stability, at least at the regional level and over a short time scale of decades.

Ecosystems can move from one state to another due to a severe disruption that acts directly on state variables [79]. State variables are quantities that change rapidly to ecologically relevant time scales, such as the density of the population [79]. In closed spruce-moss stands, the opening of stands is attributed to a poor regeneration of the main species after several disturbances [2,6,78,80] that leads a burned area unable to recover from a disruption [81]. If regeneration is good, the amount of black spruce should increase during the first 90 years and decline thereafter [52]. After 100 years but before 200 years have lapsed, the recruitment of young plants initiates the beginning of a structural change [82] and stands become irregular [52,83]. In our study area, the rate of regeneration is particularly slow and stand density is low [46,84]. Mansuy *et al.* [46] estimated that in our study area, it takes an average of 25 years for a majority of burned-over areas to reach a regenerated stage, and 45 years for stands over 7 m in height to dominate burned sites. The increase in plant density seems to happen gradually since the time of the last fire [85] and therefore, the abundance of closed stands is favoured by a low burn rate (Figure 4). The openness of the forest in connection with a temporarily higher burn rate demonstrates a potential instability of the spruce-moss forest, since a temporal change in the fire regime triggers a change in the forest mosaic, by altering the abundance of open and closed stands.

4.3. Vulnerability and Adaptation to Disturbance Regimes

The vulnerability of boreal forest ecosystems depends on the extent of their adaptive capacity and resilience to disturbances. In terms of resilience, it is assumed that the boreal forest is locally vulnerable to the loss of forest cover resulting from permafrost degradation [56], successive disturbances [78,86], or drought [87]. However, the persistence of wide landscapes of greater canopy opening caused by an increase in the burn rate could result in a long-term major change in the composition and structure of the spruce-moss forest, with a greater dominance of the less dense stands in the landscape. Consideration of species adaptation is key to effective and sustainable forest management [88]. Indeed, when assessing the risk of changes in stands or productivity losses [14,37], management decisions should be taken in a view to try to reduce the vulnerability of ecosystems. Following this study, a vulnerability threshold can be established with regard to the proportions of open stands in a landscape composed of spruce-moss stands (Figure 5), based on tolerance to change. The proportion of closed landscapes and acceptable biodiversity losses related to a decrease of closed stand abundance [89] will depend on forest management objectives.

5. Conclusions

Our objective was to evaluate if the closure of the forest was unstable over time and not in stable equilibrium with the current fire regime. As the proportion of open stands is explained mainly by the frequency of fires and the regional climate, but also by the periodic variations of the burn rate, our results showed that vegetation responds quickly to occasional changes in the fire activity. When the burn rate is higher, there is a significant increase in the abundance of open stands. The recent increase in the burn rate over the past two decades has led to a greater abundance of open stands, which can already be observed in the current landscape. It is apparent from this study that forestry planning should consider the risks associated with the temporal variability of fire regimes on the forest ecosystem, as the resulting changes can have a significant impact on biodiversity and allowable cut estimates.

Supplementary Materials: The supplementary materials of this paper are available online at www.mdpi.com/1999-4907/7/5/103/s1. File S1: Programming code used for the landscape dynamics model. SELES is required to run such code and is available upon request by contacting its author, Andrew Fall (andrew@gowlland.ca, http://www.gowlland.ca/). Furthermore, we could not share all the data used for the simulations as forest maps and inventory plots are owned by the Ministry of Forests, Wildlife and Parks of Quebec. The programming code comes however with a sample dataset. Any interested person can access the data as we did: this data is available upon request by contacting inventaires.forestiers@mffp.gouv.qc.ca (Direction des inventaires forestiers, 5700, 4e Avenue Ouest, A-108, Québec (Québec) G1H 6R1, Canada. Phone: +(1)418-627-8669).

Acknowledgments: We thank Quebec's Ministère de la Forêt, Faune et Parcs for giving us access to the data of the Nordic Ecoforest Inventory Program. This project was financially supported by the Fonds de recherche du Québec—Nature et technologies and by the scientific committee that was tasked to evaluate the northern timber allocation limit.

Author Contributions: R.R. and F.R. conceived the experiments. R.R. and F.R. analyzed the data. All authors discussed the results, and co-wrote the paper. All authors contributed to writing the paper.

Conflicts of Interest: The authors declare no conflict of interest.

References

1. Pollock, S.L.; Payette, S. Stability in the patterns of long-term development and growth of the Canadian spruce-moss forest. *J. Biogeogr.* **2010**, *37*, 1684–1697. [CrossRef]
2. Payette, S.; Bhiry, N.; Delwaide, A.; Simard, M. Origin of the lichen woodland at its southern range limit in eastern Canada: The catastrophic impact of insect defoliators and fire on the spruce-moss forest. *Can. J. For. Res.* **2000**, *30*, 288–305. [CrossRef]

3. Ministère des Ressources Naturelles et de la Faune. *Programme d'Inventaire Écoforestier Nordique. Norme de Cartographie Écoforestière. Direction des Inventaires Forestiers*; Gouvernement du Québec: Québec, QC, Canada, 2008.

4. Payette, S.; Delwaide, A. Shift of conifer boreal forest lichen-heath parkland caused by successive stand disturbances. *Ecosystems* **2003**, *6*, 540–550. [CrossRef]

5. Simard, M.; Lecomte, N.; Bergeron, Y.; Bernier, P.Y.; Paré, D. Forest productivity decline caused by successional paludification of boreal soils. *Ecol. Appl.* **2007**, *17*, 1619–1637. [CrossRef] [PubMed]

6. Girard, F.; Payette, S.; Gagnon, R. Rapid expansion of lichen woodlands within the closed-crown boreal forest zone over the last 50 years caused by stand disturbances in eastern Canada. *J. Biogeogr.* **2008**, *35*, 529–537. [CrossRef]

7. Scheffer, M.; Hirota, M.; Holmgren, M.; Nes, E.H.V.; Chapin, F.S. Thresholds for boreal biome transitions. *Proc. Natl. Acad. Sci. USA* **2012**, *109*, 21384–21389. [CrossRef] [PubMed]

8. Turner, M.G.; Romme, W.H.; Gardner, R.H.; O'Neill, R.V.; Kratz, T.K. A revised concept of landscape equilibrium: Disturbance and stability on scaled landscapes. *Landsc. Ecol.* **1993**, *3*, 213–227. [CrossRef]

9. Robitaille, A.; Saucier, J.P. *Paysages Régionaux du Québec Méridional. Ministère des Ressources Naturelles, Gouvernement du Québec*; Les publications du Québec: Québec, QC, Canada, 1998.

10. Bergeron, Y.; Dansereau, P.R. Fire history in the southern boreal forest of north-western Quebec. *Can. J. For. Res.* **1993**, *23*, 25–32.

11. Le Goff, H.; Sirois, L. Black spruce and jack pine dynamics simulated under varying fire cycles in the northern boreal forest of Quebec, Canada. *Can. J. For. Res.* **2004**, *34*, 2399–2409. [CrossRef]

12. Lavoie, L.; Sirois, L. Vegetation changes caused by recent fires in the northern boreal forest of eastern Canada. *J. Veg. Sci.* **1998**, *9*, 483–492. [CrossRef]

13. Boucher, D.; Gauthier, S.; De Grandpré, L. Structural changes in coniferous stands along a chronosequence and a productivity gradient in the northeastern boreal forest of Québec. *Ecoscience* **2006**, *13*, 172–180. [CrossRef]

14. Rapanoela, R.; Raulier, F.; Gauthier, S.; Ouzennou, H.; Saucier, J.P.; Bergeron, Y. Contrasting current and potential productivity and the influence of fire and species composition in the boreal forest: A case study in eastern Canada. *Can. J. For. Res.* **2015**, *45*, 541–552. [CrossRef]

15. Van Wagner, C.E. Age-class distribution and the forest fire cycle. *Can. J. For. Res.* **1978**, *8*, 220–227. [CrossRef]

16. Gauthier, S.; Leduc, A.; Bergeron, Y.; Le Goff, H. La fréquence des feux et l'aménagement forestier inspiré des perturbations naturelles. In *Aménagement Écosystémique en Forêt Boréale*; Gauthier, S., Vaillancourt, M.-A., Leduc, A., de Granpré, L., Kneeshaw, D., Morin, H., Drapeau, P., Bergeron, Y., Eds.; Presses de l'Université du Québec: Montréal, QC, Canada, 2008; pp. 61–78.

17. Cyr, D.; Gauthier, S.; Bergeron, Y.; Carcaillet, C. Forest management is driving the eastern North American boreal forest outside its natural range of variability. *Front. Ecol. Environ.* **2009**, *7*, 519–524. [CrossRef]

18. Liang, J.; Picard, N. Matrix model of forest dynamics: An overview and outlook. *For. Sci.* **2013**, *59*, 359–378. [CrossRef]

19. Landres, P.; Morgan, P.; Swanson, F. Overview of the use of natural variability concepts in managing ecological systems. *Ecol. Appl.* **1999**, *9*, 1179–1188.

20. Johnson, E.A.; Gutsell, S.L. Fire frequency models, methods and interpretations. *Adv. Ecol. Res.* **1994**, *25*, 239–287.

21. Bergeron, Y.; Gauthier, S.; Kafka, V.; Lefort, P.; Lesieur, D. Natural fire frequency for the eastern Canadian boreal forest: Consequences for sustainable forestry. *Can. J. For. Res.* **2001**, *31*, 384–391. [CrossRef]

22. Bergeron, Y.; Flannigan, M.; Gauthier, S.; Leduc, A.; Lefort, P. Past, current and future fire frequency in the Canadian boreal forest: Implications for sustainable forest management. *Ambio* **2004**, *33*, 356–360. [CrossRef] [PubMed]

23. Bergeron, Y.; Gauthier, S.; Flannigan, M.; Kafka, V. Fire regimes at the transition between mixedwood and coniferous boreal forest in northwestern Quebec. *Ecology* **2004**, *85*, 1916–1932. [CrossRef]

24. Stocks, B.J.; Mason, J.A.; Todd, J.B.; Bosch, E.M.; Wotton, B.M.; Amiro, B.D.; Flannigan, M.D.; Hirsch, K.G.; Logan, K.A.; Martell, D.L.; *et al.* Large forest fires in Canada, 1959–1997. *J. Geophys. Res.* **2003**, *108D1*, FFR-5. [CrossRef]

25. Mansuy, N.; Gauthier, S.; Robitaille, A.; Bergeron, Y. The effects of surficial deposit-drainage combinations on spatial variations of fire cycles in the boreal forest of eastern Canada. *Int. J. Wildland Fire* **2010**, *19*, 1083–1098. [CrossRef]
26. Metsaranta, J.M.; Kurz, W.A. Inter-annual variability of ecosystem production in boreal jack pine forests (1975–2004) estimated from tree-ring data using CBM-CFS3. *Ecol. Model.* **2012**, *224*, 111–123. [CrossRef]
27. Mouillot, F.; Field, C.B. Fire history and the global carbon budget: A 1 × 1 fire history reconstruction for the 20th century. *Glob. Chang. Biol.* **2005**, *11*, 398–420. [CrossRef]
28. Kasischke, E.S.; Turetsky, M.R. Recent changes in the fire regime across the North American boreal region—Spatial and temporal patterns of burning across Canada and Alaska. *Geophys. Res. Lett.* **2006**, *33*, L09703. [CrossRef]
29. Morris, W.; Doak, D.; Groom, M.; Kareiva, P.; Fieberg, J.; Gerber, L.; Murphy, P.; Thomson, D. *A Practical Handbook for Population Viability Analysis*; The Nature Conservancy: Arlington, VA, USA, 1999.
30. Buongiorno, J.; Gilless, J.K. *Decision Methods for Forest Resource Management*; Academic Press: Amsterdam, The Netherlands, 2003.
31. Zhou, M.; Buongiorno, J. Forest landscape management in a stochastic environment, with an application to mixed loblolly pine-hardwood forest. *For. Ecol. Manag.* **2006**, *223*, 170–182. [CrossRef]
32. Caswell, H. *Matrix Population Models: Construction, Analysis and Interpretation*; Sinauer: Sunderland, MA, USA, 1989.
33. Yemshanov, D.G.; Perera, A.H. A spatially explicit stochastic model to simulate boreal forest cover transitions: General structure and properties. *Ecol. Model.* **2002**, *150*, 189–209. [CrossRef]
34. Régnière, J.; Saint-Amant, R. *BioSIM 9—Manuel de l'Utilisateur*; Rapport d'information LAU-X-134F. Ressources Naturelles Canada, Service Canadien des Forêts, Centre de Foresterie des Laurentides: Québec, QC, Canada, 2008. Available online: ftp://ftp.cfl.scf.rncan.gc.ca/regniere/software/BioSIM/ (accessed on 11 December 2013).
35. Ministère des Ressources Naturelles et de la Faune (MRNFQ). *Limite Nordique des Forêts Attribuables*; Rapport Final du Comité. Ministère des Ressources Naturelles et de la Faune: Québec, QC, Canada, 2000. Available online: http://www.mrn.gouv.qc.ca/publications/forets/consultation/partie1.pdf (accessed on 12 November 2013).
36. Létourneau, J.P.; Matajek, S.; Morneau, C.; Robitaille, A.; Roméo, T.; Brunelle, J.; Leboeuf, A. *Norme de Cartographie Écoforestière du Programme d'Inventaire Écoforestier Nordique*; Ministère des Ressources Naturelles et de la Faune: Québec, QC, Canada, 2008.
37. Gauthier, S.; Raulier, F.; Ouzennou, H.; Saucier, J.-P. Strategic analysis of forest vulnerability to risk related to fire: An example from the coniferous boreal forest of Quebec. *Can. J. For. Res.* **2015**, *45*, 553–565. [CrossRef]
38. Fall, A.; Fortin, M.-J.; Kneeshaw, D.D.; Yamasaki, S.H.; Messier, C.; Bouthillier, L.; Smyth, C. Consequences of various landscape-scale ecosystem management strategies and fire cycles on age-class structure and harvest in boreal forests. *Can. J. For. Res.* **2004**, *34*, 310–322. [CrossRef]
39. Didion, M.; Fortin, M.-J.; Fall, A. Forest age structure as indicator of boreal forest sustainability under alternative management and fire regimes: A landscape level sensitivity analysis. *Ecol. Model.* **2007**, *200*, 45–58. [CrossRef]
40. James, P.M.A.; Fortin, M.-J.; Sturtevant, B.R.; Fall, A.; Kneeshaw, D.D. Modelling spatial interactions among fire, spruce budworm, and logging in the boreal forest. *Ecosystems* **2011**, *14*, 60–75. [CrossRef]
41. Tittler, R.; Messier, C.; Fall, A. Concentrating anthropogenic disturbance to balance ecological and economic values: Applications to forest management. *Ecol. Appl.* **2012**, *22*, 1268–1277. [CrossRef] [PubMed]
42. Fall, A.; Fall, J. A domain-specific language for models of landscape dynamics. *Ecol. Model.* **2001**, *141*, 1–18. [CrossRef]
43. Raulier, F.; Dhital, N.; Racine, P.; Tittler, R.; Fall, A. Increasing resilience of timber supply: How a variable buffer stock of timber can efficiently reduce exposure to shortfalls caused by wildfires. *For. Pol. Econ.* **2014**, *46*, 47–55. [CrossRef]
44. McRoberts, R.E.; Nelson, M.D.; Wendt, D.G. Stratified estimation of forest area using satellite imagery, inventory data, and the *k*-nearest neighbors technique. *Remote Sens. Environ.* **2002**, *82*, 457–468. [CrossRef]
45. McRoberts, R.E.; Tomppo, E.O.; Finley, A.O. Heikkinen, J. Estimating areal means and variances of forest attributes using the *k*-Nearest Neighbors technique and satellite imagery. *Remote Sens. Environ.* **2007**, *111*, 466–480. [CrossRef]

46. Mansuy, N.; Gauthier, S.; Robitaille, A.; Bergeron, Y. Regional patterns of post fire canopy recovery in the northern boreal forest of Quebec: Interactions between surficial deposit, climate, and fire cycle. *Can. J. For. Res.* **2012**, *42*, 1328–1343. [CrossRef]

47. Gower, J.C. A general coefficient of similarity and some of its properties. *Biometrics* **1971**, *27*, 857–871. [CrossRef]

48. Tuominen, S.; Fish, S.; Poso, S. Combining remote sensing, data from earlier inventories, and geostatistical interpolation in multisource forest inventory. *Can. J. For. Res.* **2003**, *33*, 624–634. [CrossRef]

49. Dunne, T.; Leopold, L.B. *Water in Environmental Planning*, 1st ed.; W.H. Freeman & Company: New York, NY, USA, 1978.

50. Sutherland, G.D.; Eng, M.; Fall, S.A. Effects of uncertainties about stand-replacing natural disturbances on forest-management projections. *J. Ecosyst. Manag.* **2004**, *4*, 1–18. Available online: http://www.forrex.org/jem/2004/vol4/no2/art5.pdf (accessed on 9 May 2016).

51. Hatcher, R.J. *A Study of Black Spruce Forests in Northern Quebec*; Department of Forests: Ottawa, ON, Canada, 1963.

52. Bouchard, M.; Pothier, D.; Gauthier, S. Fire return intervals and tree species succession in the North Shore region of eastern Quebec. *Can. J. For. Res.* **2008**, *38*, 1621–1633. [CrossRef]

53. Confais, J.; le Guen, M. *Premiers pas en Régression Linéaire avec SAS*; Revue MODULAD; SAS Institute Inc.: Cary, NC, USA, 2006.

54. Chapin, F.S., III; Callaghan, T.V.; Bergeron, Y.; Fukuda, M.; Johnstone, J.F.; Juday, G.; Zimov, S.A. Global change and the boreal forest: Thresholds, shifting states or gradual changes. *Ambio* **2004**, *33*, 361–365. [CrossRef] [PubMed]

55. Sirois, L.; Bonan, G.B.; Shugart, H.H. Development of a simulation model of forest-tundra transition zone of northeastern Canada. *Can. J. For. Res.* **1994**, *24*, 697–706. [CrossRef]

56. Price, D.T.; Alfaro, R.I.; Brown, K.J.; Flannigan, M.D.; Fleming, R.A.; Hogg, E.H.; Girardin, M.P.; Lakusta, T.; Johnston, M.; McKenney, D.W.; *et al.* Anticipating the consequences of climate change for Canada's boreal forest ecosystems. *Environ. Rev.* **2013**, *21*, 322–365. [CrossRef]

57. Johnson, E.A.; Van Wagner, C.E. The theory and use of two fire history models. *Can. J. For. Res.* **1985**, *15*, 214–220. [CrossRef]

58. Metsaranta, J.M. Potentially limited detectability of short-term changes in boreal fire regimes: A simulation study. *Int. J. Wildland Fire* **2010**, *19*, 1140–1146. [CrossRef]

59. Li, C. Estimation of fire frequency and fire cycle: A computational perspective. *Ecol. Model.* **2002**, *154*, 103–120. [CrossRef]

60. Ter-Mikaelian, M.T.; Colombo, S.J.; Chen, J. Estimating natural forest fire return interval in northeastern Ontario, Canada. *For. Ecol. Manag.* **2009**, *258*, 2037–2045. [CrossRef]

61. Boulanger, Y.; Gauthier, S.; Burton, P.J. A refinement of models projecting future Canadian fire regimes using homogeneous fire regime zones. *Can. J. For. Res.* **2014**, *44*, 365–376. [CrossRef]

62. Le Goff, H.; Flannigan, M.D.; Bergeron, Y.; Girardin, M.P. Historical fire regime shifts related to climate teleconnections in the Waswanipi area, central Quebec, Canada. *Int. J. Wildland Fire* **2007**, *16*, 607–618. [CrossRef]

63. Sirois, L. Impact of fire on *Picea mariana* and *Pinus banksiana* seedlings in subarctic lichen woodlands. *J. Veg. Sci.* **1993**, *4*, 795–802. [CrossRef]

64. Barrett, K.; Kasischke, E.S.; McGuire, A.D.; Turetsky, M.R.; Kane, E.S. Modeling fire severity in black spruce stands in the Alaskan boreal forest using spectral and non-spectral geospatial data. *Remote Sens. Environ.* **2010**, *114*, 1494–1503. [CrossRef]

65. Johnstone, J.F.; Kasischke, E.S. Stand-level effects of soil burn severity on postfire regeneration in a recently burned black spruce forest. *Can. J. For. Res.* **2005**, *35*, 2151–2163. [CrossRef]

66. Lecomte, N.; Simard, M.; Bergeron, Y. Effects of fire severity and initial tree composition on stand structural development in the coniferous boreal forest of northwestern Québec, Canada. *Ecoscience* **2006**, *13*, 152–163. [CrossRef]

67. Johnstone, J.F.; Chapin, F.S. Effects of soil burn severity on post-fire tree recruitment in boreal forest. *Ecosystems* **2006**, *9*, 14–31. [CrossRef]

68. Jin, Y.; Randerson, J.T.; Goetz, S.J.; Beck, P.S.A.; Loranty, M.M.; Goulden, M.L. The influence of burn severity on post-fire vegetation recovery and albedo change during early succession in North American boreal forests. *J. Geophys. Res.* **2012**, *117*, G01036. [CrossRef]

69. Johnson, E.A.; Miyanishi, K. Testing the assumptions of chronosequences in succession. *Ecol. Lett.* **2008**, *11*, 419–431. [CrossRef] [PubMed]

70. Taylor, A.R.; Chen, H.Y.H. Multiple successional pathways of boreal forest stands in central Canada. *Ecography* **2011**, *34*, 208–219. [CrossRef]

71. Frelich, L.E.; Reich, P.B. Neighborhood effects, disturbance severity, and community stability in forests. *Ecosystems* **1999**, *2*, 151–166. [CrossRef]

72. Miller, J.D.; Skinner, C.N.; Safford, H.D.; Knapp, E.E.; Ramirez, C.M. Trends and causes of severity, size, and number of fires in northwestern California, USA. *Ecol. Appl.* **2012**, *22*, 184–203. [CrossRef] [PubMed]

73. Gauthier, S.; Bernier, P.; Kuuluvainen, T.; Shvidenko, A.Z.; Schepaschenko, D.G. Boreal forest health and global change. *Science* **2015**, *349*, 819–822. [CrossRef] [PubMed]

74. Rowe, J.S. *Forest Regions of Canada. Bulletin 123*; Government of Canada, Department of Northern Affairs and National Resources, Forestry Branch, Headquarters: Ottawa, ON, Canada, 1959.

75. Hare, F.K.; Ritchie, J.C. The boreal Bioclimates. *Geogr. Rev.* **1972**, *62*, 333–65. [CrossRef]

76. Heinselman, M.L. The natural role of fire in northern conifer forests. In Proceedings of the Fire in the Northern Environment: A symposium, Fairbanks, AK, USA, 13–14 April 1971; Slaughter, C.W., Barney, R.J., Hansen, G.M., Eds.; USDA, Forest Service, Pacific Northwest Research Station: Portland, OR, USA; pp. 61–72.

77. Payette, S. Fire as a controlling process in the North American boreal forest. In *Systems Analysis of the Global Boreal Forest*; Shugart, H.H., Leemans, R., Bonan, G.B., Eds.; Cambridge University Press: Cambridge, UK, 1992; pp. 144–169.

78. Jasinski, J.P.P.; Payette, S. The creation of alternative stable states in the southern boreal forest, Québec, Canada. *Ecol. Monogr.* **2005**, *75*, 561–583. [CrossRef]

79. Beisner, B.E.; Haydon, D.T.; Cuddington, K. Alternative stable states in ecology. *Front. Ecol. Environ.* **2003**, *1*, 376–382. [CrossRef]

80. Gagnon, R.; Morin, H. Les forêts d'épinette noire au Québec: Dynamique, perturbations et biodiversité. *Nat. Can.* **2001**, *125*, 26–35.

81. Holling, C.S. Resilience and stability of ecological systems. *Ann. Rev. Ecol. Syst.* **1973**, *4*, 1–23. [CrossRef]

82. Groot, A.; Horton, B.J. Age and size structure of natural and second-growth peatland *Picea mariana* stands. *Can. J. For. Res.* **1994**, *24*, 225–233. [CrossRef]

83. Garet, J.; Raulier, F.; Pothier, D.; Cumming, S.G. Forest age class structures as indicators of sustainability in boreal forest: Are we measuring them correctly? *Ecol. Indic.* **2012**, *23*, 202–210. [CrossRef]

84. Van Bogaert, R.K.; Gauthier, S.; Raulier, F.; Saucier, J.P.; Boucher, D.; Robitaille, A.; Bergeron, Y. Exploring forest productivity at an early age after fire: A case study at the northern limit of commercial forests in Quebec. *Can. J. For. Res.* **2015**, *45*, 579–593. [CrossRef]

85. Irulappa Pillai Vijayakumar, D.B.; Raulier, F.; Bernier, P.; Paré, D.; Gauthier, S.; Bergeron, Y.; Pothier, D. Cover density recovery after fire disturbance controls landscape aboveground biomass carbon in the boreal forest of eastern Canada. *For. Ecol. Manag.* **2016**, *360*, 170–180. [CrossRef]

86. Girard, F.; Payette, S.; Gagnon, R. Origin of the lichen-spruce woodland in the closed-crown forest zone of eastern Canada. *Glob. Ecol. Biogeogr.* **2009**, *18*, 291–303. [CrossRef]

87. Hogg, E.H.; Bernier, P.Y. Climate change impacts on drought-prone forests in western Canada. *For. Chron.* **2005**, *81*, 675–682. [CrossRef]

88. Gauthier, S.; Bernier, P.; Burton, P.J.; Edwards, J.; Isaac, K.; Isabel, N.; Jayen, K.; Le Goff, H.; Nelson, E.A. Climate change vulnerability and adaptation in the managed Canadian boreal forest. *Environ. Rev.* **2014**, *22*, 256–285. [CrossRef]

89. Imbeau, L.; St-Laurent, M.-H.; Marzell, L.; Brodeur, V. Current capacity to conduct ecologically sustainable forest management in northeastern Canada reveals challenges for conservation of biodiversity. *Can. J. For. Res.* **2015**, *45*, 567–578. [CrossRef]

© 2016 by the authors. Licensee MDPI, Basel, Switzerland. This article is an open access article distributed under the terms and conditions of the Creative Commons Attribution (CC BY) license (http://creativecommons.org/licenses/by/4.0/).

![forests logo] *forests*

MDPI

Article

How Time since Forest Fire Affects Stand Structure, Soil Physical-Chemical Properties and Soil CO_2 Efflux in Hemiboreal Scots Pine Forest Fire Chronosequence?

Kajar Köster [1,2], Egle Köster [2], Argo Orumaa [1], Kristi Parro [1], Kalev Jõgiste [1], Frank Berninger [2], Jukka Pumpanen [3] and Marek Metslaid [1,*]

[1] Institute of Forestry and Rural Engineering, Estonian University of Life Sciences, Tartu 51014, Estonia; kajar.koster@emu.ee (K.K.); argoorumaa@hotmail.com (A.O.); kristi.parro@gmail.com (K.P.); kalev.jogiste@emu.ee (K.J.)

[2] Department of Forest Sciences, University of Helsinki, Helsinki FI-00014, Finland; egle.koster@helsinki.fi (E.K.); frank.berninger@helsinki.fi (F.B.)

[3] Department of Environmental and Biological Sciences, University of Eastern Finland, Kuopio FI-70211, Finland; jukka.pumpanen@uef.fi

* Correspondence: marek.metslaid@emu.ee; Tel.: +372-731-3193

Academic Editors: Yves Bergeron and Sylvie Gauthier
Received: 7 July 2016; Accepted: 6 September 2016; Published: 12 September 2016

Abstract: We compared the changes in aboveground biomass and initial recovery of C pools and CO_2 efflux following fire disturbances in Scots pine (*Pinus sylvesteris* L.) stands with different time since stand-replacing fire. The study areas are located in hemiboreal vegetation zone, in north-western Estonia, in Vihterpalu. Six areas where the last fire occurred in the year 1837, 1940, 1951, 1982, 1997, and 2008 were chosen for the study. Our results show that forest fire has a substantial effect on the C content in the top soil layer, but not in the mineral soil layers. Soil respiration showed a chronological response to the time since the forest fire and the values were lowest in the area where the fire was in the year 2008. The respiration values also followed seasonal pattern being highest in August and lowest in May and November. The CO_2 effluxes were lowest on the newly burned area through the entire growing season. There was also a positive correlation between soil temperature and soil respiration values in our study areas.

Keywords: fire disturbance; *Pinus sylvestris*; recovery; soil respiration

---

1. Introduction

Disturbances are an important factor influencing forest structure formation, composition and forest functioning [1,2]. Forest fires and the long-term recovery from them are important for regional carbon (C) storage because C lost in fires makes a substantial difference to regional C budgets [3]. Boreal forests are a crucial part of the climate system since they contain about 60% of the C (703 billion tons) bound in global forest biomes [4]. It is expected, that the average temperature increase predicted for the future climate will be most pronounced in the boreal region and the fire frequency, intensity and severity in boreal forests will increase as a result of prolonged drought periods [5].

Wildfires strongly influence boreal forest structure and function as they can cause losses of 15%–35% of the above-ground biomass and 37%–70% of the ground layer due to combustion [6,7]. Since both high severity (stand replacing) and intermediate severity fires are common in Eurasia [4,8], it is important to understand how these ecosystems respond to the different disturbances. In the short term, increases in disturbance will lead to a net release of C and thus contribute to global warming, but the amount of C released is also linked to the age distribution of the forests. Thus, an integrated approach studying C accumulation and energy fluxes (CO_2 effluxes) across a fire chronosequence is needed to understand the role of boreal forests in global warming [9].

Soils play a major role in sequestering atmospheric CO_2 and in emitting trace gases (e.g., CO_2, CH_4, and N_2O) that are absorbing solar radiation and enhance the global warming [10]. About half of the C, which enters the ecosystem through photosynthesis, is allocated belowground [11,12]. The turnover of soil organic matter (SOM) in boreal region is slow, with a turnover time of several decades [13]. During the fire, SOM, mostly C, is released from the forest biomass rapidly to the atmosphere through combustion and simultaneously, mineralized nitrogen (N) is released in the soil favoring the re-establishment of vegetation during the first years of succession [14]. Fires directly affect the C cycle via CO_2 emissions from biomass combustion and indirectly via long-term changes in ecosystem C dynamics through forest recovery and succession [15].

In this study, we characterize the responses of soil CO_2 efflux and soil C content to a forest fire. The changes occurring in the soil C dynamics were assessed along a fire chronosequence in hemiboreal Scots pine (*Pinus sylvestris* L.) stands of similar soil type and climatic conditions. We hypothesized that the changes in post-fire ecosystems affect C content and CO_2 emissions from forest soils across a fire chronosequence. One of the aims of this study was also to investigate the changes in post-fire soil temperatures and soil moisture content and how these factors are affecting the soil CO_2 emissions. We assumed that the recovery of C stocks in soil and CO_2 emissions is associated with the recovery of aboveground plant biomass. We expect that our chronosequence study approach will bring new quantitative information on changes in soil C dynamics after forest fires and during the forest succession in the hemiboreal forest zone, which may be useful for global C-cycle modelers.

2. Materials and Methods

2.1. Study Sites

The study area (fire chronosequence) is located in hemiboreal vegetation zone, in north-western Estonia, in Vihterpalu and Nõva [2]. The area is flat with no elevation differences and covered with pure Scots pine (*Pinus sylvestris* L.) stands on sandy soils, regenerated at a different time since forest fires. The areas belong to the *Calluna* and *Vaccinium uliginosum* site types (Table 1) [16]. The average annual temperature in the area is +5.2 °C. The coldest month is February, with an average temperature of −5.7 °C, and the warmest month is July, with an average temperature of +16.4 °C [2].

Six areas (with extensive fires 200 ha and more) have been chosen for the study: fire in 1837, 1940, 1951, 1982, 1997, and 2008. All the study areas are located within 145 km^2. The total area of the forest fire in 2008 (59°11' N 23°46' E) was about 800 ha and it started at the end of May. The forest stands selected for the current study were 70 years old at the time fire occurred [2]. In August 1997 (59°12' N 23°49' E) about 700 ha of forest were burned and the forest stands selected for the current study were 45 years old when the fire occurred. The fire in 1982 (59°12' N 23°48' E) occurred in May and about 200 ha of forest were burned and the forest stands selected for the study were 30 years old at that time. A huge fire occurred in 1951 (59°14' N 23°49' E), when more than 2000 ha of forest were burned, and the forest stands selected for the study were about 35–40 years old at that time. The total area of the forest fire in 1940 (59°10' N 23°42' E) was more than 200 ha and a forest fire of similar size occurred in 1837 (although the exact area of the forest fire in 1837 (59°13' N 23°36' E) is unknown). All areas had been exposed to stand replacing fires where all (or most) of the stand was destroyed by fire. The time since last fire was first chosen from old inventory data, and later fire occurrence dates were confirmed by taking increment cores at each selected stand. In all areas, three sample plots were established (all together 18 sample plots), that were randomly located in the study areas and the distance between sample plots was at least 100 m. Although the normal practice in Estonia following large-scale disturbances in managed forests is to intervene immediately and clear the stand regardless of whether it will be regenerated naturally or planted [2], we tried to locate our sample plots in areas where the material was not removed after disturbance, thus no management actions (also no planting after disturbance) were carried out.

Table 1. Stand and soil characteristics of the study areas. Pi—Scots pine, Bi—Birch. Geographical coordinate represent the location of middle sample plot in a fire chronosequence.

Study Area (Geographical Coordinate)	Site Type	Tree Species Composition (%)	Living Trees/ha	$D_{1.3}$ (cm)	H (m)	Soil Texture	Average Thickness of: O-/E-/BHF-/BCg-/Cg- Horizon in Soil (cm)	Soil pH (O-/E-/Mineral Horizons)	Average Soil Temperature (Growing Season) (°C)
Fire 2008 (59°11′ N 23°46′ E)	*Calluna/Cladina*	56 Pi, 44 Bi	1422	1.9	1.1	Loamy sand	3.3/10.3/12.9/11.2/11.5	4.0/4.1/4.7	13.4
Fire 1997 (59°12′ N 23°49′ E)	*Calluna/Vaccinium uliginosum*	91 Pi, 9 Bi	2683	3.9	2.9	Loamy sand	8.8/7.8/12.1/11.3/15.3	4.0/4.2/4.5	11.4
Fire 1982 (59°12′ N 23°48′ E)	*Vaccinium uliginosum*	100 Pi	3167	7.2	5.5	Loamy sand	8.7/5.5/7.9/9.1/20.8	4.0/3.9/4.4	11.5
Fire 1951 (59°14′ N 23°49′ E)	*Calluna*	93 Pi, 7 Bi	1583	12.5	11.1	Loamy sand	13.2/7.2/11.1/9.4/10.3	3.8/3.9/4.3	11.5
Fire 1940 (59°10′ N 23°42′ E)	*Calluna*	100 Pi	3117	10.4	9.4	Loamy sand	9.4/10.4/11.7/10.1/10.2	3.6/3.7/4.6	11.1
Fire 1837 (59°13′ N 23°36′ E)	*Calluna/Cladina*	100 Pi	558	21.8	13.4	Loamy sand	8.9/9.9/12.6/10.7/17.5	3.7/4.0/4.5	11.4

To characterize the stands, circular sample plots with a radius of 11.28 m (400 m^2) were established in all areas of fire chronosequence. Basic tree characteristics were measured for tree biomass calculations (diameter at breast height, tree height, crown length, crown diameter) and for characterizing the stand (stand age, the number of trees per ha, time since last fire) (Table 1). Tree ages were determined from increment cores taken from sample trees and analyzed with WinDENDRO (Regent Instruments Canada Inc., Quebec, Canada). For tree biomass calculations the formulas of Repola [17] and Repola [18] were used. Also, all dead wood (all material longer than 1.3 m and with a diameter of at least 10 cm) was measured in all sample plots for dead wood biomass calculations.

In every sample plot there were two 0.5 × 0.5 m ground vegetation squares for species composition and recovery measurements and two 0.2 × 0.2 squares for ground vegetation biomass measurements. Ground vegetation was classified into mosses, lichens and shrubs/grasses and oven dried at 60 °C until constant mass was reached.

2.2. Soil C Content and CO$_2$ Efflux

The soil was classified as a gleyic podzol [19], with loamy sand. Its profile (O–E–BHF–BCg–Cg) consists of the organic (O) horizon (material in different decomposition stages) (1–15 cm), discontinuous bleached sandy podzolic (E) horizon of varying thickness (2–14 cm), iron-illuvial loamy sand (BHF) horizon with an average thickness of 14 cm and with a gradual transition towards an unevenly colored (from grey to yellowish brown color) sandy parent material.

Soil respiration was measured manually from all sample plots (measuring interval of two weeks). Manual chamber measurements were performed on 5 collars (transect north—south orientated and the distance between collars was 5 m) in each sample plot (all together 90 collars) from May till October, to determine the CO$_2$ efflux from soil to atmosphere with diffusion gradient method [20]. The collars (diameter 0.22 m, height 0.05 m) were placed at 0.02 m depth in the organic soil layer above the rooting zone to avoid damage to roots. The collars were sealed with sand placed around the collars to reduce the air leakage from below the collar. The vegetation inside the chamber was not damaged during the measurements. For CO$_2$ efflux measurements the portable chamber (0.24 m height and 0.22 m in diameter) made of plexiglass and covered with non-transparent plastic was used. All chamber measurements were carried out during the daylight. The CO$_2$ concentration was recorded during a 5 min chamber deployment time with a diffusion type CO$_2$ probe (GMP343, Vaisala Oyj, Vantaa, Finland) and air humidity and temperature inside the chamber with relative humidity and temperature sensor (HM70, Vaisala Oyj, Vantaa, Finland). The CO$_2$ fluxes were calculated based on the change in the CO$_2$ concentration (F) in the chamber headspace in time as follows:

$$F = \frac{\Delta\left(V_c C_i\right)}{\Delta t} \tag{1}$$

were V_c is the volume of the chamber, C_i is the CO$_2$ concentration inside the chamber and t is the time.

Simultaneously with soil CO$_2$ efflux measurements also soil temperature and soil moisture content (TRIME-PICO 64, IMKO GmbH, Ettlingen, Germany) were measured.

One iButton temperature sensor was placed in each sample plot to register soil temperature changes over the year.

To characterize the soil C and N content 5 soil cores (0.5 m long and 0.05 m in diameter) were taken from each sample plot. The soil cores were divided according to morphological soil horizons to litter and humus layers, mineral layers to eluvial and illuvial horizons, and sieved. All visible roots were separated (bigger roots by sieving the soil through a 2-mm sieve and smaller roots by picking) for root biomass calculations. The roots were identified as tree roots and ground vegetation (mainly dwarf shrubs and grasses) roots and rhizomes based on morphology and color [13]. The soil pH of different horizons was determined with glass electrode in 35 mL soil suspension, consisting of 10 mL of the soil sample and 25 mL of demineralised water, which had been left overnight to stand after mixing. The soil C and N content were determined with an elemental analyzer (vario MAX CN

Elementaranalysator, Elementar Analysensysteme GmbH, Hanau, Germany) after oven-drying the samples at 105 °C for 24 h.

2.3. Statistical Analyses

Data was checked for normality using the Shapiro-Wilk test and a logarithmic transformation was made for the recorded CO_2 fluxes to approximate the residual distribution of this variable to the normal distribution. Mixed models (PROC MIXED) was used to test the different factor effects behind CO_2 fluxes from the soil. CO_2 flux was treated as dependent variables in these models, while age since the last fire disturbance was treated as fixed continuous variable, plot as random factor. A Tukey's HSD test was used for comparison of differences within factors. All calculations and statistical analyses used the plot as the experimental unit and a significance level of $p < 0.05$. All the statistical analyses were performed with SAS version 9.3 (SAS Institute Inc., Cary, NC, USA).

3. Results

3.1. Soil Physical-Chemical Properties and Above- and Belowground Biomasses

Soil pH was not significantly different between areas with different time since fire. Soil pH in mineral soil was similar in all areas, ranging between 4.3 and 4.7 (Table 1). Soil pH in humus layer was slightly higher in areas where the fire occurred in 2008, 1997 and 1982 compared to other areas (Table 1), but the differences between the areas were not statistically significant ($p > 0.05$). The average thickness of the O-horizon (3.3 cm) was significantly lower ($p < 0.05$) in the area where the fire occurred in 2008 compared to other areas (Table 1). The thickest O-horizon was in the area where fire occurred in 1951 (Table 1). The thicknesses of the other horizons within the taken samples were not statistically different between the areas, and they ranged between 5.5 and 20.8 cm (Table 1). There was a significant difference ($p < 0.05$) between the areas when growing season temperatures (May–October) were used. The average soil temperatures in the area where the fire occurred in 2008 were 13.4°C (Table 1), while on the other areas the average soil temperatures stayed around 11 °C (Table 1). The daily average soil temperatures in the area where the fire occurred in 2008 were clearly higher from the middle of the May until the end of August (Figure 1). In September and October, there was no clear difference between the areas.

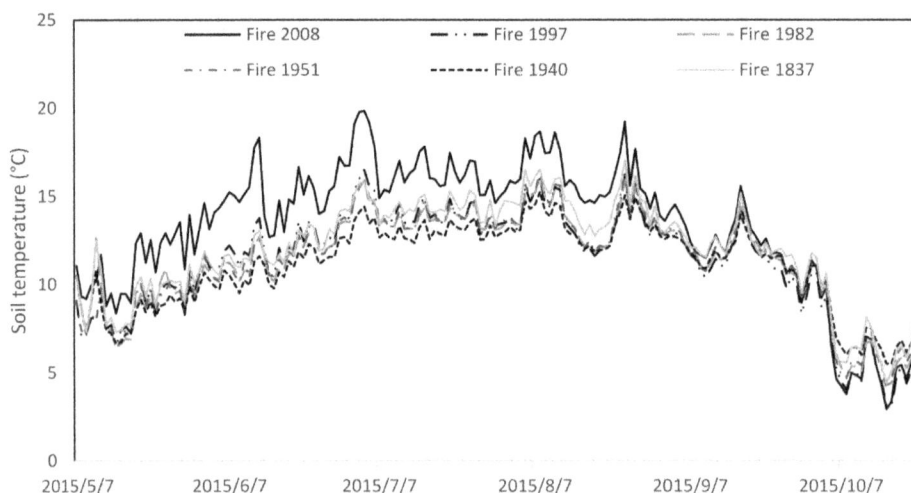

Figure 1. Daily average soil temperatures (24 h average, °C) during the measurement period in a fire chronosequence.

The living tree biomass, as well as the biomass of the ground vegetation and root biomass, increased during the post-fire succession (Figure 2). The total aboveground biomass (including tree and ground vegetation biomass) was smallest at the youngest fire area (0.53 kg m^{-2}), increased through post-fire succession and reached the maximum in the areas where the fire occurred in 1951 (12.43 kg·m^{-2}) (Figure 2). Same tendency was observed also with living root biomass (including tree and ground vegetation root biomass) being smallest at the youngest fire area (0.39 kg·m^{-2}) and highest (4.18 kg·m^{-2}) in the oldest fire area (Figure 2).

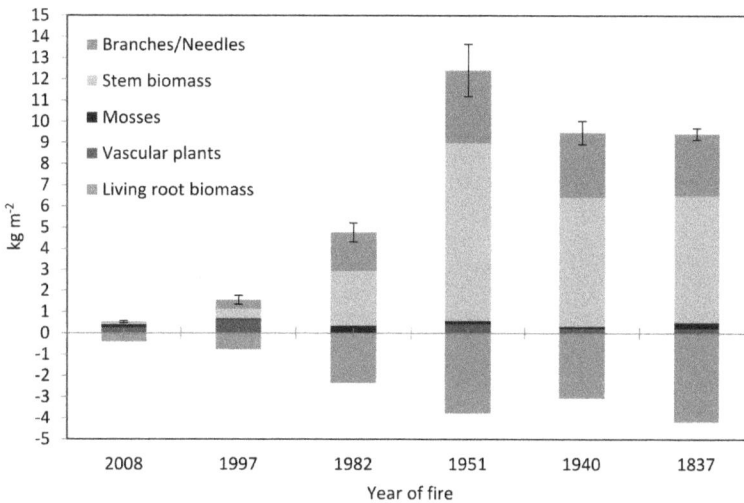

Figure 2. Average aboveground (branches/needles, tree stems, mosses and vascular plants) and belowground biomass (kg·m^{-2}) in a fire chronosequence.

In the area where the fire occurred in 2008, there was a lot of standing dead wood biomass (Figure 3), meaning that seven years after the fire most of the trees that died during fire disturbance were still standing. In the area where the fire occurred in 1997 (18 years after fire) there was almost no standing dead trees in the area and there was the highest amount of lying dead wood (Figure 3). The amount of dead wood in the study areas stabilized around 65 years after the fire (Figure 3).

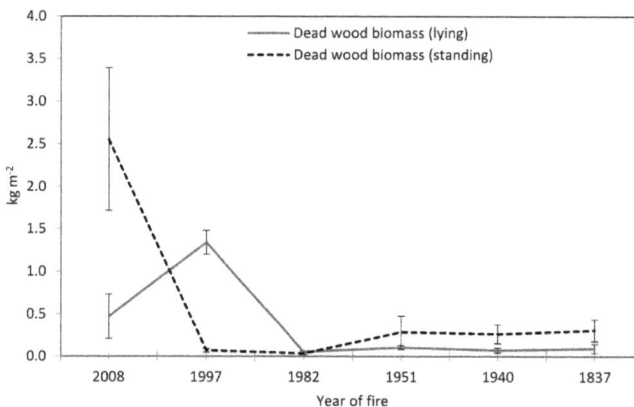

Figure 3. Average standing and lying dead wood biomass (kg·m^{-2}) in a fire chronosequence.

Total soil C stock was significantly lower ($p < 0.05$) in the area where the fire occurred in 2008 compared to other areas (Figure 4). The difference in soil total C stocks originated from the top layer (O-horizon), as the C stock in the O-horizon was much lower (only 729.2 $g \cdot C \cdot m^{-2}$) in the most recently burned area (Figure 4).

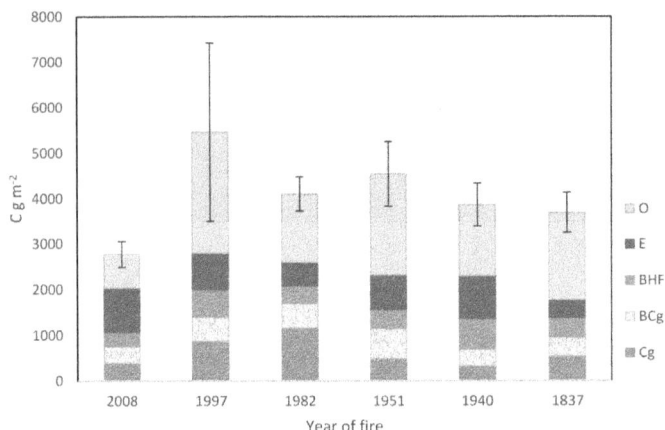

Figure 4. Average soil C content ($C \cdot g \cdot m^{-2}$) in different horizons (O-/E-/BHF-/BCg-/Cg-horizon) in a fire chronosequence.

3.2. Soil CO₂ Efflux

The results of this study revealed that factors affecting the soil CO_2 effluxes from post-fire areas were time since fire, soil temperature, time of measurement (month) and root biomass (Table 2). Factors like soil water content, soil C and N content and ground vegetation biomass had no effect on soil CO_2 efflux (Table 2). The post-fire soil CO_2 efflux increased with time since the fire in our study (Figure 5). The CO_2 efflux was lowest ($p < 0.05$) in the area where the fire occurred in 2008 (0.0747 mg CO_2 $m^{-2} \cdot s^{-1}$) and was already stable (compared to other older areas) in the area where the fire occurred in 1997 (0.1295 mg CO_2 $m^{-2} \cdot s^{-1}$), thus 18 years after the fire disturbance (Figure 5). There was also a clear correlation between soil CO_2 efflux and soil temperature in the studied fire areas($R = 0.44$, $p < 0.05$). When each year of the chronosequence was analyzed separately we found highest correlation between soil CO_2 efflux and soil temperature in the most recently burned area ($R = 0.95$, $p < 0.05$) and lowest in the area where the fire occurred in 1982 ($R = 0.59$, $p < 0.05$). In all other areas the correlation between soil CO_2 efflux and soil temperature were in the range 0.82–0.91.

Table 2. Analysis of logarithmically transformed soil CO_2 efflux: ANOVA type 3 test results for factors.

Factor	Complex Model			*p*-Value
	NDF	DDF	F	
Time since fire (year of fire)	5	924	116.99	<0.001
Time of measurement (month)	6	924	31.60	<0.001
Soil water content	1	924	0.87	0.3508
Soil temperature	1	924	141.89	<0.001
Soil C content	1	924	0.14	0.8929
Soil N content	1	924	3.36	0.0672
Ground vegetation biomass	1	924	3.39	0.0701
Root biomass (tree and ground vegetation roots)	1	924	19.23	<0.001

Note: NDF = numerator degrees of freedom for the F-test; DDF = denominator degrees of freedom; F = value of the F-statistics; *p*-value tests the null hypothesis "Factor has no effect on CO_2 efflux".

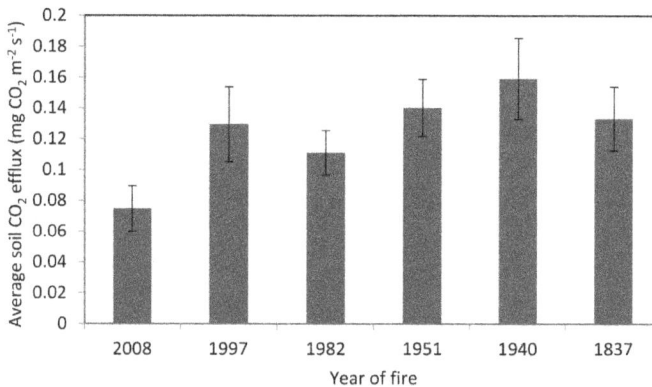

Figure 5. The average soil CO_2 efflux (mg CO_2 m^{-2}·s^{-1}) in a fire chronosequence.

4. Discussion

This study focused on stand-replacing fires in hemiboreal coniferous forests that affect C cycling and storage over large spatial scales and long time periods. Although more than 70% of the fires in Eurasia are surface fires [7], indicating that intermediate-severity fires are dominant in the area, the stand-replacing fires can also occur [4,8,21]. Fire intensity (intensity of humus reduction) is considered to be one of the most important fire-related characteristics [22], and it affects the pattern of the above- and belowground biomass recovery, community dynamics and soil processes [23].

Our results show a clear reduction of soil C stocks after the fire in the organic layer (O-horizon) on top of the soil, but fire effects were much smaller in the mineral soil. The C stocks of top soil layer had significantly lower amounts of C in the area where the last fire occurred 7 years ago in a year 2008, and the C pool started to increase significantly already 18 years after the fire in the area where fire occurred in 1997. Similarly, other studies have found that although the soil C pool in boreal forests is highly variable, the overall trend in the increase of C pool exists with increasing time since the fire [24,25]. The recovery of C pools in our study was most significant between 7 and 18 years after the fire. This is an important finding concerning the recovery of the C pool over the entire rotation period. If the fire frequency would increase from the current boreal average of approximately 100 years) between fire events, it could substantially reduce the long-term average soil C pool in the boreal forest zone [26,27]. However, these modeling studies, e.g., by Liski et al. [27] do not take into account the slow changes in soil respiration described here, and therefore overestimate the C losses experienced by forests [13].

Traditional approaches assume that the decomposition of SOM is either limited by the quality of SOM or by the environment [28]. The humus in podzol soils is usually covered with a poorly decomposed litter layer and the degree of decomposition of humus increases with depth. Fires burn most of the litter layer and often also the upper part of the humus layer. The SOM in the litter layer is easily decomposable, and weight loss from the litter approaches 10%–30% per year during the first years of decomposition [29]. The contribution of the litter layer to soil respiration has been found to be about 20% [30]. Therefore, we might assume that fires would decrease SOM quality by burning the easily decomposable litter but leaving the recalcitrant humus on the site. Direct effects of fires on the SOM quality have indeed been documented [31,32]. The amount of soil respiration is also affected by the quantity of SOM, and ground fires will cause a litter pulse because many trees are killed, but their foliage is not burnt. Therefore, litter input after a ground fire will exceed the litter input of undisturbed forests.

In previous studies it has been found that the soil CO_2 efflux is lower in recently burned areas and higher in the areas where more time has elapsed since the last fire [13,33–35]. Similarly, this study revealed that soil respiration was lowest in the area where the last fire occurred most recently in 2008.

It has been found that it may take 3–10 years for post-fire soil CO_2 efflux recovery [36,37], and the main factors affecting it are the vegetation type, vegetation coverage and post-fire biomass recovery [38,39], which contribute to the formation of new SOM. However, in our study somewhat longer recovery period was observed. Therefore, it can be assumed that in the 2008 fire area SOM has not recovered and since the main soil CO_2 efflux occurs mainly in the upper soil layers (O-horizon) the soil CO_2 efflux is lower. The main reason why the SOM has not recovered was the original site and fire severity. Pine forests in our study areas are growing on sandy soils where the organic layer of soil was thin. With stand replacing fire almost all the organic soil layer was burned and it is difficult for tree regeneration and ground vegetation to establish on pure sand. In some areas it was also noticed some movement of the sand (when the sand was exposed after fire), thus it is understandable that seven years after fire disturbance ground vegetation coverage, that is the most important source for new SOM establishment, was not completely recovered. In addition, also the root respiration (other important component of the soil CO_2 efflux) is lower as the vegetation has not fully recovered. In spring and summer, during active root growth, root respiration can account for 62% of the total soil CO_2 efflux, and in the autumn the proportion may be only 16% [40]. Thus, the soil CO_2 efflux may also be elevated due to the increase in root respiration, which in turn is caused by raise in soil temperature.

The importance of soil temperature and soil humidity on soil CO_2 efflux has been reported in several studies [33,41–43]. As temperature raises the loss of soil organic C increases and thus the soil CO_2 efflux increases [44–46]. In our study we also found that soil temperature affects the soil CO_2 efflux, while soil water content had no effect. The average soil temperature was highest in the 2008 fire area, although the soil CO_2 efflux was lowest in that area. It has been suggested that after fire the soil CO_2 efflux is lower because the vegetation is killed by fire and SOM is either damaged or destroyed [33]. Furthermore, the soil temperature rises since the sun warms the post-fire vegetation-free and darker ground (sand mixed with ash and unburned residues) more. As a result of very high temperatures the soil CO_2 efflux decreases as enzymes and microbes are deactivated [45]. Stand condition after the disturbance plays also an important role [47], because tree crowns are missing or have been damaged and it does not prevent the transmission of solar radiation on the ground [31,33]. Dead trees provide shelter from wind, which reduces evaporation that often inhibit conifer regeneration and may also act as seed catchers [2,48]. In contrast, old stand with a sparse cover of living trees allows wind to increase the evaporation. In our study there was a lot of standing dead trees in the area where the last fire occurred 7 years ago, while in the area where the last fire occurred 18 years ago there was almost no standing dead trees and there was the highest amount of lying dead wood. It is quite likely that in the 2008 fire area due to the small above- and belowground biomass the soil CO_2 efflux was lower even with the highest average soil temperature and CO_2 efflux was already stable in the 1997 fire area where the above- and belowground biomass was already recovered.

5. Conclusions

Overall, our results showed that forest fire has a substantial effect on the soil C content in the top soil layer, but not in the mineral soil layers. Soil respiration values in our study showed a chronological response to the time since the forest fire and the values were lowest in recently burned areas. The soil respiration values also followed a seasonal pattern being highest in August and lowest in May and November and there was a positive correlation between soil temperature and soil respiration values in our study areas. We also found that soil respiration follows logistic function: the recovery process happens within 10–20 years after fire and shows raising tendency.

Acknowledgments: This study was supported by the Estonian Environmental Investment Centre, the Institutional Research Funding (IUT21-4) of the Estonian Ministry of Education and Research, the Estonian Research Council (grant PUT715) and by the Academy of Finland (Projects No. 294600, 307222, 286685).

Author Contributions: K.K., M.M., K.J., F.B. and J.P. conceived and designed the experiment; K.K., M.M., A.O., K.P., K.J., F.B. and J.P. performed the experiments; K.K., A.O., and E.K. analyzed the data; all authors contributed in writing the paper.

Conflicts of Interest: The authors declare no conflict of interest.

References

1. Franklin, J.F.; Spies, T.A.; Pelt, R.V.; Carey, A.B.; Thornburgh, D.A.; Berg, D.R.; Lindenmayer, D.B.; Harmon, M.E.; Keeton, W.S.; Shaw, D.C.; et al. Disturbances and structural development of natural forest ecosystems with silvicultural implications, using Douglas-fir forests as an example. *For. Ecol. Manag.* **2002**, *155*, 399–423. [CrossRef]
2. Parro, K.; Metslaid, M.; Renel, G.; Sims, A.; Stanturf, J.A.; Jõgiste, K.; Köster, K. Impact of postfire management on forest regeneration in a managed hemiboreal forest, Estonia. *Can. J. For. Res.* **2015**, *45*, 1192–1197. [CrossRef]
3. Kashian, D.M.; Romme, W.H.; Tinker, D.B.; Turner, M.G.; Ryan, M.G. Carbon storage on landscapes with stand-replacing fires. *BioScience* **2006**, *56*, 598–606. [CrossRef]
4. Bond-Lamberty, B.; Gower, S.T.; Wang, C.; Cyr, P.; Veldhuis, H. Nitrogen dynamics of a boreal black spruce wildfire chronosequence. *Biogeochemistry* **2006**, *81*, 1–16. [CrossRef]
5. Kuzyakov, Y. Priming effects: Interactions between living and dead organic matter. *Soil Biol. Biochem.* **2010**, *42*, 1363–1371. [CrossRef]
6. Yarie, J.; Billings, S. Carbon balance of the taiga forest within Alaska: Present and future. *Can. J. For. Res.* **2002**, *32*, 757–767. [CrossRef]
7. Shorohova, E.; Kuuluvainen, T.; Kangur, A.; Jõgiste, K. Natural stand structures, disturbance regimes and successional dynamics in the Eurasian boreal forests: A review with special reference to Russian studies. *Ann. For. Sci.* **2009**, *66*, 201p1–201p20. [CrossRef]
8. Conard, S.G.; Sukhinin, A.I.; Stocks, B.J.; Cahoon, D.R.; Davidenko, E.P.; Ivanova, G.A. Determining Effects of Area Burned and Fire Severity on Carbon Cycling and Emissions in Siberia. *Clim. Chang.* **2002**, *55*, 197–211. [CrossRef]
9. Goetz, S.J.; Mack, M.C.; Gurney, K.R.; Randerson, J.T.; Houghton, R.A. Ecosystem responses to recent climate change and fire disturbance at northern high latitudes: Observations and model results contrasting northern Eurasia and North America. *Environ. Res. Lett.* **2007**, *2*, 1–9. [CrossRef]
10. Batjes, N.H. Total carbon and nitrogen in the soils of the world. *Eur. J. Soil Sci.* **1996**, *47*, 151–163. [CrossRef]
11. Högberg, P.; Read, D.J. Towards a more plant physiological perspective on soil ecology. *Trends Ecol. Evol.* **2006**, *21*, 548–554. [CrossRef] [PubMed]
12. Pumpanen, J.; Heinonsalo, J.; Rasilo, T.; Hurme, K.-R.; Ilvesniemi, H. Carbon balance and allocation of assimilated CO_2 in Scots pine, Norway spruce, and Silver birch seedlings determined with gas exchange measurements and 14C pulse labelling. *Trees* **2009**, *23*, 611–621. [CrossRef]
13. Köster, K.; Berninger, F.; Lindén, A.; Köster, E.; Pumpanen, J. Recovery in fungal biomass is related to decrease in soil organic matter turnover time in a boreal fire chronosequence. *Geoderma* **2014**, *235–236*, 74–82. [CrossRef]
14. Wan, S.; Hui, D.; Luo, Y. Fire effects on nitrogen pools and dynamics in terrestrial ecosystems: A meta-analysis. *Ecol. Appl.* **2001**, *11*, 1349–1365. [CrossRef]
15. Goulden, M.L.; McMillan, A.M.S.; Winston, G.C.; Rocha, A.V.; Manies, K.L.; Harden, J.W.; Bond-Lamberty, B.P. Patterns of NPP, GPP, respiration, and NEP during boreal forest succession. *Glob. Chang. Biol.* **2011**, *17*, 855–871. [CrossRef]
16. Parro, K.; Köster, K.; Jõgiste, K.; Vodde, F. Vegetation dynamics in a fire damaged forest area: The response of major ground vegetation species. *Balt. For.* **2009**, *15*, 206–215.
17. Repola, J. Biomass equations for birch in Finland. *Silva Fenn.* **2008**, *42*, 605–624. [CrossRef]
18. Repola, J. Biomass equations for Scots pine and Norway spruce in Finland. *Silva Fenn.* **2009**, *43*, 625–647. [CrossRef]
19. Food and Agriculture Organization of the United Nations (FAO). *World Reference Base for Soil Resources 2014. International Soil Classification System for Naming Soils and Creating Legends for Soil Maps*; FAO: Rome, Italy, 2014.
20. Beck, T.; Joergensen, R.G.; Kandeler, E.; Makeschin, F.; Nuss, E.; Oberholzer, H.R.; Scheu, S. An inter-laboratory comparison of ten different ways of measuring soil microbial biomass C. *Soil Biol. Biochem.* **1997**, *29*, 1023–1032. [CrossRef]

21. Gromtsev, A. Natural disturbance dynamics in the boreal forests of European Russia: A review. *Silva Fenn.* **2002**, *36*, 41–55. [CrossRef]
22. Schimmel, J.; Granström, A. Fire severity and vegetation response in the boreal Swedish forest. *Ecology* **1996**, *77*, 1436–1450. [CrossRef]
23. Ruokolainen, L.; Salo, K. The effect of fire intensity on vegetation succession on a sub-xeric heath during ten years after wildfire. *Ann. Bot. Fenn.* **2009**, *46*, 30–42. [CrossRef]
24. Pregitzer, K.S.; Euskirchen, E.S. Carbon cycling and storage in world forests: Biome patterns related to forest age. *Glob. Chang. Biol.* **2004**, *10*, 2052–2077. [CrossRef]
25. Bormann, B.T.; Homann, P.S.; Darbyshire, R.L.; Morrissette, B.A. Intense forest wildfire sharply reduces mineral soil C and N: The first direct evidence. *Can. J. For. Res.* **2008**, *38*, 2771–2783. [CrossRef]
26. Metsaranta, J.M.; Kurz, W.A.; Neilson, E.T.; Stinson, G. Implications of future disturbance regimes on the carbon balance of Canada's managed forest (2010–2100). *Tellus B* **2010**, *62*, 719–728. [CrossRef]
27. Liski, J.; Ilvesniemi, H.; Mäkelä, A.; Starr, M. Model analysis of the effects of soil age, fires and harvesting on the carbon storage of boreal forest soils. *Eur. J. Soil Sci.* **1998**, *49*, 407–416. [CrossRef]
28. Jenkinson, D.S.; Adams, D.E.; Wild, A. Model estimates of CO_2 emissions from soil in response to global warming. *Nature* **1991**, *351*, 304–306. [CrossRef]
29. Berg, B.; Berg, M.P.; Bottner, P.; Box, E.; Breymeyer, A.; Calvo de Anta, R.; Couteaux, M.; Escudero, A.; Gallardo, A.; Kratz, W.; et al. Litter mass loss rates in pine forests of Europe and Eastern United States: Some relationships with climate and litter quality. *Biogeochemistry* **1993**, *20*, 127–159. [CrossRef]
30. Berryman, E.M.; Marshall, J.D.; Kavanagh, K. Decoupling litter respiration from whole-soil respiration along an elevation gradient in a Rocky Mountain mixed-conifer forest. *Can. J. For. Res.* **2014**, *44*, 432–440. [CrossRef]
31. Certini, G. Effects of fire on properties of forest soils: A review. *Oecologia* **2005**, *143*, 1–10. [CrossRef] [PubMed]
32. Certini, G.; Nocentini, C.; Knicker, H.; Arfaioli, P.; Rumpel, C. Wildfire effects on soil organic matter quantity and quality in two fire-prone Mediterranean pine forests. *Geoderma* **2011**, *167–168*, 148–155. [CrossRef]
33. Luo, Y.; Zhou, X. *Soil Respiration and the Environment*; Elsevier: New York, NY, USA, 2006; p. 320.
34. Czimczik, C.I.; Trumbore, S.E.; Carbone, M.S.; Winston, G.C. Changing sources of soil respiration with time since fire in a boreal forest. *Glob. Chang. Biol.* **2006**, *12*, 957–971. [CrossRef]
35. Köster, K.; Berninger, F.; Heinonsalo, J.; Lindén, A.; Köster, E.; Ilvesniemi, H.; Pumpanen, J. The long-term impact of low-intensity surface fires on litter decomposition and enzyme activities in boreal coniferous forests. *Int. J. Wildland Fire* **2016**, *25*, 213–223. [CrossRef]
36. Kulmala, L.; Aaltonen, H.; Berninger, F.; Kieloaho, A.-J.; Levula, J.; Bäck, J.; Hari, P.; Kolari, P.; Korhonen, J.F.J.; Kulmala, M.; et al. Changes in biogeochemistry and carbon fluxes in a boreal forest after the clear-cutting and partial burning of slash. *Agric. For. Meteorol.* **2014**, *188*, 33–44. [CrossRef]
37. Köster, E.; Köster, K.; Berninger, F.; Pumpanen, J. Carbon dioxide, methane and nitrous oxide fluxes from podzols of a fire chronosequence in the boreal forests in Värriö, Finnish Lapland. *Geoderma Reg.* **2015**, *5*, 181–187. [CrossRef]
38. Raich, J.W.; Tufekcioglu, A. Vegetation and soil respiration: Correlation and controls. *Biogeochemistry* **2000**, *48*, 71–90. [CrossRef]
39. Hart, S.C.; DeLuca, T.H.; Newman, G.S.; MacKenzie, M.D.; Boyle, S.I. Post-fire vegetative dynamics as drivers of microbial community structure and function in forest soils. *For. Ecol. Manag.* **2005**, *220*, 166–184. [CrossRef]
40. Widén, B.; Majdi, H. Soil CO_2 efflux and root respiration at three sites in a mixed pine and spruce forest: Seasonal and diurnal variation. *Can. J. For. Res.* **2001**, *31*, 786–796. [CrossRef]
41. Raich, J.W.; Schelesinger, W.H. The global carbon dioxide flux in soil respiration and its relationship to vegetation and climate. *Tellus B* **1992**, *44*, 81–99. [CrossRef]
42. Buchmann, N. Biotic and abiotic factors controlling soil respiration rates in Picea abies stands. *Soil Biol. Biochem.* **2000**, *32*, 1625–1635. [CrossRef]
43. Laganière, J.; Paré, D.; Bergeron, Y.; Chen, H.Y.H. The effect of boreal forest composition on soil respiration is mediated through variations in soil temperature and C quality. *Soil Biol. Biochem.* **2012**, *53*, 18–27. [CrossRef]
44. Schlesinger, W.H.; Andrews, J.A. Soil respiration and the global carbon cycle. *Biogeochemistry* **2000**, *48*, 7–20. [CrossRef]

45. Fang, C.; Moncrieff, J.B. The dependence of soil CO_2 efflux on temperature. *Soil Biol. Biochem.* **2001**, *33*, 155–165. [CrossRef]
46. Köster, K.; Püttsepp, Ü.; Pumpanen, J. Comparsion of soil CO_2 flux between uncleared and cleared windthrow areas in Estonia and Latvia. *For. Ecol. Manag.* **2011**, *262*, 65–70. [CrossRef]
47. Köster, K.; Voolma, K.; Jõgiste, K.; Metslaid, M.; Laarmann, D. Assessment of tree mortality after windthrow using photo-derived data. *Ann. Bot. Fenn.* **2009**, *46*, 291–298. [CrossRef]
48. Moser, B.; Temperli, C.; Schneiter, G.; Wohlgemuth, T. Potential shift in tree species composition after interaction of fire and drought in the Central Alps. *Eur. J. For. Res.* **2010**, *129*, 625–633. [CrossRef]

© 2016 by the authors. Licensee MDPI, Basel, Switzerland. This article is an open access article distributed under the terms and conditions of the Creative Commons Attribution (CC BY) license (http://creativecommons.org/licenses/by/4.0/).

forests

MDPI

Article

Estimates of Wildfire Emissions in Boreal Forests of China

Kunpeng Yi [1],* and Yulong Bao [2]

[1]   Institute of Remote Sensing and Digital Earth (RADI), Chinese Academy of Sciences (CAS), Beijing 100101, China

[2]   College of Geography Science, Inner Mongolia Normal University, Hohhot 101022, China; baoyulong@imnu.edu.cn

*   Correspondence: yikp@radi.ac.cn; Tel.: +86-6484-2375; Fax: +86-6485-8721

Academic Editors: Yves Bergeron and Sylvie Gauthier

Received: 29 March 2016; Accepted: 21 July 2016; Published: 1 August 2016

Abstract: Wildfire emissions in the boreal forests yield an important contribution to the chemical budget of the troposphere. To assess the contribution of wildfire to the emissions of atmospheric trace species in the Great Xing'an Mountains (GXM), which is also the most severe fire-prone boreal forest region in China, we estimated various wildfire activities by combining explicit spatio-temporal remote sensing data with fire-induced emission models. We observed 9998 fire scars with 46,096 km^2 in the GXM between the years 1986 and 2010. The years 1987 and 2003 contributed 33.2% and 22.9%, respectively, in burned area during the 25 years. Fire activity is the strongest in May. Most large fires occurred in the north region of the GXM between 50° N and 54° N latitude due to much drier weather and higher fire danger in the northern region than in the southern region of the study domain. Evergreen and deciduous needleleaf forest and deciduous broadleaf forest are the main sources of emissions, accounting for 84%, 81%, 84%, 87%, 89%, 86%, 85% and 74% of the total annual CO_2, CH_4, CO, PM_{10}, $PM_{2.5}$, SO_2, BC and NO_x emissions, respectively. Wildfire emissions from shrub, grassland and cropland only account for a small fraction of the total emissions level (approximately 4%–11%). Comparisons of our results with other published estimates of wildfire emissions show reasonable agreement.

Keywords: wildfire; emissions; satellite; China; burned area

1. Introduction

Wildfire is a critical disturbance factor in the boreal forests, which acts as a double-edged sword in the natural context. On the one hand, fire is traditionally used as a tool to aid in many land use and related changes, including the clearing of forests for agriculture and for shifting agricultural practices. Wildfire is an important part of ecosystem services, providing nutrients and recycling material. On the other hand, wildfire always releases some emissions of gases and aerosols to the atmosphere [1]. It is considered a major source of aerosol that affects air quality, atmospheric composition and the Earth's radiation budget [2,3]. Smoke emitted by fires is composed of volatile organic compounds [4], particulate matter (PM) and numerous trace gases, including carbon monoxide (CO), carbon dioxide (CO_2), methane (CH_4) and nitrogen oxides [5]. Globally, wildfire contributes approximately 50% of the total direct CO emissions and approximately 15% of surface NO_x emissions [6]. Most of these particular matters and trace gases can have significant effects, not only on human health but also affect the climate, with potential feedback on air quality. As gaseous and aerosol emissions from fires are transported through the atmosphere, they degrade air quality by reducing visibility, creating unhealthy levels of PM, and reacting to create harmful tropospheric trace gases, such as ozone (O_3) [7]. Some species have significant and far-reaching consequences due to their long lifetime (e.g., N_2O: ~ 150 years,

CO_2: ~100 years, CH_4: ~10 years) [8]. In addition, smoke and PM influences precipitation processes, resulting in delayed, suppressed or invigorated rainfall, which changes cloud albedo [9], scatters and absorbs solar radiation [10,11], affecting atmospheric warming or cooling and contributing to climate change [12–14].

With increasing scientific and political concern regarding the carbon cycle, there is a strong impetus to better understand wildfire carbon emissions on both a global and regional scale. For several decades, researchers have made great efforts to estimate burned biomass emissions from ground-based and in situ measurements. Both spatial and temporal coverage of these studies is severely limited in pre satellite era [15–17]. However, during the past two decades, major advances have occurred in the detection of atmospheric pollution from space. The generation of satellite instruments launched since 1995 has proven to be capable of observing a wide range of chemical species at increasingly high spatial and temporal resolutions [18–20]. In addition, the transformation of raw satellite retrievals to user-friendly, archived products has considerably progressed, such that the application of satellite observations to a wide range of atmospheric problems is no longer a daunting prospect. Space-based observations, such as the Along-Track Scanning Radiometer (ATSR) [21] and the Moderate Resolution Imaging Spectroradiometer (MODIS), provide information concerning global burning hotspots at a spatial resolution of 1 km, which have allowed a better identification of wildfire and emissions [22].

Previous studies have attempted to estimate the amount of burned biomass and the fire-induced emissions in China [23–25]. The total amount of carbon emitted per year via the burning of terrestrial biomass in China has been estimated at 11.31 Tg from 1950 to 2000 [25]. This amount of carbon emissions has resulted from the atmospheric emissions of three trace gases: 40.6 Tg year^{-1} CO_2, 27.1 Tg year^{-1} CO and 0.112 Tg year^{-1} CH_4, besides 0.113 Tg year^{-1} NMHC (non-methane hydrocarbons) [25]. Black carbon (BC) emissions have increased at an average annual rate of 25.54%, from 0.014 Tg in 1990 to 0.067 Tg in 2005 [23]. Nevertheless, current approaches depend on factors such as scale, accuracy requirements and information availability, among others. Therefore, a longer-term, higher-resolution study into the emissions from wildfire on both a global and regional scale is required.

Local estimates are necessary to understand micro-scale emission mechanisms, whereas regional and global modeling is essential to assess the net effects of emissions on the atmosphere and on global climate change [26]. Some articles argue that China should take more responsibility for climate change mitigation than other counties due to much CO_2 emissions by fossil fuel consumption and by wildfire [27–29]. In China, the most severe fire-prone area is the northeastern region, particularly the GXM region. Wildfire is inevitable and is ecologically important in forests throughout much of the GXM because of the fuels, ignition sources and variable climatic conditions. It has been roughly estimated that 1.2×10^7 Mg carbon emissions were released from forest fires, with approximately 1.0×10^6 hm^2 of forest burned in the GXM during the period from 1980 to 2005 [30]. To assess the atmospheric impact of biomass burning quantitatively, accurate emissions estimates of trace gases and aerosols are required. The purpose of this work is to provide estimates of total direct emissions from wildfire by combining explicit spatio-temporal remote-sensing data with fire-induced emission models. Another major objective of this investigation is to estimate the potential range and spatial-temporal patterns of wildfire events in the GXM area. We also attempt to identify gaps and limitations in existing data and methods that must be studied in the future to improve our understanding of the role of wildfire in regional carbon dynamics.

2. Materials and Methods

2.1. Study Area

The Great Xing'an Mountains (GXM), which lie on China's northern border frontier and neighbor Russia in the north and Mongolia in the west, are commonly defined as stretching from the Heilongjiang (or Amur) River in the north to the Silas Moron River in the south. This study focuses on the Hulun Buir Plateau and the majority of the GXM (Figure 1). This region is primarily a hilly mountainous region that ranges from 450 m to 1500 m in elevation. The climate is terrestrial monsoon, with long, severe winters (mean January temperature −28.5 °C) and short, mild summers (mean July temperature 17 °C). Precipitation, which peaks in summer, is 420 mm annually and is unevenly distributed throughout the year, with more than 60% of precipitation occurring between June and August. Vegetation in this region falls within the cool temperate coniferous forests, which occur at the southern extension of the eastern Siberian light coniferous forest [31].

The species composition is relatively simple, and the forest area covers over 75% of the study area. The most dominant tree species is larch, which accounts for 80% of the study area. The second most dominant species is birch, which covers 10% of the study area. Other species, including pine, spruce, aspen and willow cover approximately 10% of the study area. Figure 1 show the land cover distribution in the study area, which was acquired from the Advanced Very High Resolution Radiometer (AVHRR) Global Land Cover Map (http://www.glcf.umd.edu/data/landcover).

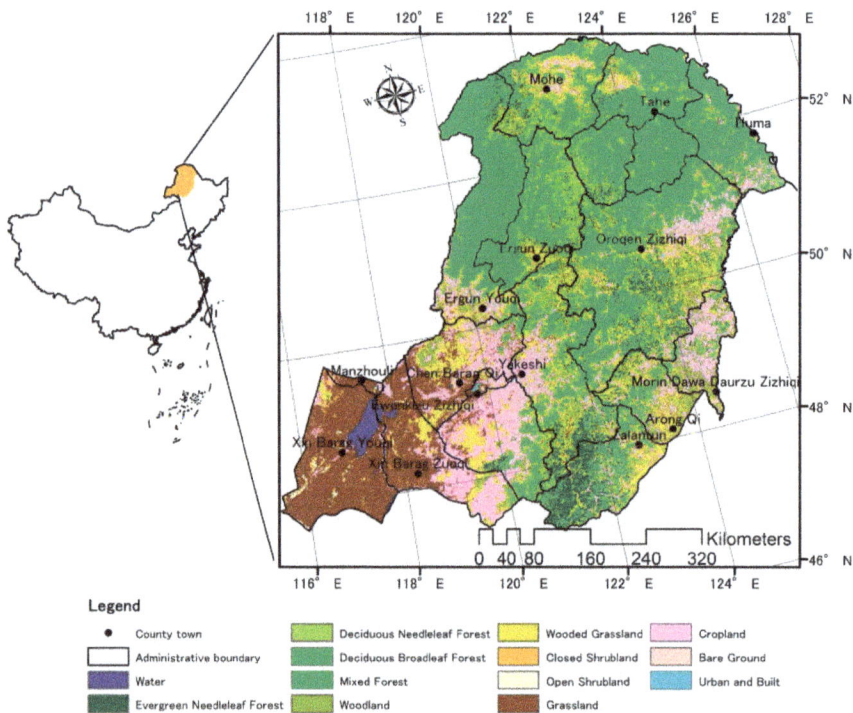

Figure 1. Land cover distribution in the study area. This distribution was derived from the AVHRR Global Land Cover Map.

2.2. Methodology:Carbon Emission Calculations

Over the past several decades, substantial efforts have been devoted to evaluating fire-induced carbon emissions, primarily by use of models. Specifically, remote-sensing technologies are assessed as a feasible way to estimate emissions from both a direct approach (i.e., smoke measurements) and an indirect approach (i.e., model emissions driving variables, such as fuel availability and combustion efficiency) [26].

The most general way to compute gas emissions is deterministic, where the trace gas emissions are linked to the amount and type of fuel consumed and to the combustion characteristics [32]. The amount of a specific gas or aerosol species, x, emitted as follows [32]:

$$M_x = MB_{consumed} \times EF_x \tag{1}$$

where Mx is the mass of species x emitted from fire EF$_x$ is the mass of species x released per kilogram of the biomass burned (emission factor); and MB$_{consumed}$ is the mass of the dry biomass burned. MB$_{consumed}$ can be estimated using the following equation:

$$MB_{consumed} = BA \times BD \times BE \tag{2}$$

where BA (ha) is the area burned by fire; BD (kg ha^{-1}) is the density of the dry biomass in the area; and BE (%) is burn efficiency or combustion completeness, i.e., the percentage of biomass consumed by fire. The model used in our study integrates a series of biophysical variables that can be estimated based on various sources from remote sensing imagery and from the literature.

2.3. DeterMination of Variables

2.3.1. Burned Area (BA)

The Great Xing'an Mountain (GXM) is a typical fire-prone region, where many species have a recognized ability to regenerate after fire. Fire has been a primary disturbance in most forests of this region and has shaped their plant communities for millions of years [33–35]. The National Forestry Bureau (NFB) of China reported information on significant fires across China since 1950, including information on fire numbers, burnt areas, and fire locations. Historically, fire regimes in this region have been characterized by frequent, low intensity surface fires mixed with sparse stand-replacing fires in relatively small areas [36]. The historical burned areas used in this study were derived from NOAA (National Oceanic and Atmospheric Administration) AVHRR for 1986–2000. We can obtain burned date and fire location from national fire inventory. Firstly, we buffer a region as a rough burned area by using fire location. Then we compare AVHRR NDVI images of pre burn date and post burn date. If AVHRR NDVIpre is larger than two times NDVIpost, then the pixels were considered as burnt area. If not, the pixel will be flagged as unburnt area (Figure 2). The AVHRR NDVI datasets from 1986–2000 provided by National Aeronautics and Space Administration (NASA) (URL: http://daac.gsfc.nasa.gov/ DATASET _DOCS /avhrr_dataset.html).

For the period from 2000 to 2010, the burned area was estimated using the satellite MODIS MCD45 Burned Area Level3 product, which provides the most comprehensive data concerning fire-affected areas in remote boreal regions. MCD45A1 is a monthly, 500-m resolution product and has been available online since April 2000. In this study, MCD45A1 burned area dataset from 2001 to 2010 were obtained from USGS (URL:http://e4ftl01.cr.usgs.gov/MOTA/MCD45A1.005/). Uncertainty of MODIS burnt area is still substantial. On the one hand, overestimation of burnt area may occur as a consequence of including unburnt patches within pixels classified as burnt. On the other hand, underestimation of area burnt may occur in pixels where only a small fraction was burned. In this case, the satellite senor cannot capture such weak signals. Therefore, the confidence of MODIS burnt pixels detection was divided into four values (QA = 1 (most confident) to QA = 4 (least confident)). In this study, if burnt pixels QA = 1, then those pixels were flag as burnt; for those pixels with less confident (QA = 2, 3 or 4), we redefined them by comparing SPOT (Satellite for Observation of Earth) NDVI images of pre burn date and post burn date. A different satellite product, SPOT vegetation, was used to check the MODIS results. If evidence were found by both of them, then the results are right. Especially, if SPOT NDVIpre is larger than two times SPOT NDVIpost, then the pixels was considered as burnt area, otherwise, the pixels were considered as unburnt area (Figure 2).

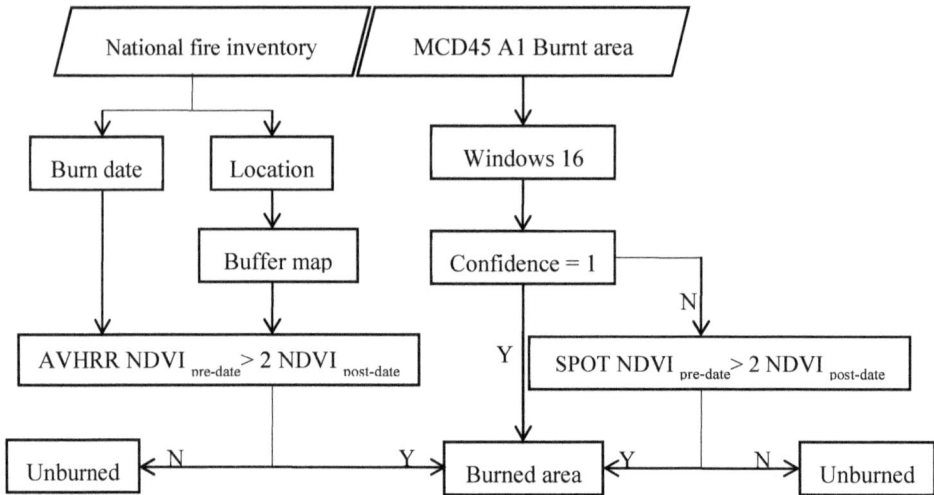

Figure 2. Flow chart used to derive burned area.

2.3.2. Biomass Density (BD) and Burning Efficiency (BE)

Biomass density and burning efficiency are two uncertain parameters in biomass burning estimates due to the high-spatial variability of the burning process and of fuel availability, which are related not only to general ecosystem characteristics but also to the micro-scale environmental conditions. Trace gas emissions are directly related to biomass and land cover through the amount and composition of the fuel [26]. Simply stated, the higher the biomass density is, the higher the amount available for burning and, therefore, the greater the total quantity of carbon that can be released as trace gases [37]. In areas of low precipitation and in regions with dry periods of high temperature, biomass consumption is higher than in more humid climates in Brazil [38]. A linear relation between the annual area burned and the fraction of biomass consumed was revealed in previous study [39], they found that in warmer years, when a higher number of fires occur, burning efficiency is higher as well in boreal forests. Thus, in warmer years, when a higher number of fires occur, the burning efficiency is also higher [26]. It is infeasible to identify the specific meteorological conditions of each fire event. Most plants live in areas with very specific climate conditions. Phytogeography tells us that the geographic distribution of plant species is primarily decided by those regional climate characteristics, such as temperature and rainfall patterns. In this sense, the land cover map can also commendably reflect the site climate characteristics. Therefore, in this study, the biomass densities (BD) and burning efficiencies (BE) were derived from values published [40–42] using the AVHRR Global Land Cover Map (AVHRR GLC-2000) at a 1 km resolution. In this study, we presume that biomass below ground does not burn, although fires can burn deep into the ground under certain conditions, such as peat fire. Table 1 shows the determination of biomass densities (BD) and burning efficiencies (BE), whose values are derived from published literature based on specific sites across the globe. Thus, we can get biomass density and burned efficiency on pixel level (Figure 3).

Table 1. Determination of emission model parameters for each bioclimatic zone against the AVHRR Global Land Cover Map.

Code	Description	BD (Mg/ha)	(BE%)
1, 3	Evergreen and deciduous needleleaf forest	140	40
4	Deciduous broadleaf forest	95	40
5	Mixed forest	12	40
6	Woodland	95	40
7	Wooded grassland	11	90
8, 9	Scrubland	43	50
10	Grassland	11	90
11	Cropland	5	90
0, 12–14	Water bodies, barren, built-up, undefined	0	0

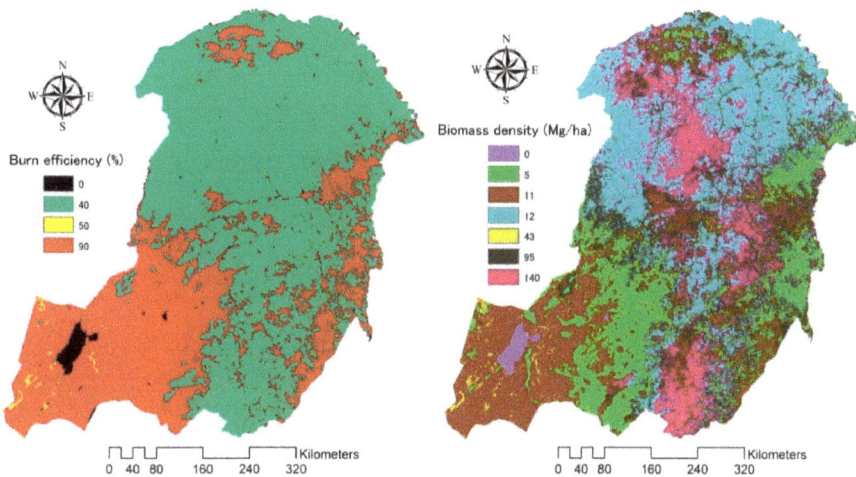

Figure 3. Biomass density (**left**) and burned efficiency (**right**) distribution in the study.

2.3.3. Determination of Emission Factors (EF)

Emission factors, usually defined as the amount of certain trace gas and aerosol species released per amount of fuel consumed, which are expressed in grams of a gas compound per kilogram of dry matter [26,43]. Emission factors are based on the correlation between the concentration of a certain gas species and the concentration of the reference species emitted and are estimated from experiments for specific conditions [26]. Emission factors for various ecosystems and environmental conditions have been derived either in natural conditions or in controlled laboratory experiments [7,41,42] where the rest of the variables were known (i.e., the amount of the compound released, the amount of fuel burned and the concentration of the element in the fuel) [26]. In the present study, emission factors were also assigned for each land cover classification in the AVHRR GLC-2000 data. These emission factors for each emitted species, which are given in Table 2, were based on previously published studies [7,42]. The average of relevant emission factors for each gaseous or particulate species was applied when more than one emission factor was available in the literature. Figure 4 shows the determination of CO_2 emission factors in China. The emission factors of other emitted species (CH_4, CO, PM_{10}, $PM_{2.5}$, SO_2, BC and NO_x) were also determined in the same way.

Table 2. Emission factors assigned to fires in each of the vegetation types of the AVHRR GLC-2000 data.

Description	CO_2 *	CH_4	CO	PM_{10}	$PM_{2.5}$	SO_2	BC	NO_x
Evergreen and deciduous needleleaf forest	1700	4.8	89	13.1	12.7	0.6	0.77	3.1
Deciduous broadleaf forest	1750	4.5	94	15	12.3	0.5	0.82	2.5
Mixed forest	1670	4.5	84	12.5	7.9	0.6	0.8	3
Woodland	1652	4.5	90	8.5	7.3	0.5	0.5	6.5
Wooded grassland	1642	3.1	80	8.5	7	0.5	0.52	3.2
Scrubland	1632	3.1	85	8.5	7	0.5	0.52	6.5
Grassland	1588	3.1	80	9.9	6.3	0.5	0.48	4
Cropland	1353	2.2	80	9.9	6.3	0.4	0.63	2.9
Water bodies, barren, urban **	0	0	0	0	0	0	0	0

* Units are gram species per kilogram of dry biomass burned (g/kg). The values given in this table were derived from published studies [7,41,42]. ** Emission factors for water bodies, barren area and urban area were assigned a value of zero.

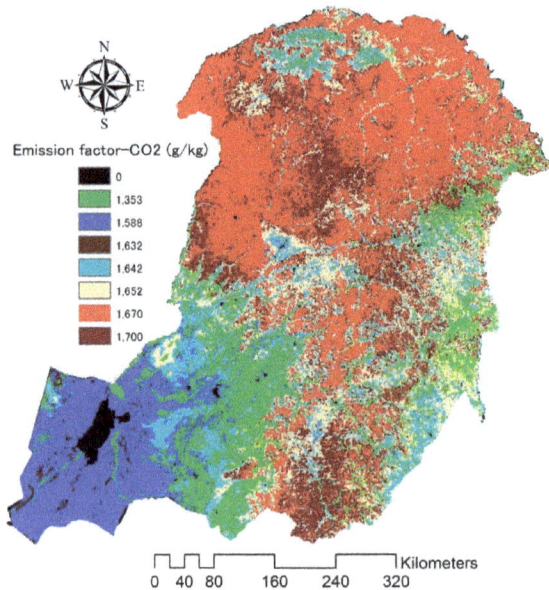

Figure 4. The determination of CO_2 emission factors in China.

3. Results and Discussion

3.1. Wildfire Distributionin the Great Xing'an Mountains

Figure 5 shows the historical burned area distribution by month from the years 1986 to 2010. In total, 9998 fire scars were observed in the GXM between the years 1986 and 2010. Particularly during the spring fire season, there are always serious large forest fire events. For example, on 6th May 1987, a catastrophic fire occurred on the northern slopes of the GXM, which burned a total area of 1.3×10^6 ha, with disastrous effects on the forest composition and structure, ecosystem processes and the landscape pattern [44]. The wildfire that broke out in Jinhe and Genhe counties on 5 May 2003 burned 7.9×10^4 hectares, including 6.3×10^4 hectares of forest areas, with an economic loss of more than 198 million RMB [44].

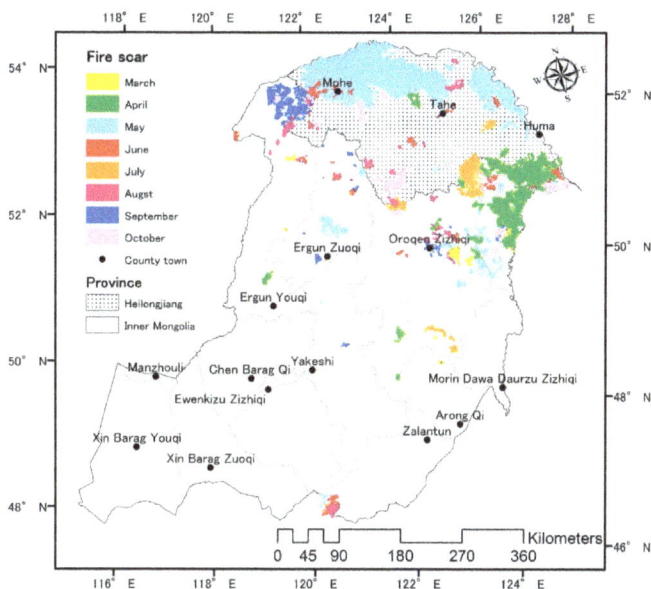

Figure 5. Burned area distribution from 1986 to 2010. Only the fire scars larger than 100 ha are shown in this figure. Note that the size of the burn scars increases slightly for the display purpose.

Spring and autumn are two primary fire seasons in the GXM. Almost 80% burned area are burned in three months (May: 52%; April: 15%; October: 12%). Normally, there are no large fire events due to high precipitation and high fuel moisture in summer (July and August) and to low temperatures accompanied by the accumulation of snow and frozen land in winter (late November to February) (Figure 6). Thus, fire events occur in spring from mid-March to mid-June and from mid-September to mid-November in fall. Large fires occurred more often in the northern region of the GXM between 50° N and 54° N latitude than in the southern region (part of Inner Mongolia Province) (Figure 5) due to much drier weather and higher fire danger in the northern region (part of Heilongjiang Province). The vegetation in this region is a cold-temperature mixed coniferous forest, in which fire is likely to occur. Burn scars and the associated vegetation succession lead to a mosaic of landscape patches. These spatial patterns of regenerating vegetation in various succession stages are important considerations for carbon budget studies in boreal forests [45].

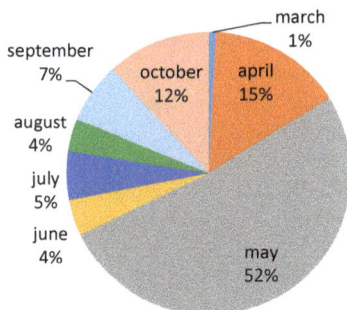

Figure 6. Monthly pattern of wildfire in the Great Xing'an Mountains.

Figure 7 shows the burned area shift from 1986 to 2010. The years 1987 and 2003 contributed 33.2% and 22.9%, respectively, in burned area during the 25 years. In total, 1.53 million hectares were charred by fires in 1987, which was the most destructive fire year in China's history, with a specific larger fire in May that burned 1.33 million hectares alone. Extreme meteorological conditions, in combination with the lack of accessibility, resulted in large, high-intensity crown fires throughout China's boreal forests in 1987 [44]. Snowfall over the northern China was typically light during the winter of 1986–1987, and the snow cover had disappeared by early April southeast of Lake Baikal and along the Amur River. Low temperatures and relative humidity prevailed throughout April, with only a light scattering of precipitation. The cumulative result was an extremely dry forest fuel situation in this region by the beginning of May 1987, which combined with increasing temperatures and strong winds to produce uncontrollable forest fires. Another severe fire year in the history is 2003, during which approximately 1.1 million hectares was burned. Dozens of fires burned in the GXM region from Inner Mongolia to Heilongjiang Province in May 2003. As the neighbor of the GXM, West Siberia's largest forest fires on record also occurred during the same period in 2003, claiming approximately 20 million hectares of land and emitting heat-trapping emissions equal to the total cuts in emissions that the European Union pledged under the Kyoto Protocol [46]. Higher temperatures and thawing permafrost are most likely contributing to the rising frequency and severity of forest fires in West Siberia [47].

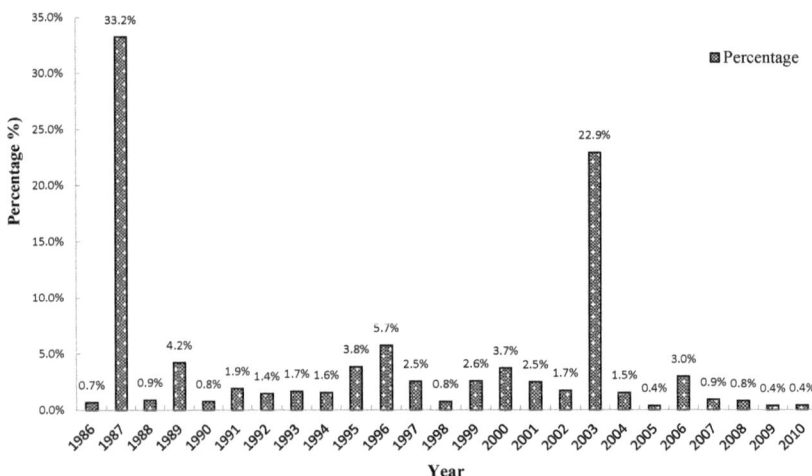

Figure 7. Yearly burned area shift in the Great Xing'an Mountains.

3.2. Trends in Biomass Burning Emissions

Yearly and monthly emissions of CH_4, CO, PM_{10}, $PM_{2.5}$, SO_2, BC and NO_x from biomass burning in the GXM have been estimated from 1986 to 2010. The spatial and temporal distribution of fire emissions is similar to that of the burned area because the burned pixels were defined as 1, whereas unburned pixels were defined as 0. Table 3 summarizes the average annual emissions of trace gases and particulate matter for different land cover types. Evergreen and deciduous needleleaf forest and deciduous broadleaf forest are the main sources of emissions, accounting for 84%, 81%, 84%, 87%, 89%, 86%, 85% and 74% of the total annual CO_2, CH_4, CO, PM_{10}, $PM_{2.5}$, SO_2, BC and NO_x emissions, respectively. Fire emissions from shrub, grassland and cropland only account for a small fraction of the total emissions level (approximately 4%–11%). The two extreme fire years of 1987 and 2003 contributed approximately 60% of total biomass burning emission, according to this study from 1986 to 2010. Cahoon et al. estimated that 14.5 million ha burned and that approximately 36 Tg CO were emitted during the catastrophic fire burned on 6 May 1987 in China using AVHRR burn scars [44]. This result suggests that the inclusion of fire emissions that are specific to a particular period and region is essential for realistically simulating air quality [7].

Table 3. Emissions of estimated trace gases and particulate matter (Tg/year) from fires for the average annual value of the period from 1986 to 2010.

Description	CO_2	CH_4	CO	PM_{10}	$PM_{2.5}$	SO_2	BC	NO_x
Evergreen and deciduous needle leaf forest	52.66	0.15	27.57	0.41	0.39	0.02	0.02	0.10
Deciduous broadleaf forest	30.65	0.08	16.47	0.26	0.22	0.01	0.01	0.04
Mixed forest	2.22	0.01	1.12	0.02	0.01	7.97×10^{-4}	1.06×10^{-3}	3.98×10^{-3}
Woodland	5.79	0.02	3.15	0.03	0.03	1.75×10^{-3}	1.75×10^{-3}	0.02
Wooded grassland	1.50	2.83×10^{-3}	0.73	0.01	0.01	4.56×10^{-4}	4.75×10^{-4}	0.00
Scrubland	3.23	0.01	1.68	0.02	0.01	9.91×10^{-4}	1.03×10^{-3}	1.29×10^{-3}
Grassland	2.90	0.01	1.46	0.02	0.01	9.13×10^{-4}	8.76×10^{-4}	0.01
Cropland	0.56	9.13×10^{-4}	0.33	4.11×10^{-3}	2.61×10^{-3}	1.66×10^{-4}	2.61×10^{-4}	1.20×10^{-3}
Total	99.51	0.28	52.51	0.77	0.68	0.04	0.04	0.19

Several published papers have also estimated the biomass burning emissions in different regions for recent years [7,48–50]. The emissions from wildfire in this study are compared with four other previous study results (Table 4). Our study period includes two extreme fire years, 1987 and 2003, which play an important role in the final average annual result from 1986–2010. Thus, to make a reasonable comparison with other studies, we calculated another average annual emission result for a normal period (2005–2010) without any extreme fire events. The areas of these study domains differ from our study area. Thus, we list all estimated regions and their area ratios in Table 4 for comparison (United States: 962.9×10^4 km^2; Canada: 998.5×10^4 km^2; China: 962.9×10^4 km^2; Our study area: 32.7×10^4 km^2). Wiedinmyer et al. estimated the entire fire-induced emissions across North America and Central America for the calendar year 2004 [7]. These authors estimated 356 Tg and 227 Tg yearly total CO_2 emissions for the United States and Canada, respectively. The yearly total CO_2 emissions for the United States are only 3.6 times and 14.7 times our resulting annual CO_2 emissions of the GXM for 1986–2010 and for 2005–2010, respectively. However, the area of the United States is 29.4 times that of our study domain. Table 4 indicates that the yearly total CO_2 emissions in the United States are much higher than in China, by approximately 3–9-fold, according to different estimates [25,49,50]. Other emission species in the United States also show a similar relation with China. Estimating trace gas emissions from biomass burning is a complex issue due to the many variables that are involved in the process. The level of uncertainty includes several factors directly related to the working scale and to each of the variables involved in the models [26]. Notably, there was a significant difference in the emissions results among different studies, even within the same study domain [25,49,50]. Compared with the CO_2 emission for all of China's, the average annual CO_2 emission (2005–2010) accounts for 24%, according to Song et al. [50], 30.3%, according to Yan et al. [49] and 59.3%, according to Lu et al. [25].

Table 4. Comparison of our biomass burning estimations with several other published estimations (Tg/year).

Literature Cited	Regions	Area ratio [1]	CO_2	CH_4	CO	PM_{10}	$PM_{2.5}$	SO_2	BC	NO_x
[7] [2]	United states	29.4:1	356	1.0	19.8	2.7	2.4	0.16	-	0.68
	Canada	30.5:1	227	0.64	12.7	1.8	1.6	0.11	-	0.43
[50] [3]	China	29.5:1	79.7	0.24	5.4	-	0.66	0.05	0.03	0.2
[23] [4]	China	29.5:1	40.7	0.11	2.7	-	-	-	-	-
[25] [5]	China	29.5:1	102	0.27	6.8	-	0.69	0.05	0.06	0.2
This study [6]	GXM	1:1	99.51	0.26	52.51	0.76	0.68	0.03	0.04	0.19
This study [7]	GXM	1:1	24.15	0.06	1.27	0.18	0.16	0.05	0.02	0.05

[1] The area ratio of other study domains and the Great Xing'an Mountains (GXM); [2] calendar year 2004 estimated by [7]; [3] calendar year 2000 estimated by [49]; [4] average from calendar years 1950 to 2000 estimated by [23]; [5] fire year 2000 (April 2000 to March 2001) estimated by [47]; [6] average annual value of the period from 1986 to 2010; [7] average annual value of the period from 2005 to 2010.

4. Estimates Challenges

Modeling methodologies are the most realistic way to accomplish emissions estimations, and remote-sensing data acquisition seems to be a feasible way to estimate those parameters required for the models. These approaches depend on factors such as the scale, accuracy requirements and information availability, among others. The variables involved in most of the models are the burned area, fuel availability, burning efficiency, and emission factors. In this study, we assumed that the emission model coefficients depend only on the type of dominating vegetation, which, in turn, is the function of geographic coordinates. In reality, the intensity of emissions during biomass burning is determined by a wider set off actors and can vary in the course of the fire risk season [51]. For example, the completeness of combustion is determined by the moisture content in combustible vegetation, which, in turn, depends on meteorological conditions (air temperature and precipitation) and seasonal factors (the time of snow cover thawing and moistening regime) in a given region [51,52]. Comparison of burned area or hotspot products often reveals factor of ten or larger disagreements [53]. Burning efficiency (BE) is also affected by instantaneous meteorological conditions on the burning days, such as wind and precipitation. Concerning the emission factors, data are available only for specific sites across the globe or for laboratory experiments. Although remote sensing may provide helpful insight into their temporal and spatial evolution in biomass burning, the accuracy of burned area extract from satellite image are need to further improve by using spectral data and vegetation properties. The use of common emission factors and biomass available for highly diverse environmental conditions introduces a high level of uncertainty into the calculations. In this sense, uncertainty is highly associated with the land cover inside class variability; areas with high variability show higher levels of uncertainty due to the difficulty in applying precise coefficients [26]. Now there are two satellites monitor Earth's greenhouse gas emissions from space, the one is NASA's Orbiting Carbon Observatory-2 (OCO-2) [54], which launched in 2014 and measures CO_2, and the other one is Japan's Greenhouse Gases Observing Satellite (GOSAT) [55], which launched in 2009 and tracks CO_2 and methane. However, it is very hard for those measurements to calculate the total trace gas emissions from specific fire events. This is because a polar orbiting satellite has a strict running period and orbit so only a fire in the area right under the satellite can be monitored, and it cannot make continuous observation on a single fire event. Satellites can monitor only the instantaneous values and cannot record the complete and entire values in a single fire event.

5. Conclusions and Recommendations

This study improves strategies that optimize input information by combining explicit spatio-temporal remote-sensing data with models to obtain emissions information. Biomass burning emissions in the boreal region yield an important contribution to the chemical budget of the troposphere. To assess the contribution of biomass burning to the emissions of atmospheric trace species in the GXM, which is the most severe fire-prone area in China, we estimated various fire activities by combining explicit spatio-temporal remote sensing data with models. We observed 9998 fire scars with a total area of 46,096 km^2 in the GXM between the years 1986 and 2010. The years 1987 and 2003 contributed 33.2% and 22.9%, respectively, in burned area during the 25 years. Fire activity is the strongest in May. Most large fires occurred in the northern region of the GXM between 50 °N and 54° N latitude due to much drier weather and higher fire danger in the northern region than in the southern region. Evergreen and deciduous needle leaf forest and deciduous broadleaf forest are the main sources of emissions, accounting for 84%, 81%, 84%, 87%, 89%, 86%, 85% and 74% of the total annual CO_2, CH_4, CO, PM_{10}, $PM_{2.5}$, SO_2, BC and NO_x emissions, respectively. Fire emissions from shrub land, grassland and cropland only account for a small fraction of the total emissions level (approximately 4%–11%). Comparisons of our results with other published estimates of biomass burning emissions show reasonable agreement; however, substantial uncertainties remain concerning the modeling parameters. The variability in emission factors greatly contributed to the uncertainty. There is an urgent requirement to obtain more accurate biomass burning estimates because of its considerable contribution to the regional and global carbon balance and to the atmosphere.

Acknowledgments: This study would not have been possible without the financial support from the Postdoctoral Science Foundation of China (Y5T0020022) and National Nature Science Foundation of China (41561099). We would like to thank University of Maryland for providing AVHRR Global Land Cover Map (AVHRR GLC-2000) data used in this research. The authors are grateful to the anonymous reviewers for their insightful and helpful comments to improve the manuscript.

Author Contributions: Kunpeng Yi and Yulong Bao conceived and designed the research; Kunpeng Yi performed the research; Kunpeng Yi and Yulong Bao analyzed the data; Yulong Bao contributed analysis tools; Kunpeng Yi wrote the paper. All authors have read and approved the final manuscript after minor modifications.

Conflicts of Interest: The authors declare no conflict of interest.

References

1. Keywood, M.; Kanakidou, M.; Stohl, A.; Dentener, F.; Grassi, G.; Meyer, C.P.; Torseth, K.; Edwards, D.; Thompson, A.M.; Lohmann, U.; et al. Fire in the air: Biomass burning impacts in a changing climate. *Crit. Rev. Environ. Sci. Technol.* **2013**, *43*, 40–83. [CrossRef]
2. Marshall, S.; Taylor, J.A.; Oglesby, R.J.; Larson, J.W.; Erickson, D.J. Climatic effects of biomass burning. *Environ. Softw.* **1996**, *11*, 53–58. [CrossRef]
3. Jacobson, L.V.; Hacon, S.; Ignotti, E.; Castio, H.; Artaxo, P.; de Leon, A.P. Effects of air pollution from biomass burning in amazon: A panel study of schoolchildren. *Epidemiology* **2009**, *20*, S90–S90. [CrossRef]
4. Akagi, S.K.; Yokelson, R.J.; Wiedinmyer, C.; Alvarado, M.J.; Reid, J.S.; Karl, T.; Crounse, J.D.; Wennberg, P.O. Emission factors for open and domestic biomass burning for use in atmospheric models. *Atmos. Chem. Phys.* **2011**, *11*, 4039–4072. [CrossRef]
5. Knox, K.J.E.; Clarke, P.J. Fire severity, feedback effects and resilience to alternative community states in forest assemblages. *For. Ecol. Manag.* **2012**, *265*, 47–54. [CrossRef]
6. Ito, A.; Penner, J.E. Global estimates of biomass burning emissions based on satellite imagery for the year 2000. *J. Geophys. Res. Atmos.* **2004**, *109*, D14S05. [CrossRef]
7. Wiedinmyer, C.; Quayle, B.; Geron, C.; Belote, A.; McKenzie, D.; Zhang, X.; O'Neill, S.; Wynne, K.K. Estimating emissions from fires in north america for air quality modeling. *Atmos. Environ.* **2006**, *40*, 3419–3432. [CrossRef]
8. Ichoku, C.; Kahn, R.; Chin, M. Satellite contributions to the quantitative characterization of biomass burning for climate modeling. *Atmos. Res.* **2012**, *111*, 1–28. [CrossRef]
9. Zhang, X.; Hecobian, A.; Zheng, M.; Frank, N.H.; Weber, R.J. Biomass burning impact on PM2.5 over the southeastern US during 2007: Integrating chemically speciated frm filter measurements, modis fire counts and pmf analysis. *Atmos. Chem. Phys.* **2010**, *10*, 6839–6853. [CrossRef]

10. Tian, D.; Hu, Y.T.; Wang, Y.H.; Boylan, J.W.; Zheng, M.; Russell, A.G. Assessment of biomass burning emissions and their impacts on urban and regional PM2.5: A georgia case study. *Environ. Sci. Technol.* **2009**, *43*, 299–305. [CrossRef] [PubMed]

11. Barrera, V.A.; Miranda, J.; Espinosa, A.A.; Meinguer, J.; Martinez, J.N.; Ceron, E.; Morales, J.R.; Miranda, P.A.; Dias, J.F. Contribution of soil, sulfate, and biomass burning sources to the elemental composition of PM10 from Mexico city. *Int. J. Environ. Res.* **2012**, *6*, 597–612.

12. Liu, Y.; Goodrick, S.L.; Stanturf, J.A. Future US wildfire potential trends projected using a dynamically downscaled climate change scenario. *For. Ecol. Manag.* **2013**, *294*, 120–135. [CrossRef]

13. Crimmins, M.A. Wildfire and Climate Interactions Across the Southwest United States. Ph.D. Thesis, The University of Arizona, Tucson, AZ, USA, 2004.

14. Aldersley, A.; Murray, S.J.; Cornell, S.E. Global and regional analysis of climate and human drivers of wildfire. *Sci. Total Environ.* **2011**, *409*, 3472–3481. [CrossRef] [PubMed]

15. Allen, A.G.; Miguel, A.H. Biomass burning in the Amazon: Characterization of the ionic component of aerosols generated tom flaming and smoldering rain-forest and savanna. *Environ. Sci. Technol.* **1995**, *29*, 486–493. [CrossRef] [PubMed]

16. Delmas, R.; Lacaux, J.P.; Menaut, J.C.; Abbadie, L.; Leroux, X.; Helas, G.; Lobert, J. Nitrogen compound emission from biomass burning in tropical african savanna fos/decafe 1991 experiment (lamto, ivory coast). *J. Atmos. Chem.* **1995**, *22*, 175–193. [CrossRef]

17. Lacaux, J.P.; Brustet, J.M.; Delmas, R.; Menaut, J.C.; Abbadie, L.; Bonsang, B.; Cachier, H.; Baudet, J.; Andreae, M.O.; Helas, G. Biomass burning in the tropical savannas of ivory coast: An overview of the field experiment fire of savannas (fos/decafe 91). *J. Atmos. Chem.* **1995**, *22*, 195–216. [CrossRef]

18. Simoneit, B.R.T. Biomass burning—A review of organic tracers for smoke from incomplete combustion. *Appl. Geochem.* **2002**, *17*, 129–162. [CrossRef]

19. Randerson, J.T.; Chen, Y.; van der Werf, G.R.; Rogers, B.M.; Morton, D.C. Global burned area and biomass burning emissions from small fires. *J. Geophys. Res. Biogeosci.* **2012**, *117*, G04012. [CrossRef]

20. Zhang, X.; Kondragunta, S. Temporal and spatial variability in biomass burned areas across the USA derived from the goes fire product. *Remote Sens. Environ.* **2008**, *112*, 2886–2897. [CrossRef]

21. Kasischke, E.S.; Hewson, J.H.; Stocks, B.; van der Werf, G.; Randerson, J. The use of atsr active fire counts for estimating relative patterns of biomass burning—A study from the boreal forest region. *Geophys. Res. Lett.* **2003**, *18*, 1969. [CrossRef]

22. Giglio, L.; Csiszar, I.; Justice, C.O. Global distribution and seasonality of active fires as observed with the terra and aqua moderate resolution imaging spectroradiometer (modis) sensors. *J. Geophys. Res.* **2006**, *111*, G02016. [CrossRef]

23. Qin, Y.; Xie, S.D. Historical estimation of carbonaceous aerosol emissions from biomass open burning in china for the period 1990–2005. *Environ. Pollut.* **2011**, *159*, 3316–3323. [CrossRef] [PubMed]

24. Huang, X.; Li, M.M.; Friedli, H.R.; Song, Y.; Chang, D.; Zhu, L. Mercury emissions from biomass burning in China. *Environ. Sci. Technol.* **2011**, *45*, 9442–9448. [CrossRef] [PubMed]

25. Lu, A.F.; Tian, H.Q.; Liu, M.L.; Liu, J.Y.; Melillo, J.M. Spatial and temporal patterns of carbon emissions from forest fires in China from 1950 to 2000. *J. Geophys. Res. Atmos.* **2006**, *5*, D05313. [CrossRef]

26. Palacios-Orueta, A.; Chuvieco, E.; Parra, A.; Carmona-Moreno, C. Biomass burning emissions: A review of models using remote-sensing data. *Environ. Monit. Assess.* **2005**, *104*, 189–209. [CrossRef] [PubMed]

27. Oberheitmann, A. CO_2-emission reduction in China's residential building sector and contribution to the national climate change mitigation targets in 2020. *Mitig. Adapteg. Strat. Glob. Change* **2012**, *17*, 769–791. [CrossRef]

28. Oberheitmann, A. A new post-kyoto climate regime based on per-capita cumulative CO_2-emission rights—Rationale, architecture and quantitative assessment of the implication for the CO_2-emissions from china, india and the annex-i countries by 2050. *Mitig. Adapteg. Strat. Glob. Chang.* **2010**, *15*, 137–168. [CrossRef]

29. Hasanbeigi, A.; Morrow, W.; Masanet, E.; Sathaye, J.; Xu, T.F. Energy efficiency improvement and CO_2 emission reduction opportunities in the cement industry in China. *Energ. Policy* **2013**, *57*, 287–297. [CrossRef]

30. Zhang, Y.; Hu, H.Q.; Wang, Q. Carbon emissions from forest fires in great Xing'an mountains from 1980 to 2005. *Procedia Environ. Sci.* **2011**, *10*, 2505–2510. [CrossRef]

31. Li, X.; He, H.S.; Wu, Z.; Liang, Y.; Schneiderman, J.E. Comparing effects of climate warming, fire, and timber harvesting on a boreal forest landscape in northeastern China. *PLoS ONE* **2013**, *8*, e59747. [CrossRef] [PubMed]
32. Seiler, W.; Crutzen, P.J. Estimates of gross and net fluxes of carbon between the biosphere and the atmosphere from biomass burning. *Clim. Chang.* **1980**, *2*, 207–247. [CrossRef]
33. Hoffmann, W.A.; Geiger, E.L.; Gotsch, S.G.; Rossatto, D.R.; Silva, L.C.R.; Lau, O.L.; Haridasan, M.; Franco, A.C. Ecological thresholds at the savanna-forest boundary: How plant traits, resources and fire govern the distribution of tropical biomes. *Ecol. Lett.* **2012**, *15*, 759–768. [CrossRef] [PubMed]
34. Moreno, J.M.; Viedma, O.; Zavala, G.; Luna, B. Landscape variables influencing forest fires in central Spain. *Int. J. Wildland Fire* **2011**, *20*, 678–689. [CrossRef]
35. Zinck, R.D.; Johst, K.; Grimm, V. Wildfire, landscape diversity and the drossel-schwabl model. *Ecol. Model.* **2010**, *221*, 98–105. [CrossRef]
36. Liu, Z.; He, H.S.; Chang, Y.; Hu, Y. Analyzing the effectiveness of alternative fuel reductions of a forested landscape in northeastern China. *For. Ecol. Manag.* **2010**, *259*, 1255–1261. [CrossRef]
37. Prasad, V.K.; Kant, Y.; Gupta, P.K.; Elvidge, C.; Badarinath, K.V.S. Biomass burning and related trace gas emissions from tropical dry deciduous forests of India: A study using dmsp-ols data and ground-based measurements. *Int. J. Remote Sens.* **2002**, *23*, 2837–2851. [CrossRef]
38. Ward, D.; Susott, R.; Kauffman, J.; Babbitt, R.; Cummings, D.; Dias, B.; Holben, B.; Kaufman, Y.; Rasmussen, R.; Setzer, A. Smoke and fire characteristics for cerrado and deforestation burns in Brazil: Base-b experiment. *J. Geophys. Res. Atmos.* **1992**, *97*, 14601–14619. [CrossRef]
39. Kasischke, E.S.; Stocks, B.J.; O'Neill, K.; French, N.H.; Bourgeau-Chavez, L.L. Direct effects of fire on the boreal forest carbon budget. In *Biomass Burning and Its Inter-Relationships with the Climate System*; Springer: Dordrecht, The Netherlands, 2000; pp. 51–68.
40. Michel, C.; Liousse, C.; Grégoire, J.M.; Tansey, K.; Carmichael, G.; Woo, J.H. Biomass burning emission inventory from burnt area data given by the spot-vegetation system in the frame of trace-p and ace-asia campaigns. *J. Geophys. Res. Atmos.* **2005**, *110*, D09304. [CrossRef]
41. Song, Y.; Liu, B.; Miao, W.J.; Chang, D.; Zhang, Y.H. Spatiotemporal variation in nonagricultural open fire emissions in China from 2000 to 2007. *Glob. Biogeochem. Cycles* **2009**, *23*, GB2008. [CrossRef]
42. Wang, X.P.; Fang, J.Y.; Zhu, B. Forest biomass and root-shoot allocation in northeast China. *For. Ecol. Manag.* **2008**, *255*, 4007–4020. [CrossRef]
43. Vasileva, A.; Moiseenko, K. Methane emissions from 2000 to 2011 wildfires in northeast Eurasia estimated with modis burned area data. *Atmos. Environ.* **2013**, *71*, 115–121. [CrossRef]
44. Cahoon, D.R.; Stocks, B.J.; Levine, J.S.; Cofer, W.R.; Pierson, J.M. Satellite analysis of the severe 1987 forest fires in northern China and southeastern Siberia. *J. Geophys. Res. Atmos.* **1994**, *99*, 18627–18638. [CrossRef]
45. Lin, Y.; Wang, L. *Typical Cases of Forest Fire in China from 1953 to 2005*; Chinese Forestry Press: Beijing, China, 2007.
46. Balzter, H.; Gonzalez, M.C.; Gerard, F.; Riaño, D. Post-fire vegetation phenology in Siberian burn scars. *Geosci. Remote Sens. Symp.* **2007**, 4652–4655. [CrossRef]
47. Van Leeuwen, T.T.; van der Werf, G.R. Spatial and temporal variability in the ratio of trace gases emitted from biomass burning. *Atmos. Chem. Phys.* **2011**, *11*, 3611–3629. [CrossRef]
48. Sheng, Y.W.; Smith, L.C.; MacDonald, G.M.; Kremenetski, K.V.; Frey, K.E.; Velichko, A.A.; Lee, M.; Beilman, D.W.; Dubinin, P. A high-resolution GIS-based inventory of the west siberian peat carbon pool. *Glob. Biogeochem. Cycles* **2004**, *18*, GB3004. [CrossRef]
49. Roy, D.P.; Jin, Y.; Lewis, P.E.; Justice, C.O. Prototyping a global algorithm for systematic fire-affected area mapping using modis time series data. *Remote Sens. Environ.* **2005**, *97*, 137–162. [CrossRef]
50. Yan, X.Y.; Ohara, T.; Akimoto, H. Bottom-up estimate of biomass burning in mainland China. *Atmos. Environ.* **2006**, *40*, 5262–5273. [CrossRef]
51. Frey, K.E.; Smith, L.C. Amplified carbon release from vast west Siberian peatlands by 2100. *Geophys. Res. Lett.* **2005**, *32*, L09401. [CrossRef]
52. Vivchar, A.V.; Moiseenko, K.B.; Pankratova, N.V. Estimates of carbon monoxide emissions from wildfires in northern Eurasia for airquality assessment and climate modeling. *Izv. Atmo. Ocean. Phys.* **2010**, *46*, 281–293. [CrossRef]

53. Urbanski, S.P.; Hao, W.M.; Nordgren, B. The wildland fire emission inventory: Western United States emission estimates and an evaluation of uncertainty. *Atmos. Chem. Phys.* **2011**, *11*, 12973–13000. [CrossRef]

54. Taylor, T.E.; O'Dell, C.W.; Frankenberg, C.; Partain, P.T.; Cronk, H.Q.; Savtchenko, A.; Nelson, R.R.; Rosenthal, E.J.; Chang, A.Y.; Fisher, B.; et al. Orbiting carbon observatory-2 (OCO-2) cloud screening algorithms: Validation against collocated modis and caliop data. *Atmos. Meas. Tech. Discuss.* **2016**, *9*, 973–989. [CrossRef]

55. Takagi, H.; Saeki, T.; Oda, T.; Saito, M.; Valsala, V.; Belikov, D.; Saito, R.; Yoshida, Y.; Morino, I.; Uchino, O.; et al. On the benefit of gosat observations to the estimation of regional CO_2 fluxes. *Sola* **2011**, *7*, 161–164. [CrossRef]

© 2016 by the authors. Licensee MDPI, Basel, Switzerland. This article is an open access article distributed under the terms and conditions of the Creative Commons Attribution (CC BY) license (http://creativecommons.org/licenses/by/4.0/).

forests

MDPI

Article

Influence of Fuel Load Dynamics on Carbon Emission by Wildfires in the Clay Belt Boreal Landscape

Aurélie Terrier [1,*], Mathieu Paquette [1], Sylvie Gauthier [2], Martin P. Girardin [2], Sylvain Pelletier-Bergeron [3] and Yves Bergeron [1,4]

[1]   Chaire Industrielle en Aménagement Forestier Durable (NSERC-UQAT-UQAM), Université du Québec à Montréal; P.O. Box 8888, Stn. Centre-ville, Montréal, QC H3C 3P8, Canada; paqmat@gmail.com
[2]   Natural Resources Canada, Canadian Forest Service, Laurentian Forestry Centre, 1055 du PEPS, P.O. Box 10380, Stn. Sainte-Foy, Québec, QC G1V 4C7, Canada; sylvie.gauthier2@canada.ca (S.G.); martin.girardin@canada.ca (M.P.G.)
[3]   Département des Sciences du Bois et de la Forêt, Université Laval, 2405 de la Terrasse, Québec, QC G1V 0A6, Canada; Sylvain.Pelletier-Bergeron@mffp.gouv.qc.ca
[4]   Forest Research Institute, Université du Québec en Abitibi-Temiscamingue, 445 blvd de l'Université, Rouyn-Noranda, QC J9X 5E4, Canada; yves.bergeron@uqat.ca
*   Correspondence: terrier.aurelie@courrier.uqam.ca.; Tel.: +1-514-987-3000 (ext. 7608)

Academic Editor: Timothy A. Martin
Received: 27 September 2016; Accepted: 18 December 2016; Published: 24 December 2016

Abstract: Old-growth forests play a decisive role in preserving biodiversity and ecological functions. In an environment frequently disturbed by fire, the importance of old-growth forests as both a carbon stock as well as a source of emissions when burnt is not fully understood. Here, we report on carbon accumulation with time since the last fire (TSF) in the dominant forest types of the Clay Belt region in eastern North America. To do so, we performed a fuel inventory (tree biomass, herbs and shrubs, dead woody debris, and duff loads) along four chronosequences. Carbon emissions by fire through successional stages were simulated using the Canadian Fire Effects Model. Our results show that fuel accumulates with TSF, especially in coniferous forests. Potential carbon emissions were on average 11.9 t·ha^{-1} and 29.5 t·ha^{-1} for old-growth and young forests, respectively. In conclusion, maintaining old-growth forests in the Clay Belt landscape not only ensures a sustainable management of the boreal forest, but it also optimizes the carbon storage.

Keywords: boreal forest; fuel load dynamics; fire behavior; carbon emission modelling; sustainable management; mitigation management

---

1. Introduction

Forest management that optimizes carbon storage in the boreal forest could be an effective tool to mitigate anthropogenic carbon emissions [1]. By storing more than one third of the global terrestrial carbon [2,3], the boreal biome is the world's largest carbon stock. However, limited by the cold temperature and short growing season, boreal forests' annual carbon uptake (0.004 Pg·C/m^2·year) is low in comparison with the tropical forests (0.008 Pg·C/m^2·year) or the temperate forests (0.009 Pg·C/m^2·year) [4]. Fortunately, large amounts of carbon can accumulate on the boreal forest's floor since the decomposition of dead material is also limited by the cold climate conditions that prevail therein [2]. Although disturbances emit large quantities of carbon, the boreal forest is a carbon sink since regeneration and time between disturbances allows total recovery of emitted carbon [5].

In Canada, the boreal forest represents approximately 60% of the economic resources for the forest industry [6]. In addition, boreal ecosystems provide a diverse range of habitats for wildlife [6]. During the last few decades, the expansion of intensive harvesting has prompted the need to find a compromise between preserving biodiversity, maintaining forest ecosystems resilience, and providing resources for the wood and paper industries [6]. Facing these needs, interests for a sustainable forest management approach based on natural fire dynamics has increased [7–10]. Natural fires shape landscape structures and composition by creating forests of various ages and by favoring the conservation of shade-intolerant species (e.g., jack pine, trembling aspen) in areas with high fire activity and of tolerant species (e.g., balsam fir) in landscapes rarely influenced by fire [6–9]. Forest management that recreates landscapes similar to those generated by natural fire preserves biodiversity and long-term ecosystem functionality [10–13].

A major challenge in sustainable management lies in maintaining forest structures the way they occur under natural conditions. Several studies have highlighted old-growth forests' importance in preserving biodiversity and ecological functions [10,14,15], for example, by supporting the presence of specialized plant and animal species or by favoring forest resilience. While forest harvesting has resulted in the creation of younger landscapes in the eastern part of North America [14,16,17], sustainable forest management aims to promote silvicultural practices that maintain late-successional forests, or forests that have the same characteristics [10,16–18]. These strategies aim to ensure forest sustainability, but their relevance can be brought into question in a mitigation management context under climate change. In old boreal forests, carbon has had time to accumulate for several decades [17]; when disturbed by fire, 40% [19] to 60% [20] of the carbon accumulated in boreal forests is likely to be emitted back into the atmosphere [21]. This phenomenon is likely to be amplified over the next century under global warming owing to the more frequent and more severe drought conditions [22], which may directly increase wildfire activity (e.g., [23–25]) and reduce tree growth rates [26]. These changes will affect the time needed for forests to recover the carbon released [5]. Given their large carbon pools and the anticipated risks posed to these pools under climate change, maintaining old-growth forests, as opposed to restoring productivity through harvesting, could have adverse effects on the global climate. However, old-growth forests are characterized by a particular fuel structure that tends to maintain higher moisture levels in comparison with younger forest stands [27–29]. This fuel structure may provide old-growth forests with some resistance to fire regime changes.

This study estimates the potential amount of carbon released into the atmosphere by fire activity along successional stages for the dominant forests (trembling aspen (TA), jack pine (JP), and black spruce (BS) forests) of the Clay Belt boreal forest in eastern North America. The Clay Belt region constitutes one of the world's largest terrestrial carbon stocks, estimated at 201 to 250 $t \cdot ha^{-1}$, with a large proportion of forests reaching values of up to 1050 $t \cdot ha^{-1}$ [30]. Therefore, it is justified to assess carbon emissions caused by fire disturbance specifically in the Clay Belt forest. The notion of the impact of fuel accumulation with the time since the last fire (TSF) on fire behavior was first mentioned by Brown [31] for *Pinus ponderosa* ecosystems in the Interior West of the United States. The policy of fire exclusion has created a situation in which fuel accumulates, resulting into catastrophic and severe fire events [32], and induced higher carbon emissions. In 2001, Johnson et al. [33] showed that this notion is not valid for boreal ecosystems. Fire cycles (time required to burn an area equal in size to the study area) in these ecosystems are longer than those in *Pinus ponderosa* ecosystems. Fuel accumulates and does not constitute a limiting factor for fire events: large severe fires in these ecosystems occur under extreme climatic conditions whatever the forest. Studies conducted in Clay Belt BS forests show that peat mosses accumulate with the TSF, thereby inducing a decrease in tree productivity (tree fuel load) [34]. Organic layer accumulation leads to a rise in the water table [35] and creates high soil moisture conditions [34]. Depth of burn (i.e., the depth of the soil organic layer consumed during a fire) is consequently low [36–39] and the fire only burns surface fuel. However, these studies did not consider all fuel material and forest types. First, we wanted to address the hypothesis that fuel accumulates with TSF in the dominant Clay Belt boreal forests, considering all fuel material. Second, it can be postulated that the high soil moisture content and low depth of burn in BS forests may induce less carbon emissions, while the higher fuel availability in TA and JP forests could induce higher carbon emissions. From this hypothesis, we verified whether or not fuel accumulation could induce higher carbon emission by fires in late-successional TA and JP forests, and not in black spruce forests. To do so, we performed a complete fuel inventory (tree biomass, herbs and shrubs, dead woody debris (DWD), and duff loads) in 61 sites differing in their forest type and TSF. Empirical fuel loads were then used to calculate the carbon emitted by one single simulated fire at each sampling site using the Canadian Fire Effects Model (CanFIRE) [40,41].

2. Materials and Methods

2.1. Study Area

The study area (49°00′–50°00′ N, 78°30′–80°00′ W) is located in the Clay Belt boreal forest in eastern North America, stretching across the Quebec and Ontario border (Figure 1).

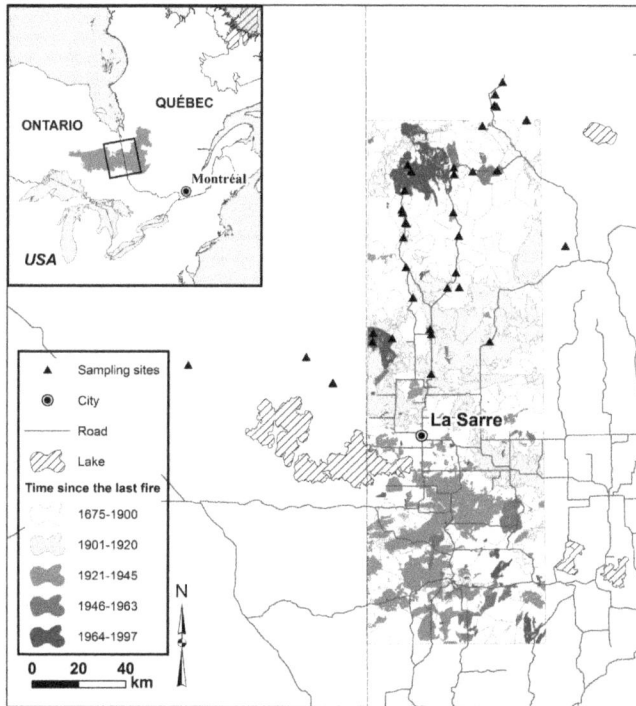

Figure 1. Geographic location of the sampling sites in the Clay Belt boreal forest, eastern North America.

A former proglacial lake (Lake Barlow-Ojibway) left a thick deposit of clay, forming the physiographic unit known today as the Clay Belt, which covers an area of approximately 145,470 km^2 [42]. The current level of fire activity is low, with a fire cycle estimated at 398 years from 1959 to 1999 [43]. The climate is subpolar and subhumid continental, characterized by long, harsh, and dry winters and, short, hot, and humid summers [44]. Average annual temperature from 1971 to 2000 recorded at the closest weather station to the north (Matagami, 49°46′ N, 77°49′ W) and to the south (La Sarre, 48°46′ N, 79°13′ W) was −0.7 °C and 0.7 °C, respectively. Mean total annual precipitation was 906 and 890 mm, respectively [44]. The poor drainage conditions induced by the presence of an impermeable clay substrate, flat topography, historical low fire activity, and cold climate facilitated the accumulation of thick layers of organic soil, a process often described as paludification [45,46]. In parts of the region, peat mosses accumulate on initially mesic soils, independently of topography or drainage, and are related to forest succession [34]. Therefore, in the prolonged absence of fire, these forests tend to convert into less productive spruce-*Sphagnum* opened forests regardless of the initial species composition [34,47–50]. Burned area and residual organic layers (i.e., layers that are not consumed by fire) jointly control forest structure and composition [51]. Shallow residual organic layers on the ground lead to the establishment of dense forests composed of black spruce (*Picea mariana* (Mill.) BSP), trembling aspen (*Populus tremuloïdes* Michx.), or jack pine (*Pinus banksiana* Lamb.) on mesic sites [48,51–53]. In contrast, thick residual organic layers favor black spruce self-replacement [53,54] and accelerate the process of paludification [34,51].

2.2. Fuel Load Dynamics with TSF

Data were collected to characterize trees, herbs, shrubs, DWD, and duff loads (t·ha^{-1}) along the successional stages of the dominant Clay Belt forests (TA, JP, BS; see Figure 5 in [48]). Black spruce forests were separated into black spruce forests originating from severe fire (BS-S) and black spruce forests originating from non-severe fire (BS-NS), since fire severity influences tree and duff load dynamics in BS forests [34]. Successional stages were determined with TSF. We used information (tree basal area, organic layer depth, TSF) that was already available for 37 sites to calculate tree and duff loads [34,48,51]. We visited these sites to complete the inventory with information on DWD, shrubs, and herbs. We selected 24 additional sites to expand TSF variability for each forest type (Table 1). TSF was determined for these additional sites by overlaying a fire reconstruction map previously published for the study area [43]. A total of 61 sites were sampled. At each site, a 30-m-sided equilateral triangle was defined in which the fuel sampling protocol was performed.

Table 1. Distribution of study sites in different successional stages and forest types.

TSF (Years)	Forest Type			
	TA	JP	BS-S	BS-NS
<90	6	7	5	5
90–150	4	1	5	5
>150	2	7	8	6
Total	**12**	**15**	**18**	**16**

Values express the number of sites. TSF: time since the last fire; TA: trembling aspen; JP: jack pine; BS-S: black spruce originating from severe fire; BS-NS: black spruce originating from non-severe fire.

Basal area for live trees larger than 3 cm in diameter at breast height (DBH) were sampled by species for the 24 additional sampling sites using the prism method (factor 2, metric). Tree loads were calculated using stand-level equations [55]. Shrub load was sampled using the method described in Brown et al. [56]. Nine 1-m^2 plots were established evenly along the triangle transect. Shrub species were identified and stem basal diameters (cm) were measured. Loads were calculated using equations linking shrub species weight in grams with the stem diameter. Previously determined equations [57] were used for *Lonicera canadensis* Bartr., *Ribes* sp., *Rosa acicularis* Lindl., and *Viburnum edule* Raf. New equations were determined with additional shrub samples for *Chamaedaphne calyculata* (L.) Moench, *Kalmia angustifolia* L., *Rhododendron groenlandicum* (Oeder) K.A. Kron & Judd, and *Vaccinium myrtilloïd*es Michx. (see details in Appendix A, Table A1). We measured herb loads (all surface vegetation that is not woody (e.g., forbs and graminoids)) using a weighted-estimate approach [56]. We established four plot rectangles (0.5 m^2), three at the center of each side and one in the middle of the triangle. We selected the plot with the most weight, which was identified as the base plot. Herbaceous vegetation was collected in this base plot and oven dried at 95 °C for 24 h for dry weight measurements. The weight of the three other plots was estimated as a fraction of the base plot and fuel load was calculated using Brown et al.'s equation [56]. Fuel load of DWD was sampled using the line intersect method [58] along the triangle sides. Pieces smaller than 7 cm were measured with a "go-no-go" gauge, and counted according to five diameter sizes (class I: 0–0.49 cm; class II: 0.5–0.99 cm, class III: 1–2.99 cm, class IV: 3–4.99 cm; class V: 5–6.99 cm). For pieces bigger than 7 cm, we measured the diameter size. DWD fuel loads were calculated for each size classes using McRae et al.'s equations [58]. Size-class specific coefficients values were extracted from Hély et al. [57] for white cedar, and from McRae et al. [58] for the other species (BS, JP, TA, WB). Finally, duff loads were calculated by multiplying total organic layer depth by bulk density. We used organic layer depths data from Simard et al. [34] for the 37 previously sampled sites. Additional organic layer depth was measured for the other 24 sites in parallel with shrub sampling in each of the nine 1-m^2 plots and averaged by sites. We used previously published mean bulk density (kg·m^{-3}) information [36] to convert organic layer depth into duff load (kg·m^{-2}). Duff data were missing for old BS-NS sites because of technical problems.

The relationship between TSF and (i) the fuel load of each material separately and (ii) total fuel load was investigated by forest type. We performed polynomial regressions using the R freeware [59]. Herb and shrub fuel loads were combined into fine aerial vegetation. DWD corresponded to the sum of all classes. Loads were logarithmically transformed (ln + 1) to linearize the relationship. First- to third-order polynomial regressions were tested, and we promoted the significant minimum-order polynomial regressions (p-values $\leq$ 0.05) in accordance with the parsimony principle.

2.3. Simulation of Carbon Emissions by Fire

Potential carbon emission (tons/ha) was investigated using the Canadian Fire Effects Model (CanFIRE, formerly the Boreal Fire Effects Model, BORFIRE) [40,41], including the modified depth of burn equations for BS forests dominated by *Sphagnum* spp. in the Quebec Clay Belt boreal landscape [36,60]. CanFIRE is a collection of Canadian fire behavior models that are used to estimate first-order fire effects on physical characteristics, and to estimate ecological effects at the stand level. For any given fire, CanFIRE calculates fire behavior information (e.g., fire intensity, rate of spread) based on the pre-fire amount of fuel and components of the Canadian Forest Fire Weather Index (FWI) System [61]. Tree fuel consumption takes place only when CanFIRE predicts a crown fire. Values are estimated as the sum of foliage and bark using tree biomass algorithms [62] and are comparable to overstory fuel consumption data recorded on experimental burns in the Canadian Forest Fire Behavior Prediction (FBP) System database [63]. DWD and duff consumption follows McRae et al.'s [64] and de Groot et al.'s [65] equations, respectively. These algorithms are driven by the Buildup Index (BUI) and Drought Code (DC) values of the FWI System for DWD and duff consumption, respectively, and were built from empirical pre-burn and post-burn fuel data collected in prescribed fires.

Since simulation objectives were to analyze differences in carbon emission with fuel structure variations under similar climatic conditions, we simulated the potential carbon emitted by one single simulated fire, using the same fire weather conditions at each sampling site. Sensitivity analyses were made to ensure that fire weather values above average did not modify our conclusions (see details in Appendix B). We determined FWI System component values by averaging daily FWI System components for each natural fire start point encountered during the interval 1971–2000 period in the study area (FWI = 18) [60]. Forest cover was determined based on species' empirical basal area as described in the previous section. Since basal area was sampled for trees with DBH larger than 3 cm, basal area, and, consequently, cover were 0 for all species in young sites. In this case, we attributed a value of 100% to the dominant species of the forest type (e.g., cover: 100% of jack pine in JP). Tree fuel load was directly estimated by CanFIRE using species' basal area. To do so, species' site index and age were used to calculate basal area with Plonski's yield tables [65]. We used TSF for species age and the same site indexes as those used by Terrier et al. [60]. Prior to running simulations, we ensured that the use of Plonski's yield tables instead of our empirical basal area data would not bring a bias to our conclusions (see details in Figure C1). DWD and duff loads were provided by our empirical data. Total duff was separated into duff-fibric and duff-humus using the fibric layer equations described in Terrier et al. [60]. Since the database was not complete for all sites (technical problems in duff sampling in old BS-NS sites), a subset of 55 sites was used for the simulations. Simulation results were averaged by forest types and successional stage classes (TSF < 90 years; TSF between 90 and 150 years; TSF > 150 years).

Some additional changes were also made to the model. In our study area, high soil moisture conditions in BS forests dominated by *Sphagnum* spp. resulted in lower depth of burn, and, consequently, in lower forest floor fuel consumption (FFFC) in comparison with other boreal forest types [36]. In the CanFIRE model, FFFC is a function of duff load and climate conditions [37], and does not consider the lower depth of burn for calculations. Therefore, FFFC equations had to be modified to reflect specific carbon emission in BS forests dominated by *Sphagnum* spp. We defined FFFC as a function of the potential depth of burn as Equation (1):

$$FFFC\left(\frac{kg}{m^2}\right) = \text{depth of burn (m)} \times \text{bulk density } (\frac{kg}{m^3}) \tag{1}$$

Since bulk density varies with the organic layer depth [37], we came up with a model that linked measured bulk density with organic layer depth. To do so, we used a published dataset that comprised 103 peat measurements of bulk density at different organic layer depths sampled in 11 sites [34] (details are presented in Appendix D).

3. Results

3.1. Fuel Load Dynamics with TSF

TSF was significantly related to all fuel loads materials (trees, fine aerial vegetation, DWD, and duff) and total fuel loads for TA, JP, BS-S, and BS-NS boreal forests in the Clay Belt. Exceptions were observed for fine aerial vegetation in coniferous forests, duff load in TA forests, and DWD of JP forests (Figure 2, Table 2).

Tree fuel loads (Table 2; Figure 2A) were predicted to approximate 0 (t·ha^{-1}) just after fire (exponential of the polynomial regression intercept ~0), except for BS-NS, where the intercept equals 0.79. Tree fuel loads for coniferous forests increased gradually before decreasing after 150 and 200 years in JP and BS forests, respectively (Figure 2A). Maxima reached 5.47, 4.75, and 4.39 for JP, BS-S, and BS-NS, respectively. In the case of TA forests, tree fuel loads increased particularly rapidly at early successional stages until around 75 years (logarithmic maximum = 6.46), followed by a slight decrease. Finally, tree fuel loads increased again at late successional stages.

Fine aerial vegetation load did not vary significantly with TSF in coniferous forests (Figure 2B); logarithmic values varied near 0 along the successional stages (Table E1). In contrast, the amount of fine aerial vegetation was particularly high at early successional stages in TA forests. A value of 570 tons/ha was observed for the younger site (TSF = 11 years) (Table E1). Fuel loads decreased significantly, reached 0 at 120 years old (Table E1), and increased rapidly at late-successional stages.

DWD load dynamics for BS forests, represented in Figure 2C, followed the same trend as those observed for the tree load: an increase was predicted at early successional stages, maxima logarithmic values of 2.79 and 1.86 for BS-S and BS-NS, respectively, were reached at 200 years, and were followed by a decrease in values. DWD load trend in TA forests was the opposite of tree load: a value of 90 t·ha^{-1} was estimated in the youngest site (TSF = 11 years) (Table E1). Values decreased for 75 years, then increased slightly, and finally decreased gradually at late successional stages. DWD load for JP forests showed no significant relationship with TSF. Values ranged from 1 to 86 t·ha^{-1} (Table E1).

Duff load showed a significant gradual increase with TSF in coniferous forests. Higher values were predicted for BS-NS (Figure 2D). Values were on average 70, 125, 292 t·ha^{-1} for JP, BS-S, and BS-NS, respectively, in sites younger than 50 years old (Table E1). The increase in duff load was faster in BS forests in comparison with JP forests (higher equation slope in Table 2). Variation in duff load was not significantly related to TSF in TA forests. Values averaged 150 all along the chronosequence.

Finally, variation in total fuel load with TSF was significant for all forest types (Figure 2E). Total fuel load gradually increased along successional stages in coniferous forests. Values varied from

57 t·ha^{-1} to 382 t·ha^{-1} in JP, from 77 t·ha^{-1} to 767 t·ha^{-1} in BS-S, and from 160 t·ha^{-1} 1084 t·ha^{-1} in BS-NS (Table E1). Total fuel in TA forests decreased slightly at early successional stages, reaching a minimum at 140 years, and increased at late successional stages (Figure 2E). Values of total fuel ranged from 241 t·ha^{-1} to 730 t·ha^{-1} in TA forests (Table E1).

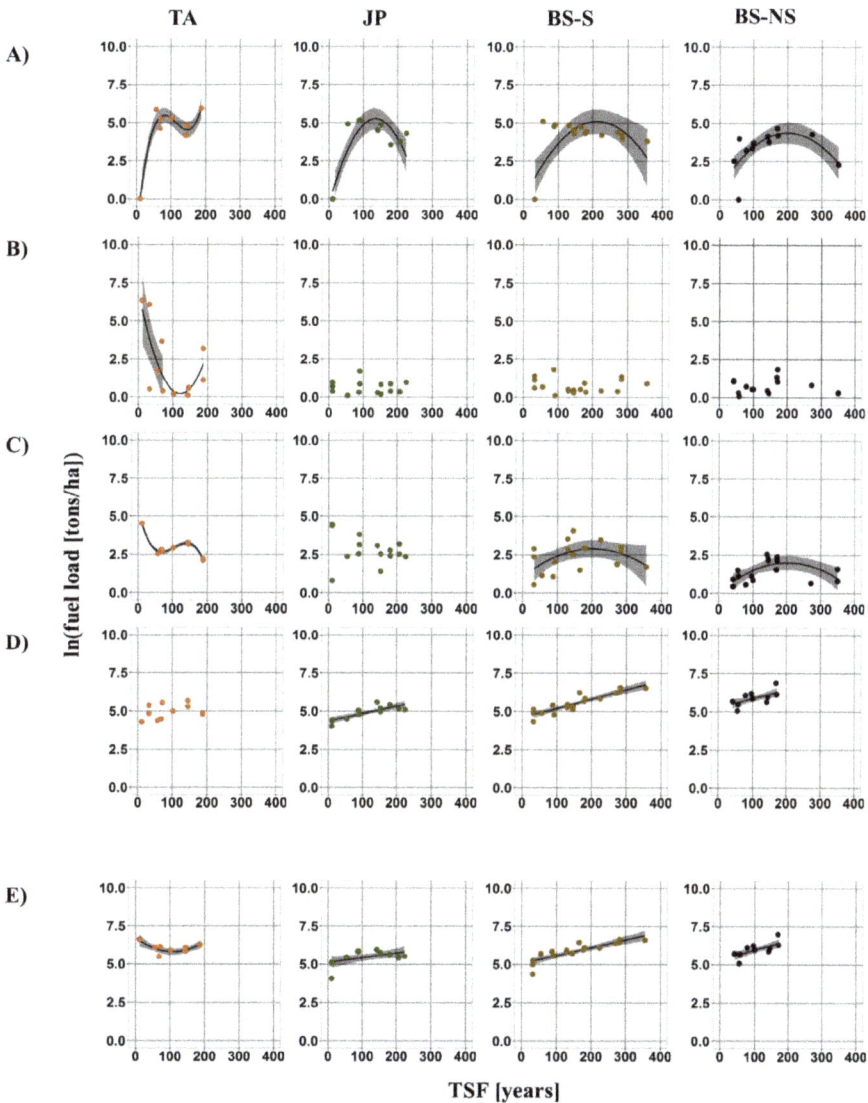

Figure 2. Changes in loads of (**A**) tree; (**B**) fine aerial vegetation (herbs and shrubs); (**C**) dead woody debris (DWD); (**D**) duff; and (**E**) total fuel (**A** + **B** + **C** + **D**) with TSF for TA, JP, BS-S, and BS-NS dominated forests. Fuel loads were logarithmically transformed and are represented by color points. The black line expresses the significant relationship (*p*-value ≤ 0.05) in the polynomial regression analysis and the grey shape corresponds to the regression standard error.

Table 2. *p*-value, adjusted R^2 and equations extracted from the regression polynomial analysis of tree, fine aerial vegetation (shrubs + herbs), DWD, duff, and total fuel loads with TSF for TA, JP, BS-S, and BS-NS dominated forests. NS stands for a non-significant relationship.

	Fuel	*p*-Value	R^2	Model
TA	Trees	≤0.005	0.92	$\ln(load) = -2.21 + 0.23 \times TSF - 0.002 \times TSF^2 + 0.000006 \times TSF^3$
	Fine aerial	≤0.05	0.47	$\ln(load) = 6.91 - 0.11 \times TSF + 0.00045 \times TSF^2$
	DWD	≤0.001	0.84	$\ln(load) = 5.31 - 0.093 \times TSF + 0.00099 \times TSF^2 - 0.000003 \times TSF^3$
	Duff	NS	NS	NS
	Total	≤0.05	0.49	$\ln(load) = 6.72 - 0.017 \times TSF + 0.00008 \times TSF^2$
JP	Trees	≤0.005 [1]	0.78	$\ln(load) = -0.38 + 0.084 \times TSF - 0.0003 \times TSF^2$
	Fine aerial	NS	NS	NS
	DWD	NS	NS	NS
	Duff	≤0.001	0.65	$\ln(load) = 4.35 + 0.0049 \times TSF$
	Total	≤0.05	0.19	$\ln(load) = 5.1 + 0.003 \times TSF$
BS-S	Trees	≤0.001 [1]	0.52	$\ln(load) = -0.049 + 0.048 \times TSF - 0.00012 \times TSF^2$
	Fine aerial	NS	NS	NS
	DWD	≤0.05 [1]	0.16	$\ln(load) = 1.03 + 0.018 \times TSF - 0.000046 \times TSF^2$
	Duff	≤0.001	0.81	$\ln(load) = 4.59 - 0.006 \times TSF$
	Total	≤0.001	0.71	$\ln(load) = 5.06 + 0.0052 \times TSF$
BS-NS	Trees	≤0.001 [1]	0.48	$\ln(Fuel\ load) = 0.79 + 0.036 \times TSF - 0.00009 \times TSF^2$
	Fine aerial	NS	NS	NS
	DWD	≤0.05 [1]	0.34	$\ln(Fuel\ load) = 0.06 + 0.019 \times TSF - 0.00005 \times TSF^2$
	Duff	≤0.05	0.38	$\ln(Fuel\ load) = 5.3 + 0.0055 \times TSF$
	Total	≤0.005	0.48	$\ln(Fuel\ load) = 5.34 + 0.006 \times TSF$

[1] Non-significant intercept.

3.2. Simulation of Carbon Emission by Fire

The CanFIRE model was used to simulate carbon emissions during a single fire along the different successional stages of the four dominant forest types (TA, JP, BS-N, and BS-NS) of the Clay Belt boreal forest. Simulation results are presented in Figure 3.

Mean carbon emissions by fire were estimated to range from 14.83 to 21.37 t·ha^{-1} in TA, from 17.97 to 24.7 t·ha^{-1} in JP, from 12.48 to 44.3 t·ha^{-1} in BS-S, and from 1.92 to 52.31 t·ha^{-1} in BS-NS. Simulated carbon emissions were on average lower for sites older than 150 years, except for JP forests, where carbon emission estimates averaged 16 t·ha^{-1} for forests younger than 90 years and 19 t·ha^{-1} for forests older than 150 years. A maximum average of 53 t·ha^{-1} of carbon emitted was simulated for BS-NS with TSF < 90 years, while a particularly low value (2 tons/ha) was simulated for older BS-NS. Aside from BS old-growth forests, potential carbon emission values simulated were lower for TA. Table 3 presents the contribution in percentage of each fuel material to carbon emission by fire. Trees barely contribute to emission, which is mostly generated by duff and DWD combustion.

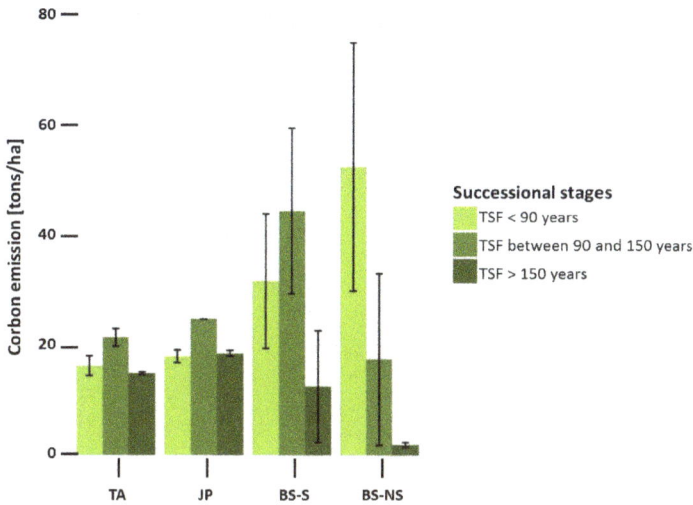

Figure 3. CanFIRE potential carbon emissions averaged by early (TSF < 90 years), intermediate (TSF between 90 and 150 years), and late (TSF > 150 years) successional stages, and by dominant forest types of the Clay Belt boreal landscape (trembling aspen (TA), jack pine (JP), black spruce originating from severe fire (BS-S), and black spruce originating from non-severe fire (BS-NS) forests). Error bars represent the mean standard error.

Table 3. Values of total potential carbon emission by forest type and successional stage, and contribution of each fuel category to the total emission.

Forest Type	Successional Stage	Total (t/ha)	Contribution of Fuel Category		
			Tree (%)	DWD (%)	Duff (%)
	TSF < 90 years	16.2	0	20	80
TA	TSF between 90 and 150 years	21.4	0	11	89
	TSF > 150 years	14.8	0	10	90
	TSF < 90 years	17.9	14	21	65
JP	TSF between 90 and 150 years	24.7	0	9	91
	TSF > 150 years	18.5	0	6	94
	TSF < 90 years	31.6	6	3	91
BS-S	TSF between 90 and 150 years	44.3	6	5	89
	TSF > 150 years	12.5	2	12	86
	TSF < 90 years	52.3	3	1	96
BS-NS	TSF between 90 and 150 years	17.4	3	6	91
	TSF > 150 years	1.9	0	53	47

4. Discussion

4.1. Fuel Load Dynamics with TSF

Our results confirmed our first hypothesis that fuel accumulates with TSF in Clay Belt coniferous forests, despite variations in the dynamics of the different fuel materials. These variations were confirmed and explained by previous studies on fuel load dynamics.

Tree dynamics strongly affect variations in fuel material with TSF [66–69]. It is therefore important to first understand Clay Belt tree dynamics to discuss fine aerial, DWD, and duff dynamics. Light, nutrients, and space are abundant just after a fire event, and vegetation can rapidly colonize disturbed sites [70]. Trees grow more or less rapidly depending on the composition [65] or depth of the residual organic layer [51,71]. Consequently, it was not surprising to predict in our study an increase in tree biomass at early successional stages for all forest types. This increase was faster for TA and JP forests, because growth rates are faster for these trees [65,72] and the low residual organic layers required for TA and JP postfire recruitment lead to the establishment of productive and dense forests [49,52–54]. In contrast, BS-NS forests showed the slowest increase due to the thick residual organic layers left after fire which favor unproductive open BS self-replacement [34,48,52–54]. Following the growth influx, a decrease in tree loads [34,73] was predicted at 80, 110, and 200 years for TA, JP, and BS forests, respectively. Over time, tree mortality starts, gaps are created, and shade-tolerant species appear in the canopy [74,75]. Our predictions concur with previous observations: Pothier et al. [76] estimated tree productivity loss after about 60 years for TA forest in Quebec; Belleau et al. [50] estimated composition changes to occur in the Clay Belt region at around 70 to 120 years for JP and TA forests, respectively, and at around 100 to 200 years for BS forests. In our analysis, TA experienced a recovery in tree biomass after 150 years. Low organic layers observed at late successional stages in TA forests probably made this productivity recovery possible.

Fine aerial vegetation establishment and growth rates are primarily limited by the understory light availability that is created by the overstory canopy [67–69]. The shady understory environment provided by coniferous species [77] results in less abundant fine aerial vegetation in coniferous forests [68,70]. Forest openings induced by the establishment of the second cohort was not associated with the development of fine aerial vegetation in old-growth coniferous forests because the cold soil temperature and nitrogen limitation induced by organic layer accumulation at the late successional stage [34] probably acted to prevent herbs and shrubs development. Herbs and shrubs dynamics in TA forests reflect higher light transmission by the deciduous canopy. At early successional stages, light availability decreases in response to tree growth, and fine aerial vegetation declines in abundance and growth [64]. Tree mortality at intermediate successional stages increases light availability, allowing understory recolonization by herbs and shrubs [70,78,79]. Late successional stages were consequently characterized by an increase in fine aerial vegetation.

DWD variations with TSF can be explained by coarse woody debris (CWD) dynamics (pieces > 7 cm) since CWD constitute most of the DWD loads [29,66]. CWD dynamics in TA forests followed the traditional U-shaped distribution summarized by Brassard and Chen [68]. Pre-disturbance and disturbance-generated woody debris that were not consumed generated a high CWD quantity at early successional stages. As TSF increased this material decomposed and a decrease in CWD occurred. Tree mortality at intermediate successional stages provided high loads of woody debris. Old-growth forests were characterized by a low quantity of CWD, as new cohort growth did not generate dead woody supply [68]. DWD debris dynamics in BS forests corresponded closely to tree fuel load dynamics, with an absence of DWD high value at early successional stages, compared with what was observed in TA forests. Our data showed an absence or low quantity of CWD in young BS forests.Such a feature in coniferous forests has already been mentioned in previous studies [66,80,81]. It may be induced by a faster decay rate after fire and the absence of pre-fire large-diameter trees [66]. CWD could also be buried in the moss layer in BS-NS due to the rapid growth of *Sphagnum* species [31]. No trend in DWD was observed in JP forests, although Brais et al. [82] found that DWD dynamics followed a 'U-shaped' temporal pattern in pure Clay Belt JP forests. We explain these differences in observations by the facts that our data cover a larger variability of TSF and our forests were not pure JP forests. Few studies have compared CWD dynamics in different cover types [69]. Differences in DWD dynamics between coniferous forests and TA forests may be induced by the lower pre-fire tree productivity (Figure 2) and the presence of woody debris with higher flammability [83–86] in coniferous forests.

Decomposition rates were higher in TA than in coniferous forests [80,86]; consequently, organic layer accumulation did not occur in TA forests. Conversely, duff loads increased gradually in coniferous choronosequences, since the lower decomposition rates associated with the paludification process favor organic layer accumulation [34,45,46]. The most interesting fact is that despite a tree and DWD load loss with TSF in coniferous forests, total fuel loads increased with TSF.

Scharlemann et al. [30] provided a global map representing carbon stocks in tree biomass and soils. As mentioned in the introduction, they estimated a higher carbon amount in the Clay Belt boreal forests, with values ranging from 201 to 250 t·ha^{-1} (402–500 t·ha^{-1} of fuel load), and forests reaching values up to 1050 t·ha^{-1} (2100 t·ha^{-1} of fuel load). Our values are in accordance with the authors: we observed fuel loads ranging from around 60 to 1080 t·ha^{-1}, with the majority of sites containing 200 to 500 t·ha^{-1}. This study confirmed the importance of the region for global carbon stocks.

4.2. Simulation of Carbon Emission by Fire

This research is the first that we are aware of to use empirical data that include both spatial and temporal variations in fuel loads to quantify fire carbon emission in boreal ecosystems. Our approach differs from other studies [20,27,28,87,88], as fuel loads were approximated by fuel type and did not consider that the temporal changes [20,27] or amount of carbon emitted were a function of the burned area [28,88] or of the landscape's total tree loads [87]. We estimated that mean carbon emission by fire ranged from 2 to 53 t·ha^{-1}, depending on the forest type and age category. These results are consistent with values provided by previous studies. Harden et al. [87] estimated long-term average for C exchange in boreal forests near Thompson, Manitoba, Canada. The values ranged from 11 to 25 t·ha^{-1} for JP forests and from 14 to 70 t·ha^{-1} for BS-feathermoss forests. No observed data were offered for BS-*Sphagnum* forests, but the authors simulated an average of 20 t·ha^{-1}. Simulated carbon emissions using the CanFIRE model in forests of Alberta, Canada, ranged from 11.3 to 42.6 t·ha^{-1}, depending on the month during which fire occurred [19]. Amiro et al. [28] estimated a mean of 20.1 t·ha^{-1} of fuel consumed by individual fire (carbon emissions of around 100 (t·ha^{-1})) in the boreal east shield from 1959 to 1999.

The most interesting point to emerge from our simulations is that despite fuel accumulation with TSF, old-growth forests emitted less carbon by fire on average, whatever the forest type. Pre-fire fuel structure determined the amount of carbon emitted by fire [27–29] since total fuel consumption in CanFIRE was a function of the amount of each fuel material. More specifically, as previously observed [27], more than 85% of the carbon emissions were generated by duff and DWD combustion. It was not surprising to simulate a low tree contribution to carbon emission, since fire does not consume trees, it kills them [89]. Emissions are consequently generated after the fire, through dead wood decomposition [90]. However, this particularity does not influence the conclusions of this study since tree fuel load is as low in late-successional than in early-successional forests. Variations in carbon emission through BS successional stages were mainly influenced by duff dynamics. As mentioned previously, organic layer accumulation leads to a rise in the water table [35] and creates high soil moisture conditions [35]. Depth of burn is consequently low [36–39]. Duff loads in TA and JP forests present little variation across successional stages, as carbon emission was determined by DWD dynamics. Old-growth forests emitted less carbon because DWD loads were lower (even if DWD variations with TSF were not significant in JP forests, values were lower in old-growth forests).

These results have important implications for forest management in the context of climate change mitigation. Managers should consider practices that favor mature forests for harvesting but also increase potential forest vulnerability to higher carbon emission and decrease long-term carbon storage [36]. In fact, although the climatic conditions expected by the end of the 21st century may induce an increase in fire activity in the Clay Belt boreal region (e.g., [22–24]), expansion of old-growth BS forests in the landscape is projected [60]. If fire activity over the next three to four decades remains similar to current levels, an increase in the proportion of forests characterized by high soil moisture conditions will occur [60]. These moisture conditions may provide landscape resilience to increased fires. Fuel should consequently continue to accumulate without generating more emissions during fires. Peatland protection (such as reducing or preventing peatland drainage) could be an alternative in order to increase forest resistance to fire and reduce future fire carbon emissions in eastern Canadian forests.

Finally, it is important to note that the CanFIRE model does not include fine aerial consumption in its calculations. However, this omission would not change the findings of this study. Fine aerial load is low for coniferous forests all along the successional stages. High fine aerial load in TA forests at early and late successional stages may amplify our results as moisture contents of this fuel reduce fire intensity and, therefore, carbon emissions.

5. Conclusions

Empirical fuel structure dynamics were examined and potential carbon emissions by wildfire were simulated along four dominant chronosequences of the Clay Belt boreal forest in eastern North America. Fuel structure was an important factor influencing carbon emission by the simulated fire, while fuel availability was not a determining factor. Given our results, we argue that maintaining old-growth forests in the Clay Belt landscape not only promotes a sustainable management of the boreal forest, it also optimizes carbon sink.

A considerable effort was made in this study to sample covering a wide variability in TSF, and our results showed a low value for TA in the TSF > 150 years category and for JP in the TSF between 90 and 150 years category. However, these conclusions may apply only to the region studied, since the paludification process brings specific high moisture dynamics in old-growth forests [34]. Direct impacts of climatic change on fuel loads should also be investigated. This includes the potential increase in tree growth with the lengthening of the growing season [91,92] or decomposition [93]. However, the particularly high organic layer thickness and high soil moisture content may prevent such impacts by limiting tree growth and decomposition.

Acknowledgments: This project was financially supported by the Natural Sciences and Engineering Research Council of Canada (NSERC; Strategic Project; Discovery Grants Program), the Canada Chair in Forest Ecology and Management (Yves Bergeron), the Canadian Forest Service Tembec Inc. Danielle Charron and Marie-Hélène Longpré contributed to the field work logistics. We are particularly grateful to Mélanie Desrochers for her help with ARCGIS maps, to Rémi St-Amant for providing climate data and help with the BioSIM software, to Martin Simard and Nicolas Lecomte for providing their data and, finally, to Christelle Hély and Samira Ouarmim for their availability and helpful advice on fuel load fieldwork and calculations. We extend our thanks to Samira Ouarmim for reviewing the paper and providing helpful comments and to Isabelle Lamarre for technical editing. Finally, we are grateful to two anonymous reviewers for their helpful comments on an earlier version of the manuscript.

Author Contributions: S.G. and Y.B. conceived and coordinated the research project. M.P. and S.G. designed data collection. M.P. and S.P.-B. did the field work. M.P., A.T., and S.P.-B. performed fuel load calculations. A.T. and M.P.G. conceived modeling and simulations analyses. A.T. and M.P. analyzed the data. A.T. wrote the paper and all the authors revised it.

Conflicts of Interest: The authors declare no conflict of interest. The founding sponsors had no role in the design of the study; in the collection, analyses, or interpretation of data; in the writing of the manuscript; and in the decision to publish the results.

Abbreviations

The following abbreviations are used in this manuscript:

BS	Black spruce
BS-S	Black spruce forests originating from severe fire
BS-NS	Black spruce forests originating from non-severe fire
CWD	Coarse woody debris
DBH	Diameter at breast height
DWD	Dead woody debris
FFFC	Forest floor fuel consumption
FWI	Fire weather index
JP	Jack pine
TA	Trembling aspen
TSF	Time since the last fire

Appendix A. Shrub Load Equations

Previously determined equations were used for load calculations of *Lonicera canadensis* Bartr., *Ribes* sp., *Rosa acicularis* Lindl., and *Viburnum edule* Raf [57].

New equations were determined with additional shrub samples for load calculations of *Chamaedaphne calyculata* (L.) Moench, *Kalmia angustifolia* L., *Rhododendron groenlandicum* (Oeder) K.A. Kron & Judd, and *Vaccinium myrtilloides* Michx. (Table A1). A total of 90 additional shrub samples were randomly collected in various triangle transects to determine equations linking species' weight (g) with stem diameter (cm) for *Chamaedaphne calyculata* (sample size = 19), *K. angustifolia* (sample size = 28), *R. groenlandicum* (sample size = 28), and *V. myrtilloides* (sample size = 15). Efforts were made to collect a high variability in diameter size. Each sample was weighted and linked with basal diameter using linear regressions. Weight and diameter values were logarithmically transformed to normalize the relationship. Linear regressions were performed using *R* freeware [59].

Table A1. Equations and adjusted R^2 from linear regressions analysis linking shrub weight (g) with stem basal diameter (cm) by species. All equations and coefficients were significant with a *p*-value ≤ 0.05.

Species	Equations	Adjusted R^2
Kalmia angustifolia [2]	$\ln(weight) = -2.13 + 2.5 \ln(D)$	0.91
Lonicera canadensis [1]	$\ln(weight) = -2.33 + 2.64 \ln(D)$	0.82
Chamaedaphne calyculata [2]	$\ln(weight) = -3.25 + 3.25 \ln(D)$	0.93
Rhododendron groenlandicum [2]	$\ln(weight) = -2.4 + 2.52 \ln(D)$	0.94
Ribes sp. [1]	$\ln(weight) = -2.18 + 2.33 \ln(D)$	0.79
Rosa acicularis [1]	$\ln(weight) = -2.09 + 2.41 \ln(D)$	0.82
Vaccinium myrtilloides [1]	$\ln(weight) = -1.93 + 1.92 \ln(D)$	0.65
Viburnum edule [2]	$\ln(weight) = -2.55 + 2.62 \ln(D)$	0.84

[1] Hély's equations [57]; [2] New equations from additional shrub sample.

Appendix B. Sensitivity Analysis of the Influence of Fire Weather Variations on Potential Carbon Emissions across Successional Stages

Sensitivity analyses were made to ensure that fire weather values above average did not modify our conclusions. To do so, we performed simulations using daily FWI System components for each natural fire start point during the interval 1971–2000 period in the study area [60] (Table B1). A total of eight fires occurred during this period, burning a total of 56,080 ha.

Table B1. Daily FWI System components for each natural fire start point encountered during the interval 1971–2000 period.

Name	FFMC	DMC	DC	ISI	BUI	FWI
Fire 1	89	22	168	15	33	24
Fire 2	93	42	85	17	41	30
Fire 3	88	22	167	9	33	17
Fire 4	90	50	125	15	49	29
Fire 5	83	26	99	5	32	10
Fire 6	82	24	101	4	30	8
Fire 7	88	30	99	8	34	15
Fire 8	88	31	88	6	33	13

FFMC: Fine Fuel Moisture Code; DMC: Duff Moisture Code; DC: Drought Code; ISI: Initial Spread Index; BUI: Buildup Index; FWI: Fire Weather Index.

Simulation results showed that potential carbon emission was the lowest for forests older than 150 years, regardless of the FWI System components (Figure B1).

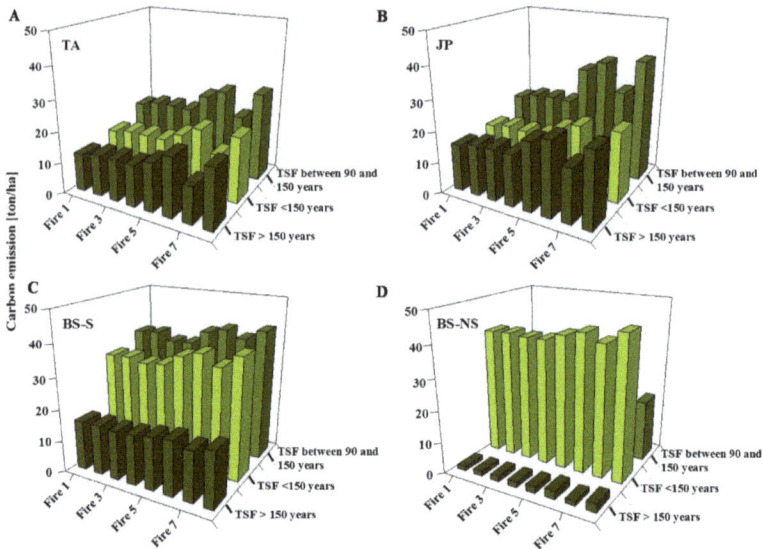

Figure B1. CanFIRE potential carbon emissions averaged by early (TSF < 90 years), intermediate (TSF between 90 and 150 years) and late (TSF > 150 years) successional stages, and by dominant forest types of the Clay Belt boreal landscape ((**A**) trembling aspen (TA), (**B**) jack pine (JP), (**C**) black spruce originating from severe fire (BS-S), and (**D**) black spruce originating from non-severe fire (BS-NS) forests). Simulations were performed using FWI System components for each natural fire start from 1971 to 2000.

Appendix C. Comparison of Empirical Basal Area and Simulated Basal Area with Plonski's Yield Table

Prior to simulations of carbon emission by fire, we ensured that the use of Plonski's yield table [65] instead of our empirical basal area data would not bring a bias to our conclusions. We calculated Pearson's correlation between tree basal areas extracted from Plonski's tables and our empirical basal areas. Comparisons showed good predictive skills, with a Pearson correlation coefficient of 0.72 (Figure C1). Plonski's tables, however, overestimated the basal area, but this difference does not change the conclusions of our study because trees contribute only slightly to the amount of carbon emitted during fire (Table 3) and simulated values are significantly similar to observed values.

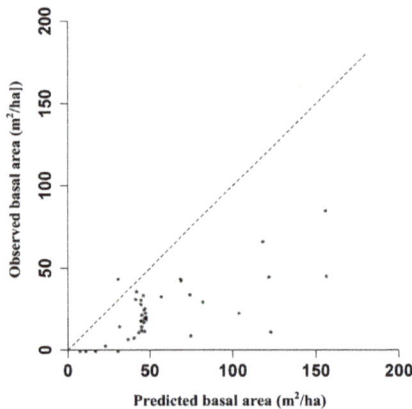

Figure C1. Comparison of empirical basal area and simulated basal area using Plonski's yield table.

Appendix D. Calibration of Bulk Density with Duff Fuel Loads

We calibrated a model linking measured bulk density (kg·m^{-3}) with organic layer depth (m) to consider heterogeneity in organic bulk density with the organic layer depth [37]. We made use of peat datasets comprising 103 measurements of bulk density at different organic layer depths selected from 11 sites in the Quebec Clay Belt boreal forest [34]. Bulk density and organic layer depths were logarithmically transformed to linearize the relationship. Linear regressions were performed using R freeware [59].

The resulting model explained 45% of the variance in bulk density (adjusted R^2 = 0.45, *p*-value ≤ 0.001). The model took on the following form:

$$\ln(Bulk\ Density + 1) = 2.72 + 0.57 \times \ln(Organic\ Layer\ Depth + 1) \tag{D1}$$

Therein bulk density progressively increases with organic layer depth (Figure D1A). Comparison of observed and predicted values indicated relatively good predictive skills, with a Pearson correlation coefficient of 0.73 (Figure D1B).

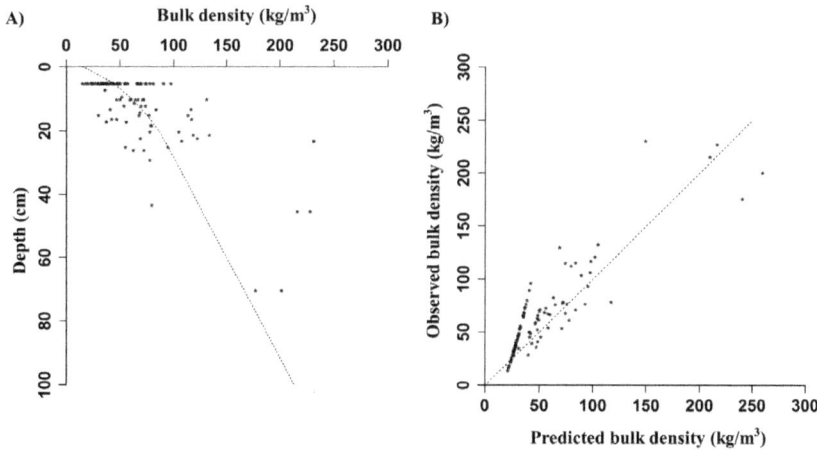

Figure D1. (**A**) Bulk density (kg/m^3) variation with organic layer depth for forested peats in the Clay Belt boreal forests. Stars represent sampling data collected by Simard et al. [34]. The dotted line represents the regression relationship; (**B**) Predicted bulk density with regression analysis versus observed values.

Appendix E. Calibration of Bulk Density with Duff Fuel Loads

Table E1. Empirical TSF and fuel loads (tree, fine aerial vegetation, DWD, duff, and total) for TA, JP, BS-S, and BS-NS dominated forests.

			Fuel Material			
Forest Type	TSF (Years)	Tree (t/ha)	Fine Aerial (t/ha)	DWD (t/ha)	Duff (t/ha)	Total (t/ha)
	11	0	569	90	70	730
	57	346	4	12	78	440
	68	100	37	16	88	241
	71	190	0	14	257	460
TA	102	200	0	17	145	362
	145	120	1	22	292	435
	145	120	1	26	197	343
	187	375	23	8	117	524
	187	375	2	7	129	514
	11	0	0	86	81	168
	11	0	2	79	77	158
	54	137	0	10	87	234
	87	173	0	12	154	339

Table E1. *Cont.*

Forest Type	TSF (Years)	Tree (t/ha)	Fine Aerial (t/ha)	DWD (t/ha)	Duff (t/ha)	Total (t/ha)
			Fuel Material			
	11	0	1	1	54	57
	89	175	4	22	152	353
	89	175	1	44	115	335
	142	91	0	21	270	382
JP	152	132	0	3	182	317
	152	132	1	12	140	285
	180	34	1	15	225	276
	180	34	0	10	225	270
	207	42	0	12	171	225
	207	42	0	23	173	238
	225	74	2	10	163	248
	32	0	1	17	170	188
	32	0	3	10	131	144
	32	0	2	1	74	77
	55	164	1	2	131	298
	87	118	5	2	223	348
	91	134	0	7	117	258
	129	127	1	33	190	350
	129	127	1	16	232	376
	145	72	1	58	169	299
BS-S	145	94	0	11	204	310
	165	115	1	3	506	625
	177	74	2	18	289	383
	181	85	0	17	350	453
	225	66	1	31	338	436
	272	79	0	6	496	581
	283	52	2	14	698	767
	283	66	3	19	541	629
	356	45	1	4	682	733
	41	11	2	1	292	306
	41	11	2	2	292	307
	41	11	2	1	292	306
	55	0	0	4	157	160
	57	53	0	2	239	295
	78	24	1	1	435	461
	96	29	1	2	483	515
BS-NS	100	40	1	1	351	393
	142	63	1	12	281	356
	146	43	0	7	389	440
	146	43	0	9	389	441
	170	106	3	4	972	1084
	172	66	5	8	464	543
	172	66	2	10	464	542

References

1. Lemprière, T.C.; Kurz, W.A.; Hogg, E.H.; Schmoll, C.; Rampley, G.J.; Yemshanov, D.; McKenney, R.; Gilsenan, A.; Beatch, D.; Blain, J.S.; et al. Canadian boreal forests and climate change mitigation 1. *Environ. Rev.* **2013**, *21*, 293–321. [CrossRef]
2. Lorenz, K.; Lal, R. *Carbon Sequestration in Forest Ecosystems*; Springer: Dordrecht, The Netherlands, 2010; pp. 5–11.
3. Dixon, R.K.; Solomon, A.M.; Brown, S.; Houghton, R.A.; Trexler, M.C.; Wisniewski, J. Carbon pools and flux of global forest ecosystems. *Science* **1994**, *263*, 185–190. [CrossRef] [PubMed]

4. Luyssaert, S.; Inglima, I.; Jung, M.; Richardson, A.D.; Reichstein, M.; Papale, D.; Piao, S.L.; Schulze, E.-D.; Wingate, L.; Matteucci, G.; et al. CO_2 balance of boreal, temperate, and tropical forests derived from a global database. *Glob. Chang. Biol.* **2007**, *13*, 2509–2537. [CrossRef]

5. Terrier, A.; Girardin, M.; Bergeron, Y. Les réservoirs de carbone en forêt boréale à l'est du Canada: Acquis et incertitudes dans la modélisation face aux changements climatiques. *VertigO-la Revue Électronique Sci. L'environnement* **2012**, *11*, 3. [CrossRef]

6. Burton, P.J.; Messier, C.; Weetman, G.F.; Prepas, E.E.; Adamowicz, W.L.; Tittler, R. The current state of boreal forestry and the drive for change. In *Towards Sustainable Management of the Boreal Forest*; Burton, P.J., Messier, C., Smith, D.W., Adamowicz, W.L., Eds.; NRC Research Press: Ottawa, ON, Canada, 2003; pp. 1–40.

7. Attiwill, P.M. The disturbance of forest ecosystems: The ecological basis for conservative management. *For. Ecol. Manag.* **1994**, *63*, 247–300. [CrossRef]

8. Bergeron, Y.; Harvey, B. Basing sylviculture on natural ecosystems dynamics: An approach applied to the southern boreal mixedwood forest of Quebec. *For. Ecol. Manag.* **1997**, *92*, 235–242. [CrossRef]

9. Angelstam, P.K. Maintaining and restoring biodiversity in European boreal forests by developing natural disturbance regimes. *J. Veg. Sci.* **1998**, *9*, 593–602. [CrossRef]

10. Gauthier, S.; Vaillancourt, M.-A.; Leduc, A.; de GrandPré, L.; Kneeshaw, D.; Morin, H.; Drapeau, P.; Bergeron, Y. *Ecosystem Management in the Boreal Forest*; Presses de l'Uniersité du Québec: Québec, QC, Canada, 2009.

11. Franklin, J.F. Preserving biodiversity: Species, ecosystems, or landscapes? *Ecol. Appl.* **1993**, *3*, 202–205. [CrossRef] [PubMed]

12. Gauthier, S.; Leduc, A.; Bergeron, Y. Forest dynamics modelling under natural fire cycles: A tool to define natural mosaic diversity for forest management. *Environ. Monit. Assess.* **1996**, *39*, 417–434. [CrossRef] [PubMed]

13. Perera, A.H.; Buse, L.J. Emulating Natural Disturbance in Forest Management. In *Emulating Natural Forest Landscape Disturbances*; Perera, A.H., Buse, L.J., Weber, M.G., Eds.; Columbia University Press: New York, NY, USA, 2004; pp. 3–7.

14. Niemela, È.J. Management in relation to disturbance in the boreal forest. *For. Ecol. Manag.* **1999**, *115*, 127–134. [CrossRef]

15. Bergeron, Y.; Fenton, N.J. Boreal forests of eastern Canada revisited: Old growth, nonfire disturbances, forest succession, and biodiversity. *Botany* **2012**, *90*, 509–523. [CrossRef]

16. Bergeron, Y.; Harvey, B.; Leduc, A.; Gauthier, S. Forest management guidelines based on natural disturbances dynamics: Stand- and forest-level considerations. *For. Chron.* **1999**, *75*, 49–54. [CrossRef]

17. Cyr, D.; Gauthier, S.; Bergeron, Y.; Carcaillet, C. Forest management is driving the eastern North American boreal forest outside its natural range of variability. *Front. Ecol. Environ.* **2009**, *7*, 519–524. [CrossRef]

18. Luyssaert, S.; Schulze, E.D.; Börner, A.; Knohl, A.; Hessenmöller, D.; Law, B.E.; Ciais, P.; Grace, J. Old-growth forests as global carbon sinks. *Nature* **2008**, *455*, 213–215. [CrossRef] [PubMed]

19. Amiro, B.D.; Cantin, A.; Flannigan, M.D.; de Groot, W.J. Future emissions from Canadian boreal forest fires. *Can. J. For. Res.* **2009**, *39*, 383–395. [CrossRef]

20. De Groot, W.J.; Cantin, A.S.; Flannigan, M.D.; Soja, A.J.; Gowman, L.M.; Newbery, A. A comparison of Canadian and Russian boreal forest fire regimes. *For. Ecol. Manag.* **2013**, *294*, 23–34. [CrossRef]

21. Chapin, F.; Woodwell, G.; Randerson, J.; Rastetter, E.; Lovett, G.; Baldocchi, D.; Clark, D.; Harmon, M.; Schimel, D.; Valentini, R.; et al. Reconciling carbon-cycle concepts, terminology, and methods. *Ecosystems* **2006**, *9*, 1041–1050. [CrossRef]

22. Yang, G.; Di, X.-Y.; Guo, Q.-X.; Shu, Z.; Zeng, T.; Yu, H.-Z.; Wang, C. The impact of climate change on forest fire danger rating in China's boreal forest. *J. For. Res.* **2011**, *22*, 249–257. [CrossRef]

23. Flannigan, M.D.; Logan, K.A.; Amiro, B.D.; Skinner, W.R.; Stocks, B.D. Future area burned in Canada. *Clim. Chang.* **2005**, *72*, 1–16. [CrossRef]

24. Bergeron, Y.; Cyr, D.; Girardin, M.P.; Carcaillet, C. Will climate change drive 21st century burn rates in Canadian boreal forest outside of its natural variability: Collating global climate model experiments with sedimentary charcoal data. *Int. J. Wildl. Fire* **2010**, *19*, 1127–1139. [CrossRef]

25. Boulanger, Y.; Gauthier, S.; Burton, P.J.; Vaillancourt, M.A. An alternative fire regime zonation for Canada. *Int. J. Wildl. Fire* **2012**, *21*, 1052–1064. [CrossRef]

26. Girardin, M.P.; Hogg, E.H.; Bernier, P.Y.; Kurz, W.A.; Guo, X.J.; Cyr, G. Negative impacts of high temperatures on growth of black spruce forests intensify with the anticipated climate warming. *Glob. Chang. Boil.* **2016**, *22*, 627–643. [CrossRef] [PubMed]

27. De Groot, W.J. Modeling Canadian wildland fire carbon emissions with the Boreal Fire Effects (BORFIRE) model. *For. Ecol. Manag.* **2006**, *234*, S224. [CrossRef]

28. Amiro, B.D.; Todd, J.B.; Wotton, B.M.; Logan, K.A.; Flannigan, M.D.; Stocks, B.J.; Mason, J.A.; Martell, K.G.; Hirsch, K.G. Direct carbon emissions from Canadian forest fires, 1959–1999. *Can. J. For. Res.* **2001**, *31*, 512–525. [CrossRef]

29. Johnston, D.C.; Turetsky, M.R.; Benscoter, B.W.; Wotton, B.M. Fuel load, structure, and potential fire behaviour in black spruce bogs. *Can. J. For. Res.* **2015**, *45*, 888–899. [CrossRef]

30. Scharlemann, J.P.; Hiederer, R.; Kapos, V.; Ravilious, C. *Updated Global Carbon Map*; UNEP-WCRC & EU-JRC: Cambridge, UK, 2009.

31. Brown, J.K. Unatural Fuel Buildup Issue. In Proceedings of the Symposium and Workshop on Wilderness Fire, Missoula, MT, USA, 15–18 November 1983.

32. Ryan, K.C.; Knapp, E.E.; Varner, J.M. Prescribed fire in North American forests and woodlands: History, current practice, and challenges. *Front. Ecol. Environ.* **2013**, *11*, e15–e24. [CrossRef]

33. Johnson, E.A.; Miyanishi, K.; Bridge, S.R.J. Wildfire regime in the boreal forest and the idea of suppression and fuel buildup. *Conserv. Biol.* **2001**, *15*, 1554–1557. [CrossRef]

34. Simard, M.; Lecomte, N.; Bergeron, Y.; Bernier, P.Y.; Paré, D. Forest productivity decline caused by successional paludification of boreal soils. *Ecol. Appl.* **2007**, *17*, 1619–1637. [CrossRef] [PubMed]

35. Fenton, N.J.; Bergeron, Y. Facilitative succession in a boreal bryophyte community driven by changes in available moisture and light. *J. Veg. Sci.* **2006**, *17*, 65–76. [CrossRef]

36. Terrier, A.; de Groot, W.J.; Girardin, M.P.; Bergeron, Y. Dynamics of moisture content in spruce-feather moss and spruce–*Sphagnum* organic layers during an extreme fire season and implications for future depths of burn in Clay Belt black spruce forests. *Int. J. Wildl. Fire* **2014**, *23*, 490–502. [CrossRef]

37. Benscoter, B.W.; Thompson, D.K.; Waddington, J.M.; Flannigan, M.D.; Wotton, B.M.; de Groot, W.J.; Turetsky, M.R. Interactive effects of vegetation, soil moisture and bulk density on depth of burning of thick organic soils. *Int. J. Wildl. Fire* **2011**, *20*, 418–429. [CrossRef]

38. Benscoter, B.W.; Wieder, R.K. Variability in organic matter lost by combustion in a boreal bog during the 2001 Chisholm fire. *Can. J. For. Res.* **2003**, *33*, 2509–2513. [CrossRef]

39. Shetler, G.; Turetsky, M.R.; Kane, E.; Kasischke, E. *Sphagnum* mosses limit total carbon consumption during fire in Alaskan black spruce forests. *Can. J. For. Res.* **2008**, *38*, 2328–2336. [CrossRef]

40. De Groot, W.J.; Bothwell, P.M.; Carlsson, D.H.; Logan, K.A. Simulating the effects of future fire regimes on western Canadian boreal forests. *J. Veg. Sci.* **2003**, *14*, 355–364. [CrossRef]

41. De Groot, W.J.; Landry, R.; Kurz, W.A.; Anderson, K.R.; Englefield, P.; Fraser, R.H.; Hall, R.J.; Banfield, E.; Raymond, D.A.; Decker, V.; et al. Estimating direct carbon emissions from Canadian wildland fires. *Int. J. Wildl. Fire* **2007**, *16*, 593–606. [CrossRef]

42. Vincent, J.S.; Hardy, L. L'évolution et l'extension des lacs glaciaires Barlow et Ojibway en territoire québécois. *Géogr. Phys. Quat.* **1977**, *31*, 357–372. [CrossRef]

43. Bergeron, Y.; Gauthier, S.; Flannigan, M.D.; Kafka, V. Fire regimes at the transition between mixedwoods and coniferous boreal forest in northwestern Quebec. *Ecology* **2004**, *85*, 1916–1932. [CrossRef]

44. Environment Canada. National Climate Data and Information Archive. 2016. Available online: http://climate. weatheroffice.gc.ca/ (accessed on 7 June 2016).

45. Fenton, N.; Lecomte, N.; Légaré, S.; Bergeron, Y. Paludification in black spruce (*Picea mariana*) forests of eastern Canada: Potential factors and management implications. *For. Ecol. Manag.* **2005**, *213*, 151–159. [CrossRef]

46. Lavoie, M.; Paré, D.; Fenton, N.; Groot, A.; Taylor, K. Paludification and management of forested peatlands in Canada: A literature review. *Environ. Rev.* **2005**, *13*, 21–50. [CrossRef]

47. Harper, K.; Boudreault, C.; de Grandpré, L.; Drapeau, P.; Gauthier, S.; Bergeron, Y. Structure, composition and diversity of old-growth black spruce boreal forest of the Clay Belt region in Québec and Ontario. *Environ. Rev.* **2003**, *11*, S79–S98. [CrossRef]

48. Lecomte, N.; Bergeron, Y. Successional pathways on different surficial deposits in the coniferous boreal forest of the Quebec Clay Belt. *Can. J. For. Res.* **2005**, *35*, 1984–1995. [CrossRef]

49. Fenton, N.J.; Béland, C.; de Blois, S.; Bergeron, Y. *Sphagnum* establishment and expansion in black spruce (*Picea mariana*) boreal forests. *Can. J. Bot.* **2007**, *85*, 43–50. [CrossRef]
50. Belleau, A.; Leduc, A.; Lecomte, N.; Bergeron, Y. Forest succession rate and pathways on different surface deposit types in the boreal forest of northwestern Quebec. *Ecoscience* **2011**, *18*, 329–340. [CrossRef]
51. Lecomte, N.; Simard, M.; Bergeron, Y. Effects of fire severity and initial tree composition on stand structural development in the coniferous boreal forest of northwestern Québec, Canada. *Ecoscience* **2006**, *13*, 152–163. [CrossRef]
52. Greene, D.F.; Macdonald, S.E.; Haeussler, S.; Domenicano, S.; Noël, J.; Jayen, K.; Charron, I.; Gauthier, S.; Hunt, S.; Gielau, E.T.; et al. The reduction of organic-layer depth by wildfire in the North American boreal forest and its effect on tree recruitment by seed. *Can. J. For. Res.* **2007**, *37*, 1012–1023. [CrossRef]
53. Johnstone, J.F.; Chapin, F.S., III; Hollingsworth, T.N.; Mack, M.C.; Romanovsky, V.; Turetsky, M. Fire, climate change, and forest resilience in interior Alaska. *Can. J. For. Res.* **2010**, *40*, 1302–1312. [CrossRef]
54. Van Cleve, K.; Oliver, L.; Schlentner, R.; Viereck, L.A.; Dyrness, C. Productivity and nutrient cycling in taiga forest ecosystems. *Can. J. For. Res.* **1983**, *13*, 747–766. [CrossRef]
55. Paré, D.; Bernier, P.; Lafleur, B.; Titus, B.D.; Thiffault, E.; Maynard, D.G.; Guo, X. Estimating stand-scale biomass, nutrient contents, and associated uncertainties for tree species of Canadian forests. *Can. J. For. Res.* **2013**, *43*, 599–608. [CrossRef]
56. Brown, J.K.; Oberheu, R.D.; Johnston, C.M. *Handbook for Inventorying Surface Fuels and Biomass in the Interior West*; General Technical Report INT-GTR-129; U.S. Department of Agriculture, Forest Service, Intermountain Forest and Range Experimental Station: Ogden, UT, USA, 1982; p. 48.
57. Hély, C.; Université de Montpellier 2, Montpellier, France. Personal communication, 2001.
58. McRae, D.J.; Alexander, M.E.; Stocks, B.J. *Measurement and Description of Fuels and Fire Behavior on Prescribed Burns: A Handbook*; Great Lakes Forest Research Centre: Sault Ste Mary, ON, Canada, 1979.
59. R Development Core Team. *R: A Language and Environment for Statistical Computing*, R Development Core Team: Vienna, Austria, 2010.
60. Terrier, A.; Girardin, M.P.; Cantin, A.; Groot, W.J.; Anyomi, K.A.; Gauthier, S.; Bergeron, Y. Disturbance legacies and paludification mediate the ecological impact of an intensifying wildfire regime in the Clay Belt boreal forest of eastern North America. *J. Veg. Sci.* **2015**, *26*, 588–602. [CrossRef]
61. Van Wagner, C.E. *Development and Structure of the Canadian Forest Fire Weather Index System*; Technical Report No. 35; Canadian Forestry Service: Ottawa, ON, Canada, 1987.
62. Lambert, M.C.; Ung, C.H.; Raulier, F. Canadian national tree aboveground biomass equations. *Can. J. For. Res.* **2005**, *35*, 1996–2018. [CrossRef]
63. Forestry Canada Fire Danger Group. *Development and Structure of the Canadian Forest Fire Behavior Prediction System*; Information Report ST-X-3; Forestry Canada Fire Danger Group: Ottawa, ON, Canada, 1992.
64. De Groot, W.J.; Pritchard, J.M.; Lynham, T.J. Forest floor fuel consumption and carbon emissions in Canadian boreal forest fires. *Can. J. For. Res.* **2009**, *39*, 367–382. [CrossRef]
65. Plonski, W.L. *Normal Yield Tables (Metric) for Major Forest Species of Ontario*; Ontario Ministry of Natural Resources: Toronto, ON, Canada, 1974; p. 40.
66. Hély, C.; Bergeron, Y.; Flannigan, M.D. Coarse woody debris in the southeastern Canadian boreal forest: Composition and load variations in relation to stand replacement. *Can. J. For. Res.* **2000**, *30*, 674–687. [CrossRef]
67. Légaré, S.; Bergeron, Y.; Paré, D. Influence of forest composition on understory cover in boreal mixedwood forests of western Quebec. *Silva Fenn.* **2002**, *36*, 353–366. [CrossRef]
68. Brassard, B.W.; Chen, H.Y. Stand structural dynamics of North American boreal forests. *Crit. Rev. Plant Sci.* **2006**, *25*, 115–137. [CrossRef]
69. Hart, S.A.; Chen, H.Y. Understory vegetation dynamics of North American boreal forests. *Crit. Rev. Plant Sci.* **2006**, *25*, 381–397. [CrossRef]
70. Greene, D.F.; Zasada, J.C.; Sirois, L.; Kneeshaw, D.; Morin, H.; Charron, I.; Simard, M.-J. A review of the regeneration dynamics of North. American boreal forest tree species. *Can. J. For. Res.* **1999**, *29*, 824–839. [CrossRef]
71. Johnstone, J.F.; Chapin, F.S., III. Effects of soil burn severity on post-fire tree recruitment in boreal forest. *Ecosystems* **2006**, *9*, 14–31. [CrossRef]

72. Pothier, D.; Savard, F. *Actualisation des Tables de Production Pour les Principales Espèces du Québec*; RN98-3054; Gouvernement du Québec, Ministère des Ressources Naturelles, Bibliothèque Nationale du Québec: Québec, QC, Canada, 1998.

73. Wardle, D.A.; Walker, L.R.; Bardgett, R.D. Ecosystem properties and forest decline in contrasting long-term chronosequences. *Science* **2004**, *305*, 509–513. [CrossRef] [PubMed]

74. Kneeshaw, D.D.; Bergeron, Y. Canopy gap characteristics and tree replacement in the southeastern boreal forest. *Ecology* **1998**, *79*, 783–794. [CrossRef]

75. Chen, H.Y.; Popadiouk, R.V. Dynamics of North American boreal mixedwoods. *Environ. Rev.* **2002**, *10*, 137–166. [CrossRef]

76. Pothier, D.; Raulier, F.; Riopel, M. Ageing and decline of trembling aspen stands in Quebec. *Can. J. For. Res.* **2004**, *34*, 1251–1258. [CrossRef]

77. Messier, C.; Parent, S.; Bergeron, Y. Effects of overstory and understory vegetation on the understory light environment in mixed boreal forests. *J. Veg. Sci.* **1998**, *9*, 511–520. [CrossRef]

78. Grandpré, L.; Gagnon, D.; Bergeron, Y. Changes in the understory of Canadian southern boreal forest after fire. *J. Veg. Sci.* **1993**, *4*, 803–810. [CrossRef]

79. Pham, A.T.; Grandpré, L.D.; Gauthier, S.; Bergeron, Y. Gap dynamics and replacement patterns in gaps of the northeastern boreal forest of Quebec. *Can. J. For. Res.* **2004**, *34*, 353–364. [CrossRef]

80. Pedlar, J.H.; Pearce, J.L.; Venier, L.A.; McKenney, D.W. Coarse woody debris in relation to disturbance and forest type in boreal Canada. *For. Ecol. Manag.* **2002**, *158*, 189–194. [CrossRef]

81. Bond-Lamberty, B.; Gower, S.T. Decomposition and fragmentation of coarse woody debris: Re-visiting a boreal black spruce chronosequence. *Ecosystems* **2008**, *11*, 831–840. [CrossRef]

82. Brais, S.; Paré, D.; Lierman, C. Tree bole mineralization rates of four species of the Canadian eastern boreal forest: Implications for nutrient dynamics following stand-replacing disturbances. *Can. J. For. Res.* **2006**, *36*, 2331–2340. [CrossRef]

83. Päätalo, M.-L. Factors influencing occurrence and impacts of fires in northern European forests. *Silva Fenn.* **1998**, *32*, 185–202. [CrossRef]

84. Hély, C.; Flannigan, M.; Bergeron, Y.; McRae, D. Role of vegetation and weather on fire behavior in the Canadian mixedwood boreal forest using two fire behavior prediction systems. *Can. J. For. Res.* **2001**, *31*, 430–441. [CrossRef]

85. Arienti, M.C.; Cumming, S.G.; Boutin, S. Empirical models of forest fire initial attack success probabilities: The effects of fuels, anthropogenic linear features, fire weather, and management. *Can. J. For. Res.* **2006**, *36*, 3155–3166. [CrossRef]

86. Krawchuk, M.A.; Cumming, S.G.; Flannigan, M.D.; Wein, R.W. Biotic and abiotic regulation of lightning fire initiation in the mixedwood boreal forest. *Ecology* **2006**, *87*, 458–468. [CrossRef] [PubMed]

87. Harden, J.W.; Trumbore, S.E.; Stocks, B.J.; Hirsch, A.; Gower, S.T.; O'neill, K.P.; Kasischke, E.S. The role of fire in the boreal carbon budget. *Glob. Chang. Biol.* **2000**, *6*, 174–184. [CrossRef]

88. Turetsky, M.R.; Kane, E.S.; Harden, J.W.; Ottmar, R.D.; Manies, K.L.; Hoy, E.; Kasischke, E.S. Recent acceleration of biomass burning and carbon losses in Alaskan forests and peatlands. *Nat. Geosci.* **2011**, *4*, 27–31. [CrossRef]

89. Whelan, R.J. *The Ecology of Fire*; Cambridge University Press: Cambridge, UK, 1995.

90. Fleming, R.A.; Candau, J.-N.; McAlpine, R.S. Landscape-Scale Analysis of Interactions between Insect Defoliation and Forest Fire in Central Canada. *Clim. Chang.* **2002**, *55*, 251–272. [CrossRef]

91. Gignac, L.D.; Vitt, D.H. Responses of northern peatlands to climate change: Effects on bryophytes. *Hattori Shokubutsu Kenkyujo Hokoku* **1994**, *75*, 119–132.

92. Breeuwer, A.; Robroek, B.J.; Limpens, J.; Heijmans, M.M.; Schouten, M.G.; Berendse, F. Decreased summer water table depth affects peatland vegetation. *Basic Appl. Ecol.* **2009**, *10*, 330–339. [CrossRef]

93. Wickland, K.P.; Neff, J.C. Decomposition of soil organic matter from boreal black spruce forest: Environmental and chemical controls. *Biogeochemistry* **2008**, *87*, 29–47. [CrossRef]

© 2016 by the authors. Licensee MDPI, Basel, Switzerland. This article is an open access article distributed under the terms and conditions of the Creative Commons Attribution (CC BY) license (http://creativecommons.org/licenses/by/4.0/).

forests

MDPI

Article

Rating a Wildfire Mitigation Strategy with an Insurance Premium: A Boreal Forest Case Study

Georgina Rodriguez-Baca [1,*], Frédéric Raulier [1] and Alain Leduc [2]

[1] Centre d'Étude de la Forêt, Faculté de foresterie, de géographie et de géomatique, Université Laval, 2405 rue de la Terrasse, Québec, QC G1V 0A6, Canada; frederic.raulier@sbf.ulaval.ca

[2] Centre d'étude de la forêt, Université du Québec à Montréal, C. P. 8888, Succursale centre-ville, Montréal, QC H3C 3P8, Canada; alain.leduc@uqam.ca

* Correspondence: georgina-renee.rodriguez-baca.1@ulaval.ca; Tel.: +1-418-656-2131

Academic Editors: Yves Bergeron and Sylvie Gauthier
Received: 10 March 2016; Accepted: 10 May 2016; Published: 13 May 2016

Abstract: Risk analysis entails the systematic use of historical information to determine the frequency, magnitude and effects of unexpected events. Wildfire in boreal North America is a key driver of forest dynamics and may cause very significant economic losses. An actuarial approach to risk analysis based on cumulative probability distributions was developed to reduce the adverse effects of wildfire. To this effect, we developed spatially explicit landscape models to simulate the interactions between harvest, fire and forest succession over time in a boreal forest of eastern Canada. We estimated the amount of reduction of timber harvest necessary to build a buffer stock of sufficient size to cover fire losses and compared it to an insurance premium estimated in units of timber volume from the probability of occurrence and the amount of damage. Overall, the timber harvest reduction we applied was much more costly than the insurance premium even with a zero interest rate. This is due to the fact that the insurance premium is directly related to risk while the timber harvest reduction is not and, as a consequence, is much less efficient. These results, especially the comparison with a standard indicator such as an insurance premium, have useful implications at the time of choosing a mitigation strategy to protect timber supplies against risk without overly diminishing the provision of services from the forest. They are also promoting the use of insurance against disastrous events in forest management planning.

Keywords: planning; financial risk; fire; timber supply; sustainable forest management

1. Introduction

Human activities depend on the sustainability of natural resources and proving sustainability requires making forecasts. In forest management, uncertainty is an important issue in the support of any planning decision and in evaluating the consequences of alternative strategies [1]. Uncertainty stems from known variability (risk), lack of knowledge (uncertainty), ignorance and indeterminacy [2]. Ignorance and indeterminacy are difficult to account for, cannot be anticipated and require scenario planning [3] or adaptive methods [2]. Risk and uncertainty are somewhat easier to evaluate beforehand with risk analysis methods [4]. Such analyses are required when there is a possibility that the outcome of an event can deviate from expectations and have a negative effect on an objective [5]. For instance, the negative effect of a disturbance on the profits from timber harvesting provides the cost of that disturbance [6]. Situations where risk and uncertainty are at the core of the problem—as it is in risk management—require different strategies and coherent risk measures [7]. The development of methods to account for risk and uncertainty has made considerable progress and they already play a role in environmental decision-making, particularly in cases of severe uncertainty due to extremely long planning horizons [8]. Although the application of risk analysis in forest planning remains rare [9], attention to risk analysis in forestry should grow even more in the coming years [10,11].

Forest managers should account for many different sources of risk and uncertainty, one of them being wildfire. Fire is a critical component of terrestrial and atmospheric dynamics [12] and is a primary driver of forest dynamics across the boreal forest region of North America. Fire is also a major source of risk and uncertainty that can cause important damages to timber resources [13]. Fire-dominated forests present challenges when designing forest management plans that maximize sustained and constant harvest volume flows because of the wide spatial and temporal variation in the frequency and severity of fire events [14]. In Canada, each year, fire burns large portions of the forest area which causes significant losses to management agencies [15]. Despite the uncertainty that characterizes forest management planning, most planning models used for strategic planning remain deterministic in North America [16,17]. Linear mathematical programming (LP) is the approach most often applied in practice to such planning problems [18,19], despite the fact that many other techniques exist [16], the assumption that all data are assumed to be known exactly and the fact that decisions made today with optimal solutions will probably be suboptimal in the future [20]. Incorporating a fire regime into timber harvest-level determination procedures leads to reductions in harvest levels when desiring a sustainable timber harvest [15,21,22]. Such reductions help implement a timber buffer stock, providing a contingency inventory in the case of unexpected timber losses. The implementation of such measures therefore implies losses of short-term revenues that must be thoroughly justified and understood.

Successful methods of dealing with uncertainty and risk need to be simple and comprehensible enough to be useful in planning and decision-making in forestry practice [23]. The best strategy for dealing with uncertainty depends on the risk preference of the decision makers, how much risk they are willing to face, and the degree of uncertainty involved. One example of a successful method is the cost-plus-loss analysis, which estimates the cost of sub-optimal decisions. It has been used effectively to justify the costs implied by sampling intensity in forest inventory [24,25], the cost of fire-fighting (examples provided in [14]) or the cost of forest planning [26]. In practice, the minimization of risk exposures and potential losses involves risk processes with one or more techniques considered in the context of financial and nonfinancial exposures [27]. Financial risk modelling refers to the use of formal econometric techniques to determine the aggregate risk of a financial portfolio that depends on the probability distributions of losses that can arise from damage. Actuaries combine the likelihood and size information to provide average, or expected losses [28]. For instance, Value at Risk (VaR) is a widely used risk measure [29], is easy to explain and easy to estimate [30,31]. In portfolio management, Bagajewicz and Barbaro [32] defined VaR as the worst expected loss under normal market conditions over a specific time interval and at a given confidence level. VaR has become a popular risk measure used by both regulated banks as well as investment practitioners. Although specific indicators such as VaR cannot guarantee the identification of the best risk-reduced solution, in many instances the use of different risk measures help identify potentially robust solutions.

With financial risk management, the expected loss is expressed in monetary terms. One loss-adaptation option is insurance [33]. Insurance transfers the cost of financing losses in exchange for a premium. For instance, a forest manager may seek to protect his planning decisions against wildfire and he can purchase an insurance policy from an insurer by paying a premium to receive a compensatory payment that should cover the loss generated by fire [27]. The determination of the premium to pay for the cover requires evaluating risk as an actuarial process of valuing the insurable risk, *i.e.*, by summing the values at risk compounded at the start of the planning horizon. Such an insurance strategy covering losses caused by fires may be interesting if the interest rate is higher than the mean volume increment rate of the forest under study.

The idea of insurance in forestry is not new, it was proposed decades ago by Shepard [34,35] who noticed that the proper valuation of forest properties is a necessary prerequisite to any successful fire-insurance undertaking. Holecy and Hanewinkel [33] proposed an actuarial model calculating appropriate probabilities to estimate insurance premiums. Lankoande *et al.* [36] evaluated efficient wildfire insurance in the presence of government intervention through a subsidy for risk. Chen *et al.* [37] proposed an insurance instrument to protect timber owners against wildfire risks as a management instrument. Although insurance is an effective mechanism to lessen the burden of loss by wildfire and is simple to explain, studies in wildfire insurance still remain limited.

The main objective of this study was therefore to provide a comparative analysis of the alternative advantages produced by two different risk management strategies: insurance premium and timber harvest reduction to build a buffer stock of timber. In the context of planning and scheduling forest harvesting, the first aim of our study was thus to quantify potential harvest losses due to wildfire under an ecosystem-based management scenario in an eastern Canadian boreal forest. We used a linear programming (LP)-based timber harvest scheduling model to determine the maximum even-flow harvest volume a forest area can sustain over the planning horizon. Interaction between fire and harvest was simulated with a landscape dynamics model to evaluate harvest losses, insurance premium and amount of buffer stock required to cover such loss.

2. Materials and Methods

2.1. Study Area

The study area corresponds to the Forest Management Unit 085-51 located between 48°50' N and 50°09' N latitude, and between 78°05' W and 79°31' W longitude in western Quebec, Canada (Figure 1). It belongs to the bioclimatic domain of balsam fir-white birch to the south (14%) and black spruce-feather mosses to the north (86%) [38]. Mean annual temperature varies from −2.5 °C to 0 °C, and total precipitation from 700 to 800 mm. The area covers 1.08 million ha, of which 542,000 hectares are timber productive. Black spruce (*Picea mariana* (Mill.) B.S.P.) and jack pine (*Pinus banksiana* Lamb.) are the most abundant tree species and also the most economically important ones. Hardwoods such as trembling aspen (*Populus tremuloides* Michx) and white birch (*Betula papyrifera* Marsh.), and to a lesser extent, balsam poplar (*Populus balsamifera*) can also occur in mixture with black spruce. The forest dynamics in the region may be simplified into three main successional pathways either dominated by black spruce, jack pine or trembling aspen [39,40]. Fire dominates the natural disturbance regime in the study area [40,41]. Current (1920–2000) and past fire cycles (1850–1920) were estimated to be around 398 and 135 years [1]. Forest management planning should account for climate change as it should affect fire regimes in the boreal forest of North America [42]. For our study area, fire burn rate is projected to increase gradually over the period 2001–2100. Bergeron *et al.* [43] estimated that under B1 (2 × CO_2) and A2 (3 × CO_2) climate scenarios, fire cycles should lower to around 254 and 79 years respectively, values lying either in between the current and historical fire cycles [44] or below the historical fire cycle.

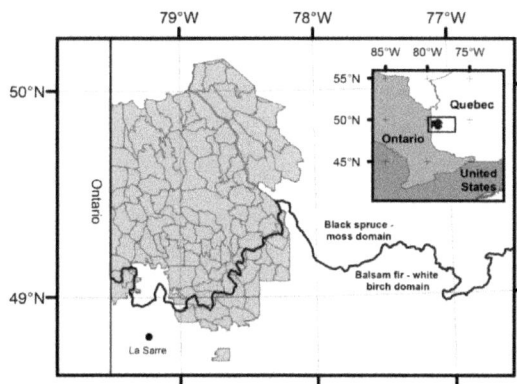

Figure 1. Location of study area Forest Management Unit 085-51. Grey polygons correspond to operating areas [45].

2.2. Timber Supply Model

We formulated the timber supply model as an optimization problem solved with linear programming, as it is the current practice in Quebec. No mitigation strategies were included at first against potential fire losses. The planning horizon was set to 150 years and divided into 30 periods of 5 years. The objective function of this model maximized harvest volume (*i.e.*, Mm^3/period) (Equation (1)). The first constraint provided an even flow of harvest volume over time (Equation (2)). For harvest planning purposes, the study area was divided into different spatially organized compartments (operating areas between 30 km^2 and 150 km^2) as a function of canopy closure and species composition [45] to emulate fire size distribution [46]. These operating areas are open to harvest when more than 30% of their timber productive area is eligible to harvest (Equation (3), [47]). Planting of jack pine after a clear-cut was limited to less than the actual plantation level (7500 ha per period (Equation (4))). A forest age structure was also targeted with a minimal abundance of three age classes (0–150 years: 63%, 150–275 years: 21% and more than 275 years: 16%) (Equation (5)) [41]. Two harvesting systems were implemented, careful logging around advanced regeneration [48] and irregular shelter-wood cuts (50% removal of merchantable volume; [49]). The areas planned to be harvested must be positive (Equation (6)).

Let

o : operational area $(1 \ldots 107)$

s : successional pathway $(1 \ldots 3)$

p : period $(1 \ldots 30)$

a : stand age

h : harvest type $(1 \ldots 2)$

c : cohort number $(1 \ldots 3)$

T_C : Timber production area belonging to cohort c, $\forall$ c, following the targeted forest structure

Variables

$$X_{op} = \begin{cases} 1, & \text{if operating area is open to harvest} \\ 0, & \text{otherwise} \end{cases} \quad \forall \, o, p$$

A_{ashop} : Area harvested (ha), $\forall \, a, s, h, o, p$

e_{ashop} : Area eligible for harvest (ha), $\forall \, a, s, h, o, p$

C_{cp} : Area belonging to cohort c (ha), $\forall \, c, p$

Parameters

$$V_{ashop} : \text{Volume yield } \left(\text{m}^3 \, \text{ha}^{-1} \right), \, \forall \, a, s, h, o, p$$

Objective function

$$Z = \sum_{a=1}^{30} \sum_{s=1}^{3} \sum_{o=1}^{107} \sum_{h=1}^{2} \sum_{p=1}^{30} V_{asohp} A_{asohp}, \tag{1}$$

Subject to

$$\sum_{a=1}^{30} \sum_{s=1}^{3} \sum_{o=1}^{107} \sum_{h=1}^{2} V_{asoh(p-1)} A_{asoh(p-1)} - \sum_{a=1}^{30} \sum_{s=1}^{3} \sum_{o=1}^{107} \sum_{h=1}^{2} V_{asohp} A_{asohp} = 0, \tag{2}$$

$$A_{asohp} \leqslant e_{asohp} X_{op} \, \forall \, a, s, o, h, p, \tag{3}$$

$$A_{hp} <= 7500, \, h \text{ being a clear cut followed by a jack pine plantation}, \, \forall \, p, \tag{4}$$

$$C_{cp} \geqslant T_c, \, \forall \, c; \, p \in [11; \, 30], \tag{5}$$

$$A_{asop} \geqslant 0 \,. \tag{6}$$

To develop the timber supply model (Model II formulation—([50], pp. 608–611)), we used the Remsoft Spatial Planning System (version 2013.12, Remsoft, Fredericton, NB, Canada) and solved it with Mosek 5.0. (Mosek ApS, Copenhagen, Denmark).

2.3. Interaction between Harvest Scheduling and Stochastic Processes

We simulated the interaction existing between harvest, fire and forest succession by adapting pre-existing modules of harvest, fire and succession already developed in the Spatially Explicit Landscape Event Simulator (SELES) [51]. Inputs are spatial rasters (forest type, stand age, operating areas), data tables (e.g., yield curves, harvested area planned by harvesting systems per operating area and per period, matrix of succession probabilities, other parameters such as a mean burn rate and a mean fire size). Fire was modeled as a percolation process [52] parameterized from historical fire occurrence data [41] to reproduce basic characteristics of a fire regime [53]. The model uses a negative exponential distribution to determine the number of fires and a Weibull distribution to determine fire sizes [54]. Simulated fires burn independently of terrain, and there is equal forest flammability regardless of stand age [21,55]. The harvest module prioritized the harvest of salvageable volume (30% of pre-fire standing volume) and subsequently the harvest scheduled by the timber supply model. If the harvest module was not able to find the harvest volume planned in designed operating areas, then it selected productive stands (with a volume greater than 50 $m^3 \cdot ha^{-1}$) not prescribed in the harvest plan until it reached the targeted timber supply level. Disturbance-specific changes in forest composition and age structure drive the interactions between fire, succession and harvest. Natural succession was modelled as a semi-Markov process [53] with probabilities of transition estimated from the proportions of each stratum by stand age class (20-year interval) observed in the forest map. The spatial resolution of the model was 10 ha per pixel and the temporal resolution five years. We performed 100 replications of each scenario, which provided stable estimates of indicators, especially VaR [56]. We used the technique of common random numbers [57] to reduce the variability generated by random effects between the scenarios [58]. Simulation outputs allowed us to quantify loss likelihood distributions and estimate insurance premiums as detailed below.

2.4. Risk Management

Simulation results with the landscape dynamics model served to estimate loss distributions (frequency distributions of differences between planned and harvested volumes per period). Value at Risk served to assess risk was estimated with the 5th percentile (α) of the loss (L) distribution for a given period p (Equation (7)):

$$\text{Prob}\ (L_p \leqslant \text{VaR}_p\ ) = \alpha. \tag{7}$$

VaR computation was performed using the R statistical software environment [59].

2.4.1. Risk Characterization

If fire risk is indeed part of the risk of timber supply disruptions, another part results from the inadequate consideration of fire risk in the timber supply planning process [15,60]. We have assumed with the timber supply model used in the present study (Section 2.2) that fire suppression is totally effective (*i.e.*, no fire risk), which is not true [12] and, consequently, overly optimistic. Risk can therefore be subdivided into two different risk types, effective risk, when effectively implementing planned forest management strategies despite fire risk, and planning risk caused by the optimism of the planning procedure. To distinguish both sources of risk, we fixed a planned timber harvest (PTH) threshold below which PTH is equal to the median realized harvest level that has been simulated with the landscape dynamics model [60] and above which PTH cannot fully be implemented anymore because of fire disturbances. Below this threshold, risk of losses (difference between median and 5th percentile) is caused by fire risk only. Above this threshold, risk of losses is a compound of risks caused by fire and planning optimism.

To characterize risk, we therefore looked for three PTH values, one for which no risk exists (*i.e.*, disruptions not occurring anymore, such that $VaR_p = 0$, $\forall p$), one for which risk is the highest while respecting all the constraints of the timber supply model and one for which PTH is equal to the median realized harvest level. We used the landscape simulation model for this purpose by decreasing the PTH originally estimated with the timber supply model by steps of 10% until no risk occurred anymore for three fire cycles. We then estimated the parameters of a piecewise linear model with one knot between realized harvest and planned timber supply values (Equation (8)):

$$\tilde{h} = PTH \text{ if } PTH \leqslant PTH_{th}, \tilde{h} = PTH_{th} + \beta \left(PTH - PTH_{th}\right) \text{ if } PTH > PTH_{th}, \tag{8}$$

where $\tilde{h}$ is the periodic median realized harvest implemented with the landscape simulation model, PTH is the timber harvest planned with the timber supply model and β and PTH_{th} are parameters estimated with the MODEL procedure (SAS Institute Inc., Cary, NC, USA).

2.4.2. Insurance Premium

A loss function proportional to the forest value may serve to characterize wildfire risk in a forest. Forest managers may seek to be protected against wildfire damages to timber supply by taking out an insurance contract. We therefore calculated the insurance premium with probabilities of the potential losses provided with periodic VaRs. Putting this into a formula, one needs to find the value of a periodic premium (P) such that (Equation (9)):

$$\sum_{p=1}^{30} \left(P - VaR_p\right) \frac{1}{(1+i)^{5 \times p - 2.5}} = 0, \tag{9}$$

where p corresponds to a number of five-year periods, and i is an interest rate. Harvest is assumed to take place in the middle of the period (hence the term -2.5). We selected different interest rates (0%, 1%, 2%, and 4%) used for discount rates for public investment [61]. Statistical computations were performed using the R statistical software environment [59].

2.4.3. Timber Supply Reduction

To prevent operational disruptions, a reduction in periodic wood harvest can be used to build a buffer stock of timber [15] serving as a back-up plan in the event that a supply disruption occurs. A supply disruption occurred whenever realized timber harvest was below 90% of the planned timber harvest volume [62]. We were interested in estimating the harvest target reduction that helped deal with timber losses caused by only fire and therefore used the difference between the maximum PTH value equal to the median realized harvest level (PTH_{th} in Equation (8)) and the one for which disruptions do not occur anymore (*i.e.*, $VaR_p = 0$, $\forall p$).

2.5. Comparison of Risk Management Strategies

At first, we estimated harvest loss distributions by simulating the implementation of the timber supply solution with the landscape simulation model for three fire cycles (100, 200 and 400 years). Simulations were then redone with the landscape simulation model for the three fire cycles by reducing the PTH value by steps of 10% until we found a PTH value that could be implemented with no risks. Periodic VaR_p and median VaR values were computed from these loss distributions. We also used these simulation results to estimate the parameters of piecewise linear models (Equation (8)) in order to find the PTH value equal to the median realized harvest level for each considered fire cycle. Premium insurance was then computed (Equation (9)) for a range of interest rates used for public investments (0%, 1%, 2% and 4% [61]) at the threshold PTH value equal to the median realized harvest level. Finally, we compared the timber supply reductions required to cancel risk to insurance premiums for each fire cycle.

3. Results

3.1. Risk Assessment

Periodic timber harvest with an ecosystem management strategy may reach values up to approximately 3.8 Mm^3 period^{-1} when not considering fire risk (Equation (1)). However, a blind implementation of such a strategy will not enable the procurement of expected timber levels (Figure 2) and timber supply disruptions caused by fire are expected (Figure 3). Despite the likely occurrence of such disruptions, the median rate of planning success reaches 97% (3.7 Mm^3 period^{-1}) of the optimal solution provided by the timber supply model with a fire cycle of 400 years and decreases only up to 73% (2.8 Mm^3 period^{-1}) with a fire cycle of 100 years (Figure 2). The chances of obtaining such a rate of success are, however, threatened by infrequent but possibly very significant disruptions. Timber supply disruptions may start to occur as soon as the 6th planning period (30 years) and, depending on the considered fire cycle, either tend to disappear after 50 years (and occur again approximately after one mean stand rotation) or maintain themselves for the rest of the planning horizon (Figure 3).

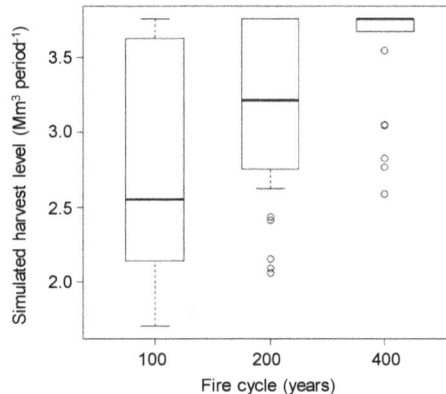

Figure 2. Box and whiskers plots representing the probability distributions of the simulated implementation of the optimal solution (3.8 Mm^3 period^{-1}) provided by the timber supply model (Equations (1)–(6)) under current (400 years) and probable interval for future fire regimes (100 and 200 years).

Maximum periodic VaRs, which provide an indication of the expected vulnerability of timber supplies to wildfire, are substantial in our study area (2.1 to 2.8 Mm^3 period^{-1}) and represent 55% to 74% of the periodic timber harvest, depending on the fire regime that is considered. Median VaR across the planning horizon with the longest fire cycle (400) years is substantially lower than the maximum VaR (0.9 *vs.* 2.2 Mm^3 period^{-1}) when compared to that resulting from a fire cycle of 100 years (2.4 *vs.* 2.9 Mm^3 period^{-1}), indicating more frequent occurrences of important timber supply disruption throughout the planning horizon with a higher burn rate (Figure 3c).

Successive implementation of a portion (30% to 90%) of the optimized timber supply solution helped find a median realized harvest level equal to PTH values for timber harvest levels up of 2.8 to 3.3 Mm^3 period^{-1}, depending on the considered fire cycle (Figure 4, Table 1). At the threshold PTH value beyond which the implementation success decreases, median VaR values (0.02 to 1.12 Mm^3 period^{-1} depending on the fire cycle) are much lower than those induced by the implementation of the entire optimized timber supply solution. They are in fact reduced by a factor varying between two and 20. Maximum VaR values are less reduced, by a factor between 1.6 (for a fire cycle of 100 years) and

1.9 (for a fire cycle of 400 years). This means that ignoring fire in the timber supply model and assuming that fire risk is totally controlled (planning optimism) increased the risk of supply disruptions by almost one order of magnitude, even with a fire cycle of 400 years. Such increased risk is, however, accompanied by an increase in realized harvest level, the rate of which varies between 0% and 60% (= slope of the second segment of the piecewise regression) (Figure 4, Table 1). This increase is only significant with a fire cycle of 400 years (Table 1).

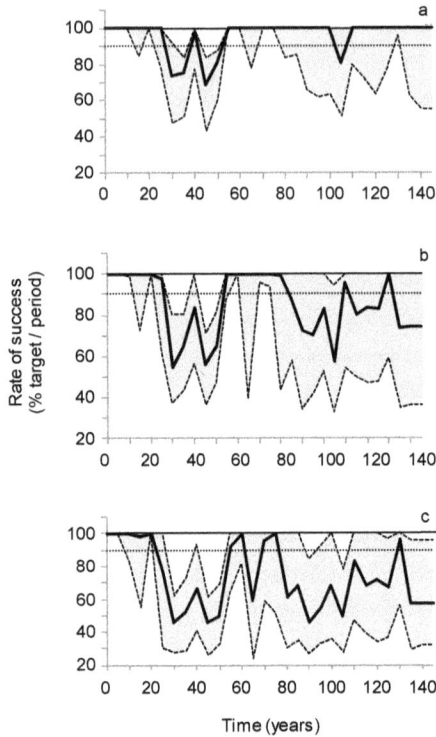

Figure 3. Probability distributions of the success rate of the simulated harvest schedule implementation under current ((**a**) 400 years) and probable future fire regimes ((**b**) 200 years; (**c**) 100 years). From bottom up, broken and bold lines represent the 5th, median and 95th percentiles. One hundred percent represents the target (continuous line) and ninety percent correspond to a cutoff value below which a timber supply disruption was considered to occur [57].

Table 1. Parameter values of piecewise linear models with one knot (threshold planned timber harvest (PTH)) (Equation (8)) relating PTH and periodic median realized harvest levels implemented with the landscape simulation model under current (400 years) and probable interval for future fire regimes (100 and 200 years).

Fire Cycle	Threshold PTH (Mm³ Period⁻¹)	β
100	2.77 (0.18)	0.04 (0.26) [a]
200	3.23 (0.15)	−0.01 (0.39) [a]
400	3.34 (0.18)	0.60 (0.25)

[a] Not significantly different from 0 at $\alpha = 0.05$ ($p = 0.78$ and $p = 0.96$, respectively). Numbers in parentheses represent a half-confidence interval.

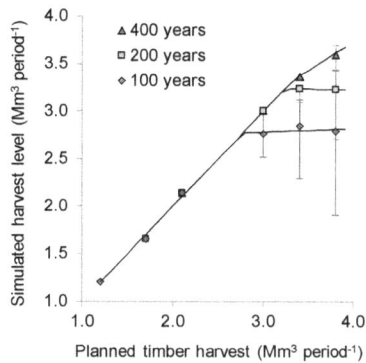

Figure 4. Relationship between planned timber harvest and its simulated implementation (median value) when considering the risk of fire for three fire cycles (100, 200 and 400 years). Parameters of the segmented linear models are provided in Table 1. Error bars represent the 5th and 95th percentiles of the probability distribution of simulated harvest levels.

3.2. Risk Management Strategies

Timber harvest reductions are required to deal with wildfire risk throughout the planning horizon: according to a timber harvest reduction strategy, a 35% of harvest reduction is necessary to avoid significant disruptions (*i.e.*, $VaR_p = 0, \forall p$) with the current fire regime (400 years, Figure 3), and such reductions increase to 48% and 56% for fire cycles of 200 and 100 years, respectively. Harvest reductions therefore seem to increase non-linearly with an increase of the fire cycle (*i.e.*, +6%/100 years between 200 and 400 years and +8%/100 years between 100 and 200 years). In fact, the sensitivity of maximum VaR to a timber harvest reduction decreases as maximum VaR tends to zero (Figure 5). Depending on the interest rate and the fire cycle, we looked at the changes in the amount of insurance premium an insurer should hold against unexpected losses as a function of fire risk (Figure 6). Insurance premiums represent between 5% and 7% of the level of supply for a fire cycle of 400 years (Figure 6), which are noticeably lower than for a timber harvest reduction strategy. With a change of fire cycle between 200 and 400 years, premium increases are also lower than those of a timber supply reduction strategy (between 5% and 11%/100 years depending on the interest rate), and increase less between 100 and 200 (between 2% and 7%/100 years). Such premium increases are more directly related to an increase in median VaR rather than to an increase in fire cycle (Figure 7).

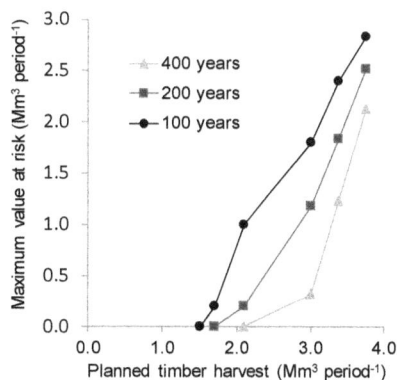

Figure 5. Relationship between planned timber harvest and maximum value at risk for three fire cycles.

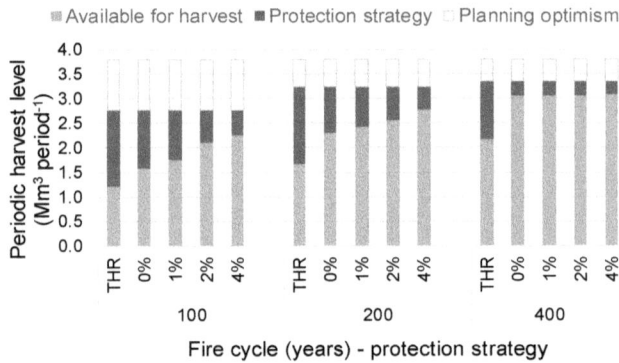

Figure 6. Distribution of the planned timber harvest into: a part that is not entirely feasible (planning optimism, in white, see Figure 4), a part that should be used to build a buffer stock of timber (**dark gray**) (with a timber harvest reduction—THR, which should not be harvested, or with an insurance premium, which should be harvested and set apart, with an interest rate between 0% and 4%) (protection strategy), and a part available for harvest (**light grey**), considering three possible fire cycles.

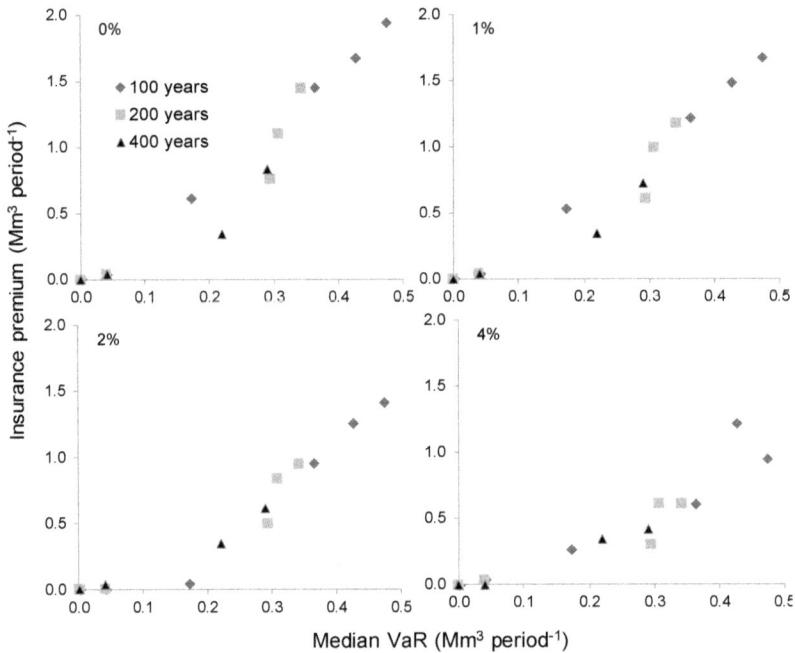

Figure 7. Relationship between median value-at-risk (VaR) and insurance premium as a function of interest rates (0% to 4%) and present (400 years) or probable fire cycles (100 and 200 years).

4. Discussion

We have evaluated two strategies to protect timber supply against disruptions with the objective of achieving over the planning horizon at least 90% of the planned harvest level: an insurance policy based on probabilities of potential losses and a reduction of timber harvest. Both mitigation strategies help build a buffer stock of available timber as a back-up plan. Our results show, however, that a constant reduction of timber harvest is costlier than an insurance policy (Figure 6) and is therefore less efficient.

This result is linked to the fact that the premium insurance is directly related to risk (Equation (9)), contrary to a timber harvest reduction strategy, which provides only an indirect way of managing a buffer stock [56]. The insurance premium also acts as an asset protected from fire. The consideration of fire impact and level of planned timber supply are the factors that most influence the planning success rate when implementing the optimized plan with a landscape dynamics model [56]. These two factors were varied in a systematic fashion in the present study. Both mitigation strategies were evaluated at the maximum PTH value that could be implemented in interaction with fire with the landscape simulation model, as proposed by [60]. We therefore succeeded in differentiating two types of risk, one due to fire only and another one due to the planning method used for the dimensioning of a sustainable timber supply.

Analysis of the risk related to the use of a specific planning method was discarded in the present study but deserves more consideration, especially in a real decision-making process: for the highest fire cycle that we have considered (400 years), the implementation of the timber supply optimal solution ($3.8 \text{ Mm}^3 \text{ period}^{-1}$) in interaction with fire had a success rate of 97%. The maximum PTH value that could be totally implemented with the landscape simulation model was 2.8 Mm^3 period. This means that a timber harvest reduction of 36% would be required to increase the success rate up to 100%, which is clearly very expensive [15,63]. In fact, since absolute protection against losses cannot be guaranteed, some level of acceptable loss expressed as a risk tolerance must be established, which can widely vary based on knowledge of exposures and proposed risk management solutions [27]. This points to the importance of choosing a level of tolerance to risk when facing a relatively low fire cycle, as already noted by [60]. Increasing tolerance to risk requires the availability of other mitigation strategies, such as the diversification of procurement sources [64], when supply disruptions occur.

Manley and Watt [65] mentioned that the possible reasons why uncertainty might have been ignored in the design of optimal forest management strategies is that it has often been assumed that forest management is based upon purely risk-neutral preferences. Brumelle *et al.* [66] made a survey of the literature on optimal forest management that took into account the presence of risk and found that 70% assumed risk neutral preferences and only 10% openly used risk averse preferences. A risk neutral forest manager would prefer adopting a strategy of passive acceptance whereas a risk adverse manager might prefer to adopt a risk mitigation strategy and continually revise his strategy in a dynamic replanning process [67]. Clearly, adopting a risk neutral strategy in the present case is an unsustainable strategy since important timber supply disruptions are to be expected (Figure 3), even with the present fire cycle of 400 years (Figure 3a) and despite an expected success rate of 97% of the harvest plan implementation. Furthermore, Gauthier *et al.* [68] showed that the increase in burn rates expected in the future, especially for the 2071 to 2100 period will impact the vulnerability of the harvest in most of western Quebec where our study area is located.

Inclusion of fire risk into the timber supply planning process has already been the subject of numerous research studies and different techniques are available for this purpose [10,11]. However, the implementation of these techniques in a real decision-making situation still remains limited in North America. For instance, inclusion of fire impact on timber supply models with linear programming requires a model structure seldom available in software designed for timber supply modeling [19,69]. Other approaches [15,22,63] remain too complex to implement with the typical problem size required to solve optimization problems of real timber supply models [19,69]. Heuristic optimization methods may overcome these limitations but do not guarantee optimality, which restrains their use in practice [16,17]. To the opposite, landscape simulation models are designed and therefore suited to analyze the interaction between harvest, natural succession and disturbances [53,70,71]. Clearly, such an approach offered two main advantages. First, simulation of the implementation of the optimized solution of a timber supply problem with such a landscape dynamics model helped assess the feasibility of the optimal solution, which conducted us to reduce the planned harvest down to a level where it was feasible at least 50% of the time. Second, at that level, we were able to estimate the amount of risk caused by fire only (and not by fire and planning method combined) and to express the simulated risk into an insurance premium, which is a common standard used for risk assessment and protection. The central view of any insurance scheme is an understanding of risk probabilities to inform the decision-making process.

5. Conclusions

Wildfire events impact optimal forest management decisions because such stochastic events may disturb the planned solution. In this paper, we used VaR as a tool to measure risk to characterize expected losses caused by fire during the implementation of a timber supply model solution in a boreal forest. We did not aim to investigate all possible mitigation strategies but rather to focus on two simple strategies: the use of an insurance premium and of a reduction of timber harvest. These strategies assume that decision-makers do not have prior information on which to base their weighting of the opinions and decisions. At the moment, the prospects around climate change are hardly encouraging and it is probable that the forest economy will diminish but decision-makers should consider shifting their attention to other promising potential schemes of strategies that could be used to deal with risk, and, maybe, only then will the risk of fire lower significantly.

Acknowledgments: This research was financially supported by the Fonds québécois de la recherche sur la nature et les technologies. We thank the Bureau du Forestier en Chef (Ministère de la Forêt, Faune et Parc du Québec) for giving us access to their timber supply analysis for FMU 085-51 (2008–2013). We also sincerely thank Hakim Ouzennou for his statistical help.

Author Contributions: G.R. and F.R. conceived the experiments and analyzed the data. All authors discussed the results, and co-wrote the paper. All authors contributed to writing the paper.

Conflicts of Interest: The authors declare no conflict of interest.

References

1. Kangas, J.; Kangas, A. Multiple criteria decision support in forest management—The approach, methods applied, and experiences gained. *For. Ecol. Manag.* **2005**, *207*, 133–143. [CrossRef]
2. Wynne, B. Uncertainty and environmental learning: Reconceiving science and policy in the preventive paradigm. *Glob. Environ. Change* **1992**, *2*, 111–127. [CrossRef]
3. Peterson, G.D.; Cumming, G.S.; Carpenter, S.R. Scenario planning: A tool for conservation in an uncertain world. *Conserv. Biol.* **2003**, *17*, 358–366. [CrossRef]
4. Hoffman, F.O.; Hammonds, J.S. Propagation of uncertainty in risk assessments: The need to distinguish between uncertainty due to lack of knowledge and uncertainty due to variability. *Risk Anal.* **1994**, *14*, 707–712. [CrossRef] [PubMed]
5. Bagajewicz, M.; Uribe, A. *Financial Risk Management in Refinery Operations Planning Using Commercial Software*; XIV Latin Ibero-American Congress on Operations Research: Cartagena de Indias, Colombia, 2008.
6. Armstrong, G.W.; Cumming, S.G. Estimating the cost of land base changes due to wildfire using shadow prices. *For. Sci.* **2003**, *49*, 719–730.
7. Bertsimas, D.; Lauprete, G.J.; Samarov, A. Shortfall as a risk measure: Properties, optimization and applications. *J. Econ. Dyn. Control* **2004**, *28*, 1353–1381. [CrossRef]
8. Hildebrandt, P.; Knoke, T. Investment decisions under uncertainty—A methodological review on forest science studies. *For. Policy Econ.* **2011**, *13*, 1–15. [CrossRef]
9. Von Gadow, K.; Hui, G. *Modelling Forest Development*; Kluwer Academic Publishers: Dordrecht, The Netherlands, 2001; pp. 1–23.
10. Yousefpour, R.; Jacobsen, J.B.; Thorsen, B.J.; Meilby, H.; Hanewinkel, M.; Oehler, K. A review of decision-making approaches to handle uncertainty and risk in adaptive forest management under climate change. *Ann. For. Sci.* **2012**, *69*, 1–15. [CrossRef]
11. Pasalodos-Tato, M.; Mäkinen, A.; Garcia-Gonzalo, J.; Borges, J.G.; Lämas, T.; Eriksson, L.O. Assessing uncertainty and risk in forest planning and decision support systems: Review of classical methods and introduction of innovative approaches. *For. Syst.* **2013**, *22*, 282–303. [CrossRef]
12. Flannigan, M.D.; Krawchuk, M.A.; de Groot, W.J.; Wotton, M.; Gowman, L.M. Implications of changing climate for global wildland fire. *Int. J. Wildl. Fire* **2009**, *18*, 483–547. [CrossRef]

13. Taylor, P.D.; Fahrig, L.; Kimberly, A. Landscape connectivity: A return to the basics. In *Connectivity Conservation*; Crooks, K.R., Sanjayan, M., Eds.; Cambridge University Press: Cambridge, UK; 2006; pp. 22–43.

14. Martell, D.; Gunn, E.; Weintraub, A. Forest management challenges for operational researchers. *Eur. J. Oper. Res.* **1998**, *104*, 1–17. [CrossRef]

15. Boychuk, D.; Martell, D.L. A multistage stochastic programming model for sustainable forest-level timber supply under risk of fire. *For. Sci.* **1996**, *42*, 10–26.

16. Kaya, A.; Bettinger, P.; Boston, K.; Akbulut, R.; Ucar, Z.; Siry, J.; Merry, K.; Cieszewski, C. Optimisation in forest management. *Curr. For. Rep.* **2016**, *2*, 1–17. [CrossRef]

17. Bettinger, P.; Chung, W. The key literature of, and trends in, forest-level management planning in North America, 1950–2001. *Int. For. Rev.* **2004**, *6*, 40–50. [CrossRef]

18. Siry, J.P.; Bettinger, P.; Merry, K.; Grebner, D.L.; Boston, K.; Cieszewski, C. *Forest Plans of North America*; Academic Press: London, UK, 2015.

19. Gunn, E.A. Models for strategic forest management. In *Handbook of Operations Research in Natural Resources*; Weintraub, A., Romero, C., Bjørndal, T., Epstein, R., Eds.; Springer: New York, 2007; pp. 317–341.

20. Acuna, M.A.; Palma, C.D.; Cui, W.; Martell, D.L.; Weintraub, A. Integrated spatial fire and forest management planning. *Can. J. For. Res.* **2010**, *40*, 2370–2383. [CrossRef]

21. Van Wagner, C.E. Simulating the effect of forest fire on long-term annual timber supply. *Can. J. For. Res.* **1983**, *13*, 451–457. [CrossRef]

22. Reed, W.J.; Errico, D. Optimal harvest scheduling at the forest level in the presence of the risk of fire. *Can. J. For. Res.* **1986**, *16*, 266–278. [CrossRef]

23. Armstrong, G.W. Sustainability of timber supply considering the risk of wildfire. *For. Sci.* **2004**, *50*, 626–639.

24. Eid, T. Use of uncertain inventory data in forestry scenario models and consequential incorrect harvest decisions. *Silva Fenn.* **2000**, *34*, 89–100. [CrossRef]

25. Borders, B.E.; Harrison, W.M.; Clutter, M.L.; Shiver, B.D.; Souter, R.A. The value of timber inventory information for management planning. *Can. J. For. Res.* **2008**, *38*, 2287–2294. [CrossRef]

26. Duvemo, K.; Lämas, T.; Eriksson, L.O.; Wikström, P. Introducing cost-plus-loss analysis into a hierarchical forestry planning environment. *Ann. Oper. Res.* **2014**, *219*, 415–431. [CrossRef]

27. Banks, E. *Risk and Financial Catastrophe*; Finance and Capital Markets Series; Palgrave Macmillan: Hampshire, Great Britain, 2009; p. 213.

28. Baranoff, E.Z.; Baranoff, E.Z. *Risk Management and Insurance*; John Wiley and sons: Hoboken, NJ, USA, 2004; pp. 48–52.

29. Jorion, P. How informative are value-at-risk disclosures? *Account. Rev.* **2002**, *77*, 911–931. [CrossRef]

30. Ardia, D. *Financial Risk Management with Bayesian Estimation of GARCH Models*; Springer: Heidelberg, Germany, 2008; p. 161.

31. Hoogerheide, L.; Van Dijk, H.K. Bayesian forecasting of value at risk and expected shortfall using adaptive importance sampling. *Int. J. Forecast.* **2010**, *26*, 231–247. [CrossRef]

32. Bagajewicz, M.J.; Barbaro, A.F. Financial risk management in the planning of energy recovery in the total site. *Ind. Eng. Chem. Res.* **2003**, *42*, 5239–5248. [CrossRef]

33. Holecy, J.; Hanewinkel, M. A forest management risk insurance model and its application to coniferous stands in southwest Germany. *For. Policy Econ.* **2006**, *8*, 161–174. [CrossRef]

34. Shepard, H.B. Forest fire insurance in the Pacific Coast states. *J. For.* **1935**, *33*, 111–116.

35. Shepard, H.B. Fire insurance for forests. *J. Land Pub. Util. Econ.* **1937**, *13*, 111–115. [CrossRef]

36. Lankoande, M.; Yoder, J.; Wandschneider, P. Optimal wildfire insurance in the wildland-urban interface in the presence of a government subsidy for fire risk mitigation. Available online: http://faculty.ses.wsu.edu/WorkingPapers/Yoder/LankoandeEtAl_InsuranceSubsidiesWildfire_2005.pdf (accessed on 11 May 2016).

37. Chen, H.; Cummins, J.D.; Viswanathan, K.S.; Weiss, M.A. Systemic risk and the interconnectedness between banks and insurers: An econometric analysis. *J. Risk Insur.* **2014**, *81*, 623–652. [CrossRef]

38. Robitaille, A.; Saucier, J.P. *Paysages Régionaux du Québec Méridional*; Les publications du Québec: Ste-Foy, QC, Canada, 1998; p. 213.

39. Nguyen-Xuan, T. Développement d'une stratégie d'aménagement forestier s'inspirant de la dynamique des perturbations naturelles pour la région nord de l'Abitibi. Available online: http://chaireafd.uqat.ca/pdf/nguyen1.pdf (accessed on 11 May 2016).

40. Bergeron, Y.; Leduc, A.; Harvey, B.D.; Gauthier, S. Natural Fire Regime: A Guide for Sustainable Management of the Canadian boreal forest. *Silva Fenn.* **2002**, *36*, 81–95. [CrossRef]

41. Gauthier, S.; Nguyen, T.; Bergeron, Y.; Leduc, A.; Drapeau, P.; Grondin, P. Developing forest management strategies based on fire regimes in northwestern Quebec. In *Emulating Natural Forest Landscape Disturbances: Concepts and Applications*; Perera, A., Buse, L.J., Weber, M.G., Eds.; Columbia University Press: New York, NY, USA, 2004; pp. 219–229.

42. Bergeron, Y.; Cyr, D.; Drever, C.R.; Flannigan, M.; Gauthier, S.; Kneeshaw, D.; Logan, K. Past, current, and future fire frequencies in Quebec's commercial forests: Implications for the cumulative effects of harvesting and fire on age-class structure and natural disturbance-based management. *Can J. For. Res.* **2006**, *36*, 2737–2744. [CrossRef]

43. Bergeron, Y.; Cyr, D.; Girardin, M.P.; Carcaillet, C. Will climate change drive 21st century burn rates in Canadian boreal forest outside of its natural variability: Collating global climate model experiments with sedimentary charcoal data. *Int. J. Wildl. Fire* **2010**, *19*, 1127–1139. [CrossRef]

44. Cyr, D.; Gauthier, S.; Bergeron, Y.; Carcaillet, C. Forest management is driving the eastern North American boreal forest outside its natural range of variability. *Front. Ecol. Environ.* **2009**, *7*, 519–524. [CrossRef]

45. Belleau, A.; Légaré, S. Project Tembec: Towards the implementation of a forest management strategy based on the natural disturbance dynamics of the northern Abitibi region. In *Ecosystem Management in the Boreal Forest*; Gauthier, S., Vaillancourt, M.-A., Leduc, A., de Grandpré, L., Kneeshaw, D., Morin, H., Drapeau, P., Bergeron, Y., Eds.; Presses de l'Université du Québec: Québec, Canada, 2009; pp. 479–499.

46. Bergeron, Y.; Gauthier, S.; Flannigan, M.; Kafka, V. Fire regimes at the transition between mixedwood and coniferous boreal forest in northwestern Quebec. *Ecology* **2004**, *85*, 1916–1932. [CrossRef]

47. Dhital, N.; Raulier, F.; Asselin, H.; Imbeau, L.; Valeria, O.; Bergeron, Y. Emulating boreal forest disturbance dynamics: Can we maintain timber supply, aboriginal land use, and woodland caribou habitat? *For. Chron.* **2013**, *89*, 54–65. [CrossRef]

48. Groot, A.; Lussier, J.-M.; Mitchell, A.K.; MacIsaac, D.A. A silvicultural systems perspective on changing Canadian forestry practices. *For. Chron.* **2005**, *81*, 50–55. [CrossRef]

49. Raymond, P.; Bédard, S.; Roy, V.; Larouche, C.; Tremblay, S. The irregular shelterwood system: Review, classification, and potential application to forests affected by partial disturbances. *J. For.* **2009**, *107*, 405–413.

50. Davis, L.S.; Johnson, K.N.; Bettinger, P.S.; Howard, T.E. *Forest Management to Sustain Ecological, Economic, and Social Values*; McGraw Hill: New York, NY, USA, 2001; p. 804.

51. Fall, A.; Fall, J. A domain-specific language for models of landscape dynamics. *Ecol. Modell.* **2001**, *141*, 1–18. [CrossRef]

52. Cumming, S.G. A parametric model of the fire-size distribution. *Can. J. For. Res.* **2001**, *31*, 1297–1303. [CrossRef]

53. Fall, A.; Fortin, M.-J.; Kneeshaw, D.D.; Yamasaki, S.H.; Messier, C.; Bouthillier, L.; Smith, C. Consequences of various landscape scale ecosysteme management strategies and fire cycles on age-class structure and harvest in boreal forest. *Can. J. For. Res.* **2004**, *34*, 310–322. [CrossRef]

54. Bouchard, M.; Pothier, D. Long-term influence of fire and harvesting on boreal forest age structure and forest composition in eastern Québec. *For. Ecol. Manag.* **2011**, *261*, 811–820. [CrossRef]

55. Gauthier, S.; Raulier, F.; Ouzennou, H.; Saucier, J.P. Strategic analysis of forest vulnerability to risk related to fire: An example from the coniferous boreal forest of Quebec. *Can. J. For. Res.* **2015**, *45*, 553–565. [CrossRef]

56. Raulier, F.; Dhital, N.; Racine, P.; Tittler, R.; Fall, A. Increasing resilience of timber supply: How a variable buffer stock of timber can efficiently reduce exposure to shortfalls caused by wildfires. *For. Policy Econ.* **2014**, *46*, 47–55. [CrossRef]

57. Schruben, L.W. Designing Correlation Induction Strategies for Simulation. Current Issues in Computer Simulation. *J. Am. Stat. Assoc.* **1979**, *73*, 504–525. [CrossRef]

58. Law, A.M.; Kelton, W.D. *Simulation Modelling and Analysis*; McGraw-Hill, Inc.: Boston, MA, USA, 1982.

59. R Foundation for Statistical Computing. *R: A Language and Environment for Statistical Computing*; R Foundation for Statistical Computing: Vienna, Austria, 2014.

60. Leduc, A.; Bernier, P.Y.; Mansuy, N.; Raulier, F.; Gauthier, S.; Bergeron, Y. Using salvage logging and tolerance to risk to reduce the impact of forest fires on timber supply calculations. *Can. J. For. Res.* **2015**, *45*, 480–486. [CrossRef]

61. Moore, M.A.; Boardman, A.E.; Vining, A.R.; Weimer, D.L.; Greenberg, D.H. "Just give me a number!" Practical values for the social discount rate. *J. Policy Anal. Manag.* **2004**, *23*, 789–812. [CrossRef]
62. Peter, B.; Nelson, J. Estimating harvest schedules and profitability under the risk of fire disturbance. *Can. J. For. Res.* **2005**, *35*, 1378–1388. [CrossRef]
63. Gassmann, H.I. Optimal harvest of a forest in the presence of uncertainty. *Can. J. For. Res.* **1989**, *19*, 1267–274. [CrossRef]
64. Tomlin, B. On the value of mitigation and contingency strategies for managing supply chain disruption risks. *Manag. Sci.* **2006**, *52*, 639–657. [CrossRef]
65. Manley, B.; Watt, R. *Forestry Insurance, Risk Pooling and Risk Minimization Options*; Ministry of Agriculture and Forestry: Wellington, New Zealand, 2009. Available online: http://maxa.maf.govt.nz/climatechange/reports/forestry-insurance-risk-pooling-and-minimisation.pdf (accessed on 11 May 2016).
66. Brumelle, S.L.; Stanbury, W.T.; Thompson, W.A.; Vertinsky, I.B.; Wehrund, D.A. A framework for the analysis of risks in forest management and silvicultural investments. *For. Ecol. Manag.* **1990**, *35*, 279–299. [CrossRef]
67. Savage, D.W.; Martell, D.L.; Wotton, B.M. Evaluation of two risk mitigation strategies for dealing with fire-related uncertainty in timber supply modelling. *Can. J. For. Res.* **2010**, *40*, 1136–1154. [CrossRef]
68. Gauthier, S.; Bernier, P.Y.; Boulanger, Y.; Guo, J.; Beaudoin, A.; Boucher, D. Vulnerability of timber supply to projected changes in fire regime in Canada's managed forests. *Can. J. For. Res.* **2015**, *45*, 1439–1447. [CrossRef]
69. Bettinger, P.; Boston, K.; Siry, J.P.; Grebner, D.L. *Forest Management and Planning*; Academic Press: Burlington, MA, USA, 2009; p. 331.
70. Sturtevant, B.R.; Miranda, B.R.; Yang, J.; He, H.S.; Gustafson, E.J. Studying fire mitigation strategies in multi-ownership landscapes: Balancing the management of fire-dependent ecosystems and fire risk. *Ecosystems* **2009**, *12*, 445–461. [CrossRef]
71. James, P.M.A.; Fortin, M.J.; Fall, A.; Kneeshaw, D.; Messier, C. The Effects of Spatial Legacies following Shifting Management Practices and Fire on Boreal Forest Age Structure. *Ecosystems* **2007**, *10*, 1261–1277. [CrossRef]

© 2016 by the authors. Licensee MDPI, Basel, Switzerland. This article is an open access article distributed under the terms and conditions of the Creative Commons Attribution (CC BY) license (http://creativecommons.org/licenses/by/4.0/).